FOURIER TRANSFORMS

FOURIER TRANSFORMS
Principles and Applications

ERIC W. HANSEN
Thayer School of Engineering, Dartmouth College

Cover Image: ©iStockphoto/olgaaltunina

Copyright © 2014 by John Wiley & Sons, Inc. All rights reserved.

Published by John Wiley & Sons, Inc., Hoboken, New Jersey.
Published simultaneously in Canada.

No part of this publication may be reproduced, stored in a retrieval system, or transmitted in any form or by any means, electronic, mechanical, photocopying, recording, scanning, or otherwise, except as permitted under Section 107 or 108 of the 1976 United States Copyright Act, without either the prior written permission of the Publisher, or authorization through payment of the appropriate per-copy fee to the Copyright Clearance Center, Inc., 222 Rosewood Drive, Danvers, MA 01923, (978) 750-8400, fax (978) 750-4470, or on the web at www.copyright.com. Requests to the Publisher for permission should be addressed to the Permissions Department, John Wiley & Sons, Inc., 111 River Street, Hoboken, NJ 07030, (201) 748-6011, fax (201) 748-6008, or online at http://www.wiley.com/go/permission.

Limit of Liability/Disclaimer of Warranty: While the publisher and author have used their best efforts in preparing this book, they make no representations or warranties with respect to the accuracy or completeness of the contents of this book and specifically disclaim any implied warranties of merchantability or fitness for a particular purpose. No warranty may be created or extended by sales representatives or written sales materials. The advice and strategies contained herein may not be suitable for your situation. You should consult with a professional where appropriate. Neither the publisher nor author shall be liable for any loss of profit or any other commercial damages, including but not limited to special, incidental, consequential, or other damages.

For general information on our other products and services or for technical support, please contact our Customer Care Department within the United States at (800) 762-2974, outside the United States at (317) 572-3993 or fax (317) 572-4002.

Wiley also publishes its books in a variety of electronic formats. Some content that appears in print may not be available in electronic formats. For more information about Wiley products, visit our web site at www.wiley.com.

Library of Congress Cataloging-in-Publication Data is available.

ISBN: 978-1-118-47914-8

Printed in the United States of America.

10 9 8 7 6 5 4 3 2 1

To my family

CONTENTS

PREFACE — xi

CHAPTER 1   REVIEW OF PREREQUISITE MATHEMATICS   1

1.1   Common Notation   1
1.2   Vectors in Space   3
1.3   Complex Numbers   8
1.4   Matrix Algebra   11
1.5   Mappings and Functions   15
1.6   Sinusoidal Functions   20
1.7   Complex Exponentials   22
1.8   Geometric Series   24
1.9   Results from Calculus   25
1.10   Top 10 Ways to Avoid Errors in Calculations   33
     Problems   33

CHAPTER 2   VECTOR SPACES   36

2.1   Signals and Vector Spaces   37
2.2   Finite-dimensional Vector Spaces   39
2.3   Infinite-dimensional Vector Spaces   64
2.4   ★ Operators   86
2.5   ★ Creating Orthonormal Bases–the Gram–Schmidt Process   94
2.6   Summary   99
     Problems   101

CHAPTER 3   THE DISCRETE FOURIER TRANSFORM   109

3.1   Sinusoidal Sequences   109
3.2   The Discrete Fourier Transform   114
3.3   Interpreting the DFT   117
3.4   DFT Properties and Theorems   126
3.5   Fast Fourier Transform   152
3.6   ★ Discrete Cosine Transform   156
3.7   Summary   164
     Problems   165

CHAPTER 4   THE FOURIER SERIES   177

4.1   Sinusoids and Physical Systems   178
4.2   Definitions and Interpretation   178
4.3   Convergence of the Fourier Series   187

4.4	Fourier Series Properties and Theorems	199
4.5	The Heat Equation	215
4.6	The Vibrating String	223
4.7	Antenna Arrays	227
4.8	Computing the Fourier Series	233
4.9	Discrete Time Fourier Transform	238
4.10	Summary	256
	Problems	259

CHAPTER 5 THE FOURIER TRANSFORM 273

5.1	From Fourier Series to Fourier Transform	274
5.2	Basic Properties and Some Examples	276
5.3	Fourier Transform Theorems	281
5.4	Interpreting the Fourier Transform	299
5.5	Convolution	300
5.6	More about the Fourier Transform	310
5.7	Time–bandwidth Relationships	318
5.8	Computing the Fourier Transform	322
5.9	★ Time–frequency Transforms	336
5.10	Summary	349
	Problems	351

CHAPTER 6 GENERALIZED FUNCTIONS 367

6.1	Impulsive Signals and Spectra	367
6.2	The Delta Function in a Nutshell	371
6.3	Generalized Functions	382
6.4	Generalized Fourier Transform	404
6.5	Sampling Theory and Fourier Series	414
6.6	Unifying the Fourier Family	429
6.7	Summary	433
	Problems	436

CHAPTER 7 COMPLEX FUNCTION THEORY 454

7.1	Complex Functions and Their Visualization	455
7.2	Differentiation	460
7.3	Analytic Functions	466
7.4	exp z and Functions Derived from It	470
7.5	Log z and Functions Derived from It	472
7.6	Summary	489
	Problems	490

CHAPTER 8 COMPLEX INTEGRATION 494

8.1	Line Integrals in the Plane	494
8.2	The Basic Complex Integral: $\oint_\Gamma z^n dz$	497
8.3	Cauchy's Integral Theorem	502

8.4	Cauchy's Integral Formula 512
8.5	Laurent Series and Residues 520
8.6	Using Contour Integration to Calculate Integrals of Real Functions 531
8.7	Complex Integration and the Fourier Transform 543
8.8	Summary 556
	Problems 557

CHAPTER 9   LAPLACE, Z, AND HILBERT TRANSFORMS   563

9.1	The Laplace Transform 563
9.2	The Z Transform 607
9.3	The Hilbert Transform 629
9.4	Summary 652
	Problems 654

CHAPTER 10   FOURIER TRANSFORMS IN TWO AND THREE DIMENSIONS   669

10.1	Two-Dimensional Fourier Transform 669
10.2	Fourier Transforms in Polar Coordinates 684
10.3	Wave Propagation 696
10.4	Image Formation and Processing 709
10.5	Fourier Transform of a Lattice 722
10.6	Discrete Multidimensional Fourier Transforms 731
10.7	Summary 736
	Problems 737

BIBLIOGRAPHY   743

INDEX   747

PREFACE

The mathematical techniques known as "transform methods" have long been a basic tool in several branches of engineering and science, and no wonder. Fourier's simple idea, radical in its time, that a function can be expressed as a sum of sine waves, is ubiquitous. It underlies fields as diverse as communications, signal and image processing, control theory, electromagnetics, and acoustics. Electrical engineers typically encounter the rudiments of Fourier transforms in undergraduate systems and circuits courses, for modeling the spectral content of signals and designing frequency selective circuits (filters). The Laplace transform, a close cousin of the Fourier transform, enables the efficient analytical solution of ordinary differential equations and leads to the popular "S plane" and "root locus" methods for analyzing linear systems and designing feedback controllers. Discrete-time versions of the Fourier and Laplace transforms model spectra and frequency responses for digital signal processing and communications. Physics and engineering students meet the Fourier series when learning about harmonic motion or solving partial differential equations, for example, for waves and diffusion. The Fourier transform also models wave propagation from acoustics to radio frequencies to optics to X-ray diffraction. The widespread dissemination of the fast Fourier transform algorithm following its publication in 1965 added a computational dimension to all of these applications, from everyday consumer electronics to sophisticated medical imaging devices.

My purpose in writing this textbook is to pull these threads together and present a unified development of Fourier and related transforms for seniors and graduate students in engineering and physics—one that will deepen their grasp of how and why the methods work, enable greater understanding of the application areas, and perhaps motivate further pursuit of the mathematics in its own right. Drafts of the book have been used by myself and others for a 10-week course in Fourier transforms and complex variable theory at the Thayer School of Engineering, Dartmouth College. The prerequisites are an introductory course in lumped parameter systems (including the Laplace transform) or differential equations. Our course is itself prerequisite to courses in signal and image processing and a more advanced course in applied analysis taken by engineering and physics graduate students.

Philosophy and Distinctives

The book is more mathematically detailed and general in scope than a sophomore or junior level signals and systems text, more focused than a survey of mathematical methods, and less rigorous than would be appropriate for students of advanced mathematics. In brief, here is the approach I have taken.

1. The four types of Fourier transform on discrete and continuous domains—discrete Fourier transform (DFT), Fourier series, discrete-time Fourier transform (DTFT), and Fourier transform—are developed as orthogonal expansions within a vector space framework. They are introduced sequentially, starting with the DFT and working up to the continuous-time Fourier transform. The same important properties and theorems are revisited for each transform in turn, reinforcing the basic ideas as each new transform is introduced. This is in contrast with an approach that either begins with the continuous-time transform and works down to the others as special cases, or develops all four in parallel.

2. The early presentation of the DFT makes it immediately available as a tool for computing numerical approximations to the Fourier series and Fourier transform. Several homework problems give the student practice using numerical tools. MATLAB® is used throughout the book to demonstrate numerical methods and to visualize important ideas, but whether to use MATLAB or some other computational tool in the course is up to the instructor.

3. The fundamentals of complex analysis and integration are included as a bridge to a more thorough understanding of the Laplace and Z transforms, and as an additional way to calculate Fourier transforms.

4. Physical interpretations and applications are emphasized in the examples and homework problems. My hope is that the student will cultivate intuition for how the mathematics work as well as gain proficiency with calculation and application.

5. Starred sections, which may be skipped on a first reading, give brief introductions to more advanced topics and references for further reading.

6. Each chapter has a table of key results, which should be particularly helpful for reference after the course is completed.

Any author of an applied mathematics book must decide the extent to which the development of the material will be supported by proofs. The level of rigor required by a mathematician generally exceeds what is needed to justify the trustworthiness of a result to an engineer. Moreover, to prove all the key theorems of Fourier analysis requires a facility with real analysis and even functional analysis that exceeds the usual mathematical background of an undergraduate engineer. The approach taken here, for the most part, is to include proofs when they build intuition about how the mathematics work or contribute to the student's ability to make calculations and apply the transforms. Otherwise, I will usually substitute informal plausibility arguments, derivations of weaker results either in the text or in the end-of-chapter problems, or computational illustrations of the principles involved. Footnotes refer the interested reader to detailed treatments in more advanced texts.

Flow of the Book

The book has 10 chapters, which are described briefly here to show how the book's main ideas are developed. *Chapter 1* is a review of the topics from geometry, trigonometry, matrix algebra, and calculus that are needed for this course. *Chapter 2* then develops some fundamentals of vector spaces, particularly the generalizations of the geometric ideas of norm, inner product, orthogonality and orthogonal expansion from vectors to functions. This provides the unifying framework for the Fourier family and acquaints the student with concepts of broad importance in engineering mathematics.

Chapters 3–5 introduce, in sequence, the DFT, the Fourier series, the DTFT, and the (continuous-time) Fourier transform. The DFT has the easiest vector space interpretation of the four transforms, since it expands finite-dimensional vectors in terms of orthogonal finite-dimensional vectors. Some basic Fourier theorems (linearity, shift, energy conservation, convolution) are first presented here, then reappear later for the other transforms. Chapter 3 includes a derivation of the fast Fourier transform (FFT) algorithm and the discrete cosine transform (DCT), a close relative of the DFT that is the mathematical foundation of JPEG image compression.

The Fourier series, *Chapter 4*, is a representation of a periodic function as an infinite series of orthogonal sines and cosines. The appearance of the infinite series raises the question of convergence and leads to the important connections among convergence of the series, asymptotic behavior of the spectrum, and smoothness of the original function. The chapter includes applications to the diffusion and wave equations and to antenna arrays, and shows how to use the DFT to compute Fourier coefficients and partial sums of Fourier series. Swapping the time and frequency domains, the Fourier series becomes the DTFT, the basic tool for discrete-time system analysis and signal processing.

The Fourier transform, *Chapter 5*, expands an aperiodic function as a continuum of orthogonal sines and cosines rather than a set of discrete oscillatory modes. Despite the additional mathematical complication, it has many of the same properties as the DFT and the Fourier series and intuition developed earlier for these transforms will carry over to the Fourier transform. The chapter emphasizes using Fourier theorems for modeling systems (impulse response and transfer function) and performing calculations. It also shows how to use the DFT to compute transforms and convolutions. A brief introduction to time-frequency transforms and wavelet transforms concludes the chapter.

Chapter 6 begins by placing the impulse (delta) function on a more secure footing than the informal notion of "infinite height, zero width, unit area" that students sometimes bring with them from earlier classes. This is followed by development of a common, generalized framework for understanding ordinary functions together with impulses and other singular functions. Sampling theory, introduced informally in Chapter 3, is studied here in depth. It is also used to unify the four Fourier transforms, via the observation that sampling in the time domain produces periodicity in the frequency domain, and vice versa.

Chapters 7 and 8 are devoted to the theory of complex functions and methods of complex integration, with a focus on ultimately applying the theory to understanding and calculating transforms. Numerical calculations of the basic complex integral $\oint z^n \, dz$ on different closed contours are used to help students visualize why the integral evaluates either to $2\pi i$ or to zero, before formally introducing the fundamental results, the Cauchy integral theorem and integral formula. The traditional subjects of conformal mapping and potential theory are omitted, but the complex variable introduction here is, I believe, sufficient preparation for subsequent courses, for example, in electromagnetism or fluid mechanics, where complex potentials may be useful.

Chapter 9 moves beyond the Fourier transform to the Laplace, Z, and Hilbert transforms. The Laplace transform is motivated by the need to handle functions, in particular ones that grow exponentially, that are beyond the reach of the Fourier transform. The familiar Laplace theorems, used to solve initial value problems, are derived and compared with their Fourier counterparts. The well-known partial fraction expansion method for Laplace inversion is connected with complex integration and extended beyond the rational functions encountered in linear system theory. The Z transform appears via the Laplace transform of a sampled function, and analogies between the transforms are emphasized. The Hilbert transform, which describes a special property of the Fourier transform of a one-sided function, is developed and applied to various problems in signal theory.

Chapter 10 concludes the book by revisiting the Fourier transform in two and three dimensions, with applications to wave propagation and imaging. The closely related Hankel and Radon transforms are introduced. Multidimensional transforms of arrays of impulses are developed and applied to sampling theory and X-ray crystallography.

Suggested Use

Most of Chapters 2–5, 7 and 8, and selected parts of Chapters 6 and 9, are covered in my 10-week (30-hour) course. In a full semester course, additional material from Chapters 6, 9, and 10 could be added. If students have already had a course in complex analysis, or if time does not permit, Chapters 7 and 8 may be skipped, with the caveat that portions of Chapter 9 are inaccessible without complex integration. However, this would permit a thorough coverage of Chapters 2–6 and 10 with selected topics from Chapter 9. While I naturally prefer the sequence of Chapters 2–4, they may be approached in a different order, with the Fourier series before the DFT, and with the vector space material presented "just in time" as the Fourier methods are introduced. End-of-chapter problems cover basic and more complex calculations, drawn from the theory itself and from many physical applications. I hope that instructors will find sufficient variety to suit their particular approaches to the material.

Acknowledgments

Ulf Österberg and Markus Testorf taught from various drafts at Dartmouth, and Ron June used portions of the text at University of California, Davis. Paul Hansen also read and commented on several chapters. My hearty thanks to all of them for being generous with their time and constructive ideas. Likewise, the anonymous comments of early reviewers both encouraged me that I was on the right track and challenged me in valuable ways. I also thank George Telecki and Kari Capone at Wiley for taking on this project. Finally, to my students at the Thayer School, with whom it has been my privilege to teach and to learn over many years, thank you.

ERIC W. HANSEN

Hanover, New Hampshire
July 2014

CHAPTER 1

REVIEW OF PREREQUISITE MATHEMATICS

This chapter reviews, mostly without proof, a number of mathematical topics that you should have encountered before—including sets, vectors and matrices, complex numbers, sinusoidal and other functions, and some results from calculus. The material may be read rapidly, and referred to later as you have need. For a comprehensive treatment of any of these items, consult your previous texts.

1.1 COMMON NOTATION

The following symbols denote mathematical objects and operations you have probably seen in your prior coursework. Additional new notation will be introduced throughout the book as new concepts are introduced.

Common operations and relations

$+, -$	Addition and subtraction
$\cdot, \times$	Multiplication (numbers) or dot and cross product, respectively (vectors)
$\div, /$	Division
$\lvert x \rvert$	Absolute value of number x
$\lVert \mathbf{x} \rVert$	Norm, or length, of vector $\mathbf{x}$
$=, \neq, \approx$	Equal, not equal, approximately equal
$>, \geq, <, \leq$	Greater than, greater than or equal, less than, less than or equal
$\gg, \ll$	Much greater than, much less than
$\sum_{k=M}^{N} x_k$	The sum $x_M + x_{M+1} + \cdots + x_N$

Defining sets

$\mathbb{R}$	The real numbers
$\mathbb{C}$	The complex numbers
$\mathbb{Z}$	The integers
$\mathbb{N}$	The natural numbers (positive integers)
$\mathbb{Q}$	The rational numbers
$\emptyset$	The empty set, { }

Fourier Transforms: Principles and Applications, First Edition. Eric W. Hansen.
© 2014 John Wiley & Sons, Inc. Published 2014 by John Wiley & Sons, Inc.

$x \in A$	x is an element of the set A, for example, $0 \in \{-1, 0, 1\}$
$x \notin A$	x is not an element of the set A, for example, $2 \notin \{-1, 0, 1\}$
(a, b)	Open interval on the real line: $x \in (a, b)$ means $a < x < b$. (a, b) also denotes an ordered pair of numbers, for example, $(1, 2)$ is the point in the xy plane with coordinates $x = 1$ and $y = 2$. Which usage applies will be clear from context.
$[a, b]$	Closed interval on the real line: $x \in [a, b]$ means $a \leq x \leq b$.
$(a, b], [a, b)$	Half-open intervals on the real line
$\{x \mid condition\}$	Set-builder notation, for example, $\{x \in \mathbb{R} \mid x > a\}$ denotes the set of all real x such that x is greater than a (i.e., the open interval (a, ∞)). $\{f \mid \int \lvert f(x) \rvert dx < \infty\}$ denotes the set of functions f which are absolutely integrable.

Relationships between sets

$A = B$	The sets A and B are equal, that is, they have the same elements.
$A \subset B$	The set A is a subset of the set B: every element of A is also an element of B, for example, $\mathbb{N} \subset \mathbb{Z}$. $A \subset B$ includes the possibility that $A = B$. Sets A and B are equal if $A \subset B$ and $B \subset A$.

Operations on sets

$A \cap B$	The intersection of sets A and B, $A \cap B = \{x \mid x \in A \text{ and } x \in B\}$, for example, $\{0, 1\} \cap \{1, 2\} = \{1\}$. Sets A and B are *disjoint* if $A \cap B = \emptyset$.
$A \cup B$	The union of sets A and B, $A \cup B = \{x \mid x \in A \text{ or } x \in B\}$, for example, $\{0, 1\} \cup \{1, 2\} = \{0, 1, 2\}$.
$A \times B$	The cartesian product of sets A and B, the set of ordered pairs (a, b) with a drawn from set A and b drawn from set B, for example, $\{0, 1\} \times \{2, 3\} = \{(0, 2), (0, 3), (1, 2), (1, 3)\}$; $\mathbb{R} \times \mathbb{R}$, also known as $\mathbb{R}^2$, is the real plane; $[a, b] \times [c, d]$ is a rectangular region in the plane, a subset of $\mathbb{R}^2$.

Functions

$f : A \to B$	A mapping f, which assigns each element of A to one or more elements of B, that is, for each $x \in A$, there is $f(x) \in B$.
$f, f(x)$	f is the function itself, $f(x)$ is the value of the function for a particular input x
$\lim_{x \to a} f(x)$	The limiting value of f as its input x approaches a
Δx	A small change in x
dx	An infinitesimal change in x
f', df/dx	The first derivative of f (differentiate $f(x)$ once)
f'', d^2f/dx^2	The second derivative of f (differentiate $f(x)$ twice)
$f^{(n)}$, d^nf/dx^n	The function formed by differentiating f n times with respect to x
$\partial f/\partial x$, $\partial f/\partial y$	Partial derivatives of a function of two variables $f(x, y)$ with respect to x, or with respect to y
$\partial^2 f/\partial x \partial y$, $\partial^2 f/\partial x^2$	Partial derivatives of $f(x, y)$ formed with respect to y, then x, or twice with respect to x
$F(x) = \int f(x)\,dx$	F is the indefinite integral, or antiderivative, of f, the function such that $f = F'$

$\int_a^b f(x)\,dx$ The definite integral, or integral, of f over the interval (a, b). If F is the antiderivative of f, then $\int_a^b f(x)\,dx = F(b) - F(a)$ (fundamental theorem of calculus)

$\begin{bmatrix} a & b & c \\ d & e & f \end{bmatrix}$ An array, or matrix, of the numbers $a, b, \ldots f$

1.2 VECTORS IN SPACE

In everyday life, we can measure the distance between two points in space. This distance is nonnegative, and zero only if the two points are identical. For two points a and b, it makes sense that the distance d from a to b is the same as the distance from b to a: $d(a, b) = d(b, a)$. Three points in space, a, b, c, constitute the vertices of a triangle. The distances between pairs of points, $d(a, b)$, $d(b, c)$, and $d(a, c)$, are the lengths of the sides of the triangle. It is always true that one side of a triangle is no longer than the sum of the lengths of the other two sides, for example,

$$d(a, c) \leq d(a, b) + d(b, c) \tag{1.1}$$

(with equality if the three points are collinear). This relationship is called the *triangle inequality* (Figure 1.1).

Distance enables us to partially orient ourselves in space relative to other objects. We can say if an object is near or far, and specify sets of objects that are within a particular distance (radius) of us. But distance does not tell direction; we can only say "how far," not "which way."

Vectors specify "which way" as well as "how far." Each point in space is uniquely positioned at the tip of a vector whose tail is fixed at a common reference point, or origin. The vector represents the displacement of the point from the origin. The length of a vector **v** is a nonnegative real number called the *norm*, denoted $\|\mathbf{v}\|$. The norm of a vector is equal to zero only if **v** is the zero vector. A *unit vector* is a vector whose norm is one. A vector is *normalized*, made into a unit vector, by dividing it by its norm: $\mathbf{v}/\|\mathbf{v}\|$. Multiplying a vector by an ordinary real number, or *scalar*, c changes (scales) its length: $\|c\mathbf{v}\| = |c|\,\|\mathbf{v}\|$. If $c > 0$, the direction of the vector is unchanged, but if $c < 0$, the direction is reversed: **v** and $-\mathbf{v} = -1\mathbf{v}$ have the same length, but point in opposite directions. Any vector **v** can be represented as the product of a nonnegative scalar equal to **v**'s norm, and a unit vector pointing in **v**'s direction.

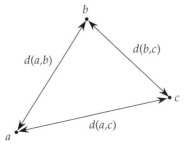

FIGURE 1.1   The triangle inequality: $d(a, c) \leq d(a, b) + d(b, c)$.

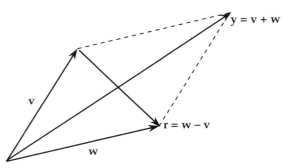

FIGURE 1.2 Addition and subtraction of vectors. The vector **y** is the sum of **v** and **w**. The vector **r** is the difference of **v** and **w**. The vector **w** is the sum of **v** and **r**. The lengths of **v**, **w**, and **r** obey the triangle inequality.

Two vectors **v** and **w** are added by translating **w** parallel to itself so that its tail is located at the tip of **v**, then constructing a new vector from the tail of **v** to the tip of **w** (Figure 1.2). This new vector (call it **y**) is the sum, or *resultant*, of **v** and **w**: **y** = **v** + **w**. You can easily show that the same resultant is obtained if **v** is translated parallel to itself so that its tail is located at the tip of **w**—that is, vector addition is commutative, **v** + **w** = **w** + **v**. In fact, vector addition has all the algebraic properties of ordinary addition. For vectors **u**, **v**, and **w**, and scalar c,

$$\mathbf{v} + \mathbf{w} = \mathbf{w} + \mathbf{v} \quad \text{(commutative)} \quad (1.2a)$$

$$(\mathbf{u} + \mathbf{v}) + \mathbf{w} = \mathbf{u} + (\mathbf{v} + \mathbf{w}) \quad \text{(associative)} \quad (1.2b)$$

$$c(\mathbf{u} + \mathbf{v}) = c\mathbf{u} + c\mathbf{v} \quad \text{(distributive)} \quad (1.2c)$$

$$\mathbf{v} + \mathbf{0} = \mathbf{v} \quad \text{(identity element)} \quad (1.2d)$$

The identity element **0** is called the zero vector, and its norm is zero: $\|\mathbf{0}\| = 0$.

If I am located in space at the tip of vector **v** and you are located at the tip of vector **w**, our respective distances from the origin are $\|\mathbf{v}\|$ and $\|\mathbf{w}\|$. We may define a third vector **r** that runs from the tip of **v** to the tip of **w** and directs me to you. Adding **r** to my position puts me at your position: **w** = **v** + **r**, and so **r** = **w** − **v**. The path from you to me is the opposite vector **v** − **w** = −**r**. The norm $\|\mathbf{r}\| = \|-\mathbf{r}\| = \|\mathbf{w} - \mathbf{v}\|$ is the distance between us. The vectors **v**, **w**, and **w** − **v** form a triangle, and their lengths, which are distances between points, obey the triangle inequality: $\|\mathbf{w} - \mathbf{v}\| \leq \|\mathbf{v}\| + \|\mathbf{w}\|$ (Figure 1.2).

The *dot product* of two vectors is a scalar quantity defined

$$\mathbf{v} \cdot \mathbf{w} = \|\mathbf{v}\| \|\mathbf{w}\| \cos \theta \quad (1.3)$$

where θ is the angle between **v** and **w**. The dot product of two nonzero vectors is zero if the angle θ is $\frac{\pi}{2}$; the vectors are then said to be *orthogonal*. Orthogonal unit vectors are said to be *orthonormal*. The dot product of a vector with itself is the square of its norm,

$$\mathbf{v} \cdot \mathbf{v} = \|\mathbf{v}\| \|\mathbf{v}\| \cos(0) = \|\mathbf{v}\|^2. \quad (1.4)$$

1.2 VECTORS IN SPACE

Algebraically, the dot product behaves like multiplication. For vectors **u**, **v**, and **w**, and scalar c,

$$\mathbf{v} \cdot \mathbf{w} = \mathbf{w} \cdot \mathbf{v} \quad \text{(commutative)} \quad (1.5a)$$

$$c(\mathbf{v} \cdot \mathbf{w}) = (c\mathbf{v}) \cdot \mathbf{w} = \mathbf{v} \cdot (c\mathbf{w}) \quad \text{(associative)} \quad (1.5b)$$

$$\mathbf{u} \cdot (\mathbf{v} + \mathbf{w}) = \mathbf{u} \cdot \mathbf{v} + \mathbf{u} \cdot \mathbf{w} \quad \text{(distributive)} \quad (1.5c)$$

A vector **v** in a plane may be expressed as the sum of two orthogonal vectors, which are called *components* of **v**. This is also called an orthogonal *decomposition* of the vector. These orthogonal vectors could be aligned with the coordinate directions (i.e., x and y) of a Cartesian system. Define unit vectors $\mathbf{e}_x$ and $\mathbf{e}_y$ pointing in the x and y directions. Then, $\mathbf{v} = v_x \mathbf{e}_x + v_y \mathbf{e}_y$, where v_x and v_y are scalar coefficients. To calculate them, take the dot product of this expression with $\mathbf{e}_x$ and $\mathbf{e}_y$, respectively.

$$\mathbf{v} \cdot \mathbf{e}_x = v_x(\mathbf{e}_x \cdot \mathbf{e}_x) + v_y(\mathbf{e}_y \cdot \mathbf{e}_x)$$
$$\mathbf{v} \cdot \mathbf{e}_y = v_x(\mathbf{e}_x \cdot \mathbf{e}_y) + v_y(\mathbf{e}_y \cdot \mathbf{e}_y).$$

Because the unit vectors are orthonormal, $\mathbf{e}_x \cdot \mathbf{e}_x = 1$, $\mathbf{e}_y \cdot \mathbf{e}_y = 1$, and $\mathbf{e}_x \cdot \mathbf{e}_y = \mathbf{e}_y \cdot \mathbf{e}_x = 0$. Hence,

$$v_x = \mathbf{v} \cdot \mathbf{e}_x, \qquad v_y = \mathbf{v} \cdot \mathbf{e}_y. \quad (1.6)$$

The coefficients v_x and v_y are called *orthogonal projections* of **v** along $\mathbf{e}_x$ and $\mathbf{e}_y$, respectively. Using the projection formulas (Equation 1.6), any vector **v** in the plane can be written as some *linear combination* $v_x \mathbf{e}_x + v_y \mathbf{e}_y$ of the same two orthonormal vectors $\mathbf{e}_x$ and $\mathbf{e}_y$. These unit vectors are said to *span* the plane, and are called an orthonormal *basis* for the plane.

Physically, orthogonal vectors represent noninteracting actions or motions. For example, in the parabolic motion of a projectile, gravity acts to decelerate/accelerate the vertical component of motion, so $\dfrac{dv_y}{dt} = -g$, but has no effect on the x component of velocity (Figure 1.3).

Expressing vectors in terms of their components relative to a common basis greatly simplifies vector calculations. Let **v** and **w** be two vectors in the plane,

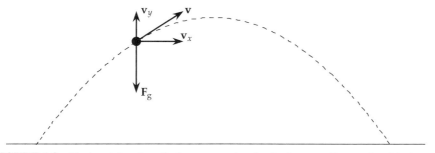

FIGURE 1.3 Horizontal and vertical components of velocity are orthogonal and independent. The gravitational force $\mathbf{F}_g$ acts only upon the vertical component of the motion, $\mathbf{v}_y$.

expressed in terms of their orthogonal components, $\mathbf{v} = v_x \mathbf{e}_x + v_y \mathbf{e}_y$ and $\mathbf{w} = w_x \mathbf{e}_x + w_y \mathbf{e}_y$, and let c be a scalar. Then,

$$c\mathbf{v} = c(v_x \mathbf{e}_x + v_y \mathbf{e}_y) = (cv_x)\mathbf{e}_x + (cv_y)\mathbf{e}_y \tag{1.7a}$$

$$\begin{aligned}\mathbf{v} + \mathbf{w} &= v_x \mathbf{e}_x + v_y \mathbf{e}_y + w_x \mathbf{e}_x + w_y \mathbf{e}_y \\ &= (v_x + w_x)\mathbf{e}_x + (v_y + w_y)\mathbf{e}_y\end{aligned} \tag{1.7b}$$

$$\begin{aligned}\mathbf{v} \cdot \mathbf{w} &= (v_x \mathbf{e}_x + v_y \mathbf{e}_y) \cdot (w_x \mathbf{e}_x + w_y \mathbf{e}_y) \\ &= v_x w_x (\mathbf{e}_x \cdot \mathbf{e}_x) + v_x w_y (\mathbf{e}_x \cdot \mathbf{e}_y) + v_y w_x (\mathbf{e}_y \cdot \mathbf{e}_x) + v_y w_y (\mathbf{e}_y \cdot \mathbf{e}_y) \\ &= v_x w_x + v_y w_y\end{aligned} \tag{1.7c}$$

$$\mathbf{v} \cdot \mathbf{v} = v_x^2 + v_y^2 = \|\mathbf{v}\|^2. \tag{1.7d}$$

Basis vectors are not unique. Instead of $\mathbf{e}_x$ and $\mathbf{e}_y$, we could use orthonormal vectors $\mathbf{e}_{x'}$ and $\mathbf{e}_{y'}$, which are rotated by 45° from $\mathbf{e}_x$ and $\mathbf{e}_y$ (Figure 1.4). The coefficients $v_{x'}$ and $v_{y'}$ will be different from v_x and v_y, because $\mathbf{e}_{x'}$ and $\mathbf{e}_{y'}$ point in different directions than $\mathbf{e}_x$ and $\mathbf{e}_y$ (Figure 1.4).

However, the resultant vector $\mathbf{v}$ is the same with either basis, and vector calculations (sum, difference, dot product, norm) carried out with components relative to either basis will yield the same results, for example, $v_x^2 + v_y^2 = v_{x'}^2 + v_{y'}^2 = \|\mathbf{v}\|^2$. For modelling a physical quantity, one basis may be preferred over others. A good example is the case of a body moving in a circular path under the influence of a central force, for example, a satellite orbiting a planet (Figure 1.5). The "polar" form, based on $\mathbf{e}_r$ and $\mathbf{e}_\theta$, is more natural than the Cartesian form, and the equations of motion are simpler than in Cartesian coordinates.

Vectors in the plane are generalized to vectors in three-dimensional (3D) space by adding a third basis vector orthogonal to the plane and a third component along that basis vector.

$$\mathbf{v} = v_x \mathbf{e}_x + v_y \mathbf{e}_y + v_z \mathbf{e}_z$$

(Figure 1.6).

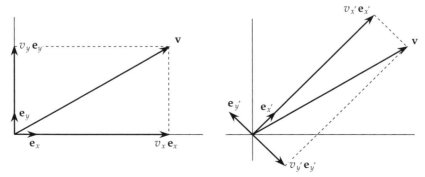

FIGURE 1.4 The decomposition of a vector depends on the choice of basis vectors.

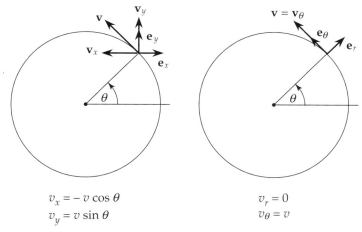

$$v_x = -v\cos\theta$$
$$v_y = v\sin\theta$$

$$v_r = 0$$
$$v_\theta = v$$

FIGURE 1.5 Circular motion is best described in polar coordinates.

The length of the vector is

$$\|\mathbf{v}\| = \sqrt{v_x^2 + v_y^2 + v_z^2},$$

generalizing the Pythagorean formula.

We shall see that, in Fourier analysis, an arbitrary function (signal or waveform) is expressed as a sum of simple (sine and cosine) functions, which behave, in a generalized sense, like orthogonal basis vectors. The next chapter will lay the foundation for this important concept.

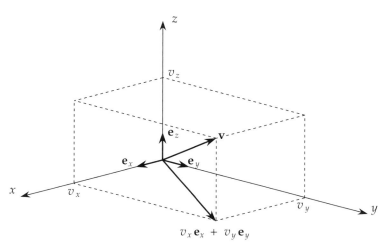

FIGURE 1.6 A vector in 3D space.

1.3 COMPLEX NUMBERS

Complex numbers arose initially in the study of roots of certain algebraic equations. You know that the solutions of the quadratic equation $x^2 - 1 = 0$ are the two square roots of 1, $+1$ and -1. The equation $x^2 + 1 = 0$, or $x^2 = -1$, on the other hand, has no real-valued solution because it requires us to take the square root of a negative number. The "imaginary number" i is defined to be $\sqrt{-1}$. It has the properties $i^2 = -1$ and $-i \cdot -i = (-1)(-1)i^2 = -1$. With this invention, the equation $x^2 + 1 = 0$ has two roots, $x = +i$ and $-i$, which are the two square roots of -1.

In mathematics and physics, the symbol i is used for the square root of -1. In engineering, j is frequently used instead of i. This is because electrical engineers in particular are accustomed to using i to denote electric current. Physicists, on the other hand, use j to stand for current density, and i for $\sqrt{-1}$. We shall use i in this book.

The product of i with a real number y is an imaginary number iy. The combination of a real number x and an imaginary number iy is called a *complex number*, and is written as a sum, $z = x + iy$. The quantities x and y are called the real and imaginary parts of z, respectively, denoted $x = \mathcal{R}e\, z$ and $y = \mathcal{I}m\, z$. (Be careful here. The imaginary part of a complex number is a *real* number. It is a common error to include the i in the imaginary part, writing $\mathcal{I}m\, z = iy$.)

A complex number $z = x + iy$ defines a point (x, y) in a plane. It is convenient to think of this point as the tip of a vector extending from the origin (Figure 1.7). The length of this vector, $\sqrt{x^2 + y^2}$, is called the *modulus*, or *magnitude*, of the complex number, and is denoted $|z|$. If z is purely real, then $|z|$ is simply an absolute value. The angle from the real (x) axis to the vector is called the *argument* of the complex number, written $\arg z$. A complex number may be specified either by its real and imaginary parts, $z = x + iy$, or by its modulus and argument, $z = r\angle\theta$, where $r = |z|$ and $\theta = \arg z$. These are known, respectively, as the rectangular (or cartesian) and polar forms.

By elementary trigonometry, we see $\tan \arg z = y/x$, or $\arg z = \arctan(y/x)$. We must be careful, however, to locate the angle in the proper quadrant of the complex plane. For the complex number $1 + i$, the ratio y/x is 1, and the angle is, by inspection, $\pi/4$. However, the number $-1 - i$, which lives in the third quadrant and has angle $-3\pi/4$, also has $y/x = (-1)/(-1) = 1$. (Figure 1.8). When asked to

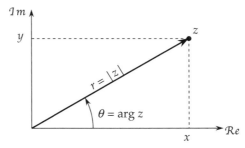

FIGURE 1.7 A complex number may be visualized as a vector in a plane.

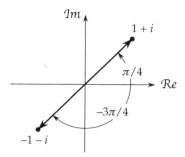

FIGURE 1.8 Although $\arg z = \arctan(\mathcal{I}m\, z/\mathcal{R}e\, z)$, one must be careful to locate the arctangent in the proper quadrant. The numbers $1 + i$ and $-1 - i$ both yield $\mathcal{I}m\, z/\mathcal{R}e\, z = 1$, yet their arguments are different.

calculate an arctangent, a pocket calculator produces a result between $-\pi/2$ and $\pi/2$, the so-called *principal value* of the arctangent. To properly calculate the argument, however, requires an arctangent function that respects the signs of the real and imaginary parts. Many calculators and computer languages have such a function; in MATLAB,[1] for example, it is called atan2. MATLAB also has a function called angle, which takes a complex number directly, so you don't have to separate the imaginary and real parts. The MATLAB functions are illustrated below.

```
z = [1+i, -1-i];
>atan(imag(z)./real(z))    % basic arctan, principal value
ans =
    0.7854       0.7854

>atan2(imag(z), real(z))   % two argument arctan, preserves quadrant
ans =
    0.7854      -2.3562

>angle(z)                  % angle of a complex number, uses atan2
ans =
    0.7854      -2.3562
```

A complex number in polar form, $z = r\angle\theta$, is converted to Cartesian form by the equations (see Figure 1.7):

$$x = r\cos\theta$$
$$y = r\sin\theta. \qquad (1.8)$$

---

[1] MATLAB is a registered trademark of The Mathworks, Inc. (http://www.mathworks.com). MATLAB is a computing and graphics system popular among engineers and segments of the scientific community. It is available for all popular operating systems, including Macintosh, Linux, and Windows. Occasionally in this text, numerical calculations will be expressed in MATLAB syntax. All numerical calculations resulting in graphs were performed with MATLAB. It is not necessary, however, for the reader to be familiar with MATLAB.

The rectangular form is more convenient for adding and subtracting complex numbers,

$$z_1 + z_2 = (x_1 + x_2) + i(y_1 + y_2) \tag{1.9}$$

while the polar form is better for multiplication and division,

$$z_1 z_2 = r_1 r_2 \angle (\theta_1 + \theta_2). \tag{1.10}$$

(The proof of this is easy using complex exponentials—see Section 1.7.) Both forms have physical interpretations in engineering and science.

The argument is multivalued (see Section 1.5). If $z = r\angle\theta$, then it is also true that $z = r\angle(\theta + 2\pi k)$, where k is any integer. We define the *principal value* of the argument, *Arg* z (with a capital "A"), to be the one between $-\pi$ and π. All other values of arg z (with a lower case "a") are obtained by adding integer multiples of 2π to the principal value.

$$\text{Arg } z \in (-\pi, \pi]$$
$$\arg z = \text{Arg } z + 2\pi k, \quad k = 0, \pm 1, \pm 2, \ldots. \tag{1.11}$$

The complex number $z^* = x - iy$, obtained by changing i to $-i$, is called the *complex conjugate* of z. The product $zz^* = (x+iy)(x-iy) = x^2 + y^2$ is $|z|^2$, the squared modulus. Because of the change of sign in z^*, $\arg z^* = \arctan[(-y)/x] = -\arg z$. The sum $z + z^* = x + iy + x - iy = 2x$, and the difference $z - z^* = 2iy$. This leads to the relationships:

$$x = \mathcal{R}e\, z = \frac{z + z^*}{2} \quad \text{and} \quad y = \mathcal{I}m\, z = \frac{z - z^*}{2i}. \tag{1.12}$$

These are particularly useful in calculating the real and imaginary parts of complex-valued functions.

It is almost always advisable to simplify a complex fraction by "rationalizing the denominator" with the complex conjugate:

$$\frac{z_1}{z_2} = \frac{z_1 z_2^*}{z_2 z_2^*} = \frac{z_1 z_2^*}{|z_2|^2}. \tag{1.13}$$

It follows that

$$\arg(1/z) = \arg\left(z^*/|z|^2\right) = \arg z^* = -\arg z. \tag{1.14}$$

As an example, consider the complex numbers $z_1 = 1 + \sqrt{3}i$ and $z_2 = 1 - i$. In polar form, they are

$$z_1 = \sqrt{1+3}\angle \arctan \frac{\sqrt{3}}{1} = 2\angle \frac{\pi}{3}$$

$$z_2 = \sqrt{1+1}\angle \arctan \frac{-1}{1} = \sqrt{2}\angle\left(-\frac{\pi}{4}\right)$$

and their sum, product, and quotient are

$$z_1 + z_2 = 2 + (\sqrt{3} - 1)i$$

$$z_1 z_2 = (1 + \sqrt{3}i)(1 - i) = 1 + \sqrt{3}i(-i) + \sqrt{3}i - i = (\sqrt{3} + 1) + i(\sqrt{3} - 1)$$

$$= 2\angle\frac{\pi}{3} \cdot \sqrt{2}\angle\left(-\frac{\pi}{4}\right) = 2\sqrt{2}\angle\frac{\pi}{12}$$

$$\frac{z_1}{z_2} = \frac{z_1 z_2^*}{|z_2|^2}$$

$$= \frac{(1 + \sqrt{3}i)(1 + i)}{2} = \frac{(1 - \sqrt{3}) + i(1 + \sqrt{3})}{2}$$

$$= \frac{2}{\sqrt{2}}\angle\left(\frac{\pi}{3} + \frac{\pi}{4}\right) = \sqrt{2}\angle\frac{7\pi}{12}$$

$$\frac{1}{z_1} = \frac{1}{2\angle\frac{\pi}{3}} = \frac{1}{2}\angle\left(-\frac{\pi}{3}\right) = \frac{1}{2}\left(\frac{1 - \sqrt{3}i}{2}\right) = \frac{1}{4} - \frac{\sqrt{3}}{4}i.$$

1.4 MATRIX ALGEBRA

A *matrix* is an array of numbers, which may be real or complex.

$$\mathbf{X} = \begin{bmatrix} x_{11} & x_{12} & \cdots & x_{1c} \\ x_{21} & x_{22} & \cdots & x_{2c} \\ \vdots & \vdots & \ddots & \vdots \\ x_{r1} & x_{r2} & \cdots & x_{rc} \end{bmatrix}$$

The dimensions of the matrix are expressed "$r \times c$" (read "r by c"), where r is the the number of rows and c is the number of columns. Particularly important special cases are

- $1 \times n$, called a row vector

$$\begin{bmatrix} x_1 & x_2 & \cdots & x_n \end{bmatrix}$$

- $n \times 1$, called a column vector

$$\begin{bmatrix} x_1 \\ x_2 \\ \vdots \\ x_n \end{bmatrix}$$

- $n \times n$, a square matrix

$$\begin{bmatrix} x_{11} & x_{12} & \cdots & x_{1n} \\ x_{21} & x_{22} & \cdots & x_{2n} \\ \vdots & \vdots & \ddots & \vdots \\ x_{n1} & x_{n2} & \cdots & x_{nn} \end{bmatrix}$$

- A diagonal matrix is a square matrix in which the elements x_{ij}, $i \neq j$, are zero:

$$\begin{bmatrix} x_{11} & & & 0 \\ & x_{22} & & \\ & & \ddots & \\ 0 & & & x_{nn} \end{bmatrix}$$

- **I**, the identity matrix, is a diagonal matrix whose diagonal elements, x_{ii}, are all equal to one.

$$\begin{bmatrix} 1 & & & 0 \\ & 1 & & \\ & & \ddots & \\ 0 & & & 1 \end{bmatrix}$$

The familiar 3D vectors from physics (Section 1.2), expressed in terms of orthogonal components relative to a basis, are compactly written as arrays. The following are equivalent representations:

$$\mathbf{v} = v_1 \mathbf{e}_1 + v_2 \mathbf{e}_2 + v_3 \mathbf{e}_3$$
$$\mathbf{v} = \begin{bmatrix} v_1 & v_2 & v_3 \end{bmatrix}$$
$$\mathbf{v} = \begin{bmatrix} v_1 \\ v_2 \\ v_3 \end{bmatrix}.$$

The presence of an underlying basis is implicit when a vector is written as an array.

The *transpose* of an array, denoted $\mathbf{X}^T$, is obtained by exchanging the rows and columns:

$$\begin{bmatrix} a & b & c \\ d & e & f \end{bmatrix}^T = \begin{bmatrix} a & d \\ b & e \\ c & f \end{bmatrix}.$$

The transpose of a row vector is a column vector, and vice versa. The complex conjugate of a matrix is made by taking the complex conjugate of each element in the array. The *adjoint* of an array, denoted $\mathbf{X}^\dagger$, is the complex conjugate of the transpose,

$$\mathbf{X}^\dagger = (\mathbf{X}^T)^* = (\mathbf{X}^*)^T$$

$$\begin{bmatrix} 1 & 2i \\ 1-3i & 4 \end{bmatrix}^\dagger = \begin{bmatrix} 1 & 1+3i \\ -2i & 4 \end{bmatrix}.$$

If $\mathbf{X}$ is real, then $\mathbf{X}^\dagger = \mathbf{X}^T$.[2]

Arrays may be added and subtracted if they are of the same dimension:

$$\begin{bmatrix} 1 \\ 2 \end{bmatrix} + \begin{bmatrix} 2 \\ 3 \end{bmatrix} = \begin{bmatrix} 3 \\ 5 \end{bmatrix}$$

is a valid operation, but

$$\begin{bmatrix} 1 \\ 2 \end{bmatrix} + \begin{bmatrix} 2 & 3 \end{bmatrix}$$

is undefined.

A matrix may be multiplied by a scalar,

$$c\mathbf{x} = c \begin{bmatrix} x_1 \\ x_2 \\ \vdots \\ x_n \end{bmatrix} = \begin{bmatrix} cx_1 \\ cx_2 \\ \vdots \\ cx_n \end{bmatrix}, \quad c\mathbf{A} = c \begin{bmatrix} a_{11} & a_{12} \\ a_{21} & a_{22} \end{bmatrix} = \begin{bmatrix} ca_{11} & ca_{12} \\ ca_{21} & ca_{22} \end{bmatrix}.$$

Two arrays may be multiplied if they have compatible dimensions. Let $\mathbf{X}$'s dimensions be $r_x \times c_x$ and $\mathbf{Y}$'s be $r_y \times c_y$. If $c_x = r_y$, then the product $\mathbf{XY}$ may be calculated. The dimensions c_x and r_y are called the inner dimensions of $\mathbf{XY}$, and the resulting matrix has the outer dimensions, r_x and c_y. Here is a way to remember this:

$$\mathbf{XY} = \begin{bmatrix} r_x \times c_x \end{bmatrix} \begin{bmatrix} r_y \times c_y \end{bmatrix} = \begin{bmatrix} r_x \times c_y \end{bmatrix}.$$
$$\qquad\qquad\quad \underbrace{\qquad\qquad\qquad}_{\text{inner}} \quad \underbrace{\qquad\qquad}_{\text{outer}}$$

In the special case of a row vector $(1 \times n)$ times a column vector $(n \times 1)$, the result is a 1×1 matrix that we take to be a scalar, by analogy with the idea of the dot product. For two column vectors $\mathbf{x}$ and $\mathbf{y}$, the *inner product* is the scalar defined by

$$\mathbf{x}^\dagger \mathbf{y} = x_1^* y_1 + x_2^* y_2 + \cdots + x_n^* y_n. \qquad (1.15)$$

The norm of a vector is given by its inner product with itself,

$$\|\mathbf{x}\| = \sqrt{\mathbf{x}^\dagger \mathbf{x}} = \sqrt{|x_1|^2 + |x_2|^2 + \cdots + |x_n|^2}.$$

---

[2] Some mathematical software, like MATLAB, separate the conjugate and transpose operations. In MATLAB, for example, the command x' computes the adjoint of x. If you just want a transpose without complex conjugation, use x.' instead. If x is real, then x' and x.' give the same result.

Notice how these definitions generalize the dot product and norm for real-valued vectors in 2D and 3D space.

The product of a matrix and a vector is calculated by repeated row–column products. The rth element of the result is the product of the rth row or column of the matrix and the vector:

$$\begin{bmatrix} \cdot & \cdot & \cdot \\ * & * & * \\ \cdot & \cdot & \cdot \end{bmatrix} \begin{bmatrix} * \\ * \\ * \end{bmatrix} = \begin{bmatrix} \cdot \\ * \\ \cdot \end{bmatrix}$$

or

$$\begin{bmatrix} * & * & * \end{bmatrix} \begin{bmatrix} \cdot & * & \cdot \\ \cdot & * & \cdot \\ \cdot & * & \cdot \end{bmatrix} = \begin{bmatrix} \cdot & * & \cdot \end{bmatrix}.$$

The product of two matrices is just more of the same. The product of the rth row and c^{th} column is the (r, c) element of the result:

$$\begin{bmatrix} \cdot & \cdot & \cdot & \cdot \\ * & * & * & * \\ \cdot & \cdot & \cdot & \cdot \end{bmatrix} \begin{bmatrix} \cdot & \cdot & * \\ \cdot & \cdot & * \\ \cdot & \cdot & * \\ \cdot & \cdot & * \end{bmatrix} = \begin{bmatrix} \cdot & \cdot & \cdot \\ \cdot & \cdot & * \\ \cdot & \cdot & \cdot \end{bmatrix}.$$

Consider the matrices $\mathbf{A} = \begin{bmatrix} 0 & 2 & 4 \\ 1 & 3 & 5 \end{bmatrix}$ and $\mathbf{B} = \begin{bmatrix} 0 & 1 \\ 2 & 3 \end{bmatrix}$. The product $\mathbf{AB}$ is not defined, because the inner dimensions (3 and 2, respectively) do not agree. On the other hand, $\mathbf{BA}$ is defined, and is

$$\mathbf{BA} = \begin{bmatrix} 0 & 1 \\ 2 & 3 \end{bmatrix} \begin{bmatrix} 0 & 2 & 4 \\ 1 & 3 & 5 \end{bmatrix} = \begin{bmatrix} 1 & 3 & 5 \\ 3 & 13 & 23 \end{bmatrix}.$$

In general, matrix multiplication does not commute, even for square matrices: $\mathbf{AB} \neq \mathbf{BA}$. For example, $\begin{bmatrix} 0 & 1 \\ 1 & 2 \end{bmatrix} \begin{bmatrix} 1 & 0 \\ 2 & 3 \end{bmatrix} = \begin{bmatrix} 2 & 3 \\ 5 & 6 \end{bmatrix}$ but $\begin{bmatrix} 1 & 0 \\ 2 & 3 \end{bmatrix} \begin{bmatrix} 0 & 1 \\ 1 & 2 \end{bmatrix} = \begin{bmatrix} 0 & 1 \\ 3 & 8 \end{bmatrix}$.

The identity matrix generalizes the scalar multiplicative identity element, 1. For any matrix $\mathbf{A}$ and identity matrices of appropriate dimension, $\mathbf{IA} = \mathbf{A}$ and $\mathbf{AI} = \mathbf{A}$.

An $m \times n$ matrix $\mathbf{A}$ transforms an n-dimensional vector $\mathbf{x}$ into an m-dimensional vector $\mathbf{y}$ through the product $\mathbf{y} = \mathbf{Ax}$. When $\mathbf{A}$ is square, $\mathbf{x}$ and $\mathbf{y}$ have the same dimensions. If, in addition, $\mathbf{y} = \mathbf{Ax} = \lambda \mathbf{x}$, where λ is a complex scalar, $\mathbf{x}$ is called an *eigenvector* of $\mathbf{A}$, and λ is the *eigenvalue* associated with $\mathbf{x}$. Eigenvalues and eigenvectors have numerous applications in engineering and physics, and there are good numerical methods for computing them.

The combination $\mathbf{x}^\dagger \mathbf{Ax}$, where $\mathbf{A}$ is a square matrix, is called a *quadratic form*. If $\mathbf{A}$ is a diagonal matrix, then

$$\mathbf{x}^\dagger \mathbf{Ax} = a_{11}|x_1|^2 + a_{22}|x_2|^2 + \cdots + a_{nn}|x_n|^2,$$

which appears to be an n-dimensional generalization of the simple quadratic $a|x|^2$. The most general quadratic form has nondiagonal $\mathbf{A}$ and includes cross-terms such as $a_{12}x_1^*x_2$. If a is positive in the scalar case, $a|x|^2$ is also positive, for all nonzero x. In the matrix case, if $\mathbf{x}^\dagger \mathbf{A} \mathbf{x} > 0$ for all nonzero $\mathbf{x}$, then $\mathbf{A}$ is said to be *positive definite*.

Matrix division is defined in a very restricted sense. If $\mathbf{A}$ is square, and $\mathbf{A}\mathbf{x} = \mathbf{y}$, then we may be able to solve for $\mathbf{x}$. In the scalar algebraic equation $ax = y$, we can calculate $x = y/a$ if a is nonzero. Otherwise, the division is undefined. In the matrix situation, the *determinant*, denoted $\det \mathbf{A}$ or $|\mathbf{A}|$, must be nonzero. The determinant will be nonzero if the rows of $\mathbf{A}$ are linearly independent (no row can be expressed as a nontrivial linear combination of the other rows). If this is the case, then $\mathbf{A}\mathbf{x} = \mathbf{y}$ represents n simultaneous equations in n unknowns that have a unique solution. So, if $\det \mathbf{A} \neq 0$, the matrix inverse $\mathbf{A}^{-1}$ may be calculated, and the unique solution to the equation $\mathbf{A}\mathbf{x} = \mathbf{y}$ is $\mathbf{x} = \mathbf{A}^{-1}\mathbf{y}$. Practical computational algorithms for solving $\mathbf{A}\mathbf{x} = \mathbf{y}$ are readily available. A matrix for which $\det \mathbf{A} = 0$ (or, as a practical matter, numerically very close to zero) cannot be inverted and is called *singular*.

If $\mathbf{A}$ is nonsingular, then $\mathbf{A}^{-1}$ exists and the products $\mathbf{A}^{-1}\mathbf{A}$ and $\mathbf{A}\mathbf{A}^{-1}$ exist and are both equal to $\mathbf{I}$, the identity matrix (analogous to the scalar case $a \times 1/a = 1$). If, in addition to being linearly independent, the rows of a square matrix are also orthogonal, then it is called an *orthogonal* matrix. The product of an orthogonal matrix and its adjoint, $\mathbf{A}\mathbf{A}^\dagger$, is a diagonal matrix. If, further, the rows of the matrix are orthonormal, then $\mathbf{A}\mathbf{A}^\dagger = \mathbf{I}$, and $\mathbf{A}$ is called a *unitary* matrix. For a unitary matrix, $\mathbf{A}^\dagger = \mathbf{A}^{-1}$.

1.5 MAPPINGS AND FUNCTIONS

A *mapping* is a rule that assigns to every point x in a set X a point y in a set Y. The set X is called the *domain* of the mapping. Frequently, we write $f : X \to Y$ to say "f is a mapping from X to Y". We can write $f : \mathbb{R} \to \mathbb{R}$ as shorthand for "f is a real-valued mapping of a real variable," and $f : \mathbb{R} \to \mathbb{C}$ to say "f is a complex-valued mapping of a real variable." For each point $x \in X$, the corresponding point $y \in Y$ is called the *image* of x, and is denoted $y = f(x)$. If each point x in the domain has only one image point, then the mapping f is called a *function*. That is, a function is a *single-valued* mapping. In this text, when a function's domain is a subset of the integers, $X \subset \mathbb{Z}$, we will denote the image of n by $f[n]$ rather than $f(n)$. A function whose domain is a set of successive integers, for example, $X = \{1, 2, \ldots, N\}$, is also called a *sequence*.

The set of all image points, $f(X) = \{y \in Y \mid y = f(x), x \in X\}$, is called the *range* of f. The range of f is a subset of Y; there may be points in Y that are not "f of something in X." If a point $y \in Y$ is the image of a point $x \in X$, we call x the *preimage* of y. An *inverse* f^{-1} can be defined using preimages. For a point $y \in Y$, $f^{-1}(y) = \{\text{preimages of } y \text{ in } X\} = \{x \in X \mid f(x) = y\}$. The inverse may or may not be a function, in that y may have more than one preimage.

If every y in the range of f has a unique preimage, we say f is *one-to-one* or *injective*. If the range of f is identically Y, that is, if every $y \in Y$ has a preimage in X, we say that f is *onto* or *surjective*. A function that is both one-to-one and onto is

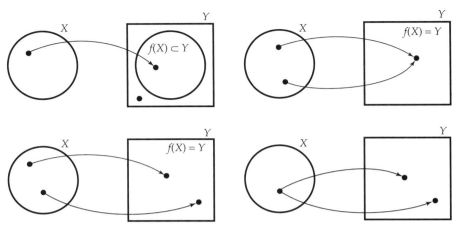

FIGURE 1.9 Mappings f from points in a set X to points in a set Y. Clockwise, from top left: One-to-one, but not onto; onto, but not one-to-one; multivalued; one-to-one and onto.

also called *bijective*. It creates a one-to-one correspondence between the points in X and the points in Y. Then, the inverse $f^{-1} : Y \to X$ is a function, with the property $f^{-1}(f(x)) = x$ and $f\left(f^{-1}(y)\right) = y$ (Figure 1.9).

Here are some examples.

- The function $f : \mathbb{R} \to \mathbb{R}$ defined by $y = 2x + 3$ is both one-to-one and onto. The inverse is $x = \dfrac{y-3}{2}$.
- The function $f : \mathbb{R} \to \mathbb{R}$ defined by $y = x^2$ is not one-to-one, because each positive real y has two preimages, $+\sqrt{y}$ and $-\sqrt{y}$, the positive and negative square roots of y. Neither is the function onto, because there is no $x \in \mathbb{R}$ for which $f(x)$ is negative.
- The logarithm function,[3] $y = \log x$, is undefined for $x = 0$, and real-valued only for $x > 0$. We restrict the domain to the positive reals $\mathbb{R}^+$, and write $f : \mathbb{R}^+ \to \mathbb{R}$ to indicate that the function maps positive reals into the reals. With this restriction of the domain, the range is all of $\mathbb{R}$, and f is one-to-one and onto. The inverse is $x = \exp y$.
- The function $f : \mathbb{R} \to \mathbb{R}$ defined by $f(x) = \cos x$ is not onto because its range is just the interval $[-1, 1]$ rather than all of $\mathbb{R}$. It is not one-to-one because there are many x values that map to the same y, for example, $\cos 2\pi k = 1$, $k \in \mathbb{Z}$. If the domain is restricted to the interval $[0, \pi]$, then cosine is one-to-one, but still not onto. If we further restrict the sets and define $f : [0, \pi] \to [-1, 1]$, cosine is both one-to-one and onto, and the inverse, arccos, is a well-defined function.

The inverse of $y = x^2$ is not a function because it assigns two values, $\pm\sqrt{y}$. By restricting the range of the square root to either the nonnegative or nonpositive real

---

[3]In this text, we use log rather than ln to denote natural logarithm. When necessary, the base-10 logarithm is denoted $\log_{10}$.

numbers, we can make it single-valued; these restrictions are called the positive and negative *branches* of the square root. Engineers and scientists, including this author, usually call the square root a *multivalued function* rather than a mapping, although the former term is, strictly speaking, an oxymoron. We shall have much more to say about multivalued functions in a later chapter.

An *even* function is one whose graph is symmetric across the origin, $f(-x) = f(x)$. An *odd* function's graph is antisymmetric across the origin, $f(-x) = -f(x)$. A function can always be expressed as the sum of an even part and an odd part, $f(x) = f_e(x) + f_o(x)$, where

$$f_e(x) = \frac{f(x) + f(-x)}{2}$$

$$f_o(x) = \frac{f(x) - f(-x)}{2}. \tag{1.16}$$

If f is an odd function, its integral on a symmetric interval $(-a, a)$ is zero:

$$\int_{-a}^{a} f(x)dx = 0. \tag{1.17}$$

and if f is an even function,

$$\int_{-a}^{a} f(x)dx = 2\int_{0}^{a} f(x)dx. \tag{1.18}$$

A complex-valued function, $f : \mathbb{R} \to \mathbb{C}$, is *Hermitian* if $f(-x) = f^*(x)$. The real part of a Hermitian function is even, and its imaginary part is odd. To see this, write

$$f(-x) = f_r(-x) + if_i(-x) = f^*(x) = f_r(x) - if_i(x),$$

and equate the real and imaginary parts,

$$f_r(-x) = f_r(x), \quad f_i(-x) = -f_i(x).$$

A function f is *bounded above* if there is a finite real number M such that $f(x) \leq M$ for all x in f's domain. A function is *bounded below* if there is a finite real number m such that $f(x) \geq m$ for all x in f's domain. If a function is bounded above, then there is a particular value of M that is the smallest possible upper bound. This value is called the *supremum* of f, denoted $\sup f$. Likewise, if a function is bounded below, then there is greatest lower bound, which is called the *infimum* of f, denoted $\inf f$.

Sometimes the supremum and infimum are just the largest and smallest values that f takes on in its domain. For example, $\sup \cos = 1$ and $\inf \cos = -1$ on the interval $(-\pi, \pi]$. Other functions never attain their bounds. The "saturating exponential" $f(x) = 1 - \exp(-x)$ approaches one as $x \to \infty$, so $\sup f = 1$. A sequence of numbers can also have a supremum and an infimum. The sequence $x = \{\frac{1}{2}, \frac{1}{4}, \ldots, \frac{1}{2^n}, \ldots\}$ has $\sup x = 1/2$ and $\inf x = 0$. The sequence attains its supremum, but not its infimum.

If a function $f : \mathbb{R} \to \mathbb{C}$ (because $\mathbb{R} \subset \mathbb{C}$, this includes both real- and complex-valued functions) is finite at $x = x_0$, and if

$$\lim_{x \to x_0^+} f(x) = \lim_{x \to x_0^-} f(x) = f(x_0),$$

where $x \to x_0^+$ and $x \to x_0^-$ mean that x approaches x_0 from above ($x > x_0$) and below ($x < x_0$), respectively, we say that f is *continuous* at x_0. (Informally, you can draw the graph of a continuous function without lifting your pen.) If f is continuous for all $x_0 \in (a, b)$, then we say that f is continuous on (a, b). The parabolic function $f(x) = x^2$ is continuous for all finite x. The slightly different function

$$g(x) = \begin{cases} 1, & x = 0 \\ x^2, & x \neq 0 \end{cases}$$

is not continuous at $x = 0$. Even though $\lim_{x \to 0} g(x) = 0$ from both sides, the limit is not equal to $g(0)$.

A function $f : (a, b) \to \mathbb{C}$, with a and b finite, is said to be *piecewise continuous* if it is continuous everywhere in the interval (a, b) except perhaps on a finite set of points $\{x_k\}_{k=1}^N \subset (a, b)$, and at these points $f(x_k^-)$ and $f(x_k^+)$ are finite. That is, f is piecewise continuous if it has at most a finite number of finite jump discontinuities on the interval (a, b). The unit step function $U(x)$, defined

$$U(x) = \begin{cases} 1, & x > 0 \\ \frac{1}{2}, & x = 0 \\ 0, & x < 0 \end{cases}.$$

is piecewise continuous (Figure 1.10).

If f is finite and continuous at x, and the limit

$$\lim_{\Delta x \to 0} \frac{f(x + \Delta x) - f(x)}{\Delta x} \qquad (1.19)$$

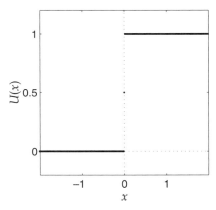

FIGURE 1.10 The unit step function is piecewise continuous. It has a finite jump at the origin.

exists, we say that f is *differentiable* at x. The limit is the derivative, $f'(x)$. For the parabolic function $f(x) = x^2$, which is continuous everywhere,

$$\frac{f(x + \Delta x) - f(x)}{\Delta x} = \frac{x^2 + 2x\Delta x + (\Delta x)^2 - x^2}{\Delta x}$$
$$= 2x + \Delta x.$$

As $\Delta x \to 0$ from above or below, the limit is $2x$, which we recognize as the derivative of x^2.

On the other hand, the step function $U(x)$ is discontinuous at the origin. The limits as the origin is approached from below and above are 0 and 1, respectively. They are not equal to each other, nor are they equal to $U(0)$, which we have defined to be $\frac{1}{2}$. If we attempt to evaluate the limit (Equation 1.19) we obtain, on the one hand,

$$\lim_{\Delta x \to 0^-} \frac{0 - \frac{1}{2}}{\Delta x} = \lim_{\Delta x \to 0^-} \frac{-1}{2\Delta x}$$

and, on the other hand,

$$\lim_{\Delta x \to 0^+} \frac{1 - \frac{1}{2}}{\Delta x} = \lim_{\Delta x \to 0^+} \frac{1}{2\Delta x},$$

neither of which exist.

A function $f : (a, b) \to \mathbb{C}$ is *piecewise smooth* if it is piecewise continuous and its derivative f' is also piecewise continuous. The absolute value function, $f(x) = |x|$, is one example. It is everywhere continuous, even at the origin, where it has a "corner" (Figure 1.11). For $x < 0$, the derivative is -1, and for $x > 0$, the derivative is $+1$. The derivative at $x = 0$ does not exist according to the definition given in Equation 1.19. But this is a single point, and because $f'(0^-) = -1$ and $f'(0^+) = 1$ are finite, the function is piecewise smooth.

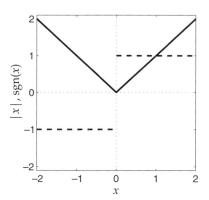

FIGURE 1.11 The absolute value function, $f(x) = |x|$, is piecewise linear and piecewise smooth. The derivative does not exist in the ordinary sense at $x = 0$. If $f'(0)$ is defined to be zero, then $f'(x) = \text{sgn } x$.

Even though the derivative of $f(x) = |x|$ is undefined at the origin according to Equation 1.19 we could assign it some reasonable value, say $f'(0) = 0$ (the average of -1 and 1). If you imagine that f is infinitesimally rounded at $x = 0$ rather than a sharp corner, this definition even makes intuitive sense. In a later chapter, we will see how to make it precise, so that $f'(x)$ is defined everywhere by the *signum* function,

$$\mathrm{sgn}(x) = \begin{cases} 1, & x > 0 \\ 0, & x = 0 \\ -1, & x < 0 \end{cases}.$$

1.6 SINUSOIDAL FUNCTIONS

The basic sinusoid, or sine wave function of time is

$$f(t) = \sin \omega t = \sin 2\pi \nu t.$$

The parameter ω is the angular frequency of the wave, expressed in radians per second, or rads/sec. Equivalently, we may use the frequency ν, expressed in cycles per second, or hertz (Hz). The two forms of frequency are related by $\omega = 2\pi\nu$.

The period T of the wave is the amount of time it takes for the wave to go through one complete cycle, that is, from $\omega t = 0$ to $\omega t = 2\pi$. Let $\omega T = 2\pi$, and the period is seen to be related to the frequency by $T = 2\pi/\omega = 1/\nu$.

Among the many trigonometric identities, two of the most useful are the sum formulae,

$$\sin(A + B) = \sin A \cos B + \cos A \sin B$$
$$\cos(A + B) = \cos A \cos B - \sin A \sin B. \quad (1.20)$$

For example, using Equation 1.20,

$$\sin(2\pi\nu t + \pi/2) = \sin 2\pi\nu t \cos \pi/2 + \cos 2\pi\nu t \sin \pi/2 = \cos 2\pi\nu t,$$

and, similarly,

$$\cos(2\pi\nu t + \pi/2) = -\sin 2\pi\nu t.$$

With this, we see that sine and cosine are related by a phase shift of 90°, or $\pi/2$ radians (Figure 1.12).

A sinusoid of arbitrary amplitude and phase, say $f(t) = C\cos(2\pi\nu t + \varphi)$, can always be written as the sum of a pure sine and cosine of the same frequency, but different amplitudes,

$$C\cos(2\pi\nu t + \varphi) = A\cos 2\pi\nu t + B\sin 2\pi\nu t. \quad (1.21)$$

Using the formula for the cosine of the sum of two angles (Equation 1.20),

$$C\cos(2\pi\nu t + \varphi) = C(\cos\varphi \cos 2\pi\nu t - \sin\varphi \sin 2\pi\nu t)$$
$$= (C\cos\varphi)\cos 2\pi\nu t + (-C\sin\varphi)\sin 2\pi\nu t,$$

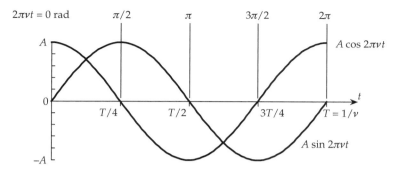

FIGURE 1.12 The sine and cosine functions.

from which it follows that

$$A = C\cos\varphi, \quad B = -C\sin\varphi$$
$$C = \sqrt{A^2 + B^2}, \quad \tan\varphi = -B/A. \tag{1.22}$$

(annotation: *magnitude* pointing to C)

This is illustrated in Figure 1.13.

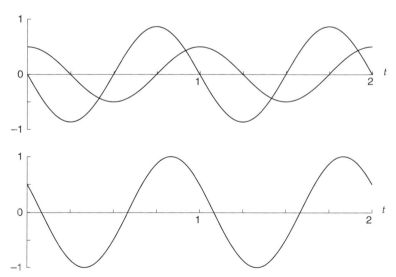

FIGURE 1.13 A cosine of arbitrary amplitude and phase may be formed from the sum of a pure sine and cosine. *Top:* The functions $\frac{1}{2}\cos 2\pi t$ and $-\frac{\sqrt{3}}{2}\sin 2\pi t$. *Bottom:* Their sum, $\cos(2\pi t + \pi/3)$.

1.7 COMPLEX EXPONENTIALS

The complex exponential $e^{i\theta}$ combines sine and cosine in a single function:

$$e^{i\theta} = \cos\theta + i\sin\theta. \tag{1.23}$$

The complex conjugate is $e^{-i\theta} = \cos\theta - i\sin\theta$. Combining these two expressions leads to the *Euler equations*,

$$\cos\theta = \frac{e^{i\theta} + e^{-i\theta}}{2} \quad \text{and} \quad \sin\theta = \frac{e^{i\theta} - e^{-i\theta}}{2i}. \tag{1.24}$$

It is often advantageous to reduce the sum and difference of two complex exponentials, $e^{i\theta_1} \pm e^{i\theta_2}$, by creating a symmetric form through factorization:

$$e^{i\theta_1} + e^{i\theta_2} = 2\, e^{i(\theta_1+\theta_2)/2} \cos\left(\frac{\theta_1 - \theta_2}{2}\right)$$

$$e^{i\theta_1} - e^{i\theta_2} = 2i\, e^{i(\theta_1+\theta_2)/2} \sin\left(\frac{\theta_1 - \theta_2}{2}\right). \tag{1.25}$$

In particular,

$$1 - e^{-i\theta} = 2i e^{-i\theta/2} \sin(\theta/2), \tag{1.26}$$

a result that will repeatedly come in handy.

Because the real part of a complex number $z = r\angle\theta$ is $r\cos\theta$ and the imaginary part is $r\sin\theta$, the complex exponential provides another way to write a complex number in polar form:

$$z = x + iy = r\cos\theta + ir\sin\theta = re^{i\theta}.$$

It also leads to a useful interpretation of sine and cosine waves. Consider

$$e^{i\omega t} = \cos\omega t + i\sin\omega t.$$

Recalling that the real and imaginary parts of a complex number define a vector in the complex plane, we may interpret $e^{i\omega t}$ as a rotating unit vector ($|e^{i\omega t}|^2 = e^{i\omega t}e^{-i\omega t} = 1$). The projection of this vector onto the real axis gives a cosine function, and the projection onto the imaginary axis gives a sine function (Figure 1.14).

And using the Euler equations (Equation 1.24),

$$\cos\omega t = \frac{e^{i\omega t} + e^{-i\omega t}}{2}, \quad \sin\omega t = \frac{e^{i\omega t} - e^{-i\omega t}}{2i}.$$

This way, the sine and cosine functions are interpreted as the sum and difference of vectors of length $\frac{1}{2}$ rotating in opposite directions with angular velocities $+\omega$ and $-\omega$ (you are invited to make a sketch like Figure 1.14 to illustrate this).

The complex exponential simplifies the solution of certain ordinary differential equations (ODEs). For example, consider the following linear ODE with constant coefficients:

$$\frac{dy}{dt} + cy = \cos\omega t.$$

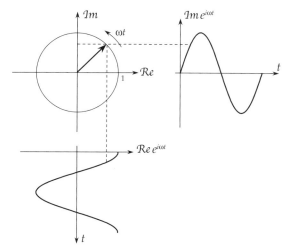

FIGURE 1.14 The complex exponential $e^{i\omega t}$ is a unit vector rotating counterclockwise with angular velocity ω. The projections of this vector along the real and imaginary axes trace out $\cos \omega t$ and $\sin \omega t$, respectively.

The steady-state solution (output of the system) in response to a sinusoidal driving function is a sinusoid having the same frequency, but with a different amplitude and phase, $y(t) = A\cos(\omega t + \varphi)$. Following the usual approach, we substitute this into the equation (using the alternative form, $a \cos \omega t + b \sin \omega t$), obtaining:

$$\frac{dy}{dt} + cy = \frac{d}{dt}(a\cos\omega t + b\sin\omega t) + c(a\cos\omega t + b\sin\omega t)$$

$$= -\omega a \sin\omega t + \omega b \cos\omega t + ca\cos\omega t + cb\sin\omega t$$

$$= (-\omega a + cb)\sin\omega t + (\omega b + ca)\cos\omega t$$

$$= \cos\omega t.$$

From here we solve the simultaneous equations

$$-\omega a + cb = 0, \qquad \omega b + ca = 1,$$

which yield

$$a = \frac{c}{\omega^2 + c^2}, \qquad b = \frac{\omega}{\omega^2 + c^2},$$

from which we obtain the more useful form (see Equation 1.22),

$$A = \frac{1}{\sqrt{\omega^2 + c^2}}, \qquad \tan\varphi = -\omega/c.$$

We will now present a simpler approach using the complex exponential. Were we to drive the equation instead with $\sin \omega t$ instead of $\cos \omega t$, the solution would be $A\sin(\omega t + \varphi)$. And if we drive the differential equation with the complex sum,

$\cos \omega t + i \sin \omega t = e^{i\omega t}$, then by the linearity property of the differential equation, the solution is $y = A \cos(\omega t + \varphi) + iA \sin(\omega t + \varphi) = A \exp(i(\omega t + \varphi))$.

Now, $e^{i\omega t}$ has the convenient property[4] $\frac{d}{dt} e^{i\omega t} = i\omega e^{i\omega t}$. This transforms the differential equation into a simple algebraic equation:

$$\frac{dy}{dt} + cy = \frac{d}{dt} A \exp(i(\omega t + \varphi)) + cA \exp(i(\omega t + \varphi))$$
$$= i\omega A e^{i\varphi} e^{i\omega t} + cA e^{i\varphi} e^{i\omega t}$$
$$= e^{i\omega t}.$$

Collecting terms,

$$(i\omega A e^{i\varphi} + cA e^{i\varphi} - 1) e^{i\omega t} = 0,$$

and because $|e^{i\omega t}| = 1$, the only way this can be true for all values of time is if

$$i\omega A e^{i\varphi} + cA e^{i\varphi} - 1 = 0.$$

The solution of this algebraic equation is

$$A e^{i\varphi} = \frac{1}{c + i\omega}.$$

(You may recognize this as the transfer function of first-order dynamic system like a resistor–capacitor or a mass-damper.)

The amplitude A and phase φ of the actual solution, $y(t) = A \cos(\omega t + \varphi)$, are given by calculating the modulus and argument of the complex function $\frac{1}{c + i\omega}$.

$$A = \left| \frac{1}{c + i\omega} \right| = \left(\frac{1}{(c + i\omega)(c - i\omega)} \right)^{1/2} = \frac{1}{\sqrt{\omega^2 + c^2}}$$

$$\tan \varphi = -\omega/c.$$

The solution may also be written:

$$y(t) = \mathcal{Re} \left\{ A e^{i\varphi} e^{i\omega t} \right\} = \mathcal{Re} \left\{ \frac{1}{c + i\omega} e^{i\omega t} \right\}.$$

1.8 GEOMETRIC SERIES

A series of the form $\sum_{n=0}^{N-1} x^n$ is called a *geometric* series. The sum of this series is an important result, easily derived, which should be committed to memory:

$$\sum_{n=0}^{N-1} x^n = \frac{1 - x^N}{1 - x}. \tag{1.27}$$

---

[4] Using the language of linear algebra, we say that $e^{i\omega t}$ is an *eigenfunction* of the differential operator $\frac{d}{dt}$, with eigenvalue $i\omega$. We will have more say about this in later chapters.

If $|x| < 1$, the sequence of partial sums $S_N = \sum_{n=0}^{N-1} x^n$ has a limit as $N \to \infty$—the series converges to

$$\sum_{n=0}^{\infty} x^n = \lim_{N \to \infty} \frac{1 - x^N}{1 - x} = \frac{1}{1 - x}, \quad |x| < 1. \tag{1.28}$$

This resembles the result of the integral,

$$\int_0^T e^{at} dt = \frac{e^{aT} - 1}{a},$$

where, if $a < 0$, the limit as $T \to \infty$ exists and is equal to

$$\int_0^\infty e^{at} dt = \lim_{T \to \infty} \frac{e^{aT} - 1}{a} = -\frac{1}{a}, \quad a < 0.$$

1.9 RESULTS FROM CALCULUS

Here are some more items from elementary calculus, which we shall need frequently.

Asymptotic behavior
Often we are interested in the asymptotic behavior of a function f as its argument x becomes very large ($x \to \infty$) or very small ($x \to 0$). For large values of x, the function $f(x) = \frac{1}{1+x}$ is just slightly smaller than $\frac{1}{x}$. We say that f is "of the order of x^{-1}," abbreviated $O(x^{-1})$ (read: "big-oh of x^{-1}"). Formally, a function f is $O(g)$ as $x \to \infty$ if there are numbers x_0 and M such that

$$|f(x)| < M|g(x)|, \quad x > x_0.$$

For small x ($x \to 0$), just change the $x > x_0$ to $x < x_0$. We generally are not interested in the actual values x_0 or M, just the dominating function g. The polynomial $p(x) = 1 + x + 2x^2$ is less than $4x^2$ for $x > 1$, but it is also less than $\frac{11}{4}x^2$ for $x > 2$. For most purposes it is sufficient to know that the asymptotic behavior is quadratic: p is $O(x^2)$. The function $f(x) = \frac{x}{1 + x^2}$ is $O(x^{-1})$ for large x, and $O(x)$ for small x.

The big-oh notation is also used to describe the error incurred in a series approximation to a function. For example, the series for e^x,

$$e^x = 1 + x + \frac{x^2}{2!} + \frac{x^3}{3!} + \cdots$$

can be abbreviated

$$e^x = 1 + x + O(x^2)$$

to indicate that when e^x is approximated by $1 + x$ for small x, deleted terms are of order x^2 and higher. When x is small, the x^2 term is the most consequential of the deleted terms, and will dominate the error.

Taylor series

In the vicinity of a point $x = a$, a function $f(x)$ may be approximated by a polynomial, by keeping the first few terms of the function's Taylor series:

$$f(x) = f(a) + f'(a)(x-a) + \frac{f''(a)}{2!}(x-a)^2 + \frac{f'''(a)}{3!}(x-a)^3 + O\{(x-a)^4\},$$

provided, of course, that the necessary derivatives exist at $x = a$. For example, the Taylor series for $\sin x$ about $x = 0$ is $x - x^3/3! + x^5/5! - \cdots$. Keeping the first term, the well-known "small angle approximation" approximates the function by a line, and keeping the next term approximates the sine by a cubic polynomial. Both approximations are tangent to the graph of the sine at the origin (Figure 1.15).

L'Hospital's (L'Hôpital's) rule

When evaluating $\lim\limits_{x \to a} \dfrac{f(x)}{g(x)}$, you may find that $\lim\limits_{x \to a} f(x)$ and $\lim\limits_{x \to a} g(x)$ are both zero or both infinite. It is still possible that the limit of the quotient will exist, however, if both numerator and denominator are decreasing or increasing at the same rate. If $|\lim\limits_{x \to a} f'(x)| < \infty$ and $0 < |\lim\limits_{x \to a} g'(x)| < \infty$, then L'Hospital's rule says

$$\lim_{x \to a} \frac{f(x)}{g(x)} = \frac{\lim\limits_{x \to a} f'(x)}{\lim\limits_{x \to a} g'(x)}.$$

L'Hospital's rule is helpful for resolving the value of functions where, at first glance, they are undefined. For example, the function $f(x) = \dfrac{\sin(\pi x)}{\pi x}$ appears to be

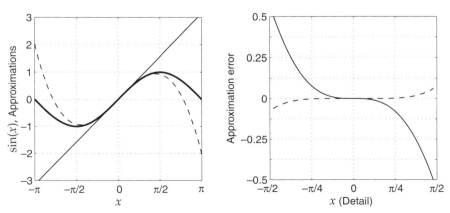

FIGURE 1.15 Low-order polynomial approximations to $\sin x$ created by truncating the Taylor series. *Left:* x (solid line) and $x - x^3/6$ (dashed line). *Right:* Approximation error over the range $\pi/2 > x > -\pi/2$. $\sin(x) - x$ (solid line) and $\sin(x) - (x - x^3/6)$ (dashed line).

singular (infinite) at the origin because the denominator is zero there. But $\sin \pi x$ is also zero at the origin, so we are not so sure. Applying L'Hospital's rule,

$$\lim_{x \to 0} \frac{\sin(\pi x)}{\pi x} = \frac{\lim_{x \to 0} \pi \cos(\pi x)}{\lim_{x \to 0} \pi} = 1. \tag{1.29}$$

We therefore define $f(0) = 1$, which removes the singularity at the origin.

L'Hospital's rule does not always work, at least not the first time. If you find that both limits are still zero or infinite, keep differentiating until one of them is finite. For example,

$$\lim_{x \to 0} \frac{\sin^2 x}{1 - \cos x} = \frac{\lim_{x \to 0} \sin^2 x}{\lim_{x \to 0} (1 - \cos x)} = \frac{0}{0}.$$

Differentiate the numerator and denominator again,

$$\frac{\lim_{x \to 0} 2 \sin x \cos x}{\lim_{x \to 0} \sin x} = \frac{0}{0},$$

and again,

$$\frac{\lim_{x \to 0}(2 \cos^2 x - 2 \sin^2 x)}{\lim_{x \to 0} \cos x} = \frac{2}{1} = 2.$$

Of course, some situations are hopeless. Consider $f(x) = \frac{\sin \pi x}{(\pi x)^2}$. Applying L'Hospital's rule,

$$\lim_{x \to 0} \frac{\sin(\pi x)}{(\pi x)^2} = \frac{\lim_{x \to 0} \pi \cos(\pi x)}{\lim_{x \to 0} 2\pi^2 x} = \frac{1}{0}.$$

The limit does not exist.

The Taylor series gives insight into what is going on with these limits. In the first example, dividing the expansion for $\sin(\pi x)$ by πx gives

$$\frac{\sin(\pi x)}{\pi x} = \frac{\pi x - \frac{(\pi x)^3}{6} + \frac{(\pi x)^5}{120} - \cdots}{\pi x} = 1 - \frac{(\pi x)^2}{6} + \frac{(\pi x)^4}{120} \cdots .$$

This series tends to 1 as $x \to 0$, in agreement with L'Hospital's rule. On the other hand, dividing by $(\pi x)^2$ results in

$$\frac{\pi x - \frac{(\pi x)^3}{6} + \frac{(\pi x)^5}{120} - \cdots}{(\pi x)^2} = \frac{1}{\pi x} - \frac{\pi x}{6} + \frac{(\pi x)^3}{120} \cdots .$$

The leading term blows up as $x \to 0$. The denominator is going to zero faster than the numerator, as x^2 vs. x; the numerator fails to neutralize the denominator and the function blows up.

Chain rule
The chain rule tells how to differentiate a function of a function:
$$\frac{d}{dx}f(u(x)) = f'(u(x))\frac{du}{dx} = \frac{df}{du}\frac{du}{dx}.$$

Integrability
It is useful to know whether the integral of a function exists before attempting to calculate the integral. A few basic cases are all that is needed.

- First, it is clear that if a positive function f is bounded, $f < M < \infty$, on a bounded interval (a, b) (i.e., $-\infty < a < b < \infty$), then the integral $\int_a^b f(x)\, dx$ has a finite value, which is bounded by $M(b - a)$.

- A positive function that grows without bound on a bounded interval (a, b) may nonetheless have a finite integral if it blows up sufficiently slowly. For example, the function $1/x$ blows up as $x \to 0$, and the integral
$$\int_a^1 \frac{dx}{x} = -\log a \to \infty \quad \text{as } a \to 0,$$
so $1/x$ is not integrable on $(0, 1)$. On the other hand, the function $1/x^{1-r}$, $r > 0$, grows more slowly than $1/x$ as x approaches 0, and the integral
$$\int_a^1 \frac{dx}{x^{1-r}} = \frac{1}{r}(1 - a^r) \to \frac{1}{r} \quad \text{as } a \to 0,$$
so $1/x^{1-r}$ is integrable on $(0, 1)$.

- A positive function on an unbounded interval, for example, $(1, \infty)$, must decay sufficiently rapidly as $x \to \infty$ in order to be integrable. Again, consider $1/x$.
$$\int_1^\infty \frac{dx}{x} = \lim_{A \to \infty} \int_1^A \frac{dx}{x} = \lim_{A \to \infty} \log A$$
and this limit does not exist because $\log A$ grows without bound. On the other hand, $1/x^{1+r}$, $r > 0$, decays faster than $1/x$, and its integral on $(1, \infty)$ is
$$\lim_{A \to \infty} \int_1^A \frac{dx}{x^{1+r}} = \lim_{A \to \infty} \frac{1}{r}\left[1 - \frac{1}{A^r}\right] = \frac{1}{r}.$$

- If a function takes on both positive and negative values, the integral of the negative portions will partially cancel the integral of the positive portions. The integral will be smaller in magnitude than the integral of $|f|$,
$$\left|\int_a^b f(x)\, dx\right| \le \int_a^b |f(x)|\, dx.$$

A function f is said to be *absolutely integrable* if the integral of its absolute value exists. Because the integral of f is bounded above by the integral of $|f|$, a sufficient, though not necessary, condition for the integrability of f is the absolute integrability of f. For example, the function $\sin(\pi x)/(\pi x)$ is not absolutely integrable on $(-\infty, \infty)$—it decays only as fast as $1/x$ for large x. However, we will see in a later chapter that it is integrable and its integral is 1.

In sum: in order to be integrable, a function must decay faster than $1/x$ for large x and grow more slowly than $1/x$ for small x (or more slowly than $1/(x-c)$ as $x \to c$). If a function is absolutely integrable, then it is integrable, but there are important cases where a function may be integrable even though it is not absolutely integrable.

Improper integrals and the Cauchy principal value

The integral $\int_a^b f(x)\, dx$ has a finite value when a and b are finite and $f(x)$ is bounded on the interval $[a, b]$. When one of the limits of integration is infinite, the integral is called *improper* and is defined

$$\int_{-\infty}^b f(x)\, dx = \lim_{a \to -\infty} \int_a^b f(x)\, dx \quad \text{or} \quad \int_a^\infty f(x)\, dx = \lim_{b \to \infty} \int_a^b f(x)\, dx$$

when the limit exists. If both limits of integration are infinite, the improper integral is defined by

$$\int_{-\infty}^\infty f(x)\, dx = \lim_{a \to \infty} \int_{-a}^c f(x)\, dx + \lim_{b \to \infty} \int_c^b f(x)\, dx$$

(typically, $c = 0$) when both the indicated limits exist. An alternative definition that is frequently useful is the *Cauchy principal value*, defined

$$\mathcal{P} \int_{-\infty}^\infty f(x)\, dx = \lim_{a \to \infty} \int_{-a}^a f(x)\, dx, \qquad (1.30)$$

when the limit exists. The Cauchy principal value often exists where the integral defined in the ordinary way fails.

Example 1.1 (sgn x has zero area). The signum function, defined

$$\mathrm{sgn}(x) = \begin{cases} -1, & x < 0 \\ 0, & x = 0 \\ 1, & x > 0 \end{cases}$$

has odd symmetry, and consequently should have zero area,

$$\int_{-\infty}^{\infty} \operatorname{sgn}(x)\, dx = 0.$$

According to the standard rules of calculus, however,

$$\int_{-\infty}^{\infty} \operatorname{sgn}(x)\, dx = -\lim_{a \to \infty} \int_{-a}^{0} dx + \lim_{b \to \infty} \int_{0}^{b} dx$$
$$= \lim_{a \to \infty} -a + \lim_{b \to \infty} b.$$

As a and b increase, the integrals separately grow without bound, and their "sum," $-\infty + \infty$, is undefined. The idea of the Cauchy principal value is to let the areas under the positive and negative portions of the integrand accumulate symmetrically, in such a way that the integral remains finite in the limit. In this example,

$$\mathcal{P}\int_{-\infty}^{\infty} \operatorname{sgn}(x)\, dx = \lim_{A \to \infty} \int_{-A}^{A} \operatorname{sgn}(x)\, dx$$
$$= \lim_{A \to \infty} |x|\Big|_{-A}^{A} = \lim_{A \to \infty} (A - A) = 0. \quad \blacksquare$$

If f is unbounded at a point c within an interval, $c \in (a, b)$, then by the conventional definition

$$\int_{a}^{b} f(x)\, dx = \lim_{\epsilon \to 0^+} \int_{a}^{c-\epsilon} f(x)\, dx + \lim_{\epsilon \to 0^+} \int_{c+\epsilon}^{b} f(x)\, dx$$

where both limits must independently exist. There is also a Cauchy principal value,

$$\mathcal{P} \int_{a}^{b} f(x)\, dx = \lim_{\epsilon \to 0^+} \left[\int_{a}^{c-\epsilon} f(x)\, dx + \int_{c+\epsilon}^{b} f(x)\, dx \right]$$

when the limit exists.

Example 1.2 (x^{-1} has zero area). Like signum, the function x^{-1} has odd symmetry and its area should be zero:

$$\int_{-\infty}^{\infty} \frac{dx}{x} = 0.$$

The integrand is singular at the origin. Under the ordinary rules of integration, the singularity is approached by a limiting process, as are the infinite limits of the integral itself.

$$\int_{-\infty}^{\infty} \frac{dx}{x} = \lim_{A \to \infty} \lim_{a \to 0} \int_{-A}^{-a} \frac{dx}{x} + \lim_{B \to \infty} \lim_{b \to 0} \int_{b}^{B} \frac{dx}{x}$$
$$= \lim_{A \to \infty} \lim_{a \to 0} \log |x|\Big|_{-A}^{-a} + \lim_{B \to \infty} \lim_{b \to 0} \log |x|\Big|_{b}^{B}$$
$$= \lim_{A \to \infty} \lim_{a \to 0} \log\left(\frac{a}{A}\right) + \lim_{B \to \infty} \lim_{b \to 0} \log\left(\frac{B}{b}\right)$$
$$= \log(0) + \log(\infty).$$

Neither limit is finite. Using the Cauchy principal value instead, the singularity is approached symmetrically, with the expectation that positive and negative areas will combine and give a finite result in the limit.

$$\mathcal{P}\int_{-\infty}^{\infty} \frac{dx}{x}\,dx = \lim_{A\to\infty}\lim_{a\to 0}\left[\int_{-A}^{-a}\frac{dx}{x} + \int_{a}^{A}\frac{dx}{x}\right]$$

$$= \lim_{A\to\infty}\lim_{a\to 0}\left[\log|x|\Big|_{-A}^{-a} + \log|x|\Big|_{a}^{A}\right]$$

$$= \lim_{A\to\infty}\lim_{a\to 0}\left[\log\left(\frac{a}{A}\right) + \log\left(\frac{A}{a}\right)\right]$$

$$= \lim_{A\to\infty}\lim_{a\to 0}\log(1) = 0.$$

∎

If an integral exists in the ordinary sense, then it also has a Cauchy principal value and, as there are no singularities for the Cauchy principal value to avoid, the ordinary value and principal value are the same. For example, the Cauchy principal value of the integral $\int_{-\infty}^{\infty}\left(1+x^2\right)^{-1}dx$ is

$$\mathcal{P}\int_{-\infty}^{\infty}\frac{dx}{1+x^2} = \lim_{A\to\infty}\int_{-A}^{A}\frac{dx}{1+x^2} = \lim_{A\to\infty}\left[\arctan(A) - \arctan(-A)\right]$$

$$= \lim_{A\to\infty} 2\arctan(A) = \pi.$$

The integral taken in the ordinary sense is

$$\int_{-\infty}^{\infty}\frac{dx}{1+x^2} = \lim_{A\to\infty}\int_{-A}^{0}\frac{dx}{1+x^2} + \lim_{B\to\infty}\int_{0}^{B}\frac{dx}{1+x^2}$$

$$= \lim_{A\to\infty}\left[\arctan(0) - \arctan(-A)\right] + \lim_{B\to\infty}\left[\arctan(B) - \arctan(0)\right]$$

$$= 0 - \left(-\frac{\pi}{2}\right) + \frac{\pi}{2} - 0 = \pi.$$

Because of this agreement, one commonly sees improper integrals on $(-\infty, \infty)$ performed as Cauchy principal values even when an ordinary integral would suffice.

Integration by parts
Integration by parts is the inverse of the product rule for differentiation, $d(uv) = u\,dv + v\,du$.

$$\int u\,dv = uv - \int v\,du.$$

As an example, consider the integral $\int x\sin x\,dx$. Let $u = x$ and $dv = \sin x\,dx$, then $du = dx$ and $v = -\cos x$. Plugging these into the formula,

$$\int x\sin x\,dx = -x\cos x - \int -\cos x\,dx = -x\cos x + \sin x.$$

It is important when integrating by parts to choose u and v in such a way that the remaining integral is simpler than the one you started with. Were you to choose instead $u = \sin x$ and $dv = x\,dx$, then $du = \cos x\,dx$ and $v = \frac{1}{2}x^2$. Then

$$\int x \sin x\,dx = \frac{1}{2}x^2 \sin x - \int \frac{1}{2}x^2 \cos x\,dx,$$

which is not an improvement.

Differentiating under the integral sign

If $f(x, y)$ is continuous on the region $[a, b] \times [c, d]$, and $\dfrac{\partial f}{\partial y}$ exists and is continuous on this region, then the integral

$$\int_a^b f(x, y)\,dx$$

is differentiable for all $y \in (c, d)$, and

$$\frac{d}{dy}\int_a^b f(x, y)\,dx = \int_a^b \frac{\partial f}{\partial y}\,dx.$$

Double integrals

If the function $f(x, y)$ is integrable on the region $[a, b] \times [c, d]$, the function $f(x_0, y)$ is integrable on $[c, d]$ for all $x_0 \in (a, b)$, and the function $f(x, y_0)$ is integrable on $[a, b]$ for all $y_0 \in (c, d)$, then the double integral may be written in terms of iterated single integrals (Fubini's theorem):

$$\int_c^d \int_a^b f(x, y)\,dx\,dy = \int_c^d \left[\int_a^b f(x, y)\,dx\right] dy = \int_a^b \left[\int_c^d f(x, y)\,dy\right] dx.$$

Unfortunately, the converse is not true, that the existence of either of the iterated integrals guarantees the existence of the double integral. A separate condition takes care of this (Tonelli's theorem):[5] If $\int_c^d \left[\int_a^b |f(x, y)|\,dx\right] dy$ or $\int_a^b \left[\int_c^d |f(x, y)|\,dy\right] dx$ exists, then the double integral $\int_c^d \int_a^b f(x, y)\,dx\,dy$ exists and Fubini's theorem applies.

A particularly important form of this, which we shall use later, is

$$\int_c^d \int_a^b f(x)\,g(y)\,h(x, y)\,dx\,dy = \int_c^d g(y) \left[\int_a^b f(x)\,h(x, y)\,dx\right] dy \quad (1.31a)$$

$$= \int_a^b f(x) \left[\int_c^d g(y)\,h(x, y)\,dy\right] dx \quad (1.31b)$$

---

[5]The Fubini and Tonelli theorems are discussed in Champeney (1987, pp. 18–19), Gasquet and Witomski (1999, p. 124), and Kolmogorov and Fomin (1975, pp. 359–362).

1.10 TOP 10 WAYS TO AVOID ERRORS IN CALCULATIONS

These should be reviewed frequently and applied as you solve the problems in this book and make calculations in your professional work.

1. Plan a strategy for solving the problem rather than diving right in and expecting the right answer to appear at the end. Make sketches to gain insight.
2. Identify your variables—time/space, frequency—and distinguish them from parameters and constants. When you integrate $f(x)dx$ between a and b, x should not appear in your answer.
3. Substitute letters for numerical parameters, for example, change $110\sin(2\pi 60t)$ to $V_o \sin(2\pi v_o t)$; at the end of the calculation, substitute the numerical values back in.
4. Divide calculations into small pieces.
5. Narrate the steps of your solution, explaining them to yourself and/or to your reader.
6. Use previously established results (e.g., theorems) to minimize calculations, but make sure you use them correctly—check their conditions and make sure they are applicable to the problem at hand.
7. Copy accurately. Double check plus and minus signs, especially as you go from line to line in a calculation. Write comfortably large and leave space between lines.
8. Watch dimensions.
   - Arguments of functions (like cos, sin, exp, log) must be dimensionless. If you have $\cos t$ (where t is understood to be time), there must be an implicit $\omega = 1$. If you have $\cos 2\pi t/T$, you're OK. If you have $\cos Tt$, you made a mistake.
   - Terms in a sum must have the same dimensions: $v + 1/T$ is OK (both are frequencies in Hz), $v + T$ is not.
   - Integrals have dimensions too. dt, dx, dv carry units of time, space, frequency (cycles/unit time or cycles/unit length), respectively.
9. Interpret your results. Look for symmetry and simplicity. Rewrite your solutions into algebraic forms that convey insight about the physical problem (mathematical software is not proficient at this). Make sketches to illustrate behavior.
10. Demand that your results make physical sense. Check limiting cases for extreme values of variables or parameters.

PROBLEMS

1.1. An alternative derivation of the dot product in terms of the components of the vectors (Equation 1.2) follows from the Law of Cosines. In the plane defined by vectors **v** and **w**, consider a triangle made from **v**, **w**, and their difference, **v** − **w**. (Figure 1.16). By the

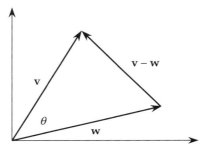

FIGURE 1.16 Illustrating Problem 1.1.

Law of Cosines,

$$\|\mathbf{v}\|^2 + \|\mathbf{w}\|^2 - 2\|\mathbf{v}\|\,\|\mathbf{w}\|\cos\theta = \|\mathbf{v}-\mathbf{w}\|^2.$$

Note that the last term on the left-hand side contains the dot product. Express the squared norms in terms of the components of $\mathbf{v}$ and $\mathbf{w}$, and complete the calculation.

1.2. Let a vector $\mathbf{v}$ be expressed in terms of the orthonormal basis $\{\mathbf{e}_1, \mathbf{e}_2\}$: $\mathbf{v} = v_1\mathbf{e}_1 + v_2\mathbf{e}_2$. Consider a different orthonormal basis $\{\mathbf{e}'_1, \mathbf{e}'_2\}$ and the expression of $\mathbf{v}$ in terms of this basis, $\mathbf{v} = v'_1\mathbf{e}'_1 + v'_2\mathbf{e}'_2$.

(a) Show that the coefficients v'_1 and v'_2 are related to the coefficients v_1 and v_2 by a linear transformation

$$\begin{bmatrix} v'_1 \\ v'_2 \end{bmatrix} = \begin{bmatrix} a_{11} & a_{12} \\ a_{21} & a_{22} \end{bmatrix} \begin{bmatrix} v_1 \\ v_2 \end{bmatrix}$$

and derive expressions for the coefficients a_{ij} in terms of the basis vectors.

(b) Show that the dot product is *invariant* to change of basis, that is, for two vectors $\mathbf{v}$ and $\mathbf{w}$ expressed in terms of the two bases, that

$$\mathbf{v} \cdot \mathbf{w} = v_1 w_1 + v_2 w_2 = v'_1 w'_1 + v'_2 w'_2.$$

By setting $\mathbf{v} = \mathbf{w}$, this also shows that the norm is invariant to change of basis.

(c) For the same vectors and bases as in the previous part, show that the vector sum is invariant to change of basis.

(d) Consider a scalar c and calculate its value c' in the new basis.

In advanced treatments of vectors in physics, vectors and scalars are defined by their transformation properties under change of coordinates, which is the same as change of basis.[6]

1.3. Derive the formulas for even and odd functions, Equations 1.16, 1.17, and 1.18.

1.4. Calculate the even and odd parts of the following functions:

(a) $f(x) = U(x-1)$

(b) $f(x) = e^{-x}U(x)$

(c) $f(x) = x[U(x) - U(x-1)]$.

---

[6]See, for example, Arfken (1985, Section 1.2).

1.5. Show that the derivative of an odd function is even and the derivative of an even function is odd.

1.6. Beginning with the sum formulae (Equation 1.20), derive the usual trigonometric identities for

(a) $\sin 2A$ and $\cos 2A$

(b) $\sin(A - B)$ and $\cos(A - B)$

(c) $\sin^2 A$ and $\cos^2 A$.

1.7. Derive a formula similar to Equation 1.22 for $C\sin(2\pi v t + \varphi)$.

1.8. One way to appreciate the identity $\exp(i\theta) = \cos\theta + i\sin\theta$ is via the Taylor series. Derive the Taylor series for $\exp(i\theta)$, separate the real and imaginary parts, and show that these are equal to the Taylor series for $\cos\theta$ and $\sin\theta$, respectively.

1.9. Show that $|\exp(i\theta)| = 1$ using the Euler equations.

1.10. Establish the geometric series identity (Equation 1.27). To prove by induction, show that the identity is correct for $N = 1$ and $N = 2$. Then, assuming it is correct for $N = K - 1$ (the inductive hypothesis), prove that it is correct for $N = K$.

1.11. Using L'Hospital's rule, evaluate the following limits:

(a) $\dfrac{1 - e^{2x}}{x}$, as $x \to 0$

(b) $\dfrac{1 - \cos \pi x}{\pi x}$, as $x \to 0$.

1.12. Calculate the following integrals, by parts:

(a) $\displaystyle\int xe^{ax}\,dx$

(b) $\displaystyle\int x^2 \cos bx\,dx$.

1.13. Calculate the following improper integrals:

(a) $\displaystyle\int_0^1 \dfrac{dx}{x^{1/2}}$

(b) $\displaystyle\int_{-\infty}^{\infty} \dfrac{dx}{1 + x^2}$.

1.14. Calculate the Cauchy principal values of the following integrals:

(a) $\displaystyle\int_{-1}^{1} \dfrac{dx}{x^2}$

(b) $\displaystyle\int_{-1}^{1} \dfrac{dx}{x^3}$

(c) $\displaystyle\int_{-1}^{2} \dfrac{dx}{x^3}$.

CHAPTER 2

VECTOR SPACES

Quantities in the physical world are modeled mathematically by functions of one or more independent variables. These functions are often called *signals*, particularly when they result from measurements or are purposely designed to convey information. An audio signal is a function of one independent variable, time. A photographic image is a function of two spatial variables. A video image is a function of one temporal and two spatial variables. The sound wave emanating from a loudspeaker is a function of time and three spatial variables.

For mathematical analysis, it is convenient to group signals into classes with common properties, such as: "all speech signals," "all 256×256 pixel images," "all signals with amplitude less than one volt," "all continuous functions". This chapter introduces several important signal classes and the mathematical structures that model them. With the right set of rules, these signals, be they finite-dimensional vectors, infinite sequences, or functions, can be collected into families that display behavior strikingly similar to physical vectors. These special families are called *linear spaces*, or *vector spaces*.

The properties of physical vectors are helpfully understood in geometric terms. The magnitude of a vector is expressed by its norm. Two vectors representing, say, force or velocity can be added to produce a new vector that combines the effects of the individual vectors. The dot product of two vectors expresses their degree of interaction or dependence. Vectors whose dot product is zero are orthogonal; they are noninteracting or represent independent actions. The dot product of a vector with a unit coordinate vector is the projection of the vector along that coordinate direction. A vector is decomposed into the sum of noninteracting components by projecting it along mutually orthogonal unit vectors.

Within the general framework introduced in this chapter, signals of various kinds are seen to act like vectors. They can be added and subtracted. Norms are defined, so that the closeness of one signal to another, or the degree to which one signal approximates another, can be quantified. The dot product is generalized, and with it some functions are seen to be orthogonal; an arbitrary function can be decomposed into a sum of simple orthogonal functions. Consequently, your geometric intuition about physical vectors will carry over to studying signals and, in particular, understanding and using Fourier analysis.

Fourier Transforms: Principles and Applications, First Edition. Eric W. Hansen.
© 2014 John Wiley & Sons, Inc. Published 2014 by John Wiley & Sons, Inc.

2.1 SIGNALS AND VECTOR SPACES

Signals may be classified in several ways, some of which are more powerful mathematically than others. One important classification is based on the independent variable, which for our purposes here will be taken to be time, but could be space or something else. If the domain of a signal is the set of real numbers $\mathbb{R}$ or a real interval $(a, b) \in \mathbb{R}$, it is called a *continuous-time* signal. If the domain is a discrete set of values $\{t_1, t_2, \ldots\}$, then f is a *discrete-time* signal. When the time values are integers, for example, $t_k = k$, the discrete-time signal f is also called a *sequence*. The term *signal* will be used interchangeably with *sequence* and *function* as the context permits. In this book, the continuous-time signal f evaluated at time $t \in \mathbb{R}$ is denoted $f(t)$, as usual. The nth element of the discrete-time signal f will be denoted $f[n]$ or f_n, and the entire sequence will be written $f = (f[1], f[2], \ldots) = (f[n])_{n=1}^{\infty}$, or $(f_n)_{n=1}^{\infty}$. We will also use sequences whose index sets are the nonnegative integers, $n = 0, 1, 2, \ldots$, and all integers, $n = \ldots, -2, -1, 0, 1, 2, \ldots$. Continuous- and discrete-time signals are drawn in the manner shown in Figure 2.1.

The most useful classifications of signals are those that possess certain algebraic properties, for example, that if signals u and v belong to a set of signals V, then their sum $u + v$, defined appropriately, also belongs to V. The idea of a *vector space* (or *linear space*) generalizes the vector ideas familiar from geometry and physics to objects that behave algebraically like physical vectors. Although we will call the elements of a vector space *points* or *vectors*, they may, in fact, not be vectors in the physical sense but any mathematical objects, such as functions, for which the addition and multiplication operations can be defined. In this book, the vectors will be real or complex valued, the scalars will correspondingly be real or complex numbers, and the space itself will be called a *real vector space* or *complex vector space*, respectively.

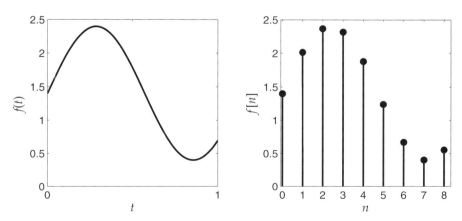

FIGURE 2.1 A continuous-time signal (*left*), and a discrete-time signal, or sequence (*right*).

The following definition is suitable to our purposes.

Definition 2.1. A *vector space* V is a nonempty set with two operations, called vector addition and scalar multiplication. For elements $u, v, w \in V$ and scalars[1] a, b, vector addition $(u + v)$ and scalar multiplication (au) have the properties:

(a) Closure: $u + v \in V$, $au \in V$
(b) Commutative: $u + v = v + u$
(c) Associative: $(u + v) + w = u + (v + w)$, $(ab)u = a(bu)$ (Note that ab is the product of two scalars, while bu is the product of a scalar and a vector.)
(d) Distributive: $a(u + v) = au + av$, $(a + b)u = au + bu$ (Note that $a + b$ is the sum of two scalars, while $u + v$ is the sum of two vectors.)
(e) Additive identity element: There exists an element $0 \in V$ such that $0 + u = u$ for every $u \in V$.
(f) Additive inverse elements: For each $u \in V$ there exists an element $w \in V$ such that $u + w = 0$. (It may conveniently be written $w = -u$.)
(g) Multiplicative identity: There is a scalar 1 with the property $1u = u$.

Example 2.1 (Examples of real and complex vector spaces).

(a) The real numbers $\mathbb{R}$ and the complex numbers $\mathbb{C}$. (Here, the vectors and scalars are one and the same.)
(b) The set of all n-tuples[2] $\{(x_1, x_2, \ldots, x_n) \mid x_k \in \mathbb{R}\}$, known as $\mathbb{R}^n$, with addition and multiplication defined in the usual way: for $x, y \in \mathbb{R}^n$,

$$x + y = (x_1 + y_1, x_2 + y_2, \ldots, x_n + y_n)$$
$$ax = (ax_1, ax_2, \ldots, ax_n), a \in \mathbb{R}.$$

These n-tuples, or n-dimensional vectors, arise in numerous practical applications. N observations of an experimental quantity or, equivalently, N values of a discrete-time signal, constitute a vector in an N-dimensional space. An image, which is an $N \times N$ array of picture elements, or pixels, may be regarded as a single vector in an N^2-dimensional space by stringing the rows of the image together.

(c) The set of all infinite sequences $(x_n)_{n=1}^{\infty}$ whose sums converge absolutely:

$$\sum_{n=1}^{\infty} |x_n| < \infty.$$

This vector space has the name ℓ^1 (read *ell one*).

---

[1] In this book, the elements of a vector space are either real or complex valued, and the scalars are either real or complex numbers, respectively.

[2] We will usually dispense with boldfaced notation for vectors, in keeping with the more general definition above. Whether x_k refers to the kth component of a vector x or the kth element of a set will be clear from context.

(d) The set of continuous functions $f : [0, 1] \to \mathbb{R}$, denoted $C[0, 1]$, with addition and scalar multiplication defined,

$$(f + g)(x) = f(x) + g(x)$$
$$(af)(x) = af(x).$$

Because the sum of two continuous functions is continuous, the closure property (c) is satisfied. That the other properties hold is not hard to show. ∎

2.2 FINITE-DIMENSIONAL VECTOR SPACES

A vector space is *finite dimensional* if the vectors are n-tuples $(v_1, v_2, \ldots, v_n), n < \infty$. A vector space is *infinite dimensional* if it is not finite dimensional. The vector space $\mathbb{R}^n$ is finite dimensional, while ℓ^1 and $C[0, 1]$ are infinite dimensional. In this section several important vector space principles are developed for finite-dimensional spaces. These will be familiar from geometry in two and three dimensions; what is new is just the extension to complex numbers and higher dimensions. Following this, in Section 2.3 we will see how the same geometric intuition extends naturally and powerfully to infinite-dimensional spaces. Finite-dimensional vector spaces are the foundation for Chapter 3, and Chapters 4 and 5 are based on infinite-dimensional spaces.

2.2.1 Norms and Metrics

The minimum requirements for a vector space are laid out in Definition 2.1. In addition to these, it will be useful to have a definition of length for the elements of a vector space. The usual definition of length for a vector $v = (v_1, v_2)$ in a plane is $\|v\| = \sqrt{v_1^2 + v_2^2}$. It has four properties: (a) it is a nonnegative real number; (b) it is zero only for the zero vector $v = (0, 0)$; (c) multiplying a vector by a scalar c multiplies the length by $|c|$; and (d) for two vectors u and v, $\|u + v\| \leq \|u\| + \|v\|$ (the triangle inequality). Properties (a) through (c) are obvious. The triangle inequality makes intuitive sense (draw a picture, or see Figure 1.1); to verify it mathematically requires some algebra:

$$\|u + v\| = \sqrt{(u_1 + v_1)^2 + (u_2 + v_2)^2} \stackrel{?}{\leq} \|u\| + \|v\| = \sqrt{u_1^2 + u_2^2} + \sqrt{v_1^2 + v_2^2}$$

Square both sides and collect terms,

$$u_1^2 + 2u_1v_1 + v_1^2 + u_2^2 + 2u_2v_2 + v_2^2 \stackrel{?}{\leq} u_1^2 + u_2^2 + v_1^2 + v_2^2 + 2\sqrt{(u_1^2 + u_2^2)(v_1^2 + v_2^2)}$$

$$u_1v_1 + u_2v_2 \stackrel{?}{\leq} \sqrt{(u_1^2 + u_2^2)(v_1^2 + v_2^2)}.$$

Square both sides again and collect terms,

$$u_1^2v_1^2 + u_2^2v_2^2 + 2u_1u_2v_1v_2 \stackrel{?}{\leq} u_1^2v_1^2 + u_2^2v_2^2 + u_1^2v_2^2 + u_2^2v_1^2$$

$$2u_1u_2v_1v_2 \stackrel{?}{\leq} u_1^2v_2^2 + u_2^2v_1^2.$$

Collect terms on the right hand side,

$$(u_1 v_2)^2 - 2(u_1 v_2)(u_2 v_1) + (u_2 v_1)^2 \stackrel{?}{\geq} 0,$$

then factor,

$$(u_1 v_2 - u_2 v_1)^2 \geq 0.$$

This is certainly true for all u and v.

The formula $\|v\| = \sqrt{v_1^2 + v_2^2}$ is not the only way to define the length of a vector, nor is the idea of length confined to real-valued vectors. As we encounter vectors of more general type—real or complex, finite- or infinite-dimensional, or even functions of one or more variables—we will want to have an idea of length. It turns out that the way to do this is simply to invoke the same four properties listed above. We will introduce it here in the context of finite-dimensional spaces, then revisit it when we get to infinite-dimensional spaces.

Definition 2.2. A *norm* $\|\cdot\|$ on a vector space V is a mapping $\|\cdot\| : V \to \mathbb{R}$, which satisfies the following conditions for vectors $v, w \in V$ and scalars c:

(a) Nonnegativity: $\|v\| \geq 0$
(b) Nondegeneracy: $\|v\| = 0$ if and only if $v = 0$
(c) Scaling: $\|cv\| = |c| \|v\|$
(d) Triangle inequality: $\|v + w\| \leq \|v\| + \|w\|$

A vector space on which a norm is defined is called a *normed vector space*, or just *normed space*.

Example 2.2. Here are some norms for finite-dimensional vector spaces.

(a) On the real line, the vectors are the real numbers and the norm is identical with the absolute value, $\|x\| = |x|$.
(b) In the complex plane, the vectors are the complex numbers and the norm is identical with the complex magnitude, $\|z\| = |z| = \sqrt{zz^*}$.
(c) For a vector $v = (v_1, v_2, \ldots, v_n) \in \mathbb{R}^n$, the familiar *euclidean norm* is

$$\|v\|_2 = \left(\sum_{k=1}^{n} v_k^2 \right)^{1/2}, \tag{2.1}$$

and $\mathbb{R}^n$ with this norm is called *euclidean space*. We can also define the following noneuclidean norms:

$$\|v\|_1 = \sum_{k=1}^{n} |v_k| \tag{2.2}$$

$$\|v\|_\infty = \max_{k=1,2,\ldots,n} |v_k|. \tag{2.3}$$

(For simplicity, we will henceforth write $\max_k$ instead of $\max_{k=1,2,\ldots,n}$ when the range of the index is understood.) Properties (a)–(c) are straightforward to verify for each of these. Verification of the triangle inequality is more difficult. We shall prove it here for the euclidean norm, and leave the other two to the problems.

The straightforward algebraic approach used to verify the triangle inequality in two dimensions is obviously too complicated to generalize to arbitrary n. Instead, we use the *Schwarz inequality*, which shall be explained later in a more general form. For two points in $\mathbb{R}^n$, $u = (u_1, u_2, \ldots, u_n)$ and $v = (v_1, v_2, \ldots, v_n)$, the Schwarz inequality says

$$\sum_{i=1}^{n} |u_i v_i| \leq \left(\sum_{i=1}^{n} u_i^2 \right)^{1/2} \left(\sum_{i=1}^{n} v_i^2 \right)^{1/2} = \|u\|_2 \|v\|_2.$$

So consider $u, v \in \mathbb{R}^n$, and write

$$\|u + v\|^2 = \sum_{k=1}^{n} (u_k + v_k)^2 = \sum_{k=1}^{n} u_k^2 + 2 u_k v_k + v_k^2$$

$$= \sum_{k=1}^{n} u_k^2 + 2 \sum_{k=1}^{n} u_k v_k + \sum_{k=1}^{n} v_k^2$$

$$= \|u\|^2 + 2 \sum_{k=1}^{n} u_k v_k + \|v\|^2.$$

Taking absolute values of both sides, and using the fact that the triangle inequality holds for real numbers,[3]

$$\|u + v\|^2 = \left| \|u\|^2 + 2 \sum_{k=1}^{n} u_k v_k + \|v\|^2 \right| \leq \|u\|^2 + 2 \left| \sum_{k=1}^{n} u_k v_k \right| + \|v\|^2$$

$$\leq \|u\|^2 + 2 \sum_{k=1}^{n} |u_k v_k| + \|v\|^2.$$

By the Schwarz inequality, the middle term is bounded by $2 \|u\| \|v\|$, so

$$\|u + v\|^2 \leq \|u\|^2 + 2\|u\|\|v\| + \|v\|^2 = (\|u\| + \|v\|)^2.$$

Taking the square root of both sides completes the proof. ∎

Later we will consider norms for infinite-dimensional vector spaces, such as the generalization of the absolute value norm for infinite sequences,

$$\|v\| = \sum_{k=1}^{\infty} |v_k|,$$

and for functions, for example, defined on the interval $(0, 1)$,

$$\|f\| = \int_0^1 |f(x)|\, dx.$$

---

[3] That is, $|x + y| \leq |x| + |y|$ for real x and y; see Rosenlicht (1968, p. 34).

When a vector in three dimensions represents the displacement of a point in space from the origin, the norm of the vector gives the distance of the point from the origin. If two points in space are displaced from the origin by vectors u and v, the norm of the difference of the displacement vectors, $\|u - v\|$, is the distance $d(u, v)$ between the two points. We say that the norm *induces a metric* (provides a measure of distance) on the space. The metric d has properties very similar to the norm.

Definition 2.3. A *metric* on a space V is a function $d : V \times V \to \mathbb{R}$ having the following properties for $u, v, w \in V$:

(a) Nonnegativity: $d(u, v) \geq 0$
(b) Nondegeneracy: $d(u, v) = 0$ if and only if $u = v$
(c) Symmetry: $d(u, v) = d(v, u)$
(d) Triangle inequality: $d(u, w) \leq d(u, v) + d(v, w)$.

If a space V has a metric, then it is called a *metric space*.

The formula $d(u, v) = \|u - v\|$ is a valid metric, no matter how the norm is defined. Properties (a)–(c) are obvious. To show that it satisfies the triangle inequality, write $u - w = (u - v) + (v - w)$. This is the sum of two vectors, $u - v$ and $v - w$, and because norms obey the triangle inequality, we have

$$\|u - w\| \leq \|u - v\| + \|v - w\|,$$

that is,

$$d(u, w) \leq d(u, v) + d(v, w).$$

Although we have presented normed spaces first, a metric space is actually more general (less restricted) than a normed space. A metric space is specified solely by a distance measure. It need not have the algebraic properties of a vector space (addition of points, multiplication of points by scalars). That is, all normed spaces are metric spaces, but not vice-versa. All of the vector spaces we will routinely use in this book have norms; hence, they are also metric spaces with $d(u, v) = \|u - v\|$.

Example 2.3 (Some metrics). Here are some metrics for finite-dimensional spaces. Corresponding metrics for infinite-dimensional spaces will be defined later.

(a) *Absolute value metric.* On the real line, the customary distance measure is the absolute value, $d(x, y) = |x - y|$. The first three metric properties are easy to verify. For the triangle inequality, write $x - z = (x - y) + (y - z)$, then

$$d(x, z) = |x - z| = |(x - y) + (y - z)|$$
$$\leq |x - y| + |y - z| = d(x, y) + d(y, z).$$

(b) *Euclidean metric.* The euclidean norm induces a metric on $\mathbb{R}^n$ or $\mathbb{C}^n$,

$$d_2(u, v) = \left[\sum_{k=1}^{n} |u_k - v_k|^2\right]^{1/2} = \|u - v\|_2. \qquad (2.4)$$

Again, this is just the generalization of the Pythagorean theorem to n dimensions. In our familiar $\mathbb{R}^3$ world, it is the distance "as the crow flies" between points u and v.

(c) *Taxicab metric.* The absolute value metric on $\mathbb{R}$ formally extends to $\mathbb{R}^n$ and $\mathbb{C}^n$ like this:

$$d_1(u, v) = \sum_{k=1}^{n} |u_k - v_k| = \|u - v\|_1. \qquad (2.5)$$

This is sometimes called the *taxicab metric* or the *Manhattan metric*. Suppose you are in a city with a rectangular grid of streets and you want to travel from point u to point v. If you can fly directly from u to v, the distance you travel is given by the euclidean metric. However, if you travel on the streets, you use the taxicab metric to compute the distance.

(d) *Maximum metric.* For two points $u, v \in \mathbb{R}^n$ or $\mathbb{C}^n$, the *maximum metric* is

$$d_\infty(u, v) = \max_k |u_k - v_k| = \|u - v\|_\infty. \qquad (2.6)$$

The maximum metric is a worst-case measure. Suppose you are traveling between two cities u and v with coordinates (u_1, u_2) and (v_1, v_2) on opposite corners of a rectangle. Your highway goes along one edge of the rectangle from $u = (u_1, u_2)$ to the intermediate point $w = (v_1, u_2)$, then turns 90° and continues to $v = (v_1, v_2)$. Further suppose that your gas tank does not hold enough fuel to cover the entire distance from u to w and then w to v. However, there is a filling station at w. If your tank can hold enough fuel for the longer of the two distances you can refill at w and complete the trip. The critical distance is the maximum metric, $\max\{|u_1 - v_1|, |u_2 - v_2|\}$. ■

A more general norm is defined

$$\|u\|_p = \left(\sum_{k=1}^{n} |u_k|^p\right)^{1/p}, \quad p \geq 1 \qquad (2.7)$$

with an associated metric

$$d_p(u, v) = \|u - v\|_p = \left(\sum_{k=1}^{n} |u_k - v_k|^p\right)^{1/p}, \quad p \geq 1. \qquad (2.8)$$

The taxicab and euclidean metrics are particular cases ($p = 1$ and 2, respectively). It can be shown that these general definitions have the necessary properties. The maximum metric can also be shown to be a limiting case of this general metric,

$$d_\infty(u, v) = \lim_{p \to \infty} d_p(u, v). \qquad (2.9)$$

Example 2.4 (Different metrics give different distances). Let $x = (0, 1)$ and $y = (1, -2)$. Then we have

$$d_1(x, y) = |0 - 1| + |1 + 2| = 4$$
$$d_2(x, y) = \sqrt{|0 - 1|^2 + |1 + 2|^2} = \sqrt{10}$$
$$d_\infty(x, y) = \max\{|0 - 1|, |1 + 2|\} = 3$$

Observe that $d_\infty < d_2 < d_1$. This ordering holds in general for $\mathbb{R}^n$ (see the problems). ∎

Having a metric permits you to identify regions in a space whose points are close to one another. On the real line a *neighborhood* of a point x is defined to be an open interval $(x - r, x + r)$ for some $r > 0$. A point y is within this neighborhood if $d(x, y) < r$. With the euclidean metric, a neighborhood of x on the plane is an open disk of radius r, the set $\{y \mid \|x - y\|_2 < r\}$. In $\mathbb{R}^3$ with the euclidean metric, a neighborhood of a point is an open sphere. Depending on the choice of metric, neighborhoods can have different shapes.

Finally, suppose V is a metric space and U is a subset of V. We say that U is *bounded* if $d(x, y) < \infty$ for every $x, y \in U$. The greatest distance between two points in U, $\sup_{x,y \in U} d(x, y)$ is called the *diameter* of U.

2.2.2 Inner Products

Two vectors in the plane, $u = (u_1, u_2)$ and $v = (v_1, v_2)$, have a dot product defined by

$$u \cdot v = u_1 v_1 + u_2 v_2 = \|u\| \|v\| \cos \theta,$$

where θ is the angle between u and v. The dot product has algebraic properties that are analogous to those of multiplication: (a) commutative, $u \cdot v = v \cdot u$; (b) associative with a scalar, $(cu) \cdot v = c(u \cdot v)$; (c) distributive with addition, $(u + v) \cdot w = u \cdot w + v \cdot w$; and (d) "squaring," $u \cdot u = \|u\|^2 > 0$, unless $u = 0$. The dot product has physical significance, for example, the calculation of work, and is also the vehicle for decomposing vectors into orthogonal components (see Section 1.2).

If the elements of u and v are complex-valued, we still want $u \cdot u = \|u\|^2$; to achieve this we redefine the dot product as

$$u \cdot v = u_1 v_1^* + u_2 v_2^*.$$

The commutativity property changes to $u \cdot v = (v \cdot u)^*$, but the other properties stay the same.[4] The complex definition includes real vectors as a special case.

Just as the familiar geometric distance between two points (metric) and length of a vector (norm) may be extended to more general vector spaces, we may extend the dot product. Once again, the approach is simply to use the properties of the dot product as the definition for the more general operation, which is called the *inner product*.

---

[4]In quantum mechanics, the definition is $u \cdot v = u_1^* v_1 + u_2^* v_2$, which is the complex conjugate of the definition used in this text. MATLAB also follows the quantum mechanics convention in its dot() function.

2.2 FINITE-DIMENSIONAL VECTOR SPACES

Definition 2.4. An *inner product* $\langle \cdot, \cdot \rangle$ on a vector space V is a mapping $\langle \cdot, \cdot \rangle : V \times V \to \mathbb{C}$ such that, for $u, v, w \in V$ and c a constant,

(a) $\langle v, w \rangle = \langle w, v \rangle^*$
(b) $\langle cv, w \rangle = c \langle v, w \rangle$
(c) $\langle u + v, w \rangle = \langle u, w \rangle + \langle v, w \rangle$
(d) $\langle v, v \rangle > 0$, when $v \neq 0$.

A vector space with an inner product is called an *inner product space*.

Some straightforward consequences of this definition are

$$\langle v, cw \rangle = c^* \langle v, w \rangle \tag{2.10}$$
$$\langle u, v + w \rangle = \langle u, v \rangle + \langle u, w \rangle \tag{2.11}$$
$$\langle v, 0 \rangle = \langle 0, v \rangle = 0 \tag{2.12}$$
$$\text{If } \langle u, v \rangle = \langle u, w \rangle \text{ for all } u \in V, \text{ then } v = w. \tag{2.13}$$

Moreover, the quantity $\sqrt{\langle u, u \rangle}$ can be shown to have all the properties of a norm. Thus, an inner product space is a normed space, and the norm of the space is defined $\|u\| = \sqrt{\langle u, u \rangle}$. The proofs of these are left as problems.

Example 2.5 (Inner products in $\mathbb{R}^n$ and $\mathbb{C}^n$). For vectors in our familiar euclidean space $\mathbb{R}^3$, the dot product is an inner product,

$$\langle u, v \rangle = u_1 v_1 + u_2 v_2 + u_3 v_3,$$

which readily generalizes to $\mathbb{R}^n$,

$$\langle u, v \rangle = \sum_{k=1}^{n} u_k v_k. \tag{2.14}$$

For complex vectors $u, v \in \mathbb{C}^n$,

$$\langle u, v \rangle = \sum_{k=1}^{n} u_k v_k^* \tag{2.15}$$

is an inner product; when u and v are real, this is the same as Equation 2.14. We will examine each of the four properties in turn. Let $u, v, w \in \mathbb{C}^n$ and $c \in \mathbb{C}$. Then,

(a) $\langle u, v \rangle = \sum_{k=1}^{n} u_k v_k^* = \left[\sum_{k=1}^{n} u_k^* v_k \right]^* = \left[\sum_{k=1}^{n} v_k u_k^* \right]^* = \langle v, u \rangle^*$

(b) $\langle cu, v \rangle = \sum_{k=1}^{n} c u_k v_k^* = c \sum_{k=1}^{n} u_k v_k^* = c \langle u, v \rangle$

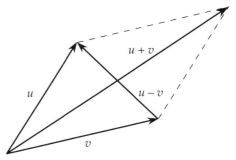

FIGURE 2.2 Illustrating the parallelogram law for vectors in the plane. The diagonals of the parallelogram are the vectors $u - v$ and $u + v$. The parallelogram law says that the sum of the squared lengths of the diagonals is the sum of the squared lengths of the four sides, $\|u+v\|^2 + \|u-v\|^2 = 2\|u\|^2 + 2\|v\|^2$.

(c)
$$\langle u+v, w \rangle = \sum_{k=1}^{n}(u_k + v_k)w_k^* = \sum_{k=1}^{n} u_k w_k^* + v_k w_k^*$$
$$= \sum_{k=1}^{n} u_k w_k^* + \sum_{k=1}^{n} v_k w_k^* = \langle u, w \rangle + \langle v, w \rangle.$$

(d) We have
$$\langle u, u \rangle = \sum_{k=1}^{n} u_k u_i^* = \sum_{k=1}^{n} |u_k|^2 = \|u\|_2^2$$

and by the definition of the norm, $\langle u, u \rangle = \|u\|^2 \geq 0$, with $\langle u, u \rangle = 0$ only if $u = 0$. ∎

There is a cluster of interesting results connecting the inner product and the norm. These include the *parallelogram law* (Figure 2.2):

$$\|u+v\|^2 + \|u-v\|^2 = 2\|u\|^2 + 2\|v\|^2, \tag{2.16}$$

and the *polarization identity*:

$$4\langle u, v \rangle = \|u+v\|^2 - \|u-v\|^2 + i\|u+iv\|^2 - i\|u-iv\|^2. \tag{2.17}$$

These are considered in more detail in the problems. The most important relationship is the *Cauchy–Schwarz inequality*. For vectors in a plane, the dot product $u \cdot v = \|u\|\|v\|\cos\theta$ is bounded in magnitude by $\|u\|\|v\|$, because $|\cos\theta| \leq 1$. The Cauchy–Schwarz inequality generalizes this bound to any inner product space.

Theorem 2.1 (Cauchy–Schwarz inequality). Let V be an inner product space, and $u, v \in V$. Then,

$$|\langle u, v \rangle| \leq \|u\|\|v\| \tag{2.18}$$

with equality only if $u = cv$, c constant.

Proof: If $u = cv$, then we have directly

$$|\langle u, v \rangle| = |\langle cv, v \rangle| = |c|\|v\|^2 = \|cv\|\|v\|$$
$$= \|u\|\|v\|.$$

On the other hand, if $u \neq cv$, then $\langle u - cv, u - cv \rangle = \|u - cv\|^2 > 0$, and

$$\langle u - cv, u - cv \rangle = \langle u, u \rangle - \langle u, cv \rangle - \langle cv, u \rangle + \langle cv, cv \rangle$$
$$= \|u\|^2 - c^* \langle u, v \rangle - c \langle u, v \rangle^* + |c|^2 \|v\|^2$$
$$= \|u\|^2 - 2\mathcal{R}e\left(c^* \langle u, v \rangle\right) + |c|^2 \|v\|^2 > 0.$$

This relationship, $\|u\|^2 - 2\mathcal{R}e\left(c^* \langle u, v \rangle\right) + |c|^2 \|v\|^2 > 0$, holds for all c. In particular, it holds for

$$c = t \frac{\langle u, v \rangle^*}{|\langle u, v \rangle|}, t \geq 0$$

(note $|c| = t$), which gives

$$\|u\|^2 - 2t|\langle u, v \rangle| + t^2 \|v\|^2 > 0.$$

The left-hand side is a quadratic expression in t, of the form $\alpha t^2 - 2\beta t + \gamma$, with $\alpha, \gamma > 0$. Its graph is a convex-upward parabola, whose minimum value is $\gamma - \beta^2/\alpha = \|u\|^2 - |\langle u, v \rangle|^2 / \|v\|^2$. If this minimum is positive, then the left hand side is positive and the inequality is true; hence, make

$$\|u\|^2 > \frac{|\langle u, v \rangle|^2}{\|v\|^2}$$
$$\Rightarrow |\langle u, v \rangle| < \|u\|\|v\|. \blacksquare$$

This proof made use only of the general properties of inner product and norm. It will also hold for infinite-dimensional vectors and functions, as long as we can define meaningful inner products and norms. Specialized to vectors in euclidean space $\mathbb{R}^n$, the Cauchy–Schwarz inequality is the same as the Schwarz inequality we used earlier to show that the euclidean metric (Equation 2.1) satisfies the triangle inequality. That proof, in turn, can be generalized (see the problems) to show that in any inner product space, not just $\mathbb{R}^n$ or $\mathbb{C}^n$, the norm $\|u\| = \langle u, u \rangle^{1/2}$ satisfies the triangle inequality:

$$\|u + v\| \leq \|u\| + \|v\|. \tag{2.19}$$

These results anticipate the move to more general vector spaces in the next section.

Example 2.6 (Correlation receivers). Practical applications of inner products abound. Here is one. In radar, a target object is located by transmitting bursts of electromagnetic radiation from an antenna, and detecting the waves that are reflected from the target back to the antenna. The range to the target is computed from the time delay between transmission and reception, divided by the speed of wave propagation (approximately the speed of light). The reflected wave is much weaker than the transmitted wave; its signal strength decreases with the fourth power of the distance to the target.

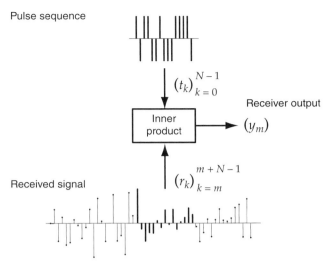

FIGURE 2.3 Correlation receiver. As the received signal sequence (r_k) flows through, an N-length subsequence is isolated and the inner product with the pulse sequence (t_k) is computed. A peak in the output sequence (y_m) indicates that an instance of the input sequence has been detected in the received signal.

The energy in a radar pulse v is proportional to $\|v\|_2^2$, while the peak power is proportional to $\|v\|_\infty^2$. In early radars, the transmitted pulse was a single intense impulse, and the effective range of the system was limited by the peak power the transmitter could produce. Modern radars use pulse sequences rather than single impulses. A sequence of smaller pulses (v_n) can have high total energy $\sum_n |v_n|^2$ at a lower peak value $\max_n |v_n|$. The following is a much-simplified description of the signal processing that takes place in such a system.[5]

Detection of instances of the pulse sequence in the received signal sequence is based on an operation called *cross-correlation*, defined

$$y_m = \sum_{k=0}^{N-1} t_k r_{k+m},$$

where $(t_k)_{k=0}^{N-1}$ is the basic pulse sequence and (r_k) is the received signal sequence consisting of pulse echoes from the target, weakened by propagation over distance and obscured by noise. As the received signal sequence flows through the receiver in time, successive N-length subsequences $(r_k)_{k=m}^{m+N-1} = (r_{k+m})_{k=0}^{N-1}$ are pulled out and the inner product with the original sequence $(t_k)_{k=0}^{N-1}$ is computed (Figure 2.3). This inner product is the cross-correlation of (t_k) and (r_k). The system is called a *correlation receiver* or *matched filter* (because the receiver is *matched* to a particular pulse sequence).

---

[5]In a real pulse-coded radar system, the pulse sequence modifies, or *modulates*, the frequency or phase of a sinusoidal wave. Moreover, because the target is frequently moving, the effect of Doppler shift on this wave must be taken into account. For a detailed discussion, see Levanon and Mozeson (2004).

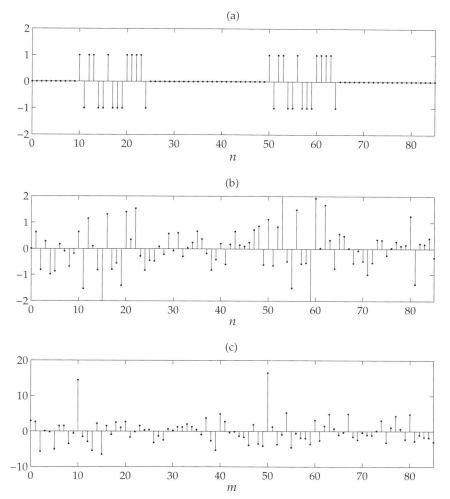

FIGURE 2.4 Inputs and outputs of the correlation receiver. (*a*) A transmitted signal consisting of two pulse sequences at $n = 10$ and $n = 50$. (*b*) Received signal. The pulse echoes from the target are obscured by noise. (*c*) Cross-correlation of pulse sequence with received signal. The peaks at $m = 10$ and $m = 50$ indicate pulse sequence detection.

According to the Schwarz inequality, the inner product is maximized when the two vectors are proportional to each other. Thus, we expect to see a peak in the output of the receiver at times m when the subsequence $(r_k)_{k=m}^{m+N-1}$ is proportional to the pulse sequence $(t_k)_{k=0}^{N-1}$. Because of noise, exact proportionality is not possible, but with proper pulse sequence design, a high degree of discrimination can be achieved (Figure 2.4). ∎

Finally, we extend the geometric idea of orthogonality. Perpendicular vectors are easily visualized, at least in the real spaces $\mathbb{R}^2$ and $\mathbb{R}^3$. Mathematically, they are orthogonal if and only if their dot product is zero. The generalization to arbitrary inner product spaces is straightforward, even if we cannot so easily picture it.

Definition 2.5 (Orthogonality). Let V be an inner product space, and $u, v \in V$. The vectors u and v are orthogonal if and only if their inner product $\langle u, v \rangle$ is zero.

Theorem 2.2 (Pythagorean theorem). Let V be an inner product space and $u, v \in V$. If the vectors u and v are orthogonal, then

$$\|u + v\|^2 = \|u\|^2 + \|v\|^2. \tag{2.20}$$

If u and v are real and $\|u + v\|^2 = \|u\|^2 + \|v\|^2$, then they are orthogonal.

Proof is left to the problems.

2.2.3 Orthogonal Expansion and Approximation

From prior experience with vectors in physics, you know that it is often useful to express an arbitrary vector, say in a plane or in three dimensions, as a sum of two or three orthogonal components (e.g., Figure 1.3). We will now generalize this idea to vectors in spaces of arbitrary (but finite) dimension, and consider the important problem of approximating a vector by one with fewer components.

Basis vectors

We begin with some definitions, which may be familiar to you from linear algebra.

Definition 2.6. Let V be a vector space, $\{v_k\}_{k=1}^n \subset V$ be a set of vectors in V, and $\{c_k\}_{k=1}^n$ be scalars. Then

(a) The vectors $\{v_k\}$ are *linearly independent* if the only coefficients $\{c_k\}$ that result in $c_1 v_1 + c_2 v_2 + \cdots + c_n v_n = 0$ are $c_1 = c_2 = \cdots = c_n = 0$. Otherwise, the vectors are linearly *dependent*, and one of the v_k can be written in terms of the others, for example,

$$v_1 = -\frac{c_2}{c_1} v_2 - \cdots - \frac{c_n}{c_1} v_n.$$

(Figure 2.5).

(b) The set of all linear combinations $c_1 v_1 + c_2 v_2 + \cdots + c_n v_n$ is called the *linear span* of the set $\{v_k\}$.

(c) If the linear span of $\{v_k\}$ is identical to V, then the vectors $\{v_k\}$ are called a *basis* for V. Any vector in V is expressible as a linear combination of basis vectors. The $\{v_k\}$ are said to *span* V. The number of basis vectors, n, is the *dimension* of V.

(d) A set of n mutually orthogonal unit vectors is called an *orthonormal set* of vectors, which we will denote $\{e_k\}_{k=1}^n$. If this orthonormal set spans V, it is an *orthonormal basis* for V.

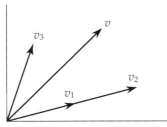

FIGURE 2.5 Vectors v_1 and v_2 are not linearly independent. The vector v cannot be expressed as a linear combination of v_1 and v_2. However, v_3 is independent of v_1 and v_2, and v can be written as a linear combination of v_1 and v_3 or v_2 and v_3.

Any vector $v \in \mathbb{R}^n$ can be written as a linear combination of n orthonormal basis vectors:

$$v = \sum_{k=1}^{n} c_k e_k.$$

To calculate the coefficients, take the inner product of both sides with each basis vector in turn:

$$\langle v, e_k \rangle = \left\langle \sum_{j=1}^{n} c_j e_j, e_k \right\rangle = \sum_{j=1}^{n} c_j \langle e_j, e_k \rangle = c_k.$$

Therefore, the expansion of v in the basis is

$$v = \sum_{k=1}^{n} \langle v, e_k \rangle e_k. \tag{2.21}$$

The coefficients $c_k = \langle v, e_k \rangle$ represent projections of v along each of the basis vectors.

A vector ϕ is normalized by dividing it by its norm, that is, $e_k = \dfrac{\phi_k}{\|\phi_k\|}$. If a set of vectors $\{\phi_k\}$ is orthogonal but not normalized, then an expansion in terms of the $\{\phi_k\}$ may be shown to be

$$v = \sum_{k=1}^{n} \frac{\langle v, \phi_k \rangle}{\|\phi_k\|^2} \phi_k. \tag{2.22}$$

Example 2.7. Here are some orthogonal bases for finite-dimensional spaces.

(a) The *standard basis* in $\mathbb{R}^n$ consists of vectors e_k with 1 in the kth position and 0 in the other positions:

$$e_1 = (1, 0, 0, \ldots, 0)$$
$$e_2 = (0, 1, 0, \ldots, 0)$$
$$\vdots$$
$$e_n = (0, 0, 0, \ldots, 1).$$

(b) The so-called Haar basis in $\mathbb{R}^4$ is

$$\phi_1 = (1, 1, 1, 1)$$
$$\phi_2 = (1, 1, -1, -1)$$
$$\phi_3 = (1, -1, 0, 0)$$
$$\phi_4 = (0, 0, 1, -1).$$

These vectors are orthogonal but not normalized.

(c) The complex exponential basis in $\mathbb{C}^N$ is the set $\{\phi_k\}_{k=0}^{N-1}$, where

$$\phi_k = \left(\exp\left(\frac{i2\pi kn}{N}\right)\right)_{n=0}^{N-1} = \left(1, e^{i2\pi k/N}, e^{i4\pi k/N}, \ldots, e^{i2(N-1)\pi k/N}\right).$$

By convention, the indexing for this basis is from 0 to $N-1$ rather than from 1 to N. These vectors are orthogonal, but not normalized. The orthogonal projection of a vector $f = (f[0], f[1], \ldots, f[N-1]) \in \mathbb{C}^n$ onto this basis,

$$F[k] = \langle f, \phi_k \rangle = \sum_{n=0}^{N-1} f[n] \exp\left(-\frac{i2\pi kn}{N}\right), \quad k = 0, 1, \ldots, N-1, \quad (2.23)$$

is called the *discrete Fourier transform* (DFT) of f. The DFT is the subject of Chapter 3. ∎

Orthogonal approximation

Deleting one or more terms from the orthogonal expansion (Equation 2.21) results in a lower-dimensional approximation, which turns out to be optimal in a certain sense. Many problems in data analysis and signal processing may be expressed as approximations of a vector by another vector of lower dimension. Here are some examples:

- To display a three-dimensional (3D) structure on a computer screen, the 3D data must be mapped to a 2D array of pixels—assigning each point in $\mathbb{R}^3$ to a point in $\mathbb{R}^2$.
- In linear regression, data points $\{(x_1, y_1), (x_2, y_2), \ldots, (x_n, y_n)\}$ are approximated by a straight line,

$$y_k \approx ax_k + b, \quad k = 1, 2, \ldots, n.$$

The slope a and intercept b parameterize the line. We go from n pairs of values to two coefficients, or from $\mathbb{R}^{2n}$ to $\mathbb{R}^2$.

- In the original form of JPEG image compression, an $N \times N$ pixel picture is divided into 8×8 pixel subimages. Each subimage, considered to be a vector in $\mathbb{R}^{64}$, is expanded into a linear combination of predetermined basis images, $g = \sum_{k=0}^{63} c_k v_k$ (Example 2.12, below). Only the P most important of the coefficients c_k are kept, effectively reducing a 64-dimensional vector to P dimensions.

- In image restoration, a noisy $N \times N$ pixel picture is represented as a vector in N^2-dimensional space. It is decomposed into a sum of orthogonal basis vectors with the goal of identifying and removing the components that contain most of the noise. The denoised image is reconstituted from the remaining components. This process is also called *filtering*.

We begin by defining a *subspace* of a vector space.

Definition 2.7 (Subspace). Let V be a vector space. A subset $U \subset V$ is a *subspace* of V if U is also a vector space under the same definition of addition and multiplication as V.

Example 2.8 (Subspaces). Consider the 3D real space $\mathbb{R}^3$.

(a) The set $U = \{(x, y, 0) \mid x, y \in \mathbb{R}\}$ (the xy plane) is a 2D subspace of $\mathbb{R}^3$. It is easy to check, in particular, that the sum of two vectors in the plane is also in the plane.

(b) The set $U = \{(x, y, 1) \mid x, y \in \mathbb{R}\}$ (a plane passing through $z = 1$) is a subset of $\mathbb{R}^3$ but not a subspace of $\mathbb{R}^3$. For a counterexample, let $u = (0, 1, 1)$ and $v = (1, 0, 1)$; the sum $u + v = (1, 1, 2)$ is not in U. Moreover, a subspace, being a vector space in its own right, must always have an additive identity element; in this case that element is $(0, 0, 0)$, which is not in U.

(c) The set $U = \{(x, ax, 0) \mid a, x \in \mathbb{R}\}$ (a line in the xy plane, passing through the origin with slope a) is a 1D subspace of $\mathbb{R}^3$.

(d) The set $U = \{0\}$ is a trivial (zero-dimensional) subspace of $\mathbb{R}^3$. ∎

Subspaces of an inner product space may readily be constructed. Let V be an n-dimensional inner product space with an orthonormal basis $\{e_k\}_{k=1}^n$. Select m of these basis vectors—without loss of generality, $e_1 \ldots e_m$. The linear span of these vectors is an m-dimensional subspace U. The linear span of the remaining basis vectors, $e_{m+1} \ldots e_n$, is also a subspace, which we will call $U^\perp$. Every vector in $U^\perp$ is orthogonal to every vector in U, because every basis vector of $U^\perp$ is orthogonal to every basis vector of U. $U^\perp$ is called the *orthogonal complement* of U. We say that V is the *direct sum* of U and $U^\perp$, written $V = U \oplus U^\perp$.[6] Every vector in V can be written as the sum of a vector in U and an orthogonal vector in $U^\perp$,

$$v = \sum_{k=1}^{n} a_k e_k = \underbrace{\sum_{k=1}^{m} a_k e_k}_{\in U} + \underbrace{\sum_{k=m+1}^{n} a_k e_k}_{\in U^\perp}.$$

Note that the vectors in U have n elements, but the subspace is m-dimensional; each vector in U is built from only m of the n-element basis vectors.

---

[6]For more about sums and direct sums of subspaces, see Axler (1997, pp. 14–18) and Young (1988, pp. 39–42).

Example 2.9. Returning to the subspaces of $\mathbb{R}^3$ from the previous example,

(a) $U = \{(x, y, 0) \mid x, y \in \mathbb{R}\}$ (the xy plane) is a 2D subspace of $\mathbb{R}^3$. The orthogonal complement is $U^\perp = \{(0, 0, z) \mid z \in \mathbb{R}\}$ (the z-axis).

(b) $U = \{(x, ax, 0) \mid a, x \in \mathbb{R}\}$ (a line in the xy plane, passing through the origin with slope a) is a 1D subspace of $\mathbb{R}^3$. The orthogonal complement is $U^\perp = \{(-ay, y, z) \mid y, z \in \mathbb{R}\}$, a plane perpendicular to U.

(c) $U = \{0\}$ is a zero-dimensional subspace of $\mathbb{R}^3$. The orthogonal complement is $U^\perp = \mathbb{R}^3$. ∎

Now we consider how to approximate a vector in an inner product space V by a vector in a lower-dimensional subspace U. To begin, let v be a vector in the real plane (Figure 2.6). We seek an approximate vector $\hat{v}$ constrained to the x-direction: $\hat{v} = ae_x$. The difference between v and $\hat{v}$ is an error vector, $\tilde{v} = v - \hat{v}$. We seek $\hat{v}$ to minimize the norm of the error, $\|\tilde{v}\|$, that is, to make the error vector as "short" as possible.

From the figure, it appears that the error vector $\tilde{v}$ will be shortest when it is perpendicular to the approximation $\hat{v}$, making the approximation $\hat{v}$ the orthogonal projection of v onto the x-axis. For this simple example, the approximating subspace U is the x-axis, and its orthogonal complement $U^\perp$ is the y-axis; $\hat{v} \in U$ and $\tilde{v} \in U^\perp$.

To formally calculate the best approximation, define $\mathcal{E}^2$ to be the square of the euclidean norm of the error:

$$\mathcal{E}^2 = \|v - \hat{v}\|_2^2 = \langle v - \hat{v}, v - \hat{v} \rangle$$
$$= \|v\|^2 + \|\hat{v}\|^2 - 2\langle v, \hat{v} \rangle.$$

Substitute $\hat{v} = ae_x$,

$$\mathcal{E}^2 = \|v\|^2 + a^2 - 2a\langle v, e_x \rangle$$

and find a, which minimizes $\mathcal{E}^2$:

$$\frac{\partial \mathcal{E}^2}{\partial a} = 2a - 2\langle v, e_x \rangle = 0 \quad \Rightarrow \quad a = \langle v, e_x \rangle,$$

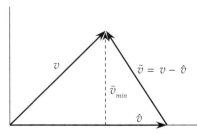

FIGURE 2.6 Approximating a 2D vector v by a 1D vector $\hat{v}$. We seek the $\hat{v}$ that is "closest" to v, in the sense that the norm of the error vector $\tilde{v}$ is minimized. Under the euclidean norm, this occurs when the error $\tilde{v}$ is perpendicular to the approximation $\hat{v}$.

then
$$\hat{v} = \langle v, e_x \rangle e_x \quad \text{and} \quad \mathcal{E}_{\min}^2 = \|v\|^2 - \langle v, e_x \rangle^2.$$

The approximation that minimizes the 2-norm of the error is indeed the orthogonal projection of v onto the basis vector e_x, represented by the inner product $\langle v, e_x \rangle$. The minimized error is given by the Pythagorean formula, which makes sense since v is the hypotenuse of a right triangle and $\hat{v}$ and $\tilde{v}_{\min}$ are the other two sides.

Let us now make a modest extension to three dimensions. If $v \in \mathbb{R}^3$, the minimum 2-norm approximation of v on the x-axis is still $\hat{v} = \langle v, e_x \rangle e_x$, with squared error $\mathcal{E}_{\min}^2 = \|v\|^2 - \langle v, e_x \rangle^2$. Now add a dimension, approximating v by a vector in the xy plane, which we have seen is a 2D subspace of $\mathbb{R}^3$. Again we use

$$\mathcal{E}^2 = \|v\|^2 + \|\hat{v}\|^2 - 2\langle v, \hat{v} \rangle,$$

this time with $\hat{v} = ae_x + be_y$. We have

$$\mathcal{E}^2 = a^2 + b^2 - 2(a\langle v, e_x \rangle + b\langle v, e_y \rangle)$$
$$\frac{\partial \mathcal{E}^2}{\partial a} = 2a - 2\langle v, e_x \rangle = 0 \quad \Rightarrow \quad a = \langle v, e_x \rangle$$
$$\frac{\partial \mathcal{E}^2}{\partial b} = 2b - 2\langle v, e_y \rangle = 0 \quad \Rightarrow \quad b = \langle v, e_y \rangle$$

so
$$\hat{v} = \langle v, e_x \rangle e_x + \langle v, e_y \rangle e_y$$
$$\text{and } \mathcal{E}_{\min}^2 = \|v\|^2 - (\langle v, e_x \rangle^2 + \langle v, e_y \rangle^2).$$

In going from a 1D approximation to a 2D approximation, the coefficient for e_x is unchanged. *The best 2D approximation simply adds a component to the best 1D approximation.* You can think of the projection onto the plane as happening in two steps. First, the vector is projected onto the x-axis, leaving behind a residual vector $v - \langle v, e_x \rangle e_x$, which is perpendicular to the x-axis. This residual is then projected onto the y-axis, yielding

$$\langle v - \langle v, e_x \rangle e_x, e_y \rangle = \langle v, e_y \rangle - \langle \langle v, e_x \rangle e_x, e_y \rangle = \langle v, e_y \rangle - \langle v, e_x \rangle \underbrace{\langle e_x, e_y \rangle}_{=0}$$
$$= \langle v, e_y \rangle.$$

It is easy to see how to generalize this to an m-dimensional approximation of an n-dimensional vector. Each successive projection onto one of the subspace's m orthonormal basis vectors extracts new information about v without changing the coefficients that have already been calculated. The approximation may, more generally, be constructed from any set of independent vectors spanning the subspace, but only when these vectors are orthonormal are the coefficients given by simple inner products $\langle v, e_k \rangle$. For this reason, we generally prefer to work with orthonormal bases.[7]

---

[7] It is also possible to formulate the approximation problem in terms of the 1-norm and ∞-norm, but the euclidean norm is much easier to work with.

In general, because the approximation $\hat{v}$ is constructed from the basis vectors for the subspace U, the error vector $\tilde{v} = v - \hat{v}$ is perpendicular to U, that is, for a basis vector $e_j \in U$,

$$\langle \tilde{v}, e_j \rangle = \langle v - \sum_{k=1}^{m} \langle v, e_k \rangle e_k, e_j \rangle = \langle v, e_j \rangle - \sum_{k=1}^{m} \langle v, e_k \rangle \underbrace{\langle e_k, e_j \rangle}_{=1, j=k}$$

$$= \langle v, e_j \rangle - \langle v, e_j \rangle = 0.$$

Thus, $\tilde{v} \in U^\perp$, and $\hat{v}$ is the best (euclidean) approximation to v.

These results are collected in the following important theorem.

Theorem 2.3 (Orthogonality principle). Let V be an n-dimensional inner product space and U an m-dimensional subspace of V. If $v \in V$, then

(a) the approximation $\hat{v} \in U$ that minimizes $\|v - \hat{v}\|_2$ is the orthogonal projection of v onto U:

$$\hat{v} = \sum_{k=1}^{m} \langle v, e_k \rangle e_k,$$

where $\{e_k\}_{k=1}^{m}$ is an orthonormal set in U.

(b) the residual vector $\tilde{v} = v - \hat{v}$ is orthogonal to U, and hence to the approximation: $\langle \tilde{v}, \hat{v} \rangle = 0$ and $\tilde{v} \in U^\perp$.

(c) the minimized approximation error is

$$\mathcal{E}_{\min}^2 = \|v - \hat{v}\|^2 = \|v\|^2 - \sum_{k=1}^{m} |\langle v, e_k \rangle|^2.$$

In the event that the subspace U is identical to V, then $\hat{v} = v$ and $\mathcal{E}_{\min}^2 = 0$, that is, $\|v\|^2 = \sum_{k=1}^{m} |\langle v, e_k \rangle|^2$, which is just the Pythagorean theorem: the squared norm of a vector is the sum of squares of its orthogonal components. This is, in fact, a special case of the following more general relationship for orthogonal expansions.

Theorem 2.4 (Parseval's formula). Let V be an n-dimensional inner product space and $\{e_k\}_{k=1}^{n}$ be an orthonormal basis for V. For $u, v \in V$,

$$\langle u, v \rangle = \sum_{k=1}^{n} \langle u, e_k \rangle \langle v, e_k \rangle^*. \tag{2.24}$$

and if $u = v$,

$$\|v\|^2 = \sum_{k=1}^{n} |\langle v, e_k \rangle|^2. \tag{2.25}$$

2.2 FINITE-DIMENSIONAL VECTOR SPACES

Proof: Express u in terms of the basis, then

$$\langle u, v \rangle = \left\langle \sum_{k=1}^{n} \langle u, e_k \rangle e_k, v \right\rangle = \sum_{k=1}^{n} \langle u, e_k \rangle \langle e_k, v \rangle = \sum_{k=1}^{n} \langle u, e_k \rangle \langle v, e_k \rangle^*.$$

∎

If we think of the coefficient sequences $\{\langle u, e_k \rangle\}_{k=1}^{n}$ and $\{\langle v, e_k \rangle\}_{k=1}^{n}$ as n-vectors, then Equation 2.24 says that the inner product of the coefficient sequences is equal to the inner product of the original vectors. *Orthogonal expansion preserves inner products.* Moreover, interpreting the squared norm as energy, Equation 2.25 shows that the coefficients $\langle v, e_k \rangle$ "capture" all of v's energy. *Orthogonal expansion conserves energy.*

Equations 2.24 and 2.25 extend to infinite-dimensional inner product spaces. In the literature the equations are variously called *Parseval's formula* (or theorem), *Plancherel's formula, Rayleigh's formula*, and the *power formula*. Historically, different versions of these relationships were derived in different contexts: by Parseval for the Fourier series, by Plancherel and Rayleigh for the Fourier transform. However, they all say the same thing—orthogonal expansions preserve inner products and conserve energy—so little seems to be gained by making distinctions. With apologies to history, I will call them all "Parseval's formula" in this book.

Example 2.10. We will expand the vector $v = (1, 2, 3, 4)$ in the Haar basis (Example 2.7). First, we normalize the basis vectors to make them into an orthonormal set:

$$e_1 = \frac{\phi_1}{\|\phi_1\|_2} = \frac{(1,1,1,1)}{\sqrt{4}} = \frac{1}{2}(1,1,1,1)$$

$$e_2 = \frac{\phi_2}{\|\phi_2\|_2} = \frac{1}{2}(1,1,-1,-1)$$

$$e_3 = \frac{\phi_3}{\|\phi_3\|_2} = \frac{1}{\sqrt{2}}(1,-1,0,0)$$

$$e_4 = \frac{\phi_4}{\|\phi_4\|_2} = \frac{1}{\sqrt{2}}(0,0,1,-1).$$

Then do the projections,

$$c_1 = \langle v, e_1 \rangle = \left\langle (1,2,3,4), \frac{1}{2}(1,1,1,1) \right\rangle = \frac{1}{2}(1+2+3+4) = 5$$

$$c_2 = \langle v, e_2 \rangle = \left\langle (1,2,3,4), \frac{1}{2}(1,1,-1,-1) \right\rangle = \frac{1}{2}(1+2-3-4) = -2$$

$$c_3 = \langle v, e_3 \rangle = \left\langle (1,2,3,4), \frac{1}{\sqrt{2}}(1,-1,0,0) \right\rangle = \frac{1}{\sqrt{2}}(1-2+0+0) = -\frac{1}{\sqrt{2}}$$

$$c_4 = \langle v, e_4 \rangle = \left\langle (1,2,3,4), \frac{1}{\sqrt{2}}(0,0,1,-1) \right\rangle = \frac{1}{\sqrt{2}}(0+0+3-4) = -\frac{1}{\sqrt{2}}.$$

Check Parseval's formula:
$$\|v\|^2 = 1^2 + 2^2 + 3^2 + 4^2 = 30$$
$$\sum_{k=1}^{4} |c_k|^2 = 5^2 + (-2)^2 + \left(-\frac{1}{\sqrt{2}}\right)^2 + \left(-\frac{1}{\sqrt{2}}\right)^2 = 30.$$

Now suppose we want to construct the best 2D approximation to v in this basis. We select c_1 and c_2, the two coefficients with the greatest absolute value. The subspace U is the linear span of $\{e_1, e_2\}$ and its orthogonal complement $U^\perp$ is the linear span of $\{e_3, e_4\}$. The best approximation is

$$\hat{v} = 5e_1 - 2e_2 = 5 \cdot \frac{1}{2}(1,1,1,1) - 2 \cdot \frac{1}{2}(1,1,-1,-1)$$
$$= \left(\frac{3}{2}, \frac{3}{2}, \frac{7}{2}, \frac{7}{2}\right).$$

The error vector is

$$\tilde{v} = v - \hat{v} = (1,2,3,4) - \left(\frac{3}{2}, \frac{3}{2}, \frac{7}{2}, \frac{7}{2}\right)$$
$$= \left(-\frac{1}{2}, \frac{1}{2}, -\frac{1}{2}, \frac{1}{2}\right) = -\frac{1}{2}e_3 - \frac{1}{2}e_4.$$

Check that $\tilde{v}$ is orthogonal to the approximation:

$$\langle \hat{v}, \tilde{v} \rangle = \left\langle \left(\frac{3}{2}, \frac{3}{2}, \frac{7}{2}, \frac{7}{2}\right), \left(-\frac{1}{2}, \frac{1}{2}, -\frac{1}{2}, \frac{1}{2}\right) \right\rangle$$
$$= \frac{1}{4}(-3 + 3 - 7 + 7) = 0.$$

The squared norm of the error is, by direct calculation,

$$\|\tilde{v}\|^2 = \left(-\frac{1}{2}\right)^2 + \left(\frac{1}{2}\right)^2 + \left(-\frac{1}{2}\right)^2 + \left(\frac{1}{2}\right)^2 = 1.$$

Using the Pythagorean formula instead, the squared norm of the error is the sum of the squares of the coefficients that were dropped:

$$\|\tilde{v}\|^2 = |c_3|^2 + |c_4|^2 = \left(-\frac{1}{\sqrt{2}}\right)^2 + \left(-\frac{1}{\sqrt{2}}\right)^2 = 1.$$

∎

Example 2.11. Repeat the previous example using MATLAB. It is efficient to collect all the normalized basis vectors into a matrix,

```
% Normalized Haar basis vectors in a matrix
H = [   [1 1 1 1]/2;
        [1 1 -1 -1]/2;
        [1 -1 0 0]/sqrt(2);
        [0 0 1 -1]/sqrt(2)  ]
```

2.2 FINITE-DIMENSIONAL VECTOR SPACES

```
H =
        0.5         0.5         0.5         0.5
        0.5         0.5        -0.5        -0.5
    0.70711    -0.70711           0           0
          0           0     0.70711    -0.70711
```

The inner products of the vector $v = (1, 2, 3, 4)$ with the basis vectors are computed by multiplying the row vector on the right by the transpose of the basis vector matrix:

```
% The vector to be decomposed
v = [1 2 3 4];

% Decomposition
c = v * H'
c =
        5          -2    -0.70711    -0.70711
```

(annotation: transpose of basis vector)

To demonstrate the "conservation of energy" $\|v\|^2 = \sum_k |c_k|^2$, calculate the dot products of v and c with themselves.

```
% Verifying "conservation of energy"
v * v'
ans =
    30

c * c'
ans =
    30
```

The reconstruction of v from its coefficients, $v = \sum_k c_k e_k$, is computed by multiplying the coefficient vector on the right by the basis vector matrix (not its transpose).

```
% Reconstructing v
c * H
ans =
    1    2    3    4
```

To approximate v in a lower-dimensional subspace, perform the same calculation with the unwanted coefficients set to zero.

```
% Two-dimensional approximation - keep only the two largest
% coefficients
vhat = [c(1:2) 0 0] * H
vhat =
       1.5         1.5         3.5         3.5
```

Calculate the dot product of the approximation and the error to show that they are orthogonal.

```
% The error is orthogonal to the approximation
vhat * (v-vhat)'
ans =
     0
```

The size of the approximation error is computed in two ways, as before.

```
% Approximation error, calculated two ways
(v-vhat)*(v-vhat)'
ans =
     1
c(3:4)*c(3:4)'
ans =
     1
```
∎

Again, be careful not to confuse the dimensionality of the basis vectors with the dimensionality of the subspace and the approximation. In these examples, the basis vectors span 4D real space, $\mathbb{R}^4$. The approximating subspace is 2D, a plane spanned by the two 4D basis vectors e_1 and e_2. The approximation $\hat{v}$ is 2D, requiring only two coordinates c_1 and c_2 to locate it in the plane.

Example 2.12 (Expansion in the JPEG basis). The set of basis images used in JPEG image compression is shown in Figure 2.7.

We will use it to decompose and reconstruct the 8×8 pixel block shown in Figure 2.8. There are 64 pixels, so we may think of the image as a vector in 64D real space, $\mathbb{R}^{64}$. The inner product of the original image, v, and the nth basis image e_n, is the double sum

$$c_n = \sum_{i=1}^{8} \sum_{j=1}^{8} v[i,j] e_n[i,j], \quad n = 0, 1, \ldots, 63,$$

where i and j are row and column indices, respectively.

Figure 2.9 shows the partial reconstructions of the image from a subset of the basis images,

$$\hat{v}_N = \sum_{n=0}^{N-1} c_n e_n.$$

The Nth approximation $\hat{v}_N$ is a projection of the image into an N-dimensional subspace of $\mathbb{R}^{64}$. Instead of the 64 original pixel values, the image is represented by N of the expansion coefficients.

The figure shows the progressive refinement of the reconstruction as more basis images are added. For this particular image, the improvement of the reconstruction is not steady but occurs in jumps. There are big changes at $N = 4, 11, 13, \ldots$. Why this happens may be understood by looking at the values of the coefficients c_n

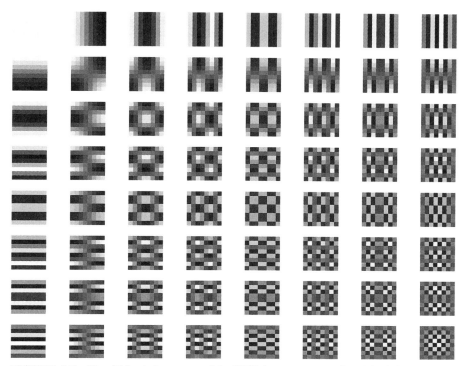

FIGURE 2.7 The 64 basis images used in JPEG image compression. Black is -1, white is 1.

(Figure 2.10(a)). The first four coefficients, c_0 through c_3, are zero or very nearly zero. The original image is effectively orthogonal to basis images e_0 through e_3 (the reader is invited to compare the original image with the basis images to see that this makes sense). The fourth basis image has the same gross symmetry as the original; the projection is strong and the coefficient is large. There is a big drop in the approximation error norm $\|v - \hat{v}_N\|$ at this point (Figure 2.10(b)). The

FIGURE 2.8 An 8×8 pixel image. Black is -1, white is 1.

62 CHAPTER 2 VECTOR SPACES

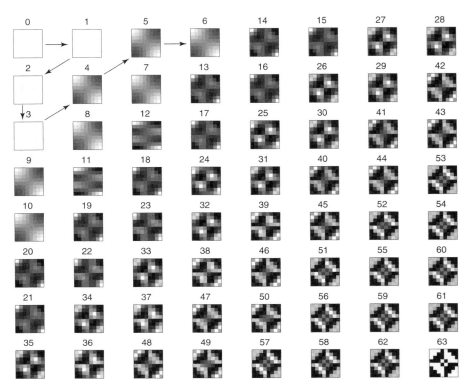

FIGURE 2.9 The image of Figure 2.8, progressively reconstructed from its projections in the JPEG basis. The images are arranged in an 8×8 array, numbered in a zigzag fashion as shown. Image 0 uses only the 0th basis image, image 1 uses basis images 0 and 1, etc., until all 64 basis images are included in image 63.

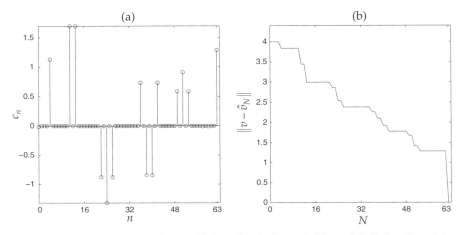

FIGURE 2.10 (a) The expansion coefficients for the image in Figure 2.8. Only a few of the 64 components are significant. (b) The approximation error as basis images are successively included in the reconstruction of Figure 2.9.

2.2 FINITE-DIMENSIONAL VECTOR SPACES 63

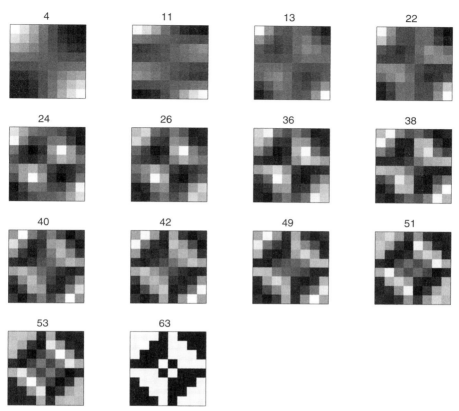

FIGURE 2.11 Approximating the image of Figure 2.8, using only the strongest components. Left to right, from top: Using only the fourth basis image; using the fourth and eleventh; using basis images 4, 11, and 13; etc.

subsequent jumps in the quality of the reconstructed image all occur at points where the coefficients are large.

With so many small components, it is reasonable to think that the image could be accurately represented by fewer than 64 numbers, by projecting it onto the basis images and keeping only the strongest components. This is shown in Figure 2.11.

Real JPEG compression algorithms break the image up into 8×8 blocks and project each block onto the basis images. Components are then kept or discarded according to their strength, relative to the known sensitivity of the human visual system to each basis image; that is, a component is discarded if its absence would not be noticed by a human observer. The selected coefficients can be stored with less memory than the original image, or transmitted in less time. The image is reconstructed for the end user from the coefficients and the basis images.[8] ∎

---

[8] For more about JPEG, see Mallat (1999, pp. 561–566) and Pennebaker and Mitchell (1993). The mathematical form of the JPEG basis images will be derived in Chapter 3.

2.3 INFINITE-DIMENSIONAL VECTOR SPACES

In principle, the extension of the norm, inner product, and orthogonal approximation from finite-dimensional vectors to infinite-dimensional vectors (infinite sequences) is natural: One simply makes the vectors (much) longer, and changes the sums to infinite series. Let v and w be infinite sequences,

$$v = (v_n)_{n=1}^{\infty} \quad \text{and} \quad w = (w_n)_{n=1}^{\infty},$$

then the inner product and norm would be

$$\langle v, w \rangle = \sum_{n=1}^{\infty} v_n w_n^* \quad \text{and} \quad \|v\|^2 = \sum_{n=1}^{\infty} |v_n|^2.$$

Now consider two functions f and g, $x \in (0, 1)$ (or, more generally, (a, b), where a and b may both be infinite). Imagine them to be vectors too, with x as the index and $f(x), g(x)$ as the respective components. By analogy with the finite-dimensional case, we replace sums by integrals and have

$$\langle f, g \rangle = \int_0^1 f(x) g^*(x) \, dx \quad \text{and} \quad \|f\|^2 = \int_0^1 |f(x)|^2 \, dx.$$

These generalizations will be justified and explained in this section, and applied in subsequent chapters.

In order to be useful, the infinite series defining our more general norm and inner product must converge. Recall that an infinite series $\sum_{n=1}^{\infty} x_n$ is convergent if the sequence of partial sums, $S_N = \sum_{n=1}^{N} x_n$, converges as $N \to \infty$. We begin our discussion, then, with a review of convergent sequences.

2.3.1 Convergent Sequences

A *sequence* is an ordered set of points drawn from some space, denoted $\{x_1, x_2, \ldots, x_N\}$, $\{x_n\}_{k=1}^{N}$, or $(x_n)_{k=1}^{N}$. In some applications sequences are indexed with nonnegative integers, $(x_n)_{n=0}^{N-1}$, or even with negative integers, $(x_n)_{n=-N}^{N}$. The number of terms in the sequence can be finite or infinite. The points themselves may be numbers, for example, $\{0, 1, 2, 3, \ldots\}$, geometric points in a plane, for example, $\{(1, 1), (1, \frac{1}{2}), (\frac{1}{2}, \frac{1}{2}), (\frac{1}{2}, \frac{1}{4}), \ldots\}$, or N-dimensional vectors in $\mathbb{R}^N$. Significantly, for what comes later, they may also be functions, for example, $\{1, e^{-x}, e^{-2x}, \ldots\} = (e^{-nx})_{n=0}^{\infty}$.

If the points x_n of an infinite sequence belong to a metric space, do they draw successively closer to one particular point as n increases?

Definition 2.8 (Convergent sequence). An infinite sequence, $(x_n)_{n=1}^{\infty}$, of elements in a metric space V *converges* in the space if, for some $x \in V$ and any $\epsilon > 0$, there is an N such that $d(x_n, x) < \epsilon$ for $n > N$. We say that $\lim_{n \to \infty} x_n = x$, and x is called the *limit point* of the sequence.

FIGURE 2.12 The convergent sequence $(\frac{1}{n})_{n=1}^{\infty}$. The points accumulate in a neighborhood of 0, the limit of the sequence.

In other words, if you put a neighborhood of arbitrarily small radius ϵ around x, you can find an N such that x_N and all the terms following it are inside the neighborhood. No matter how small the neighborhood becomes, there will always be more points inside the neighborhood than outside it.

Example 2.13 (Convergent sequences on the real line).

(a) The sequence $\{1, \frac{1}{2}, \frac{1}{3}, \ldots, \frac{1}{n}, \ldots\}$ converges to zero (Figure 2.12). Consider the neighborhood $(-\epsilon, \epsilon)$ around $x = 0$ and take $N > \frac{1}{\epsilon}$. Then, for any $n > N$,

$$d(x_n, 0) = \frac{1}{n} < \frac{1}{N} < \epsilon.$$

(b) The sequence $\{0, \frac{1}{2}, \frac{3}{4}, \frac{7}{8}, \ldots, 1 - 2^{-n}, \ldots\}$ converges to one. Consider the neighborhood $(1 - \epsilon, 1 + \epsilon)$. We want the points $1 - 2^{-n}$ to be in the neighborhood for n sufficiently large. So we look for N such that $1 - \epsilon < 1 - 2^{-N} < 1 + \epsilon$. This works out to $2^{-N} < \epsilon$. Take $N > -\log_2 \epsilon$. Then, for any $n > N$,

$$d(x_n, 1) = |(1 - 2^{-n}) - 1| = 2^{-n} < 2^{-N} < \epsilon. \qquad \blacksquare$$

This definition of convergence is useful if we already have an idea of what the limit of the sequence is. In the two cases above, it was easy to guess the limit, because $\frac{1}{n} \to 0$ and $1 - 2^{-n} \to 1$ as $n \to \infty$. When the limit is not obvious, the following definition is helpful in assessing convergence. Instead of considering the distance between each point and a hypothetical limit, it looks at the distance between points within the sequence.

Definition 2.9 (Cauchy sequence). A sequence $(x_k)_{k=1}^{\infty}$ in a metric space V is a *Cauchy sequence* if, for any $\epsilon > 0$, there is an N such that $d(x_n, x_m) < \epsilon$ for $n, m > N$.

Example 2.14. The two sequences we considered above are Cauchy sequences.

(a) With $x_n = \frac{1}{n}$, we have

$$d(x_n, x_m) = \left|\frac{1}{n} - \frac{1}{m}\right| \leq \left|\frac{1}{n}\right| + \left|\frac{1}{m}\right| \quad \text{(Triangle inequality)}.$$

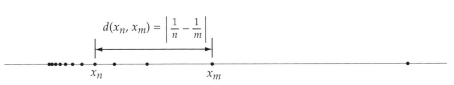

FIGURE 2.13 The sequence $(\frac{1}{n})_{n=1}^{\infty}$ is a Cauchy sequence. Pairs of points $(\frac{1}{n}, \frac{1}{m})$ get closer as m, n increase.

We seek an N such that, for $n, m > N$, $|\frac{1}{n}| + |\frac{1}{m}| < \epsilon$. Choosing $N > 2/\epsilon$ will do the trick. Then $\frac{1}{m}, \frac{1}{n}$ are each less than $\frac{\epsilon}{2}$ and so

$$d(x_n, x_m) < \frac{\epsilon}{2} + \frac{\epsilon}{2} < \epsilon.$$

(b) With $x_n = 1 - 2^{-n}$, we have

$$d(x_n, x_m) = |(1 - 2^{-n}) - (1 - 2^{-m})|$$
$$= |2^{-n} - 2^{-m}| \leq 2^{-n} + 2^{-m} \qquad \text{(Triangle inequality)}.$$

Take $N > -\log_2(\frac{\epsilon}{2})$, then for $n, m > N$,

$$d(x_n, x_m) \leq 2^{-n} + 2^{-m} < 2 \cdot 2^{-N} < 2 \cdot \frac{\epsilon}{2} = \epsilon. \qquad \blacksquare$$

The successive points in a convergent sequence become arbitrarily close to a limit point. Successive points in a Cauchy sequence, on the other hand, become arbitrarily close to each other (Figure 2.13).

It is evident that convergent sequences are Cauchy; if the points are approaching a limit, they must also be crowding together. The converse also seems reasonable, that a Cauchy sequence should be convergent, but it is not always true because the limit may not exist in the same space as the sequence. For example, the decimal approximations to π, $\{3, 3.1, 3.14, 3.142, 3.1416, \ldots\}$ are a Cauchy sequence in the space of rational numbers. However, the limit, π, is irrational, so the sequence of rational approximations to π does not converge in the space of rational numbers. If we consider the larger space of real numbers, which contains π as well as all rational numbers, then the sequence is convergent.

Suppose a vector v is the limit of a Cauchy sequence (v_n). The Nth term of this sequence is an approximation to v at some level of accuracy that may be satisfactory in practice and easier to compute than the actual vector v. Examples include polynomial approximations to functions, and, as we shall see in a later chapter, Fourier series. To say that the approximating sequences always converge, it is important that both the vectors and their approximations be in the same space. Thus, in addition to requiring that a vector space have a norm (and hence a metric), we impose the following restriction:

Definition 2.10 (Complete metric space). A metric space is *complete* if every Cauchy sequence in the space converges (has a limit) in the space.

The rational numbers are not complete (the above sequence of decimal approximations to π is a counterexample), but the space of real numbers is complete,[9] as is its n-dimensional generalization, $\mathbb{R}^n$. A Cauchy sequence in $\mathbb{R}^n$ is guaranteed to be convergent.

2.3.2 Infinite Sequences and the ℓ^p Spaces

Now let us consider in more detail the generalization of vectors in $\mathbb{R}^n$ to vectors with countably infinite members, $x = (x_1, x_2, \ldots)$, that is, infinite sequences or discrete-time signals. The set of all such real-valued sequences is called $\mathbb{R}^\infty$. The analogous space of complex sequences is called $\mathbb{C}^\infty$. It is clear that $\mathbb{R}^\infty$ and $\mathbb{C}^\infty$ are vector spaces, with addition and scalar multiplication defined in the usual way,

$$x + y = (x_1, x_2, \ldots) + (y_1, y_2, \ldots) = (x_1 + y_1, x_2 + y_2, \ldots)$$
$$cx = c(x_1, x_2, \ldots) = (cx_1, cx_2, \ldots).$$

It is natural to define norms for infinite sequences using infinite sums,

$$\|x\|_1 = \sum_{n=1}^{\infty} |x_n|$$

$$\|x\|_2 = \left(\sum_{n=1}^{\infty} |x_n|^2 \right)^{1/2}$$

and generalize the maximum norm by the supremum,

$$\|x\|_\infty = \sup_n |x_n|.$$

These definitions meet the requirements for norms that were spelled out earlier (Definition 2.2), *provided that the series converge*. There are vectors for which the infinite sums diverge, for example, constant vectors $x, y = (1, 1, \ldots)$. In order to have useful norms, we must restrict attention to subsets of $\mathbb{R}^\infty$ or $\mathbb{C}^\infty$ containing vectors for which the sums converge.

Definition 2.11. Let $x \in \mathbb{R}^\infty$ or $\mathbb{C}^\infty$. Then, x is *absolutely summable* if $\sum_{n=1}^{\infty} |x_n| < \infty$; x is *square summable* if $\sum_{n=1}^{\infty} |x_n|^2 < \infty$; and x is *bounded* if $\sup_n |x_n| < \infty$.

With these definitions, the following can be shown:

1. The set of all absolutely summable sequences is a vector space (which we call ℓ^1, read "ell one") with norm $\|x\|_1 = \sum_{n=1}^{\infty} |x_n|$.
2. The set of all square summable sequences is a vector space, ℓ^2 ("ell two"), with norm $\|x\|_2 = \left(\sum_{n=1}^{\infty} |x_n|^2 \right)^{1/2}$.

---

[9] See, for example, Rosenlicht (1968, p. 52) and Folland (2002, p. 28).

68 CHAPTER 2 VECTOR SPACES

3. The set of all bounded sequences is a vector space, ℓ^∞ ("ell infinity"), with norm $\|x\|_\infty = \sup_n |x_n|$.

If $x, y \in \ell^2$, then by an application of the Cauchy–Schwarz inequality, the infinite sum $\sum_{n=1}^\infty x_n y_n^*$ can be shown to be convergent. It also satisfies the requirements for an inner product. Thus ℓ^2 is an inner product space with

$$\langle x, y \rangle = \sum_{n=1}^\infty x_n y_n^*, \tag{2.26}$$

and just as with a finite-dimensional euclidean space, $\|x\|_2 = \sqrt{\langle x, x \rangle}$. Orthogonality, the parallelogram law, and the polarization identity also carry over from finite-dimensional spaces to ℓ^2, since they depend only on the defining properties of norm and inner product, and not on the dimensionality of the space.

In general, for $1 \le p \le \infty$, ℓ^p is the space of sequences x with finite *p-norm*, defined

$$\|x\|_p = \left(\sum_{n=1}^\infty |x_n|^p \right)^{1/p}. \tag{2.27}$$

The following may be shown:

- The ℓ^∞ norm is the limit, as $p \to \infty$, of the p-norm.
- Of all the ℓ^p spaces, only ℓ^2 is an inner product space.
- The ℓ^p spaces are nested: if $1 \le p < q \le \infty$, then $\ell^p \subset \ell^q$. In particular, $\ell^1 \subset \ell^2 \subset \ell^\infty$. If a sequence is absolutely summable, then it is also square summable and bounded. On the other hand, if a sequence is unbounded, then it is not summable in any sense.

Example 2.15. Here are some illustrative examples of infinite-dimensional vectors.

(a) The sequence $x = (1 - \frac{1}{n})_{n=1}^\infty = (0, \frac{1}{2}, \frac{2}{3}, \frac{3}{4}, \ldots)$ has supremum norm $\|x\|_\infty = 1$, and belongs to ℓ^∞. It does not, however, belong to ℓ^1 or ℓ^2.

(b) The sequence $x = (\frac{1}{n})_{n=1}^\infty = (1, \frac{1}{2}, \frac{1}{3}, \ldots)$ belongs to ℓ^∞ and ℓ^2, but not ℓ^1.

(c) The sequence $x = (\frac{1}{2^{n-1}})_{n=1}^\infty = (1, \frac{1}{2}, \frac{1}{4}, \ldots)$ belongs to ℓ^1, ℓ^2, and ℓ^∞.

(d) The sequence $x = (2^{n-1})_{n=1}^\infty = (1, 2, 4, \ldots)$ does not belong to any ℓ^p space. ∎

For completeness, one may also define the set ℓ^0 of sequences, which are zero except for a finite number of finite values. Vectors in ℓ^0 behave like finite-dimensional vectors, and each of the ℓ^p spaces contains the set ℓ^0. For any vector $v \in \ell^p, p \ge 1$, there is a vector $u \in \ell^0$ arbitrarily close to v, in the sense that $\|v - u\|_p < \epsilon$—simply take u to be the truncation of v, $u = (v_1, v_2, \ldots, v_N, 0, \ldots)$, with N as large as needed to achieve the desired error ϵ.

2.3.3 Functions and the L^p Spaces

Vector spaces may also be composed of real- or complex-valued functions of a real variable, for example, continuous-time signals. They are then called *function spaces*. Norms and inner products in these spaces are defined by integrals rather than sums. Requiring the integrals to be convergent restricts the functions that occupy these spaces.

Definition 2.12. Let $Q = [a, b] \in \mathbb{R}$ (Q can also be $(-\infty, b]$, $[a, \infty)$, or $(-\infty, \infty)$) and $f : Q \to \mathbb{C}$ be a function. Then, f is *absolutely integrable* on Q if $\int_Q |f(x)| dx < \infty$; f is *square integrable* on Q if $\int_Q |f(x)|^2 dx < \infty$; and f is *bounded* on Q if $\sup_Q |f| < \infty$.

With these definitions, it can be shown that:

1. The set of all absolutely integrable functions on Q is a vector space, denoted $L^1(Q)$ ("ell one") or simply L^1, if the domain is understood, with norm

$$\|f\|_1 = \int_Q |f(x)| dx. \tag{2.28}$$

2. The set of all square-integrable functions on Q is a vector space, denoted $L^2(Q)$ ("ell two"), with norm

$$\|f\|_2 = \left(\int_Q |f(x)|^2 dx \right)^{1/2}. \tag{2.29}$$

3. The set of all bounded functions on Q is a vector space with norm

$$\|f\|_u = \sup_Q |f|. \tag{2.30}$$

The subscript u denotes "uniform." The supremum norm is also called the *uniform norm*.[10] With a slight modification, to be discussed later, the vector space of bounded functions is denoted L^∞.

In general, the space $L^p(Q)$, where $1 \leq p < \infty$, is the space of functions f on Q having finite *p-norm*, defined

$$\|f\|_p = \left(\int_Q |f(x)|^p dx \right)^{1/p}. \tag{2.31}$$

Examples of functions in L^1, L^2, and L^∞

The choice of norm places restrictions on the functions that populate a space, as the next few examples show.

Example 2.16. The "sinc" function:

$$\operatorname{sinc} x = \frac{\sin \pi x}{\pi x},$$

---

[10] See Folland (1999, p. 121). If a sequence of functions $(f_n)_{n=1}^\infty$ converges to a function f in such a way that $\|f_n - f\|_u \to 0$ as $n \to \infty$, we say that the sequence converges *uniformly* to f. Uniform convergence is an important concept in the theory of Fourier series, Chapter 4.

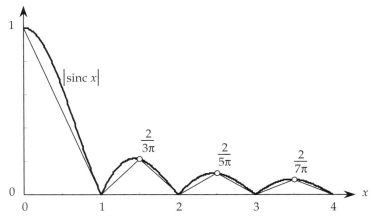

FIGURE 2.14 The function | sinc | is bounded below by triangles. The sum of the areas of these triangles diverges.

which will figure prominently in later chapters, is bounded ($|\operatorname{sinc} x| \leq \operatorname{sinc} 0 = 1$ by L'Hospital's rule), so it belongs to L^∞. Its 1-norm,

$$\int_{-\infty}^{\infty} |\operatorname{sinc} x|\, dx = 2 \int_{0}^{\infty} \left|\frac{\sin \pi x}{\pi x}\right| dx,$$

however, is unbounded. To show this, we devise another function that is everywhere less than or equal to sinc, but which has an unbounded 1-norm (Figure 2.14). Between $x = 0$ and $x = 1$, | sinc | is bounded below by a triangle with area $\frac{1}{2}$. On the kth interval $(k, k + 1)$, it is bounded below by a triangle of unit base and height $\frac{|\sin \pi(k+1/2)|}{\pi(k+1/2)} = \frac{1}{\pi(k+1/2)}$, whose area is $I_k = 1/\pi(2k + 1)$.

The norm is bounded below by the sum of these triangular areas,

$$\|\operatorname{sinc}\|_1 > 2 \left(\frac{1}{2} + \sum_{k=1}^{\infty} I_k \right) = 1 + \frac{2}{\pi} \sum_{k=1}^{\infty} \frac{1}{2k+1},$$

which is a divergent series. ∎

Example 2.17. Sinc does not belong to $L^1(\mathbb{R})$ because it does not decay fast enough as $x \to \infty$. We will show that the 2-norm $\|\operatorname{sinc}\|_2$ *is* finite, so sinc belongs to $L^2(\mathbb{R})$. Using a table of integrals, we can calculate $\int_{-\infty}^{\infty} |\operatorname{sinc} x|^2 dx = 1$. We can also carry out another bounding argument. This time, we seek to bound sinc^2 *above* by a function whose integral is finite (Figure 2.15). On the interval $[0,1]$, $\operatorname{sinc}^2 x \leq 1$, and for $x > 1$, $\operatorname{sinc}^2 x < 1/(\pi x)^2$. The integral is bounded by $2 + 2\int_1^\infty \frac{dx}{(\pi x)^2} = 2 + 2/\pi^2 < \infty$; therefore $\|\operatorname{sinc}\|_2 < \infty$. The dominating function $1/x^2$ dies fast enough that sinc x belongs to the space of functions with bounded 2-norm. In fact, this argument can be extended to show that sinc $\in L^p(\mathbb{R})$ for all $p > 1$. ∎

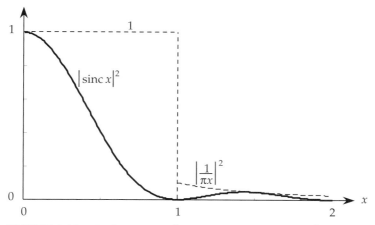

FIGURE 2.15 The function sinc² is bounded above by $1/(\pi x)^2$ for $x > 1$.

Example 2.18. Whereas sinc is in $L^2(\mathbb{R})$ but not in $L^1(\mathbb{R})$, the function $f : [0, \infty) \to \mathbb{R}$ defined by $f(x) = \frac{1}{\sqrt{x}(1+x^2)}$ is in $L^1[0, \infty)$ but not in $L^2[0, \infty)$ or $L^\infty[0, \infty)$ (Figure 2.16). Clearly, $|f| \to \infty$ as $x \to 0$, so $f \notin L^\infty[0, \infty]$. It can be shown that

$$\int_0^\infty |f(x)|dx = \int_0^\infty \frac{dx}{\sqrt{x}(1+x^2)} = \frac{\pi}{\sqrt{2}}$$

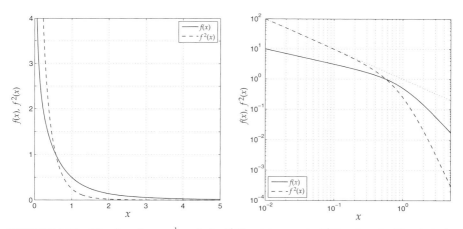

FIGURE 2.16 The function $\frac{1}{\sqrt{x}(1+x^2)}$ is in $L^1[0, \infty)$ but not in $L^2[0, \infty)$ or $L^\infty[0, \infty)$. *Left:* Linear plot. *Right:* Log–log plot. A graph of x^{-1} (dotted line) is superimposed on the log–log plot to show the asymptotic properties of the functions. Both $|f|$ and $|f|^2$ decay faster than x^{-1} for large x, which is fast enough for integrability. As x decreases toward zero, $|f|$ grows more slowly than x^{-1} so it is integrable, but $|f|^2$ grows like x^{-1} and is unintegrable. Both functions are unbounded as $x \to 0$.

so $f \in L^1[0, \infty]$. However,

$$\int_\epsilon^\infty |f(x)|^2 dx = \int_\epsilon^\infty \frac{dx}{x(1+x^2)^2} = \int_\epsilon^\infty \left(\frac{1}{x} - \frac{x}{1+x^2} - \frac{x}{(1+x^2)^2}\right) dx$$

is unbounded as $\epsilon \to 0^+$ (note $\int \frac{dx}{x} = \log x$), so $f \notin L^2[0, \infty]$. ∎

Example 2.19. The sine function, $f(x) = \sin x$, belongs to $L^1[-\pi, \pi]$, $L^2[-\pi, \pi]$, and $L^\infty[-\pi, \pi]$:

$$\int_{-\pi}^\pi |\sin x|\, dx = 2\int_0^\pi \sin x\, dx = 2$$

$$\int_{-\pi}^\pi |\sin x|^2\, dx = 2\int_0^\pi \sin^2 x\, dx = \int_0^\pi (1 - \cos 2x)\, dx = \pi$$

$$|\sin x| \le 1, x \in [-\pi, \pi].$$

However, while $\sin x$ is in $L^\infty(\mathbb{R})$, it is not in $L^1(\mathbb{R})$ or $L^2(\mathbb{R})$. This will be important in Chapters 5 and 6, when we study Fourier transforms. ∎

Physical signals and function spaces

The L^2 norm has a physical interpretation in terms of the energy or power in a signal. In an electrical system, for example, the instantaneous power dissipated in a resistor of resistance R with a current $I(t)$ flowing through it is $P(t) = I^2(t)R$. By analogy, we may regard the squared magnitude of any function f as a generalized instantaneous power. The integral of power is energy, so the integral of $|f|^2$—the square of the L^2 norm—is the total energy of the function:

$$E = \int_{-\infty}^\infty |f(t)|^2 dt = \|f\|_2^2.$$

Saying a function has finite energy is another way of saying that it is in $L^2(\mathbb{R})$. We expect actual physical signals to have finite energy as well as bounded amplitude, although certain convenient idealizations, like the unit step function, do not.

Averaging instantaneous power $|f|^2$ over a finite time interval, say $(-\frac{T}{2}, \frac{T}{2})$, gives the mean-square value, or average power, over the interval. Of particular interest in applications is the long-time average as $T \to \infty$.

$$P_{\text{avg}} = \lim_{T \to \infty} \frac{1}{T} \int_{-T/2}^{T/2} |f(t)|^2 dt. \tag{2.32}$$

The square root of this integral is called the *root mean square*, or *rms*, value of f:

$$f_{\text{rms}} = \left(\lim_{T \to \infty} \frac{1}{T} \int_{-T/2}^{T/2} |f(t)|^2 dt\right)^{1/2}. \tag{2.33}$$

Example 2.20. In electrical engineering, one distinguishes between signals with finite energy and those with finite average power.

(a) We have already seen that sinc t is square integrable, so it has finite energy.

(b) The sine function $f(t) = A \sin \omega t$ does not have finite energy, because the integral

$$\int_{-\infty}^{\infty} |A \sin \omega t|^2 dt$$

is not finite. However, we can calculate the average power over $(-\frac{T}{2}, \frac{T}{2})$,

$$\frac{1}{T} \int_{-T/2}^{T/2} |A \sin \omega t|^2 dt = \frac{A^2}{2}\left(1 - \frac{\sin \omega T}{\omega T}\right).$$

Taking the limit as $T \to \infty$, we find $P_{\text{avg}} = \frac{A^2}{2}$, which is finite. The rms value is $f_{\text{rms}} = \frac{A}{\sqrt{2}}$.

The well-known formula for DC power, $P = V^2/R$, also holds for AC circuits by considering average power and rms voltage, $P_{\text{avg}} = V_{\text{rms}}^2/R$. ∎

⋆ A subtle point about function spaces

Everything that has been said so far about function spaces is true for functions that are at least piecewise continuous, which are the ones encountered in most, if not all, practical situations. The full picture requires some adjustments, which are described briefly in the following paragraphs and may be skipped on a first reading.

One of the requirements for a norm is that $\|f\| = 0$ if and only if $f = 0$. Taking the L^1 norm as an example, $\int_Q |f(x)| dx$ is certainly zero if $f = 0$ for all $x \in Q$. But let $x_0 \in Q$ and define a function f_0 by

$$f_0(x) = \begin{cases} 1, & x = x_0 \\ 0, & \text{otherwise} \end{cases}.$$

The integral of this function is also zero (the single point contributes zero area to the integral). This violates the requirement that $\|f\| = 0$ *only if* $f = 0$. Worse, there is an infinite number of such f with any countable number of isolated points different from zero, whose integrals are also zero. The difference between any two of these functions, $f - g$, has a countable number of isolated points different from zero, and the norm of the difference, $\|f - g\|$, is zero even though $f \neq g$. We would like for $\|f - g\| = 0$ to mean that $f = g$ for all $x \in Q$, but we cannot have it if the norm is blind to point differences. Only the uniform norm (Equation 2.30) will reliably give $\|f - g\| \neq 0$ for all $f \neq g$.

The issue calls for a more refined idea of equality. If a function f is zero at all but a countable number of points $\{x_1, x_2, \dots\}$, we say that f is zero *almost everywhere* (a.e.). The set of points where $f(x) \neq 0$ is called a *null set*. It is also called a *set of measure zero*. The *measure* of a set X is a function $m: X \to [0, \infty]$ that generalizes the

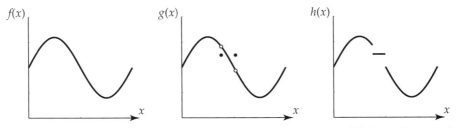

FIGURE 2.17 The functions f and g are equal almost everywhere, differing only a set of measure zero. The functions f and h, on the other hand differ on an interval, and are not equal.

idea of the length of an interval. If $X = [a, b]$ is an interval, then its measure is $m(X) = b - a$. If X is the countable union of non-overlapping intervals, then $m(X)$ is the sum of the lengths of the intervals, for example, $m([\frac{1}{2}, 1] \cup [\frac{1}{8}, \frac{1}{4}] \cup \ldots) = \frac{1}{2} + \frac{1}{8} + \cdots = \frac{2}{3}$. A single point has measure zero, as does a finite set of isolated points, $\{a_1, a_2, \ldots, a_n\}$. In fact, a countable set of points, like the rational numbers, can be shown to have measure zero, and there are exotic sets constructed from intervals rather than points that can also be shown to have measure zero.[11]

Definition 2.13. Let $Q \subset \mathbb{R}$ and $f : Q \to \mathbb{C}$ be a function. Any property of f is said to hold *almost everywhere* (a.e.) in Q if the set

$$\{x \in Q \mid \text{the property does not hold for } f(x)\}$$

has measure zero.

The integral of a bounded function on a set of measure zero is zero (this is obvious for a finite set of isolated points, and can be shown for more general sets). If f and g are equal almost everywhere then $f - g$ is zero except on a set of measure zero, so they are also equal in norm, $\|f - g\| = 0$. On the other hand, if f and g differ on an interval (a, b), then $f \neq g$ pointwise and $\|f - g\| \neq 0$. (Figure 2.17).

Unless we say otherwise, in a function space all properties will be assumed to hold in the almost everywhere sense.

The uniform norm $\|\cdot\|_u$ is not compatible with the idea of equality almost everywhere. It only gives $\|f\|_u = 0$ when $f = 0$ everywhere. We will keep this norm around, because it is useful in certain applications, but we also need to find a supremum norm that will give $\|f\| = 0$ when $f = 0$ (a.e.). Following the above definition, we will say that a function is *bounded almost everywhere* by $M > 0$ if the set

$$\{x \mid |f(x)| > M\}$$

has measure zero. We then define another norm,

$$\|f\|_\infty = \inf\{M > 0 \mid |f| \leq M \text{ a.e.}\}. \qquad (2.34)$$

---

[11] Measure is a fundamental topic in advanced real analysis texts. An introduction that is not too technical may be found in Wilcox and Myers (1978, Chapters 2 and 3).

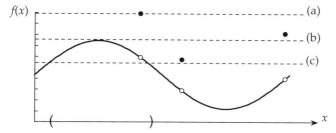

FIGURE 2.18 Supremum and essential supremum norm. The dashed lines represent candidate values for the upper bound. f is less than or equal to bound (a) everywhere; this is the supremum (uniform) norm, $\|f\|_u$. The value (b) is exceeded by f only on a set of measure zero (two isolated points). This is the essential supremum norm, $\|f\|_\infty$. The value (c) excludes all the isolated points, but f exceeds this value on an interval.

This norm is the smallest M such that $|f| > M$ only on a set of measure zero. It is called the *essential supremum norm*, and is illustrated in Figure 2.18. It is also the limit, as $p \to \infty$, of the general p-norm (Equation 2.31). It is this norm that is properly the defining property of the space L^∞.

Example 2.21. Let $f : (-2, 2) \to \mathbb{R}$ be defined by $f(x) = x$. The p-norm is

$$\|f\|_p = \left(\int_{-2}^{2} |x|^p dx \right)^{1/p} = \left(2 \int_{0}^{2} x^p dx \right)^{1/p} = \left(2 \left. \frac{x^{p+1}}{p+1} \right|_0^2 \right)^{1/p}$$

$$= \left(\frac{2^{p+2}}{p+1} \right)^{1/p}$$

Letting $p = 1$,

$$\|f\|_1 = \frac{2^3}{2} = 4.$$

Letting $p = 2$,

$$\|f\|_2 = \left(\frac{2^4}{3} \right)^{1/2} = \frac{4}{\sqrt{3}}$$

The ∞-norm is, by inspection, $\|f\|_\infty = 2$. To check that the integral gives the same result, use the fact that $x^a = \exp(a \log x)$:

$$\|f\|_\infty = \lim_{p \to \infty} \exp\left(\frac{1}{p} \log \left(\frac{2^{p+2}}{p+1} \right) \right)$$

$$= \lim_{p \to \infty} \exp\left(\frac{(p+2) \log 2}{p} - \frac{\log(p+1)}{p} \right).$$

The limit of the first term in the exponent is $\log 2$, and the limit of the second term is 0, because p grows faster than $\log(p+1)$. Both limits are finite, so

$$\|f\|_\infty = \exp(\log 2 - 0) = 2.$$

as expected. ∎

Inner products and orthogonal functions

We may define an inner product for functions $f, g : Q \to \mathbb{C}$ by analogy with the finite-dimensional case:

$$\langle f, g \rangle = \int_Q f(x)\, g^*(x)\, dx \qquad (2.35)$$

is an inner product when the integral exists. This is a common definition, but not the only possible one (see the problems for another). It satisfies the four requirements for an inner product (Definition 2.4):

(a) $\langle f, g \rangle = \int_Q f(x)g^*(x)\,dx = \left(\int_Q g(x)f^*(x)\,dx\right)^* = \langle g, f \rangle^*$

(b) $\langle cf, g \rangle = \int_Q cf(x)g^*(x)\,dx = c\int_Q f(x)g^*(x)\,dx = c\langle f, g \rangle$

(c) $\langle f + g, h \rangle = \int_Q (f(x) + g(x))h^*(x)\,dx = \int_Q f(x)h^*(x)\,dx + \int_Q g(x)h^*(x)\,dx = \langle f, h \rangle + \langle g, h \rangle$

(d) $\langle f, f \rangle = \int_Q f(x)f^*(x)\,dx = \int_Q |f(x)|^2\,dx > 0$ unless $f = 0$ a.e.

Given an inner product, we can talk about *orthogonal functions*. Following the earlier definition of orthogonality (2.5), we will say that two functions f and g are orthogonal on an interval Q if their inner product integral is zero,

$$\int_Q f(x)g^*(x)\,dx = 0.$$

It is important to note that orthogonality depends on the choice of interval. Two functions which are orthogonal on one interval may fail to be orthogonal on a different interval. We will illustrate the ideas with a few examples.

Example 2.22 (Orthogonal polynomials). The inner product of the polynomial functions $f(x) = x$ and $g(x) = 1 - 2x^2$ is the integral $\int_Q x(1 - 2x^2)\,dx$. They are orthogonal on an interval Q where this integral is zero. By inspection of the graph (Figure 2.19(a)), the integrand $x(1 - 2x^2)$ is an odd function and will integrate to zero

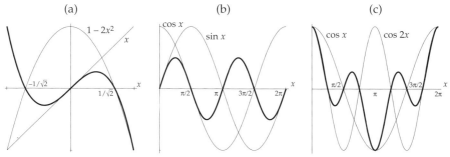

FIGURE 2.19 Pairs of orthogonal functions. The heavier lines denote their products. (a) x and $1 - 2x^2$ are orthogonal on $(-1, 0)$, $(0, 1)$, and any even interval, such as $(-\frac{1}{\sqrt{2}}, \frac{1}{\sqrt{2}})$. (b) $\cos x$ and $\sin x$ are orthogonal on $(0, \pi)$ but not on $(0, \frac{\pi}{2})$. (c) $\cos x$ and $\cos 2x$ are orthogonal on $(0, \pi)$ but not on $(\frac{\pi}{2}, \frac{3\pi}{2})$.

on any even interval $Q = (-a, a)$. More generally, the integral will be zero on any interval where the integrand has as much area below the x axis as above, such as $(0, b)$ or $(-b, 0)$ for an appropriate value of b:

$$\int_0^b x(1 - 2x^2)dx = \left[\frac{1}{2}x^2 - \frac{1}{2}x^4\right]_0^b = \frac{b^2 - b^4}{4},$$

which is zero for $b = 1$. ∎

Example 2.23 (Orthogonal trigonometric functions). Orthogonal sets of trigonometric functions are the cornerstone of Fourier analysis.

(a) Consider the trigonometric functions $\sin x$ and $\cos x$. By inspection of the graph (Figure 2.19(b)), their product, $\sin x \cos x = \frac{1}{2} \sin 2x$, has period π and will integrate to zero on any interval of length π, such as $(0, \pi)$ or $(-\frac{\pi}{2}, \frac{\pi}{2})$. The integral is also zero for intervals of length $n\pi$. The functions are not orthogonal on any other interval, for example, the interval $(0, \frac{\pi}{2})$:

$$\int_0^{\pi/2} \sin x \cos x \, dx = \frac{1}{2}.$$

Graphically, we see that on this interval the positive and negative areas under the curve are not equal, so the integral is not zero.

(b) An even and an odd function, like $\sin x$ and $\cos x$, will always be orthogonal for some interval, because the product of an even and an odd function is odd. For an example with two even functions, consider $\cos x$ and $\cos 2x$ (Figure 2.19(c)). Their product, $\cos x \cos 2x$, is periodic and oscillatory, and by inspection of the graph has odd symmetry about the point $x = \frac{\pi}{2}$. We expect, therefore, that these functions will be orthogonal on the interval $(0, \pi)$. Carrying out the calculation,

$$\int_0^{\pi} \cos x \cos 2x \, dx = \left[\frac{1}{6} \sin 3x + \frac{1}{2} \sin x\right]_0^{\pi} = 0.$$

Unlike the previous example, however, they are not orthogonal on arbitrary intervals of length π, for consider $(\frac{\pi}{2}, \frac{3\pi}{2})$:

$$\int_{\pi/2}^{3\pi/2} \cos x \cos 2x \, dx = \left[\frac{1}{6} \sin 3x + \frac{1}{2} \sin x\right]_{\pi/2}^{3\pi/2} = \frac{2}{3}.$$

The question of orthogonality on more general intervals is considered in the problems. ∎

Complete function spaces

Our ultimate goal is to approximate functions by sequences of other functions— in particular, by sums of sines and cosines, which are continuous functions. We will seek to construct Cauchy sequences of continuous functions, (g_n), such that $\|f - g_n\| \to 0$ for some target function f. Now, it is easy to see that if two continuous functions agree everywhere on an interval, that is, $f - g = 0$, $x \in (a,b)$, then the norm of their difference is zero, $\int_a^b |f(x) - g(x)|^2\, dx = 0$; conversely, if f and g are continuous on (a,b) and $\|f - g\| = 0$, then $f = g$ everywhere on (a,b). So it would seem that continuous functions can only approximate other continuous functions. As the following example shows, however, there are Cauchy sequences of continuous functions whose limits are not continuous.

Example 2.24. The functions f_k, defined

$$f_k(x) = \begin{cases} 0, & 1 - 2^{-k} > x > 0 \\ 2^k(x - 1) + 1, & 1 > x \geq 1 - 2^{-k} \\ 1, & 2 > x \geq 1 \end{cases}$$

belong to the space $C(0,2)$ of functions that are continuous on $(0,2)$. By inspection of the graphs (Figure 2.20), we expect the sequence to be convergent.

Without loss of generality, assume $n > m$. The norm $\|f_n - f_m\|_2$ can be shown to be

$$\|f_n - f_m\|_2 = \frac{2^{-m/2}}{\sqrt{3}}\left(1 - 2^{-(n-m)}\right),$$

which approaches zero as $n, m \to \infty$. Thus, (f_k) is a Cauchy sequence.

Again with reference to the graph, it appears that the sequence is converging to the step function denoted f_∞, and indeed, it can be shown that $\|f_\infty - f_k\| \to 0$ as $k \to \infty$. ∎

Recall that a complete space contains the limits of all its Cauchy sequences. This example showed a sequence of continuous functions with discontinuous limit,

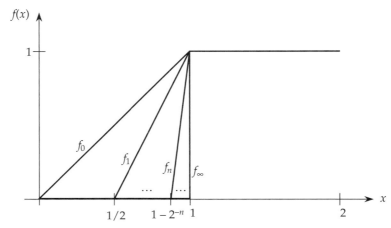

FIGURE 2.20 A sequence of continuous functions converging to a discontinuous function. The vector space $C(0,2)$ with euclidean norm is not complete.

and demonstrates that $C(0, 2)$ (with this choice of norm) is not a complete space.[12] The larger space containing piecewise continuous as well as continuous functions is also not complete. It can be shown that there are Cauchy sequences of Riemann integrable piecewise continuous functions whose limits are discontinuous everywhere and not Riemann integrable.[13] However, such functions are arguably pathological and will be encountered rarely, if at all, in practical applications.

The spaces L^1 and L^2, indeed, all L^p for $\infty > p \geq 1$, can be shown to be complete. They contain continuous and piecewise continuous functions, functions that deviate from piecewise continuity at isolated points, and functions that defy conventional Riemann integration.[14] Significantly, however, it can be shown that a function f in L^p (including any of the wildly discontinuous ones) can be approximated arbitrarily well by a continuous function g in L^p, and there exists a Cauchy sequence of continuous functions (g_n) that converges to f in norm, $\|f - g_n\| \to 0$. We say that the continuous functions are *dense* in L^p.[15] This opens the door to using sequences of continuous functions, like the partial sums of Fourier series, to approximate arbitrary functions in L^p.

The spaces ℓ^2 and L^2 of square-summable sequences and square-integrable functions can be shown to be complete inner product spaces. A complete inner product space is called a *Hilbert space*. Hilbert spaces generalize familiar finite-dimensional euclidean spaces like $\mathbb{R}^n$ and $\mathbb{C}^n$ to infinite dimensions. (Some authors consider $\mathbb{R}^n$ and $\mathbb{C}^n$ to be Hilbert spaces; others confine the term to infinite-dimensional spaces.) Many problems of practical interest in engineering and physics, including Fourier analysis, are elegantly solved through the use of Hilbert space theory.[16]

The other ℓ^p and L^p spaces ($p \neq 2$) may be shown to be complete normed spaces—also known as *Banach spaces*—but they are not inner product spaces.[17] The lack of an inner product makes Banach space theory more difficult and less useful for our purposes than Hilbert space theory. Although they do not have an inner product, they do possess a generalization of the Cauchy–Schwarz inequality, called *Hölder's inequality*.[18] For sequences $x \in \ell^p$ and $y \in \ell^q$, with $1/p + 1/q = 1$,

$$\left| \sum_{n=1}^{\infty} x_n y_n^* \right| \leq \|x\|_p \|y\|_q \qquad (2.36)$$

---

[12] Whether a function space is complete or incomplete can depend critically on the choice of norm. The space $C(0, 2)$, shown in Example 2.24 to be incomplete under the L^2 norm, is also incomplete under the L^1 norm but complete under the supremum norm. See Oden and Demkowicz (1996, pp. 346–348).

[13] Folland (1992, p. 75).

[14] The full theory of L^p spaces depends on the so-called Lebesgue integral, a standard topic in advanced real analysis. For an introduction that is not too technical, see Gasquet and Witomski (1999, pp. 95–130). Fortunately, if a function is Riemann integrable, then it is also Lebesgue integrable, and the two integrals agree. Thus, in all of our practical calculations we may employ results of the more general theory while continuing to use our familiar Riemann integral.

[15] For more about the functions in L^p, see Gasquet and Witomski (1999, pp. 137–139) and also Stade (2005, pp. 163–179), Dym and McKean (1972, p. 22), and Folland (1992, pp. 72–75).

[16] See Young (1988) for a survey of the theory and some of its principal applications. Axler (1997) is an excellent exposition of the theory for finite-dimensional spaces.

[17] For completeness proofs for ℓ^p and L^p, see Oden and Demkowicz (1996, pp. 346–350).

[18] For proofs, see Oden and Demkowicz (1996, pp. 282, 335).

and for functions $f \in L^p(Q)$ and $g \in L^q(Q)$, with $1/p + 1/q = 1$,

$$\left| \int_Q f(x)g^*(x)dx \right| \leq \|f\|_p \|g\|_q. \tag{2.37}$$

In particular, if $f \in L^1$, then the integral $\int_Q f(x)g^*(x)dx$ is guaranteed to be finite only if $g \in L^\infty$ ($\frac{1}{1} + \frac{1}{\infty} = 1$). Hölder's inequality also says that in this case, $fg^* \in L^1$. When $p = 2$, Hölder's inequality is identical to the Cauchy–Schwarz inequality.

Relationships among the various spaces discussed in this chapter are diagrammed in Figure 2.21.

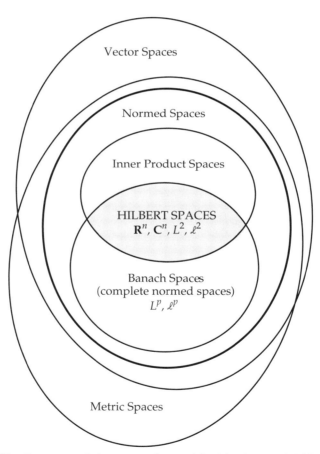

FIGURE 2.21 Taxonomy of the spaces discussed in this chapter. A Hilbert space is a complete inner product space. The finite-dimensional spaces $\mathbb{R}^n$ and $\mathbb{C}^n$, with euclidean norm, are Hilbert spaces, as are the infinite-dimensional spaces ℓ^2 and L^2. The spaces ℓ^1, L^1, ℓ^∞, and L^∞ are Banach spaces.

2.3.4 Orthogonal Expansions in ℓ^2 and L^2

Now that we have seen that inner products and orthogonality can be defined for infinite sequences and functions, it is natural to extend the idea of orthogonal decomposition beyond finite-dimensional vectors. In particular, the subjects of the next three chapters are the four basic types of Fourier expansion, each of which is an orthogonal decomposition in some vector space.

1. The discrete Fourier transform, for finite sequences:

$$f[n] = \frac{1}{N} \sum_{k=0}^{N-1} F[k] e^{i2\pi kn/N}, \quad n = 0, 1, \ldots, N-1.$$

2. The discrete-time Fourier transform, for infinite sequences:

$$f[n] = \frac{1}{2\pi} \int_{-\pi}^{\pi} F_d(\theta) e^{in\theta} d\theta, \quad n = \ldots, -2, -1, 0, 1, 2, \ldots.$$

3. The Fourier series, for functions on bounded intervals. On the interval $[-\pi, \pi]$,

$$f(x) = \sum_{k=-\infty}^{\infty} c_k e^{ikx}, \quad x \in [-\pi, \pi].$$

4. The Fourier transform, for functions on the real line:

$$f(x) = \int_{-\infty}^{\infty} F(\nu) e^{i2\pi \nu x} d\nu, \quad x \in \mathbb{R}.$$

All four expansions are built on complex exponentials, $e^{i2\pi kn/N}$, $e^{in\theta}$, e^{ikx}, or $e^{i2\pi\nu x}$, which will be shown to constitute four orthogonal sets. The coefficients in the linear combinations are $F[k]$, $F_d(\theta)$, c_k, or $F(\nu)$, respectively.

So far, we have seen that the *finite*-dimensional vector spaces $\mathbb{R}^n$ and $\mathbb{C}^n$ possess orthonormal basis sets $\{e_k\}_{k=1}^n$. Any vector v in a finite-dimensional space V can be written as a linear combination of basis vectors by taking inner products, $v = \sum_{k=1}^n \langle v, e_k \rangle e_k$. Parseval's formula says that all the "energy" in v is captured by the expansion coefficients, $\|v\|^2 = \sum_{k=1}^n |\langle v, e_k \rangle|^2$. The truncated series $\hat{v} = \sum_{k=1}^m \langle v, e_k \rangle e_k$ is an approximation to v in the m-dimensional subspace U spanned by $\{e_k\}_{k=1}^m$. The remaining basis vectors $\{e_k\}_{k=m+1}^n$ span the orthogonal complement $U^\perp$, and the residual vector $\tilde{v} = v - \hat{v}$ is perpendicular to $\hat{v}$ and belongs to $U^\perp$.

We next consider the expansion of a vector v in an *infinite*-dimensional inner product space V on an infinite orthonormal set contained in V,

$$\hat{v} = \sum_{k=1}^{\infty} c_k e_k, \quad c_k = \langle v, e_k \rangle. \tag{2.38}$$

Example 2.25. The following are infinite orthonormal sets.

(a) The standard basis in $\mathbb{R}^\infty$,

$$\begin{aligned} e_1 &= (1, \quad 0, \quad 0, \quad \ldots) \\ e_2 &= (0, \quad 1, \quad 0, \quad \ldots) \\ &\vdots \end{aligned}$$

Projecting a vector $v = (v_1, v_2, \ldots) \in \ell^2$ onto this basis returns the elements of v:

$$\langle v, e_k \rangle = v_k.$$

(b) The Legendre polynomials on $[-1, 1]$, the first three of which are

$$p_0(x) = \frac{1}{\sqrt{2}}$$

$$p_1(x) = \sqrt{\frac{3}{2}} x$$

$$p_2(x) = \frac{1}{2}\sqrt{\frac{5}{2}}(3x^2 - 1).$$

(c) Complex exponentials on $[-\pi, \pi]$,

$$e_k(x) = \frac{1}{\sqrt{2\pi}} e^{ikx}, \quad k = 0, \pm 1, \pm 2, \ldots. \tag{2.39}$$

To verify the orthonormality of this set, calculate

$$\langle e_n, e_m \rangle = \int_{-\pi}^{\pi} e_n(x) e_m^*(x) dx = \frac{1}{2\pi} \int_{-\pi}^{\pi} e^{i(n-m)x} dx$$

$$= \frac{1}{2\pi} \frac{e^{i(n-m)\pi} - e^{-i(n-m)\pi}}{i(n-m)} = \frac{\sin(n-m)\pi}{(n-m)\pi}$$

$$= \begin{cases} 1, & n = m \quad \text{(Equation 1.29)} \\ 0, & \text{otherwise.} \end{cases}$$

Expanding a function $f \in L^2[-\pi, \pi]$ on this set leads to the complex Fourier series, which is the subject of Chapter 4.

$$f(x) = \sum_{k=-\infty}^{\infty} \left\langle f, \frac{1}{\sqrt{2\pi}} e^{ikx} \right\rangle \frac{1}{\sqrt{2\pi}} e^{ikx} = \sum_{k=-\infty}^{\infty} \frac{1}{2\pi} \langle f, e^{ikx} \rangle e^{ikx}$$

$$= \sum_{k=-\infty}^{\infty} c_k e^{ikx}, \tag{2.40a}$$

where $c_k = \frac{1}{2\pi} \langle f, e^{ikx} \rangle = \frac{1}{2\pi} \int_{-\pi}^{\pi} f(\xi) e^{-ik\xi} d\xi.$ \hfill (2.40b)

∎

The question that concerns us now is whether, and under what conditions, the infinite series $\sum_{k=1}^{\infty} c_k e_k$ with $c_k = \langle v, e_k \rangle$ (Equation 2.38) is actually equivalent to the vector v. There are three questions to answer:

1. Does the series converge at all?
2. If it does converge, does it converge to a vector in V (is V a complete space)?
3. If the limit is in the same space as v, is it, in fact, equal to v?

Our first concern is whether the infinite series converges. The nth partial sum of the series is $\hat{v}_n = \sum_{k=1}^{n} c_k e_k$. It is contained in V because it is a finite linear combination of the $\{e_k\}$, which are in V. The sequence of partial sums will converge if it is a Cauchy sequence, so check to see if the sequence is Cauchy. The distance between two partial sums is

$$\|\hat{v}_n - \hat{v}_m\|^2 = \left\|\sum_{k=m+1}^{n} c_k e_k\right\|^2 = \sum_{k=m+1}^{n} |c_k|^2,$$

by the orthonormality of the $\{e_k\}$. If the whole infinite series $\sum_{k=1}^{\infty} |c_k|^2$ converges, then we know that the coefficients $|c_k|$ must decay to 0 as $k \to \infty$. Consequently, a sufficient condition for the finite sum $\sum_{k=m+1}^{n} |c_k|^2$ to go to zero as $m, n \to \infty$ is that the entire series $\sum_{k=1}^{\infty} |c_k|^2 = \sum_{k=1}^{\infty} |\langle v, e_k \rangle|^2$ converges. Assurance of this is supplied by the following theorem.

Theorem 2.5 (Bessel's inequality). Let $\{e_k\}_{k=1}^{\infty}$ be an orthonormal set in an inner product space V. Then, for a vector $v \in V$,

$$\sum_{k=1}^{\infty} |\langle v, e_k \rangle|^2 \leq \|v\|^2. \tag{2.41}$$

Proof: Make a finite-dimensional approximation, $\hat{v} = \sum_{k=1}^{n} \langle v, e_k \rangle e_k$. There is an error vector $\tilde{v} = v - \hat{v}$, which is orthogonal to $\hat{v}$. The Pythagorean theorem gives $\|v\|^2 = \|\hat{v}\|^2 + \|\tilde{v}\|^2$. Hence,

$$\|\hat{v}\|^2 = \left\|\sum_{k=1}^{n} \langle v, e_k \rangle e_k\right\|^2 = \sum_{k=1}^{n} |\langle v, e_k \rangle|^2 = \|v\|^2 - \|\tilde{v}\|^2 \leq \|v\|^2,$$

which holds for all n. Taking the limit as $n \to \infty$ yields Equation 2.41. ∎

In a finite-dimensional space, the approximation and the error, $\hat{v}$ and $\tilde{v}$, are two sides of a right triangle, with v the hypotenuse. So $\hat{v}$ can be no longer than v. Bessel's inequality says that the relationship continues to hold in an infinite-dimensional space.

This answers the first question: the sequence of approximations $(\hat{v}_n)$ is a Cauchy sequence, and has a limit. As for the second question, if V is a complete inner product space—a Hilbert space—the sum converges to a vector that is also in V:

$$\lim_{n\to\infty} \sum_{k=1}^{n} \langle v, e_k \rangle e_k \in V.$$

In particular, the spaces ℓ^2 and L^2 are Hilbert spaces. This leaves the third question: is the limit $\lim_{n\to\infty} \sum_{k=1}^{n} \langle v, e_k \rangle e_k$ equal to the vector v? In an n-dimensional space, we know that v is identically $\sum_{k=1}^{n} \langle v, e_k \rangle e_k$, the error $\tilde{v}$ is 0, and $\|v\|^2 = \sum_{k=1}^{n} |\langle v, e_k \rangle|^2$ (Parseval's formula). The best we have been able to say so far about the infinite-dimensional case is Bessel's inequality,

$$\sum_{k=1}^{\infty} |\langle v, e_k \rangle|^2 \leq \|v\|^2,$$

which leaves open the possibility that the approximation error, $\tilde{v} = v - \sum_{k=1}^{\infty} \langle v, e_k \rangle e_k$, could be nonzero. After projecting v onto all the e_k, there could still be something left over, indicating that the orthonormal set $\{e_k\}$, though infinite, is not big enough.

Suppose, in an infinite-dimensional inner product space, we have an orthonormal set $\{e_k\}$ that *is* big enough. Then it will not be possible to find any more vectors in that space that are orthogonal to the $\{e_k\}$ (for if such a vector could be found, we would have to add it to the set). Such a set is said to be *complete*.[19]

Definition 2.14 (Complete orthonormal set). An orthonormal set $\{e_k\}$ in an inner product space V is *complete* if the only vector in V orthogonal to each of the e_k is the zero vector.

Now consider the projection of a vector v onto a complete orthonormal set $\{e_k\}$. The approximation error using N of the vectors is $\tilde{v} = v - \sum_{k=1}^{N} \langle v, e_k \rangle e_k$. Its projections onto the $\{e_k\}$ are

$$\langle \tilde{v}, e_k \rangle = \left\langle v - \sum_{k=1}^{N} \langle v, e_k \rangle e_k, e_n \right\rangle = \langle v, e_n \rangle - \sum_{k=1}^{N} \langle v, e_k \rangle \underbrace{\langle e_k, e_n \rangle}_{=1, k=n} = \begin{cases} \langle v, e_n \rangle & n > N \\ 0 & n \leq N \end{cases}.$$

In the limit as $N \to \infty$, $\langle \tilde{v}, e_k \rangle = 0$ for all k. The error $\tilde{v}$ is orthogonal to each of the e_n, and by the definition, $\tilde{v} = 0$. It follows that $v = \sum_{k=1}^{\infty} \langle v, e_k \rangle e_k$. This answers the third question: the expansion of a vector v on a complete orthonormal set is equal to v in the 2-norm. The complete orthonormal set spans the space V and is a basis for the space. We can think of the vectors $\{e_k\}$ as establishing a coordinate system in V, for which the inner products $\langle v, e_k \rangle$ are the coordinates.

---

[19]This is a different use of "complete" than when we speak of a complete vector space.

Parseval's formula, the statement that orthogonal expansion preserves norms (conserves energy), extends to infinite-dimensional space. To derive the relationship, express the error norm in terms of the inner product, obtaining

$$\left\| v - \sum_{k=1}^{N} \langle v, e_k \rangle e_k \right\|^2 = \langle v, v \rangle - \left\langle v, \sum_{m=1}^{N} \langle v, e_m \rangle e_m \right\rangle - \left\langle \sum_{k=1}^{N} \langle v, e_k \rangle e_k, v \right\rangle$$

$$+ \left\langle \sum_{k=1}^{N} \langle v, e_k \rangle e_k, \sum_{m=1}^{N} \langle v, e_m \rangle e_m \right\rangle$$

$$= \|v\|^2 - \sum_{m=1}^{N} \langle v, e_m \rangle^* \langle v, e_m \rangle - \sum_{k=1}^{N} \langle v, e_k \rangle \langle v, e_k \rangle^*$$

$$+ \sum_{k=1}^{N} \langle v, e_k \rangle \sum_{m=1}^{N} \langle v, e_m \rangle^* \underbrace{\langle e_k, e_m \rangle}_{=1, k=m}$$

$$= \|v\|^2 - \sum_{k=1}^{N} |\langle v, e_k \rangle|^2.$$

Then, as $N \to \infty$,

$$\|\tilde{v}\|^2 = \lim_{N \to \infty} \left\| v - \sum_{k=1}^{N} \langle v, e_k \rangle e_k \right\|^2 = \|v\|^2 - \sum_{k=1}^{\infty} |\langle v, e_k \rangle|^2 = 0$$

because, as we have seen, $\tilde{v} \to 0$. Thus,

$$\|v\|^2 = \sum_{k=1}^{\infty} |\langle v, e_k \rangle|^2. \tag{2.42}$$

It can also be shown that just as in the finite-dimensional case, orthogonal expansion preserves inner products as well as norms (Parseval),

$$\langle u, v \rangle = \sum_{k=1}^{\infty} \langle u, e_k \rangle \langle v, e_k \rangle^*. \tag{2.43}$$

These results are summarized in the following theorem.

Theorem 2.6. Let $\{e_k\}$ be a complete orthonormal set in a Hilbert space V. Then, for vectors $u, v \in V$:

(a) $v = \sum_{k=1}^{\infty} \langle v, e_k \rangle e_k$ in norm.
(b) $\|v\|^2 = \sum_{k=1}^{\infty} |\langle v, e_k \rangle|^2$, preserving norms.
(c) $\langle u, v \rangle = \sum_{k=1}^{\infty} \langle u, e_k \rangle \langle v, e_k \rangle^*$, preserving inner products.

Proving that a particular orthonormal set is complete is often quite technical, and we will simply take the completeness of some key sets for granted. In particular, the complex exponential set, $\{e^{ikx}\}_{k=-\infty}^{\infty}$, which we know is orthogonal on $[-\pi, \pi]$, is complete.[20] It is a basis for the Hilbert space $L^2[-\pi, \pi]$, and the foundation for our study of Fourier series in Chapter 4.

★ 2.4 OPERATORS

An *operator*, also called a *transformation*, is a mapping between two vector spaces (the spaces are always assumed to have the same set of scalars, i.e., real or complex numbers). We write $T: V \to W$ to indicate that an operator T maps vectors in V to vectors in W, and write $w = T(v)$ or just $w = Tv$ to say that w is the result of T operating on v. The branch of mathematics known as *functional analysis* is the study of operators on linear spaces. It has deep connections to Fourier analysis, system theory, and physics, especially quantum mechanics.[21] In this section we will set out a few elementary ideas about operators. It may be skipped on a first reading.

Linearity
The most useful operators for our purposes are *linear*. An operator $T: V \to W$ is *linear* if, for all vectors $f, g \in V$ and scalars c,

$$T(f + g) = T(f) + T(g) \tag{2.44a}$$
$$T(cf) = cT(f). \tag{2.44b}$$

The idea of linearity extends easily to a finite linear combination of vectors,

$$T\left(\sum_{k=1}^{n} c_k f_k\right) = \sum_{k=1}^{n} c_k T(f_k).$$

This is the same definition of linearity you may be familiar with from system theory, that is, a system is linear if its response to a composite input $ax(t) + by(t)$ is the sum of a times the response to $x(t)$ plus b times the response to $y(t)$.

If we have two linear operators, $S, T: V \to W$, their sum may be defined in a natural way,

$$(S + T)v = Sv + Tv.$$

The sum of two linear operators is also linear. (In fact, it can be shown that the set of all linear operators mapping V to W is a vector space.) Operators may also be composed: if $S: V \to W$ and $T: U \to V$ are operators, then $ST: U \to W$ is defined

$$STu = S(T(u)).$$

First, T maps $u \in U$ to a vector $T(u) \in V$. Then, S maps $T(u)$ to the vector $S(T(u)) \in W$. Composition does not, in general, commute: $ST \neq TS$ (e.g., matrix multiplication). The composition of two linear operators is also linear. Using addition and composition, other linear operators may be constructed from simpler ones. For example, the derivative operator $\frac{d}{dx}$ is linear. The composition with itself is the

---

[20] For proofs of the completeness of the complex exponential functions e^{ikx}, see Folland (1992, pp. 78–79), Young (1988, pp. 45–52), and Dym and McKean (1972, pp. 30–36).

[21] An introduction to functional analysis at the level of this text is Oden and Demkowicz (1996).

second derivative operator, $\frac{d^2}{dx^2}$, and the linear differential equation $\frac{d^2 f}{dx^2} + p(x)f = u$ is an operator equation, $Lf = \left(\frac{d^2}{dx^2} + p(x)\right)f = u$. Linear operators are ubiquitous in mathematics, science, and engineering.

One-to-one and onto operators; null space
Many properties of mappings and functions in $\mathbb{R}$ (Section 1.5) carry over to operators. For an operator $T: V \to W$, the vector space V is the *domain* of T, and the set $T(V) = \{Tv \mid v \in V\} \subseteq W$ is the *range* of T. If $T(V) = W$, we say that T is *onto* W (or *surjective*)—every point in W is the image under T of some point in V. We say that T is *one-to-one* (or *injective*) if $T(v_1) = T(v_2)$ only if $v_1 = v_2$, that is, each point in V maps to a unique point in W. An operator that is both one-to-one and onto is also called *bijective*. A bijection establishes a one-to-one correspondence between the points in V and the points in W (recall Figure 1.9).

An operator $T: V \to W$ is not one-to-one if there is more than one point in V mapping to some point in W. Of particular interest are those points in V that map to the zero vector in W (remember, every vector space has a zero vector, to serve as the additive identity). They comprise the *null space* of T, $N(T) = \{v \in V \mid Tv = 0\}$. The null space is a subspace of the domain V, representing information that is lost (or obliterated) by T. In signal processing, for example, an ideal filter is a linear operator, and its stopband (those frequencies that are blocked by the filter) is the operator's null space.

Example 2.26 (Operators on finite-dimensional spaces). The most familiar examples of linear operators on vector spaces are matrices. All linear operators on finite-dimensional vector spaces have matrix representations. Here are some examples of matrix operators.

Consider the mapping $\mathcal{A}: \mathbb{R}^3 \to \mathbb{R}^2$ represented by the matrix $\mathbf{A} = \begin{bmatrix} 1 & 0 & 0 \\ 0 & 1 & 0 \end{bmatrix}$. It operates on a vector $\begin{bmatrix} a \\ b \\ c \end{bmatrix}$ to produce the vector $\begin{bmatrix} 1 & 0 & 0 \\ 0 & 1 & 0 \end{bmatrix} \begin{bmatrix} a \\ b \\ c \end{bmatrix} = \begin{bmatrix} a \\ b \end{bmatrix}$. This operator is onto but not one-to-one, because every vector in $\mathbb{R}^2$ is the image of an infinite number of vectors in $\mathbb{R}^3$ (just drop the third component). The null space of $\mathcal{A}$ consists of all vectors of the form $\begin{bmatrix} 0 \\ 0 \\ c \end{bmatrix}$.

On the other hand, consider the mapping $\mathcal{B}: \mathbb{R}^2 \to \mathbb{R}^3$ defined by the matrix $\mathbf{B} = \begin{bmatrix} 1 & 0 \\ 0 & 1 \\ 0 & 0 \end{bmatrix}$. It operates on a vector $\begin{bmatrix} a \\ b \end{bmatrix}$ to produce the vector $\begin{bmatrix} 1 & 0 \\ 0 & 1 \\ 0 & 0 \end{bmatrix} \begin{bmatrix} a \\ b \end{bmatrix} = \begin{bmatrix} a \\ b \\ 0 \end{bmatrix}$. This operator is one-to-one but not onto. There is no vector in $\mathbb{R}^2$ that maps to a vector in $\mathbb{R}^3$ having a nonzero third element. The null space is trivial, consisting solely of the zero vector $\begin{bmatrix} 0 \\ 0 \end{bmatrix}$.

Finally, the operator $\mathcal{C}: \mathbb{R}^2 \to \mathbb{R}^2$ with square matrix $\mathbf{C} = \begin{bmatrix} 1 & 0 \\ 0 & 2 \end{bmatrix}$ maps a vector $\begin{bmatrix} a \\ b \end{bmatrix} \in \mathbb{R}^2$ to another vector in $\mathbb{R}^2$, namely $\begin{bmatrix} a \\ 2b \end{bmatrix}$. This operator is both one-to-one and onto. Again, the null space consists solely of the zero vector. Not every square matrix represents a bijection, however: consider $\mathbf{D} = \begin{bmatrix} 1 & 0 \\ 2 & 0 \end{bmatrix}$ and $\mathbf{E} = \begin{bmatrix} 1 & 2 \\ 0 & 0 \end{bmatrix}$. Are these one-to-one? Onto? What are their null spaces? ∎

Example 2.27 (Operators on infinite-dimensional spaces). Here are a few operators that work on infinite-dimensional spaces.

1. Differentiation: The operator $\frac{d}{dx}: C^{(1)}[0,1] \to C[0,1]$ maps functions $f(x)$ that are continuously differentiable on $[0,1]$ to their derivatives $f'(x)$. It is a linear operator. It is not one-to-one, because all functions of the form $g(x) = f(x) + c$ have the same derivative, namely, $f'(x)$; its null space is the set of all constant functions. It is, however, onto, because every continuous function is the derivative of some continuously differentiable function.

2. Forward and backward shift: For sequences $v = (v_0, v_1, v_2, \ldots) \in \ell^2$, define forward shift $\mathcal{T}_f$ and backward shift $\mathcal{T}_b$ by $\mathcal{T}_f v = (0, v_0, v_1, \ldots)$ and $\mathcal{T}_b v = (v_1, v_2, \ldots)$. Both operators are linear. The forward shift operator is one-to-one, but not onto. The backward shift operator is onto, but not one-to-one (what is its null space)?

3. Inner product: Let V be a complex inner product space. The inner product operator $\langle \cdot, \cdot \rangle : V \times V \to \mathbb{C}$ is linear in the first operand: $\langle af_1 + bf_2, g \rangle = a \langle f_1, g \rangle + b \langle f_2, g \rangle$. If V is a real inner product space, $\langle \cdot, \cdot \rangle$ is also linear in the second operand, but not if V is a complex inner product space.

4. Norm: Let V be a normed space. The norm operator, $\|\cdot\| : V \to \mathbb{R}$, is not linear. ∎

Operator norm and bounded operators
In general, an operator $\mathcal{T}: V \to W$ will change the norm of a vector, making it "longer" or "shorter." Let the spaces V and W have norms $\|\cdot\|_V$ and $\|\cdot\|_W$, respectively. They need not be the same norm, for example, we could have $V = L^1$ and $W = L^\infty$. If a linear operator always maps a vector $v \in V$ with finite norm $\|v\|_V$ into a vector $\mathcal{T}v \in W$ with finite norm $\|\mathcal{T}v\|_W$, we say that the operator is *bounded*. The amount of "stretch" for a particular vector v is $\frac{\|\mathcal{T}v\|_W}{\|v\|_V}$; the supremum of this ratio over all (nonzero) vectors v is the maximum stretch that the operator can produce. It can be shown that this supremum has all the properties of a norm, and so it is called the *operator norm*,

$$\|\mathcal{T}\| = \sup \left\{ \frac{\|\mathcal{T}v\|_W}{\|v\|_V} \,\bigg|\, v \neq 0 \right\}. \tag{2.45a}$$

An equivalent definition is

$$\|\mathcal{T}\| = \sup \left\{ \|\mathcal{T}v\|_W \,\big|\, \|v\|_V \leq 1 \right\}. \tag{2.45b}$$

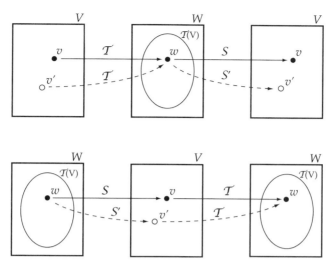

FIGURE 2.22 Inverse operators. *Top:* Left inverse. If T is one-to-one, there is only one point v in V that maps to the point w in W, and there is an inverse S such that $STv = v$ for every $v \in V$. If T is not one-to-one, there is no left inverse. More than one point in V maps to w (e.g., the dashed line), and it is not possible to "look back" through T to know whether Sw should be v or v'. *Bottom:* Right inverse. If T is not one-to-one, then there is more than one possible right inverse: $TSw = TS'w = w$. But if T is one-to-one, there is only one path from V to w, and only one right inverse. Moreover, if T is one-to-one, the right and left inverse are the same operator.

Clearly, a bounded operator has finite norm. It can be shown that a bounded linear operator is also continuous: vectors that are close to each other in V are mapped by a bounded T to vectors that are close to each other in W.[22]

Inverse operators

If an operator is one-to-one, so that every point in the range $T(V)$ comes from a unique point in the domain V, we may define an inverse operator that maps every point in the range back to where it came from in the domain. That is, the composition of T with its inverse is the *identity operator* that maps a vector to itself. However, because composition does not commute, the inverse may in principle be applied before or after T, that is, there are two inverses. Let $S : W \to V$ and $T : V \to W$. In the composition ST, a vector $v \in V$ is mapped to $Tv \in W$, then mapped back to $STv \in V$. If $STv = v$, we say that S is a *left inverse* of T. In the other composition, TS, a vector $w \in T(V)$ is first mapped to $v = Sw \in V$, then back to $TSw \in T(V)$. If $TSw = w$, we say that S is a *right inverse* of T (Figure 2.22).

Consider the existence (or not) of a left inverse. If T is not one-to-one, then there are (at least) two points, call them v and v', that are mapped by T to the same image point w. The left inverse S can be defined so that Sw is either v or v', but not

---

[22] For a precise statement of the continuity and boundedness of linear operators, see Oden and Demkowicz (1996, pp. 405–407).

both. To have $STv = v$, there can be only one preimage in V for each input to S. This includes the null space: if $T(v) = 0$ for some $v \neq 0$, then T is not one-to-one, because there are two points, 0 and v, that map to 0. Therefore, T will have one (and only one) left inverse *only* if it is one-to-one.

For the right inverse, if T is not one-to-one, then for some $w \in T(V)$ there is more than one identity path from w through V back to w. The right inverse must be defined so that only one of the paths is followed. Either one is satisfactory—an operator may have more than one right inverse.

So, if T is one-to-one, then there is one left inverse and one right inverse, and Figure 2.22 suggests that these inverses are the same S. If T is onto as well as one-to-one, then the domain of S is the whole space W, and we have an appealing symmetry: $T: V \to W$ and $S: W \to V$. Finally, if the (left or right) inverse S is bounded, we say that T is *invertible* and define $T^{-1} = S$. The inverse T^{-1} is also a linear operator.

Example 2.28 (Inverse operators on finite-dimensional spaces). Consider again the finite-dimensional operators discussed in Example 2.26. The matrix $\mathbf{A} = \begin{bmatrix} 1 & 0 & 0 \\ 0 & 1 & 0 \end{bmatrix}$ performs the mapping $\begin{bmatrix} a \\ b \\ c \end{bmatrix} \mapsto \begin{bmatrix} a \\ b \end{bmatrix}$. It is not one-to-one. It has no left inverse, because there is no way to recover the third element c once it is lost. On the other hand, it has an infinite number of right inverses, mapping $\begin{bmatrix} a \\ b \end{bmatrix} \mapsto \begin{bmatrix} a \\ b \\ c \end{bmatrix}$ with c arbitrary. A matrix form of the right inverse is $\begin{bmatrix} 1 & 0 \\ 0 & 1 \\ r & s \end{bmatrix}$, and the product with $\mathbf{A}$ gives

$$\begin{bmatrix} 1 & 0 & 0 \\ 0 & 1 & 0 \end{bmatrix} \begin{bmatrix} 1 & 0 \\ 0 & 1 \\ r & s \end{bmatrix} = \begin{bmatrix} 1 & 0 \\ 0 & 1 \end{bmatrix},$$

the identity for $\mathbb{R}^2$. But this is not a left inverse, because

$$\begin{bmatrix} 1 & 0 \\ 0 & 1 \\ r & s \end{bmatrix} \begin{bmatrix} 1 & 0 & 0 \\ 0 & 1 & 0 \end{bmatrix} = \begin{bmatrix} 1 & 0 & 0 \\ 0 & 1 & 0 \\ r & s & 0 \end{bmatrix},$$

which is not an identity for $\mathbb{R}^3$.

The matrix $\mathbf{B} = \begin{bmatrix} 1 & 0 \\ 0 & 1 \\ 0 & 0 \end{bmatrix}$ performs the mapping $\begin{bmatrix} a \\ b \end{bmatrix} \mapsto \begin{bmatrix} a \\ b \\ 0 \end{bmatrix}$. It is one-to-one but not onto. It has a left inverse that maps $\begin{bmatrix} a \\ b \\ 0 \end{bmatrix} \mapsto \begin{bmatrix} a \\ b \end{bmatrix}$. One of the matrix forms of the left inverse is $\begin{bmatrix} 1 & 0 & 0 \\ 0 & 1 & 0 \end{bmatrix}$. Multiplying this with **B**,

$$\begin{bmatrix} 1 & 0 & 0 \\ 0 & 1 & 0 \end{bmatrix} \begin{bmatrix} 1 & 0 \\ 0 & 1 \\ 0 & 0 \end{bmatrix} = \begin{bmatrix} 1 & 0 \\ 0 & 1 \end{bmatrix},$$

the identity for $\mathbb{R}^2$. It is also a right inverse, for

$$\begin{bmatrix} 1 & 0 \\ 0 & 1 \\ 0 & 0 \end{bmatrix} \begin{bmatrix} 1 & 0 & 0 \\ 0 & 1 & 0 \end{bmatrix} = \begin{bmatrix} 1 & 0 & 0 \\ 0 & 1 & 0 \\ 0 & 0 & 0 \end{bmatrix}.$$

This is not an identity for $\mathbb{R}^3$, but it is an identity for the range of **B**, the set of all vectors $\begin{bmatrix} a \\ b \\ 0 \end{bmatrix}$, which is all it has to be. (The composition TS maps $T(V) \to T(V)$; only if T is bijective does TS map $W \to W$.)

Finally, the matrix $\mathbf{C} = \begin{bmatrix} 1 & 0 \\ 0 & 2 \end{bmatrix}$ represents the operation $\begin{bmatrix} a \\ b \end{bmatrix} \mapsto \begin{bmatrix} a \\ 2b \end{bmatrix}$. It is one-to-one and onto, and has a left inverse represented by the matrix $\begin{bmatrix} 1 & 0 \\ 0 & \frac{1}{2} \end{bmatrix}$. This is also a right inverse, mapping any vector $\begin{bmatrix} c \\ d \end{bmatrix} \in \mathbb{R}^2$ to the vector $\begin{bmatrix} c \\ d/2 \end{bmatrix}$, which is then mapped by **C** back to $\begin{bmatrix} c \\ d \end{bmatrix}$. ∎

Inverses for the forward and backward shift operators on ℓ^2 are considered in the problems.

In the application to signal processing, an operator T often represents some loss of information, and inverting T amounts to trying to recover the lost information. For example, consider the representation of a signal vector v as an orthogonal expansion,

$$v = \sum_k c_k e_k, \quad c_k = \langle v, e_k \rangle.$$

After passing v through a system characterized by a linear operator T, the resulting vector is

$$w = \sum_k c_k \, T e_k.$$

Now, if for some k, $\|Te_k\|$ is small, or worse, if $Te_k = 0$ (e_k is in the null space of T), the basis vector e_k will be absent, or nearly so, from w. This attenuation could be by design, for example, T is a lowpass filter, or it could represent some sort of degradation, like the blurring of an image. Recovering e_k's original contribution to v, if desired, requires amplification of the attenuated component, and if $\|Te_k\|$ is very small or zero, $\|T^{-1}\|$ could be very large or even unbounded. In practice, perfect signal recovery is usually not possible, and one constructs approximate inverses (called *pseudoinverses*) that are computationally tractable.[23]

Isometric and unitary operators; isomorphic spaces

Parseval's formula says that orthogonal projection preserves inner products and norms. If f and g are vectors in a Hilbert space and $F = (F_n)$ and $G = (G_n)$ are their orthogonal projections onto a basis for the space, then $\langle f, g \rangle = \langle F, G \rangle$ and $\|f\| = \|F\|$. We will see now that Parseval's formula is just one instance of a general property of certain linear operators.

Let V and W be normed spaces, and $T: V \to W$ be an operator. If the norm of the operator's output is always the same as the norm of its input, $\|Tv\|_W = \|v\|_V$, we say that T is an *isometry* or a *norm-preserving* linear operator. This is a stronger statement than $\|T\| = 1$. The latter says only that $\|Tv\| \le \|v\|$, and in fact, since the operator norm is a supremum, there might not be any $v \in V$ such that $\|Tv\|$ is exactly equal to $\|v\|$. On the other hand, if T is an isometry, $\|Tv\|$ is identically $\|v\|$ for all $v \in V$.

Going further, suppose V and W are Hilbert spaces. If $T: V \to W$ is invertible and preserves inner products, $\langle f, g \rangle = \langle Tf, Tg \rangle$, it is called a *unitary* operator. Setting $f = g$, we have $\langle f, f \rangle = \langle Tf, Tf \rangle$, showing that a unitary operator is also isometric.

We may gain some insight into the significance of unitary operators by returning to vectors in a plane. Recall that the dot product of two vectors has the property $u \cdot v = \|u\| \|v\| \cos \theta$, where θ is the angle between the vectors. So, the angle θ may be calculated from

$$\theta = \cos^{-1} \frac{u \cdot v}{\|u\| \|v\|}.$$

Any linear transformation of the vectors u and v that preserves their inner product, that is, a unitary operator, preserves the angle between them. For example, consider a rotation in the plane, represented by the matrix

$$\mathbf{T} = \begin{bmatrix} \cos \varphi & -\sin \varphi \\ \sin \varphi & \cos \varphi \end{bmatrix}.$$

Suppose $u = \begin{bmatrix} 1 \\ 0 \end{bmatrix}$ and $v = \begin{bmatrix} 1 \\ 1 \end{bmatrix}$. By inspection, their dot product is 1 and their norms are 1 and $\sqrt{2}$, respectively. The angle between them is $\cos^{-1} \frac{1}{\sqrt{2}} = \frac{\pi}{4}$. After rotation, the vectors become $\mathbf{T}u = \begin{bmatrix} \cos \varphi \\ \sin \varphi \end{bmatrix}$ and $\mathbf{T}v = \begin{bmatrix} \cos \varphi - \sin \varphi \\ \sin \varphi + \cos \varphi \end{bmatrix}$. Their norms are still 1 and $\sqrt{2}$, and their dot product is still 1. Thus, the angle between them is unchanged. The operator simply rotates all vectors by the same amount, preserving their norms and the angles between. Another operation is reflection through the y axis,

$$\mathbf{T} = \begin{bmatrix} -1 & 0 \\ 0 & 1 \end{bmatrix}.$$

---

[23] For the application of linear operators to image recovery from noisy, blurred data, see Andrews and Hunt (1977) and Barrett and Myers (2004).

With the same u and v, we have $\mathbf{T}u = \begin{bmatrix} -1 \\ 0 \end{bmatrix}$ and $\mathbf{T}v = \begin{bmatrix} -1 \\ 1 \end{bmatrix}$. Their norms are still 1 and $\sqrt{2}$, and their dot product is, again, 1. Their angle is preserved.

An operator that maps a finite-dimensional vector space to itself, $\mathcal{T}: V \to V$, is represented by a square matrix $\mathbf{T}$. The inner product $\langle \mathcal{T}u, \mathcal{T}v \rangle$ is

$$\langle \mathcal{T}u, \mathcal{T}v \rangle = (\mathbf{T}v)^\dagger \mathbf{T}u = v^\dagger \mathbf{T}^\dagger \mathbf{T}u,$$

where $\dagger$ denotes the adjoint, or complex conjugate transpose, $\mathbf{T}^\dagger = (\mathbf{T}')^*$. If $\mathcal{T}$ is unitary, then

$$v^\dagger \mathbf{T}^\dagger \mathbf{T}u = v^\dagger u \Rightarrow \mathbf{T}^\dagger \mathbf{T} = \mathbf{I}.$$

That is, $\mathbf{T}^\dagger = \mathbf{T}^{-1}$. A matrix $\mathbf{T}$ having this property is called, appropriately, a *unitary matrix*. The product $\mathbf{T}^\dagger \mathbf{T}$ consists of dot products between the rows of $\mathbf{T}$. Thus, the rows of a unitary matrix are mutually orthogonal. We verify this for the preceding examples. The product $\mathbf{T}^\dagger \mathbf{T}$ for the rotation operator is

$$\begin{bmatrix} \cos\varphi & \sin\varphi \\ -\sin\varphi & \cos\varphi \end{bmatrix} \begin{bmatrix} \cos\varphi & -\sin\varphi \\ \sin\varphi & \cos\varphi \end{bmatrix}$$
$$= \begin{bmatrix} \cos^2\varphi + \sin^2\varphi & -\cos\varphi\sin\varphi + \cos\varphi\sin\varphi \\ \cos\varphi\sin\varphi - \cos\varphi\sin\varphi & \sin^2\varphi + \cos^2\varphi \end{bmatrix} = \begin{bmatrix} 1 & 0 \\ 0 & 1 \end{bmatrix},$$

and for the reflection operator,

$$\begin{bmatrix} -1 & 0 \\ 0 & 1 \end{bmatrix} \begin{bmatrix} -1 & 0 \\ 0 & 1 \end{bmatrix} = \begin{bmatrix} 1 & 0 \\ 0 & 1 \end{bmatrix}.$$

We may now, with an exercise of imagination, think of the angle between two vectors f and g in a Hilbert space, and define it in a similar way,

$$\theta = \cos^{-1} \frac{\langle f, g \rangle}{\|f\| \|g\|}.$$

We operate on f and g with a unitary operator $\mathcal{T}$. By virtue of being unitary, $\mathcal{T}$ is also isometric. The angle between $\mathcal{T}f$ and $\mathcal{T}g$ is

$$\cos^{-1} \frac{\langle \mathcal{T}f, \mathcal{T}g \rangle}{\|\mathcal{T}f\| \|\mathcal{T}g\|} = \cos^{-1} \frac{\langle f, g \rangle}{\|f\| \|g\|},$$

the same as the angle between f and g. In this sense, the two spaces V and W that are connected by $\mathcal{T}$ have the same "shape." We say that they are *isomorphic*.

The Hilbert space analogy to coordinate rotation is multiplication by a complex number $ae^{i\varphi}$ ($a > 0$). This operator maps a complex Hilbert space V to itself. We have $\|\mathcal{T}f\| = \|ae^{i\varphi}f\| = a\|f\|$ and similarly for g. By taking $a = 1$ we make $\mathcal{T}$ an isometry. Then, the inner product is $\langle \mathcal{T}f, \mathcal{T}g \rangle = \langle e^{i\varphi}f, e^{i\varphi}g \rangle = e^{i\varphi}e^{-i\varphi}\langle f, g \rangle = \langle f, g \rangle$. The inner product between f and g, and hence the angle between them, is preserved under $\mathcal{T}$. The mapping is invertible, with $\mathcal{T}^{-1} = e^{-i\varphi}$; thus it is unitary.

It can be shown that all *separable* Hilbert spaces (those that have a countable basis) are isomorphic either to $\mathbb{C}^n$ (if finite dimensional) or to ℓ^2 (if infinite dimensional). The operator that accomplishes this bit of magic is the orthogonal expansion developed in the previous section. For example, let f be a vector in a Hilbert space V with an orthonormal basis $\{e_n\}_{n=1}^{\infty}$ and calculate the orthogonal projections $F_n = \langle f, e_n \rangle$. We then can show the following:

- Projection onto the basis is an operator $\mathcal{P}$ that maps the vector f to the infinite sequence $F = (F_n)_{n=1}^{\infty}$. It is a linear operator by the linearity of the inner product.
- $F \in \ell^2$ because, by Parseval's formula, $\|F\|^2 = \sum_{n=1}^{\infty} |F_n|^2 = \|f\|^2 < \infty$. Thus, $\mathcal{P}$ is a linear operator from V to ℓ^2.
- $\mathcal{P}$ is one-to-one. If it were not, then there would be vectors $f \neq g$ such that $\mathcal{P}f = \mathcal{P}g$. But, by linearity, we would have $\mathcal{P}f - \mathcal{P}g = \mathcal{P}(f-g) = 0$ with $f - g \neq 0$. This means that $f - g$ must be orthogonal to every basis vector, $\langle f - g, e_k \rangle = 0$ for all k, which cannot be, since the basis is a complete set.
- $\mathcal{P}$ preserves inner products: $\langle f, g \rangle = \langle F, G \rangle = \langle \mathcal{P}f, \mathcal{P}g \rangle$ (Parseval's formula). Consequently, $\mathcal{P}$ is isometric, hence bounded, and continuous.
- $\mathcal{P}$ is invertible, as we saw in the previous section: $\mathcal{P}^{-1} F = \sum_{n=1}^{\infty} F_n e_n = f$ (a.e.). Thus $\mathcal{P}$ is unitary.

★ 2.5 CREATING ORTHONORMAL BASES — THE GRAM–SCHMIDT PROCESS

A basis for a Hilbert space can be any complete set of independent vectors, and there are realistic situations where the natural basis is not orthonormal. On the other hand, orthonormal bases are considerably easier to work with. Given a nonorthogonal basis, it is possible to construct an orthonormal basis that spans the same space. The method used to do this is called the *Gram–Schmidt process*.

First consider vectors v_1 and v_2 in the plane ($\mathbb{R}^2$) (Figure 2.23). Arbitrarily select $e_1 = v_1 / \|v_1\|$ to be the first basis vector. Then, express v_2 in terms of e_1 and a (yet unknown) residual vector v_2', which is orthogonal to e_1:

$$v_2 = v_2' + \langle v_2, e_1 \rangle e_1.$$

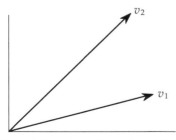
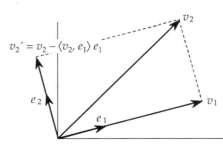

FIGURE 2.23 The Gram–Schmidt process for constructing orthonormal basis vectors. The first vector, v_1, defines the direction of the first basis vector, e_1. The second basis vector is constructed from the residual after v_2 is projected onto e_1.

2.5 CREATING ORTHONORMAL BASES—THE GRAM–SCHMIDT PROCESS

Solving for v'_2,
$$v'_2 = v_2 - \langle v_2, e_1 \rangle e_1.$$

Normalized, v'_2 becomes the other basis vector,
$$e_2 = \frac{v'_2}{\|v'_2\|}.$$

This can be taken to three dimensions by projecting a third independent vector v_3 onto e_1 and e_2:
$$v'_3 = v_3 - (\langle v_3, e_2 \rangle e_2 + \langle v_3, e_1 \rangle e_1)$$
$$e_3 = \frac{v'_3}{\|v'_3\|}.$$

Generalizing to n dimensions, if $\{v_1, v_2, \ldots v_n\}$ are linearly independent vectors in $\mathbb{R}^n$, then an orthonormal basis $\{e_1, e_2, \ldots e_n\}$ for $\mathbb{R}^n$ is constructed recursively as follows:

$$e_k = \frac{v'_k}{\|v'_k\|}, k = 1, 2, \ldots n$$
$$v'_k = v_k - \sum_{i=1}^{k-1} \langle v_k, e_i \rangle e_i. \qquad (2.46)$$

Example 2.29. We will create an orthonormal basis from the vectors
$$v_1 = (1, \quad 2, \quad 2)$$
$$v_2 = (1, \quad 0, \quad 1)$$
$$v_3 = (1, \quad -1, \quad 2).$$

Take the first unit vector to be
$$e_1 = \frac{v_1}{\|v_1\|} = \frac{1}{3}(1, \quad 2, \quad 2).$$

The second vector is
$$v'_2 = v_2 - \langle v_2, e_1 \rangle e_1 = (1, \quad 0, \quad 1) - \left\langle (1, \quad 0, \quad 1), \frac{1}{3}(1, \quad 2, \quad 2) \right\rangle \cdot \frac{1}{3}(1, \quad 2, \quad 2)$$
$$= \frac{1}{3}(2, \quad -2, \quad 1)$$
$$e_2 = \frac{v'_2}{\|v'_2\|} = \frac{1}{3}(2, \quad -2, \quad 1).$$

The third vector is
$$v'_3 = v_3 - \langle v_3, e_2 \rangle e_2 - \langle v_3, e_1 \rangle e_1 = \frac{1}{3}(-2, \quad -1, \quad 2)$$
$$e_3 = \frac{v'_3}{\|v'_3\|} = \frac{1}{3}(-2, \quad -1, \quad 2).$$

These vectors are shown in Figure 2.24. ∎

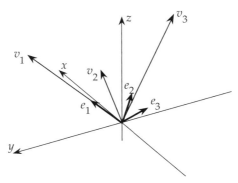

FIGURE 2.24 Result of Gram–Schmidt construction (Example 2.29).

Example 2.30 (Dependent vectors).   Let us see what happens if the vectors to be orthogonalized are not linearly independent. In the previous example, change v_3 to $(0,\ 2,\ 1)$. The resulting v'_3 turns out to be

$$v'_3 = v_3 - \langle v_3, e_2 \rangle e_2 - \langle v_3, e_1 \rangle e_1$$
$$= (0,\ 2,\ 1) + \left(\tfrac{2}{3},\ -\tfrac{2}{3},\ \tfrac{1}{3}\right) - \left(\tfrac{2}{3},\ \tfrac{4}{3},\ \tfrac{4}{3}\right)$$
$$= 0.$$

As one might expect, with only two independent vectors Gram–Schmidt gives back only two orthogonal vectors. A general verification of this is left to the problems. ∎

Example 2.31 (Orthogonal polynomials).   The monomials $\{1, x, x^2 \ldots\}$ are not orthogonal. We will use the Gram–Schmidt process to construct from them a set of polynomials orthonormal on $[-1, 1]$. These are called the Legendre polynomials. The first three members of the family are

$$p_0(x) = \frac{1}{\|1\|} = \frac{1}{\sqrt{2}}$$

$$p_1(x) = \frac{x - \left(\int_{-1}^{1} \frac{x}{\sqrt{2}} dx\right) \frac{1}{\sqrt{2}}}{\|\cdots\|} = \sqrt{\frac{3}{2}} x$$

$$p_2(x) = \frac{x^2 - \left(\int_{-1}^{1} \sqrt{\frac{3}{2}} x^3 dx\right) \sqrt{\frac{3}{2}} x - \left(\int_{-1}^{1} \frac{x^2}{\sqrt{2}} dx\right) \frac{1}{\sqrt{2}}}{\|\cdots\|} = \frac{1}{2}\sqrt{\frac{5}{2}}(3x^2 - 1).$$

They are illustrated in Figure 2.25. ∎

Finally, here is an extended example that brings together many of the ideas of this chapter.

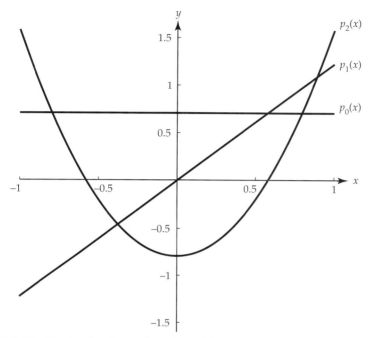

FIGURE 2.25 The first three Legendre polynomials, obtained by orthogonalizing the monomials $\{1, x, x^2\}$ on the interval $[-1, 1]$.

Example 2.32 (Linear regression). In an experiment, observations of a dependent variable y have been made at n values of the independent variable x, yielding n pairs $(x_1, y_1), (x_2, y_2), \ldots, (x_n, y_n)$. We wish to model the relationship between x and y by a straight line,

$$y = ax + b - \tilde{y},$$

where $\tilde{y}$ is the residual, or fitting error. For each pair $(x_i, y_i), i = 1, 2, \ldots n,$

$$\hat{y}_i = ax_i + b \quad \text{(model)}$$
$$\tilde{y}_i = \hat{y}_i - y_i \quad \text{(residual)}$$

Gathering all the points into n-vectors, we have

$$\hat{y} = ax + bu$$
$$\text{where } \hat{y} = (\hat{y}_1 \quad \hat{y}_2 \quad \ldots \quad \hat{y}_n)$$
$$x = (x_1 \quad x_2 \quad \ldots \quad x_n)$$
$$u = (1 \quad 1 \quad \ldots \quad 1)$$

that is, we are approximating y by a linear combination of two vectors, x and u.

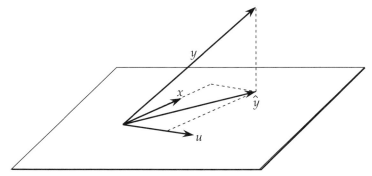

FIGURE 2.26 Illustrating linear regression as a projection onto the plane.

We wish to minimize $\|\tilde{y}\|^2 = \|y - \hat{y}\|^2$. By the orthogonality principle, $\|\tilde{y}\|^2$ will be minimized by projecting y onto the subspace (in this case, a plane) spanned by x and u (Figure 2.26).

We cannot project y onto the plane simply by projecting onto x and u, however. They are not (in general) orthonormal, though they will be guaranteed to be linearly independent by choosing at least two of the x values to be different (Why?). In order to carry out the projection, we will construct an orthogonal basis by Gram–Schmidt. We take the first basis vector to be

$$e_1 = \frac{u}{\|u\|} = \frac{u}{\sqrt{n}}$$

and construct e_2:

$$x' = x - \langle x, e_1 \rangle e_1 = x - \frac{\langle x, u \rangle}{n} u$$
$$e_2 = \frac{x'}{\|x'\|}.$$

But $\langle x, u \rangle = \sum_{i=1}^{n} x_i$, so $\langle x, u \rangle / n$ is the average of all the x-values, which we shall denote $\bar{x}$. Hence,

$$x' = x - \bar{x} u$$
$$\|x'\|^2 = \langle x, x \rangle - 2\bar{x} \langle x, u \rangle + \bar{x}^2 \langle u, u \rangle$$
$$= \sum_{i=1}^{n} x_i^2 - 2n\bar{x}^2 + n\bar{x}^2.$$

The sum $\sum_{i=1}^{n} x_i^2$ is n times the average of the squares of the x-values, which we will write $n\overline{x^2}$. (Be careful not to confuse $\overline{x^2}$ with $\bar{x}^2$!). So,

$$\|x'\|^2 = n\overline{x^2} - n\bar{x}^2 = ns_{xx},$$

where $s_{xx} = \overline{x^2} - \overline{x}^2$ is known in statistics as the *sample variance*. With these simplifications, we have an orthonormal basis,

$$e_1 = \frac{u}{\sqrt{n}}$$

$$e_2 = \frac{x - \overline{x}u}{\sqrt{ns_{xx}}}.$$

The projections of the data onto the basis are

$$\langle y, e_1 \rangle = \frac{\langle y, u \rangle}{\sqrt{n}} = \sqrt{n}\,\overline{y}$$

$$\langle y, e_2 \rangle = \frac{\langle y, x \rangle - \overline{x}\langle y, u \rangle}{\sqrt{ns_{xx}}}$$

$$= \frac{\sum_{i=1}^n y_i x_i - n\overline{x}\,\overline{y}}{\sqrt{ns_{xx}}} = \frac{n(\overline{xy} - \overline{x}\,\overline{y})}{\sqrt{n}\sqrt{s_{xx}}} = \sqrt{n}\,\frac{s_{xy}}{\sqrt{s_{xx}}}$$

where $s_{xy} = \overline{xy} - \overline{x}\,\overline{y}$ is called the *sample covariance*.

Putting all this together, the approximation is

$$\hat{y} = \langle y, e_1 \rangle e_1 + \langle y, e_2 \rangle e_2$$

$$= \overline{y}u + \frac{s_{xy}}{s_{xx}}(x - \overline{x}u),$$

or, in terms of individual points,

$$\hat{y}_i = ax_i + b,$$
$$\text{where } a = \frac{s_{xy}}{s_{xx}}, b = \overline{y} - \frac{s_{xy}}{s_{xx}}\overline{x},$$

which is the classic result from statistics.

The approximation to y came out in terms of (orthogonal, non-unit) vectors u and $x - \overline{x}u$, constructed by Gram–Schmidt from the nonorthogonal vectors u and x. Physically, the vector $\overline{y}u$ is the part of y that can be modelled as a constant, that is, the mean value. The other basis vector, $x - \overline{x}u$, picks up that part of y which differs from the mean. Finally, the vector $\tilde{y} = \hat{y} - y$ is the part of y that cannot be accounted for by a linear model—due either to noise or some additional nonlinear relationship between x and y. ∎

2.6 SUMMARY

This is a quick reference for the main results of this chapter, which will be used in subsequent chapters.

An inner product space is also a normed space and a normed space is also a metric space. In a complete metric space, all Cauchy sequences $(f_k)_{k=1}^\infty$ converge to limits f in the space.

A *Hilbert space* (e.g., $\mathbb{R}^n, \mathbb{C}^n, \ell^2, L^2$) is a complete inner product space. All the familiar geometric properties of vectors in finite-dimensional euclidean space are preserved in Hilbert space. For f, g in a Hilbert space, and $\{e_k\}$ a complete orthonormal set (basis),

$$\langle f, g \rangle = \sum_k f_k g_k^* \quad \text{or} \quad \int_Q f(x) g^*(x) dx$$

$$\|f\| = \langle f, f \rangle^{1/2} = \left(\sum_k |f_k|^2 \right)^{1/2} \quad \text{or} \quad \left(\int_Q |f(x)|^2 dx \right)^{1/2}$$

$$|\langle f, g \rangle| \leq \|f\| \|g\| \quad \text{(Cauchy–Schwarz)}$$

$$f = \sum_k \langle f, e_k \rangle e_k \quad \text{(Orthogonal expansion)}$$

$$\langle f, g \rangle = \sum_k \langle f, e_k \rangle \langle g, e_k \rangle^* \quad \text{(Parseval)}$$

$$\|f\|^2 = \sum_k |\langle f, e_k \rangle|^2.$$

A truncated orthogonal expansion $\sum_{k=1}^m \langle f, e_k \rangle e_k$ gives an m-dimensional approximation to f that minimizes the euclidean norm

$$\left\| f - \sum_{k=1}^m \langle f, e_k \rangle e_k \right\|_2.$$

A *Banach space* (e.g., ℓ^p, L^p) is a complete normed space. A general Banach space lacks an inner product and the nice properties that come with it. If a Banach space does have an inner product, it is a Hilbert space. For particular Banach spaces $\ell^1, \ell^\infty, L^1, L^\infty$,

$$\|f\|_1 = \sum_k |f_k| \quad \text{or} \quad \int_Q |f(x)| dx$$

$$\|f\|_\infty = \lim_{p \to \infty} \|f\|_p$$

$$= \sup_k (|f_k|) \quad \text{or} \quad \operatorname{ess\,sup}_{x \in Q} |f(x)|.$$

For $f \in \ell^1$ (or L^1) and $g \in \ell^\infty$ (or L^∞), Hölder's inequality gives

$$\left. \begin{array}{l} \sum_k f_k g_k^* \\ \int_Q f(x) g^*(x) dx \end{array} \right\} \leq \|f\|_1 \|g\|_\infty.$$

In all the L^p spaces, two functions f, g are equal almost everywhere if they differ only on a set of measure zero. If $f = g$ (a.e.), then $\|f - g\| = 0$. Pointwise equality is a special case of equality almost everywhere.

An *operator* T is a mapping between two vector spaces. For a linear operator, $T(af + bg) = aTf + bTg$. An operator is bounded if $\|Tf\|$ is finite for all f with finite norm. An operator S is the inverse of an operator T if S is bounded, if $STv = v$, and if $TSw = w$. An isometric operator preserves norms, $\|Tv\| = \|v\|$, and a unitary operator preserves inner products, $\langle Tu, Tv \rangle = \langle u, v \rangle$. A unitary operator on a finite-dimensional vector space is represented by an orthogonal matrix, $\mathbf{TT}^\dagger = \mathbf{I}$. Projection onto an orthogonal basis is a unitary operator.

PROBLEMS

2.1. *Signal models and vector spaces*

Which of the following signal sets comprise a vector space? Explain why or why not. Unless otherwise specified, addition and scalar multiplication are defined in the natural way, with $c \in \mathbb{R}$.

(a) All nonnegative signals ($v(t) \geq 0$).

(b) All 60 Hz sine waves.

(c) All 512×512 pixel images. Pixels may take on integer values between 0 and 255. Scalars c likewise may take on integer values between 0 and 255.

(d) Same as (c), but addition and multiplication are defined modulo-256.

(e) All signals that are one-sided ($f(t) = 0$ when $t < 0$).

(f) All signals that are one-sided ($f(t) = 0$ when $t < 0$) and decay exponentially in time (e.g., damped harmonic oscillation).

(g) All piecewise-constant functions (e.g., the voltage output of a digital-to-analog converter).

(h) All waves on a 1 m string with fixed ends (i.e., solutions of the classic vibrating string wave equation).

(i) All signals with zero average value, $\frac{1}{T}\int_0^T f(t)dt = 0$.

(j) All functions with unit area, $\int_{-\infty}^{\infty} f(t)dt = 1$.

2.2. *Vector spaces*

Determine which of the following sets are vector spaces. For those which are not, give a counterexample. For each vector space, determine a basis and the dimension of the space.

(a) $\{(x_1, x_2, x_3) \in \mathbb{R}^3 \mid x_1 + x_2 + x_3 = 0\}$

(b) $\{(x_1, x_2, x_3) \in \mathbb{R}^3 \mid x_1 + x_2 + x_3 = 1\}$

(c) {Real $m \times n$ matrices A}

(d) {Real $n \times n$ matrices A with $\det A = 0$}

(e) {Real $n \times n$ matrices A with $A = A'$}

(f) {Real polynomials $p(x)$ of degree n}

(g) {Real polynomials $p(x)$ of degree $\leq n$ with $p(0) = 1$}

(h) {Real polynomials $p(x)$ of degree $\leq n$ with $p(1) = 0$}.

2.3. *Vector spaces*

Complete the proof, begun in Example 2.1, that $C[0, 1]$ is a vector space.

2.4. *Norms*

For a *complex* vector $v \in \mathbb{C}^2$, ($v = (v_1, v_2)$, $v_1, v_2 \in \mathbb{C}$), it is reasonable to define the norm

$$\|v\| = \sqrt{|v_1|^2 + |v_2|^2} = \sqrt{v_1 v_1^* + v_2 v_2^*}.$$

Following the derivation in Section 2.2.1, show that this definition satisfies the four criteria for a norm (in particular, that the triangle inequality holds).

2.5. *Norms*

Consider the complex vector $x = (1, 2i)$, and calculate the norms $\|x\|_1$, $\|x\|_2$, and $\|x\|_\infty$.

2.6. *Norms*

One of the criteria for a norm is that it satisfies the triangle inequality, $\|x + y\| \leq \|x\| + \|y\|$. Show that this inequality is also true:

$$\|x - y\| \geq \Big|\|x\| - \|y\|\Big|.$$

2.7. Show that the metrics $d_1(x, y) \geq d_2(x, y) \geq d_\infty(x, y)$ for all $x, y \in \mathbb{R}^n$.

2.8. In the real plane, a neighborhood is a disk and the boundary of a closed neighborhood is a circle. Draw the "unit circles" $\{x \mid d_p(0, x) = 1\}$ for $p = 1, 2$, and ∞. Relate your drawing to the solution of Problem 2.7.

2.9. Prove the following inner product relationships. These are supposed to hold for *any* inner product. Therefore, you cannot base a proof on a particular form like $\langle u, v \rangle = u_1 v_1 + u_2 v_2$. Instead, use the general properties in Definition 2.4.

(a) $\langle v, cw \rangle = c^* \langle v, w \rangle$
(b) $\langle u, v + w \rangle = \langle u, v \rangle + \langle u, w \rangle$
(c) $\langle v, 0 \rangle = \langle 0, v \rangle = 0$
(d) If $\langle u, v \rangle = \langle u, w \rangle$ for all $u \in V$, then $v = w$.

2.10. Prove the parallelogram law geometrically for vectors in the plane. *Hint:* Use the law of cosines.

2.11. *Parallelogram law*

(a) Prove the parallelogram law for arbitrary vectors. Begin by expressing $\|u + v\|^2$ and $\|u - v\|^2$ in terms of inner products.

(b) Then prove this corollary,

$$\|u + v\|^2 \leq 2\|u\|^2 + 2\|v\|^2. \tag{2.47}$$

2.12. *Polarization identity*

(a) Derive the polarization identity (Equation 2.17).

(b) Then prove the corollaries,

$$|\langle u, v \rangle| \leq \|u\|^2 + \|v\|^2 \quad \text{complex } u, v \tag{2.48}$$

and

$$|\langle u, v \rangle| \leq \frac{\|u\|^2 + \|v\|^2}{2} \quad \text{real } u, v. \tag{2.49}$$

2.13. In an inner product space, the norm is calculated from the inner product. The polarization identity says that, in an inner product space, the inner product can also be calculated from the (2-)norm. Show that this relationship does not hold for the absolute value (1-)norm in $\mathbb{R}^n$, $\|u\|_1 = \sum_{k=1}^n |u_k|$. Hence, the absolute value norm cannot be calculated as an inner product. *Hint:* Consider the particular vectors $u = (a, 0, 0, \ldots, 0)$ and $v = (0, b, 0, \ldots, 0)$.

2.14. Let u and v be orthogonal, and let a and b be scalars.
  (a) Show that au and bv are orthogonal.
  (b) Show that $u + v$ and $u - v$ are orthogonal if $\|u\| = \|v\|$.

2.15. Let $u = (2, 1 + i, i, 1 - i)$ and $v = (1, i, -1, i)$. Calculate the following:
  (a) $\|u\|_1, \|u\|_2,$ and $\|u\|_\infty$
  (b) $\|v\|_1, \|v\|_2,$ and $\|v\|_\infty$
  (c) $\langle u, v \rangle$
  (d) $\langle v, u \rangle$.

2.16. Prove the Pythagorean theorem (Theorem 2.2) for general inner product spaces.

2.17. Let vectors e_1 and e_2 be orthonormal. Calculate the euclidean distance between them, $\|e_1 - e_2\|_2$.

2.18. For the orthonormal functions $\{e_k\}_{k=-\infty}^{\infty}$, $e_k(x) = \frac{1}{\sqrt{2\pi}} e^{ikx}$, calculate the distance between them in $L^2[-\pi, \pi]$, that is, $\|e_m - e_n\| = \langle e_m - e_n, e_m - e_n \rangle^{1/2}$. Compare this to the distance between $\mathbf{e}_x$ and $\mathbf{e}_y$, the Cartesian basis vectors for $\mathbb{R}^2$. You should get the same answer for both calculations, reinforcing the idea that orthonormal functions in an infinite-dimensional space are, in a very important sense, just like orthonormal vectors in a finite-dimensional space.

2.19. Suppose you have a set of six vectors, each of which belongs to the space $\mathbb{R}^4$.
  (a) How many vectors do you need to make a basis for the space?
  (b) If you select this many vectors from the set, will you have a basis?

2.20. Derive the following alternative form for the Cauchy–Schwarz inequality:
$$|\langle f, g \rangle + \langle g, f \rangle| \leq 2\|f\|_2 \|g\|_2. \tag{2.50}$$

2.21. Prove that the norm in an inner product space, $\|x\| = \langle x, x \rangle^{1/2}$, satisfies the triangle inequality.

2.22. Show that ℓ^1 and ℓ^∞ are normed vector spaces. (As usual, the hard part is verifying the triangle inequality.)

2.23. Show that ℓ^2 is an inner product space.
  (a) Show that the infinite series Equation 2.26 converges. *Hint*: Because $\mathbb{C}$ is an inner product space with $\langle x, y \rangle = xy^*$ (Example 2.5), the inequality (Equation 2.48) holds for complex numbers: $|xy^*| \leq |x|^2 + |y|^2$.
  (b) Having shown that the series converges, verify that it satisfies the specifications for an inner product.

2.24. *Nested spaces.*
Show that $\ell^1 \subset \ell^2 \subset \ell^\infty$.

2.25. Calculate the 1-, 2-, and ∞-norms of the geometric sequence, $x = (a^{n-1})_{n=1}^{\infty}$. For what values of a are the norms finite?

2.26. *Nested spaces.*
For a function f defined on a bounded interval $[a, b]$,
  (a) Show that if f is bounded a.e., then it is absolutely integrable and square integrable.
  (b) Show that if f is square integrable, then it is also absolutely integrable.
  (c) Hence, show that $L^\infty[a, b] \subset L^2[a, b] \subset L^1[a, b]$.

2.27. Show that if $f \in L^1$ and $f \in L^\infty$, then $f \in L^2$.

2.28. Give an example of a function f that is *not* square integrable on the interval $[0, L]$, and hence cannot be well approximated by an orthogonal function expansion.

2.29. Show that the L^1-norm satisfies the conditions required of a norm, including the triangle inequality.

2.30. Show that the L^2-norm satisfies the conditions required of a norm, including the triangle inequality.

2.31. Show that the L^∞-norm satisfies the conditions required of a norm, including the triangle inequality. Additionally show that L^∞ is complete (a Banach space).

2.32. Let $f, g \in L^p$, $1 \leq p < \infty$. Prove *Minkowski's inequality*,

$$\|f + g\|_p \leq \|f\|_p + \|g\|_p. \tag{2.51}$$

Hint: Use Hölder's inequality. Minkowski's inequality shows that the norms in L^p satisfy the triangle inequality, and hence that L^p are normed spaces. The same inequality holds for infinite sequences, $f, g \in \ell^p$, and their respective norms.

2.33. Let a sequence of functions $(f_n)_{n=0,1,\ldots}$ be defined by

$$f_n = \frac{1}{nx + 1}, \quad x \in [0, 1].$$

(a) What is the limiting function f of this sequence as $n \to \infty$?

(b) Does f_n converge pointwise or uniformly to f?

2.34. What is $\|f - g\|_1$ when $f(x) = \sin x$ and $g(x) = \cos x$?

2.35. In the design of feedback control systems, a classic problem is to create a system whose response to a step input follows that input as closely as possible. The design problem is posed in terms of minimizing the error between the ideal and actual responses. Let $f(t)$ be the error, and J a single number that measures the total error. Four different error measures are typically considered:

Integrated Absolute Error (IAE)	$J = \int_0^\infty \|f(t)\|\, dt$
Integrated Square Error (ISE)	$J = \int_0^\infty \|f(t)\|^2\, dt$
Integrated Time-Absolute Error (ITAE)	$J = \int_0^\infty \|f(t)\|\, t\, dt$
Integrated Time-Square Error (ITSE)	$J = \int_0^\infty \|f(t)\|^2\, t\, dt.$

IAE and ISE are recognizable as a 1-norm and squared 2-norm, respectively. Are ITAE and the square root of ITSE also valid norms? Give proofs.

2.36. The *root mean square* (rms) value of a function is defined (Equation 2.33)

$$f_{\text{rms}} = \left(\lim_{T \to \infty} \frac{1}{T} \int_{-T/2}^{T/2} |f(t)|^2 dt \right)^{1/2}$$

(a) Is the set of functions with finite rms value a vector space?

(b) Is the rms value a norm? Why or why not?

2.37. Consider the following functions, whose common domain is the interval $(0, 1)$:
$$f_1(x) = 1$$
$$f_2(x) = x + a.$$
Find the constant a such that the two functions are orthogonal.

2.38. Let $u, v : [0, 1] \to \mathbb{R}$ be given by $u(x) = 1$, $v(x) = x$. Calculate the projections of u on v and v on u.

2.39. Using Parseval's formula (Equation 2.42), show that the only vector orthogonal to every vector in the complete orthonormal set $\{e_k\}$ is the zero vector. This enables Theorem 2.6a or 2.6b to be used to verify the completeness of an orthonormal set.

2.40. Show that orthogonal expansion in an infinite-dimensional space preserves inner products, (Equation 2.43).

2.41. Find a relationship between a and b such that the functions $\cos x$ and $\cos 2x$ are orthogonal on the interval (a, b). Use graphs to make a conjecture, then prove it.

2.42. Derive the general expansion
$$v = \sum_{k=1}^{n} \frac{\langle v, \phi_k \rangle}{\|\phi_k\|^2} \phi_k,$$
where $\{\phi_k\}_{k=1}^{n}$ is a set of mutually orthogonal vectors.

2.43. Which of the operators in Example 2.27 are bounded? Are any of them isometries?

2.44. *Matrix representations*

Consider a linear operator $T: V \to W$, where V and W are finite-dimensional spaces. Let $(d_j)_{j=1}^{n}$ and $(e_k)_{k=1}^{m}$ be orthonormal bases for V and W, respectively. Thus, any vector $v \in V$ may be written $v = \sum_{j=1}^{n} v_j d_j$, and similarly for $w \in W$, $w = \sum_{k=1}^{m} w_k e_k$. Show that the operation $w = Tv$ may be written in matrix form,
$$\begin{bmatrix} w_1 \\ w_2 \\ \vdots \\ w_m \end{bmatrix} = \mathbf{T} \begin{bmatrix} v_1 \\ v_2 \\ \vdots \\ v_n \end{bmatrix}$$
and give explicit expressions for the elements T_{kj} of the matrix $\mathbf{T}$.

2.45. Show that the following operators are linear:
(a) $Tu = u$.
(b) $Tu = 0$.

2.46. Consider the forward and backward shift operators on ℓ^2, $T_f(v_0, v_1, v_2, \ldots) = (0, v_0, v_1, \ldots)$ and $T_b(v_0, v_1, v_2, \ldots) = (v_1, v_2, \ldots)$, respectively.
(a) Are they bounded?
(b) Is either one an isometry?
(c) Are they one-to-one and/or onto?
(d) Discuss the invertibility of T_f and T_b.

2.47. The convolution integral,
$$g(t) = \int_{-\infty}^{\infty} f(\tau) h(t - \tau) \, d\tau,$$

describes many continuous-time signal processing systems. The function h is often called the *convolution kernel*.

(a) Show that convolution is a linear operator, $g = Tf$.

(b) If $f \in L^1$, what properties must h have in order for T to be a bounded linear operator, that is, $|g(t)| < \infty$ for all t?

(c) Normally, operators do not commute. But consider two convolution operators T_1 and T_2 with kernels h_1 and h_2, respectively. Show that $T_2 T_1 = T_1 T_2$.

2.48. The sum

$$y[n] = \sum_{k=0}^{N-1} a^k x[n-k],$$

where a is a complex constant, describes the input–output properties of a class of discrete-time signal processing systems.

(a) Show that it is a linear operator, $y = Tx$.

(b) If $x \in \ell^\infty$, what properties must a have in order for T to be a bounded linear operator, that is, $|y[n]| < \infty$ for all n?

(c) How do your answers to (a) and (b) change if $N \to \infty$ so that the sum is an infinite series?

2.49. Consider a system represented by the operator $g(t) = Tf(t) = [1 + mf(t)] \cos \omega t$, where $|f(t)| \leq 1$ and $m \in (0, 1]$ is a constant.

(a) Is T a linear operator?

(b) Is T bounded?

2.50. Let V, W be vector spaces and consider the set $\mathcal{L}(V, W)$ of all linear operators $T : V \to W$. Define operator addition in the natural way, $(T_1 + T_2)v = T_1 v + T_2 v$. Show that $\mathcal{L}(V, W)$ is a vector space.

2.51. Let $S : V \to W$ and $T : U \to V$ be linear operators with operator norms $\|S\|$ and $\|T\|$. Show that $\|ST\| \leq \|S\| \|T\|$.

2.52. The following vectors are 4D *Walsh functions*. Walsh functions of higher dimension have applications in signal processing, information coding, cryptography, and statistics.

$$\begin{aligned}
\phi_0 &= (1 \quad 1 \quad 1 \quad 1) \\
\phi_1 &= (1 \quad 1 \quad -1 \quad -1) \\
\phi_2 &= (1 \quad -1 \quad -1 \quad 1) \\
\phi_3 &= (1 \quad -1 \quad 1 \quad -1).
\end{aligned}$$

(a) Verify that they are an orthogonal basis for $\mathbb{R}^4$.

(b) Calculate their norms.

(c) Express the vector $v = (1, 2, 3, 4)$ as a linear combination of the Walsh functions, that is, calculate the coefficients c_k in $v = \sum_{k=0}^{3} c_k \phi_k$. Check your answer by summing the series and obtaining v.

(d) Find the best 3D approximation to v. This is the linear combination $\hat{v}$ of *three* Walsh functions that minimizes the 2-norm $\|\hat{v} - v\|_2$.

2.53. Write a MATLAB program to do Gram–Schmidt orthogonalization of a set of independent vectors.

2.54. Consider two vectors in the plane, $x_1 = (3, 2)$ and $x_2 = (0, 5)$.
  (a) Are they linearly independent? Show why or why not.
  (b) Do they span $\mathbb{R}^2$?
  (c) If they are a basis, use Gram–Schmidt to obtain an orthonormal basis.

2.55. Given the following nonorthogonal vectors,
$$x_1 = (8, 2, 7), \quad x_2 = (-7, -6, 4), \quad x_3 = (-7, -8, -5),$$
  (a) use the Gram–Schmidt process to create an orthonormal set of basis vectors, $\{e_1, e_2, e_3\}$. Verify that your basis vectors are indeed orthonormal.
  (b) expand the vector $y = (-3, -2, 4)$ in terms of $\{e_1, e_2, e_3\}$. Verify that your expansion is correct.
  (c) what is the minimum 2-norm approximation of y in terms of e_1 and e_2? What is the norm of the approximation error?
  Hint: Use MATLAB for these calculations.

2.56. Three functions, ϕ_1, ϕ_{2a}, and ϕ_{2b}, are shown in Figure 2.27.
  (a) Which function, ϕ_{2a} or ϕ_{2b}, is orthogonal to ϕ_1?
  (b) Construct an orthonormal set $\{e_1, e_2\}$ from ϕ_1 and the function you selected in (a).
  (c) Make an orthogonal expansion of the function f in the functions e_1, e_2. Make an accurate sketch of your approximation.
  (d) Is the orthonormal set $\{e_1, e_2\}$ a basis for $L^2[-1, 1]$? Why or why not?

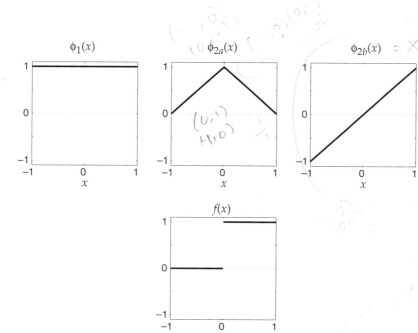

FIGURE 2.27 For Problem 2.56.

2.57. Show that the integral

$$2\pi \int_0^1 |f(r)|^2 \, r \, dr$$

is a valid norm. Then show that

$$2\pi \int_0^1 f(r) \, g^*(r) \, r \, dr$$

is a valid inner product. The additional factor of r in the integrand is called a *weight function*.

2.58. The *Zernike polynomials*, an orthogonal family important in optics, are obtained by orthogonalizing the monomials $\{1, r^2, r^4, \ldots\}$ on the unit disk, that is, using the inner product defined in the previous problem. Derive the first three Zernike polynomials.

CHAPTER 3

THE DISCRETE FOURIER TRANSFORM

In the last chapter, it was shown how an arbitrary vector in an N-dimensional space can be represented as a linear combination of N mutually orthogonal basis vectors in that space. In this chapter we consider $\mathbb{C}^N$, the space of N-dimensional complex vectors, and a basis $\{\phi_0, \phi_1, \ldots \phi_{N-1}\}$ of complex exponential vectors:

$$\phi_m = \left(\exp\left(\frac{i2\pi mn}{N}\right)\right)_{n=0}^{N-1}.$$

The expansion in this basis of a vector $f = (f[0], f[1], \ldots f[N-1]) \in \mathbb{C}^N$ is

$$f = \frac{1}{N} \sum_{m=0}^{N-1} F[m] \phi_m,$$

$$\text{where } F[m] = \langle f, \phi_m \rangle = \sum_{n=0}^{N-1} f[n] \exp\left(-\frac{i2\pi mn}{N}\right),$$

and is called the *discrete Fourier transform* (DFT) of f.

The DFT is the principal computational tool of Fourier analysis and a logical place to begin our study. Being a mapping between finite-dimensional spaces, its mathematics are uncomplicated by questions about convergence of series or integrals. We will derive the basic properties of the DFT, including several theorems that facilitate its application, and develop the fast Fourier transform (FFT) algorithm for computing the DFT. The chapter concludes by introducing a close relative of the DFT, the discrete cosine transform (DCT), which is widely applied in signal compression algorithms such as JPEG.

3.1 SINUSOIDAL SEQUENCES

Recall that a sequence is an ordered set of points belonging to a space. Sequences are commonly indexed by positive integers, $x = (x[n])_{n=1}^{N}$, or nonnegative integers, $x = (x[n])_{n=0}^{N-1}$. The sequence length N may be finite or infinite. In a later chapter we shall also see infinite sequences indexed by positive and negative integers,

---

Fourier Transforms: Principles and Applications, First Edition. Eric W. Hansen.
© 2014 John Wiley & Sons, Inc. Published 2014 by John Wiley & Sons, Inc.

110 CHAPTER 3 THE DISCRETE FOURIER TRANSFORM

$x = (x[n])_{n \in \mathbb{Z}}$. A finite sequence of real or complex numbers is equivalent to a vector in N-dimensional space. In many applications, a sequence models the result of observing, or *sampling*, a function of a continuous variable at regular intervals. If, for example, $v(t)$ is a time-varying voltage, then the sequence of values (samples) produced by an analog-to-digital converter is modeled by the sequence $v[n] = v(n\Delta t)$, where Δt is the time interval between observations.

We are interested here in the properties of sinusoidal sequences: the complex exponential sequence $(\exp(i\theta n))$, and its real and imaginary parts, $(\cos \theta n)$ and $(\sin \theta n)$. The parameter θ is the *frequency* or, as it is sometimes called in the signal processing literature, *discrete frequency* or *digital frequency*. The "units" of digital frequency are radians/sample.[1]

Indistinguishable sequences and aliasing

A sinusoidal function $f(t) = \cos 2\pi v t$ can have a frequency v that is any nonnegative real number, but something different happens with sinusoidal sequences. Two complex exponential sequences $(\exp(i\theta n))$ and $(\exp(i\theta' n))$ are indistinguishable if their frequencies differ by an integer multiple of 2π, $\theta' = \theta + 2\pi k$:

$$\exp[i(\theta + 2\pi k)n] = \exp(i\theta n)\exp(i2\pi k n) = \exp(i\theta n).$$

The set of *unique* complex exponential sequences is thus restricted to a 2π frequency range: by convention, $\theta \in [-\pi, \pi)$. All other frequencies produce sequences identical to the ones in this set. This is a fundamental issue in signal processing, as the following example shows.

Example 3.1 (A glimpse of sampling theory). Let $f(t) = \cos 2\pi v t$ be a sinusoidal signal with frequency v (Hz). A sequence $f[n]$ is generated by sampling $f(t)$ at regular intervals, $t = n\Delta t$. Then,

$$f[n] = f(n\Delta t) = \cos 2\pi v (n\Delta t)$$
$$= \cos \theta n$$

where $\theta = 2\pi v \Delta t$. The reciprocal of the sampling interval is called the *sampling frequency* or *sampling rate*, $v_s = 1/\Delta t$. In terms of the sampling rate, $\theta = 2\pi v/v_s$. Digital frequency expresses the analog frequency relative to the sampling rate: the sequence resulting from sampling a 10 Hz sinusoid at a 100 Hz rate has the same digital frequency, $\theta = \frac{\pi}{5}$ rads/sample, as the sequence produced by sampling a 20 Hz sinusoid at a 200 Hz rate.

Now, consider another signal $g(t) = \cos 2\pi v' t$, whose samples are $g[n] = \cos \theta' n$, where $\theta' = 2\pi v' \Delta t$. The samples $g[n]$ and $f[n]$ will be identical if $\theta' = \theta + 2\pi k$, or

$$v' = v + \frac{k}{\Delta t} = v + k v_s.$$

---
[1] In digital signal processing, continuous-time angular frequency (rads/sec) and discrete-time (digital) frequency (rads/sample) are typically represented by the variables Ω and ω, respectively. In this text we use ω and θ for these quantities.

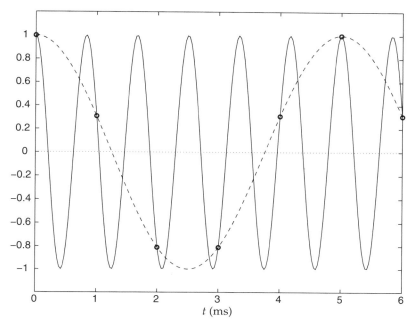

FIGURE 3.1 Aliasing. Sampling a high frequency signal with a sampling interval that is too large produces samples indistinguishable from those of a lower frequency signal. Here, the high frequency signal is 1200 Hz, the sampling rate is 1000 Hz, and the aliased frequency is 200 Hz.

To be specific, let $v_s = 1000$ Hz and $v' = 1200$ Hz. With $k = 1$, the above equation is satisfied by $v = 200$ Hz. At a 1 kHz sampling rate, the samples of a 1200 Hz sinusoid are the same as those of a 200 Hz sinusoid. This phenomenon, in which a high frequency signal's samples are indistinguishable from those of a lower frequency signal, is called *aliasing*. It occurs when the sampling rate v_s is too small (the sampling interval Δt is too large) to capture the high frequency signal's variations (Figure 3.1).

To avoid aliasing, the sampling rate must always be chosen so that the digital frequency of the sample sequence, $\theta = 2\pi v/v_s$, is less than π, or

$$v_s > 2v. \tag{3.1}$$

The sampling rate must be at least twice the frequency of the signal, or "two samples per cycle." For a complex signal containing sinusoids of many frequencies, for example, a music recording, the sampling rate must be twice the highest frequency present in the signal. This minimum allowable sampling rate is called the *Nyquist rate*.[2] At the time of this writing, compact disc and MP3 recording technology use a sampling rate of 44.1 kHz, just over twice the 22 kHz limit of human hearing. Other digital audio recording and editing formats use sampling rates from 48 kHz to 192 kHz. ∎

---

[2] Nyquist (1928, reprinted 2002); Hartley (1928).

Periodic, even, and odd sequences

A function f of a continuous variable is *periodic* with period $T > 0$ if $f(t + T) = f(t)$ for all values of t. Otherwise, it is *aperiodic*. A periodic function is completely described by its values over one period, for example, $t \in [0, T)$. Similarly, a sequence $(x[n])$ is periodic with period $N \in \mathbb{N}$ if $x[n + N] = x[n]$ for all n. If a sequence $(x[n])$ has period N, it is completely described by a vector of its samples over one period, $(x[n])_{n=0}^{N-1}$.

A function f is even if $f(t) = f(-t)$, odd if $f(t) = -f(-t)$, and Hermitian if $f(t) = f^*(-t)$. If f is periodic as well as even, then symmetry is observed within a single period: $f(t) = f(T - t)$. Likewise, $f(t) = -f(T - t)$ if f is odd and periodic, and $f(t) = f^*(T - t)$ if f is Hermitian and periodic. The same definitions hold for sequences. We will say that an infinite sequence $(x[n])$ with period N is

$$\left.\begin{array}{c}\text{even}\\ \text{odd}\\ \text{Hermitian}\end{array}\right\} \text{ if } x[n] = \left\{\begin{array}{c}x[N-n]\\ -x[N-n]\\ x^*[N-n]\end{array}\right\}, \quad n = 0, 1, 2, \ldots, N-1. \tag{3.2}$$

Sinusoidal functions of a real variable are always periodic: $\cos 2\pi \nu t = \cos 2\pi \nu(t + T)$, where the period T is the reciprocal of the frequency (in Hz), $T = 1/\nu$. This is not the case for sinusoidal sequences. Apply the periodicity condition $x[n] = x[n + N]$ to the complex exponential sequence:

$$\exp(i\theta n) = \exp(i\theta(n + N)) = \exp(i\theta n) \exp(i\theta N).$$

It is periodic if and only if $\exp(i\theta N) = 1$. This occurs only when the frequency θ is of the form

$$\theta = 2\pi \frac{m}{N}, \tag{3.3}$$

a *rational* multiple of 2π. When you consider how few rational numbers there are (compared with the real numbers), you can anticipate that a sinusoidal sequence obtained, for example, by sampling a physical signal $f(t) = \cos 2\pi \nu t$, will rarely be periodic.

Orthogonal sinusoidal sequences

Now consider the vector $\phi_m = (\phi_m[n])_{n=0}^{N-1}$ defined by

$$\phi_m[n] = \exp\left(\frac{i2\pi mn}{N}\right), \quad n = 0, 1, \ldots, N-1,$$

which is one period of a complex exponential sequence with frequency $\theta = 2\pi m/N$. The set $\{\phi_m\}$ of all such vectors has five important properties.

1. The complex exponential $\exp(i2\pi mn/N)$ is periodic in m as well as n:

$$\phi_m = \phi_{m+N}. \tag{3.4}$$

Thus, there are only N unique vectors in the set: $\phi_0, \phi_1, \ldots, \phi_{N-1}$.

3.1 SINUSOIDAL SEQUENCES

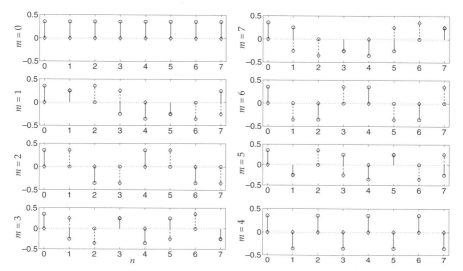

FIGURE 3.2 Normalized complex exponential basis vectors $\{e_m\} = \{\phi_m/\|\phi_m\|\}$ for $N = 8$. Real part is indicated by solid lines and circular markers, imaginary part by dashed lines and diamond markers. Note the Hermitian symmetry in m and n, for example, $\phi_1 = \phi_7^*$, $\phi_1[n] = \phi_1^*[8-n]$.

2. The complex exponential $\exp(i2\pi mn/N)$ is Hermitian in m and n:

$$\phi_m[N-n] = \exp\left(\frac{i2\pi m(N-n)}{N}\right) = \exp\left(\frac{-i2\pi mn}{N}\right) = \left[\exp\left(\frac{i2\pi mn}{N}\right)\right]^*$$
$$= \phi_m^*[n] \tag{3.5}$$

$$\phi_{N-m}[n] = \exp\left(\frac{i2\pi (N-m)n}{N}\right) = \exp\left(\frac{-i2\pi mn}{N}\right) = \left[\exp\left(\frac{i2\pi mn}{N}\right)\right]^*$$
$$= \phi_m^*[n] \tag{3.6}$$

3. The vectors ϕ_m and ϕ_{N-m} have the same digital frequency, $\theta = 2\pi m/N$, because $\phi_{N-m} = \phi_m^*$. If N is even, there are only $\frac{N}{2} + 1$ unique digital frequencies ($\frac{N+1}{2}$ if N is odd), and each is an integer multiple, or harmonic, of $\frac{2\pi}{N}$. Excepting $\theta_0 = 0$ (and $\theta_{N/2} = \pi$ if N is even), each digital frequency is represented by a complex conjugate pair of vectors, ϕ_m and ϕ_{N-m} (Figure 3.2).

4. The N vectors $\{\phi_m\}_{m=0}^{N-1}$ are orthogonal, hence they span $\mathbb{C}^N$. Verify orthogonality by calculating the inner product

$$\langle \phi_k, \phi_m \rangle = \sum_{n=0}^{N-1} \exp\left(\frac{i2\pi kn}{N}\right) \exp\left(\frac{-i2\pi mn}{N}\right) = \sum_{n=0}^{N-1} \left[\exp\left(\frac{i2\pi(k-m)}{N}\right)\right]^n$$
$$= \frac{1 - \exp(i2\pi(k-m))}{1 - \exp(i2\pi(k-m)/N)} \quad \text{(using Equation (1.27))}.$$

The fraction $\frac{1-\exp(i2\pi k)}{1-\exp(i2\pi k/N)}$ is zero except when $\exp(i2\pi k)$ and $\exp(i2\pi k/N)$ are both one, which happens when k is an integer multiple of N. Applying L'Hospital's rule gives

$$\frac{1-\exp(i2\pi k)}{1-\exp(i2\pi k/N)} = \begin{cases} N, & k = rN, r \in \mathbb{Z} \\ 0, & \text{otherwise} \end{cases}.$$

This result reappears sufficiently often that it is worth introducing a more compact notation. Define the *unit-sample sequence* δ,

$$\delta[n] = \begin{cases} 1, & n = 0 \\ 0, & \text{otherwise} \end{cases}. \tag{3.7}$$

Also define the *comb sequence* with period N,

$$\text{III}_N[k] = \sum_{r=-\infty}^{\infty} \delta[k - rN] = \begin{cases} 1, & k = rN, r \in \mathbb{Z} \\ 0, & \text{otherwise} \end{cases}. \tag{3.8}$$

We then have

$$\frac{1-\exp(i2\pi k)}{1-\exp(i2\pi k/N)} = N\,\text{III}_N[k]$$

and

$$\langle \phi_k, \phi_m \rangle = \frac{1-\exp(i2\pi(k-m))}{1-\exp(i2\pi(k-m)/N)} = N\,\text{III}_N[k-m].$$

The indices k and m both range from 0 to $N-1$. Their difference, $k-m$, ranges from $-(N-1)$ to $N-1$. The comb sequence III_N has only one nonzero sample in this range, at $k-m=0$. Thus, we may express the orthogonality of the vectors $\{\phi_m\}$ simply by

$$\langle \phi_k, \phi_m \rangle = N\delta[k-m]. \tag{3.9}$$

The unit sample $\delta[k-m]$ is also written δ_{km}, and called the *Kronecker delta*.

5. The norm $\|\phi_m\| = \sqrt{\langle \phi_m, \phi_m \rangle} = \sqrt{N}$. The orthogonal ϕ_m are made into an ortho*normal* set by dividing each one by its norm,

$$e_m = \frac{\phi_m}{\sqrt{N}} = \left(\frac{1}{\sqrt{N}} \exp\left(\frac{i2\pi mn}{N}\right)\right)_{n=0}^{N-1}. \tag{3.10}$$

3.2 THE DISCRETE FOURIER TRANSFORM

The vectors $\{e_m\}$ are an orthonormal basis for $\mathbb{C}^N$ (Figure 3.2). A vector (i.e., a finite-length sequence) $f \in \mathbb{C}^N$ may be represented by an orthogonal expansion in these basis vectors,

$$f = \sum_{m=0}^{N-1} \langle f, e_m \rangle e_m = \sum_{m=0}^{N-1} \left\langle f, \frac{\phi_m}{\sqrt{N}} \right\rangle \frac{\phi_m}{\sqrt{N}}.$$

(Of course, all the elements of f must have finite magnitude.) By convention, the two factors of $1/\sqrt{N}$ are combined and brought to the front,

$$f = \frac{1}{N} \sum_{m=0}^{N-1} \langle f, \phi_m \rangle \, \phi_m,$$

that is,

$$f[n] = \frac{1}{N} \sum_{m=0}^{N-1} \underbrace{\left[\sum_{k=0}^{N-1} f[k] \exp\left(-\frac{i2\pi mk}{N}\right) \right]}_{F[m]} \exp\left(\frac{i2\pi mn}{N}\right), \quad n = 0, 1, \ldots N-1.$$

The inner sum is called the *discrete Fourier transform* (DFT) of f, denoted F. This establishes the following result.

Theorem 3.1 (Discrete Fourier transform). Let $f \in \mathbb{C}^N$ be a vector. Then

$$f[n] = \frac{1}{N} \sum_{m=0}^{N-1} F[m] \exp\left(\frac{i2\pi mn}{N}\right), \quad n = 0, 1, \ldots, N-1, \tag{3.11a}$$

where $F \in \mathbb{C}^N$ is the discrete Fourier transform of f,

$$F[m] = \sum_{n=0}^{N-1} f[n] \exp\left(-\frac{i2\pi mn}{N}\right), \quad m = 0, 1, \ldots, N-1. \tag{3.11b}$$

The vectors f and F are a *DFT pair*. Equation 3.11b is called the *forward* (or *analysis*) transform, and Equation 3.11a is called the *inverse* (or *synthesis*) transform.

We may write $F = DFT\{f\}, f \xrightarrow{DFT} F$, or $f \longmapsto F$ as shorthand for the forward transform, and $f = DFT^{-1}\{F\}$ or $F \xrightarrow{DFT^{-1}} f$ for the inverse transform. Likewise, we write $f \longleftrightarrow F$ to indicate that f and F are a DFT *pair*.

The DFT expresses the vector f as a combination of harmonically related sinusoids with digital frequencies $\theta = 2\pi m/N$. The forward transform F tells how much of each sinusoidal component is contained in f. It gives the "recipe" for constructing f from these sinusoidal components, according to the inverse DFT formula.

The DFT is principally a computational tool, and for computational reasons (explained in Section 3.5), N is usually taken to be a power of two, $N = 2^p$. We shall assume this here for convenience, but you should note that the mathematical properties of the DFT are virtually unaffected by the choice of N. Sometimes we will make reference to $F[N/2]$ or $f[N/2]$. If N is odd, these terms do not exist and any statement that uses them is not true. Other than that, the evenness or oddness of N makes no difference in the theory.

⋆ Matrix form of the DFT

Occasionally it is useful to represent the DFT as a matrix-vector multiplication. For convenience, define

$$W_N = \exp\left(\frac{i2\pi}{N}\right), \tag{3.12}$$

so that the mth DFT basis vector is

$$\phi_m = \left(W_N^0,\ W_N^m,\ W_N^{2m},\ \ldots\ W_N^{(N-1)m}\right).$$

Collect the complex conjugates of the basis vectors as the rows of a matrix,

$$\mathbf{D} = \begin{bmatrix} \phi_0^* \\ \phi_1^* \\ \phi_2^* \\ \vdots \\ \phi_{N-1}^* \end{bmatrix} = \begin{bmatrix} W_N^0 & W_N^0 & W_N^0 & \cdots & W_N^0 \\ W_N^0 & W_N^{-1} & W_N^{-2} & \cdots & W_N^{-(N-1)} \\ W_N^0 & W_N^{-2} & W_N^{-4} & \cdots & W_N^{-2(N-1)} \\ \vdots & \vdots & \vdots & \ddots & \vdots \\ W_N^0 & W_N^{-(N-1)} & W_N^{-2(N-1)} & \cdots & W_N^{-(N-1)(N-1)} \end{bmatrix}, \tag{3.13}$$

that is, $D_{mn} = W_N^{-mn}$. The matrix $\mathbf{D}$ is seen to be symmetric, so $\mathbf{D}^\dagger = (\mathbf{D}^*)' = \mathbf{D}^*$. Moreover,

$$\mathbf{D}^\dagger \mathbf{D} = \mathbf{D}^* \mathbf{D}' = \begin{bmatrix} \phi_0 \\ \phi_1 \\ \phi_2 \\ \vdots \\ \phi_{N-1} \end{bmatrix} \begin{bmatrix} \phi_0^\dagger & \phi_1^\dagger & \phi_2^\dagger & \cdots & \phi_{N-1}^\dagger \end{bmatrix}$$

$$= \begin{bmatrix} \langle \phi_0, \phi_0 \rangle & \langle \phi_0, \phi_1 \rangle & \langle \phi_0, \phi_2 \rangle & \cdots & \langle \phi_0, \phi_{N-1} \rangle \\ \langle \phi_1, \phi_0 \rangle & \langle \phi_1, \phi_1 \rangle & \langle \phi_1, \phi_2 \rangle & \cdots & \langle \phi_1, \phi_{N-1} \rangle \\ \langle \phi_2, \phi_0 \rangle & \langle \phi_2, \phi_1 \rangle & \langle \phi_2, \phi_2 \rangle & \cdots & \langle \phi_2, \phi_{N-1} \rangle \\ \vdots & \vdots & \vdots & \ddots & \vdots \\ \langle \phi_{N-1}, \phi_0 \rangle & \langle \phi_{N-1}, \phi_1 \rangle & \langle \phi_{N-1}, \phi_2 \rangle & \cdots & \langle \phi_{N-1}, \phi_{N-1} \rangle \end{bmatrix}$$

$$= \begin{bmatrix} N & 0 & 0 & \cdots & 0 \\ 0 & N & 0 & \cdots & 0 \\ 0 & 0 & N & \cdots & 0 \\ \vdots & \vdots & \vdots & \ddots & \vdots \\ 0 & 0 & 0 & \cdots & N \end{bmatrix} = N\mathbf{I}.$$

Thus $\mathbf{D}$ is invertible, and $\mathbf{D}^{-1} = \frac{1}{N}\mathbf{D}^\dagger$.

Now define the column vectors

$$\mathbf{f} = \begin{bmatrix} f[0]\ f[1]\ f[2]\ \ldots\ f[N-1] \end{bmatrix}'$$
$$\tilde{\mathbf{f}} = \begin{bmatrix} F[0]\ F[1]\ F[2]\ \ldots\ F[N-1] \end{bmatrix}'$$

(we denote the DFT by $\tilde{\mathbf{f}}$ rather than $\mathbf{F}$ to maintain the convention that vectors are named by lowercase letters and matrices by uppercase letters). Then, the DFT is compactly written

$$\tilde{\mathbf{f}} = \mathbf{D}\mathbf{f} \tag{3.14}$$

and the inverse DFT is

$$\mathbf{f} = \frac{1}{N}\mathbf{D}^{\dagger}\tilde{\mathbf{f}} = \frac{1}{N}\mathbf{D}^{*}\tilde{\mathbf{f}}. \tag{3.15}$$

The invertibility of $\mathbf{D}$ guarantees that the DFT is one-to-one and onto—every vector $f \in \mathbb{C}^N$ has one and only one DFT vector $F \in \mathbb{C}^N$, and this F is the DFT of no other vector in $\mathbb{C}^N$.

The matrix $\frac{1}{\sqrt{N}}\mathbf{D}$ is unitary, $\left(\frac{1}{\sqrt{N}}\mathbf{D}\right)^{\dagger}\left(\frac{1}{\sqrt{N}}\mathbf{D}\right) = \mathbf{I}$. Sometimes, although it departs from the standard definition (Equation 3.11), one sees the DFT written using the unitary matrices,

$$\tilde{\mathbf{f}} = \frac{1}{\sqrt{N}}\mathbf{D}\mathbf{f}$$

$$\mathbf{f} = \frac{1}{\sqrt{N}}\mathbf{D}^{\dagger}\tilde{\mathbf{f}} = \frac{1}{\sqrt{N}}\mathbf{D}^{*}\tilde{\mathbf{f}}. \tag{3.16}$$

Example 3.2. For $N = 4$, the DFT matrices are

$$\mathbf{D} = \begin{bmatrix} W_4^0 & W_4^0 & W_4^0 & W_4^0 \\ W_4^0 & W_4^{-1} & W_4^{-2} & W_4^{-3} \\ W_4^0 & W_4^{-2} & W_4^{-4} & W_4^{-6} \\ W_4^0 & W_4^{-3} & W_4^{-6} & W_4^{-9} \end{bmatrix} = \begin{bmatrix} 1 & 1 & 1 & 1 \\ 1 & -i & -1 & i \\ 1 & -1 & 1 & -1 \\ 1 & i & -1 & -i \end{bmatrix}$$

$$\mathbf{D}^{\dagger} = \begin{bmatrix} 1 & 1 & 1 & 1 \\ 1 & i & -1 & -i \\ 1 & -1 & 1 & -1 \\ 1 & -i & -1 & i \end{bmatrix}.$$

You can work out the algebra (or use MATLAB) and verify that $\mathbf{D}^{\dagger}\mathbf{D} = 4\mathbf{I}$. ∎

3.3 INTERPRETING THE DFT

A common use of the DFT is computational spectrum analysis: given a portion of a sampled signal, $f[n] = f(n\Delta t)$, $n = 0, 1, \ldots N - 1$, model f as a sum of sinusoids and identify the prominent frequencies. Spectrum analysis is a sophisticated branch of signal processing beyond the scope of this text.[3] What we shall do here is calculate a few DFTs of simple signals "by hand" in order to develop some basic intuition, see how to interpret the DFT, and expose some common pitfalls in its application.

---

[3] Good introductions to DFT-based spectrum analysis include Porat (1997, Chapters 6 and 13), Oppenheim and Schafer (2010, Chapter 10), Kay (1988, Chapter 4), and Percival and Walden (1993, Chapter 6).

We begin with the simple complex exponential sequence $f = \left(\exp i\theta_0 n\right)_{n=0}^{N-1}$, where the digital frequency θ_0 can take any value between $-\pi$ and π. Its DFT is

$$F[m] = \sum_{n=0}^{N-1} \exp\left(i\theta_0 n\right) \exp\left(-\frac{i2\pi mn}{N}\right) = \sum_{n=0}^{N-1} [\exp(-i(m\Delta\theta - \theta_0))]^n$$

$$= \frac{1 - \exp(-iN(m\Delta\theta - \theta_0))}{1 - \exp(-i(m\Delta\theta - \theta_0))}, \quad \text{(geometric series)}$$

where $\Delta\theta = 2\pi/N$. Then, using Equation 1.26,

$$F[m] = \exp\left[-\frac{i(N-1)(m\Delta\theta - \theta_0)}{2}\right] \frac{\sin\left(\frac{N(m\Delta\theta - \theta_0)}{2}\right)}{\sin\left(\frac{m\Delta\theta - \theta_0}{2}\right)}. \quad (3.17)$$

In-bin sinusoid

We first consider a special case, when the digital frequency θ_0 is an integer multiple of $\Delta\theta$. These N special digital frequencies, $k\Delta\theta = 0, \Delta\theta, 2\Delta\theta, \ldots, (N-1)\Delta\theta$, are often called DFT *bins*. A signal at one of these frequencies is said to be "in a bin."

Substituting $\theta_0 = k\Delta\theta = 2\pi k/N$ ($k = 0, 1, \ldots, N-1$) into Equation 3.17,

$$F[m] = \exp\left[-i\pi\left(\frac{N-1}{N}\right)(m-k)\right] \frac{\sin(\pi(m-k))}{\sin\left(\frac{\pi(m-k)}{N}\right)} = \begin{cases} N, & m = k \\ 0, & \text{otherwise} \end{cases}$$

$$= N\delta[m-k], \quad m = 0, 1, \ldots N-1. \quad (3.18)$$

This can also be seen by noting that with $\theta_0 = 2\pi k/N$, the vector f is $\left(\exp(i2\pi kn/N)\right)$, which is ϕ_k, the kth DFT basis vector. Then, by orthogonality (Equation 3.9),

$$F[m] = \langle \phi_k, \phi_m \rangle = N\delta[m-k], \quad m = 0, 1, \ldots, N-1.$$

An in-bin cosine sequence $c = \left(\cos\frac{2\pi kn}{N}\right)_{n=0}^{N-1}$ is the sum of two in-bin complex exponentials:

$$c = \left(\cos\frac{2\pi kn}{N}\right)_{n=0}^{N-1} = \frac{1}{2}\left(e^{i2\pi kn/N}\right)_{n=0}^{N-1} + \frac{1}{2}\left(e^{-i2\pi kn/N}\right)_{n=0}^{N-1}$$

$$= \frac{1}{2}\phi_k + \frac{1}{2}\phi_{N-k},$$

where we have used the fact that $\phi_{-k} = \phi_{N-k}$ (Equation 3.4). The DFT is

$$C[m] = \left\langle \frac{1}{2}\phi_k + \frac{1}{2}\phi_{N-k}, \phi_m \right\rangle$$

$$= \frac{1}{2}\langle \phi_k, \phi_m \rangle + \frac{1}{2}\langle \phi_{N-k}, \phi_m \rangle \quad \text{(linearity of the inner product)}$$

$$= \frac{N}{2}\delta[m-k] + \frac{N}{2}\delta[m-(N-k)], \quad m = 0, 1, \ldots, N-1. \quad (3.19)$$

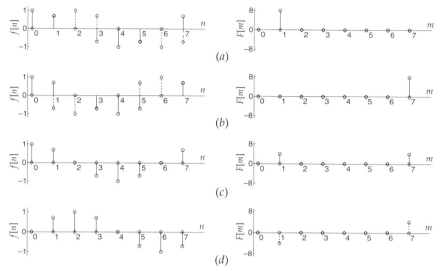

FIGURE 3.3 In-bin trigonometric sequences and their DFTs. (a) Complex exponential $f[n] = e^{i2\pi n/8}$ (left) and DFT $F[m] = 8\delta[m-1]$ (right). Solid stems and circular markers denote real part, dashed stems and diamond markers denote imaginary part. (b) Complex exponential $f[n] = e^{i2\pi\,7n/8} = e^{-i2\pi n/8}$ and DFT $F[m] = 8\delta[m-7]$. (c) Cosine $f[n] = \cos(2\pi n/8)$ and DFT $F[m] = 4\delta[m-1] + 4\delta[m-7]$. (d) Sine $f[n] = \sin(2\pi n/8)$ and DFT $F[m] = -i4\delta[m-1] + 4i\delta[m-7]$.

The DFT of an in-bin sine sequence $s = \left(\sin\frac{2\pi kn}{N}\right)_{n=0}^{N-1}$ is calculated in the same manner:

$$s = \left(\sin\frac{2\pi kn}{N}\right)_{n=0}^{N-1} = \frac{1}{2i}\left(e^{i2\pi kn/N}\right)_{n=0}^{N-1} - \frac{1}{2i}\left(e^{-i2\pi kn/N}\right)_{n=0}^{N-1}$$

$$= -\frac{i}{2}\phi_k[n] + \frac{i}{2}\phi_{N-k}[n]$$

$$S[m] = -\frac{iN}{2}\delta[m-k] + \frac{iN}{2}\delta[m-(N-k)], \quad m = 0, 1, \ldots, N-1. \quad (3.20)$$

The complex exponential, sine, and cosine sequences and their DFTs are compared in Figure 3.3.

The special frequencies $\theta_0 = 2\pi k/N, k = 0, 1, \ldots, N-1$ produce periodic sinusoidal sequences. They project onto only one (if a complex exponential) or two (if a sine or cosine) DFT basis vectors, and the DFT $F[m]$ is zero except for one or two values of m.

Out-of-bin sinusoids—spectral leakage

In contrast to these N special frequencies, all other values of θ_0 between $-\pi$ and π generate sinusoidal sequences that are *not* periodic and that project onto *all* DFT basis vectors. They are said to be *out-of-bin* or *between bins*.

Consider, for example, the 32-point sequences $f = \left(\cos\frac{2\pi 4n}{32}\right)$ and $g = \left(\cos\frac{2\pi 4.5n}{32}\right)$, shown with their DFTs in Figure 3.4.

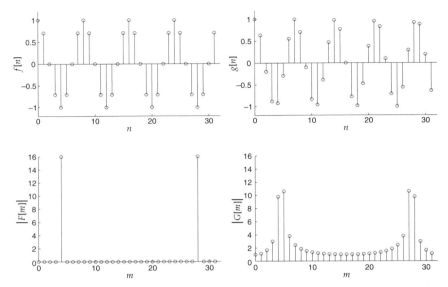

FIGURE 3.4 DFT magnitudes $|F|$ and $|G|$ of an in-bin sinusoid (*left*) and an out-of-bin sinusoid (*right*).

The DFT of f consists of two nonzero values, at $m = 4$ and $m = 32 - 4 = 28$. On the other hand, g's DFT, while it peaks near $m = 4$ and $m = 28$, is spread across all the DFT bins. In the signal processing literature, this phenomenon is called *spectral leakage*. The signal's energy "leaks" into bins adjacent to the peaks because there is no bin located exactly at the signal's frequency. Signals encountered in practice are invariably between bins, and consequently spectral leakage is common.

To explore this phenomenon further, return to Equation 3.17. The DFT of the complex exponential $f[n] = e^{i\theta_0 n}$ is

$$F[m] = \exp\left[-\frac{i(N-1)(m\Delta\theta - \theta_0)}{2}\right] \frac{\sin\left(\frac{N(m\Delta\theta - \theta_0)}{2}\right)}{\sin\left(\frac{m\Delta\theta - \theta_0}{2}\right)},$$

with $\Delta\theta = 2\pi/N$, the DFT bin spacing. The ratio of sines in this expression is a version of a function called the *Dirichlet kernel*, which is also important in the theory of the Fourier series (Chapter 4). The Dirichlet kernel is defined[4]

$$D_N(x) = \frac{\sin \pi N x}{\sin \pi x}. \quad (3.21)$$

Graphs of D_N, showing its salient features, are shown in Figure 3.5.

---

[4]In the theory of Fourier series, the Dirichlet kernel is traditionally defined $D_N(x) = \sum_{n=-N}^{N} e^{i2\pi n x} = \frac{\sin(2N+1)\pi x}{\sin \pi x}$. The definition used in this book is common in signal processing. In terms of this alternative definition of D_N, the traditional Dirichlet kernel is $D_{2N+1}(x)$.

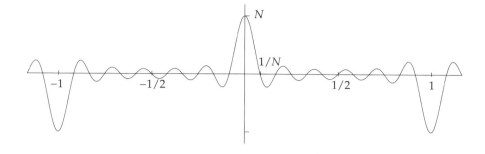

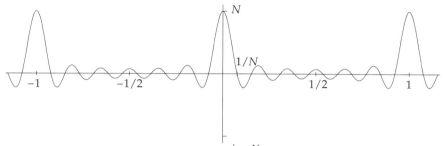

FIGURE 3.5 The Dirichlet kernel, $D_N(x) = \dfrac{\sin \pi N x}{\sin \pi x}$. $D_N(\text{integer}) = \pm N$. There are $N-1$ zero crossings between each peak, evenly spaced at $\frac{1}{N}$. *Top:* $N = 12$. When N is even, the peaks alternate sign, and $D_N(\pm \frac{1}{2}) = 0$. *Bottom:* $N = 13$. When N is odd, the peaks are the same sign, and $D_N\left(\pm \frac{1}{2}\right) = (-1)^{(N-1)/2} \neq 0$.

In terms of the Dirichlet kernel, Equation 3.17 is

$$F[m] = \exp\left[-\frac{i(N-1)(m\Delta\theta - \theta_0)}{2}\right] D_N\left(\frac{m\Delta\theta - \theta_0}{2\pi}\right)$$

$$= \exp\left[-\frac{i(N-1)(\theta - \theta_0)}{2}\right] D_N\left(\frac{\theta - \theta_0}{2\pi}\right)\bigg|_{\theta = m\Delta\theta}.$$

Except for the leading phase factors, the DFT is proportional to a Dirichlet kernel, centered at the digital frequency $\theta = \theta_0$, and sampled at the bin frequencies. If θ_0 is a bin frequency, then all the samples except $m = \theta/\Delta\theta$ are at zero crossings. This is the in-bin case. In the out-of-bin case, θ_0 is not an integer multiple of $\Delta\theta$. None of the samples occur at zero crossings, and the characteristic spread associated with spectral leakage is observed (Figure 3.6).

The amelioration of leakage effects, an important issue in signal processing, is explored in the problems.

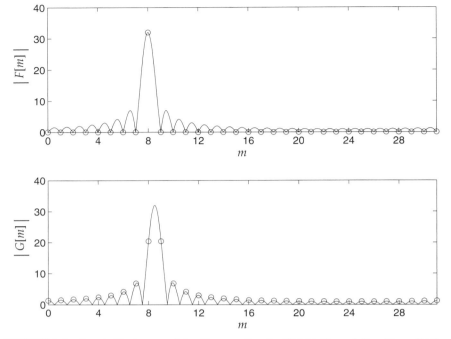

FIGURE 3.6 Spectral leakage explained by sampling the Dirichlet kernel. *Top:* The solid line is the absolute value of the Dirichlet kernel centered at $\theta = 2\pi \cdot 8/32$. The DFT magnitude $|F|$ of the in-bin function $\cos 2\pi n/32$ is identical to samples of the Dirichlet kernel at frequencies $\theta = 2\pi m/32$ (circles), which happen to be zero crossings, except for $m = 8$. *Bottom:* The solid line is the absolute value of the Dirichlet kernel, centered at $\theta = 2\pi \cdot 8.5/32$. The DFT magnitude $|G|$ of the out-of-bin function $\cos 2\pi 8.5n/32$ is identical to samples of the Dirichlet kernel at frequencies $\theta = 2\pi m/32$ (circles); none of the samples are zero.

Rectangle sequence
Another simple but important sequence is the rectangle,

$$f[n] = \begin{cases} 1, & n = 0, 1, \ldots P-1 \\ 0, & \text{otherwise} \end{cases}.$$

The DFT is

$$F[m] = \sum_{n=0}^{P-1} \exp\left(-\frac{i 2\pi m n}{N}\right) = \frac{1 - e^{-i2\pi m P/N}}{1 - e^{-i2\pi/N}} = e^{-i\pi m(P-1)/N} D_P\left(\frac{m}{N}\right), \quad (3.22)$$

another instance of the Dirichlet kernel. The rectangle sequence and the magnitude of its DFT are plotted in Figure 3.7.

There are two important features to note from the figure. First, the height of the DFT is P, the same as the width (number of unit samples) of the rectangle. Second, the width of the DFT, measured by the distance from the origin to the first zero crossing of the underlying Dirichlet kernel, is N/P, inversely proportional to

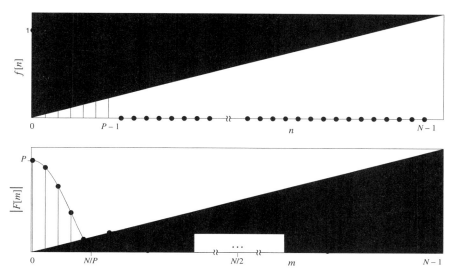

FIGURE 3.7 A P-sample rectangle sequence (*top*) and the magnitude of its DFT (*bottom*). The width of the DFT, as measured by the distance to the first zero crossing of the underlying Dirichlet kernel, is N/P, inversely proportional to the width of the rectangle. The height of the DFT is $F[0] = P$, proportional to the rectangle's width.

the rectangle's width. Thus, making the rectangle wider causes its DFT to become higher and narrower, and vice versa. *This reciprocal relationship between widths in the time and frequency domains is a property of all Fourier transforms.*

Two limiting cases of the rectangle sequence are of interest. First, let $P = 1$, which shrinks the sequence back to just a unit sample at the origin, that is, $\delta[n]$. Second, let $P = N$, which stretches the sequence out to a constant, 1 for all n. The DFTs of these two special cases are

$$\delta[n] \longmapsto D_1(m/N) = \frac{\sin(\pi m/N)}{\sin(\pi m/N)} = 1 \qquad (3.23)$$

and

$$1 \longmapsto e^{-i\pi m(N-1)/N} D_N(m/N) = e^{-i\pi m(N-1)/N} \underbrace{\frac{\sin(\pi m)}{\sin(\pi m/N)}}_{=N\delta[m]}$$

$$= N\delta[m]. \qquad (3.24)$$

Physical frequencies, digital frequencies, and bins

To apply the DFT to real-world signals, we must map physical ("analog") frequency to digital frequency. Suppose we take N samples of a signal at sampling interval Δt. To avoid aliasing, the signal's frequency cannot exceed $\nu = 1/2\Delta t = \nu_s/2$; the corresponding digital frequency is $\theta = \pi$ (or $-\pi$). A signal of this frequency appears in the $N/2$ bin of the DFT. A constant signal (also called DC, by analogy with a direct current electrical signal) appears in bin 0. The other bin frequencies are evenly distributed between these extremes, with frequency interval $\Delta v = 1/N\Delta t = v_s/N$. Bins 1 through $N/2 - 1$ have frequencies $v_m = m\Delta v = mv_s/N$ and cover the frequency range from 0 to half the sampling rate, $v_s/2$. The remaining bins, numbered $N/2 + 1$ through $N - 1$, cover negative digital frequencies between $-\pi$ and 0. Their corresponding analog frequencies are also negative, ranging between $-v_s/2$ and 0 with the analog frequency interval $\Delta v = v_s/N$.

Negative frequencies are interpreted by remembering that basis vectors ϕ_m and ϕ_{N-m} are complex conjugates of one another,

$$\phi_m[n] = e^{i2\pi mn/N}$$
$$\phi_{N-m}[n] = e^{i2\pi(N-m)n/N} = e^{-i2\pi mn/N},$$

and actually have the same frequency ($2\pi m/N$ radians/sample). The apparent negative frequency of the vector ϕ_{N-m} only means that it is the complex conjugate of another vector (namely, ϕ_m) with a positive sign in the exponent. Likewise, a negative analog frequency simply corresponds to a complex exponential $e^{-i2\pi vt}$, which is the complex conjugate of a complex exponential $e^{i2\pi vt}$ having a positive sign in its exponent. A real sine or cosine signal is the combination of a positive and a negative frequency component, a basis vector and its complex conjugate.

The correspondences among bin number, digital frequency, and analog frequency are shown in Figure 3.8 in two ways, on a linear scale and wrapped around a unit circle. The unit circle is particularly helpful for visualizing aliasing. If the analog frequency exceeds the Nyquist limit ($1/2\Delta t$), the digital frequency exceeds π. The angle θ on the unit circle flips to a negative value between 0 and $-\pi$, and this new value is the aliased frequency.

Example 3.3. A 1 kHz sinusoid is sampled at 8 kHz, and 512 samples are taken. The peaks corresponding to this signal will appear at bin $m = 1$ kHz / $\Delta v = 1$ kHz ÷ (8 kHz/512) = 64, and also at bin $512 - 64 = 448$. ∎

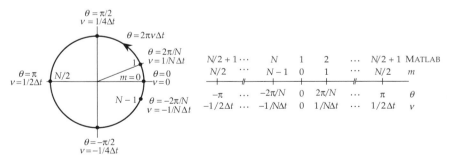

FIGURE 3.8 The relationship among analog frequency v (in Hz), digital frequency $\theta = 2\pi v\Delta t$ (in radians/sample), and DFT bin m diagrammed using the unit circle (*left*) and a conventional frequency axis (*right*). The conventional axis also shows the conversion between "0-based" array indexing used in mathematics and by some computer languages (e.g., C) and the "1-based" array indexing used by MATLAB and other computer languages.

Example 3.4. A 5 kHz sinusoid is sampled at 8 kHz. The signal frequency exceeds 4 kHz, so there will be aliasing. We shall calculate the lower, alias frequency. The digital frequencies of the signal's spectral peaks are $\theta = \pm 2\pi \cdot 5\,\text{kHz}/8\,\text{kHz} = \pm 1.25\pi$, which exceed $\pm\pi$, as expected. The point at 1.25π on the unit circle is the same as the point at -0.75π, and -1.25π is the same as $+0.75\pi$. The analog frequency corresponding to $\theta = \pm 0.75\pi$ is $\nu = \frac{0.75\pi}{2\pi} \cdot 8\,\text{kHz} = 3\,\text{kHz}$. After sampling, the 5 kHz signal is indistinguishable from a 3 kHz signal. ∎

Example 3.5. To analyze, with the DFT, signals of bandwidth up to 10 kHz, with a 25 Hz bin spacing (or less), one needs at least $2 \times 10{,}000/25 = 800$ bins. Rounding 800 up to a power of 2 gives $N = 1024$. The resulting bin spacing is $20{,}000/1024 \approx 19.5$ Hz. ∎

The interpretation of a DFT spectrum is frequently aided by displaying the frequency components on a conventional axis, with negative frequencies to the left of the origin and positive frequencies to the right. MATLAB provides a function, fftshift, which performs this task by swapping the first and second halves of the DFT vector:

before fftshift: $\left(F[0], F[1], \ldots, F\left[\frac{N}{2}-1\right], F\left[\frac{N}{2}\right], \ldots, F[N-1] \right)$

after fftshift: $\left(F\left[\frac{N}{2}\right], \ldots, F[N-1], F[0], F[1], \ldots, F\left[\frac{N}{2}-1\right] \right).$

Another common way to display a DFT, when one wants to know how much energy or power is present at each frequency, is the power spectrum. Because the DFT is an orthogonal expansion, there is a version of Parseval's formula (Theorem 2.4). In the next section we will derive Parseval's formula for the DFT,

$$\sum_{n=0}^{N-1} |f[n]|^2 = \frac{1}{N} \sum_{m=0}^{N-1} |F[m]|^2.$$

It can be rewritten

$$\sum_{n=0}^{N-1} |f[n]|^2 = \frac{1}{N}|F[0]|^2 + \frac{1}{N}\sum_{m=1}^{N/2-1} \left(|F[m]|^2 + |F[N-m]|^2 \right) + \frac{1}{N}\left|F\left[\frac{N}{2}\right]\right|^2.$$

In this form, positive and negative frequency components are grouped together. We then define the power spectrum, for $m = 0, 1, \ldots, N/2$, by

$$P_f[m] = \begin{cases} \frac{1}{N}|F[0]|^2, & m = 0 \\ \frac{1}{N}\left(|F[m]|^2 + |F[N-m]|^2\right), & m = 1, 2, \ldots, \frac{N}{2}-1 \\ \frac{1}{N}\left|F\left[\frac{N}{2}\right]\right|^2, & m = \frac{N}{2} \end{cases} \quad (3.25)$$

These three ways of displaying DFT data—normal, shifted, and power spectrum—are shown in Figure 3.9.

Further applications to signal processing are taken up in the problems.

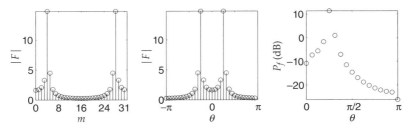

FIGURE 3.9 The DFT of an out-of-bin sinusoid, displayed in different forms. (*a*) The absolute value of the DFT, $|F|$, vs. frequency index m. (*b*) The DFT has been arranged (e.g., by `fftshift`) so that the positive and negative frequency components are symmetrically displayed about the origin. The horizontal axis is digital frequency, $\theta = m\Delta\theta$. (*c*) The DFT power spectrum P_f. The values of the spectrum are displayed in decibels, $10\log_{10} P_f$. The horizontal axis is digital frequency.

3.4 DFT PROPERTIES AND THEOREMS

Calculations with the DFT are made easier by certain theorems. Some of these are direct applications of general results for orthogonal expansions. Others are unique to the choice of a trigonometric basis, and appear, in various forms, throughout the Fourier family.

Linearity of the DFT
The DFT is an inner product, $F[m] = \langle f, \phi_m \rangle$, and the inner product is linear, $\langle af + bg, \phi_m \rangle = a \langle f, \phi_m \rangle + b \langle g, \phi_m \rangle$. Consequently, the DFT is linear.

Theorem 3.2 (Linearity). Let $f, F \in \mathbb{C}^N$ and $g, G \in \mathbb{C}^N$ be DFT pairs, and let $a, b \in \mathbb{C}$ be constants. Then

$$af + bg \longmapsto aF + bG. \tag{3.26}$$

It is a simple exercise to show that the inverse DFT is also linear.

Example 3.6 (Raised cosine sequence). Tapered sequences such as $f[n] = \frac{1}{2} + \frac{1}{2}\cos(2\pi n/N)$ are important in signal processing. To derive the DFT of this sequence, use the linearity theorem. From previous calculations (Equations 3.18 and 3.19),

$$1 = \exp(i2\pi 0 \cdot n) \longmapsto N\delta[m]$$
$$\cos(2\pi n/N) \longmapsto \frac{N}{2}\delta[m-1] + \frac{N}{2}\delta[m-(N-1)].$$

Combining these, we obtain

$$F[m] = \frac{N}{2}\delta[m] + \frac{N}{4}\delta[m-1] + \frac{N}{4}\delta[m-(N-1)].$$

There is a peak at $m = 0$ for the constant component, and peaks at $m = \pm 1$ for the cosine. ∎

Parseval's formula

Parseval's formula says that the DFT preserves inner products and norms.

Theorem 3.3 (Parseval's formula). Let $f, F \in \mathbb{C}^N$ and $g, G \in \mathbb{C}^N$ be DFT pairs. Then
$$\langle f, g \rangle = \frac{1}{N} \langle F, G \rangle \quad \text{and} \quad \|f\|^2 = \frac{1}{N} \|F\|^2,$$
that is,
$$\sum_{n=0}^{N-1} f[n]g^*[n] = \frac{1}{N} \sum_{m=0}^{N-1} F[m]G^*[m] \tag{3.27a}$$
and
$$\sum_{n=0}^{N-1} |f[n]|^2 = \frac{1}{N} \sum_{m=0}^{N-1} |F[m]|^2. \tag{3.27b}$$

Proof: Begin with the general form (Equation 2.24), specialized to N dimensions,
$$\langle f, g \rangle = \sum_{m=0}^{N-1} \langle f, e_m \rangle \langle g, e_m \rangle^*$$
and substitute $e_m = \phi_m / \sqrt{N}$, where $\phi_m[n] = e^{i2\pi mn/N}$ (Equation 3.10). Then,
$$\langle f, g \rangle = \sum_{m=0}^{N-1} \frac{1}{\sqrt{N}} \langle f, \phi_m \rangle \frac{1}{\sqrt{N}} \langle g, \phi_m \rangle^* = \frac{1}{N} \sum_{m=0}^{N-1} F[m]G^*[m] = \frac{1}{N} \langle F, G \rangle.$$
Set $f = g$ and $F = G$ to obtain the corresponding formula for the norm. ■

Example 3.7. The four-element vectors $f = (1, 1, 1, 1)$ and $g = (1, 1, -1, -1)$ are orthogonal. Their respective DFTs are
$$F[m] = \sum_{n=0}^{3} f[n]e^{-i2\pi mn/4} = \sum_{n=0}^{3} e^{-i2\pi mn/4} = \frac{1 - e^{-i2\pi m}}{1 - e^{-i2\pi m/4}}, \quad m = 0, 1, 2, 3$$
$$\Rightarrow F = (4, 0, 0, 0)$$
$$G[m] = \sum_{n=0}^{3} g[n]e^{-i2\pi mn/4} = 1 + e^{-i\pi m/2} - e^{-i\pi m} - e^{-i3\pi m/2}, \quad m = 0, 1, 2, 3$$
$$\Rightarrow G = (0, 2(1-i), 0, 2(1+i)).$$

By inspection, the vectors F and G are also orthogonal. As for the norms,
$$\|f\|^2 = 4, \quad \|F\|^2 = 16$$
$$\|g\|^2 = 4, \quad \|G\|^2 = |2(1-i)|^2 + |2(1+i)|^2 = 16$$
and, with $N = 4$, we have
$$\|f\|^2 = \frac{1}{4}\|F\|^2 \quad \text{and} \quad \|g\|^2 = \frac{1}{4}\|G\|^2$$
as expected, by Parseval's theorem. ■

Sum of samples

The values $f[0]$ and $F[0]$ have a special relationship to the sequences F and f, deriving from the fact that $\phi_0[n] = 1$ and $\phi_m[0] = 1$ for all m, n.

Theorem 3.4 (Area theorem). Let $f, F \in \mathbb{C}^N$ be a DFT pair. Then,

$$F[0] = \sum_{n=0}^{N-1} f[n] \tag{3.28a}$$

$$f[0] = \frac{1}{N} \sum_{m=0}^{N-1} F[m]. \tag{3.28b}$$

It is called the area theorem because, if the values $f[n]$ are regarded as samples of a continuous function $f(t)$ ($f[n] = f(n\Delta t)$), then $\sum_{n=0}^{N-1} f[n]\Delta t = F[0]\Delta t$ approximates the area $\int_0^{N\Delta t} f(t)dt$.

Example 3.8. The DFT of the unit sample sequence $\delta[n] = (1, 0, \ldots, 0)$ was found to be (Equation 3.23)

$$\sum_{n=0}^{N-1} \delta[n] e^{-i2\pi mn/N} = 1, \quad m = 0, 1, \ldots, N-1$$
$$= (1, 1, \ldots, 1).$$

With this pair $f = (1, 0, \ldots, 0)$ and $F = (1, 1, \ldots, 1)$, we observe

$$f[0] + f[1] + \cdots + f[N-1] = 1 = F[0]$$

and

$$\frac{1}{N}(F[0] + F[1] + \cdots + F[N-1]) = 1 = f[0],$$

as predicted by the area theorem. ∎

Example 3.9. The DFT of the P-sample rectangle sequence was found to be (Equation 3.22)

$$F[m] = e^{-i\pi m(P-1)/N} D_P\left(\frac{m}{N}\right),$$

where D_P is the Dirichlet kernel. The sum of the rectangle's samples is P, as is $F[0]$, in agreement with the area theorem. The sum of the DFT's samples are computed directly from the expression for the DFT,

$$\frac{1}{N}\sum_{m=0}^{N-1} F[m] = \sum_{m=0}^{N-1}\left[\frac{1}{N}\sum_{n=0}^{P-1}\exp\left(-j\frac{2\pi mn}{N}\right)\right] = \sum_{n=0}^{P-1}\underbrace{\frac{1}{N}\sum_{m=0}^{N-1}\exp\left(-j\frac{2\pi mn}{N}\right)}_{=\delta[n]}$$

$$= \sum_{n=0}^{P-1}\delta[n] = 1$$

and this is the value of the rectangle sequence at $n = 0$. ∎

Periodicity of the DFT

The DFT inherits an important periodicity property from the periodicity of its basis vectors.

Theorem 3.5 (Periodicity). Let $f, F \in \mathbb{C}^N$ be a DFT pair. Then f and F are each one period of an infinite sequence with period N.

Proof: Define the doubly infinite sequence $\tilde{F} : \mathbb{Z} \to \mathbb{C}$ by the formula

$$\tilde{F}[m] = \sum_{n=0}^{N-1} f[n] e^{-i2\pi mn/N}, \quad m = \ldots -2, -1, 0, 1, 2, \ldots.$$

$\tilde{F}$ has period N,

$$\tilde{F}[m+N] = \sum_{n=0}^{N-1} f[n] e^{-i2\pi(m+N)n/N} = \sum_{n=0}^{N-1} f[n] e^{-i2\pi mn/N}$$
$$= \tilde{F}[m],$$

and F is identically the subsequence $(\tilde{F}[0], \tilde{F}[1], \ldots, \tilde{F}[N-1])$. The proof for f is identical. ∎

DFT symmetries

Because f and F are, implicitly, single periods of underlying periodic sequences, we may apply the earlier definitions of even, odd, and Hermitian symmetry: A vector $f \in \mathbb{C}^N$ is even if $f[N-n] = f[n]$, odd if $f[N-n] = -f[n]$, and Hermitian if $f[N-n] = f^*[n]$. An arbitrary vector may be written as the sum of an even part and an odd part, $f = f_e + f_o$, where

$$f_e[n] = \frac{f[n] + f[N-n]}{2}$$
$$f_o[n] = \frac{f[n] - f[N-n]}{2}, \quad n = 0, 1, \ldots, N-1. \quad (3.29)$$

Symmetries in f are mirrored in F (Figure 3.10).

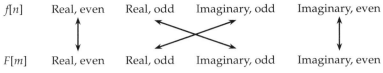

FIGURE 3.10 Symmetry properties of the discrete Fourier transform.

Theorem 3.6 (DFT symmetries). Let $f, F \in \mathbb{C}^N$ be a DFT pair. These statements and their converses are true:

(a) If f is real, then F is Hermitian: $F[N - m] = F^*[m]$.
(b) If f is even (odd), then F is even (odd).
(c) If f is real and even (real and odd), then F is real and even (imaginary and odd).

Proof:

(a) The DFT basis functions have Hermitian symmetry in the frequency index m, $\phi_{N-m} = \phi_m^*$ (Equation 3.6). Hence,
$$F[N - m] = \langle f, \phi_{N-m} \rangle = \langle f, \phi_m^* \rangle.$$
You can show that $\langle f, \phi_m^* \rangle = \langle f^*, \phi_m \rangle^*$. But, because f is real, $f^* = f$, and so
$$F[N - m] = \langle f, \phi_m \rangle^* = F^*[m].$$
Now suppose F is Hermitian. We will show that f is real by showing that the imaginary part of f is zero. The imaginary part of f is
$$\mathcal{I}m\, f = \frac{f - f^*}{2i}.$$
The DFT of f^* is
$$f^* \longmapsto \langle f^*, \phi_m \rangle = \langle f, \phi_m^* \rangle^* = \langle f, \phi_{N-m} \rangle^* = F^*[N - m]$$
because the basis vectors are Hermitian. But because F is Hermitian, $F^*[N - m] = F[m]$, that is, f and f^* have the same DFT, hence they are equal, and $f - f^* = 0$.

(b) By definition,
$$F[m] = \sum_{n=0}^{N-1} f[n] \phi_m[n];$$
but also,
$$F[N - m] = \sum_{n=0}^{N-1} f[n] \phi_{N-m}[n] = \sum_{n=0}^{N-1} f[n] \phi_m^*[n] \quad \text{(using Equation 3.6)}$$
$$= \sum_{n=0}^{N-1} f[N - n] \phi_m^*[N - n] = \sum_{n=0}^{N-1} f[N - n] \phi_m[n] \quad \text{(using Equation 3.5)}.$$

Now, the even part of F is (Equation 3.29)

$$F_e[m] = \frac{1}{2}(F[m] + F[N-m]) = \frac{1}{2}\sum_{n=0}^{N-1}(f[n]+f[N-n])\phi_m[n]$$

$$= \sum_{n=0}^{N-1} f_e[n]\phi_m[n]$$

and the odd part is, similarly,

$$F_o[m] = \sum_{n=0}^{N-1} f_o[n]\phi_m[n].$$

If f is even, then $f_o = 0$; consequently, $F_o = 0$ and F is even. Likewise, if f is odd, then $f_e = 0$ and $F_e = 0$, so F is odd. The converse follows by a symmetric derivation.

(c) If f is real, then F is Hermitian (part (a)). If additionally f is even, we know that F is even (part (b)), and this even part is real because F is Hermitian. Likewise, if f is odd as well as real, then F is odd, and it is also imaginary because F is Hermitian. The converse follows by a symmetric derivation. ∎

Example 3.10. The DFTs of in-bin cosine and sine sequences were previously calculated. Now we consider their symmetry properties.

(a) The cosine sequence $c = \left(\cos\frac{2\pi kn}{N}\right)_{n=0}^{N-1}$, $k = 0, 1, \ldots, N-1$, is even:

$$\cos\frac{2\pi k(N-n)}{N} = \underbrace{\cos\frac{2\pi kN}{N}}_{=1}\cos\frac{2\pi kn}{N} + \underbrace{\sin\frac{2\pi kN}{N}}_{=0}\sin\frac{2\pi kn}{N}$$

$$= \cos\frac{2\pi kn}{N}.$$

Its DFT is (Equation 3.19):

$$C[m] = \frac{N}{2}\delta[m-k] + \frac{N}{2}\delta[m-(N-k)], \quad m = 0, 1, \ldots, N-1.$$

This sequence is also even, in keeping with the DFT symmetry theorem. The cosine sequence and its DFT are shown, for $N = 8$, in Figure 3.11.

(b) The sine sequence $s = \left(\sin\frac{2\pi kn}{N}\right)_{n=0}^{N-1}$, $k = 0, 1, \ldots, N-1$, is odd:

$$\sin\frac{2\pi k(N-n)}{N} = \sin\frac{2\pi kN}{N}\cos\frac{2\pi kn}{N} - \cos\frac{2\pi kN}{N}\sin\frac{2\pi kn}{N}$$

$$= -\sin\frac{2\pi kn}{N}.$$

The DFT is (Equation 3.20):

$$S[m] = -\frac{iN}{2}\delta[m-k] + \frac{iN}{2}\delta[m-(N-k)], \quad m = 0, 1, \ldots, N-1.$$

It is odd and imaginary, as predicted by the symmetry theorem. The sine function and its DFT are shown, for $N = 8$, in Figure 3.12. ∎

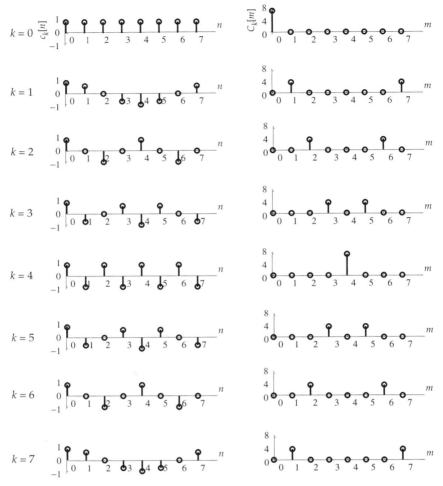

FIGURE 3.11 Cosine sequences $c_k[n] = \cos(2\pi kn/8)$ (*left*) and their DFTs $C_k[m] = 4\delta[m - k] + 4\delta[m - (8 - k)]$ (*right*). Both c and C are real valued and have even symmetry. As the frequency index k increases (*top to bottom*), $\delta[m - k]$ moves to a higher bin (higher positive frequency) and $\delta[m - (8 - k)]$ moves to a lower bin (higher negative frequency). The samples of c_k and c_{8-k} are indistinguishable, and their DFTs are identical.

Knowing the symmetry properties of the DFT is valuable in computation. Suppose you need to compute the product, FG, of two Hermitian DFTs. By symmetry, the values of FG in the negative frequency bins can be computed as the complex conjugates of the values in the positive frequency bins; thus, only $N/2$ actual complex products need be calculated. The symmetries are also useful for error checking. If you find that the inverse DFT of a purportedly Hermitian sequence has a non-negligible imaginary part, this is a clue that the sequence is actually not Hermitian (or, alternatively, that your DFT algorithm is incorrect).

3.4 DFT PROPERTIES AND THEOREMS

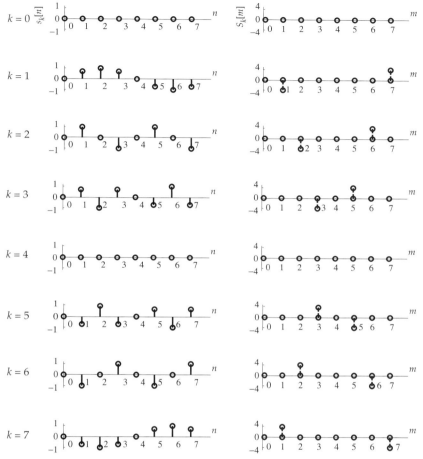

FIGURE 3.12 Sine sequences $s_k[n] = \sin(2\pi k n/8)$ (*left*) and their DFTs $S_k[m] = -4i\delta[m - k] + 4i\delta[m - (8 - k)]$ (*right*). s is real valued and has odd symmetry; S is imaginary valued and has odd symmetry. As the frequency index k increases (*top to bottom*), $\delta[m - k]$ moves to a higher bin (higher positive frequency), and $\delta[m - (8 - k)]$ moves to a lower bin (higher negative frequency). At $k = 4$ they overlap and cancel: $\sin(\pi n) = 0$. The samples of s_k and s_{8-k} are indistinguishable except for a sign flip; their DFTs are likewise identical except for a sign flip.

Example 3.11 (Rectangle sequence). Previously we calculated the DFT of a rectangle sequence (Equation 3.22):

$$F[m] = e^{-i\pi m(P-1)/N} D_P\left(\frac{m}{N}\right).$$

The rectangle sequence is real but not symmetric; thus, we expect the DFT to be Hermitian. Check this:

$$F^*[N - m] = e^{+i\pi(N-m)(P-1)/N} D_P\left(\frac{N-m}{N}\right).$$

The Dirichlet kernel is even or odd depending on whether P is odd or even, that is, $D_P(\frac{N-m}{N}) = (-1)^{P-1} D_P(\frac{m}{N})$ (Figure 3.5). As for the complex exponential,

$$e^{+i\pi(N-m)(P-1)/N} = e^{+i\pi(P-1)} e^{-i\pi m(P-1)/N} = (-1)^{P-1} e^{-i\pi m(P-1)/N}.$$

Putting these together,

$$F^*[N-m] = (-1)^{P-1} e^{-i\pi m(P-1)/N} \cdot (-1)^{P-1} D_P\left(\frac{m}{N}\right)$$
$$= e^{-i\pi m(P-1)/N} D_P\left(\frac{m}{N}\right) = F[m],$$

as expected. ∎

Shifting

The DFT's periodicity is also important when we consider shifts, or translations, of a sequence f. A shift is expressed by writing $f[n-r]$, where r can be positive or negative. Because f has N elements, we must be clear about the meaning of $f[n-r]$ when $n-r$ is less than 0 or greater than $N-1$. Taking $r=2$, for example, a two-sample shift of the sequence $(f[0], f[1], f[2], \ldots, f[7])$ would result in $(f[-2], f[-1], f[0], \ldots, f[5])$. What meaning do we attach to $f[-2]$ and $f[-1]$?

Since f is one period of a periodic function $\tilde{f}$, $\tilde{f}[-2]$ is the same as $\tilde{f}[6]$, and $\tilde{f}[-1]$ is the same as $\tilde{f}[7]$. Thus, the two-sample shift of f is $(f[6], f[7], f[0], f[1], \ldots, f[5])$. That is, the shifted index $n-r$ is interpreted modulo-N, and the notation $f[n-r]$ denotes the $(n-r) \bmod N$ element of the vector f: $f[-2 \bmod 8] = f[6]$, and $f[-1 \bmod 8] = f[7]$.

Imagine that the values $f[0] \ldots f[N-1]$ are equispaced around a circle. A positive shift ($r > 0$) corresponds to a counterclockwise rotation, and a negative shift to a clockwise rotation of the values around the circle. For this reason, the shift is said to be *circular* or *cyclic* (Figure 3.13). When working with finite sequences and the DFT, all shifts are cyclic.

The DFT of a shifted sequence has a particularly simple form.

Theorem 3.7 (Shift theorem). Let $f, F \in \mathbb{C}^N$ be a DFT pair. Then,

$$\left(f[n-r]\right)_{n=0}^{N-1} \longleftrightarrow \left(e^{-i2\pi rm/N} F[m]\right)_{m=0}^{N-1}. \tag{3.30}$$

Proof: Simply calculate the inverse transform,

$$\frac{1}{N} \sum_{m=0}^{N-1} e^{-i2\pi rm/N} F[m] e^{i2\pi mn/N} = \frac{1}{N} \sum_{m=0}^{N-1} F[m] e^{i2\pi m(n-r)/N}$$
$$= f[n-r], \quad n = 0, 1, \ldots, N-1.$$

∎

3.4 DFT PROPERTIES AND THEOREMS 135

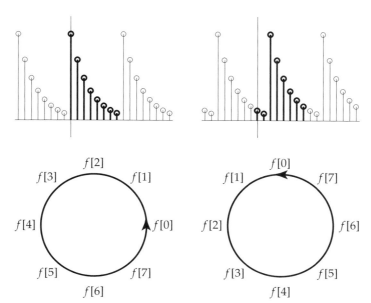

FIGURE 3.13 Shifting functions in the DFT. *Left:* A periodic sequence (*top*) and the cyclic representation of one period (*bottom*). *Right:* A shift of the periodic sequence (*top*) is equivalent to a cyclic shift of a single period (*bottom*).

Example 3.12. Consider the shifted cosine sequence

$$f[n-2] = \cos\frac{2\pi(n-2)}{8} = \sin\frac{2\pi n}{8}, \quad n = 0, 1, 2, \ldots 7.$$

Using the shift theorem, the DFT of $\cos 2\pi(n-2)/8$ is

$$\begin{aligned}
\text{DFT}\left\{\cos\frac{2\pi(n-2)}{8}\right\} &= e^{-i2\pi(2)m/8}\text{DFT}\left\{\cos\frac{2\pi n}{8}\right\} \\
&= e^{-i\pi m/2}\left(\frac{N}{2}\delta[m-1] + \frac{N}{2}\delta[m-7]\right) \\
&= e^{-i\pi/2}\frac{N}{2}\delta[m-1] + e^{-i7\pi/2}\frac{N}{2}\delta[m-7] \\
&= -\frac{iN}{2}\delta[m-1] + \frac{iN}{2}\delta[m-7],
\end{aligned}$$

which is the DFT of $\sin\frac{2\pi n}{8}$ (Figure 3.14).

It is important to note that shifting does not affect the magnitude of the DFT:

$$|e^{-i2\pi mr/N}F[m]| = |F[m]|,$$

because $|e^{-i2\pi mr/N}| = 1$. Thus, both the sine and cosine in this example have the same DFT magnitude, namely $\frac{N}{2}\delta[m-1] + \frac{N}{2}\delta[m-7]$. ∎

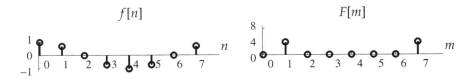
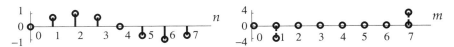

FIGURE 3.14 The sine sequence $\sin(2\pi n/8)$ is a two-sample shift of the cosine sequence: $\cos(2\pi(n-2)/8) = \sin(2\pi n/8)$ (*left*). Their DFTs have the same frequency components, $\delta[m-1]$ and $\delta[m-7]$, but different phases (*right*), according to the shift theorem.

Example 3.13 (Shifted rectangle sequence). In some applications, we want an even rectangle sequence,

$$f[n] = \begin{cases} 1, & n = 0, 1, \ldots P \text{ and } N - P \ldots N - 1 \\ 0, & \text{otherwise} \end{cases}.$$

This is a left cyclic shift by P samples of a $2P + 1$-sample rectangle sequence like the one analyzed earlier. Using the shift theorem with Equation 3.22, the DFT is

$$f[n] \longmapsto e^{-i2\pi m(-P)/N} \cdot e^{-i\pi m((2P+1)-1)/N} D_{2P+1}\left(\frac{m}{N}\right) = D_{2P+1}\left(\frac{m}{N}\right).$$

We note also that the DFT is real and even, corresponding to the symmetry of the rectangle sequence. In fact, we see that the phase factor in Equation 3.22 is due to the rectangle's lack of even symmetry. ∎

Convolution

The linearity of the DFT enables the DFT of a sum $f + h$ to be easily calculated as the sum $F + H$. We will now see how the shift theorem enables us to express the DFT of the product of two sequences. The result is important in applications of the DFT, particularly to linear system theory and signal processing.

Theorem 3.8 (Convolution theorem). Let $f, h \in \mathbb{C}^N$ have DFTs $F, H \in \mathbb{C}^N$, and define the product $fh \in \mathbb{C}^N$ by $fh[n] = f[n]h[n]$. Then, the DFT of fh is

$$fh \longmapsto \frac{1}{N} F \circledast H \tag{3.31a}$$

where

$$(F \circledast H)[m] = \sum_{k=0}^{N-1} F[k]H[m-k] \tag{3.31b}$$

is called the *cyclic convolution* of F and H (the shift $m - k$ is understood to be cyclic, or modulo-N). Similarly,

$$f \circledast h \longmapsto FH. \qquad (3.31c)$$

Proof: The DFT of fh can be written as an inner product,

$$\text{DFT}\{fh\} = \sum_{n=0}^{N-1} f[n]h[n]e^{-i2\pi mn/N} = \sum_{n=0}^{N-1} f[n]\left(h^*[n]e^{i2\pi mn/N}\right)^*.$$

Parseval's theorem then says that this inner product can be written in terms of DFTs,

$$\sum_{n=0}^{N-1} f[n]\left(h^*[n]e^{i2\pi mn/N}\right)^* = \frac{1}{N}\sum_{k=0}^{N-1} F[k]\left[\sum_{n=0}^{N-1}\left(h^*[n]e^{i2\pi mn/N}\right)e^{-i2\pi kn/N}\right]^*$$

$$= \frac{1}{N}\sum_{k=0}^{N-1} F[k]\left[\sum_{n=0}^{N-1} h[n]e^{-i2\pi(m-k)n/N}\right]$$

$$= \frac{1}{N}\sum_{k=0}^{N-1} F[k]H[m-k].$$

For the second part, write the DFT of the convolution as a double sum,

$$\text{DFT}\{f \circledast h\}[m] = \sum_{n=0}^{N-1}\left[\sum_{k=0}^{N-1} f[k]h[n-k]\right]e^{-i2\pi mn/N}$$

$$= \sum_{k=0}^{N-1} f[k]\left[\sum_{n=0}^{N-1} h[n-k]e^{-i2\pi mn/N}\right]$$

$$= \sum_{k=0}^{N-1} f[k]\left[e^{-i2\pi km/N}H[m]\right] \text{ (shift theorem)}$$

$$= \left[\sum_{k=0}^{N-1} f[k]e^{-i2\pi km/N}\right]H[m] = F[m]H[m].$$

∎

Cyclic convolution can be visualized with the aid of Figure 3.15. Imagine that the values of f and h are arranged around the circumferences of two concentric circles. The outer circle, carrying h (reversed), is rotated around the inner circle, carrying f. The value of the convolution for a particular step (n) in the rotation is the sum of products of the terms that are aligned opposite each other on the two circles.

Example 3.14. Let $f = \delta[n - a]$ and $g = \delta[n - b]$, where $a, b, n \in \{0, 1, \ldots N - 1\}$. We will calculate $f \circledast g$ using both direct summation and the convolution theorem.

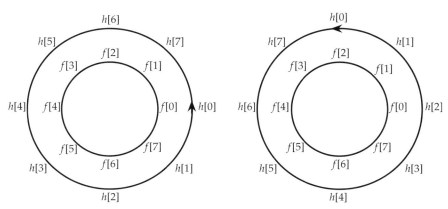

FIGURE 3.15 Cyclic convolution for two functions with $N = 8$. The values $h[-k]$ are rotated past the values $f[k]$, and the sum of products is computed. *Left:* Alignment for $n = 0$. $(f \circledast h)[0] = f[0]h[0] + f[1]h[7] + \cdots + f[7]h[1]$. *Right:* Alignment for $n = 2$. $(f \circledast h)[2] = f[2]h[0] + f[3]h[7] + \cdots + f[1]h[1]$.

Direct

$$\delta[n-a] \circledast \delta[n-b] = \sum_{k=0}^{N-1} \delta[k-a]\delta[(n-k)-b].$$

The first δ is zero except for $k = a$, so the product of the two δs is zero except for $k = a$. This gives

$$\delta[n-a] \circledast \delta[n-b] = \delta[(n-a)-b] = \delta[n-(a+b)]. \quad (3.32)$$

Convolution Theorem The DFT of $\delta[n-a]$ is

$$F[m] = \sum_{n=0}^{N-1} \delta[n-a]e^{-i2\pi mn/N} = e^{-i2\pi am/N} \quad (3.33)$$

and, likewise, the DFT of $\delta[n-b]$ is $G[m] = e^{-i2\pi bm/N}$. Their product is $FG[m] = e^{-i2\pi(a+b)m/N}$. The inverse DFT of FG is

$$(f \circledast g)[n] = \frac{1}{N}\sum_{m=0}^{N-1} e^{-i2\pi(a+b)m/N} e^{i2\pi mn/N} = \frac{1}{N}\langle \phi_n, \phi_{a+b}\rangle,$$

where we have recognized that this inverse DFT is just an inner product of DFT basis functions. Then, because the basis functions are orthogonal, the result is

$$(f \circledast g)[n] = \delta[n-(a+b)],$$

just as we calculated by the direct method. ∎

3.4 DFT PROPERTIES AND THEOREMS

Example 3.15. More generally, let $f \in \mathbb{C}^N$ be any vector, and calculate its convolution with $\delta[n-a]$. We have

$$f \circledast \delta[n-a] = \sum_{k=0}^{N-1} f[k]\delta[(n-k)-a].$$

The δ is one when $n - k - a = 0$, or $k = n - a$, and zero otherwise. It sifts out the $k = n - a$ term of the sum, giving

$$f \circledast \delta[n-a] = f[n-a]. \tag{3.34}$$

Convolution with $\delta[n-a]$ shifts f to $n = a$, where the unit sample is located. Using the convolution theorem instead, the DFT of $\delta[n-a]$ is $e^{-i2\pi am/N}$, so the convolution is the inverse DFT of $F[m]e^{-i2\pi am/N}$. But by the shift theorem, $F[m]e^{-i2\pi am/N}$ is the DFT of $f[n-a]$. ∎

Example 3.16. Let $f = (0, 1, 0, 2, 0, 0, 0, 0)$ and $h = (1, 1, 0, 0, 0, 0, 0, 1)$ and calculate $f \circledast h$. Only $f[1]$ and $f[3]$ are nonzero. The explicit sums for $g = f \circledast h$ simplify to $g[n] = f[1]h[n-1] + f[3]h[n-3] = h[n-1] + 2h[n-3]$. So we have

$$\begin{aligned}
g[0] &= h[7] + 2h[5] = 1 \\
g[1] &= h[0] + 2h[6] = 1 \\
g[2] &= h[1] + 2h[7] = 3 \\
g[3] &= h[2] + 2h[0] = 2 \\
g[4] &= h[3] + 2h[1] = 2 \\
g[5] &= h[4] + 2h[2] = 0 \\
g[6] &= h[5] + 2h[3] = 0 \\
g[7] &= h[6] + 2h[4] = 0.
\end{aligned}$$

Instead, using what we know about the unit sample sequence, we may write $f[n] = \delta[n-1] + 2\delta[n-3]$. Then, $g[n] = \delta[n-1] \circledast h + 2\delta[n-3] \circledast h = h[n-1] + 2h[n-3]$ (Equation 3.34). Now, taking cyclic shifts,

$$\begin{aligned}
h[n-1] &= (1, 1, 1, 0, 0, 0, 0, 0) \\
2h[n-3] &= (0, 0, 2, 2, 2, 0, 0, 0),
\end{aligned}$$

and adding,

$$g = (1, 1, 3, 2, 2, 0, 0, 0).$$

We may also express $h[n] = \delta[n] + \delta[n-1] + \delta[n-7]$, and use (Equation 3.32),

$$\begin{aligned}
g[n] &= \left(\delta[n-1] + 2\delta[n-3]\right) \circledast \left(\delta[n] + \delta[n-1] + \delta[n-7]\right) \\
&= \delta[n-1] + 2\delta[n-3] + \delta[n-2] + 2\delta[n-4] + \delta[n-8] + 2\delta[n-10].
\end{aligned}$$

Using $(n-8) \bmod 8 = 0$ and $(n-10) \bmod 8 = n-2$,

$$g[n] = \delta[n-1] + 2\delta[n-3] + \delta[n-2] + 2\delta[n-4] + \delta[n] + 2\delta[n-2]$$
$$= \delta[n] + \delta[n-1] + 3\delta[n-2] + 2\delta[n-3] + 2\delta[n-4].$$ ∎

Example 3.17 (Truncated sinusoid). A rectangle sequence is useful for modeling the truncation of another sequence. For example, an in-bin cosine truncated to P samples is

$$f[n] = R_P[n] \cos(2\pi kn/N)$$

where $R_P[n] = \begin{cases} 1, & n = 0, 1, \dots P-1 \\ 0, & \text{otherwise} \end{cases}$.

The (N-point) DFTs of the individual sequences are, from previous calculations,

$$R_P[n] \longmapsto e^{i\pi(P-1)/N} D_P(m/N)$$
$$\cos(2\pi kn/N) \longmapsto \frac{N}{2}\delta[m-k] + \frac{N}{2}\delta[m-(N-k)].$$

Then using the convolution theorem (Equations 3.31 and 3.34),

$$F[m] = \frac{1}{N} \cdot e^{i\pi(P-1)m/N} D_P(m/N) \circledast \left(\frac{N}{2}\delta[m-k] + \frac{N}{2}\delta[m-(N-k)] \right)$$
$$= \frac{1}{2} e^{i\pi(P-1)(m-k)/N} D_P\left(\frac{m-k}{N}\right) + \frac{1}{2} e^{i\pi(P-1)(m-(N-k))/N} D_P\left(\frac{m-(N-k)}{N}\right).$$

Ignoring the exponential phase factors, the DFT is seen to be two Dirichlet kernels of height $\frac{P}{2}$ and centered at the bins corresponding to the sinusoid's frequency. The "sharpness" of the spectrum, measured by the narrowness of the two peaks, improves with larger values of P (Figure 3.16). When $P \to N$ (no truncation), the Dirichlet kernels reduce to delta sequences, and we recover the DFT of the cosine alone,

$$F[m] \to \frac{1}{2} e^{i\pi(N-1)(m-k)/N} N\delta[m-k] + \frac{1}{2} e^{i\pi(N-1)(m-(N-k))/N} N\delta[m-(N-k)]$$
$$= \frac{N}{2}\delta[m-k] + \frac{N}{2}\delta[m-(N-k)].$$ ∎

⋆ **Matrix form of convolution**

The cyclic convolution $g = h \circledast f$ may be written as a matrix-vector product,

$$\mathbf{g} = \begin{bmatrix} g[0] \\ g[1] \\ g[2] \\ \vdots \\ g[N-1] \end{bmatrix} = \begin{bmatrix} h[0] & h[N-1] & \cdots & h[2] & h[1] \\ h[1] & h[0] & \cdots & h[3] & h[2] \\ h[2] & h[1] & \cdots & h[4] & h[3] \\ \vdots & \vdots & \ddots & \vdots & \vdots \\ h[N-1] & h[N-2] & \cdots & h[1] & h[0] \end{bmatrix} \begin{bmatrix} f[0] \\ f[1] \\ f[2] \\ \vdots \\ f[N-1] \end{bmatrix}$$

$$= \mathbf{Hf}. \tag{3.35}$$

Each row of the matrix $\mathbf{H}$ is a cyclic shift of the row above it, and the first row is a cyclic shift of the last row. A matrix with this structure is called *circulant*. Circulant

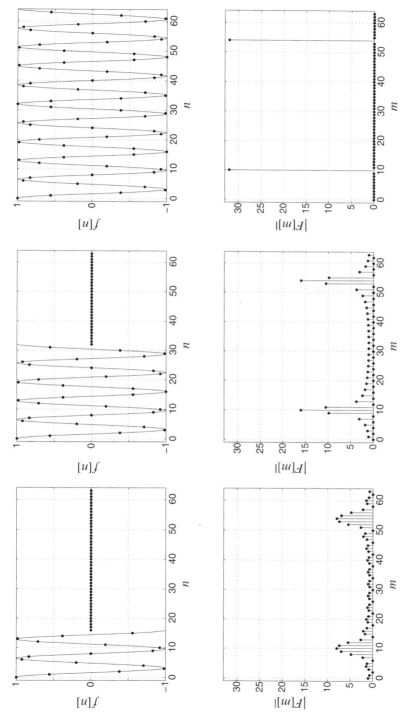

FIGURE 3.16 Truncated cosine sequences and their DFT magnitudes. The cosine has frequency $\theta_0 = 10/64$, truncated to $P = 16$ points (*left*) and $P = 32$ points (*center*), and $P = N = 64$ points (untruncated, *right*). As P increases, the spectral peaks become sharper.

141

matrices have a particular relationship to the DFT matrix (Equation 3.13). Writing the convolution theorem (Equation 3.31) in matrix-vector form,

$$\tilde{\mathbf{g}} = \begin{bmatrix} G[0] \\ G[1] \\ G[2] \\ \vdots \\ G[N-1] \end{bmatrix} = \begin{bmatrix} H[0] & & & & 0 \\ & H[1] & & & \\ & & H[2] & & \\ & & & \ddots & \\ 0 & & & & H[N-1] \end{bmatrix} \begin{bmatrix} F[0] \\ F[1] \\ F[2] \\ \vdots \\ F[N-1] \end{bmatrix}$$

$$= \tilde{\mathbf{H}}\tilde{\mathbf{f}}. \tag{3.36}$$

Note that $\tilde{\mathbf{H}}$ is a diagonal matrix. Now express $\tilde{\mathbf{g}}$ and $\tilde{\mathbf{f}}$ in terms of the unitary matrix form of the DFT (Equation 3.16),

$$\tilde{\mathbf{g}} = \sqrt{N}\left(\frac{1}{\sqrt{N}}\mathbf{D}\right)\mathbf{g}$$

$$\tilde{\mathbf{f}} = \sqrt{N}\left(\frac{1}{\sqrt{N}}\mathbf{D}\right)\mathbf{f},$$

and substitute these into Equation 3.36:

$$\sqrt{N}\left(\frac{1}{\sqrt{N}}\mathbf{D}\right)\mathbf{g} = \tilde{\mathbf{H}}\sqrt{N}\left(\frac{1}{\sqrt{N}}\mathbf{D}\right)\mathbf{f}.$$

The factors of $\sqrt{N}$ on both sides of this equation can be dropped. Then, because $(\frac{1}{\sqrt{N}}\mathbf{D})^\dagger(\frac{1}{\sqrt{N}}\mathbf{D}) = \mathbf{I}$, we can solve for $\mathbf{g}$ by multiplying both sides, on the left, by $(\frac{1}{\sqrt{N}}\mathbf{D})^\dagger$:

$$\mathbf{g} = \left(\frac{1}{\sqrt{N}}\mathbf{D}\right)^\dagger \tilde{\mathbf{H}} \left(\frac{1}{\sqrt{N}}\mathbf{D}\right)\mathbf{f}.$$

Comparing this with Equation 3.35, observe that

$$\mathbf{H} = \left(\frac{1}{\sqrt{N}}\mathbf{D}^\dagger\right)\tilde{\mathbf{H}}\left(\frac{1}{\sqrt{N}}\mathbf{D}\right),$$

which leads to the final result,

$$\tilde{\mathbf{H}} = \left(\frac{1}{\sqrt{N}}\mathbf{D}\right)\mathbf{H}\left(\frac{1}{\sqrt{N}}\mathbf{D}\right)^\dagger. \tag{3.37}$$

The circulant matrix $\mathbf{H}$ is diagonalized—transformed into the diagonal matrix $\tilde{\mathbf{H}}$—by the unitary DFT matrix. This is the matrix-vector form of the DFT relationship $h \longmapsto H$.

Zero padding

The next DFT theorem is motivated by the problem of estimating the frequency of a sinusoidal signal from the DFT of its samples. The DFT of a complex exponential signal with amplitude $A > 0$ and phase φ, $A \exp(i(\theta_0 n - \varphi))$, is a sampled Dirichlet kernel,

$$F[m] = A e^{-i\varphi} \exp\left[-\frac{i(N-1)(\theta - \theta_0)}{2}\right] D_N\left(\frac{\theta - \theta_0}{2\pi}\right)\bigg|_{\theta = m\Delta\theta},$$

and its magnitude is

$$|F[m]| = A \left|D_N\left(\frac{m\Delta\theta - \theta_0}{2\pi}\right)\right|.$$

From inspection of the graph of the DFT magnitude, it seems reasonable to pick the bin $\hat{m}$ where $|F[m]|$ is largest, and make the frequency estimate $\hat{\theta}_0 = \hat{m}\Delta\theta$. The worst case estimation error is $\frac{\Delta\theta}{2}$, occurring when θ_0 is exactly between two bin frequencies.

The frequency estimate will be improved if $\Delta\theta$, the bin spacing, can be made smaller. If we can obtain more samples of the signal, so that N is larger, then $\Delta\theta = 2\pi/N$ will be smaller and the frequency estimate will be better. However, while it is always better to have more data, it is possible to decrease $\Delta\theta$ even if we cannot increase the number of samples. Begin with the formula for the DFT,

$$F[m] = \sum_{n=0}^{N-1} f[n] e^{-i2\pi mn/N} = \sum_{n=0}^{N-1} f[n] e^{-imn\Delta\theta}.$$

Nothing formally prevents us from using a different value for $\Delta\theta$, as long as we can still calculate the sum. Assume this is possible and replace $\Delta\theta$ with $\Delta\theta' = 2\pi/N'$, where $N' \geq N$. The DFT with N' bins, denoted $F[m; N']$, is

$$F[m; N'] = \sum_{n=0}^{N-1} f[n] e^{-i2\pi mn/N'}, \quad m = 0, 1, 2, \ldots, N' - 1. \tag{3.38}$$

(Note $F[m; N] = F[m]$.) If we let $f[n] = e^{i\theta_0 n}$ and carry the calculation through,

$$F[m; N'] = \sum_{n=0}^{N-1} e^{i\theta_0 n} e^{-i2\pi mn/N'} = \sum_{n=0}^{N-1} \exp[-i(\theta - \theta_0)n]\bigg|_{\theta = 2\pi m/N'}$$

$$= \exp\left[-\frac{i(N-1)(\theta - \theta_0)}{2}\right] D_N\left(\frac{\theta - \theta_0}{2\pi}\right)\bigg|_{\theta = 2\pi m/N'}. \tag{3.39}$$

This is a more finely sampled Dirichlet kernel. An example is shown in Figure 3.17, below.

The computational implementation of Equation 3.38 requires some additional work. It is an operation that takes a vector of length N and returns a vector of length N'. As we will soon see in Section 3.5, efficient algorithms for computing the DFT require that the input and output vectors have the same length. To express

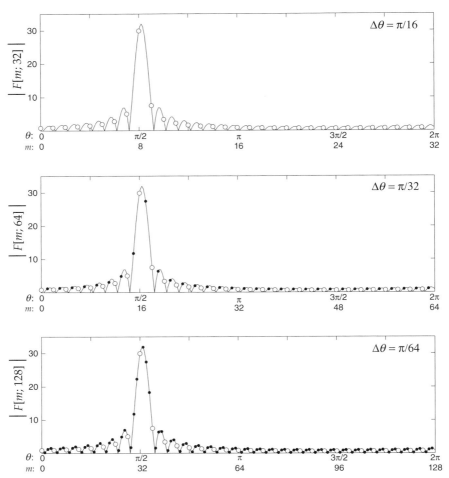

FIGURE 3.17 Improving frequency estimation by finer sampling of the Dirichlet kernel (Equation 3.39). The signal has $\theta_0 = 0.5125\pi$. *Top:* With $N' = N = 32$, the original DFT values are shown by circles. The maximum magnitude is in bin 8, $\hat{\theta}_0 = 0.5\pi$. *Middle:* The new samples are shown by dots. With $N' = 2N = 64$, $\Delta\theta = \pi/32$, the maximum magnitude is in bin 16, and again, $\hat{\theta}_0 = 0.5\pi$. *Bottom:* With $N' = 4N = 128$, $\Delta\theta = \pi/64$, and now the maximum magnitude is found in bin 33, which happens to be very close to the peak of the underlying Dirichlet kernel. The frequency estimate improves to $\hat{\theta}_0 = 33 \times \pi/64 = 0.5156\pi$.

Equation 3.38 as a "true" DFT, we can simply lengthen the sum with dummy input values:

$$F[m; N'] = \sum_{n=0}^{N-1} f[n] \exp(-i2\pi mn/N') + \sum_{n=N}^{N'-1} 0 \cdot \exp(-i2\pi mn/N').$$

That is, $F[m; N']$ is the DFT of a new vector formed by appending $N' - N$ zeros to the original vector f. This operation is called *zero padding*.

3.4 DFT PROPERTIES AND THEOREMS

Zero padding provides more bins, and thus a better chance of finding a bin close to the true frequency of a sinusoidal signal. But because we have not actually increased the amount of data, the $N' - N$ additional DFT values must be, in some way, dependent on the original N DFT values. Return to Equation 3.38 and express f as the inverse DFT of F:

$$F[m; N'] = \sum_{n=0}^{N-1} \left[\frac{1}{N} \sum_{k=0}^{N-1} F[k] \exp(+i2\pi kn/N) \right] \exp(-i2\pi mn/N').$$

Exchange the order of summation,

$$F[m; N'] = \sum_{k=0}^{N-1} F[k] \left[\frac{1}{N} \sum_{n=0}^{N-1} \exp(+i2\pi kn/N) \exp(-i2\pi mn/N') \right]$$

$$= \sum_{k=0}^{N-1} F[k] \left[\frac{1}{N} \sum_{n=0}^{N-1} \exp[i2\pi(k - mN/N')n/N] \right]$$

and defining $r = N'/N$,

$$= \sum_{k=0}^{N-1} F[k] \underbrace{\frac{1 - \exp[i2\pi(k - m/r)]}{N(1 - \exp[i2\pi(k - m/r)/N])}}_{K_N(k-m/r)}, \quad m = 0, 1, 2, \ldots, N' - 1.$$

In general, the values of $F[m; N']$ are linear combinations of the original DFT values, with coefficients supplied by a kernel function, $K_N(k - m/r)$, closely related to the Dirichlet kernel:

$$K_N(x) = \frac{1 - \exp[i2\pi x]}{N(1 - \exp[i2\pi x/N])} = \frac{1}{N} \exp\left(\frac{i\pi(N-1)x}{N}\right) D_N\left(\frac{x}{N}\right).$$

The effect of zero padding is to *interpolate* between the original values of the DFT. In the special case where the padding factor r is an integer, at those values m that are integer multiples of r (i.e., $m = rp, p = 0, 1, \ldots, N - 1$), the interpolating kernel simplifies to

$$K_N(k - p) = \frac{1 - \exp[i2\pi(k - p)]}{N(1 - \exp[i2\pi(k - p)/N])} = \delta[k - p],$$

and we have $F[rp; rN] = F[p]$. Figure 3.17 illustrates this, that every second value of $F[m; 2N]$ and every fourth value of $F[m; 4N]$ is one of the original values of $F[m]$.

We will see in Section 3.5 that the highest computational efficiency for the DFT is obtained when the number of points is a power of 2. In certain practical applications, if one has a vector of length N not a power of 2, zero padding can be used to extend the vector to a length N', which *is* a power of 2. Frequently, the computation of the DFT will be faster for the padded vector, even though more DFT values are being computed.

Just as zero padding in the time domain results in interpolation in the frequency domain, we expect zero padding in the frequency domain to result in interpolation in the time domain. But we cannot just append zeros to $F[m]$. When we zero padded f, we added null time samples and preserved the relationships among the original values of f. When we pad F we will be adding null frequency samples, but must add them in such a way that relationships among the original values of F are also preserved. The DFT values $F[m]$ and $F[N - m]$ are paired, associated with basis vectors ϕ_m and ϕ_{N-m} that have the same digital frequency. After zero padding, they must stay paired so that they continue to be associated with basis vectors of the same frequency. Thus, zero padding in the frequency domain looks like this: when N is even,

$$\left(F[0], F[1], \ldots, F\left[\frac{N}{2} - 1\right], \frac{1}{2}F\left[\frac{N}{2}\right], \underbrace{0, \ldots 0}_{N'-N-1}, \frac{1}{2}F\left[\frac{N}{2}\right], F\left[\frac{N}{2} + 1\right], \ldots, F[N - 1]\right),$$

and when N is odd,

$$\left(F[0], F[1], \ldots, F\left[\frac{N-1}{2}\right], \underbrace{0, \ldots, 0}_{N'-N}, F\left[\frac{N+1}{2}\right], \ldots, F[N - 1]\right).$$

(The reason for splitting $F[N/2]$ between two bins is explored in the problems.)

Frequency domain zero padding is illustrated in Figure 3.18 for a single cosine. As expected, new values are interpolated between the points of the original sequence. Near the ends, the interpolation does not appear to smoothly follow the expected

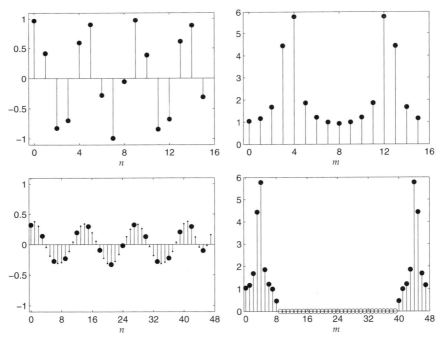

FIGURE 3.18 Effect of zero padding in the frequency domain. *Clockwise, from top left:* The original sequence $f[n] = \cos(0.44545\pi n + 1.9186\pi)$. The DFT magnitude, $|F[m]|$. The magnitude of the DFT, after inserting zeros in bins 9 through 39. The inverse DFT of the padded DFT interpolates two samples between each of the original samples (heavy dots) and scales by $1/3$.

sinusoidal shape. We also observe that the magnitude of the interpolated sequence appears to be $1/3$ the size of the original sequence for $N' = 3N$. Both of these phenomena are explained by calculating the inverse DFT of the padded DFT.

Writing out the terms of the inverse DFT explicitly,

$$f[n; N'] = \frac{1}{N'} \left[F[0] e^{i2\pi 0 n/N'} + F[1] e^{i2\pi 1 n/N'} + \cdots + F\left[\frac{N}{2}-1\right] e^{i2\pi(N/2-1)n/N'} \right.$$
$$+ \frac{1}{2} F\left[\frac{N}{2}\right] e^{i2\pi(N/2)n/N'} + 0 + \cdots + 0 + \frac{1}{2} F\left[\frac{N}{2}\right] e^{i2\pi(N'-N/2)n/N'}$$
$$\left. + F\left[N - \left(\frac{N}{2}-1\right)\right] e^{i2\pi(N'-(N/2-1))n/N'} + \cdots + F[N-1] e^{i2\pi(N'-1)n/N'} \right].$$

We will simplify this sum by exploiting the periodicity of the DFT. In the second and third lines, note that $e^{i2\pi(N'-m)n/N'} = e^{-i2\pi mn/N'}$. In the same terms, replace $F[N-k]$ by $F[-k]$. We then have

$$f[n; N'] = \frac{1}{N'} \left[F[0] e^{i2\pi 0 n/N'} + F[1] e^{i2\pi 1 n/N'} + \cdots + F\left[\frac{N}{2}-1\right] e^{i2\pi(N/2-1)n/N'} \right.$$
$$+ \frac{1}{2} F\left[\frac{N}{2}\right] e^{i2\pi(N/2)n/N'} + 0 + \cdots + 0 + \frac{1}{2} F\left[-\frac{N}{2}\right] e^{-i2\pi(N/2)n/N'}$$
$$\left. + F\left[-\left(\frac{N}{2}-1\right)\right] e^{-i2\pi(N/2-1)n/N'} + \cdots + F[-1] e^{-i2\pi n/N'} \right],$$

which is compactly written

$$f[n; N'] = \frac{1}{N'} \sum_{m=-N/2}^{N/2} F[m] R_N[m] \exp\left(\frac{i2\pi mn}{N'}\right), \quad n = 0, 1, 2, \ldots N'-1,$$

where

$$R_N[m] = \begin{cases} 1, & |m| < N/2 \\ \frac{1}{2}, & |m| = N/2 \\ 0, & N/2 < |m| \le N'/2 \end{cases}.$$

Now, express $F[m]$ as the DFT of f,

$$f[n; N'] = \frac{1}{N'} \sum_{m=-N/2}^{N/2} \left[\sum_{k=0}^{N-1} f[k] \exp\left(-\frac{i2\pi mk}{N}\right) \right] R_N[m] \exp\left(\frac{i2\pi mn}{N'}\right)$$
$$= \frac{1}{N'} \sum_{k=0}^{N-1} f[k] \sum_{m=-N/2}^{N/2} R_N[m] \exp\left(-\frac{i2\pi mk}{N}\right) \exp\left(\frac{i2\pi mn}{N'}\right)$$
$$= \frac{1}{r} \sum_{k=0}^{N-1} f[k] \underbrace{\frac{1}{N} \sum_{m=-N/2}^{N/2} R_N[m] \exp\left(-\frac{i2\pi m(k-n/r)}{N}\right)}_{K_N(k-n/r)}.$$

148 CHAPTER 3 THE DISCRETE FOURIER TRANSFORM

Using the usual geometric series manipulations and some trigonometric identities, the inner sum works out to

$$K_N(k-n/r) = \frac{1}{N} \sum_{m=-N/2}^{N/2} R_N[m] \exp\left(-\frac{i2\pi m(k-n/r)}{N}\right)$$

$$= \frac{1}{N} \cos\left(\frac{\pi(k-n/r)}{N}\right) D_N\left(\frac{k-n/r}{N}\right),$$

where D_N is the Dirichlet kernel. When N is odd, the cosine factor is absent (see the problems).

This analysis shows that the values of $f[n; N']$ are interpolated from the original values of f, and the result is scaled down by r, the padding factor:

$$f[n; rN] = \frac{1}{r} \sum_{k=0}^{N-1} f[k] K_N(k-n/r).$$

Because of the periodicity of both f and the interpolation kernel, the values near one end of the vector contribute to the interpolated values near the other end of the vector. Large differences between $f[0]$ and $f[N-1]$ show up in the interpolated values near the ends, as observed in Figure 3.18.

These results are collected in the following theorem.

Theorem 3.9 (Zero padding). Let $f, F \in \mathbb{C}^N$ be a DFT pair, and let $r = N'/N$, where $N' \geq N$.

1. In the time domain, define $f_r \in \mathbb{C}^{N'}$ by zero padding,

$$f_r = \Big(f[0] \cdots f[N-1], \underbrace{0 \cdots 0}_{N'-N}\Big).$$

The DFT of f_r, written $F[m; N']$, consists of values interpolated from F,

$$F[m; N'] = \sum_{k=0}^{N-1} F[k] K_N(k-m/r), \quad m = 0, 1, 2, \ldots N'-1, \quad (3.40)$$

where

$$K_N(x) = \frac{1}{N} \frac{1-e^{i2\pi x}}{1-e^{i2\pi x/N}} = \frac{1}{N} \exp\left[i2\pi \left(\frac{N-1}{N}\right)x\right] D_N\left(\frac{x}{N}\right)$$

and D_N is the Dirichlet kernel.

2. In the frequency domain, define $F_r \in \mathbb{C}^{N'}$ by zero padding,

$$F_r = \begin{cases} \Big(F[0] \cdots F\left[\frac{N}{2}-1\right], \frac{1}{2}F\left[\frac{N}{2}\right], \underbrace{0\cdots 0}_{N'-N-1}, \frac{1}{2}F\left[\frac{N}{2}\right], F\left[\frac{N}{2}+1\right] \cdots F[N-1]\Big) & N \text{ even} \\ \Big(F[0] \cdots F\left[\frac{N-1}{2}\right], \underbrace{0\cdots 0}_{N'-N}, F\left[\frac{N+1}{2}\right] \cdots F[N-1]\Big) & N \text{ odd} \end{cases}$$

The inverse DFT of F_r, written $f[n; N']$, consists of values interpolated from f,

$$f[n; N'] = \frac{1}{r} \sum_{k=0}^{N-1} f[k] \, K_N(k - n/r), \quad n = 0, 1, 2, \ldots N' - 1, \quad (3.41)$$

where

$$K_N(x) = \begin{cases} \dfrac{1}{N} \cos\left(\dfrac{\pi x}{N}\right) D_N\left(\dfrac{x}{N}\right), & N \text{ even} \\ \dfrac{1}{N} D_N\left(\dfrac{x}{N}\right), & N \text{ odd} \end{cases}.$$

★ Redundant basis vectors

We digress briefly to take another look at Equation 3.38, which will lead to a different interpretation of the interpolated frequency bins. Recall

$$F[m; N'] = \sum_{n=0}^{N-1} f[n] e^{-i 2\pi m n / N'}, \quad m = 0, 1, 2, \ldots, N' - 1.$$

The original signal f is N-dimensional, but the DFT F is N'-dimensional. If we calculate the inverse DFT in the usual way, we have

$$\frac{1}{N'} \sum_{m=0}^{N'-1} F[m; N'] e^{+i 2\pi m n / N'} = \frac{1}{N'} \sum_{m=0}^{N'-1} \left[\sum_{k=0}^{N-1} f[k] e^{-i 2\pi m k / N'} \right] e^{+i 2\pi m n / N'}$$

$$= \sum_{k=0}^{N-1} f[k] \left[\frac{1}{N'} \sum_{m=0}^{N'-1} e^{+i 2\pi m (n - k) / N'} \right].$$

The bracketed quantity is unity when $k = n, n \pm N', \ldots$, and zero otherwise. For $n = 0$, it sifts out the $k = 0$ term of the sum, which is $f[0]$. For $n - 1$, it sifts out $f[1]$, and so forth, to $n = N - 1$. For $n = N$ and higher, up to $n = N' - 1$, there is no corresponding term in the sum and the result is zero. Thus, we have

$$\frac{1}{N'} \sum_{m=0}^{N'-1} F[m; N'] e^{+i 2\pi m n / N'} = f[n], \quad n = 0, 1, \ldots N - 1.$$

This expression is an expansion of the N-vector f in terms of N' basis vectors. But $\mathbb{C}^N$ is spanned by only N orthogonal basis vectors. The set of DFT vectors $\{e^{i 2\pi m n / N'}\}_{n=0}^{N-1}$, $m = 0 \ldots N' - 1$, comprise a nonorthogonal set with $N' - N$ redundant members. The expansion on this redundant set is the interpolated DFT we obtained by zero padding.

Zero packing

The previous theorem gives the DFT of a sequence that has been extended by appending zeros. We may also consider what happens when a sequence is extended by replication; instead of appending zeros, $p-1$ copies of the sequence itself are appended. We define the sequence g_p by

$$g_p[n] = f[n \bmod N], n = 0, 1, 2, \ldots pN - 1.$$

The DFT of g_p is $\sum_{n=0}^{pN-1} f[n \bmod N] \exp\left(-\frac{i2\pi mn}{pN}\right)$, $m = 0, 1, \ldots, pN - 1$. Now, write $n = v + kN$, where $k = \lfloor n/N \rfloor$ and $v = n \bmod N$, and make a double sum over $v = 0, 1, \ldots N - 1$ and $k = 0, 1, \ldots p - 1$:

$$G_p[m] = \sum_{k=0}^{p-1} \sum_{v=0}^{N-1} f[v] \exp\left(-\frac{i2\pi m(v + kN)}{pN}\right)$$

$$= \sum_{k=0}^{p-1} \exp\left(-\frac{i2\pi mk}{p}\right) \sum_{v=0}^{N-1} f[v] \exp\left(-\frac{i2\pi (m/p)v}{N}\right) \quad m = 0, 1, 2, \ldots pN - 1.$$

The first sum is

$$\sum_{k=0}^{p-1} \exp\left(-\frac{i2\pi mk}{p}\right) = \frac{1 - e^{i2\pi m}}{1 - e^{i2\pi m/p}} = \begin{cases} p, & m = 0, p, 2p, \ldots \\ 0, & \text{otherwise} \end{cases}$$

and the second sum is a DFT, $F[m/p]$. Thus, we have

$$G_p[m] = \begin{cases} pF\left[\dfrac{m}{p}\right], & m = 0, p, 2p, \ldots (N-1)p \\ 0, & \text{otherwise} \end{cases}.$$

The DFT of a p-fold periodic replication of a sequence is the DFT of the original sequence, with $p - 1$ zeros inserted after each of the values of the original DFT. Also, the p-fold replication causes the DFT values to be scaled up by a factor of p. This periodic insertion of zeros is called *zero packing*. For convenience, we will define a sequence $F[m/p]$ to be zero when m is not an integer multiple of p, that is, it is obtained by taking a sequence $F[m]$ and packing $p - 1$ zeros after each sample. Then the above relationship may be compactly written

$$\left(f[n \bmod N]\right)_{n=0}^{pN-1} \longleftrightarrow \left(pF[m/p]\right)_{m=0}^{pN-1}$$

or just $f[n \bmod N] \longleftrightarrow pF[m/p]$, where we understand that the indices n and m run from 0 to $pN - 1$.

---

[5] See Kovačević and Chebira (2007a, 2007b) for an introduction to frames. Also see Mallat (1999, pp. 125–138).

The converse of this result says that if a sequence is packed with zeros, then its DFT is the periodic replication of the original DFT. Let $f[n/p]$ denote the sequence $f[n]$ with $p-1$ zeros packed after each sample. Then the DFT of $f[n/p]$ is

$$\sum_{n=0}^{pN-1} f\left[\frac{n}{p}\right] \exp\left(-\frac{i2\pi mn}{pN}\right) = \sum_{n=0,p,2p\ldots}^{(N-1)p} f\left[\frac{n}{p}\right] \exp\left(-\frac{i2\pi m(n/p)}{N}\right)$$

$$= \sum_{k=0}^{N-1} f[k] \exp\left(-\frac{i2\pi mk}{N}\right), \quad m = 0, 1, 2, \ldots pN-1.$$

This is the DFT of f, which is periodic with period N. As the index m runs from 0 to $pN-1$, we get p periods of F. We may write:

$$\left(f[n/p]\right)_{n=0}^{pN-1} \longleftrightarrow (F[m \bmod N])_{m=0}^{pN-1}$$

or just $f[n/p] \longleftrightarrow F[m \bmod N]$, where the ranges of the indices n and m are implicit. Zero packing is illustrated in Figure 3.19.

These results are summarized in the following theorem.

Theorem 3.10 (Zero packing). Let $f, F \in \mathbb{C}^N$ be a DFT pair. Define the zero-packed sequence $(f[n/p])_{n=0}^{pN-1}$ by inserting $p-1$ zeros after each sample of f,

$$(f[n/p]) = (f[0], \underbrace{0 \ldots 0}_{p-1 \text{ zeros}}, f[1], 0\ldots 0, \ldots f[N-1], 0 \ldots 0).$$

Then

$$f[n/p] \longleftrightarrow F[m \bmod N] \qquad (3.42a)$$

$$f[n \bmod N] \longleftrightarrow pF[m/p], \qquad (3.42b)$$

where $m, n = 0, 1, 2, \ldots pN-1$.

With both padding and packing, appending values to a vector in one domain causes values to be inserted between a vector's points in the other domain. One either appends zeros (zero padding) or replicas of the vector (zero packing). The inserted values are either zeros (zero packing) or interpolated from the values of the vector (zero padding). The apparent asymmetry, with a factor of p before the DFT in Equation 3.42b but not in Equation 3.42a, is considered in the problems. Packing and padding are frequently used together in signal processing.

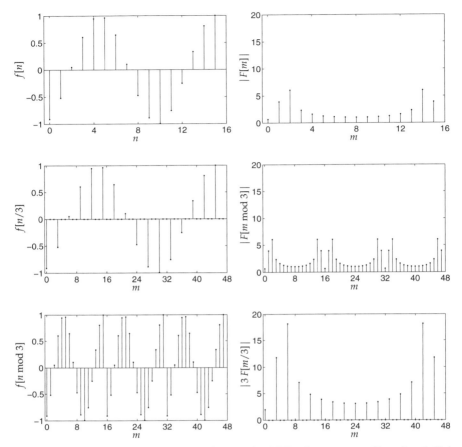

FIGURE 3.19 The effect of zero packing on the DFT of a sequence (Equation 3.42a). *Top:* A 16-point cosine sequence $f[n] = \cos(0.19022\pi n + 1.1356\pi)$ and the magnitude of its DFT, $|F[m]|$. *Middle:* The 48-point sequence $f[n/3]$ is created by inserting two zeros after every sample of the original sequence. Its DFT is $F[m \mod 3]$, a threefold replication of the original DFT. *Bottom:* The 48-point sequence $f[n \mod 3]$ is created by replicating the original sequence three times. Its DFT is $3F[m/3]$, with two zero bins packed between each of the original DFT values.

3.5 FAST FOURIER TRANSFORM

The DFT is an important practical tool because there are efficient algorithms for computing it. This section introduces the most popular method, the Cooley–Tukey fast Fourier transform.[6]

---

[6] See Cooley and Tukey (1965). The basic idea was known to Gauss, and was rediscovered several times before the widespread availability of digital computers provided the right soil for it to take root. See the historical notes in Cooley (1987).

Direct computation of the DFT using the summation

$$F[m] = \sum_{n=0}^{N-1} f[n]e^{-i2\pi mn/N}, \quad m = 0, 1, \ldots N-1$$

requires, in general, N multiplications and $N-1$ additions for each of the N elements of F. If f is complex-valued, then each multiplication itself requires four real multiplies and two real adds: $(a+ib)(c+id) = (ac-bd) + i(bc+ad)$. Each of the $N-1$ complex adds requires two real addition operations: $(a+ib) + (c+id) = (a+c) + i(b+d)$. The total operation count is $4N^2$ (real) multiplies and $4N^2 - 2N$ (real) adds. In some digital processing hardware, a multiply-add is one operation, and it is unnecessary to keep separate track of multiplies and adds. What we have, then, is $4N^2$ real operations. This number can be lowered somewhat. If f is real, then only $2N^2$ operations are needed, but in this case, F is also Hermitian, so only the values for $m = 0$ through $N/2$ need to be calculated, reducing the total by roughly half again. If f is even or odd, then further symmetries can be exploited. But the fact remains that direct calculation of the DFT is proportional to N^2. We say that it is $O(N^2)$, or "quadratic time."

On the other hand, the FFT algorithm is $O(N \log_2 N)$ when N is a power of 2. Even for a small transform, for example, $N = 32$, the savings are substantial: 170 vs. 1024. For $N = 512$, a size commonly encountered in signal analysis, the difference is huge: approximately 4600 vs. 256000! The FFT accomplishes this feat by exploiting the symmetries inherent in the DFT basis vectors.

The Cooley–Tukey algorithm

We will illustrate the approach by deriving an FFT for $N = 8$. An eight-point FFT should require $N \log_2 N = 8 \times 3 = 24$ operations. For a fixed value of m, the DFT sum is

$$F[m] = \sum_{n=0}^{7} f[n]W_8^{-mn} = f[0]W_8^0 + f[1]W_8^{-m} + f[2]W_8^{-2m} + f[3]W_8^{-3m}$$
$$+ f[4]W_8^{-4m} + f[5]W_8^{-5m} + f[6]W_8^{-6m} + f[7]W_8^{-7m},$$

where $W_N^n = e^{i2\pi n/N}$. In the folklore of the FFT, the W_N^n are called *twiddle factors*.

The first step is to recognize that $W_8^{-m} = W_8^{-m}W_8^0$, $W_8^{-3m} = W_8^{-m}W_8^{-2m}$, etc. Separating the odd- and even-indexed terms in the series, we obtain

$$F[m] = \left(f[0]W_8^0 + f[2]W_8^{-2m} + f[4]W_8^{-4m} + f[6]W_8^{-6m} \right)$$
$$+ W_8^{-m}\left(f[1]W_8^0 + f[3]W_8^{-2m} + f[5]W_8^{-4m} + f[7]W_8^{-6m} \right).$$

Observing that $W_8^2 = W_4$,

$$F[m] = \underbrace{\left(f[0]W_4^0 + f[2]W_4^{-m} + f[4]W_4^{-2m} + f[6]W_4^{-3m}\right)}_{F_{0246}[m]}$$
$$+ W_8^{-m}\underbrace{\left(f[1]W_4^0 + f[3]W_4^{-m} + f[5]W_4^{-2m} + f[7]W_4^{-3m}\right)}_{F_{1357}[m]}$$
$$= F_{0246}[m] + W_8^{-m}F_{1357}[m],$$

where F_{0246} and F_{1357} are *four*-point DFTs, of the sequences $(f[0], f[2], f[4], f[6])$ and $(f[1], f[3], f[5], f[7])$, respectively. They are combined by the twiddle factor W_8^{-m}.

Let us take stock of the operation count for this division of the calculation. The two four-point DFTs each require four multiply-adds for each of the eight values of $F[m]$. Combining them with the twiddle factor contributes another multiply-add. The total is nine operations, which looks like a step backward when compared with eight operations for the direct calculation. But the four-point DFTs are periodic with period 4, that is,

$$F_{0246}[m+4] = f[0]W_4^0 + f[2]W_4^{-(m+4)} + f[4]W_4^{-2(m+4)} + f[6]W_4^{-3(m+4)}$$
$$= f[0]W_4^0 + f[2]W_4^{-m} + f[4]W_4^{-2m} + f[6]W_4^{-3m}$$
$$= F_{0246}[m]$$
$$F_{1357}[m+4] = F_{1357}[m],$$

because $W_4^{4k} = 1$. This simplifies the eight values of F to

$$F[0] = F_{0246}[0] + W_8^0 F_{1357}[0] = F_{0246}[0] + F_{1357}[0]$$
$$F[4] = F_{0246}[0] + W_8^{-4} F_{1357}[0] = F_{0246}[0] - F_{1357}[0]$$

$$F[1] = F_{0246}[1] + W_8^{-1} F_{1357}[1]$$
$$F[5] = F_{0246}[1] + W_8^{-5} F_{1357}[1] = F_{0246}[1] - W_8^{-1} F_{1357}[1]$$

$$F[2] = F_{0246}[2] + W_8^{-2} F_{1357}[2]$$
$$F[6] = F_{0246}[2] + W_8^{-6} F_{1357}[2] = F_{0246}[2] - W_8^{-2} F_{1357}[2]$$

$$F[3] = F_{0246}[3] + W_8^{-3} F_{1357}[3]$$
$$F[7] = F_{0246}[3] + W_8^{-7} F_{1357}[3] = F_{0246}[3] - W_8^{-3} F_{1357}[3],$$

where we have used $W_8^{-4} = e^{-i2\pi 4/8} = e^{-i\pi} = -1$. Now let us do the operation count.

$$\underbrace{16 \text{ operations}}_{\text{four-point DFT}} \times \underbrace{2}_{\text{two DFTs}} + \underbrace{4}_{\text{x twiddle factors}} = 36.$$

This is just over half of the 64 operations required by the direct calculation.

3.5 FAST FOURIER TRANSFORM

The process can be repeated, expressing each of the four-point DFTs in terms of two-point DFTs:

$$F_{0246}[m] = f[0]W_4^0 + f[2]W_4^{-m} + f[4]W_4^{-2m} + f[6]W_4^{-3m}$$
$$= \left(f[0] + f[4]W_2^{-m}\right) + W_4^{-m}\left(f[2] + f[6]W_2^{-m}\right)$$
$$= F_{04}[m] + W_4^{-m}F_{26}[m]$$

and, similarly,

$$F_{1357}[m] = F_{15}[m] + W_4^{-m}F_{37}[m].$$

F_{04} is the DFT of $(f[0], f[4])$, F_{26} is the DFT of $(f[2], f[6])$, etc. These two-point DFTs are periodic with period 2, so $F_{04}[m+2] = F_{04}[m]$, etc. Hence,

$$F_{0246}[0] = F_{04}[0] + F_{26}[0]$$
$$F_{0246}[2] = F_{04}[0] - F_{26}[0]$$
$$F_{0246}[1] = F_{04}[1] + W_4^{-1}F_{26}[1]$$
$$F_{0246}[3] = F_{04}[1] - W_4^{-1}F_{26}[1]$$

and similarly for F_{1357}. Now a two-point DFT is very simple, for example:

$$F_{26}[0] = f[2]W_2^0 + f[6]W_2^0 = f[2] + f[6]$$
$$F_{26}[1] = f[2]W_2^0 + f[6]W_2^{-1} = f[2] - f[6].$$

It requires only two operations (an add and a subtract). The operation count for a four-point DFT computed this way, then, is

$$\underbrace{2 \text{ operations}}_{\text{two-point DFT}} \times \underbrace{4}_{\text{four DFTs}} + \underbrace{2}_{\times \text{ twiddle factors}} = 10$$

instead of 16 by direct calculation. The final operation count for the eight-point DFT is

$$\underbrace{10 \text{ operations}}_{\text{four-point DFT}} \times \underbrace{2}_{\text{two DFTs}} + \underbrace{4}_{\times \text{ twiddle factors}} = 24,$$

as expected.

The order of computations in the FFT is often presented as a signal-flow diagram (Figure 3.20). The shape of the signal-flow for the basic two-point DFT suggests its common name—*butterfly*. We see from the diagram that the entire FFT is just a chain of butterfly operations. The diagram also reveals where the $N \log_2 N$ performance comes from. Moving from right to left, the eight-point DFT is expressed as two four-point DFTs, combined with four butterflies (eight operations). Each of the four-point DFTs is, in turn, broken down into four two-point DFTs, which are combined with eight operations. Finally, the four two-point DFTs require a total of eight operations. The number of levels is $\log_2 8 = 3$, and each level requires eight operations, two per butterfly. For a general N, which is a power of two, there will be $\log_2 N$ levels, and each level will require N twiddle-factor operations to do its butterflies: $N \log_2 N$.

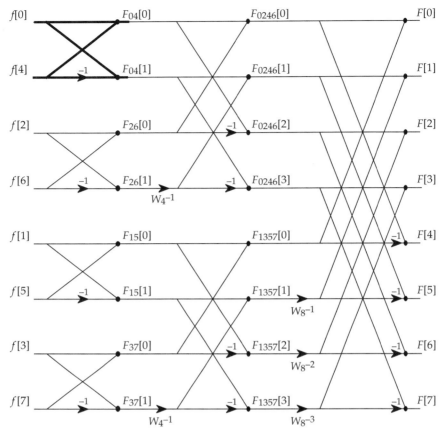

FIGURE 3.20 Signal-flow diagram for an eight-point fast Fourier transform. The basic butterfly calculation is shown in heavier lines at the upper left. Arrows correspond to multiplication operations, and dots are addition operations, for example, $F_{04}[0] = f[0] + f[4]$ and $F_{04}[1] = f[0] - f[4]$. There are $\log_2 8 = 3$ levels, with eight multiply-add operations per level. The first level (*left*) computes four two-point DFTs. The second level (*center*) combines these into two four-point DFTs. The third level (*right*) combines the four-point DFTs into the eight-point DFT.

Further aspects of using the FFT are explored in the problems. More comprehensive discussions of the FFT can be found in most signal processing texts.[7]

⋆ 3.6 DISCRETE COSINE TRANSFORM

To conclude the chapter, we introduce a transform closely related to the DFT, which is particularly useful in signal and image compression. The basic idea in image

---

[7]See, for example, Porat (1997, Chapter 4), Brigham (1988, Chapter 8), and Oppenheim and Schafer (2010, Chapter 9).

compression is to store an image using less memory or to transmit an image in less time. If the decompressed image is exactly the same as the original uncompressed image, we say that the compression is *lossless*. Frequently, some loss of accuracy can be tolerated in order to achieve higher compression. Then the compression is called *lossy*, and the objective is to achieve the highest compression possible consistent with the image fidelity required by the end user.

One of the most popular lossy compression methods at the present time is JPEG (for Joint Photographic Experts Group, the organization that published the standard). In JPEG compression, an image is divided into non-overlapping 8×8 pixel blocks. Each block is treated as a point in $\mathbb{R}^{64}$ and projected onto 64 orthogonal basis images (Example 2.12). The expansion coefficients are then processed to determine which components are perceptually important—roughly, which ones would be conspicuous by their absence in the final image; the rest are set to zero. A subsequent lossless coding step efficiently packs the coefficients for each block into a bit stream for storage or transmission. Decompression consists of unpacking the coefficients, resynthesizing the blocks, and fitting them back together into an image.

We will consider here only the basic problem of computing a Fourier expansion for a real-valued eight-point vector: one row of an 8×8 block. Statistical analysis of images has shown that a properly chosen Fourier basis is nearly optimal in the sense of packing the most signal energy into the fewest coefficients.[8]

In preparation for the mathematical derivation we present an example.

Example 3.18. Consider a ramp sequence, $f = (0, 1, 2, 3, 4, 5, 6, 7)$, which is representative of how an image might change over a relatively smooth eight-pixel run. The ramp and its DFT are shown in Figure 3.21. All the DFT coefficients are strong. The ramp sequence is not symmetric and does not bear any resemblance to one period of a sine or cosine. Consequently, it does not project well onto any particular DFT basis vector. Attempting to set the weakest DFT values to zero results in considerable error in the reconstruction.

Appending the mirror image of f as shown in Figure 3.22 converts the asymmetric ramp into a symmetric triangular sequence.

$$f_s[n] = \begin{cases} f[n], & n = 0, 1, \ldots 7 \\ f[15-n], & n = 8, 9, \ldots 15 \end{cases}. \quad (3.43)$$

This new sequence f_s is not strictly even, since $f_s[n] \neq f_s[16-n]$. However, it is almost an even periodic sequence, and so we would expect it to project more strongly onto low frequency basis vectors than did the original ramp. This is what we observe in Figure 3.22. In fact, most of the values of the DFT F_s can be set to zero with negligible effect on the reconstruction. ■

In this example, both f and f_s are real, so their DFTs are Hermitian. This means that five (complex) values of F (eight real numbers, because $F[0]$ and $F[4]$ are real) must be stored to accurately reconstruct f. On the other hand, only three values of F_s (five real numbers) are needed get a virtually error-free reconstructon of f_s, the first

---
[8] Ahmed *et al.* (1974); Unser (1984).

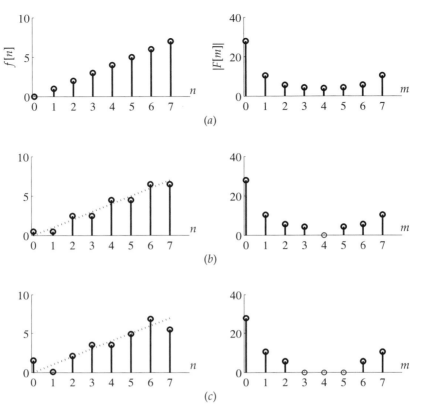

FIGURE 3.21 The DFT is not a good choice for compression. (a) A ramp sequence (*left*) and the magnitude of its DFT (*right*). There are significant contributions in all bins. (b) and (c) Attempting compression by setting the smallest DFT values to zero (light circles). (b) Setting $F[4] = 0$. (c) Setting $F[3], F[4], F[5]$ to zero. Significant error results.

half of which is the reconstructed f. If f_s actually had even symmetry, then F_s would be real-valued and three real numbers would suffice for reconstructing f—a savings of about 60% over the uncompressed version!

What makes even and odd sequences special for the DFT is the fact that the basis vectors ϕ_m are Hermitian. The real and imaginary parts of ϕ_m are even and odd, respectively. The inner product of an even sequence with an odd sequence can be shown to be zero, so when we calculate the DFT of a real, even sequence, the odd part of the DFT, which is also the imaginary part, is zero. Likewise, when we calculate the DFT of a real, odd sequence, the even part of the DFT, the real part, is zero. The trouble with using the DFT with f_s as we have constructed it is that its symmetries do not match the symmetries of the DFT basis functions. So let us see if we can construct an alternative basis that will match up correctly with f_s.

Figure 3.23 shows the symmetrized ramp sequence again, periodically extended over all n. It has a period of 16, and it is symmetric about two points, $n = -0.5$ and

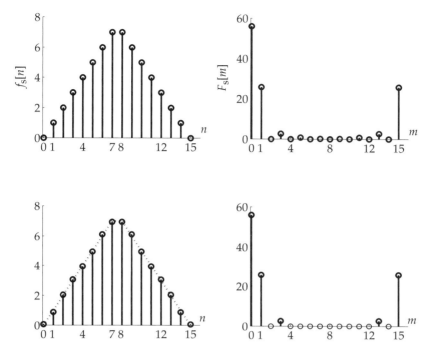

FIGURE 3.22 Making a sequence symmetric by mirror reflection. *Top:* The 16-point DFT of the symmetric sequence f_s is more concentrated in lower bins than the DFT of the original ramp. *Bottom:* All but five values of F_s are set to zero, with negligible effect on the reconstruction.

$n = 7.5$. (There are other ways to symmetrize f so that f_s is even (see the problems for examples) but the resulting sequences have lengths 14 or 15, which are not optimal for using the FFT.) In contrast, the DFT basis vector $\phi_m = \left(e^{i2\pi mn/16}\right)_{n=0}^{15}$ is symmetric about $n = 0$ and $n = 8$. The figure shows that if we shift the complex exponential left by a half sample, a new vector $\left(e^{i2\pi m(n+\frac{1}{2})/16}\right)_{n=0}^{15}$ results that has the same symmetry as f_s.

The inner products of the symmetrized ramp f_s with the DFT basis vector ϕ_1 and the shifted vector $\left(e^{i2\pi(n+\frac{1}{2})/16}\right)_{n=0}^{15}$ are, respectively,

$$\langle f_s, \phi_1 \rangle = -50.5483 + i10.0547$$

$$\left\langle f_s, \left(\exp \frac{i2\pi(n+\frac{1}{2})}{16}\right)\right\rangle = \left\langle f_s, \left(\cos \frac{i2\pi(n+\frac{1}{2})}{16}\right)\right\rangle + i\left\langle f_s, \left(\sin \frac{i2\pi(n+\frac{1}{2})}{16}\right)\right\rangle$$

$$= -51.5386 - i0.0000.$$

The imaginary part is zero in the second case because the symmetry of $\left(e^{i2\pi(n+\frac{1}{2})/16}\right)_{n=0}^{15}$ matches that of the signal f_s. We have, in effect, designed the

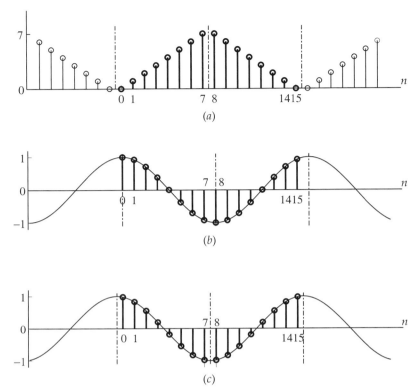

FIGURE 3.23 Designing a basis. (a) The vector f_s has even symmetry with respect to the points $n = -0.5, 7.5, 15.5$. (b) The DFT basis vectors do not share this symmetry. The real part of ϕ_1, $(\cos 2\pi n/16)$, is shown for example. (c) The vector $\left(\cos \frac{2\pi(n+\frac{1}{2})}{16}\right)$ has the same symmetry as f_s. This suggests that a basis constructed from the vectors $\left\{\left(\cos \frac{2\pi m(n+\frac{1}{2})}{16}\right)_{n=0}^{15}\right\}_{m=0}^{15}$ is better suited to representing f_s.

vectors $\left(\sin \frac{2\pi m(n+\frac{1}{2})}{16}\right)_{n=0}^{15}$ to be orthogonal to f_s. This allows us to replace the shifted complex exponential vectors with the shifted cosine vectors $\left(\cos \frac{2\pi m(n+\frac{1}{2})}{16}\right)$.

Another observation from the figure is that, in the inner product

$$\left\langle f_s, \left(\cos \frac{2\pi(n+\frac{1}{2})}{16}\right)\right\rangle = \sum_{n=0}^{15} f_s[n] \cos \frac{2\pi(n+\frac{1}{2})}{16},$$

the sum of terms 0 through 7 is identical to the sum of terms 8 through 15, because of the symmetry about $n = 7.5$. So,

$$\left\langle f_s, \left(\cos \frac{2\pi(n+\frac{1}{2})}{16}\right)\right\rangle = 2\sum_{n=0}^{7} f[n] \cos \frac{2\pi(n+\frac{1}{2})}{16}.$$

3.6 DISCRETE COSINE TRANSFORM   161

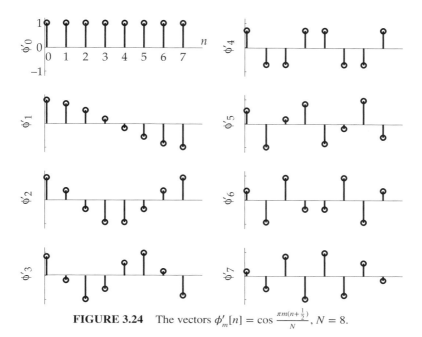

FIGURE 3.24 The vectors $\phi'_m[n] = \cos\frac{\pi m(n+\frac{1}{2})}{N}$, $N = 8$.

These symmetries should hold for the more general vectors $\phi'_m = \left(\cos\frac{2\pi m(n+\frac{1}{2})}{16}\right)_{n=0}^{7}$, which we suspect will span $\mathbb{R}^8$. But, also, $\phi'_{16-m} = \phi'_m$ and $\phi'_8 = 0$, so only the inner products for $m = 0, 1, \ldots 7$ need to be calculated. If a formal derivation confirms our intuition, we will have a real transform for f (not f_s) based on the set of vectors $\left\{\left(\cos\frac{\pi m(n+\frac{1}{2})}{8}\right)_{n=0}^{7}\right\}_{m=0}^{7}$.

Derivation of the discrete cosine transform
It is not any harder to develop the transform for a sequence of arbitrary length N. Define the vectors $\{\phi'_m\}_{m=0}^{N-1}$, where

$$\phi'_m[n] = \cos\frac{\pi m(n+\frac{1}{2})}{N}, \quad n = 0, 1, \ldots N-1$$

(Figure 3.24).

These vectors are orthogonal,

$$\langle \phi'_m, \phi'_k \rangle = \frac{N}{2\sigma_m}\delta[m-k], \qquad (3.44)$$

where

$$\sigma_m = \begin{cases} \frac{1}{2}, & m = 0 \\ 1, & \text{otherwise} \end{cases}$$

so they span $\mathbb{R}^N$. Normalizing, we have an orthonormal basis $\{e_m\}_{m=0}^{N-1}$, where

$$e_m[n] = \sqrt{\sigma_m}\sqrt{\frac{2}{N}}\cos\frac{\pi m(n+\frac{1}{2})}{N}. \qquad (3.45)$$

Using this basis, we can represent f by orthogonal projection,

$$f[n] = \sum_{m=0}^{N-1} F_c[m]e_m[n]$$

$$F_c[m] = \langle f, e_m \rangle = \sum_{n=0}^{N-1} f[n]e_m[n].$$

This proves the following theorem.

Theorem 3.11 (Discrete cosine transform). Let $f \in \mathbb{R}^N$. Then

$$f[n] = \sqrt{\frac{2}{N}} \sum_{m=0}^{N-1} F_c[m] \sqrt{\sigma_m} \cos\frac{\pi m(n+\frac{1}{2})}{N}, \quad n = 0, 1, \ldots N-1 \quad (3.46\text{a})$$

$$F_c[m] = \sqrt{\frac{2}{N}} \sum_{n=0}^{N-1} f[n] \sqrt{\sigma_m} \cos\frac{\pi m(n+\frac{1}{2})}{N}, \quad m = 0, 1, \ldots N-1 \quad (3.46\text{b})$$

$$\sigma_m = \begin{cases} \frac{1}{2}, & m = 0 \\ 1, & \text{otherwise} \end{cases}. \qquad (3.46\text{c})$$

The vector $F_c \in \mathbb{R}^N$ is called the DCT of f.

In the particular case $N = 8$,

$$f[n] = \frac{1}{2}\sum_{m=0}^{7} F_c[m]\sqrt{\sigma_m}\cos\frac{\pi m(n+\frac{1}{2})}{8}, \quad n = 0, 1, \ldots 7$$

$$F_c[m] = \frac{1}{2}\sum_{n=0}^{7} f[n]\sqrt{\sigma_m}\cos\frac{\pi m(n+\frac{1}{2})}{8}, \quad m = 0, 1, \ldots 7.$$

The DCT of the ramp sequence is shown in Figure 3.25. The DCT concentrates most of the ramp's energy in the first two bins, unlike the DFT (compare Figure 3.21). The sequence that is reconstructed using only the low frequency coefficients is very close to the original.

In JPEG image compression, the original image is divided into 8×8 pixel blocks. Each block, represented as an array $(x[i,j])_{i,j=0}^{7}$, is projected onto and

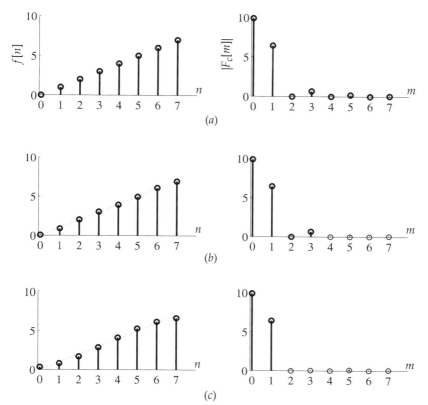

FIGURE 3.25 (a) A ramp sequence (*left*) and the magnitude of its discrete cosine transform (*right*). Unlike the DFT (compare Figure 3.21), most of the energy is concentrated in the low frequency bins. (b) and (c) Attempting compression by setting the smallest DCT values to zero (light circles). (b) Setting $F_c[4]$ through $F_c[7] = 0$. (c) Setting $F_c[2]$ through $F_c[7] = 0$. Even with only two coefficients, the reconstruction is very good.

reconstructed from basis images $\{e_{kl}\}_{k,l=0}^{7}$ constructed from the DCT basis vectors (Equation 3.45).

$$x[i,j] = \sum_{k=0}^{7} \sum_{l=0}^{7} X_c[k,l] e_{kl}[i,j] \tag{3.47a}$$

$$X_c[k,l] = \sum_{i=0}^{7} \sum_{j=0}^{7} x[i,j] e_{kl}[i,j] \tag{3.47b}$$

$$e_{kl}[i,j] = \frac{1}{4}\sqrt{\sigma_k \sigma_l} \cos\frac{\pi k(i+\frac{1}{2})}{8} \cos\frac{\pi l(j+\frac{1}{2})}{8}. \tag{3.47c}$$

These basis images (Equation 3.47c) are the ones displayed in Figure 2.7 and used in Example 2.12 of Chapter 2.

3.7 SUMMARY

Discrete Fourier transform

$$f[n] = \frac{1}{N} \sum_{m=0}^{N-1} F[m] \exp\left(\frac{i2\pi mn}{N}\right), \quad n = 0, 1, \ldots, N-1 \qquad (3.11a)$$

$$F[m] = \sum_{n=0}^{N-1} f[n] \exp\left(-\frac{i2\pi mn}{N}\right), \quad m = 0, 1, \ldots, N-1. \qquad (3.11b)$$

Interpretation

A function $f(x)$ is sampled at intervals Δx, creating the sequence $(f[n])_{n=0}^{N-1}$, $f[n] = f(n\Delta x)$. In the DFT sequence $(F[m])_{m=0}^{N-1}$, bins $m = 0, 1, \ldots, \frac{N}{2} - 1$ correspond to frequencies $\nu = 0, \Delta\nu, \ldots, \left(\frac{N}{2} - 1\right)\Delta\nu$, and bins $m = \frac{N}{2}, \frac{N}{2} + 1, \ldots, N-1$ correspond to "negative" frequencies $\nu = -\frac{N}{2}\Delta\nu, \left(-\frac{N}{2} + 1\right)\Delta\nu, \ldots, -\Delta\nu$. The bin spacing is $\Delta\nu = \dfrac{1}{N\Delta t}$.

An "in-bin" signal projects onto only one DFT bin frequency. An "out-of-bin" signal projects onto all DFT bin frequencies (spectral leakage).

Fast Fourier transform

Direct computation of the DFT sum requires $O(N^2)$ operations. The FFT algorithm requires $O(N \log_2 N)$ operations for N a power of 2.

Discrete cosine transform

$$f[n] = \sqrt{\frac{2}{N}} \sum_{m=0}^{N-1} F_c[m] \sqrt{\sigma_m} \cos\frac{\pi m(n + \frac{1}{2})}{N}, \quad n = 0, 1, \ldots N-1 \qquad (3.46a)$$

$$F_c[m] = \sqrt{\frac{2}{N}} \sum_{n=0}^{N-1} f[n] \sqrt{\sigma_m} \cos\frac{\pi m(n + \frac{1}{2})}{N}, \quad m = 0, 1, \ldots N-1 \qquad (3.46b)$$

$$\sigma_m = \begin{cases} \frac{1}{2}, & m = 0 \\ 1, & \text{otherwise} \end{cases}.$$

Basic DFT pairs derived in this chapter

$$\delta[n] \longleftrightarrow (1 \ldots 1) \qquad (3.23)$$

$$(1 \ldots 1) \longleftrightarrow N\delta[m] \qquad (3.24)$$

$$(\underbrace{1 \ldots 1}_{P}\ 0 \ldots 0) \longleftrightarrow e^{i\pi(P-1)m/N} D_P(m/N) \qquad (3.22)$$

$$e^{i2\pi kn/N} \longleftrightarrow N\delta[m-k] \qquad (3.18)$$

$$\cos(2\pi kn/N) \longleftrightarrow \frac{N}{2}\delta[m-k] + \frac{N}{2}\delta[m-(N-k)] \qquad (3.19)$$

$$\sin(2\pi kn/N) \longleftrightarrow -\frac{iN}{2}\delta[m-k] + \frac{iN}{2}\delta[m-(N-k)]. \qquad (3.20)$$

DFT Properties and Theorems

Periodicity	f and F are each one period of a sequence with period N, always interpreted as $f[n \bmod N]$ and $F[m \bmod N]$.	
Symmetry	See Figure 3.10	
Linearity	$af + bg \longmapsto aF + bG$	(3.26)
Parseval	$\sum_{n=0}^{N-1} f[n]g^*[n] = \frac{1}{N} \sum_{m=0}^{N-1} F[m]G^*[m]$ (inner products preserved)	(3.27a)
	$\sum_{n=0}^{N-1} |f[n]|^2 = \frac{1}{N} \sum_{m=0}^{N-1} |F[m]|^2$ (norms preserved)	(3.27b)
Area	$F[0] = \sum_{n=0}^{N-1} f[n], \quad f[0] = \frac{1}{N} \sum_{m=0}^{N-1} F[m]$	(3.28)
Cyclic shift	$\left(f[n-r]\right)_{n=0}^{N-1} \longmapsto \left(e^{-i2\pi rm/N} F[m]\right)_{m=0}^{N-1}$	(3.30)
Convolution	$f \circledast h[n] = \sum_{k=0}^{N-1} f[k]h[n-k]$	
	$f \circledast h \longmapsto FH \quad fh \longmapsto \frac{1}{N} F \circledast H$	(3.31)
Zero padding	Zero padding in one domain interpolates in the other domain.	(3.40) and (3.41)
Zero packing	Zero packing in one domain periodically replicates in the other domain.	(3.42)

PROBLEMS

3.1. Show that the inner product of an even sequence and an odd sequence is always zero. Then, show that the inner product of a real, even sequence with a DFT basis vector is real, and the inner product of a real, odd sequence with a DFT basis vector is imaginary.

3.2. *Twiddle factor algebra*
Verify the following twiddle-factor identities. Recall $W_N = e^{i2\pi/N}$.
(a) $W_N^N = 1$, that is, W_N is the so-called N^{th} root of unity
(b) $\left(W_N^m\right)^* = W_N^{-m}$
(c) $W_{N/2} = W_N^2$
(d) $1 - W_N^n = -2i W_{2N}^n \sin\left(\frac{\pi n}{N}\right)$.

3.3. *Discrete sinusoid calculations*
Using identities for trigonometric functions and geometric series,
(a) show that
$$\sum_{n=0}^{N-1} \cos^2 \theta n = \frac{N}{2} + \frac{1}{2} \cos\left[(N-1)\theta\right] \frac{\sin N\theta}{\sin \theta}$$

and, in particular, verify

$$\sum_{n=0}^{N-1} \cos^2 \frac{2\pi kn}{N} = \begin{cases} N, & k=0, \frac{N}{2} \\ \frac{N}{2}, & \text{otherwise} \end{cases}.$$

(b) derive the corresponding results for the sequence $\sin^2 \theta n$.

3.4. *Simple DFT exercises*
This problem is designed to help you develop a more intuitive feel for how the DFT operates.

(a) Calculate 16 samples of one period of a cosine. Make sure the frequency is such that it is in the first DFT bin (lowest nonzero frequency). Is this sample sequence even or odd? Calculate the DFT analytically, by evaluating the DFT sum. Compare this with the DFT computed by MATLAB. Do the symmetries of the DFT agree with theory?

(b) Increase the frequency of the cosine in (a), always keeping it in a bin (for simplicity), watching both the sample set and the DFT, until you observe aliasing.

(c) Repeat part (a) with a sine function having the same period.

3.5. *Between-bin behavior*
Examine carefully what happens to the DFT of a sinusoidal signal as the digital frequency is changed from one bin to the next higher bin. Let $\theta_1 = \frac{2\pi(16)}{128}$ and $\theta_2 = \frac{2\pi(17)}{128}$. Using MATLAB, calculate 128 samples of $f[n] = \cos \theta n$. Vary θ in small increments from θ_1 to θ_2, and observe the effect on the DFT magnitude $|F|$.

3.6. Consider a sinusoidal signal with $v = 20$ Hz, sampled at a rate $v_s = 1/\Delta t = 100$ Hz. Take $N = 256$ samples of this signal, compute and plot the magnitude of the DFT.

(a) At what bins do the spectral peaks appear? Verify that this agrees with theory.

(b) Repeat the experiment with $v_s = 200$ Hz and $N = 256$.

(c) Repeat the experiment with $v_s = 100$ Hz and $N = 128$.

(d) Repeat the experiment with $v = 10$ Hz, $v_s = 100$ Hz and $N = 128$.

(e) Perform additional experiments, as needed, until you understand completely and can explain concisely the relationship among v, v_s, N, and the locations of the spectral peaks.

3.7. A function $f(t)$ is sampled at rate $v_s = 1000$ Hz, and $N = 512$ samples are taken. The sampled function $f[n]$ and the magnitude of its DFT, $F[m]$, are plotted in Figure 3.26.

(a) What is the spacing, Δv (Hz), of the DFT bins?

(b) It is evident that $f[n]$ contains a number of sinusoidal components, as well as some noise. Estimate the frequencies of the sinusoids in Hz.

(c) Suppose the same function is sampled at a 500 Hz rate, again taking 512 samples. Describe the effect on the spectrum.

3.8. A signal $x(t) = \sin 2\pi v_0 t$ with frequency $v_0 = 300$ Hz is sampled with $\Delta t = 0.001$ s. $N = 64$ samples are taken.

(a) What is the digital frequency θ of the sampled signal $x[n]$?

(b) The DFT $X[m]$ is computed. At approximately what bin(s) will there be peaks in the spectrum?

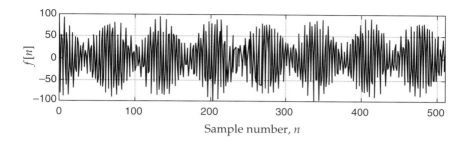

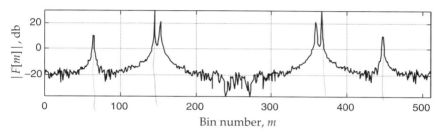

FIGURE 3.26 For Problem 3.7. A sampled waveform and its DFT spectrum, in decibels ($20 \log_{10} |F[m]|$).

(c) Given the data vector $x[n]$, $n = 0, 1, \ldots 63$, it is desired to estimate the frequency of $x(t)$ to the nearest 1 Hz. How should the data be processed to achieve this accuracy?

3.9. For the same function $x(t)$ as the previous problem,

(a) how large can Δt be without producing aliasing?

(b) what is $x[n]$ if you sample $x(t)$ with this Δt? Discuss the practical importance of your result.

3.10. A signal $f(t)$ was sampled at an 8 kHz rate for 512 samples, and the DFT $F[m]$ was computed. The power spectrum $|F[m]|^2$ is shown in Figure 3.27.

(a) From the spectrum, determine the frequency of $f(t)$.

(b) It is suspected that the sampling rate might not be high enough, and that the peaks shown are actually aliases of the correct spectrum. Describe an experiment you can perform to confirm or refute this hypothesis. You have access to the data acquisition system and are able to resample $f(t)$ at 4 kHz and at 16 kHz. You may assume that the true frequency of $f(t)$ is not "too high" for this test to succeed.

3.11. *Aliasing*

Devise a general method for determining the aliased frequency that results when a sinusoidal signal of frequency v is sampled with $v_s < 2v$.

3.12. *Linearity*

Show that the inverse DFT is linear. Begin by showing $f[n] = \frac{1}{N} \langle F, \phi_n^* \rangle$. $n = 0, 1, \ldots, N - 1$.

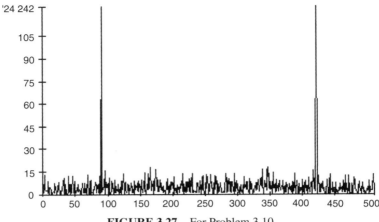

FIGURE 3.27 For Problem 3.10.

3.13. *Symmetry*

If an N-dimensional vector f is real-valued, its DFT is Hermitian, $F[N - m] = F^*[m]$; the real and imaginary parts of F are even and odd, respectively.

(a) Let F be the DFT of a real-valued vector, with N even. Show that $F[N/2]$ is real.

(b) Suppose an N-dimensional vector g is imaginary-valued. Discuss the symmetries in its DFT.

3.14. Create a vector x of $N = 16$ random numbers and compute the DFT, X (e.g., using the MATLAB commands x=rand(1,N) and X = fft(x)).

(a) Plot separately the real and imaginary parts of the DFT X and describe their symmetries.

(b) Compute the sums of the values of x and of X, and verify that the area theorem holds.

(c) Compute the 2-norms of x and X and verify that Parseval's formula holds.

3.15. Repeat the previous problem for a vector $x = (\cos \theta n)_{n=0}^{15}$, where $\theta \in [-\pi, \pi)$. Do it both for in-bin and between-bin values of θ.

3.16. *The shift theorem*

For the shifted sinusoidal sequence $f[n] = \cos\left[2\pi k(n - p)/N\right]$, where k and p are integers, calculate the DFT $F[m]$. Then, for $N = 16$ and $k = 2$, use MATLAB to compute and plot both f and F for $p = 0, 1, \ldots N - 1$. Plot the real part of F with solid stems and the imaginary part with dashed stems. Calculate the magnitude $|F[m]|$ and phase $\arg F[m]$ for each p, and compare with your analytic result.

3.17. *The convolution theorem*

Compute the convolution in Example 3.16 in MATLAB, using the DFT.

3.18. *The rectangle function*

Consider the sequence

$$f[n] = \begin{cases} 1, & n = 0, 1, \ldots P \\ 0, & n = P + 1, \ldots N - P - 1 \\ 1, & n = N - P, \ldots N - 1 \end{cases}.$$

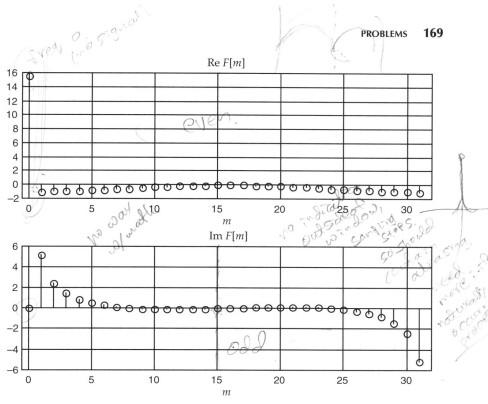

FIGURE 3.28 For Problem 3.19.

(a) Show that f has even symmetry.

(b) Calculate an expression for the DFT, $F[m]$, in terms of the Dirichlet kernel. *Hint:* Exploit the periodicity of the DFT basis vectors to write the DFT of the rectangle as a sum from $-P$ to P. Then make a change of variable so the sum runs from 0 to $2P$.

(c) With reference to the properties of the Dirichlet kernel displayed in Figure 3.5, make an accurate sketch of F, showing the effects of N and P. What do you observe about the symmetry properties of F and the effect of P on f and F?

(d) Compute the DFT using MATLAB, for $N = 32$ and $P = 1, 2, 4, 8$. Plot both f and F. What do you observe about the symmetry properties of F, and the effect of P on f and F?

(e) Alter f slightly by setting $f[P + 1] = 1$, and compute F for $N = 32$ and $P = 4$. Compare the DFT to the $P = 4$ result for part (d). Explain the difference, qualitatively and quantitatively.

3.19. A periodic function $f(t)$ was sampled with $\Delta t = 0.1$ s for $N = 32$ samples. The DFT was computed and is plotted in Figure 3.28.

Answer the following questions. Justify your answers.

(a) Is the original function f even, odd, or neither? Or is there insufficient information to make a decision?

(b) Is the original function f real, imaginary, or neither? Or is there insufficient information to make a decision?

(c) What is the fundamental frequency of f (in Hz)?

(d) How much power is at the fundamental frequency?

(e) Calculate the approximate average value of f, which is $\dfrac{1}{T}\displaystyle\sum_{n=0}^{N-1} f(n\Delta t)\,\Delta t$.

(f) Does f appear to have been sampled fast enough to avoid aliasing?

3.20. *Zero padding*

Let $f, F \in \mathbb{C}^N$ be a DFT pair. It was stated in the text that when N is even, the correct way to zero pad in the frequency domain is

$$\left(F[0] \cdots F\left[\tfrac{N}{2}-1\right], \tfrac{1}{2}F\left[\tfrac{N}{2}\right], 0 \cdots 0, \tfrac{1}{2}F\left[\tfrac{N}{2}\right], F\left[\tfrac{N}{2}+1\right] \cdots F[N-1]\right).$$

Why must we split $F[N/2]$ into two values in the padded DFT, as opposed to writing

$$\left(F[0] \cdots F\left[\tfrac{N}{2}-1\right], F\left[\tfrac{N}{2}\right], 0 \cdots 0, 0, F\left[\tfrac{N}{2}+1\right] \cdots F[N-1]\right)$$

or

$$\left(F[0] \cdots F\left[\tfrac{N}{2}-1\right], 0, 0 \cdots 0, F\left[\tfrac{N}{2}\right], F\left[\tfrac{N}{2}+1\right] \cdots F[N-1]\right)?$$

Consider the contribution of $F[N/2]$ to the N'-point inverse DFT in each of these three cases.

3.21. *Zero padding*

(a) Fill in the steps in the derivation of the interpolation kernel in Equation 3.41 for even N.

(b) Fill in the steps in the derivation of the interpolation kernel in Equation 3.41 for odd N.

(c) Show that, in either case, when the padding factor $r = N'/N$ is an integer, then every r^{th} value of the inverse DFT is proportional to one of the original values of f,
$$f[rp; rN] = \tfrac{1}{r} f[p].$$

3.22. *Zero padding and redundant basis vectors*

(a) Show that the DFT expansion, with zero padding, of an N-vector f can be written in the matrix form

$$\mathbf{F} = \underbrace{\begin{bmatrix} \phi_0^\dagger \\ \phi_1^\dagger \\ \vdots \\ \phi_{N'-1}^\dagger \end{bmatrix}}_{N' \times N} \mathbf{f}$$

where $\mathbf{F}$ is $N' \times 1$, $\mathbf{f}$ is $N \times 1$, $\dagger$ denotes conjugate transpose, and the basis vectors ϕ_m are

$$\phi_m^\dagger = \begin{bmatrix} 1 & e^{-i2\pi \cdot m/N'} & e^{-i2\pi \cdot 2m/N'} & \cdots & e^{-i2\pi(N-1)m/N'} \end{bmatrix}.$$

(b) For an illustrative example, let $N = 4$ and $N' = 5$. Show that the N' basis vectors are not an orthogonal set by calculating the matrix product with MATLAB,

$$\begin{bmatrix} \phi_0^\dagger \\ \phi_1^\dagger \\ \vdots \\ \phi_{N'-1}^\dagger \end{bmatrix} \begin{bmatrix} \phi_0 & \phi_1 & \cdots & \phi_{N'-1} \end{bmatrix}.$$

(c) The vector **f** is reconstructed from the DFT expansion by multiplying the elements of **F** by their corresponding basis vectors and summing. Show that this is written, in matrix form, as

$$\frac{1}{N'} \begin{bmatrix} \phi_0 & \phi_1 & \cdots & \phi_{N'-1} \end{bmatrix} \mathbf{F} = \frac{1}{N'} \underbrace{\begin{bmatrix} \phi_0 & \phi_1 & \cdots & \phi_{N'-1} \end{bmatrix} \begin{bmatrix} \phi_0^\dagger \\ \phi_1^\dagger \\ \vdots \\ \phi_{N'-1}^\dagger \end{bmatrix}}_{N \times N} \mathbf{f},$$

then carry out the matrix multiplication with MATLAB and show that an $N \times N$ identity matrix results. Try this with other values of N and N' (always with $N' > N$, of course).

3.23. *Zero packing*
The forward DFTs in Equations 3.42 were derived in the text. Carry out the calculations to derive the inverse DFTs.

3.24. *Zero packing*
Using the symmetry and orthogonality properties of the DFT basis vectors, explain why the DFT of a periodically replicated sequence, $f[n \bmod N]$, $n = 0, 1, 2, \ldots pN - 1$, is zero except for those bins that are integer multiples of p.

3.25. *Zero packing*
Show that the zero packing DFT equations (Equation 3.42) obey the area theorem.

3.26. *Zero packing*
This is an example of a practical application of zero packing in signal processing.

(a) Exploration with MATLAB
Begin with a 32-sample ramp function, $x[n] = an$, $n = 0, 1, \ldots 31$. Create a y vector by packing 3 zeros between each x value (this is also called "4× oversampling"). Compute X and Y, the DFTs of x and y. Check that the DFTs agree with Theorem 3.10.
Now, *filter* Y by setting certain values to zero. The transfer function of the filter is defined as follows:

$$H[m] = \begin{cases} 4, m = 0, 1, \ldots, 15 \\ 2, m = 16 \\ 0, m = 17, 18, \ldots, 111 \\ 2, m = 112 \\ 4, m = 113, 114, \ldots, 127 \end{cases}.$$

In MATLAB, multiply Y times H to get a new vector G, and compute the inverse DFT of G. If you do this right, the imaginary part of the inverse DFT should be negligible. Here is an M-file you can use.

```
function g = oversample(x, rate)
% OVERSAMPLE - intersperse zeros and filter.
% E.W. Hansen

nx=length(x);
n=0:nx-1;
y=zeros(1, rate*nx);

y(rate*n+1) = x(n+1);    % Intersperse the zeros

fy = fft(y);             % Go into the DFT domain

% Build the filter (crude rectangle)
fh = [ones(1,nx/2), 0.5, zeros(1,(rate-1)*nx-1), 0.5,
ones(1,nx/2-1)];
fg = rate*fy.*fh;        % Apply the filter and
                         % inverse DFT
g = real(ifft(fg));
end;
```

Plot (the real part of) g and compare it with y. What is the observed effect of the filtering? Pay particular attention to the central region of the plot, away from the ends. Also compare g and y numerically.

You can examine the effect of the filter on Y by calculating the DFT of g (or putting a plot statement in the M-file after fg is calculated. Notice that this is approximately the same as zero padding the DFT of x. But since filtering can be done in the time domain (according to the convolution theorem), this approach enables us to effectively zero pad in the frequency domain by packing and filtering in the time domain.

(b) Calculation

Using the convolution theorem, it is easy to figure out what is going on here. When you multiply Y times H and take the inverse DFT, you are computing the (circular) convolution of y and h, which you will recall is defined (for $4N$ sample points):

$$g[n] = \sum_{k=0}^{4N-1} y[k]h[n-k].$$

So, the crux of this problem is figuring out what h is. You know H, so you can just use the definition:

$$h[n] = \frac{1}{4N} \sum_{m=0}^{4N-1} H[m] \exp\left(\frac{+i2\pi mn}{4N}\right), n = 0, 1, 2, \ldots, 4N-1$$

and carry out the calculation, which should eventually yield the result

$$h_n = \frac{1}{4N} \cos\left(\frac{\pi n}{4N}\right) \frac{\sin\left(\frac{\pi n}{4}\right)}{\sin\left(\frac{\pi n}{4N}\right)}, \quad n = 1, 2, \ldots, 4N-1,$$

which you will note is purely real, an outcome guaranteed by the even symmetry of H.

(c) Discussion
Plot h and note in particular where its zero crossings fall. How does the convolution explain the result you observed in (a)?

What does the oversampling procedure accomplish? Assuming you could design a filter that did not introduce so much ripple, what could oversampling be used for?

3.27. Several of the DFT theorems connect operations performed on a vector with effects on its DFT. Consider cyclic shift, convolution (with fixed h), zero padding, and zero packing as operators on vectors in $\mathbb{C}^N$. Which ones are

(a) Linear?

(b) One-to-one and/or onto?

(c) Isometric?

(d) Invertible?

(e) Unitary?

(f) Commutative, either with themselves or with the other operators?

3.28. *DFT of truncated signals*

When the DFT is used for spectrum analysis, the finite amount of data leads to the spectral leakage artifact. The truncation can be modeled by multiplying a (theoretically infinite) data set by a rectangular window. The Dirichlet kernel is the DFT of this rectangular window.

But, other N-point data windows are possible, including

Bartlett(triangular window)

$$B_N[n] = \begin{cases} \frac{2n}{N}, & n = 0, 1, \ldots, N/2 \\ 2 - \frac{2n}{N}, & n = 1 + N/2, \ldots, N - 1 \end{cases}.$$

Hanning(raised cosine window)

$$H_N[n] = \frac{1}{2}\left[1 - \cos\left(\frac{2\pi n}{N}\right)\right], n = 0, 1, \ldots, N - 1.$$

Calculate the magnitudes of the DFTs of these three functions (with MATLAB). Choose a moderate value of N, say 128 or 256. If you wish, you can also zero-pad the window sequence before calculating the DFT. Plot them on the same scale, and compare them according to two criteria: width (estimated full width at half maximum, FWHM) of the mainlobe (the big bump) and maximum height of the sidelobes (the little bumps). You should use a logarithmic vertical axis to bring out the sidelobes more clearly.

3.29. *Spectral leakage*

This problem is designed to give you experience with the spectral leakage phenomenon, its practical consequences, and a way of ameliorating its effects.

Define the power spectrum of a signal $f[n]$ to be

$$P_f[m] = \begin{cases} \frac{1}{N}|F[m]|^2, & m = 0 \\ \frac{1}{N}\left(|F[m]|^2 + |F[N-m]|^2\right), & m = 1, 2, \ldots N/2 \end{cases}$$

where $F[m]$ is the DFT of $f[n]$. It is conventionally plotted in decibels, $10\log_{10} P_f[m]$, vs. frequency, $v_m = m\Delta v$.

(a) Consider sampling a 2.0 kHz cosine at a 22.050 kHz sampling rate (v_s). Calculate and plot the power spectrum of this signal for $N = 64, 128, 256, 512$. Describe your observations.

(b) Now create a sum of cosines, based on sampling the following function:

$$f(t) = \cos\left(2\pi v_1 t + \varphi_1\right) + a\cos\left(2\pi v_2 t + \varphi_2\right)$$

where $v_1 = 2$ kHz, $v_2 = 2.5$ kHz, $v_s = 22.050$ kHz, and φ_1 and φ_2 are random phases created by the MATLAB command `2*pi*rand(1)`. Plot the power spectrum of this function for $N = 512$, adjusting a until the second peak is just below the level of visibility. This weaker signal is invisible because of spectral leakage from the stronger signal. Vary N. Does this affect the visibility of the weak signal?

(c) Spectral leakage can be ameliorated somewhat by smoothly tapering the data vector to zero at the edges. This is accomplished by multiplying the data, $f[n]$, by a window function $w[n]$. One simple window function is the Hamming window, which is computed by the MATLAB function `hamming(N)`, where N is the length of the window. Using the $f[n]$ you created in part (b), with the weak signal's spectral peak just below the threshold of visibility, multiply $f[n]$ by a Hamming window and recompute the power spectrum. Plot it and compare the result with what you observed in part (b).

(d) Use MATLAB to calculate, plot, and compare the magnitudes of the DFTs of the rectangular window and the Hamming window. For convenience, use 16 samples of each window padded with zeros to a total of 128 samples. Use your plots with Figure 3.6 to explain how the Hamming window helps against spectral leakage.

3.30. There are 88 keys on a piano, producing musical notes ranging in frequency from 27.5 Hz to 4186 Hz. The frequency spacing between notes is nonuniform. Each note is higher than the preceding one by a factor of $2^{1/12}$. For example, the frequency of the next-to-lowest note is $2^{1/12} \times 27.5 = 29.14$ Hz, and the frequency of the next-to-highest note is $\frac{4186}{2^{1/12}} = 3951$ Hz.

It is desired to design a spectrum analyzer, based on the DFT, for piano music. The frequency resolution must be such that adjacent notes are always seen as distinct peaks.

(a) Specify a minimum sampling rate, in Hz.

(b) Specify a maximum bin spacing, in Hz.

(c) Because music is spectrally complex, it is advantageous to break a recording into shorter segments and observe how the spectral content changes in time. Using the lowest number of bins, which is compatible with your sampling rate and bin spacing as well as being computationally attractive for the FFT algorithm, how long (in seconds) will one of these short segments be?

PROBLEMS

3.31. *Fast Fourier transform*
Derive a Cooley–Tukey type algorithm for $N = 9$ and draw the signal-flow diagram. Begin by dividing the vector f into three vectors, $(f[0], f[3], f[6])$, $(f[1], f[4], f[7])$, and $(f[2], f[5], f[8])$. Now show how to construct F from the three-point DFTs of these three vectors.

3.32. *Fast Fourier transform*
In deriving the FFT algorithm, a four-point DFT was expressed in terms of two-point DFTs. The purpose of this exercise is to derive a fast four-point DFT that does not use two-point DFTs.

(a) Show that the four-point DFT can be written

$$F[m] = f[0] + (-i)^m f[1] + (-1)^m f[2] + i^m f[3], \quad m = 0, 1, 2, 3$$

and note that each of the powers of i is either $1, i, -1$, or $-i$.

(b) Show how to multiply a complex number $a + ib$ by a power of i *without* doing any multiplication. Hence, show how to implement the four-point DFT without multiplication and draw its signal-flow diagram—a four-point butterfly.

(c) Suppose you want to compute a 16-point FFT. Show how to do this using all four-point butterflies. This is called a *radix-4* FFT. Carefully count the operations required for the radix-4 implementation and compare with the operation count for the radix-2 FFT developed in the text.

3.33. *Fast Fourier transform*
Here is another trick used in good FFT programs to shave a factor of two off the operation count. Suppose the input $f[n]$ is real. Then we know that the DFT F is Hermitian, and we only need to compute about half the values; the rest are complex conjugates of the ones we compute. Furthermore, at the first stage, half the multiply-adds—those that operate on the imaginary part of f—are wasted, because $\mathcal{I}m\, f = 0$.

Construct a new vector g, according to the following rule:

$$g[n] = f[2n] + if[2n + 1], \quad n = 0, 1, \ldots N/2 - 1,$$

that is, the real part of g contains the even-indexed values of f, $f[0], f[2], \ldots, f[N-2]$, and the imaginary part of g contains the odd-indexed values, $f[1], f[3], \ldots f[N-1]$. (This looks like the first step in deriving an FFT.) Now, show how to construct the desired DFT, $(F[0], F[1], \ldots F[N-1])$, from the $N/2$-point DFT of g, $(G[0], G[1], \ldots G[N/2-1])$. Estimate the operation count.

3.34. *Discrete cosine transform*
Show that $\cos\left(\frac{\pi(2N-m)(n+\frac{1}{2})}{N}\right) = -\cos\left(\frac{\pi m(n+\frac{1}{2})}{N}\right)$, $m = 0, 1, \ldots 2N - 1$, and also that $\cos\left(\frac{\pi N(n+\frac{1}{2})}{N}\right) = 0$.

3.35. *Orthogonality of the DCT basis*
Prove Equation 3.44.

3.36. *Discrete cosine transform*
Is the DCT a unitary transform?

3.37. *Discrete cosine transform*

The DCT is based on a DFT whose basis vectors are naturally compatible with sequences having a certain reflection symmetry, for example, the sequence $f = (0, 1, \ldots 7)$ is symmetrized to $f_s = (0, 1, \ldots 7, 7, 6, \ldots 0)$. There are other DCTs that are derived from different symmetrizations.

(a) It is possible to symmetrize f into an even sequence by not repeating $f[0]$ or $f[7]$ in the reflection, that is,
$$f_s = (0, 1, 2, 3, 4, 5, 6, 7, 6, 5, 4, 3, 2, 1).$$
This sequence is even, but its length is 14.

(b) Another even symmetrization repeats $f[7]$ but not $f[0]$:
$$f_s = (0, 1, 2, 3, 4, 5, 6, 7, 7, 6, 5, 4, 3, 2, 1).$$
Its length is 15.

(c) The fourth way to symmetrize f is to repeat $f[0]$ but not $f[7]$:
$$f_s = (0, 1, 2, 3, 4, 5, 6, 7, 6, 5, 4, 3, 2, 1, 0).$$
It is not even, and its length is 15.

Derive discrete cosine transforms for f based on these symmetrizations.

3.38. *Computing the DCT*

Develop an expression for the DCT (Equation 3.46) in terms of the DFT, and hence show how to compute the DCT using the FFT.

3.39. *Discrete Hartley transform*

Yet another discrete Fourier-type transform is defined by using the set of basis vectors $\{\eta_m\}_{m=0}^{N-1}$, where
$$\eta_m[n] = \operatorname{cas}\left(\frac{2\pi mn}{N}\right), \quad n = 0, 1, \ldots N - 1$$
$$\operatorname{cas} x = \cos x + \sin x.$$
The transform so defined,
$$F^h[m] = \langle f, \eta_m \rangle,$$
is called the *discrete Hartley transform* (DHT).[9]

(a) Show that the vectors $\{\eta_m\}_{m=0}^{N-1}$ are an orthogonal basis for $\mathbb{C}^N$.

(b) Derive summation expressions for the forward and inverse transforms.

(c) Express the DHT in terms of the DFT, and vice-versa. Express the DFT magnitude spectrum $|F[m]|$ in terms of $F^h[m]$.

---

[9] See Bracewell (1986).

CHAPTER 4

THE FOURIER SERIES

Oscillatory phenomena are ubiquitous in physical systems. Here are some examples. When a guitar string is plucked, it vibrates. The vibration of the string sets up an acoustic wave in the air which, when it reaches your eardrum, causes it to vibrate. This vibration is mechanically transmitted to a fluid in the cochlea. The oscillating fluid stimulates sensory cells, which transmit to the brain electrical signals that are, in turn, interpreted as the sound of the guitar. The light emitted by a laser comes from an electromagnetic standing wave oscillating in the laser's resonant cavity. The wave itself is fed by atomic processes described, in the language of quantum mechanics, by waves. Numerous optical devices rely on the interactions of optical waves with periodic structures. The electromagnetic fields radiated by highly directive antenna arrays and the acoustic fields used in medical ultrasound imaging are combinations of waves. The shaking of the ground during an earthquake and the resulting motions of buildings are also periodic vibrations. Moreover, even systems that display no oscillatory behavior, like the evolution of the temperature distribution in a bar of metal as it is heated, can be described mathematically as portions of periodic functions.

The mathematical models for these diverse physical systems are based on a simple idea, proposed by Fourier in 1807, that any function defined on a finite interval $[0, L]$, including certain discontinuous functions, can be decomposed as the sum of harmonically related sinusoids. It seemed outlandish to some of Fourier's contemporaries—how could sines and cosines combine to make arbitrary smooth functions, much less unsmooth ones? Yet, over the course of the next century, investigations by numerous mathematicians proved Fourier correct and expanded the reach of his ideas.

The subject of this chapter is the *Fourier series*, which in its most concise form is

$$f(x) = \sum_{n=-\infty}^{\infty} c_n e^{inx}.$$

Like the discrete Fourier transform (DFT), it is a linear combination of orthogonal sinusoidal basis functions. Unlike the DFT, it is an *infinite* series. We shall consider what kinds of function f have a Fourier series representation, and how well the series actually represents f. We will also develop several theorems relating the function and its Fourier coefficients, analogous to the DFT theorems of the preceding chapter, and demonstrate the theory in a few physical applications. The final section of the

---

Fourier Transforms: Principles and Applications, First Edition. Eric W. Hansen.
© 2014 John Wiley & Sons, Inc. Published 2014 by John Wiley & Sons, Inc.

chapter introduces the discrete-time Fourier transform, a variation on the Fourier series important in digital signal processing.

4.1 SINUSOIDS AND PHYSICAL SYSTEMS

The Fourier series is an orthogonal expansion on a basis of complex exponentials e^{ikx}, or trigonometric functions, $\sin kx$ and $\cos kx$. Mathematically, the complex exponentials are eigenfunctions of the differential operator $\frac{d}{dx}$, and the sinusoidal functions are eigenfunctions of the differential operator $\frac{d^2}{dx^2}$. That is, when you apply $\frac{d}{dx}$ to the complex exponential, the result is also a complex exponential:

$$\frac{d}{dx} e^{ikx} = ik e^{ikx}.$$

When you apply $\frac{d^2}{dx^2}$ to sine and cosine, the results are sine and cosine, respectively:

$$\frac{d^2}{dx^2} \sin kx = -k^2 \sin kx$$

$$\frac{d^2}{dx^2} \cos kx = -k^2 \cos kx.$$

These facts make trigonometric functions the natural choice for constructing solutions to certain differential equations that describe physical phenomena, like the the damped driven harmonic oscillator,

$$\frac{d^2 y}{dt^2} + 2\zeta \omega_n \frac{dy}{dt} + \omega_n^2 y = \cos \omega_0 t,$$

the heat equation,

$$\frac{\partial^2 F(x,t)}{\partial x^2} = \frac{1}{k} \frac{\partial F(x,t)}{\partial t},$$

Laplace's equation,

$$\frac{\partial^2 F(x,t)}{\partial x^2} = 0,$$

and the wave equation,

$$\frac{\partial^2 F(x,t)}{\partial x^2} = \frac{1}{c^2} \frac{\partial^2 F(x,t)}{\partial t^2}.$$

4.2 DEFINITIONS AND INTERPRETATION

The complex exponentials on $[-\pi, \pi]$,

$$e_n(x) = \frac{1}{\sqrt{2\pi}} e^{inx} \quad n = 0, \pm 1, \pm 2, \ldots$$

4.2 DEFINITIONS AND INTERPRETATION

are a complete orthonormal set. Formally, following Equation 2.38, we can write an expansion of a function $f : [-\pi, \pi] \to \mathbb{C}$,

$$f(x) = \sum_{n=-\infty}^{\infty} \langle f, e_n \rangle e_n(x) = \sum_{n=-\infty}^{\infty} \left[\int_{-\pi}^{\pi} f(x) \frac{1}{\sqrt{2\pi}} e^{-inx} \, dx \right] \frac{1}{\sqrt{2\pi}} e^{inx}.$$

By convention, the factors of $\frac{1}{\sqrt{2\pi}}$ are collected with the integral, so we have

$$f(x) = \sum_{n=-\infty}^{\infty} c_n e^{inx}, \tag{4.1a}$$

where

$$c_n = \frac{1}{2\pi} \int_{-\pi}^{\pi} f(x) e^{-inx} \, dx, \quad n = 0, \pm 1, \pm 2, \ldots. \tag{4.1b}$$

This is called the *Fourier series* for f.

In order for this to be useful, we need to know if the integral exists and if the infinite series converges. A rule for the existence of Fourier coefficients is easily established. The integral is bounded above,

$$|c_n| = \left| \frac{1}{2\pi} \int_{-\pi}^{\pi} f(x) e^{-inx} \, dx \right| \leq \frac{1}{2\pi} \int_{-\pi}^{\pi} \left| f(x) e^{-inx} \right| \, dx \quad \text{(triangle inequality)}$$

$$= \frac{1}{2\pi} \int_{-\pi}^{\pi} |f(x)| \, dx.$$

Consequently, the Fourier coefficients c_n exist if f is absolutely integrable ($f \in L^1[-\pi, \pi]$). The question of convergence is harder. While Fourier coefficients exist for any absolutely integrable function, it has been known since the early 1900s that there are pathological L^1 functions whose Fourier series diverge at *all* points.

A better result is available if we restrict attention to the Hilbert space $L^2[-\pi, \pi]$, where the geometric ideas of basis and orthogonal expansions can be used. $L^2[-\pi, \pi]$ is a subspace of $L^1[-\pi, \pi]$ that contains nearly all functions of physical interest. On a bounded interval such as $[-\pi, \pi]$, integrability depends only on how rapidly a function grows at any singular points in the interval. If a function grows sufficiently slowly to still be integrable, its square root will grow even slower. Thus, if a function is square integrable it must also be absolutely integrable. Moreover, if a function is bounded then it will be integrable when raised to any finite power. So, $L^\infty[-\pi, \pi] \subset L^2[-\pi, \pi] \subset L^1[-\pi, \pi]$. With $f \in L^2[-\pi, \pi]$, the results of Section 2.3.4 may be applied; the complex exponentials are a basis for $L^2[-\pi, \pi]$ and the sequence of partial sums, $S_N = \sum_{|n| \leq N} c_n e^{inx}$, converges to f in the L^2 norm,[1]

$$\lim_{N \to \infty} \|f - S_N\|_2 = 0.$$

---

[1] For proofs of the norm convergence, see Folland (1992, pp. 76–79), Oden and Demkowicz (1996, pp. 545–548), Young (1988, pp. 45–52), or Dym and McKean (1972, pp. 30–36). Carleson (1966) showed, by a difficult proof, that the partial sums also converge to f almost everywhere. Convergence in norm for the broader class of L^1 functions is discussed in Dym and McKean (1972, pp. 37–43). For a brief discussion of L^1 functions whose Fourier series diverge pointwise, see Champeney (1987, pp. 36–38).

This very general result encompasses just about any function one would encounter in engineering or physics, but it is still not as useful as we would like. If we want to evaluate the Fourier series at a point, we can only sum a finite number of terms, and the mere fact of norm convergence does not say how many terms we need in order to obtain a desired accuracy or give any indication about the rate of convergence as the number of terms, N, increases. We will address these questions in a later section.

With the change of variable $x \to \frac{2\pi x}{L}$, the functions

$$e_n(x) = \frac{1}{\sqrt{L}} \exp\left(\frac{i2\pi nx}{L}\right), \quad n \in \mathbb{Z} \tag{4.2}$$

are an orthonormal basis for $L^2[0, L]$. The Fourier series with these basis functions, which we shall use from now on, is

$$f(x) = \sum_{n=-\infty}^{\infty} c_n \exp\left(\frac{i2\pi nx}{L}\right) \tag{4.3a}$$

$$c_n = \frac{1}{L} \int_0^L f(x) \exp\left(-\frac{i2\pi nx}{L}\right) dx, \quad n \in \mathbb{Z}. \tag{4.3b}$$

With appropriate changes to the limits of integration, any interval of length L will do, for example, $\left[-\frac{L}{2}, \frac{L}{2}\right]$.

The function f defined on $[0, L]$ with $f(0) = f(L)$ may be viewed as a function that exists on that interval only, or as one period of a function that extends to infinity. In the former case, the Fourier series is useful for describing systems like vibrating strings and laser resonators where the physical extent of the system is finite. In the latter, the Fourier series decomposes a periodic waveform into sinusoidal components with frequencies $v_n = n/L$. The lowest frequency, $v_1 = 1/L$, is called the *fundamental*, and the higher frequencies are called *harmonics*. The frequency $v_2 = 2/L$ is the second harmonic, $v_3 = 3/L$ is the third harmonic, etc. If $L = 2\pi$ and x is an angle φ, the Fourier series can be used to decompose the angular dependence of a function defined in polar coordinates, $f(r, \varphi)$; examples occur in optics and in the analysis of vibrating circular membranes.

The Fourier series (Equation 4.3) resembles the DFT (Equation 3.11),

$$f[n] = \frac{1}{N} \sum_{m=0}^{N-1} F[m] \exp\left(\frac{i2\pi mn}{N}\right), \quad n = 0, 1, \ldots, N-1$$

$$F[m] = \sum_{n=0}^{N-1} f[n] \exp\left(-\frac{i2\pi mn}{N}\right), \quad m = 0, 1, \ldots, N-1,$$

in two ways.

- The function $f(x)$ has period L, the sequence $f[n]$ has implicit period N.
- The frequencies of the Fourier components in the Fourier series are harmonics of the fundamental $v_1 = 1/L$; the digital frequencies of the Fourier components in the DFT are harmonics of $\Delta\theta = 2\pi/N$.

There are also two important differences.
- The Fourier series has a countable infinity of distinct frequencies, rather than the N discrete frequencies of the DFT bins.
- While the DFT sequence $(F[m])$ is implicitly periodic, $F[m] = F[m + N]$, the Fourier coefficient sequence $(c_n)_{n \in \mathbb{Z}}$ is not.

The "classical" Fourier series

The real plane $\mathbb{R}^2$, spanned by the coordinate vectors $\{\mathbf{e}_x, \mathbf{e}_y\}$, is also spanned by the same vectors rotated by 45°, $\left\{ \frac{1}{\sqrt{2}}(\mathbf{e}_x + \mathbf{e}_y), \frac{1}{\sqrt{2}}(\mathbf{e}_x - \mathbf{e}_y) \right\}$. These are linear combinations of $\mathbf{e}_x$ and $\mathbf{e}_y$. In the same way, we may construct an alternative basis for $L^2[0, L]$ from appropriate linear combinations of the complex exponential basis functions $\left\{ \exp\left(\frac{i2\pi nx}{L}\right) \right\}_{n=-\infty}^{\infty}$. In particular, consider the sine and cosine functions,

$$\sin\left(\frac{2\pi nx}{L}\right) = \frac{\exp\left(\frac{i2\pi nx}{L}\right) - \exp\left(-\frac{i2\pi nx}{L}\right)}{2i}$$

$$\cos\left(\frac{2\pi nx}{L}\right) = \frac{\exp\left(\frac{i2\pi nx}{L}\right) + \exp\left(-\frac{i2\pi nx}{L}\right)}{2}.$$

We only need to consider $n \geq 0$, because $\cos\left(-\frac{2\pi nx}{L}\right) = \cos\left(\frac{2\pi nx}{L}\right)$ and $\sin\left(-\frac{2\pi nx}{L}\right) = -\sin\left(\frac{2\pi nx}{L}\right)$. To check orthogonality on the interval $[0, L]$, we calculate inner products. For the cosines,

$$\left\langle \cos\left(\frac{2\pi nx}{L}\right), \cos\left(\frac{2\pi mx}{L}\right) \right\rangle = \int_0^L \cos\left(\frac{2\pi nx}{L}\right) \cos\left(\frac{2\pi mx}{L}\right) dx$$

$$= \frac{1}{2} \int_0^L \cos\left(\frac{2\pi(n-m)x}{L}\right) dx$$

$$+ \frac{1}{2} \int_0^L \cos\left(\frac{2\pi(n+m)x}{L}\right) dx.$$

The first integral is zero unless $n = m$, and the second integral is zero unless $n = -m$, which can occur only when $n = m = 0$, because $n, m \geq 0$. In these cases we have $\int_0^L \cos(0)\, dx = L$. Therefore, the inner product is

$$\left\langle \cos\left(\frac{2\pi nx}{L}\right), \cos\left(\frac{2\pi mx}{L}\right) \right\rangle = \begin{cases} L, & n = m = 0 \\ \frac{L}{2}, & n = m = 1, 2, \ldots \\ 0, & n \neq m \end{cases}.$$

Similarly,

$$\left\langle \sin\left(\frac{2\pi nx}{L}\right), \sin\left(\frac{2\pi mx}{L}\right) \right\rangle = \begin{cases} \frac{L}{2}, & n = m = 1, 2, \ldots \\ 0, & n \neq m \end{cases}.$$

and

$$\left\langle \cos\left(\frac{2\pi nx}{L}\right), \sin\left(\frac{2\pi mx}{L}\right)\right\rangle = 0, \quad \text{all } n, m.$$

After normalizing, we have a new orthonormal set derived from the complex exponentials, which also spans $L^2[0, L]$:

$$\left\{\frac{1}{\sqrt{L}}, \sqrt{\frac{2}{L}}\cos\left(\frac{2\pi nx}{L}\right), \sqrt{\frac{2}{L}}\sin\left(\frac{2\pi nx}{L}\right)\right\}_{n=1}^{\infty}. \tag{4.4}$$

We will call an orthogonal expansion in the sine–cosine basis the *classical* Fourier series,

$$f(x) = \frac{a_0}{2} + \sum_{n=1}^{\infty} a_n \cos\left(\frac{2\pi nx}{L}\right) + \sum_{n=1}^{\infty} b_n \sin\left(\frac{2\pi nx}{L}\right) \tag{4.5a}$$

$$a_n = \frac{2}{L}\int_0^L f(x)\cos\left(\frac{2\pi nx}{L}\right)dx, \quad n = 0, 1, 2, \ldots \tag{4.5b}$$

$$b_n = \frac{2}{L}\int_0^L f(x)\sin\left(\frac{2\pi nx}{L}\right)dx, \quad n = 1, 2, \ldots. \tag{4.5c}$$

You can show that the complex coefficients are related to the classical coefficients by

$$a_n = c_n + c_{-n}, \quad b_n = i(c_n - c_{-n})$$
$$c_n = \frac{a_n - ib_n}{2}, \quad c_{-n} = \frac{a_n + ib_n}{2}, \quad n \geq 0. \tag{4.6}$$

The case $n = 0$ is special. We have $a_0 = 2c_0$, which double counts the constant term, and $b_0 = 0$, which is trivial, because $\sin\left(\frac{2\pi(0)x}{L}\right) = 0$. Consequently, in Equation 4.5a a_0 is divided by two and b_0 is deleted.

In the classical series, one frequency component (at $\nu_n = n/L$) is represented by a pair of terms,

$$a_n \cos 2\pi \nu_n x + b_n \sin 2\pi \nu_n x,$$

the combination of a sine and cosine. The same component in the complex series also consists of two terms,

$$c_n e^{i2\pi \nu_n x} + c_{-n} e^{-i2\pi \nu_n x},$$

a complex exponential and its complex conjugate. The conjugate term $c_{-n}e^{-i2\pi\nu_n x}$ is often said to have "negative frequency," because the complex exponential appears to have frequency $\nu = -\nu_n$.

Another useful Fourier representation calculates the magnitude and phase of each frequency component. If you measure a (real) sinusoidal signal in the lab, you will usually obtain its magnitude and phase shift relative to some time origin. The Fourier component at frequency $\nu = \nu_n$ is modeled by $A_n \cos(2\pi\nu_n x + \varphi_n)$, where $A_n \geq 0$ and $\varphi_n \in [-\pi, \pi)$. The relationships between magnitude and phase and the coefficients of the complex series and the classical series are considered in the problems.

The Fourier spectrum

The Riemann–Lebesgue lemma says that the Fourier coefficients tend to zero at high frequencies.[2]

Theorem 4.1 (Riemann–Lebesgue lemma). Let $f \in L^1[0, L]$ with $f(0) = f(L)$, and let its Fourier coefficients be $(c_n)_{n \in \mathbb{Z}}$ (Equation 4.3). Then,

$$\lim_{|n| \to \infty} |c_n| = 0.$$

Moreover, when $f \in L^2$, we inherit Parseval's formula from Hilbert space theory. It will be shown later that, for the Fourier series, Parseval's formula is

$$\frac{1}{L} \int_0^L |f(x)|^2 \, dx = \sum_{n=-\infty}^{\infty} |c_n|^2.$$

If $f \in L^2$, then $(c_n) \in \ell^2$. If $P = \frac{1}{L} \int_0^L |f(x)|^2 \, dx$ is viewed as the average power in f, then Parseval's formula is a statement of *conservation of energy*. Whether you calculate average power in the "time domain," from the actual signal, or in the "frequency domain," by summing the squared magnitudes of the Fourier coefficients, you get the same result. No energy is misplaced in going to the Fourier representation.[3]

The average power can be rewritten

$$P = \frac{1}{L} \int_0^L |f(x)|^2 \, dx = |c_0|^2 + \sum_{n=1}^{\infty} \left(|c_n|^2 + |c_{-n}|^2 \right),$$

grouping positive and negative frequency components together. We then define the *power spectrum* by

$$P_f[n] = \begin{cases} |c_0|^2, & n = 0 \\ |c_n|^2 + |c_{-n}|^2, & n \geq 1 \end{cases}. \tag{4.7}$$

The terms in the power spectrum $P_f[n]$ give the portions of f's average power at zero frequency ($n = 0$) and at the fundamental and harmonic frequencies, $\nu_n = n/L$, $n \geq 1$. A plot of P_f vs. ν_n is also called a *line spectrum* (Figures 4.1 and 4.2). The Riemann–Lebesgue lemma and Parseval's formula both express the fact that the power spectra of real-world (finite average power) signals eventually "roll off" at high frequencies.

We can begin to interpret the Fourier series with the help of some examples.

Example 4.1. Consider a sinusoidal signal "riding" on a constant voltage (DC bias),

$$V(t) = A + B \cos 2\pi \nu_0 t,$$

---

[2] For functions in L^2, the Riemann–Lebesgue lemma is a simple corollary of Bessel's inequality, Theorem 2.5. For the most general (L^1) case, see Dym and McKean (1972, p. 39).

[3] In the operator language of Section 2.4, allowing for the factor of $1/L$, the Fourier series expansion is a unitary (invertible, norm-preserving) operator between two Hilbert spaces, L^2 and ℓ^2.

and calculate the Fourier series of one period of this periodic function. The period is $T = 1/\nu_0$. The Fourier coefficients are given by the integral,

$$c_n = \frac{1}{T}\int_0^T [A + B\cos 2\pi\nu_0 t]\, e^{-i2\pi nt/T}\, dt.$$

Convert the cosine to exponential form and divide the integral into three parts.

$$\begin{aligned} c_n &= \frac{A}{T}\int_0^T e^{-i2\pi nt/T}\, dt \\ &\quad + \frac{B}{2T}\int_0^T e^{+i2\pi t/T} e^{-i2\pi nt/T}\, dt + \frac{B}{2T}\int_0^T e^{-i2\pi t/T} e^{-i2\pi nt/T}\, dt \\ &= \frac{A\left[e^{-i2\pi n} - 1\right]}{-i2\pi n} + \frac{B\left[e^{-i2\pi(n-1)} - 1\right]}{-i4\pi(n-1)} + \frac{B\left[e^{-i2\pi(n+1)} - 1\right]}{-i4\pi(n+1)}. \end{aligned}$$

Now, because $e^{-i2\pi n} = 1$ for integer n, the numerator of each term is zero. But for certain values of n—0, 1, and -1, respectively—each of the denominators is also zero. We use L'Hospital's rule to determine the coefficients in these particular cases, and arrive at the final result,

$$c_n = \begin{cases} A, & n = 0 \\ B/2, & n = \pm 1 \\ 0, & \text{otherwise} \end{cases}. \tag{4.8}$$

The Fourier series is

$$f(t) = \frac{B}{2} e^{-i2\pi t/T} + A + \frac{B}{2} e^{+i2\pi t/T},$$

which is nothing more than the original function written in complex exponential form. This is not surprising, since the original signal was exactly one period of a sinusoid.

We verify by a direct calculation that Parseval's formula holds. First, in the time domain,

$$\begin{aligned} \|V\|^2 &= \int_0^T V^2\, dt = \int_0^T \left[A^2 + B^2\cos^2 2\pi\nu_0 t + 2AB\cos 2\pi\nu_0 t\right] dt \\ &= \int_0^T \left[A^2 + \frac{B^2}{2} + \frac{B^2}{2}\cos 4\pi\nu_0 t + 2AB\cos 2\pi\nu_0 t\right] dt \\ &= \left(A^2 + \frac{B^2}{2}\right) T \end{aligned}$$

and

$$P_{\text{avg}} = \frac{1}{T}\|V\|^2 = A^2 + \frac{B^2}{2}.$$

Doing the calculation in the frequency domain with the Fourier coefficients,

$$\sum_{n=-\infty}^{\infty} |c_n|^2 = \left(\frac{B}{2}\right)^2 + A^2 + \left(\frac{B}{2}\right)^2 = A^2 + \frac{B^2}{2}.$$

The function and its spectrum are shown in Figure 4.1. ∎

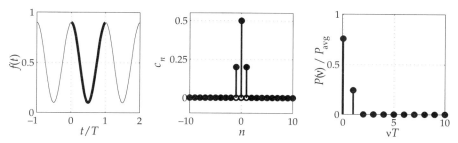

FIGURE 4.1 The function $A + B\cos 2\pi\nu_0 t$ with $A = 0.5$ and $B = 0.4$, its Fourier coefficients (*center*), and its power spectrum (*right*). The time axis and the frequency axis in the power spectrum are normalized the by the period T. The vertical axis on the right is normalized by the average power in the function, so the graph represents the fraction of the power at each frequency.

This example was simple because f was periodic with period T, just like an "in-bin" signal with the DFT. Something like the spectral leakage effect in the DFT also occurs with Fourier series, as the next example shows.

Example 4.2. Consider a raised cosine, as in the previous example, but with $\nu_0 = 5/4T$. Take the segment on $[0, T]$, assign $f(0) = f(T)$, and periodically replicate it. The Fourier coefficients are

$$c_0 = \frac{1}{T}\int_0^T \left[A + B\cos\left(\frac{5\pi t}{2T}\right)\right] dt = A + \frac{2}{5\pi}B \tag{4.9a}$$

$$c_n = \frac{1}{T}\int_0^T \left[A + B\cos\left(\frac{5\pi t}{2T}\right)\right] e^{-i2\pi nt/T} dt$$

$$= \frac{2}{\pi}\frac{5 + i4n}{25 - 16n^2} B, \quad n \neq 0. \tag{4.9b}$$

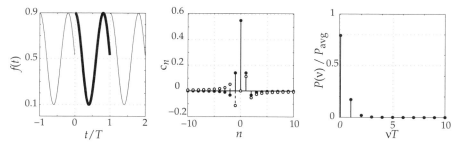

FIGURE 4.2 A discontinuous sinusoidal function (*left*), its Fourier coefficients (*center*), and its power spectrum (*right*). The real parts of the Fourier coefficients are shown with solid stems and filled circles. The imaginary parts are shown with dashed stems and open circles. Compare the DFT of an out-of-bin sinusoid (Figure 3.4).

Because of the discontinuity, energy is now distributed across all Fourier components (Figure 4.2). ∎

The next example is a square wave, useful for its ability to model on–off signals.

Example 4.3. Let f be the square wave shown in Figure 4.3. It has period T and a 50% duty cycle (the positive portion is 50% of the period). The function is specified piecewise:

$$f(x) = \begin{cases} 1, & \frac{T}{2} > x \geq 0 \\ -1, & T > x \geq \frac{T}{2} \end{cases}.$$

The complex Fourier coefficients are given by the sum of integrals over the two segments of the function.

$$c_0 = \frac{1}{T}\int_0^{T/2} dt + \frac{1}{T}\int_{T/2}^{T}(-1)dt = 0 \qquad (4.10a)$$

$$c_n = \frac{1}{T}\int_0^{T/2} e^{-i2\pi nt/T} dt + \frac{1}{T}\int_{T/2}^{T}(-1)e^{-i2\pi nt/T} dt$$

$$= -\frac{e^{-i\pi n} - 1}{i2\pi n} + \frac{1 - e^{-i\pi n}}{i2\pi n}$$

$$= \frac{1 - (-1)^n}{i\pi n} = \begin{cases} \frac{2}{i\pi n}, & n = \pm 1, \pm 3, \ldots \\ 0, & \text{otherwise (}n\text{ even)} \end{cases} \qquad (4.10b)$$

where we have used the identities $e^{-i2\pi n} = 1$ and $e^{-i\pi n} = (-1)^n$. The even-indexed coefficients are all zero; the Fourier expansion consists only of odd harmonics of the fundamental frequency $1/T$ (Figure 4.3). Every Fourier component is of the form

$$\frac{2}{i\pi n}e^{i2\pi nt/T} - \frac{2}{i\pi n}e^{-i2\pi nt/T} = \frac{4}{\pi n}\sin\frac{2\pi nt}{T}, \quad n = 1, 3, \ldots.$$

We can see why this is so with the aid of Figure 4.4. The cosine basis functions are orthogonal to the square wave for both even and odd n (hence no cosines), and the

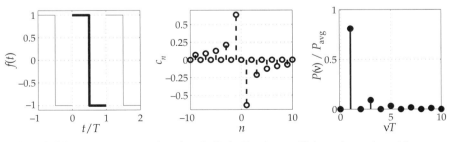

FIGURE 4.3 A square wave function (*left*), its Fourier coefficients (*center*), and its power spectrum (*right*).

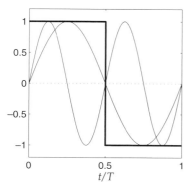
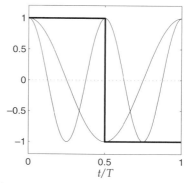

FIGURE 4.4 The square wave shown with $\sin\frac{2\pi nt}{T}$ for $n = 1, 2$ (*left*) and $\cos\frac{2\pi nt}{T}$ for $n = 1, 2$ (*right*). The square wave is orthogonal to $\sin\frac{2\pi nt}{T}$ for even n, and orthogonal to $\cos\frac{2\pi nt}{T}$ for both even and odd n. Consequently, the Fourier series for the square wave consists only of odd-indexed sine wave components.

sine basis functions are orthogonal to the square wave for even n but not for odd n (hence no sines of even harmonics). ∎

4.3 CONVERGENCE OF THE FOURIER SERIES

The sequence of partial sums of a square-integrable function f's Fourier series converges to f in the L^2 norm. Real-world signals and the functions used to model physical systems normally have some degree of continuity and smoothness (differentiability) in addition to being square integrable, and these additional attributes improve the convergence of the Fourier series. This section describes several of these relationships. We begin with some definitions, which should be familiar from calculus.

Definition 4.1 (Continuity and smoothness). Consider a function $f : [a, b] \to \mathbb{C}$, with finite a and b. By $f(c^-)$ and $f(c^+)$ we mean the limits of f as $x = c$ is approached from below (through values less than c) and from above (through values greater than c), respectively.

- **(a)** f is *continuous* on $[a, b]$ if, for $x_0 \in (a, b)$, $f(x_0^-)$ and $f(x_0^+)$ are finite and are both equal to $f(x_0)$, $f(a^+)$ is finite and equal to $f(a)$, and $f(b^-)$ is finite and equal to $f(b)$. The class of functions continuous on $[a, b]$ is denoted $C[a, b]$, or $C^{(0)}[a, b]$.

- **(b)** f is *piecewise continuous* if it is continuous everywhere in $[a, b]$ except perhaps on a finite set of points $\{x_k\}_{k=1}^N$, and at these points $f(x_k^-)$ and $f(x_k^+)$ are finite. That is, f has at most a finite number of finite jump discontinuities. The class of piecewise continuous functions on $[a, b]$ is denoted $PC[a, b]$.

- **(c)** f is *piecewise smooth* if it is piecewise continuous and its derivative f' is also piecewise continuous. The derivative is permitted to be undefined at points where f has a jump discontinuity. The class of piecewise smooth functions on $[a, b]$ is denoted $PS[a, b]$.

188 CHAPTER 4 THE FOURIER SERIES

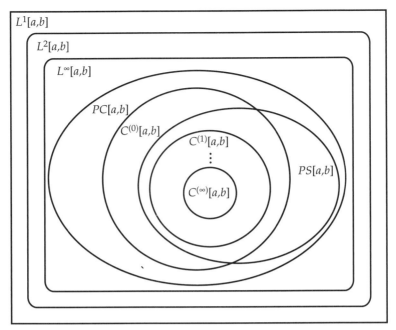

FIGURE 4.5 Inclusions of classes of functions on a bounded interval $[a, b]$. Functions possessing only boundedness or integrability (L^p) are shown with rectangles. Functions that are at least piecewise continuous or piecewise smooth are shown with ellipses. Functions having continuity or continuous differentiability are shown with circles.

- (d) f is *continuously differentiable* if f and its derivative f' are continuous. A function may be continuously differentiable more than once. The class of functions that are p-times continuously differentiable on $[a, b]$ is denoted $C^{(p)}[a, b]$. The class of infinitely continuously differentiable functions is denoted $C^{(\infty)}[a, b]$.

These classes of functions overlap (Figure 4.5).

- A continuous function is also piecewise continuous (with a single "piece"): $C[a, b] \subset PC[a, b]$.
- A piecewise smooth function is also piecewise continuous: $PS[a, b] \subset PC[a, b]$.
- A continuously differentiable function is also piecewise smooth: $C^{(1)}[a, b] \subset PS[a, b]$
- A piecewise continuous function on a bounded interval $[a, b]$ is also bounded, square integrable, and absolutely integrable: $PC[a, b] \subset L^\infty[a, b] \subset L^2[a, b] \subset L^1[a, b]$.
- A p-times continuously differentiable function is also $p - 1$ times continuously differentiable: $C^{(\infty)} \ldots \subset C^{(2)} \subset C^{(1)} \subset C^{(0)}$.

Here are some examples.

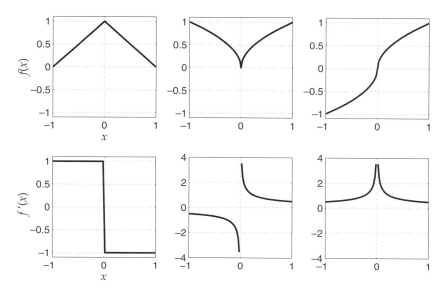

FIGURE 4.6 Three examples of functions with differing degrees of continuity and smoothness. *Left*: The triangle function is continuous and piecewise smooth on $[-1, 1]$. *Center*: The square root $\sqrt{|x|}$ is continuous on $\mathbb{R}$, but not piecewise smooth, because the derivative has an infinite jump at the origin. *Right*: The square root $f(x) = \sqrt{|x|} \, \text{sgn}(x)$ is continuous on $\mathbb{R}$, but not continuously differentiable, because the derivative blows up at the origin.

Example 4.4 (Continuity and smoothness).

(a) The raised cosine $f(t) = A + B \cos 2\pi v_0 t$ in Example 4.1 is continuous and infinitely differentiable (hence smooth) everywhere on the real line: $f \in C^{(\infty)}(\mathbb{R})$. It is absolutely integrable and square integrable on any bounded interval $[a, b]$, but not on an unbounded interval like $[0, \infty)$ or $(-\infty, \infty)$.

(b) The cosine $f(t) = A + B \cos \left(\frac{5\pi t}{2T} \right)$ (Example 4.2) is piecewise continuous on $[0, T]$ (jumps at $t = 0, T$) and smooth on $(0, T)$: $f \in PS[0, T]$.

(c) The square wave (Example 4.3) is piecewise continuous on $[0, T]$ (jumps at $t = 0, T/2, T$), and smooth on $(0, T/2) \cup (T/2, T)$: $f \in PS[0, T]$.

(d) The triangle function, $f(x) = 1 - |x|$, is piecewise smooth and continuous on $[-1, 1]$: $f \in PS \cap C[-1, 1]$ (Figure 4.6).

(e) The square root function $f(x) = \sqrt{|x|}$ is continuous on $\mathbb{R}$, but not piecewise smooth, because $f'(0^-) = -\infty$ and $f'(0^+) = \infty$: $f \in C(\mathbb{R})$ (Figure 4.6).

(f) The function $f(x) = \sqrt{|x|} \, \text{sgn}(x)$ is continuous on $\mathbb{R}$, but not continuously differentiable, because $f'(0)$ is infinite: $f \in C(\mathbb{R})$ (Figure 4.6). ∎

Just as there are different degrees of continuity and smoothness, there are different degrees of convergence.

Definition 4.2 (Convergence of series). Let $f : [a, b] \to \mathbb{C}$ be a function, $(f_n)_{n=1}^{\infty}$, where $f_n : [a, b] \to \mathbb{C}$, be a sequence of functions, and $S_N = \sum_{n=1}^{N} f_n$ denote the Nth partial sum of the infinite series $\sum_{n=1}^{\infty} f_n$.

(a) The series converges *absolutely* if $\sum_{n=1}^{\infty} |f_n(x)| < \infty$ for all $x \in [a, b]$.

(b) The series converges *pointwise* to f if $\lim_{N \to \infty} S_N(x) = f(x)$ for all $x \in [a, b]$. That is, you can make the error $|f(x) - S_N(x)|$ as small as you like, at any point x, by taking N sufficiently large.

(c) The series converges to f *in norm* if the norm of the error between the partial sums and f goes to zero,

$$\|f - S_N\| \to 0 \text{ as } N \to \infty.$$

When the norm is the L^1 or L^2 norm, the convergence is also called *convergence in mean* or *convergence in mean square*, respectively.

(d) The series converges *uniformly* to f if the supremum (uniform norm) of the error between the partial sums and f goes to zero,

$$\|f - S_N\|_u = \sup_{x \in [a,b]} |f(x) - S_N(x)| \to 0 \text{ as } N \to \infty.$$

We shall present some important connections between regularity (continuity and smoothness) and convergence in a series of theorems, without proof.[4] They impose progressively greater regularity on f, from piecewise smooth, to continuous and piecewise smooth, to continuously differentiable (smaller and smaller sets in Figure 4.5), and correspondingly reap the benefits of progressively better convergence. The theorems are followed by several examples.

Theorem 4.2. Let $f : \mathbb{R} \to \mathbb{C}$ be periodic with period L, and piecewise smooth. Let S_N be the Nth partial sum of f's Fourier series. Then $c_n = O(|n|^{-1})$ and, for all x,

$$\lim_{N \to \infty} S_N(x) = \frac{1}{2}[f(x^-) + f(x^+)].$$

At points x where f is continuous, $f(x^-) = f(x^+) = f(x)$ and the theorem says that the partial sums of the Fourier series converge exactly to $f(x)$. At points x_k where f is not continuous it can only have finite jump discontinuities, because it is piecewise continuous (by virtue of being piecewise smooth). The theorem then says that the Fourier series converges to the *average* of the left and right limits, $f(x^-)$ and $f(x^+)$. The addition of piecewise smoothness improves convergence from pointwise almost everywhere (with possible divergence at some points) to pointwise everywhere except at jumps, but with well-defined values at the jumps.

---

[4] For proofs of the convergence theorems, see Folland (1992, pp. 35–37 and 41–42) and Dym and McKean (1972, pp. 31–34).

Theorem 4.3. Let $f : \mathbb{R} \to \mathbb{C}$ be periodic with period L, continuous, and piecewise smooth. Let S_N be the Nth partial sum of f's Fourier series. Then, for all x, $S_N(x)$ converges to $f(x)$ absolutely and uniformly as $N \to \infty$.

The important thing that continuity adds is absolute and uniform convergence of the Fourier series. For the Fourier series, absolute convergence means that

$$\sum_{n=-\infty}^{\infty} |c_n e^{i2\pi nx/L}| = \sum_{n=-\infty}^{\infty} |c_n| < \infty,$$

that is, $(c_n)_{n \in \mathbb{Z}} \in \ell^1$, which means that the Fourier coefficients decay faster than $|n|^{-1}$ as $|n| \to \infty$. Not only do we have pointwise convergence everywhere, by virtue of f's continuity, we also have a sense of how rapidly the series converges, because we know what the asymptotic behavior of the Fourier coefficients is. It can also be shown that the converse is true: if the Fourier series converges absolutely and uniformly, then the function it converges to is continuous.

Theorem 4.4. Let $f : \mathbb{R} \to \mathbb{C}$ be periodic with period L, and $\infty > p \geq 1$.

(a) If $f \in C^{(p-1)}[0, L]$ and $f^{(p-1)}$ is piecewise smooth, then $|n^p c_n| \to 0$ as $|n| \to \infty$. That is, the Fourier coefficients decay faster than $|n|^{-p}$.

(b) If $f \in C^{(\infty)}[0, L]$, then the Fourier coefficients decay faster than $|n|^{-p}$ for all $p \geq 1$.

(c) If $f \in C^{(p)}[0, L]$, then the approximation error is bounded, $\|f - S_N\|_u < KN^{-p+1/2}$, where $K > 0$ is a constant (uniform convergence).

(d) If a set of coefficients (c_n) is bounded such that, except for c_0, they decay faster than $|n|^{-p}$ (i.e., $|c_n| < K|n|^{-p-\alpha}$, where $K > 0$ and $\alpha > 1$), then the Fourier series converges absolutely and uniformly to a function $f \in C^{(p)}[0, L]$.

For $p = 1$, f is continuous and piecewise smooth, and part (a) of the theorem is the same as Theorem 4.3. The smoother a function is, as measured by the degree of continuous differentiability p, the faster the Fourier series converges, as measured by both the asymptotic behavior of the Fourier coefficients (part (a)), and the decay of the approximation error (part (c)). When a function is infinitely continuously differentiable, the ultimate in smoothness, part (b) of the theorem says that the Fourier coefficients (c_n) decay faster than $|n|^{-p}$ for all $p \geq 1$. Such sequences are said to be *rapidly decreasing*. An example is a decaying exponential, $(a^{|n|})_{n \in \mathbb{Z}}$, $|a| < 1$. The most extreme example of continuous differentiability is a pure cosine, $\cos 2\pi kx$. Its coefficients are (trivially) rapidly decreasing (identically zero for $|n| > k$). Finally, part (d) gives the converse of part (a): the faster the Fourier coefficients decay, the smoother the resulting function is. All the results of this chapter on smoothness and convergence are tabulated in Table 4.1.[5]

---

[5] Additional results on smoothness and convergence of the Fourier series are given in Champeney (1987, pp. 156–162) and Gasquet and Witomski (1999, p. 46).

TABLE 4.1 Relationships between smoothness and Fourier series convergence, summarizing Theorems 4.1, 4.2, 4.3, and 4.4. As f becomes more regular, $p \geq 1$, the Fourier coefficients decay more rapidly and the convergence of the Fourier series improves.

If f is	then the coefficients	and convergence $S_N \to f$ is
L^1	$c_n \to 0$	uncertain
L^2	$(c_n) \in \ell^2$	in L^2 norm
PS	$nc_n \leq K$	pointwise, $S_N(x) \to \frac{1}{2}\left[f(x^+) + f(x^-)\right]$
$C^{(p-1)}, f^{(p-1)} \in PS$	$n^p c_n \to 0$	absolute, uniform
$C^{(p)}$		$\|S_N - f\|_u \leq K N^{-p+\frac{1}{2}}$
$C^{(\infty)}$	(c_n) rapidly decreasing	$\|S_N - f\|_u$ rapidly decreasing

In the examples that follow, we will consider the square wave, which is piecewise smooth, the triangle wave, which is piecewise smooth and continuous, and a parabolic wave, which can be continuously differentiated once and has a piecewise smooth derivative.

Example 4.5 (Square wave). The Nth partial sum of the Fourier series for the square wave is (Equation 4.10)

$$S_N(t) = \sum_{n=-N}^{N} \frac{1-(-1)^n}{i\pi n} e^{i2\pi n/T} = \sum_{n=1,3,5\ldots}^{N} \frac{4}{\pi n} \sin \frac{2\pi n t}{T}. \qquad (4.11)$$

A few of these partial sums are shown in Figure 4.7.

As more terms are added, the error norm $\|f - S_N\|_2$ decays, indicating that partial sums are converging to f in the L^2 norm. Moreover, the partial sums look more like the original square wave, evidence of pointwise convergence. Right at the points of discontinuity, $t = 0, T/2, T, \ldots$, where the square wave jumps between $+1$ and -1, the partial sums take on the average value, zero, as predicted by Theorem 4.2. That this should be so is apparent from Equation 4.11, where each sine wave is zero at $t = 0, \pm\frac{T}{2}, \pm T, \ldots$.

What is also noticeable is the overshoot in the partial sum near the jumps. As N increases and the partial sums get "squarer," the peaks narrow and crowd closer to the jumps, but their amplitudes do not seem to decrease. We will analyze this in detail later, and find that there is a finite asymptotic value for the peak of the overshoot. The Fourier series of the square wave converges in norm and pointwise, but not uniformly (the maximum error does not decrease). Neither is convergence absolute, for the sum

$$\sum_{n=-\infty}^{\infty} |c_n| = \frac{4}{\pi}\left(1 + \frac{1}{3} + \frac{1}{5} + \ldots\right)$$

diverges. ∎

4.3 CONVERGENCE OF THE FOURIER SERIES

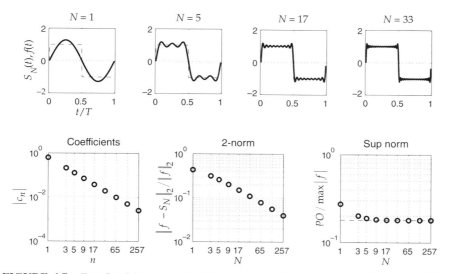

FIGURE 4.7 *Top:* Partial sums of the Fourier series of a square wave. The value of the sum at the discontinuity is the average of the extremes. *Bottom, left to right:* Logarithmic plots of the coefficient decay, $|c_n|$ vs. n, the error norm $\|f - S_N\|_2$ (as a fraction of $\|f\|_2$), and the peak overshoot (as a fraction of max $|f|$). The Fourier series exhibits pointwise and norm convergence, but not uniform convergence.

Example 4.6 (Triangle wave). One period of a triangular waveform is described by the function

$$f(t) = \begin{cases} t, & \frac{T}{2} > t \geq 0 \\ T - t, & T \geq t \geq \frac{T}{2} \end{cases}. \qquad (4.12)$$

The Fourier coefficients are

$$c_0 = \frac{1}{T} \int_0^{T/2} t \, dt + \frac{1}{T} \int_{T/2}^T (T - t) \, dt = \frac{T}{4} \qquad (4.13a)$$

$$\begin{aligned} c_n &= \frac{1}{T} \int_0^{T/2} t \, e^{-i2\pi nt/T} \, dt + \frac{1}{T} \int_{T/2}^T (T - t) \, e^{-i2\pi nt/T} \, dt \\ &= (-1)^n \frac{T}{4} \frac{1 - (-1)^n + i\pi n}{\pi^2 n^2} + (-1)^n \frac{T}{4} \frac{1 - (-1)^n - i\pi n}{\pi^2 n^2} \\ &= -\frac{T}{2} \frac{1 - (-1)^n}{\pi^2 n^2} = \begin{cases} -\frac{T}{\pi^2 n^2}, & n = \pm 1, \pm 3, \ldots \\ 0, & n = \pm 2, \pm 4, \ldots \end{cases}. \end{aligned} \qquad (4.13b)$$

As with the square wave, the even-indexed coefficients are zero. You are invited to calculate the coefficients of the classical series and compare them with those of the square wave. Several partial sums are shown in Figure 4.8.

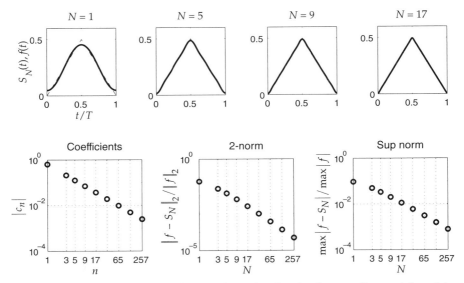

FIGURE 4.8 *Top:* Partial sums of the Fourier series of a triangle wave. *Bottom, left to right:* Logarithmic plots of the coefficient decay, $|c_n|$ vs. n, the norm $\|f - S_N\|_2$ (as a fraction of $\|f\|_2$), and the norm $\max |f - S_N|$ (as a fraction of $\max |f|$). The Fourier series converges pointwise, in norm, and uniformly.

In keeping with Theorem 4.3, the maximum error between the partial sums and the original function continues to decrease as more terms are included, indicating that the series converges uniformly. Also, $|c_n|$ is $O(|n|^{-2})$, so

$$\sum_{n=-\infty}^{\infty} |c_n| = \frac{T}{4} + \frac{2T}{\pi^2}\left(1 + \frac{1}{4} + \frac{1}{9} + \cdots\right)$$

converges, and the Fourier series converges absolutely as well as uniformly. ∎

Example 4.7 (Parabolic wave). A parabolic waveform (Figure 4.9) is described by the function

$$f(t) = \begin{cases} \frac{1}{2}\left(x - \frac{T}{4}\right)^2 - \frac{T^2}{32}, & \frac{T}{2} > x \geq 0 \\ -\frac{1}{2}\left(x\frac{3T}{4}\right)^2 + \frac{T^2}{32}, & T > x \geq \frac{T}{2} \end{cases} \quad (4.14a)$$

and its Fourier coefficients are

$$c_0 = 0 \quad (4.14b)$$

$$c_n = \frac{T^2}{4}\frac{i(1-(-1)^n)}{\pi^3 n^3} = \begin{cases} \frac{iT^2}{2\pi^3 n^3}, & n = \pm 1, \pm 3, \ldots \\ 0, & n = \pm 2, \pm 4, \ldots \end{cases}. \quad (4.14c)$$

4.3 CONVERGENCE OF THE FOURIER SERIES   195

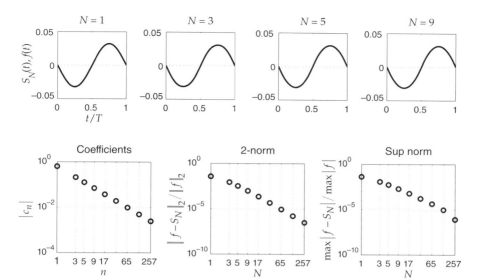

FIGURE 4.9 *Top:* Partial sums of the Fourier series of a parabolic wave. *Bottom, left to right:* Logarithmic plots of the coefficient decay, $|c_n|$ vs. n, the norm $\|f - S_N\|_2$ (as a fraction of $\|f\|_2$), and the norm $\max |f - S_N|$ (as a fraction of $\max |f|$). The Fourier series converges pointwise, in norm, and uniformly.

This function is continuously differentiable ($C^{(1)}$) and its derivative is piecewise smooth (see the problems). The Fourier series converges rapidly, as can be seen in Figure 4.9, where even the third partial sum is barely distinguishable from the original function. ∎

Figure 4.10 displays the convergence of these three series on common axes and demonstrates the connection between smoothness and convergence. From their algebraic expressions, we see that the coefficients for the square, triangle, and parabolic

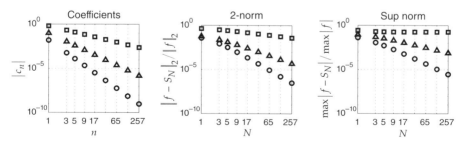

FIGURE 4.10 Comparing rates of decay of the coefficients $|c_n|$, the error norm $\|f - S_N\|_2$, (as a fraction of $\|f\|_2$), and the norm $\max |f - S_N|$ (as a fraction of $\max |f|$), for the Fourier series of the square (squares), triangle (triangles), and parabolic (circles) waves. For the square wave, peak overshoot is shown for maximum error. The smoother the function, the more rapidly the series converges.

waves decay increasingly rapidly, as $|n|^{-1}$, $|n|^{-2}$, and $|n|^{-3}$, respectively. A numerical calculation of the slopes of the asymptotes showed that the decay of the L^2 norm followed a similar pattern, $O(N^{-1/2})$ for the square wave, $O(N^{-3/2})$ for the triangle wave, and $O(N^{-5/2})$ for the parabolic wave. Theorem 4.4. does not apply to the square wave, because it is not even piecewise smooth. For the triangle wave, the uniform convergence is predicted to be at least as fast as $N^{-1/2}$. In fact, our numerical calculation gave $\|f - S_N\|_\infty = O(N^{-1})$. For the parabolic wave, the uniform convergence is predicted to be $O(N^{-3/2})$, and we observed N^{-2} in our calculation.

A closer look at jump discontinuities

We observed, in the case of the square wave, a lack of uniform convergence. The peak overshoot of the partial sum S_N near the jump appears to approach a constant value with increasing N. However, the frequency of the ripple increases and the peak squeezes toward the jump, $t = T/2$, leading one to think that as $N \to \infty$ the peak might become so narrow that it simply "evaporates" in the limit. The verification that this does, in fact, occur is the substance of the proof of Theorem 4.2. To explore the idea, consider the following form of the partial sum for an arbitrary Fourier series,

$$S_N(x) = \sum_{n=-N}^{N} \langle f, e_n \rangle e_n = \sum_{n=-N}^{N} \left[\int_0^L f(\xi) e_n^*(\xi) d\xi \right] e_n(x).$$

Exchanging the order of summation and integration (which we can always do when the integral and sum have finite limits),

$$S_N(x) = \int_0^L f(\xi) \left[\sum_{n=-N}^{N} e_n^*(\xi) e_n(x) \right] d\xi = \frac{1}{L} \int_0^L f(\xi) \left[\sum_{n=-N}^{N} e^{i2\pi n(x-\xi)/L} \right] d\xi.$$

The sum is a form of geometric series (Equation 1.27), which works out to

$$\sum_{n=-N}^{N} e^{-i2\pi n(x-\xi)/L} = e^{i2\pi N(x-\xi)/L} \sum_{n=0}^{2N} e^{-i2\pi n(x-\xi)/L}$$

$$= e^{i2\pi N(x-\xi)/L} \frac{1 - e^{-i2\pi(2N+1)(x-\xi)/L}}{1 - e^{-i2\pi(x-\xi)/L}}$$

$$= e^{i2\pi N(x-\xi)/L} \frac{e^{-i\pi(2N+1)(x-\xi)/L}}{e^{-i\pi(x-\xi)/L}} \frac{\sin \frac{\pi(2N+1)(x-\xi)}{L}}{\sin \frac{\pi(x-\xi)}{L}}$$

$$= \frac{\sin \frac{\pi(2N+1)(x-\xi)}{L}}{\sin \frac{\pi(x-\xi)}{L}} = D_{2N+1}\left(\frac{x-\xi}{L}\right), \quad (4.15)$$

a Dirichlet kernel. The Dirichlet kernel $D_{2N+1}(x)$ is plotted, for several values of N, in Figure 4.11.

It is a periodic function, of course, since it is the sum of a Fourier series. Zero crossings occur at integer multiples of $\frac{1}{2N+1}$, except when x is an integer, in which

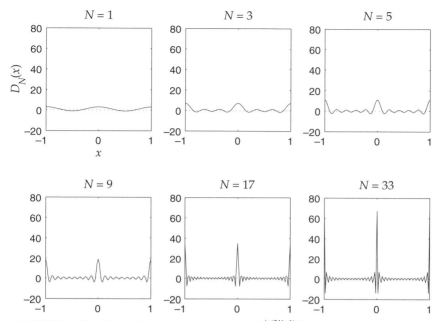

FIGURE 4.11 The Dirichlet kernel, $D_{2N+1}(x) = \frac{\sin(2N+1)\pi x}{\sin \pi x}$, for various values of N.

case L'Hospital's rule gives the value $2N + 1$. Moreover, integrating the defining series term-by-term over one period gives

$$\int_{-1/2}^{1/2} D_{2N+1}(x)dx = \int_{-1/2}^{1/2} \sum_{n=-N}^{N} e^{-i2\pi nx} dx = \sum_{n=-N}^{N} \int_{-1/2}^{1/2} e^{-i2\pi nx} dx,$$

and all the terms except $n = 0$ integrate to zero, leaving

$$\int_{-1/2}^{1/2} D_{2N+1}(x)dx = 1. \tag{4.16}$$

So, while the Dirichlet kernel becomes narrower and higher as N increases, it always has unit area.

Using the Dirichlet kernel, the partial sum is compactly written as a convolution,[6]

$$S_N(x) = \frac{1}{L} \int_0^L f(\xi) D_{2N+1}\left(\frac{x-\xi}{L}\right) d\xi. \tag{4.17}$$

The integral is not hard to visualize when f is a unit-period square wave.

---

[6] We will see in the next chapter that convolution integrals model the input–output behavior of linear, time-invariant systems, such as electronic filters. Truncating the Fourier series to make the partial sum removes higher frequency Fourier components. Convolution with the Dirichlet kernel is a lowpass filter.

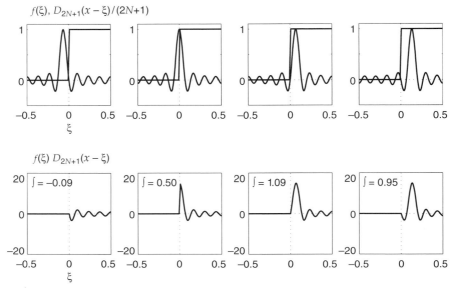

FIGURE 4.12 Graphical interpretation of the convolution of a square wave (period $L = 1$) with the Dirichlet kernel. *Top row:* The reversed and shifted kernel, $D_{2N+1}(x - \xi)$, is overlaid on the square wave. *Bottom row:* The product of the square wave and the kernel. The area under this function is the value of the convolution $f * D_{2N+1}$ for the given shift, x. Left to right: $x = -\frac{1}{2N+1}$, negative overshoot; $x = 0$, the integral is $\frac{1}{2}$, the average $\frac{f(0^-)+f(0^+)}{2}$; $x = \frac{1}{2N+1}$, the integral overshoots 1; $x = \frac{2}{2N+1}$, the integral undershoots 1. The amount of overshoot is approximately 9% of the height of the jump.

The kernel is flipped end for end, making $D_{2N+1}(-\xi/L)$, then shifted to $\xi = x$ to make $D_{2N+1}\left(\frac{x-\xi}{L}\right)$. It is then multiplied by $f(\xi)$ and the area under the product is calculated. With reference to Figure 4.12, the shifted kernel and the square wave are shown superimposed in the top row of graphs. In the bottom row are displayed the products of the square wave and kernel. As we go from left to right, the center of the kernel is to the left of the jump, over the jump, and to the right of the jump.

- When the kernel is centered on the jump discontinuity ($x = 0$), the product is just half the kernel and the integral is half the area under the kernel, $\frac{1}{2}$. This is another view of the Fourier series converging to the average $\frac{f(0^-)+f(0^+)}{2}$ at $x = 0$.
- When $x = \frac{1}{2N+1}$, the main lobe of the kernel is completely on the high side of the jump, causing the graph of the product to be unbalanced. The area (calculated using MATLAB) is 1.09—we have overshoot.
- Shifting the kernel farther, to $x = \frac{2}{2N+1}$, brings a negative lobe to the high side of the jump, which will subtract area from the product—so much, in fact, that now we have undershoot, as the area drops to 0.95.

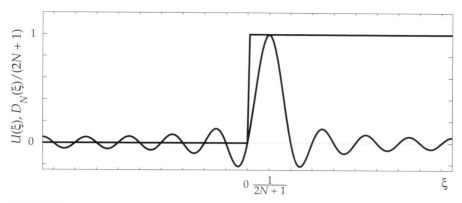

FIGURE 4.13 Setup for calculating maximum overshoot at a jump discontinuity. The Dirichlet kernel is convolved with a unit step function. The peak overshoot occurs approximately where the first zero crossing of the kernel coincides with the jump. The zero crossings for D_{2N+1} are spaced at intervals of $\frac{1}{2N+1}$.

Continuing in this way, the value of the convolution, and hence the partial sum, will oscillate between overshoot and undershoot. The lobes of the kernel become smaller as the shift increases, and the magnitude of the overshoots and undershoots will decrease as we move away from the jump, as we observed in Figure 4.7.

It remains to calculate the maximum value of the overshoot. From looking at the graph, we expect this to occur at a point $\xi = x_p \approx \frac{1}{2N+1}$, which places the main lobe just to the right of the jump, under the high side of the square wave (Figure 4.13).

Integrating over one period of the square wave, the peak overshoot PO is

$$PO = \int_{-\frac{1}{2}}^{\frac{1}{2}} U(\xi) D_{2N+1}(x_p - \xi)\, d\xi \;-\; 1 = \int_0^{\frac{1}{2}} D_{2N+1}(x_p - \xi)\, d\xi \;-\; 1,$$

where $U(x)$ is the unit step function. The overshoot is extremized by setting the derivative of PO to zero.

$$\frac{d}{dx_p} PO = \int_0^{\frac{1}{2}} \frac{d}{dx_p} D_{2N+1}(x_p - \xi)\, d\xi$$

$$= \int_0^{\frac{1}{2}} D'_{2N+1}(x_p - \xi)\, d\xi = D_{2N+1}(x_p) - D_{2N+1}(x_p - 1/2) = 0.$$

A correct value of x_p may be determined numerically, then used to numerically evaluate the peak overshoot integral. Figure 4.14 shows the result for various values of N. The peak overshoot very quickly settles down to around 9%. This is the well-known *Gibbs phenomenon*.

4.4 FOURIER SERIES PROPERTIES AND THEOREMS

The Fourier series has several properties that facilitate calculations and interpretation. We have already met similar ones in studying the DFT, and throughout this section

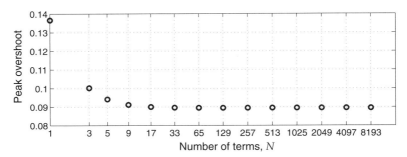

FIGURE 4.14 Gibbs phenomenon. The maximum overshoot at a jump discontinuity is approximately 9% of the height of the jump, regardless of how many terms in the series are summed.

you are invited to compare the Fourier series theorems with their DFT counterparts. We assume that a function f is periodic with period L and absolutely integrable (unless otherwise noted) over one period. We will use the notation $f \longmapsto (f_n)$ to denote the mapping from a function f to its Fourier coefficients. If the Fourier series with coefficients (f_n) converges (in any sense) to f, we write $f \longleftrightarrow (f_n)$ and say that the function f and its coefficient sequence (f_n) are a Fourier series pair.

Linearity

The Fourier coefficients are inner products, $f_n = \langle f, \phi_n \rangle$, where $\phi_n(x) = \frac{1}{L} \exp\left(\frac{i 2\pi n x}{L}\right)$, and the inner product is linear, $\langle af + bg, \phi_n \rangle = a \langle f, \phi_n \rangle + b \langle g, \phi_n \rangle$. The Fourier series, $\sum_{n \in \mathbb{Z}} f_n \phi_n$, also has the form of an inner product, and is linear. Consequently, the Fourier expansion is a linear mapping between a function and its Fourier coefficients.

Theorem 4.5 (Linearity). Let $f \longleftrightarrow (f_n)$, $g \longleftrightarrow (g_n)$ be Fourier series pairs, and let $a, b \in \mathbb{C}$ be constants. Then

$$af + bg \longleftrightarrow a(f_n) + b(g_n) = (af_n + bg_n). \qquad (4.18)$$

In Example 4.1 we calculated the Fourier coefficients of $f(t) = A + B \cos 2\pi t/T$. The function is the sum of a constant A and a scaled cosine $B \cos 2\pi t/T$. The constant has only a single Fourier coefficient, $c_0 = A$. A cosine at frequency $v = 1/T$ has two coefficients, $c_{-1} = c_1 = \frac{1}{2}$. The Fourier coefficients of the scaled cosine $B \cos 2\pi t/T$ are, by linearity, $B \cdot \frac{1}{2}$. And, by linearity, the coefficients of $A + B \cos 2\pi t/T$ are the sum of the coefficients of the two terms,

$$A + B \cos 2\pi t/T \longleftrightarrow (c_n) = \begin{cases} A, & n = 0 \\ B/2, & n = \pm 1 \\ 0, & \text{otherwise} \end{cases}.$$

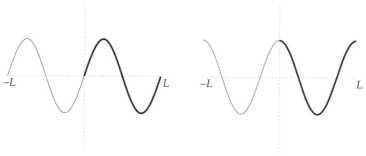

FIGURE 4.15 Evenness and oddness for a periodic function. An odd function (*left*) has $f(L-x) = -f(x)$, and an even function (*right*) has $f(L-x) = f(x)$.

Symmetries
A function f with period L is even if $f(x) = f(L-x)$ and odd if $f(x) = -f(L-x)$. (Figure 4.15). Likewise, a complex sequence $(c_n)_{n \in \mathbb{Z}}$ is even if $c_n = c_{-n}$, odd if $c_n = -c_{-n}$, and Hermitian if $c_n = c_{-n}^*$.

Theorem 4.6 (Fourier series symmetries). Let f be a function and (c_n) be its Fourier coefficients. These statements and their converses are true:

(a) If f is real, the Fourier coefficients are Hermitian: $c_n = c_{-n}^*$.

(b) If f is even (odd), then (c_n) is even (odd).

(c) If f is real and even (real and odd), then (c_n) is real and even (imaginary and odd).

Proof: The proofs are very similar to those for the DFT.

(a) Directly calculate c_{-n}^*,

$$c_{-n}^* = \left[\frac{1}{L}\int_0^L f(x) \exp\left(-\frac{i2\pi(-n)x}{L}\right) dx\right]^* = \frac{1}{L}\int_0^L f^*(x) \exp\left(-\frac{i2\pi nx}{L}\right) dx$$

but because f is real, $f = f^*$, so

$$c_{-n}^* = \frac{1}{L}\int_0^L f(x) \exp\left(-\frac{i2\pi nx}{L}\right) dx = c_n.$$

Now suppose $c_n = c_{-n}^*$. Then

$$\sum_{n=-\infty}^{\infty} c_n e^{j2\pi nx/L} = c_0 + \sum_{n=1}^{\infty} c_n e^{j2\pi nx/L} + c_{-n} e^{-j2\pi nx/L}$$

$$= c_0 + \sum_{n=1}^{\infty} c_n e^{j2\pi nx/L} + \left(c_n e^{j2\pi nx/L}\right)^*$$

$$= c_0 + \sum_{n=1}^{\infty} 2\mathrm{Re}\left\{c_n e^{j2\pi nx/L}\right\},$$

which is real.

(b) If f is even, then the odd part of f is zero. Calculate the odd part of f in terms of the Fourier coefficients:

$$f_o = \frac{1}{2}[f(x) - f(-x)]$$

$$= \frac{1}{2}\left[\sum_{n=-\infty}^{\infty} c_n \exp\left(\frac{i2\pi nx}{L}\right) - \sum_{n=-\infty}^{\infty} c_n \exp\left(\frac{i2\pi n(-x)}{L}\right)\right]$$

$$= \frac{1}{2}\sum_{n=-\infty}^{\infty}(c_n - c_{-n})\exp\left(\frac{i2\pi nx}{L}\right) = 0$$

for all x, which requires that $c_n = c_{-n}$, that is, the coefficient sequence is even. Now suppose that $c_n = c_{-n}$, then

$$c_0 + \sum_{n=1}^{\infty} c_n e^{j2\pi nx/L} + c_{-n} e^{-j2\pi nx/L} = c_0 + \sum_{n=1}^{\infty} 2c_n \cos(2\pi nx/L),$$

which is even. The proofs for odd f and an odd coefficient sequence are the same.

(c) If f is real, then (c_n) is Hermitian (part (a)). If, additionally, f is even, we know that (c_n) is even (part (b)) and therefore real because (c_n) is also Hermitian. Likewise, if f is odd as well as real, then (c_n) is odd, and it is also imaginary because (c_n) is Hermitian. The proof of the converse is the same. ∎

You are invited to consider for yourself the less common case of f being imaginary and even, or imaginary and odd. All four cases are summarized in Figure 4.16. These symmetries also hold for f on an interval $[-L/2, L/2]$, with the usual definitions of evenness ($f(x) = f(-x)$) and oddness ($f(x) = -f(-x)$).

Example 4.8. The square and parabolic waves (Examples 4.3 and 4.7) are real and odd functions, and their respective Fourier coefficients, $c_n = \frac{2}{in}$ and $c_n = \frac{iL^2}{2\pi^3 n^3}$, $n = \pm 1, \pm 3, \ldots$, are imaginary and odd. The triangle wave (Example 4.6) is real and even, and its Fourier coefficients $c_n = -\frac{T}{\pi^2 n^2}$, $n = \pm 1, \pm 3, \ldots$, are real and even. ∎

The symmetry relationships can save computation time. If you know at the outset that the coefficients are going to be even, or odd, or Hermitian, then you only need to compute, say, the positive-n coefficients and can obtain the rest by symmetry. The symmetries are also beneficial for error checking. If you calculate a set of Fourier

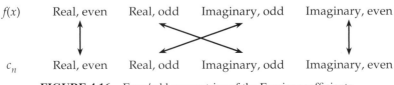

FIGURE 4.16 Even/odd symmetries of the Fourier coefficients.

coefficients and find that they violate symmetries that you know must hold, then you have caught an error and can correct it.

There are also nice symmetry relationships for the classical series: an even function has a cosine series, and an odd function has a sine series. Proof is left to the problems.

Example 4.9. We have seen that the square wave's Fourier components are of the form $\frac{4}{\pi n} \sin \frac{2\pi nt}{T}$ (Example 4.3), and its Fourier series can therefore be expressed

$$f(t) = \sum_{n=1,3,\ldots} \frac{4}{\pi n} \sin \frac{2\pi nt}{T}.$$

It can similarly be shown that the Fourier components for the triangle wave are

$$-\frac{T}{\pi^2 n^2} e^{i2\pi nt/T} - \frac{T}{\pi^2 n^2} e^{i2\pi nt/T} = -\frac{2T}{\pi^2 n^2} \cos \frac{2\pi nt}{T}, \quad n = 1, 3, \ldots,$$

so its Fourier series is

$$f(t) = \frac{T}{4} - \sum_{n=1,3,\ldots} \frac{2T}{\pi^2 n^2} \cos \frac{2\pi nt}{T}. \quad \blacksquare$$

The special forms of the Fourier series for even and odd functions can be used to derive Fourier expansions for functions on $[0, L]$, which themselves lack even or odd symmetry.

Example 4.10. The ramp function $f(x) = x/L$, $L > x \geq 0$ extends periodically to the sawtooth wave shown in Figure 4.17. The Fourier coefficients for this function,

$$c_n = \frac{1}{L} \int_0^L \frac{x}{L} e^{-i2\pi nx/L} dx = \frac{i}{2\pi n},$$

decay slowly, $O(|n|^{-1})$, owing to the jump discontinuities at $x = 0, \pm L, \pm 2L, \ldots$. We know that the Fourier series will not converge uniformly; there will be overshoot in the partial sums near the jumps. A representation with better convergence properties is obtained by symmetrically reflecting the ramp about $x = L$, as shown in Figure 4.17. This new function,

$$f_e(x) = \begin{cases} \dfrac{x}{L}, & L > x \geq 0 \\ 2 - \dfrac{x}{L}, & 2L > x \geq L \end{cases}$$

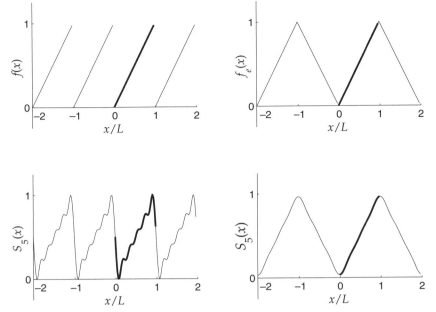

FIGURE 4.17 Deriving the Fourier series for a function on an interval. *Left:* A piecewise continuous function and its periodic extension. Because of the jump discontinuity, the partial sums (S_5 shown) display overshoot. *Right:* The function is reflected and periodically extended to a piecewise smooth function with even symmetry, equal to the original function on $[0, L]$. The resulting Fourier expansion has better convergence properties.

periodically extends to a triangle wave, which is piecewise smooth and has even symmetry. The Fourier cosine series coefficients are

$$a_n = \frac{2}{2L}\int_0^{2L} f_e(x) \cos\frac{2\pi n x}{2L} dx = \frac{1}{L} \cdot 2 \int_0^L \frac{x}{L} \cos\frac{\pi n x}{L} dx$$

$$= \begin{cases} 1, & n = 0 \\ 0, & n \text{ even} \\ -\dfrac{4}{\pi^2 n^2}, & n \text{ odd} \end{cases}.$$

With this Fourier series,

$$f_e(x) = \frac{1}{2} - \sum_{n=1,3,\ldots} \frac{4}{\pi^2 n^2} \cos\frac{n\pi x}{L}$$

we just restrict x to $[0, L]$ to obtain a series for $f(x)$. The coefficients $(a_n)_{n>0}$ decay $O(n^{-2})$, giving absolute and uniform convergence. The superiority of this representation is illustrated by the partial sums in Figure 4.17. You are also invited to compare this even symmetrization technique with the discrete cosine transform (Section 3.6). ∎

4.4 FOURIER SERIES PROPERTIES AND THEOREMS

Parseval's formula

For functions in $L^2[0, L]$, the Fourier series preserves inner products. Energy conservation follows as a special case.

Theorem 4.7 (Parseval). If $f, g \in L^2[0, L]$ have complex Fourier coefficients (f_n) and (g_n), respectively, then

$$\frac{1}{L} \int_0^L f(x) g^*(x) \, dx = \sum_{n=-\infty}^{\infty} f_n g_n^* \qquad (4.19a)$$

and taking $f = g$,

$$\frac{1}{L} \int_0^L |f(x)|^2 \, dx = \sum_{n=-\infty}^{\infty} |f_n|^2. \qquad (4.19b)$$

Proof: This is just a particular case of Equation 2.43,

$$\langle f, g \rangle = \sum_{n=-\infty}^{\infty} \langle f, e_n \rangle \langle g, e_n \rangle^*.$$

For the orthonormal basis $\{e_n\}$, use $e_n = \frac{1}{\sqrt{L}} \exp\left(\frac{i 2\pi n x}{L}\right)$. The inner products are $\langle f, e_n \rangle = \sqrt{L} f_n$ and similarly for g. Then,

$$\langle f, g \rangle = \int_0^L f(x) g^*(x) \, dx = L \sum_{n=-\infty}^{\infty} f_n g_n^*.$$

Substituting $f = g$ and $f_n = g_n$ yields the energy conservation formula. ∎

Example 4.11. Parseval's formula implies that if f and g are orthogonal functions, their respective Fourier coefficient sequences (f_n) and (g_n) will also be orthogonal. We know, for example, that $\cos 2\pi x$ and $\sin 2\pi x$ are orthogonal on $[0, 1]$. Their Fourier coefficients c_n and s_n, calculated over the same interval, are

$$c_n = \begin{cases} \frac{1}{2}, & n = -1 \\ \frac{1}{2}, & n = 1 \\ 0, & \text{otherwise} \end{cases}$$

$$s_n = \begin{cases} \frac{i}{2}, & n = -1 \\ -\frac{i}{2}, & n = 1 \\ 0, & \text{otherwise} \end{cases}$$

and their inner product is

$$\sum_{n=-\infty}^{\infty} c_n s_n^* = \frac{1}{2} \cdot \frac{i}{2} + \frac{1}{2} \cdot -\frac{i}{2} = 0.$$
∎

Area theorem

The average value of a function is proportional to the area under one period of its graph. If the function models a voltage or current waveform, then the average has a convenient physical interpretation as the constant baseline or "DC" value.

Theorem 4.8 (Area theorem). Let f be a function and (c_n) be its Fourier coefficients. Then,

$$c_0 = \frac{1}{L} \int_0^L f(x)\, dx \tag{4.20a}$$

$$f(0) = \sum_{n=-\infty}^{\infty} c_n. \tag{4.20b}$$

Proof is left to the problems.

Example 4.12. The area under one period of the triangle wave in Example 4.6 is $A = \frac{1}{2} \cdot T \cdot \frac{1}{2} = \frac{T}{4}$, the same as c_0 of its Fourier series. ∎

Shifting

For a function f defined on the real line, a shift $f(x - r)$ is a translation along the x axis. When f is defined on an interval $[0, L]$, a shift is interpreted modulo-L, similar to the interpretation for the DFT. Equivalently, it is a translation of the periodic extension of f.

Theorem 4.9 (Shift theorem). Let $f \longleftrightarrow (c_n)$ be a Fourier series pair. Then, for the shifted function $f(x - r)$,

$$f(x - r) \longleftrightarrow \left(e^{-i2\pi rn/L} c_n\right). \tag{4.21}$$

Proof is left to the problems.

Example 4.13. A particularly simple demonstration of the shift theorem uses the sine and cosine functions. The Fourier coefficients for $\cos 2\pi x$ are $c_1 = c_{-1} = \frac{1}{2}$, and for $\sin 2\pi x$, $s_1 = -s_{-1} = -\frac{i}{2}$. Now, because $\sin 2\pi x = \cos(2\pi x - \frac{\pi}{2}) = \cos 2\pi(x - \frac{1}{4})$, the shift theorem predicts

$$s_n = \exp\left(-2\pi n \frac{1}{4}\right) c_n = e^{-i\pi n/2} c_n.$$

In particular,

$$s_1 = e^{-i\pi/2} c_1 = -i\frac{1}{2}$$

$$s_{-1} = e^{-i\pi(-1)/2} c_{-1} = +i\frac{1}{2}.$$

∎

It is important to note that shifting does not affect the magnitude of the Fourier coefficients:

$$\left|e^{-i2\pi rn/L} c_n\right| = |c_n|,$$

since $\left|e^{-i2\pi rn/L}\right| = 1$. In the Fourier series,

$$\sum_{n=-\infty}^{\infty} e^{-i2\pi rn/L} c_n e^{j2\pi nx/L} = \sum_{n=-\infty}^{\infty} c_n e^{j2\pi n(x-r)/L}.$$

The mixture of basis functions—magnitudes and relative phases—specified by the coefficients (c_n) is unmodified by the shift. Rather, the phase shift $e^{-i2\pi rn/L}$ is manifested in a translation of the basis functions, and combining them with the c_n results in a translation of f.

Differentiation
A series $\sum_{n=1}^{\infty} r_n(x)$ so $r_n(x)$ that converges absolutely and uniformly to a function f can be differentiated term by term, and the derived series converges to the derivative of f,[7]

$$\sum_{n=1}^{\infty} r'_n(x) = f'(x).$$

If a function is continuous and piecewise smooth, its Fourier series is absolutely and uniformly convergent. It may be differentiated term by term to yield the Fourier series of the function's derivative.[8]

Theorem 4.10 (Derivative theorem). Let f be continuous and piecewise smooth and let its derivative f' be piecewise smooth. Let (c_n) be the Fourier coefficients of f. Then, the Fourier coefficients (c'_n) of f' are given by

$$c'_n = \frac{i2\pi n}{L} c_n \qquad (4.22)$$

and the Fourier series $\sum_{n=-\infty}^{\infty} c'_n e^{-2\pi nx/L}$ converges pointwise (Theorem 4.2) to f'.

Proof is left to the problems.

Integration
The integral $F(x) = \int_0^x f(\xi)\, d\xi$ of an absolutely integrable function f, $x \in [0, L]$, is also absolutely integrable:

$$\int_0^L \left|\int_0^x f(\xi)d\xi\right| dx < L \max_{x \in [0,L]} \left|\int_0^x f(\xi)d\xi\right| = L \int_0^L |f(\xi)|\, d\xi dx < \infty.$$

---

[7] Differentiability of series is a standard topic in real analysis. See, for example, Rosenlicht (1968, pp. 140–150).

[8] As long as f' is absolutely integrable, it has Fourier coefficients given by (Equation 4.22). See Champeney (1987, pp. 164–165) for this and other alternative statements of the derivative theorem. The conditions imposed on f in our version of the theorem guarantee pointwise convergence of the derived series to f'.

Thus, F also has a Fourier series. To find out what the Fourier coefficients of F are, formally integrate the Fourier series for f term by term:

$$F(x) = \int_0^x \sum_{n=-\infty}^{\infty} c_n e^{i2\pi n\xi/L} \, d\xi = \int_0^x c_0 \, dx + \sum_{n\neq 0} c_n \int_0^x e^{i2\pi n\xi/L} \, d\xi$$

$$= c_0 x + \sum_{n\neq 0} \frac{L}{i2\pi n} c_n \left[e^{i2\pi nx/L} - 1 \right]$$

$$= c_0 x - \sum_{n\neq 0} \frac{L}{i2\pi n} c_n + \sum_{n\neq 0} \frac{L}{i2\pi n} c_n e^{i2\pi nx/L}, \quad x \in [0, L].$$

The restriction of x to the bounded interval $[0, L]$ is important, so that the ramp term $c_0 x$ does not grow without bound. We may conveniently subtract $c_0 x$ from F, leaving

$$F(x) - c_0 x = -\sum_{n\neq 0} \frac{L}{i2\pi n} c_n + \sum_{n\neq 0} \frac{L}{i2\pi n} c_n e^{i2\pi nx/L}.$$

The first term, $-\sum_{n\neq 0} \frac{L}{i2\pi n} c_n$, converges. By the Riemann–Lebesgue lemma, the coefficients $c_n \to 0$ as $|n| \to \infty$. Thus $|c_n/n| \to 0$ faster than $O\left(|n|^{-1}\right)$, guaranteeing convergence. The sum is independent of x; it must be the constant part of $F - c_0 x$.

The second term, $\sum_{n\neq 0} \frac{L}{i2\pi n} c_n e^{i2\pi n\xi/L}$, is a Fourier series. Its coefficients also decay faster than $O(|n|^{-1})$, enabling it to converge pointwise to something at least piecewise smooth. The series lacks a constant term, so it must represent the nonconstant part of $F - c_0$.

We have essentially proved the following theorem.

Theorem 4.11 (Integral theorem). Let $f \longmapsto (c_n)$, and define the function F by

$$F(x) = \int_0^x f(\xi) \, d\xi, \quad x \in [0, L].$$

Then F has a series representation given by

$$F(x) = C_0 x + \sum_{n=-\infty}^{\infty} C_n e^{i2\pi nx/L}, \quad x \in [0, L], \tag{4.23a}$$

where

$$C_0 = \frac{1}{L} \int_0^L \left(F(x) - c_0 x \right) dx = -\sum_{n\neq 0} \frac{L}{i2\pi n} c_n = -\sum_{n\neq 0} C_n \tag{4.23b}$$

$$C_n = \frac{L}{i2\pi n} c_n, \quad n = \pm 1, \pm 2, \ldots . \tag{4.23c}$$

The Fourier series $\sum_{n=-\infty}^{\infty} C_n e^{i2\pi nx/L}$ converges pointwise to $F(x) - c_0 x$.

The theorem is stated in its most general form, for a function f that is merely absolutely integrable. If additionally f is piecewise smooth, then F is continuous and the coefficients of the integrated series decay at least $O(|n|^{-2})$, guaranteeing absolute and uniform convergence.

The relation of differentiation and integration to the smoothness of a function and the behavior of its Fourier coefficients is of fundamental importance. We have seen already that the smoothness of a function, measured by how many times it can be continuously differentiated, is directly related to the rate of the decay of its Fourier coefficients (Theorem 4.4). Here is how the derivative and integral theorems fit into this picture:

- Differentiation makes a function less smooth (think: corners become jumps). The derivative of a p times differentiable function is $p - 1$ times differentiable. In the Fourier domain, differentiation multiplies the Fourier coefficients by n, decreasing their rate of decay and boosting the high frequency components.
- On the other hand, integration makes a function smoother (jumps become corners and corners become rounded). The integral of a p times differentiable function is $p + 1$ times differentiable. In the Fourier domain, integration divides the Fourier coefficients by n, increasing their rate of decay and attenuating the high frequency components.

Unless a function is infinitely continuously differentiable, repeated differentiation will eventually render it discontinuous. With each differentiation, the Fourier coefficients decay more slowly (by a factor of n), until eventually the Fourier series loses uniform convergence. With one additional differentiation, the derivatives at the jump discontinuities blow up, the Fourier coefficients are now constant or growing with n, and the Fourier series fails to converge to any function in the ordinary sense (but see Chapter 6 for a resolution of this apparent problem).

Example 4.14. The triangle wave in Example 4.6 is the integral of the square wave in Example 4.3, as can be seen by inspection of their graphs, and the square wave is the derivative of the triangle wave. Comparing their Fourier coefficients,

$$c_n = \begin{cases} \frac{2}{in}, & n = \pm 1, \pm 3, \ldots \\ 0, & \text{otherwise} \end{cases} \quad \text{square wave}$$

$$C_n = \begin{cases} \frac{T}{4}, & n = 0 \\ -\frac{T}{\pi^2 n^2}, & n = \pm 1, \pm 3, \ldots \\ 0, & \text{otherwise} \end{cases} \quad \text{triangle wave}$$

note that

$$\frac{2}{in} = \frac{i2\pi n}{T} \cdot -\frac{T}{\pi^2 n^2}$$

in accordance with the derivative and integral theorems. Moreover, for the triangle wave,

$$-\sum_{n \neq 0} C_n = \sum_{n=\pm 1, \pm 3, \ldots} \frac{T}{\pi^2 n^2} = \frac{2T}{\pi^2} \sum_{n=1,3,\ldots} \frac{1}{n^2} = \frac{2T}{\pi^2} \frac{\pi^2}{8} = \frac{T}{4} = C_0,$$

as expected by the integral theorem. ∎

Convolution

Let f, h have period L; the convolution $f \circledast h$ is defined by

$$f \circledast h(x) = \int_0^L f(\xi) h(x - \xi) \, d\xi = \int_0^L f(x - \xi) h(\xi) \, d\xi \qquad (4.24)$$

when the integral exists. Because f and h are periodic, the integral can be taken over any interval of length L without changing the result. Like the DFT, convolution is cyclic, with the shift interpreted modulo-L. The convolution is itself periodic with period L,

$$f \circledast h(x + L) = \int_0^L f(\xi) h(x + L - \xi) \, d\xi = \int_0^L f(\xi) h(x - \xi) \, d\xi = f \circledast h(x). \qquad (4.25)$$

Example 4.15 (Convolution of square waves). Consider the square wave function,

$$f(x) = \begin{cases} 1, & \frac{L}{2} > x \geq 0 \\ 0, & L > x \geq \frac{L}{2} \end{cases}.$$

The convolution, $g = f \circledast f$, is illustrated in Figure 4.18. The function $f(x - \xi)$ is a reversed version of f, and slides across $f(\xi)$ as x is varied. Since f is defined piecewise, it is convenient to perform the integral in steps, based on the value of x. When $\frac{L}{2} > x \geq 0$, the two functions are both nonzero between $\xi = 0$ and $\xi = x$. The integral in this case is

$$\int_0^x 1 \cdot 1 \, d\xi = \xi \Big|_0^x = x.$$

When $L > x \geq \frac{L}{2}$, the functions are both nonzero between $\xi = x - \frac{L}{2}$ and $\xi = \frac{L}{2}$. The integral is

$$\int_{x-L/2}^{L/2} 1 \cdot 1 \, d\xi = \xi \Big|_{x-L/2}^{L/2} = x - L.$$

The result, combining both pieces, is

$$g(x) = \begin{cases} x, & \frac{L}{2} > x \geq 0 \\ x - L, & L > x \geq \frac{L}{2}, \end{cases}$$

which has the shape of a triangle. ∎

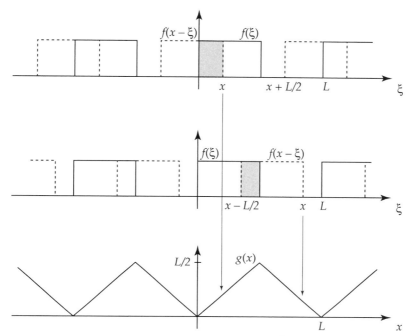

FIGURE 4.18 Convolution of two square waves, $g(x) = \int_0^L f(\xi)f(x-\xi)\,d\xi$. *Top:* Overlap of $f(\xi)$ (solid) and $f(x-\xi)$ (dashed) when $\frac{L}{2} > x \geq 0$. Integration takes place over the shaded overlap between 0 and x. *Center:* When $L > x \geq \frac{L}{2}$, the integration takes place over the shaded overlap between $x - L/2$ and $L/2$. *Bottom:* The result of the convolution is a triangle wave.

Provided that $f \circledast h$ is at least absolutely integrable ($f \circledast h \in L^1$), it has a Fourier series. Let f and h have Fourier coefficients (f_n) and (h_n), respectively. The Fourier coefficients for the convolution are formally derived by inserting the convolution integral into Equation 4.3,

$$f \circledast h \longmapsto \frac{1}{L}\int_0^L \left[\int_0^L f(\xi)h(x-\xi)\,d\xi\right] e^{-i2\pi nx/L}\,dx,$$

then reversing the order of integration (Fubini's theorem),

$$= \int_0^L f(\xi)\left[\frac{1}{L}\int_0^L h(x-\xi)e^{-i2\pi nx/L}\,dx\right]d\xi$$
$$= \left[\int_0^L f(\xi)e^{-i2\pi n\xi/L}\,d\xi\right] h_n \qquad \text{(shift theorem)}$$
$$= L f_n h_n.$$

Likewise, the product fh is periodic and, if it is at least absolutely integrable on $[0, L]$, it has a Fourier series. To calculate its Fourier coefficients, express f in terms of its

TABLE 4.2 Details about convolutions and products of functions. The norm of the convolution is bounded, $\|f \circledast h\| \le \|f\| \|h\|$, where the norms are calculated in the appropriate spaces. When h is bounded ($h \in L^\infty$), the product fh inherits the norm properties of f, that is, if $f \in L^2$ then $fh \in L^2$.

If f is	and h is	then $f \circledast h$ is	and fh is
L^1	L^1	L^1	
L^1	L^2	L^2	
L^1	L^∞	L^∞ and continuous	L^1
L^2	L^2	L^∞ and continuous	L^1

Fourier series and substitute it into Equation 4.3,

$$fh \longmapsto \frac{1}{L}\int_0^L f(x)h(x)e^{-i2\pi nx/L}\,dx$$

$$= \frac{1}{L}\int_0^L h(x)\left[\sum_{k=-\infty}^{\infty} f_k e^{i2\pi kx/L}\right]e^{-i2\pi nx/L}\,dx,$$

then reverse the order of summation and integration,

$$= \sum_{k=-\infty}^{\infty} f_k\left[\frac{1}{L}\int_0^L h(x)e^{-i2\pi(n-k)x/L}\,dx\right]$$

$$= \sum_{k=-\infty}^{\infty} f_k h_{n-k}.$$

This sum is the (noncyclic) convolution of the Fourier coefficient sequences (f_n) and (h_n), denoted $(f_n) * (h_n)$. If the inverse relationships hold as well—that is, if the Fourier series for $f \circledast h$ and fh are convergent—we have

$$f \circledast h \longleftrightarrow (Lf_n h_n) \tag{4.26a}$$
$$fh \longleftrightarrow (f_n) * (h_n). \tag{4.26b}$$

In order for the forward relationships $f \circledast h \longmapsto (Lf_n h_n)$ and $fh \longmapsto (f_n) * (h_n)$ to be true, the convolution $f \circledast h$ and product fh must be at least absolutely integrable. The table in Table 4.2 summarizes the principal cases of interest.[9] It says that the convolution of two functions is at least as well behaved as the component functions; if they have Fourier series, then so does their convolution. This is due to the smoothing effect of integration. (Consider, for example, that the convolution of two square waves, which are discontinuous, is a triangle wave, which is continuous.) On the other hand, the product is no better than its factors, and the existence of a Fourier series for the product is subject to more stringent conditions.

In the reverse direction, convergence of the Fourier series $\sum_n Lf_n h_n \exp\left(\frac{i2\pi nx}{L}\right)$ to $f \circledast h$ and $\sum_n [(f_n) * (h_n)] \exp\left(\frac{i2\pi nx}{L}\right)$ to fh is guaranteed if the coefficient

---
[9] The convolution and product relationships in Table 4.2 are similar to those in Section 5.5. References to proofs are provided there.

sequences $(f_n h_n)$ and $(f_n) * (h_n)$ are at least square summable.[10] Now, the product $(f_n h_n) \in \ell^2$ if one of the coefficient sequences is in ℓ^2 while the other is at least bounded (see Table 4.3 later). This happens if, for example, $f \in L^1$ and $h \in L^2$. The convolution $(f_n) * (h_n) \in \ell^2$ if one of the coefficient sequences is in ℓ^2 while the other is in ℓ^1. This requires either f or h to be at least piecewise smooth while the other is at least in L^2.

We can assemble all these conditions into a theorem.

Theorem 4.12 (Convolution theorem for Fourier series). Let f, h have period L and Fourier coefficients $(f_n), (g_n)$.

(a) If f and h are in $L^1[0, L]$, the convolution $g = f \circledast h$ is in $L^1[0, L]$ and its Fourier coefficients are

$$g_n = L f_n h_n.$$

If f or h is additionally in $L^2[0, L]$, then the Fourier series $\sum_{n=-\infty}^{\infty} g_n e^{i 2 \pi n x / L}$ converges in L^2 to $f \circledast h$.

(b) If f and h are in $L^2[0, L]$, then the product $g = fh$ is in $L^1[0, L]$ and its Fourier coefficients are given by the convolution sum,

$$g_n = \sum_{k=-\infty}^{\infty} f_k h_{n-k} = \sum_{k=-\infty}^{\infty} f_{n-k} h_k.$$

If f or h is additionally piecewise smooth, then the Fourier series $\sum_{n=-\infty}^{\infty} g_n e^{i 2 \pi n x / L}$ converges in L^2 to fh.

In most practical applications, the functions f and h will be bounded and at least piecewise smooth (square waves or better). Their product fh will be bounded and at least piecewise smooth, and their convolution $f \circledast h$ will be bounded and at least continuous. Both will have Fourier series expansions that converge pointwise or better, according to Theorem 4.2.

Example 4.16 (More about the convolution of square waves). In Example 4.15 we saw that the convolution of two square waves is a triangle wave. Here we will show that this result agrees with the convolution theorem. The Fourier coefficients of the square and triangle waves were calculated in Examples 4.3 and 4.6. The square wave in the earlier example had values between -1 and 1 rather than 0 and 1. The latter wave is related to the former through a scaling by $\frac{1}{2}$ followed by adding $\frac{1}{2}$. Using the linearity theorem with Equation 4.10, we calculate the coefficients of the

---

[10]Requiring square summability of $(f_n h_n)$ and $(f_n) * (h_n)$ for convergence of their respective Fourier series is less restrictive than requiring absolute summability, because $\ell^1 \subset \ell^2 \subset \ell^\infty$.

0-to-1 square wave:

$$f_0 = \frac{1}{2}$$

$$f_n = \frac{1}{2} \cdot \frac{2}{i\pi n} = \frac{1}{i\pi n}, \quad (n \text{ odd, otherwise } 0).$$

Then, according to the convolution theorem, the Fourier coefficients of the triangle wave should be $g_n = Lf_n \cdot f_n = Lf_n^2$:

$$g_0 = \frac{L}{4}$$

$$g_n = L\left(\frac{1}{i\pi n}\right)^2 = -\frac{L}{\pi^2 n^2}.$$

These agree with Equation 4.13. ∎

Example 4.17 (Harmonic distortion). The ideal electronic amplifier produces an output that is simply a scaled version of its input, $y(t) = Ax(t)$. In real amplifiers, even ones of high quality, there are nonlinearities so that the output is more closely modeled by $y(t) = A_1 x(t) + A_2 x^2(t)$. If x is a pure sinusoid, $x(t) = \cos 2\pi \nu_0 t$, then the output is

$$y(t) = A_1 \cos 2\pi \nu_0 t + A_2 \cos^2 2\pi \nu_0 t = \frac{A_2}{2} + A_1 \cos 2\pi \nu_0 t + \frac{A_2}{2} \cos 4\pi \nu_0 t.$$

The cosines have frequencies ν_0 and $2\nu_0$, so they share the common period $T = 1/\nu_0$ and the Fourier series of y can be calculated. The Fourier coefficients are

$$c_n = \begin{cases} \frac{A_2}{2}, & n = 0 \\ \frac{A_1}{2}, & n = \pm 1 \\ \frac{A_2}{4}, & n = \pm 2 \\ 0, & \text{otherwise} \end{cases}.$$

To illustrate the operation of the convolution theorem, we will recalculate the Fourier coefficients of the nonlinear term $A_2 \cos^2 2\pi \nu_0 t$. The Fourier coefficients of $\cos 2\pi \nu_0 t$ are the sequence $(c_n) = \frac{1}{2}\delta[n-1] + \frac{1}{2}\delta[n+1]$. Then, calculating the convolution $(c_n) * (c_n)$, we have

$$(c_n) * (c_n) = \sum_{k=-\infty}^{\infty} \left(\frac{1}{2}\delta[k-1] + \frac{1}{2}\delta[k+1]\right)\left(\frac{1}{2}\delta[n-k-1] + \frac{1}{2}\delta[n-k+1]\right)$$

$$= \frac{1}{4}\sum_k \delta[k-1]\delta[n-k-1] + \frac{1}{4}\sum_k \delta[k+1]\delta[n-k-1]$$

$$+ \frac{1}{4}\sum_k \delta[k-1]\delta[n-k+1] + \frac{1}{4}\sum_k \delta[k+1]\delta[n-k+1]$$

$$= \frac{1}{4}\delta[n-2] + \frac{1}{4}\delta[n] + \frac{1}{4}\delta[n] + \frac{1}{4}\delta[n+2] = \frac{1}{4}\delta[n-2] + \frac{1}{2}\delta[n] + \frac{1}{4}\delta[n+2].$$

Hence, the Fourier coefficients of $A_2 \cos^2 2\pi v_0 t$ are $\frac{A_2}{2}$, $n = 0$, and $\frac{A_2}{4}$, $n = \pm 2$, as before.

To be specific, let $A_1 = 10$, $A_2 = 0.1$, and $v_0 = 100$ Hz. There will be a spurious DC component in the output, which is routinely blocked by a capacitor, and also a spurious second harmonic at 200 Hz. This is called second-harmonic distortion, and is quantified by the ratio of second harmonic power to the total (AC) power. In this case,

$$\text{harmonic distortion} = \frac{|c_{-2}|^2 + |c_2|^2}{|c_{-1}|^2 + |c_1|^2 + |c_{-2}|^2 + |c_2|^2} = \frac{0.0025}{25.0025} \approx 0.01\%.$$ ∎

4.5 THE HEAT EQUATION

The first application of the Fourier series, described by Fourier in his seminal *Théorie Analytique de la Chaleur* (1822), was the description of heat conduction in a thin metal bar. Here, the bar is assumed to be sufficiently thin compared with its length that the temperature is uniform across its width. It is also assumed that the bar is insulated so that no heat escapes along its length, and its ends are maintained at zero temperature (Figure 4.19). The temperature, denoted $u(x, t)$, obeys a partial differential equation,

$$\frac{\partial^2 u(x,t)}{\partial x^2} = \frac{1}{D} \frac{\partial u(x,t)}{\partial t} \tag{4.27}$$

$u(0, t) = u(L, t) = 0$ boundary conditions

$u(x, 0) = f(x)$ initial condition

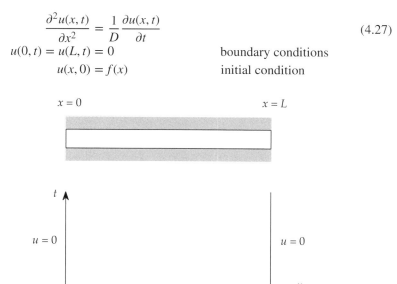

FIGURE 4.19 Solving the heat equation in a thin bar. *Top:* The sides are insulated, so no heat flows through them. The temperature profile is assumed to be uniform in the transverse direction. *Bottom:* Time–space plot showing the boundary conditions $u(0, t) = u(L, t) = 0$ and the initial condition $u(x, 0) = f(x)$.

where D is called the thermal diffusivity (units m²/s). The equation describes the evolution of temperature in the bar in space and time, beginning with an initial distribution $u(x, 0)$, and subject to specifications on the temperature and heat flux (proportional the gradient $\partial u/\partial x$) at the ends, $x = 0$ and $x = L$. The same equation describes one-dimensional (1D) diffusion processes, where u is interpreted as a concentration of material and the gradient $\partial u/\partial x$ is the flux of material.

Separation of variables

Many partial differential equations, including the present one, can be solved using a method called *separation of variables*. The solution is assumed to be separable as the product of a function of position alone and a function of time alone:

$$u(x, t) = X(x)T(t).$$

Making this substitution into the heat equation gives

$$X''T = \frac{1}{D}XT'$$

and dividing through by XT separates the variables:

$$\frac{X''}{X} = \frac{1}{D}\frac{T'}{T}.$$

The left-hand side of this equation depends on x alone while the right-hand side depends on t alone. Yet they are equal to each other, for all values of x and t. The only way this can happen is for both sides to be equal to a constant, which we denote $-\lambda$:

$$\frac{X''}{X} = \frac{1}{D}\frac{T'}{T} = -\lambda.$$

This procedure converts the partial differential equation into two ordinary differential equations,

$$X'' + \lambda X = 0$$
$$T' + \lambda DT = 0.$$

The parameter λ is called an *eigenvalue*. The corresponding solutions of the X equation are called *eigenfunctions*, denoted $\phi(x)$. There may be more than one eigenvalue–eigenfunction pair that satisfies the differential equation. By linearity, the eventual solution may be a combination of eigenfunctions.

From the theory of ordinary differential equations, we identify three possible forms for the eigenfunctions, depending on the sign of the eigenvalue:

$$\phi(x) = \begin{cases} \sinh\sqrt{-\lambda}x, \cosh\sqrt{-\lambda}x, & \lambda < 0 \\ c_1 x + c_2, & \lambda = 0 \\ \sin\sqrt{\lambda}x, \cos\sqrt{\lambda}x, & \lambda > 0 \end{cases}.$$

All of these satisfy the X equation, but the physically correct solutions are the ones that also satisfy the boundary conditions, $u(0, t) = u(L, t) = 0$. The boundary conditions must hold for all t, so they are equivalent to requiring $\phi(0) = \phi(L) = 0$. The first boundary condition, $\phi(0) = 0$, disqualifies cos and cosh, because they are nonzero at the origin. The linear solution, $c_1 x + c_2$, satisfies both boundary conditions only if $c_1 = c_2 = 0$, which is trivial. The sinh function is zero at the origin but nowhere else, so it fails to satisfy the boundary condition at $x = L$. Only $\sin\sqrt{\lambda}x$ can satisfy both boundary conditions. It is zero at $x = 0$, and will also be zero at $x = L$ if $\sqrt{\lambda} = \frac{n\pi}{L}$, where n is any positive integer (negative integers are redundant). There is an infinite number of eigenvalues and eigenfunctions,

$$\lambda_n = \frac{n^2 \pi^2}{L^2}, \quad n = 1, 2, \ldots$$

$$\phi_n(x) = \sin\sqrt{\lambda}x = \sin\frac{n\pi x}{L}.$$

Next we take up the T equation,

$$T' + \lambda D T = 0.$$

This is a first-order differential equation. We know that $\lambda > 0$ from the previous discussion, so the correct solution of this equation has the form of a decaying exponential, $\psi_n(t) = \exp(-\lambda_n D t)$. Combining this with the x-dependence, for each n there is a solution of the partial differential equation of the form $\psi_n(t)\phi_n(x) = \exp(-\lambda_n k t)\sin\frac{n\pi x}{L}$. These are called the *modes* of the system. The general solution of the partial differential equation is a linear superposition of modes,

$$u(x,t) = \sum_{n=1}^{\infty} b_n \psi_n(t) \phi_n(x)$$

$$= \sum_{n=1}^{\infty} b_n \exp\left(-\frac{n^2 \pi^2 D}{L^2} t\right) \sin\frac{n\pi x}{L}.$$

It remains to apply the initial condition, $u(x,0) = f(x)$. Substituting $t = 0$ into the general solution yields

$$u(x,0) = f(x) = \sum_{n=1}^{\infty} b_n \sin\frac{n\pi x}{L},$$

a Fourier sine series! The $\{b_n\}$ are Fourier coefficients for the initial temperature profile $f(x)$.

The eigenfunctions $\phi_n(x) = \sin\frac{n\pi x}{L}$ are orthogonal on $[0, L]$:

$$\langle \phi_n, \phi_m \rangle = \int_0^L \sin\frac{n\pi x}{L} \sin\frac{m\pi x}{L}\, dx = \begin{cases} 0, & m \neq n \\ \frac{L}{2}, & m = n \end{cases}.$$

Thus, the set $\left\{\sqrt{\frac{2}{L}}\sin\frac{n\pi x}{L}\right\}_{n=1}^{\infty}$ is orthonormal, and it can also be shown to be complete. The form of an orthogonal expansion with this basis is

$$f = \sum_{n=1}^{\infty} \left\langle f, \sqrt{\frac{2}{L}}\sin\frac{n\pi x}{L}\right\rangle \sqrt{\frac{2}{L}}\sin\frac{n\pi x}{L} = \sum_{n=1}^{\infty} \underbrace{\frac{2}{L}\left\langle f, \sin\frac{n\pi x}{L}\right\rangle}_{b_n} \sin\frac{n\pi x}{L}.$$

Finally, we have the complete solution to the heat equation,

$$u(x,t) = \sum_{n=1}^{\infty} b_n \exp\left(-\frac{n^2\pi^2 D}{L^2}t\right) \sin\frac{n\pi x}{L} \qquad (4.28)$$

$$b_n = \frac{2}{L}\int_0^L f(x) \sin\frac{n\pi x}{L}\,dx.$$

Example 4.18. For a particular case, let the initial temperature profile be rectangular with width w,

$$f(x) = \begin{cases} 0, & \frac{L}{2}-\frac{w}{2} \geq x \geq 0 \\ \frac{A}{w}, & \frac{L}{2}+\frac{w}{2} > x > \frac{L}{2}-\frac{w}{2} \\ 0, & L \geq x \geq \frac{L}{2}+\frac{w}{2} \end{cases}.$$

The Fourier coefficients are

$$b_n = \frac{2}{L}\int_{\frac{L}{2}-\frac{w}{2}}^{\frac{L}{2}+\frac{w}{2}} \frac{A}{w}\sin\frac{n\pi x}{L}\,dx = \begin{cases} 0, & n \text{ even} \\ (-1)^{\frac{n-1}{2}}\frac{2A}{L}\operatorname{sinc}\frac{nw}{2L}, & n \text{ odd} \end{cases}$$

(where $\operatorname{sinc} x = \frac{\sin \pi x}{\pi x}$). The temperature profile $u(x,t)$ is plotted in Figure 4.20 for various values of t. For computational reasons, the series is truncated at $n = 41$, and for $t = 0$ the poor convergence of the Fourier series is obvious. Within a very short time, however, the solution smooths out, and the temperature profile spreads as t increases. The ends remain at zero temperature, as specified by the boundary conditions. Heat flows out the ends and, as it is lost, the temperature of the bar decreases toward zero everywhere.

The rapid smoothing of the truncated Fourier series is due to the exponential factor. By themselves, the coefficients b_n decay $O(n^{-1})$, and the consequent bad convergence is observed at $t = 0$. For $t > 0$, the b_n are multiplied by a decaying exponential factor, so that the Fourier coefficients are effectively $b_n \exp\left(-\frac{n^2\pi^2 D}{L^2}t\right)$. Convergence for even small values of t is extremely rapid, and as t increases, the Fourier series comes to be dominated by a few low frequency terms (Figure 4.21) that are spatially spread out. ∎

Bar with insulated ends

Different boundary conditions lead to different solutions. For a second example, let the ends of the bar be insulated rather than held at a constant temperature. The boundary conditions are zero heat flow at $x = 0$ and $x = L$. Heat flow is proportional to the temperature gradient, so the boundary conditions are expressed by $\frac{\partial u}{\partial x} = 0$ at $x = 0$ and $x = L$ (Figure 4.22).

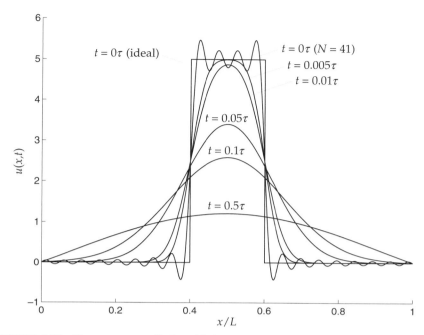

FIGURE 4.20 Temperature profile in a thin bar, computed from Equation 4.28 with $L = 1$, $A = 1$, $w = 0.2$, and $D = 1$. The constant $\tau = \frac{L^2}{\pi^2 D}$ is the time constant of the exponential factor when $n = 1$. The Fourier series is truncated at $n = 41$. Overshoot is observed for $t = 0$, but for $t \geq 0.005\tau$, the solution is monotonically decreasing away from the center of the bar. Because the ends of the bar are held at zero temperature, heat flows from the center through the ends, and the temperature of the bar approaches zero everywhere.

The solution by separation of variables proceeds as before, until we get to the possible solutions of the X equation.

$$\phi(x) = \begin{cases} \sinh\sqrt{-\lambda}x,\ \cosh\sqrt{-\lambda}x, & \lambda < 0 \\ c_1 x + c_2, & \lambda = 0 \\ \sin\sqrt{\lambda}x,\ \cos\sqrt{\lambda}x, & \lambda > 0 \end{cases}.$$

The boundary conditions are imposed on the derivative, $\frac{\partial u}{\partial x}\big|_{(0,t)} = \frac{\partial u}{\partial x}\big|_{(L,t)} = 0$, which must be true for all t. The equivalent boundary conditions on ϕ are $\phi'(0) = \phi'(L) = 0$. Differentiating the possible solutions,

$$\phi'(x) = \begin{cases} \sqrt{-\lambda}\cosh\sqrt{-\lambda}x,\ \sqrt{-\lambda}\sinh\sqrt{-\lambda}x, & \lambda < 0 \\ c_1, & \lambda = 0 \\ \sqrt{\lambda}\cos\sqrt{\lambda}x,\ -\sqrt{\lambda}\sin\sqrt{\lambda}x, & \lambda > 0 \end{cases}.$$

The sinh and cosh functions are disqualified, as before. The boundary conditions are satisfied for $\lambda = 0$ if $c_1 = 0$ (c_2 is unrestricted by the boundary conditions). For

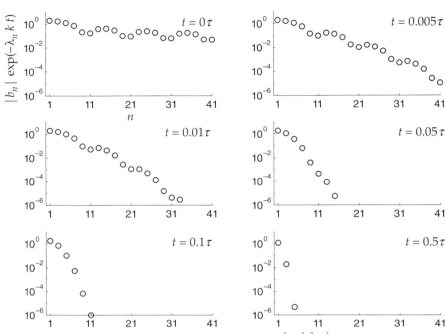

FIGURE 4.21 Decay of the Fourier coefficients $b_n \exp\left(-\frac{n^2\pi^2 D}{L^2}t\right)$ in Equation 4.28. Same parameter values as Figure 4.20. The increasingly rapid rolloff of the Fourier coefficients is the reason for the smoothing and spreading of the initial temperature profile.

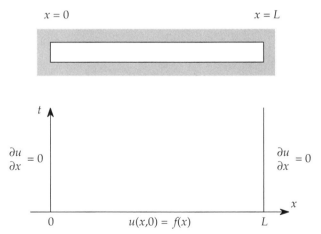

FIGURE 4.22 Solving the heat equation in a thin bar. *Top:* The sides and ends are insulated so no heat flows through them. The temperature profile is assumed to be uniform in the transverse direction. *Bottom:* Time–space plot showing the boundary conditions $\frac{\partial u}{\partial x}\big|_{(0,t)} = \frac{\partial u}{\partial x}\big|_{(L,t)} = 0$ and the initial condition $u(x,0) = f(x)$.

$\lambda > 0$, the cosine function will work. The derivative $\phi'(x) = -\lambda \sin \sqrt{\lambda} x$ is zero for $x = 0$ and also for $x = L$ if $\sqrt{\lambda} = \frac{n\pi}{L}$, $n = 1, 2, \ldots$. From here, the solution proceeds as before, until we arrive at the general solution:

$$u(x,t) = \frac{a_0}{2} + \sum_{n=1}^{\infty} a_n \exp\left(-\frac{n^2\pi^2 D}{L^2}t\right) \cos\frac{n\pi x}{L}.$$

In renaming the constant term from c_2 to $\frac{a_0}{2}$, we recognize this as a Fourier cosine series. Imposing the initial condition,

$$u(x,0) = f(x) = \frac{a_0}{2} + \sum_{n=1}^{\infty} a_n \cos\frac{n\pi x}{L}.$$

The $(a_n)_{n=0}^{\infty}$ are the Fourier cosine coefficients of the initial profile $f(x)$. They are calculated in like manner as the sine coefficients in the previous case, and the final solution of the heat equation is obtained:

$$u(x,t) = \frac{a_0}{2} + \sum_{n=1}^{\infty} a_n \exp\left(-\frac{n^2\pi^2 D}{L^2}t\right) \cos\frac{n\pi x}{L}$$

$$a_n = \frac{2}{L}\int_0^L f(x)\cos\frac{n\pi x}{L}\,dx. \tag{4.29}$$

Example 4.19. Using the same initial profile as in Example 4.18, we calculate the Fourier cosine coefficients,

$$a_n = \frac{2}{L}\int_{\frac{L}{2}-\frac{w}{2}}^{\frac{L}{2}+\frac{w}{2}} \frac{A}{w}\cos\frac{n\pi x}{L}\,dx = \begin{cases} \frac{2A}{L}, & n = 0 \\ 0, & n \text{ odd} \\ (-1)^{n/2}\frac{2A}{L}\operatorname{sinc}\frac{nw}{2L}, & n \text{ even} \end{cases}$$

The temperature profile $u(x,t)$ is plotted in Figure 4.23 for various values of t. For computational reasons, the series is truncated at $n = 40$. The solution has the same convergence behavior as the previous example.

The interesting feature of this solution is the result of insulating the ends of the rod. Because no heat can escape, the temperatures of the ends of the rod rise as the initial profile spreads out. As $t \to \infty$, all the terms in the solution decay to zero except the constant term. The steady-state temperature of the rod is given by this term, $\frac{A}{L}$. ■

General validity of the solutions

In both examples above, the solution of the heat equation takes the form of an infinite series,

$$u(x,t) = a_0/2 + \sum_{n=1}^{\infty} \frac{a_n}{b_n} \exp\left(-\frac{n^2\pi^2 D}{L^2}t\right) \frac{\cos}{\sin}\frac{n\pi x}{L}.$$

The individual terms of the series are shown to satisfy the heat equation and the boundary conditions, and at $t = 0$ the coefficients can be calculated to satisfy the initial condition. There remains the question whether the infinite series converges,

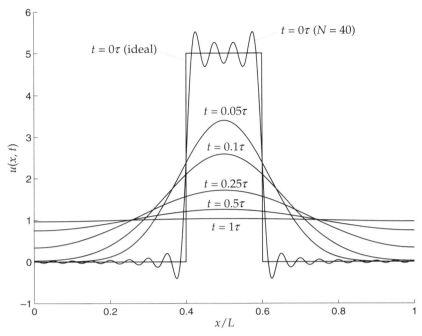

FIGURE 4.23 Temperature profile in a thin bar with insulated ends, computed from Equation 4.29 with $L = 1$, $A = 1$, $w = 0.2$, and $D = 1$. The constant $\tau = \frac{L^2}{\pi^2 D}$ is the time constant of the exponential factor when $n = 1$. The Fourier series is truncated at $n = 40$. Overshoot is observed for $t = 0$, but even for relatively small t, the solution is smooth. Because the ends of the bar are insulated, as heat flows from the center to the ends it causes the temperature of the ends to rise. For large t, the temperature of the bar approaches $\frac{A}{L} = 1$ everywhere.

and does so in such a way that it satisfies the partial differential equation. Fortunately, the answer is affirmative. The key is the exponential factor, $\exp\left(-\frac{n^2 \pi^2 D}{L^2} t\right)$, which is rapidly decreasing in n for $t > 0$. For any initial profile f that has a Fourier expansion (which includes all functions of physical interest), the Fourier coefficients (a_n) or (b_n) will exist, and by the Riemann–Lebesgue lemma, they will decay with increasing n. Multiplying them by the rapidly decreasing exponential factor renders the infinite series absolutely and uniformly convergent for $t > 0$. Consequently, it can be differentiated term by term with respect to x or t, and the derived series will also converge, to the respective partial derivatives of u, for example,

$$\frac{\partial u}{\partial x} = \sum_{n=1}^{\infty} \frac{n\pi b_n}{L} \exp\left(-\frac{n^2 \pi^2 D}{L^2} t\right) \cos \frac{n\pi x}{L}$$

$$\frac{\partial u}{\partial t} = \sum_{n=1}^{\infty} -\frac{n^2 \pi^2 D b_n}{L^2} \exp\left(-\frac{n^2 \pi^2 D}{L^2} t\right) \sin \frac{n\pi x}{L}.$$

Differentiation has multiplied the Fourier coefficients by n and by n^2, but the rapidly decreasing exponential factor counteracts this growth and continues to render the

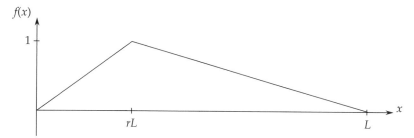

FIGURE 4.24 Initial profile of a plucked string.

series absolutely and uniformly convergent, and differentiable term by term. The series can be differentiated any number of times and the result will always converge, absolutely and uniformly, to the corresponding partial derivative of u. That is, the infinite series and its derivatives are infinitely continuously differentiable functions of x and t, for $t > 0$. This is more than is required to validate the infinite series as a solution of the heat equation.

4.6 THE VIBRATING STRING

Standing wave systems, for example, the classical equation of the vibrating string, are well-suited to Fourier series analysis. A string is stretched tight between two points separated by a distance L, and displaced away from its rest position, as shown in Figure 4.24.

You may imagine that $x = 0$ corresponds to the bridge of a guitar (on the body of the instrument), and $x = L$ corresponds to the nut (at the top of the neck, just below the tuning pegs). The string is about to be plucked at a distance rL from the bridge, where $1 > r > 0$. It is assumed to be released with zero initial velocity. The vertical displacement of the the string obeys a type of *wave equation*,

$$\frac{\partial^2 y(x,t)}{\partial x^2} = \frac{\mu}{F}\frac{\partial^2 y(x,t)}{\partial t^2}, \quad (4.30)$$

$y(0,t) = y(L,t) = 0$ boundary conditions
$y(x,0) = f(x), \ y'(x,0) = 0$ initial conditions

where μ is the mass per unit length of the string and F is the tension in the string.[11]

Like the heat equation, the wave equation for the vibrating string may be solved by separation of variables. Assume that the solution $y(x, t)$ is the product of two functions of x and t alone:

$$y(x,t) = X(x)T(t).$$

Substituting this into Equation 4.30 gives

$$X''T = \frac{1}{v^2}XT''$$

---

[11] For derivations of the equation of the vibrating string, see Churchill and Brown (1987, pp. 4–7) and Elmore and Heald (1969, pp. 1–5, 42–44).

where $v^2 = F/\mu$, for convenience. Dividing through by XT separates the variables:

$$\frac{X''}{X} = \frac{T''}{v^2 T}.$$

The left-hand side of this equation depends on x alone while the right-hand side depends on t alone. Yet they are equal to each other, for all values of x and t. The only way this can happen is for both sides to be equal to a constant, which we denote $-\lambda$:

$$\frac{X''}{X} = \frac{T''}{v^2 T} = -\lambda.$$

We now have two ordinary differential equations,

$$X'' + \lambda X = 0$$
$$T'' + \lambda v^2 T = 0,$$

with boundary and initial conditions

$$X(0) = X(L) = 0$$
$$T'(0) = 0.$$

We address the X equation first. There are three possible forms for the eigenfunctions $\phi(x)$:

$$\phi(x) = \begin{cases} \sinh \sqrt{-\lambda}x, \cosh \sqrt{-\lambda}x, & \lambda < 0 \\ c_1 x + c_2, & \lambda = 0 \\ \sin \sqrt{\lambda}x, \cos \sqrt{\lambda}x, & \lambda > 0 \end{cases}.$$

The cos and cosh functions are nonzero at $x = 0$, and fail to satisfy the boundary condition. The sinh function is zero at $x = 0$ but nowhere else, so it cannot satisfy the boundary condition at $x = L$. The linear solution, $c_1 x + c_2$, satisfies the boundary conditions only if $c_1 = c_2 = 0$, which is trivial. Finally, $\sin \sqrt{\lambda}x$ is zero at $x = 0$ and will also be zero at $x = L$ if $\sqrt{\lambda}L = n\pi$, $n = 1, 2, \ldots$. This fixes the eigenvalues and eigenfunctions of the X equation,

$$\lambda_n = \frac{n^2 \pi^2}{L^2}, \quad n = 1, 2, \ldots$$
$$\phi_n(x) = \sin \sqrt{\lambda_n} x = \sin \frac{n\pi x}{L}.$$

The solution for T proceeds in the same way. This time we know that the eigenvalues are positive, so we immediately have eigenfunctions

$$\psi_n(t) = \sin \sqrt{\lambda_n} vt, \cos \sqrt{\lambda_n} vt.$$

Only the cosine satisfies the initial condition $T'(0) = 0$.

The spatial and temporal eigenfunctions, taken together, define the modes of vibration of the string,

$$\phi_n(x)\psi_n(t) = \sin \frac{n\pi x}{L} \cos \frac{n\pi vt}{L}.$$

4.6 THE VIBRATING STRING

The general solution of the wave equation $y(x,t)$ is a linear superposition of the modes,

$$y(x,t) = \sum_{n=1}^{\infty} b_n \phi_n(x) \psi_n(t) = \sum_{n=1}^{\infty} b_n \sin \frac{n\pi x}{L} \cos \frac{n\pi v t}{L}.$$

The precise mixture of modes is specified by the coefficients $\{b_n\}$, and these are chosen to satisfy the remaining initial condition, $y(x,0) = f(x)$:

$$y(x,0) = f(x) = \sum_{n=1}^{\infty} b_n \sin \frac{n\pi x}{L}.$$

This is a Fourier sine series. As shown in the previous section, the expansion coefficients are given by the integral

$$b_n = \frac{2}{L} \int_0^L f(x) \sin \frac{n\pi x}{L} dx.$$

Substituting $v^2 = F/\mu$, the final solution of the partial differential equation is

$$y(x,t) = \sum_{n=1}^{\infty} b_n \sin\left(\frac{\pi n x}{L}\right) \cos\left(\sqrt{\frac{F}{\mu}} \frac{\pi n t}{L}\right) \quad (4.31)$$

$$b_n = \frac{2}{L} \int_0^L f(x) \sin \frac{n\pi x}{L} dx.$$

The solution of the wave equation is sinusoidal in space and in time. The spatial profile of each mode is sinusoidal, with period (wavelength) equal to $2L/n$ (Figure 4.25). The amplitude of each mode varies sinusoidally in time with frequency $v_n = \frac{n\sqrt{F/\mu}}{2L}$. We associate the fundamental frequency, $v_1 = \frac{\sqrt{F/\mu}}{2L}$, with the pitch of the sound made by the string. Making the string shorter increases the pitch (the high strings on a piano are shorter than the low strings). Tightening the string (increasing F) also increases the pitch and provides a convenient way to tune the string to a particular pitch. Length and tension are insufficient to cover the range of frequencies required of a piano, and the strings of a guitar must all be the same length. The lower strings on a guitar or piano are wound with wire to make them heavier (increase μ) and thereby lower their pitch.

The Fourier coefficients b_n give the relative strengths of the modes. The basic pitch is unaffected by this distribution (as long as $b_1 \neq 0$), but the quality of the sound, called its *timbre*, is very much dependent on the distribution of harmonic overtones.

Example 4.20 (Plucked string). Let the initial string profile $f(x)$ be described by two linear segments (Figure 4.24),

$$f(x) = \begin{cases} \dfrac{x}{rL}, & rL > x \geq 0 \\ \dfrac{1 - x/L}{1 - r}, & L \geq x \geq rL \end{cases}.$$

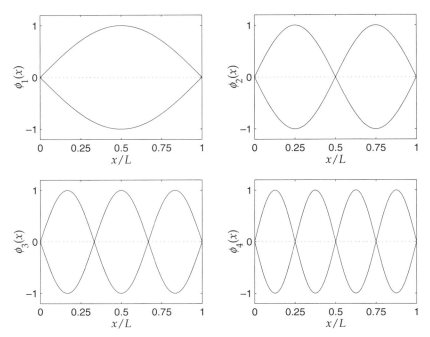

FIGURE 4.25 Low order modes of the vibrating string.

We can use the Fourier series to explore the harmonic structure, observing the effect of the parameter r on the Fourier coefficients.

The Fourier coefficients are given by the sum of two integrals,

$$b_n = \frac{2}{L} \int_0^{rL} \frac{x}{rL} \sin\left(\frac{\pi n x}{L}\right) dx + \frac{2}{L} \int_{rL}^{L} \frac{1 - x/L}{1 - r} \sin\left(\frac{\pi n x}{L}\right) dx,$$

which simplifies to

$$b_n = \frac{2 \sin r\pi n}{\pi^2 n^2 r(1-r)}, \quad n > 0. \tag{4.32}$$

Combining Equation 4.32 with Equation 4.31 yields the final result,

$$y(x,t) = \sum_{n=1}^{\infty} \frac{2 \sin \pi n r}{\pi^2 n^2 r(1-r)} \cos\left(\sqrt{\frac{F}{\mu}} \frac{\pi n t}{L}\right) \sin\left(\frac{\pi n x}{L}\right). \tag{4.33}$$

The spectrum of Fourier coefficients for different values of r is shown in Figure 4.26. The harmonics are weakest and the sound thinnest when the string is plucked in the middle, $r = 1/2$. As the string is plucked closer to the bridge, $r = 1/10$, the harmonics strengthen and the sound becomes richer. ∎

General validity of the solution

As with the heat equation, we must be concerned with the convergence properties of the solution (Equation 4.31). The modes $\phi_n(x)\psi_n(t)$ individually satisfy the wave equation and the boundary conditions, and by linearity, any finite series of modes

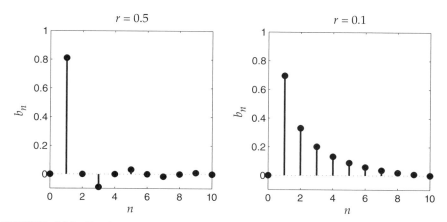

FIGURE 4.26 Fourier coefficients for the vibrating string. *Left:* Plucked at the center, $r = 1/2$. *Right:* Plucked near the bridge, $r = 1/10$.

is also a solution. Convergence of the infinite series depends on the decay rate of the coefficients, which in turn depends on the smoothness of the function being expanded—in this case, the initial profile of the string. Physically, the initial profile must be at least continuous (a jump discontinuity would break the string). The Fourier coefficients (b_n) will decay $O(n^{-2})$ or better, guaranteeing absolute and uniform convergence of the series. But the series must be differentiable twice with respect to x and with respect to t, and this imposes a more stringent constraint on the coefficients. Differentiating twice multiplies the coefficients by n^2. In order for the derived series to possess absolute and uniform convergence, the Fourier coefficients must decay $O(n^{-4})$ or better. According to Theorem 4.4, the initial profile f must belong to the class $C^{(3)}$—three times continuously differentiable.

The initial profile in the above example does not possess the required degree of smoothness, so technically, the infinite series derived there is not a solution of the wave equation. Yet, our experience tells us that a plucked string does vibrate in a sensible way. The problem is not with the physics, but with the mathematical model, which is overidealized. A real plucked string will not have a sharp corner, but will be rounded with a small radius. Mathematically, this will make f smooth enough that its Fourier coefficients have the necessary decay rate. Realistic coefficients may be more difficult to calculate, but if the radius is small there may not be a great difference numerically between the true coefficients and the ideal ones, so that the coefficients for the sharp profile can serve as a useful approximation to the true coefficients. We may also consider the behavior of such a solution as the radius of the bend approaches zero, the idealized case. Such limiting procedures are the subject of a later chapter.

4.7 ANTENNA ARRAYS

Fourier series are also applicable to the theory of antenna arrays. Many antennas, from the Very Large Array for radio astronomy to the ultrasonic transducers used for medical imaging, are constructed from multiple smaller emitters and detectors.

Initially, we will assume that the array is a 1D (line) array of point elements. The results obtained will be accurate enough to elucidate the distinctive behavior of arrays. In later chapters we will relax these assumptions to eventually model 2D arrays with finite-sized elements.

Spherical waves

The general form of the wave equation is

$$\nabla^2 u = \frac{1}{v^2} \frac{\partial^2 u}{\partial t^2}, \tag{4.34}$$

where u is a function of spatial coordinates and of time, and v has units of velocity. We are interested here in 3D waves that depend spatially only on the radius from the origin. The Laplacian in spherical coordinates without angular variation is

$$\nabla^2 u = \frac{1}{r} \frac{\partial^2}{\partial r^2}(ru),$$

so the wave equation takes the form

$$\frac{1}{r} \frac{\partial^2}{\partial r^2}(ru) = \frac{1}{v^2} \frac{\partial^2 u}{\partial t^2},$$

which may also be written

$$\frac{\partial^2}{\partial r^2}(ru) = \frac{1}{v^2} \frac{\partial^2}{\partial t^2}(ru). \tag{4.35}$$

This looks just like the 1D (vibrating string) equation, with $y = ru$. We may follow the same separation of variables method used to solve the equation of the vibrating string. We take $ru(r, t) = R(r)T(t)$ and begin with the separated equation,

$$\frac{R''}{R} = \frac{1}{v^2} \frac{T''}{T} = -k^2,$$

leading to ordinary differential equations

$$R'' + k^2 R = 0$$
$$T'' + k^2 v^2 T = 0.$$

Addressing the R equation, we again identify three cases depending on the sign of the eigenvalue k^2. Unlike the vibrating string, the spatial domain is unbounded, and a physically reasonable solution should not blow up as $r \to \infty$. Moreover, the total energy in the solution, proportional to the integral of the squared amplitude over the surface of a sphere, should not blow up:

$$E \propto \int_{\text{sphere}} \left|\frac{R(r)}{r}\right|^2 dA = 4\pi r^2 \left|\frac{R(r)}{r}\right|^2 = 4\pi |R(r)|^2 < \infty.$$

In fact, since there are no damping mechanisms in the model, $|R|$ should be constant.

$k^2 < 0$: The solutions are of the form $\exp(\pm|k|r)$. Dividing by r gives $r^{-1}\exp(\pm|k|r)$. One of these clearly blows up as $r \to \infty$, and the other decays. We can reject immediately the growing exponential. The total energy of the decaying exponential solution is $4\pi \exp(-2|k|r)$, which decays as $r \to \infty$. Because energy is not conserved, we reject this solution as well.

$k^2 = 0$: The solution is of the form $c_1 r + c_0$, and dividing by r, $c_1 + c_0 r^{-1}$. This solution is finite as $r \to \infty$, and the total energy is

$$4\pi r^2 \left(c_1 + \frac{c_0}{r}\right)^2 = 4\pi \left(c_1^2 r^2 + 2 c_1 c_0 r + c_0^2\right).$$

Taking $c_1 = 0$ yields a finite result.

$k^2 > 0$: The solutions are complex exponentials, $\exp(\pm ikr)$. Dividing by r gives $r^{-1} \exp(\pm ikr)$. Neither of these solutions blows up as $r \to \infty$, and both have total energy 4π.

The $k^2 > 0$ and $k^2 = 0$ cases can be combined into one expression for the radial dependence:

$$\frac{R(r)}{r} = \frac{\exp(\pm ikr)}{r}, \quad k \geq 0, r > 0.$$

The constant k is a spatial angular frequency (radians/distance). The spatial period, or wavelength, is $\lambda = 2\pi/k$.

Knowing $k \geq 0$ yields a complex exponential time variation, $\exp(ikvt)$. We identify $kv = \omega$, the customary angular frequency (radians/sec), which is also $2\pi\nu$, where ν is the frequency in Hertz. The spatial and temporal domains are linked by the wave velocity, $v = \lambda \nu = \omega/k$.

The radial and temporal solutions combine to give waves of the form

$$u(r,t) = \frac{A}{r} \exp(\pm i(kr - \omega t)), \quad \frac{A}{r} \exp(\pm i(kr + \omega t)),$$

where A is a constant. This is the basic spherical wave. It is called spherical because the wavefronts, loci of points of constant phase $kr \pm \omega t = \text{const}$, are spherical shells with radius $r = \mp \omega t/k + \text{const} = \mp vt + \text{const}$. Choosing the $+$ or $-$ sign selects a wave that propagates away from or toward the origin with velocity v:

$$u(r,t) = \begin{cases} \dfrac{A}{r} \exp(\pm i(kr - \omega t)) & \text{outward} \\[2mm] \dfrac{A}{r} \exp(\pm i(kr + \omega t)) & \text{inward} \end{cases} \quad (4.36)$$

or, in terms of trigonometric functions,

$$u(r,t) = \begin{cases} \dfrac{A}{r} \cos(kr - \omega t + \varphi) & \text{outward} \\[2mm] \dfrac{A}{r} \cos(kr + \omega t + \varphi) & \text{inward} \end{cases}, \quad (4.37)$$

where φ is an arbitrary constant phase (Equation 1.21).

The wave amplitude $u(r,t)$ and the energy density $|u(r,t)|^2$ are singular at the origin, indicating a concentration of all the wave's energy into a point. The wave either emanates from this point or is focused into this point. In the case of the outwardly propagating wave, we escape the singularity by noting that actual sources have finite

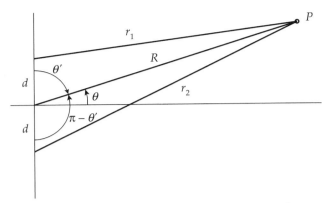

FIGURE 4.27 Geometry for calculating field due to two point sources.

size and the source energy is distributed over a finite volume. The point source is a useful model, though, particularly when we are working at distances r much greater than the actual size of the source. As for the other case, inwardly propagating spherical waves are always spatially limited in practice. One may focus a portion of a spherical wave toward a point, but never a complete sphere. The spatial limitation causes the wave to spread at the point of focus.

Two element antenna
Consider two point sources, separated by a distance $2d$, emitting spherical waves of equal amplitude A (Figure 4.27).

The total field due to the superposition of the two waves is measured at a point P, assumed to be coplanar with the source locations. The quantities r_1, r_2, and R are the distances from the two sources and the origin, respectively, to P. The outward propagating spherical waves are represented by complex exponentials,

$$\tilde{u}(r, t) = \frac{A}{r} e^{i(kr - \omega t)}.$$

The complex field amplitude measured at P is the sum of the two waves,

$$\tilde{u}_P(R, \theta, t) = \tilde{u}_1(r_1, t) + \tilde{u}_2(r_2, t) = \frac{A}{r_1} e^{i(kr_1 - \omega t)} + \frac{A}{r_2} e^{i(kr_2 - \omega t)}.$$

The actual field is obtained by taking the real part of the complex field.

The first step in deriving an expression for the total field is to express the distances r_1 and r_2 in terms of R, θ, and d. By the Law of Cosines,

$$r_1 = \left(d^2 + R^2 - 2dR \cos \theta'\right)^{1/2} = \left(d^2 + R^2 - 2dR \sin \theta\right)^{1/2}$$

$$= R\sqrt{1 - \frac{2d \sin \theta}{R} + \frac{d^2}{R^2}}$$

$$r_2 = \left(d^2 + R^2 - 2dR \cos(\pi - \theta')\right)^{1/2} = \left(d^2 + R^2 + 2dR \sin \theta\right)^{1/2}$$

$$= R\sqrt{1 + \frac{2d \sin \theta}{R} + \frac{d^2}{R^2}}.$$

Next, we assume that the observation point is very far from the sources, so that $R \gg d$. This is called the *far-field* approximation, widely used in practice. Under the radicals, there are two terms in d/R, which are much less than 1. This enables us to use the first-order approximation $\sqrt{1+\epsilon} \approx 1 + \epsilon/2$ (for $\epsilon \ll 1$), yielding

$$r_1 \approx R\left(1 - \frac{d}{R}\sin\theta + \frac{d^2}{2R^2}\right) \approx R - d\sin\theta$$

$$r_2 \approx R\left(1 + \frac{d}{R}\sin\theta + \frac{d^2}{2R^2}\right) \approx R + d\sin\theta,$$

where we have kept terms to first order in d/R. Now we have

$$\tilde{u}_P(R,\theta,t) = \frac{A}{r_1}e^{i(kR - kd\sin\theta - \omega t)} + \frac{A}{r_2}e^{i(kR + kd\sin\theta - \omega t)}.$$

The distances r_1 and r_2 also appear in the magnitudes. Using the first order approximation

$$\frac{1}{1 \pm \epsilon} \approx 1 \mp \epsilon,$$

we may write

$$\frac{1}{r_1} \approx \frac{1}{R\left(1 - \frac{d}{R}\sin\theta\right)} \approx \frac{1 + \frac{d}{R}\sin\theta}{R}$$

$$\frac{1}{r_2} \approx \frac{1}{R\left(1 + \frac{d}{R}\sin\theta\right)} \approx \frac{1 - \frac{d}{R}\sin\theta}{R}$$

and, because $d/R \ll 1$, we may further approximate

$$\frac{1}{r_1} \approx \frac{1}{r_2} \approx \frac{1}{R},$$

and obtain

$$\tilde{u}_P(R,\theta,t) \approx \frac{A}{R}e^{i(kR - \omega t)}\left(e^{-ikd\sin\theta} + e^{+ikd\sin\theta}\right)$$
$$= \frac{2A}{R}e^{i(kR - \omega t)}\cos(kd\sin\theta). \tag{4.38}$$

Finally, take the real part of $\tilde{u}_P$ to arrive at the actual field,

$$u(R,\theta,t) \approx \frac{2A}{R}\cos(kR - \omega t)\cos(kd\sin\theta). \tag{4.39}$$

Mathematically, Equation 4.38 says that in the far field, the wave amplitude is proportional to the sum of two complex exponentials, $e^{-ikd\sin\theta}$ and $e^{+ikd\sin\theta}$, like a Fourier series with two terms. We are about to see how all this connects, but first we should think about what the result, particularly Equation 4.39, says about the nature

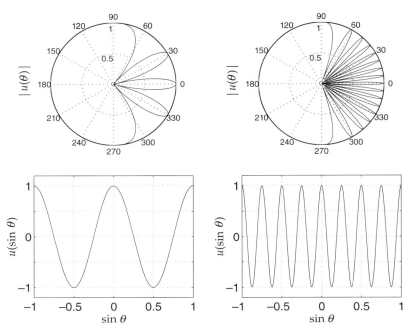

FIGURE 4.28 Angular variation of field due to two point sources. *Left:* $d/\lambda = 1$, *Right:* $d/\lambda = 4$. *Top:* The magnitude, $|u_P(\theta)|$, in a polar representation. *Bottom:* The field amplitude, $u_P(\theta)$, plotted vs. $\sin\theta$.

of the wave. First, consider the limiting case $d \to 0$, in which both sources coalesce into one source of strength $2A$. The result is what we would expect:

$$u_P = \frac{2A}{R} \cos(kR - \omega t),$$

a single diverging spherical wave. Then, when d is nonzero the amplitude of the diverging wave is modulated by the factor $\cos(kd \sin\theta) = \cos\left(2\pi \frac{d}{\lambda} \sin\theta\right)$. This function is plotted in Figure 4.28 for two values of the ratio d/λ, source position expressed in multiples of the wavelength.

The upper plots show the field strength that would be measured as you move in a semicircle around the sources. At intervals, the two waves interfere constructively and destructively, producing a characteristically lobed pattern. The lower plots show that the angular variation is sinusoidal, when plotted vs. the so-called *direction sine*, $\sin\theta$.

Multiple elements

The above analysis shows that a point source at a lateral distance d from the origin contributes a complex exponential $e^{+ikd \sin\theta}$ to the far-field antenna pattern. This can

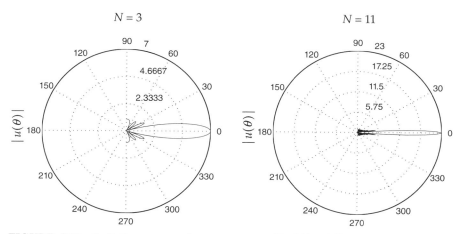

FIGURE 4.29 Radiation pattern of antenna array with $d/\lambda = 1/2$, $N = 3$ (*Left*), $N = 11$ (*Right*).

be generalized to multiple emitters, located at regular distances nd. The resulting field pattern for $2N + 1$ sources is

$$u_P(R, \theta, t) = \frac{A}{R} \cos(kR - \omega t) \sum_{n=-N}^{N} e^{inkd \sin\theta} = \frac{A}{R} \cos(kR - \omega t) D_{2N+1}\left(\frac{d}{\lambda} \sin\theta\right),$$
(4.40)

where, again, D_{2N+1} is the Dirichlet kernel. Radiation patterns are shown in Figure 4.29, for $d/\lambda = 1/2$ and two values of N. With larger N, the lobe of the array pattern is narrower—a transmitting antenna is better able to direct its energy, and a receiving antenna is better able to discriminate among sources at different angles.

The number and spacing of the elements are not all that one can do to control the antenna pattern. Later we will see how the amplitude and phase of the individual elements can be adjusted to modify the antenna's performance and how these appear in a Fourier model.

4.8 COMPUTING THE FOURIER SERIES

The Fourier series is a theoretical tool, useful for modeling natural phenomena and systems. It says that a periodic function can be broken into a sum of harmonically related sinusoidal functions, and it specifies what the proper mix of sinusoids is. It does not, however, enable you to perform a Fourier analysis of a function (signal), based on experimental data points. The DFT is the link between the theory of Fourier series and the practice of Fourier analysis. In this section we will see how to use the

DFT for Fourier analysis (computing coefficients) and Fourier synthesis (computing partial sums).

Analysis: computing Fourier coefficients

Consider an experiment in which you take N uniformly-spaced values (samples) of a function on an interval $[0, T)$,

$$f[n] = f(n\Delta t), \quad n = 0, 1, \ldots, N-1$$

where $\Delta t = T/N$ is the sampling interval. Implicitly, the observed portion of f is one period of a periodic function with period T. Expressing this period in terms of a Fourier series, the samples can be written

$$f[n] = \sum_{k=-\infty}^{\infty} c_k e^{i2\pi k(n\Delta t)/T} = \sum_{k=-\infty}^{\infty} c_k e^{i2\pi kn/N},$$

where the (c_k) are, as usual, the Fourier coefficients of f. Next, calculate the DFT of $f[n]$,

$$F[m] = \sum_{n=0}^{N-1} f[n] e^{-i2\pi mn/N} = \sum_{n=0}^{N-1} \left[\sum_{k=-\infty}^{\infty} c_k e^{i2\pi kn/N} \right] e^{-i2\pi mn/N}$$

$$= \sum_{k=-\infty}^{\infty} c_k \left[\sum_{n=0}^{N-1} e^{i2\pi(k-m)n/N} \right].$$

The inner sum is just the orthogonality equation for the DFT basis,

$$\sum_{n=0}^{N-1} e^{i2\pi(k-m)n/N} = \frac{1 - e^{i2\pi(k-m)}}{1 - e^{i2\pi(k-m)/N}},$$

which gives zero except for $k - m = rN$, where m runs from 0 to $N-1$ and k and r range over all the integers. It is equivalent to a sum of unit samples (compare the comb sequence III_N, Equation 3.8),

$$\sum_{n=0}^{N-1} e^{i2\pi(k-m)n/N} = \sum_{r=-\infty}^{\infty} N\delta[k - (m + rN)].$$

We have, finally,

$$F[m] = \sum_{k=-\infty}^{\infty} c_k \sum_{r=-\infty}^{\infty} N\delta[k - (m + rN)]$$

$$= N \sum_{r=-\infty}^{\infty} \sum_{k=-\infty}^{\infty} c_k \delta[k - (m + rN)]$$

$$= N \sum_{r=-\infty}^{\infty} c_{m+rN}, \quad m = 0, 1, \ldots, N-1. \tag{4.41}$$

4.8 COMPUTING THE FOURIER SERIES

Equation 4.41 connects the DFT with the Fourier coefficients.

$$F[0] = N(\cdots + c_{-2N} + c_{-N} + c_0 + c_N + c_{2N} + \cdots)$$
$$F[1] = N(\cdots + c_{-2N+1} + c_{-N+1} + c_1 + c_{N+1} + c_{2N+1} + \cdots)$$
$$\vdots$$
$$F[N/2] = N(\cdots + c_{-3N/2} + c_{-N/2} + c_{N/2} + c_{3N/2} + \cdots)$$
$$\vdots$$
$$F[N-2] = N(\cdots + c_{-2N-2} + c_{-N-2} + c_{-2} + c_{N-2} + c_{2N-2} + \cdots)$$
$$F[N-1] = N(\cdots + c_{-2N-1} + c_{-N-1} + c_{-1} + c_{N-1} + c_{2N-1} + \cdots).$$

Each value of the DFT is a sum of Fourier coefficients. The reason for this is the peculiar effect that sampling has on sinusoids, as we saw in Example 3.1. Take $F[1]$ as an example. The Fourier coefficient c_1 corresponds to the Fourier series basis function $e^{i2\pi(1)t/T}$. Sampling it at $t = n\Delta t$, we obtain $e^{i2\pi n \Delta t/T} = e^{i2\pi n/N}$. The next term, c_{N+1}, belongs to the basis function $e^{i2\pi(N+1)t/T}$, whose samples are also $e^{i2\pi n/N}$. All Fourier series components of the form $e^{i2\pi(1+rN)t/T}$ reduce, after sampling, to $e^{i2\pi n/N}$, the first DFT basis vector. Consequently, $F[1]$ contains contributions from all the frequencies $\nu_{1+rN} = (1+rN)/T$, each of which aliases down to $\nu_1 = 1/T$:

$$F[1] = \left\langle \left[\cdots + c_{-N+1} e^{i2\pi(-N+1)t/T} + c_1 e^{i2\pi t/T} + c_{N+1} e^{i2\pi(N+1)t/T} + \cdots \right]_{t=nT}, e^{i2\pi n/N} \right\rangle$$
$$= \left\langle \left(\cdots + c_{-N+1} e^{i2\pi n/N} + c_1 e^{i2\pi n/N} + c_{N+1} e^{i2\pi n/N} + \cdots \right), e^{i2\pi n/N} \right\rangle$$
$$= \left(\cdots + c_{-N+1} + c_1 + c_{N+1} + \cdots \right) \cdot \left\langle e^{i2\pi n/N}, e^{i2\pi n/N} \right\rangle$$
$$= N \left(\cdots + c_{-N+1} + c_1 + c_{N+1} + \cdots \right).$$

Because of aliasing the DFT will generally be a hopeless jumble of Fourier coefficients. But, suppose that f contains only the frequencies $\{0, 1/T, 2/T, \ldots, N/2T\}$. Then, its Fourier series will consist only of the coefficients $\{c_0, c_{\pm 1}, c_{\pm 2}, \ldots, c_{\pm N/2}\}$, and the DFT values will be exactly

$$F[0] = Nc_0$$
$$F[1] = Nc_1$$
$$\vdots$$
$$F[N/2] = N\left(c_{N/2} + c_{-N/2}\right)$$
$$\vdots$$
$$F[N-2] = Nc_{-2}$$
$$F[N-1] = Nc_{-1}.$$

In this case, the DFT provides exact information about the line spectrum of f. A function f that contains only frequency components below some upper limit B is said to be *bandlimited* to B. To avoid aliasing, a function must be bandlimited to $B = N/2T$, or since $T = N\Delta t$,

$$\Delta t \leq \frac{1}{2B}. \tag{4.42}$$

The reciprocal $1/\Delta t$ is the Nyquist rate, which we have seen before (compare Equation 3.1). In practice, a perfectly bandlimited signal is impossible to achieve, but

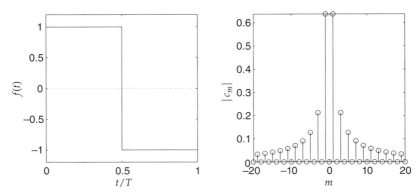

FIGURE 4.30 A square wave function and its Fourier coefficients.

aliasing can usually be reduced to tolerable levels by sampling with a small Δt, and applying a lowpass filter to the signal prior to sampling.

Figures 4.30 and 4.31 demonstrate the aliasing effect. The Fourier coefficients of a square wave were previously calculated in Equation 4.10. They are plotted in Figure 4.30. In Figure 4.31, the Fourier coefficients are estimated by computing the

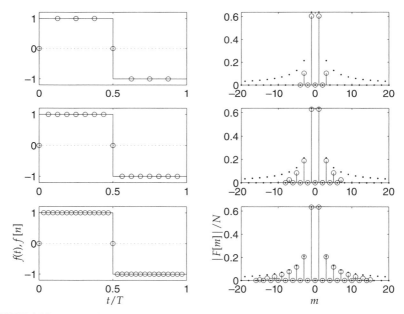

FIGURE 4.31 Using the DFT to estimate the Fourier coefficients of a square wave. The dots display the magnitudes of the true Fourier series coefficients. Aliasing causes the DFT values to be different than the true coefficients (*top, middle*), but smaller sampling intervals reduce this effect (*bottom*).

DFT of samples of the square wave, for different sampling intervals. Aliasing causes the DFT coefficients to be considerably different than the corresponding Fourier series coefficients, particularly in the first two spectra. In the third spectrum, the sampling interval is small enough that the DFT is a reasonable approximation to the true line spectrum, at least for low frequencies. Because its Fourier series coefficients only decay $O(|n|^{-1})$, the square wave's effective bandwidth is very high, and the sampling interval must be quite small in order to achieve accurate results.

Synthesis: computing partial sums

To evaluate a function described by a Fourier series, one computes values of a partial sum,

$$S_N(t) = \sum_{n=-N}^{N} c_n e^{i2\pi nt/T}.$$

Assume that S_N is to be evaluated at points evenly spaced over one period, $[0, T)$. To satisfy the Nyquist criterion, the sampling interval Δt can be at most half the period of the highest Fourier component (two samples per cycle). That frequency is N/T, so $\Delta t \leq \frac{T}{2N}$. Choosing a Δt that is much smaller than $T/2N$ will generate more closely spaced values and graphs with a smoother appearance. Sampling the partial sum with $t = m\Delta t$ gives

$$S_N(m\Delta t) = \sum_{n=-N}^{N} c_n e^{i2\pi nm\Delta t/T} = \sum_{n=-N}^{N} c_n e^{i2\pi nm/M}, \quad m = 0, 1, \ldots, M-1.$$

We could compute this sum directly, using $(2N+1)M$ multiply-adds and an equivalent number of evaluations of the complex exponential. However, except for the limits on the sum it looks like an inverse DFT, so it makes sense to try to express the sum as a DFT to exploit the $M \log_2 M$ efficiency of the FFT algorithm. To this end, embed the Fourier coefficients $(c_n)_{n=-N}^{N}$ in a vector F of length M as follows:

$$F = (c_0, c_1, \ldots, c_N, \underbrace{0, \ldots, 0}_{M-(2N+1)}, c_{-N}, \ldots, c_{-1}).$$

This vector has the form of a DFT with the Fourier coefficients c_n in the appropriate bins. There are $M - (2N+1)$ zeros inserted between c_N and c_{-N} so that F has the same number of points as S_N, as required for the DFT. The zeros represent the Fourier components with $|n| > N$ that are not included in the partial sum. With this, we have

$$S_N(m\Delta t) = \sum_{k=0}^{M-1} F[k] e^{i2\pi km/M}$$
$$= M \, DFT^{-1}\{F\}, \quad m = 0, 1, \ldots, M-1 \quad (4.43a)$$
$$\text{where } F = (c_0, c_1, \ldots, c_N, \underbrace{0, \ldots, 0}_{M-(2N+1)}, c_{-N}, \ldots, c_{-2}, c_{-1}). \quad (4.43b)$$

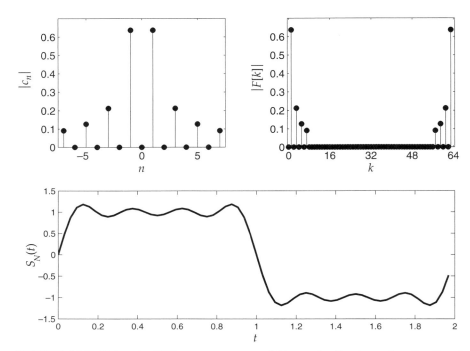

FIGURE 4.32 Using the DFT to compute a partial sum of a Fourier series, $N = 7$ and $M = 64$. *Top, left:* The magnitudes $|c_n|$ of the Fourier coefficients for $|n| \leq N$. *Top, right:* The magnitude of the vector F. *Bottom:* The partial sum S_7. DFT synthesis is about an order of magnitude faster than direct calculation of the partial sum.

In Figure 4.32 we show the result of a DFT synthesis of S_7 for a square wave of period $T = 2$. Using MATLAB, the DFT synthesis used about 2900 floating-point operations (flops) for $M = 64$ as opposed to 21,000 flops for direct synthesis.

4.9 DISCRETE TIME FOURIER TRANSFORM

The Fourier series equations (Equation 4.1) connect a function on a continuous time domain to another function defined on a discrete frequency domain. If we exchange the time and frequency domains, so that the time domain is discrete and the frequency domain is continuous, the *discrete-time Fourier transform* (DTFT) is obtained. It is defined[12]

$$F_d(\theta) = \sum_{n=-\infty}^{\infty} f[n] e^{-in\theta} \tag{4.44a}$$

$$f[n] = \frac{1}{2\pi} \int_{-\pi}^{\pi} F_d(\theta) e^{in\theta} \, d\theta. \tag{4.44b}$$

---

[12] In the signal processing literature, the discrete-time Fourier transform is usually written $F\left(e^{j\omega}\right)$, where ω is the digital frequency.

We may also indicate that F_d is the Fourier transform of f by the notation $f \longmapsto F_d$, or a transform pair by the notation $f \longleftrightarrow F_d$. The frequency variable θ is the digital frequency introduced in Chapter 3, and is measured in radians/sample, or just radians. The Fourier transform is periodic, $F_d(\theta) = F_d(\theta + 2\pi)$.

Here are three examples of Fourier transform calculations.

Example 4.21. Let f be a unit sample,

$$f[n] = \delta[n-k] = \begin{cases} 1, & n=k \\ 0, & \text{otherwise} \end{cases}.$$

The Fourier transform is

$$F_d(\theta) = \sum_{n=-\infty}^{\infty} \delta[n-k]e^{-in\theta} = e^{-ik\theta}. \tag{4.45}$$

Example 4.22. Let f be a rectangle function,

$$f[n] = \begin{cases} 1, & |n| \le N \\ 0, & \text{otherwise} \end{cases}.$$

The Fourier transform is

$$F_d(\theta) = \sum_{n=-\infty}^{\infty} f[n]e^{-in\theta} = \sum_{n=-N}^{N} e^{-in\theta} = D_{2N+1}\left(\frac{\theta}{2\pi}\right), \tag{4.46}$$

a Dirichlet kernel.

Example 4.23. Let f be a decaying exponential,

$$f[n] = a^n U[n] = \begin{cases} (1, a, a^2, \ldots), & n \ge 0 \\ 0, & n < 0 \end{cases} \quad |a| < 1.$$

The Fourier transform is

$$\begin{aligned} F_d(\theta) &= \sum_{n=0}^{\infty} a^n e^{-in\theta} \\ &= \lim_{N \to \infty} \sum_{n=0}^{N-1} (ae^{-i\theta})^n = \lim_{N \to \infty} \frac{1 - a^N e^{-iN\theta}}{1 - ae^{-i\theta}} \\ &= \frac{1}{1 - ae^{-i\theta}}. \end{aligned} \tag{4.47}$$

The function is plotted in Figure 4.33.

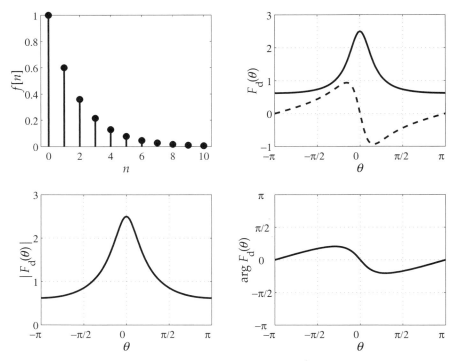

FIGURE 4.33 Discrete-time Fourier transform of the sequence $f[n] = a^n U[n]$. *Clockwise from top left:* The function f, for $1 > a > 0$; the Fourier transform F_d (real part solid, imaginary part dashed); the phase of the transform, $\arg F_d$; the magnitude of the transform, $|F_d|$.

4.9.1 Convergence Properties

The Fourier transform of a bounded, finite-duration sequence, $f \in \ell^0$, is a finite sum and will always exist. If $f \in \ell^1$, it is easy to see that the Fourier transform sum is bounded, $F_d \in L^\infty[-\pi, \pi]$:

$$\left| \sum_{n=-\infty}^{\infty} f[n] e^{-in\theta} \right| \leq \sum_{n=-\infty}^{\infty} |f[n] e^{-in\theta}| = \sum_{n=-\infty}^{\infty} |f[n]| < \infty.$$

If $f \in \ell^2$, the sequence of partial sums,

$$S_N = \sum_{|n| \leq N} f[n] e^{-in\theta},$$

converges to F_d in the L^2 norm,

$$\lim_{|N| \to \infty} \|F_d - S_N\|_2 = 0.$$

All the results for Fourier series in $L^2[-\pi, \pi]$ carry over to the DTFT. If $f \in \ell^2$, then $F_d \in L^2[-\pi, \pi]$. Beyond this, Theorem 4.4d connects the rate of decay of a sequence and the smoothness of its Fourier transform.

- If f decays faster than $|n|^{-p}$ ($p \geq 1$), then F_d is p-times continuously differentiable, $F_d \in C^{(p)}[-\pi, \pi]$.
- An exponential sequence $f[n] = a^n U[n]$, if $|a| < 1$, is rapidly decreasing and its Fourier transform $F_d(\theta) = \frac{1}{1-ae^{-i\theta}}$ is infinitely continuously differentiable ($C^{(\infty)}[-\pi, \pi]$). If $|a| \geq 1$, the exponential sequence is not in ℓ^1 and cannot be Fourier transformed.
- The transform of a finite sequence is a finite sum of complex exponentials, which is also in $C^{(\infty)}[-\pi, \pi]$.

The convergence of the inverse transform is covered by Theorem 4.4a–4.4c. Most practical applications of the DTFT involve finite sequences or combinations of decaying exponential sequences, whose transforms always exist and are infinitely continuously differentiable. On the other hand, there are some important sequences, like $f[n] = \cos \theta_0 n$, which are not ℓ^1 and whose transforms do not converge in any usual sense. This problem will be addressed in a later chapter.

4.9.2 Theorems

Many of the theorems for the Fourier series carry over to the discrete-time transform in the expected way. We list them here, mostly without proof.

Linearity (cf. Theorem 4.5)

$$af + bg \longleftrightarrow aF_d + bG_d \qquad (4.48)$$

Parseval (cf. Theorem 4.7)

If $f \in \ell^2$, then

$$\sum_{n=-\infty}^{\infty} f[n] g^*[n] = \frac{1}{2\pi} \int_{-\pi}^{\pi} F_d(\theta) G_d^*(\theta)\, d\theta \qquad (4.49a)$$

$$\sum_{n=-\infty}^{\infty} |f[n]|^2 = \frac{1}{2\pi} \int_{-\pi}^{\pi} |F_d(\theta)|^2\, d\theta. \qquad (4.49b)$$

There is no Riemann–Lebesgue lemma for the DTFT, because the frequency domain is bounded.

Area (cf. Theorem 4.8)

$$F_d(0) = \sum_{n=-\infty}^{\infty} f[n] \qquad (4.50a)$$

$$f[0] = \frac{1}{2\pi} \int_{-\pi}^{\pi} F_d(\theta)\, d\theta \qquad (4.50b)$$

for functions f and F_d for which the indicated sum and integral exist.

Symmetries (cf. Theorem 4.6)

Refer to Figure 4.16, replacing c_n with $f[n]$ and $f(\theta)$ with $F_d(\theta)$.

Shift

For infinite discrete sequences, a time shift $n - r$ is noncyclic, as contrasted with the finite sequences used with the DFT. On the other hand, a frequency $\theta - \beta$ is cyclic, because the DTFT is periodic in θ.

Theorem 4.13 (Shift). Let $f \longleftrightarrow F_d$ be a DTFT pair. Then

$$f[n - r] \longleftrightarrow e^{-ir\theta} F_d(\theta) \qquad (4.51)$$
$$e^{in\beta} f[n] \longleftrightarrow F_d(\theta - \beta). \qquad (4.52)$$

Finite-difference (derivative)

This is the discrete version of the derivative theorem for Fourier series.

Theorem 4.14 (Difference). Let $\Delta_1 f$ be the first difference of f, defined by

$$\Delta_1 f[n] = f[n] - f[n - 1].$$

The Fourier transform of the first difference is

$$\Delta_1 f \longmapsto (1 - e^{-i\theta}) F_d(\theta). \qquad (4.53)$$

Proof: Use the shift theorem. ∎

Cumulative sum (integral)

By analogy with the integral theorem for Fourier series, define the cumulative sum Σ_f of a sequence f,

$$\Sigma_f[n] = \sum_{k=-\infty}^{n} f[k].$$

This operation is the inverse of the finite difference, as you can see by taking the cumulative sum of $\Delta_1 f$:

$$\Sigma_{\Delta_1 f}[n] = \sum_{k=-\infty}^{n} \Delta_1 f[k] = \sum_{k=-\infty}^{n} (f[k] - f[k-1])$$

$$= \sum_{k=-\infty}^{n} f[k] - \sum_{k=-\infty}^{n} f[k-1] = \sum_{k=-\infty}^{n} f[k] - \sum_{k=-\infty}^{n-1} f[k]$$

$$= f[n].$$

We can write the cumulative sum as a recursion,

$$\Sigma_f[n] = \sum_{k=-\infty}^{n-1} f[k] + f[n] = \Sigma_f[n-1] + f[n].$$

Now let $\tilde{\Sigma}_f$ be the Fourier transform of Σ_f, and take the Fourier transform of both sides of the recursion (use the shift theorem):

$$\left(1 - e^{-i\theta}\right) \tilde{\Sigma}_f(\theta) = F_d(\theta).$$

Solving for $\tilde{\Sigma}_f$

$$\tilde{\Sigma}_f(\theta) = \frac{F_d(\theta)}{1 - e^{-i\theta}}.$$

Note that $\tilde{\Sigma}_f$ blows up at $\theta = 0$ unless $F_d(0) = 0$ (the integral of a constant grows without bound). With this qualification, we have

Theorem 4.15 (Integration). Let $f \in \ell^1$ and F_d be its Fourier transform, and let $\Sigma_f[n] = \sum_{k=-\infty}^{n} f[k]$ be the cumulative sum of f. If $F_d(0) = 0$, the Fourier transform $\tilde{\Sigma}_f$ of the cumulative sum is

$$\tilde{\Sigma}_f(\theta) = \frac{F_d(\theta)}{1 - e^{-i\theta}}. \tag{4.54}$$

Compare Equations 4.53 and 4.54 and observe the inverse relationship. In one you multiply by $1 - e^{-i\theta}$ and in the other you divide by $1 - e^{-i\theta}$. Compare also the derivative and integral theorems for the Fourier series, (Equations 4.22 and 4.23): In the former you multiply by $i2\pi n/L$ and in the latter, you divide by $i2\pi n/L$.

Convolution

The convolution for sequences f and h is defined

$$f * h[n] = \sum_{k=-\infty}^{\infty} f[k]h[n-k] = \sum_{k=-\infty}^{\infty} f[n-k]h[k], \tag{4.55}$$

when the sum converges. In applications of discrete convolution, for example, signal processing, it frequently happens that f and h are either of finite duration or are one sided. If h has finite duration, $h = (\ldots, 0, h[0], h[1], \ldots h[N-1], 0, \ldots)$, then the convolution is a finite sum,

$$f * h[n] = \sum_{k=0}^{N-1} f[n-k]h[k]$$

and is clearly bounded for all n if f and h are bounded. If f and h are one sided, $f[n], h[n] = 0, n < 0$, then the convolution is also a finite sum,

$$f * h[n] = \sum_{k=0}^{n} f[k]h[n-k] = \sum_{k=0}^{n} f[n-k]h[k],$$

but it is not guaranteed to be bounded as $n \to \infty$ without additional conditions on f or h.

We may formally derive convolution and product relationships like those for the Fourier series. For the convolution, substitute the convolution sum into the definition for the Fourier transform (Equation 4.44),

$$f * h \longmapsto \sum_{n=-\infty}^{\infty} \left(\sum_{k=-\infty}^{\infty} f[k]h[n-k] \right) e^{-in\theta},$$

and reverse the order of summation,

$$= \sum_{k=-\infty}^{\infty} f[k] \left(\sum_{n=-\infty}^{\infty} h[n-k] e^{-in\theta} \right)$$

$$= \sum_{k=-\infty}^{\infty} f[k] e^{-ik\theta} H_{\mathrm{d}}(\theta) \qquad \text{(shift theorem)}$$

$$= F_{\mathrm{d}}(\theta) H_{\mathrm{d}}(\theta).$$

For the product, express f as an inverse Fourier transform, $f[n] = \frac{1}{2\pi} \int_{-\pi}^{\pi} F_{\mathrm{d}}(\varphi) e^{in\varphi} d\varphi$, and insert it into the forward transform,

$$fh \longmapsto \sum_{n=-\infty}^{\infty} f[n] h[n] e^{in\theta}$$

$$= \sum_{n=-\infty}^{\infty} \left(\frac{1}{2\pi} \int_{-\pi}^{\pi} F_{\mathrm{d}}(\varphi) e^{in\varphi} d\varphi \right) h[n] e^{-in\theta},$$

then reverse the order of summation and integration,

$$= \frac{1}{2\pi} \int_{-\pi}^{\pi} F_{\mathrm{d}}(\varphi) \left(\sum_{n=-\infty}^{\infty} h[n] e^{-in(\theta-\varphi)} \right) d\varphi$$

$$= \frac{1}{2\pi} \int_{-\pi}^{\pi} F_{\mathrm{d}}(\varphi) H_{\mathrm{d}}(\theta-\varphi) d\varphi$$

$$= F_{\mathrm{d}} \circledast H_{\mathrm{d}}. \qquad \text{(recall Equation 4.24)}$$

Assuming the inverse relationships also hold, we have two transform pairs,

$$f * h \longleftrightarrow F_{\mathrm{d}} H_{\mathrm{d}} \qquad (4.56\mathrm{a})$$
$$fh \longleftrightarrow F_{\mathrm{d}} \circledast H_{\mathrm{d}}. \qquad (4.56\mathrm{b})$$

Of course, we must determine the conditions on f and h under which the convolution sum converges and the transform results are valid. These are summarized in Table 4.3, below.[13]

The ℓ^p spaces are nested, $\ell^1 \subset \ell^2 \subset \ell^\infty$ (opposite the nesting for the $L^p[0,L]$ spaces). The table says that the product of two sequences is at least as well behaved as its factors. The square summability of f and h is sufficient to guarantee that they and their product have Fourier transforms. Moreover, $F_{\mathrm{d}}, H_{\mathrm{d}} \in L^2[-\pi, \pi]$ and $F_{\mathrm{d}} \circledast H_{\mathrm{d}}$ is bounded and continuous (Table 4.2). Thus, the inverse transform exists as well.

On the other hand, the convolution $f * h$ is no better than f or h, and may be worse. It is not good enough for f and h to be square summable, for this gives a convolution that is merely bounded and not guaranteed to decay rapidly enough to have a Fourier transform. If h is no better than square summable, then f must be absolutely summable in order for f, h, and $f * h$ to be transformable. Under these

---

[13] The convolution and product relationships in Table 4.3 are the discrete versions of results in Section 5.5. Also see Gasquet and Witomski (1999, pp. 367–372).

TABLE 4.3 Details about convolutions and products of sequences. The norm of the convolution is bounded, $\|f * h\| \leq \|f\| \|h\|$, where the norms are calculated in the appropriate spaces. When h is bounded ($h \in \ell^\infty$), the product fh inherits the norm properties of f, that is, if $f \in \ell^2$ then $fh \in \ell^2$. The last two rows describe common practical cases: finite length (ℓ^0) and rapidly decreasing (r.d.) sequences.

If f is	and h is	then $f * h$ is	and fh is
ℓ^1	ℓ^1	ℓ^1	ℓ^1
ℓ^1	ℓ^2	ℓ^2	ℓ^1
ℓ^1	ℓ^∞	ℓ^∞	ℓ^1
ℓ^2	ℓ^2	ℓ^∞	ℓ^1
ℓ^∞ r.d.	$\ell^0 \cap \ell^\infty$ r.d.	ℓ^∞ r.d.	$\ell^0 \cap \ell^\infty$ r.d.

conditions, the product $F_d H_d$ will be at least absolutely integrable (Table 4.2), and the inverse transform will also exist.

These results are summarized in the following theorem.

Theorem 4.16 (Convolution). Let $f \longleftrightarrow F_d$ and $h \longleftrightarrow H_d$ be DTFT pairs.

(a) If $f \in \ell^2$ and $h \in \ell^1$, then $f * h \in \ell^2$, $F_d H_d \in L^1[-\pi, \pi]$, and

$$f * h \longleftrightarrow F_d H_d.$$

(b) If $f, h \in \ell^2$, then $fh \in \ell^1$, $F_d \circledast H_d \in L^\infty \cap C^0$, and

$$fh \longleftrightarrow F_d \circledast H_d.$$

Rapidly decreasing sequences are absolutely summable, and covered by this theorem. Additionally, when f and h are rapidly decreasing, their Fourier transforms F_d and H_d are infinitely continuously differentiable. So is their product, $F_d H_d$, and as a result the inverse transform $f * h$ exists and is rapidly decreasing. Conversely, the convolution $F_d * H_d$ is infinitely continuously differentiable and its inverse transform, fh, is rapidly decreasing.

Example 4.24. Let $f[n] = a^n U[n]$ and $g[n] = b^n U[n]$, where $|a|, |b| < 1$. Both are rapidly decreasing. We demonstrate the convolution theorem by calculating $f * g$ in two ways.

Evaluating the convolution sum directly (Figure 4.34),

$$f * g[n] = \sum_{k=-\infty}^{\infty} a^k U[k] \cdot b^{n-k} U[n-k] = \sum_{k=0}^{n} a^k b^{n-k}, n \geq 0$$

$$= b^n \frac{1 - (ab^{-1})^{n+1}}{1 - ab^{-1}} U[n] = \frac{b^{n+1} - a^{n+1}}{b - a} U[n].$$

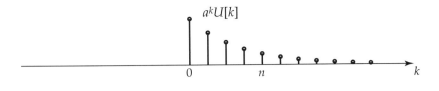

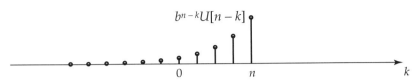

FIGURE 4.34 Illustrating the overlap of the two one-sided exponential sequences in the convolution $f * g = \sum_{k=-\infty}^{\infty} a^k U[k] \cdot b^{n-k} U[n-k]$. (*Top*) The sequence ($a^k U[k]$). (*Bottom*) The reversed and shifted sequence, ($b^{n-k} U[n-k]$). When $n < 0$, the sequences do not overlap and the value of the convolution is zero. When $n \geq 0$, the convolution is a finite sum from 0 to n.

Using the convolution theorem with Equation 4.47,

$$F_d G_d(\theta) = \frac{1}{1 - ae^{-i\theta}} \frac{1}{1 - be^{-i\theta}}$$
$$= -\frac{a}{b-a} \frac{1}{1 - ae^{-i\theta}} + \frac{b}{b-a} \frac{1}{1 - be^{-i\theta}}.$$

Therefore,

$$f * g[n] = -\frac{a}{b-a} a^n U[n] + \frac{b}{b-a} b^n U[n] = \frac{b^{n+1} - a^{n+1}}{b-a} U[n]. \qquad \blacksquare$$

Dilation: upsampling and downsampling

The linearity and shift theorems tell what happens to the Fourier transform when a sequence is scaled in amplitude and when it is translated along the time axis. It is also of practical interest to know what happens to the Fourier transform when a function is stretched or squeezed by scaling the time axis. This is called *dilation*.

The dilation of a function of a continuous variable, $f(x)$, is written $f(ax)$, where a is a constant. When $|a| < 1$, the function is stretched, and when $|a| > 1$, it is squeezed. For a discrete-time signal (sequence), the expression $f[an]$ only makes sense when an is an integer. For $|a| > 1$ (squeezing), this means that a must also be an integer, which we will henceforth call P. Consider then the sequence $f[nP]$. The $n = 0$ value is $f[0]$, the $n = 1$ value is $f[P]$, the $n = 2$ value is $f[2P]$, etc. That is, the sequence is squeezed by selecting every P^{th} sample from f, for example, for $P = 3$,

$$(\ldots\; 1\; 2\; 3\; 4\; 5\; 6\; 7\; 8\; 9\; \ldots) \longmapsto (\ldots\; 1\; 4\; 7\; \ldots).$$

This is also called *downsampling*. A downsampled sequence is denoted $f_{\downarrow P}$:

$$f_{\downarrow P}[n] = f[nP]. \tag{4.57}$$

In the opposite case, $|a| < 1$ (stretching), we can have integer an only if a is a rational number, and then only for some n. Let $a = 1/P$. We will have integer n/P for values of n that are integer multiples of P. The $n = 0$ value of the sequence $f[n/P]$ is $f[0]$, the $n = P$ value is $f[1]$, the $n = 2P$ value is $f[2]$, etc. All other values, for example, $n = 1, 2, \ldots, P - 1$, cannot be obtained from the original sequence f, and are set to zero. Stretching f is done by inserting $P - 1$ zeros between each of the original samples, for example, for $P = 3$,

$$(\ldots\ 1\ 2\ 3\ 4\ \ldots) \longmapsto (\ldots\ 0\ 0\ 1\ 0\ 0\ 2\ 0\ 0\ 3\ 0\ 0\ 4\ 0\ 0\ \ldots).$$

This is also called *upsampling*. An upsampled sequence is denoted $f_{\uparrow P}$:

$$f_{\uparrow P}[n] = \begin{cases} f[n/P], & n = rP, \quad r \in \mathbb{Z} \\ 0, & \text{otherwise} \end{cases}. \quad (4.58)$$

Another interpretation of upsampling and downsampling is frequently employed in signal processing. Suppose the original sequence f is obtained by sampling a continuous-time signal at sampling interval T, that is, $f[n] = f(nT)$. If zeros are to be inserted into f in real time, we have to maintain the time spacing of the original values. The sampling interval of the upsampled sequence $f_{\uparrow P}$ must decrease to $\frac{T}{P}$; the sampling rate increases from $\frac{1}{T}$ to $\frac{P}{T}$. If samples are to be removed from f in real time, we also have to maintain the original time spacing between the samples. The spacing of the remaining samples is PT; the sampling rate decreases from $\frac{1}{T}$ to $\frac{1}{PT}$.

The Fourier transform of an upsampled sequence $f_{\uparrow P}$ is

$$F_{\uparrow P}(\theta) = \sum_{n=-\infty}^{\infty} f_{\uparrow P}[n] e^{-in\theta} = \sum_{r=-\infty}^{\infty} f[r] e^{-irP\theta}$$
$$= F_{\mathrm{d}}(P\theta). \quad (4.59)$$

Stretching in the time domain causes a squeezing in the frequency domain. $F_{\mathrm{d}}(P\theta)$ is a squeezing of P periods of F_{d} into the interval $[-\pi, \pi]$ (Figure 4.35).

The Fourier transform of a downsampled sequence $f_{\downarrow P}$ is

$$F_{\downarrow P}(\theta) = \sum_{n=-\infty}^{\infty} f_{\downarrow P}[n] e^{-in\theta} = \sum_{n=-\infty}^{\infty} f[nP] e^{-inP(\theta/P)}.$$

Introduce the comb sequence (Equation 3.8),

$$\mathrm{III}_P[n] = \sum_{r=-\infty}^{\infty} \delta[n - rP] = \begin{cases} 1, & n = rP,\ r \in \mathbb{Z} \\ 0, & \text{otherwise} \end{cases}.$$

Then (this is subtle—expand a few terms to convince yourself),

$$F_{\downarrow P}(\theta) = \sum_{n=-\infty}^{\infty} f[n]\, \mathrm{III}_P[n]\, e^{-in\theta/P}.$$

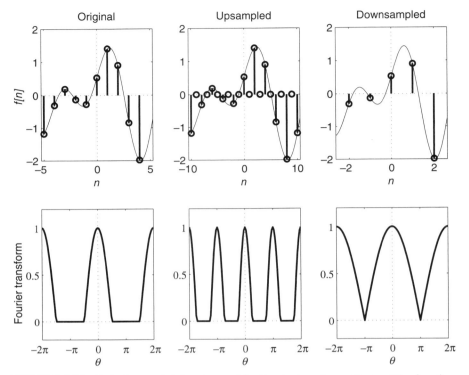

FIGURE 4.35 *Left:* A sampled function and its Fourier transform. *Center:* The function, upsampled by 2, and its Fourier transform. Upsampling causes the Fourier transform to be compressed. *Right:* The function, downsampled by 2, and its Fourier transform. Downsampling stretches the Fourier transform. Aliasing occurs unless $f[n]$ is bandlimited to $|\theta| < \pi/2$.

The discrete-time comb function has a Fourier representation (see the problems for a derivation),

$$\mathrm{III}_P[n] = \frac{1}{P} \sum_{m=0}^{P-1} e^{i2\pi mn/P}.$$

Using this,

$$F_{\downarrow P}(\theta) = \sum_{n=-\infty}^{\infty} f[n] \left[\frac{1}{P} \sum_{m=0}^{P-1} e^{i2\pi mn/P} \right] e^{-in\theta/P}.$$

The order of summation can be reversed because $f \in \ell^1$:

$$F_{\downarrow P}(\theta) = \frac{1}{P} \sum_{m=0}^{P-1} \left[\sum_{n=-\infty}^{\infty} f[n] e^{-in(\theta - 2\pi m)/P} \right]$$

$$= \frac{1}{P} \sum_{m=0}^{P-1} F_{\mathrm{d}}\left(\frac{\theta - 2\pi m}{P} \right). \qquad (4.60)$$

Downsampling by a factor of P causes a stretching in the Fourier domain by a factor of P, in such a way that the Fourier transform remains periodic with period 2π (Figure 4.35). This can lead to aliasing unless f is bandlimited to $|\theta| < \frac{\pi}{P}$. This makes sense, because downsampling by P throws away samples and effectively lengthens the sampling interval from T to PT. Correspondingly, the Nyquist frequency decreases from $1/2T$ to $1/2PT$.

The preceding derivations are summarized in the following theorem.

Theorem 4.17 (Dilation). Let $f \longleftrightarrow F_d$ be a DTFT pair. Let $P \geq 1$ be an integer. Then

$$f_{\uparrow P} \longleftrightarrow F_d(P\theta) \tag{4.61a}$$

$$f_{\downarrow P} \longleftrightarrow \frac{1}{P} \sum_{m=0}^{P-1} F_d\left(\frac{\theta - 2\pi m}{P}\right) \tag{4.61b}$$

4.9.3 Discrete-time Systems

The most common application of the DTFT is the analysis of discrete-time linear, time-invariant (LTI) systems. These are modeled by linear difference equations with constant coefficients,

$$g[n] - a_1 g[n-1] - \cdots - a_N g[n-N] = b_0 f[n] + b_1 f[n-1] + \cdots + b_M f[n-M]. \tag{4.62}$$

The sequence f is the input, or driving function, and the sequence g is the output, or response. There may be initial conditions specifying the values $g[-1], g[-2], \ldots g[-N]$. We shall be interested only in the case where initial conditions are zero.

Mathematically, there is nothing to prevent the difference equation from also having terms of the form $b_{-k} f[n+k]$ with $k > 0$, but a physical system obeying such an equation would have the ability to anticipate its input and produce a response before receiving any stimulus. Such a system is said to be *noncausal*. We shall restrict our attention to *causal* systems, described by difference equations of the form given in Equation 4.62.

Discrete-time LTI systems are classified by their responses when the the input is a unit sample $\delta[n]$ and initial conditions are zero. This is called the *unit-sample response* or *impulse response*, and is usually denoted h.

Definition 4.3 (FIR and IIR systems). Let h be the impulse response of a causal, discrete-time LTI system; that is, h is the solution of the equation

$$h[n] - a_1 h[n-1] - \cdots - a_N h[n-N] = b_0 \delta[n] + b_1 \delta[n-1] + \cdots + b_M \delta[n-M], \quad n \geq 0$$

with

$$h[-1] = h[-2] = \cdots = h[-N] = 0.$$

(a) If all the coefficients $\{a_1, a_2, \ldots a_n\}$ are zero, then $h = (b_0, b_1, \ldots b_M, 0, \ldots)$. The system is said to be *finite-impulse response* (FIR) or *nonrecursive*.

(b) If any of the coefficients $\{a_1, a_2, \ldots a_n\}$ is nonzero, then the system is said to be *infinite-impulse response* (IIR) or *recursive*.

Example 4.25 (FIR and IIR systems).

(a) The FIR sequence $h = (1, 2, 1, 0, 0, \ldots)$ satisfies the nonrecursive difference equation

$$h[n] = x[n] + 2x[n-1] + x[n-2], \ n \geq 0$$

with $x[n] = \delta[n]$.

(b) The IIR sequence $h = (1, \frac{1}{2}, \frac{1}{4}, \ldots, \frac{1}{2^n}, \ldots)$ satisfies the recursive difference equation

$$h[n] - \frac{1}{2}h[n-1] = x[n], \ n \geq 0, \ h[-1] = 0$$

with $x[n] = \delta[n]$.

An arbitrary input f may be expressed

$$f[n] = \sum_k f[k]\delta[n-k].$$ ∎

The response of an LTI system to a shifted unit sample $\delta[n-k]$ is a shifted impulse response $h[n-k]$, and by linearity, the system response g is a superposition of shifted impulse responses,

$$g[n] = \sum_k f[k]h[n-k] = f * h[n], \quad (4.63)$$

a convolution.

Example 4.26 (Simple impulse responses).

(a) A system with impulse response $h = \delta[n-1]$ is called a *unit delay*. The output of a unit delay is a replica of the input, delayed by one sample:

$$f * h = f * \delta[n-1] = \sum_{k=-\infty}^{\infty} f[k]\delta[n-k-1] = f[n-1].$$

(b) A system with impulse response $h = \frac{1}{M}(\delta[n] + \delta[n-1] + \cdots + \delta[n-(M-1)])$ is an M-sample *moving average* or *boxcar filter*. The output at time n is the average of the current sample and the preceding $M-1$ samples:

$$f * h = \sum_{k=-\infty}^{\infty} f[n-k] \cdot \frac{1}{M}(\delta[k] + \delta[k-1] + \cdots + \delta[k-(M-1)])$$
$$= \frac{1}{M}(f[n] + f[n-1] + \cdots + f[n-(M-1)]). \quad \blacksquare$$

Frequency response
Consider specifically a complex exponential input $f[n] = e^{i\theta n}$. The output is

$$g[n] = \sum_{k=-\infty}^{\infty} h[n-k]e^{i\theta k}.$$

Make a change of variable $m = n - k$, and

$$g[n] = \sum_{m=-\infty}^{\infty} h[m]e^{i\theta(n-m)} = \left[\sum_{m=-\infty}^{\infty} h[m]e^{-i\theta m}\right] e^{i\theta n} = H_\text{d}(\theta)e^{i\theta n}.$$

The output of the system is equal to the input multiplied by a complex weighting factor, $H_\text{d}(\theta)$. This quantity is the Fourier transform of the impulse response. It is called the *transfer function* of the system. Expressing the transfer function in polar form, $H_\text{d}(\theta) = |H_\text{d}(\theta)|e^{i\varphi(\theta)}$, the output is

$$g[n] = |H_\text{d}(\theta)|e^{i[\theta n + \varphi(\theta)]}.$$

If the impulse response is real (the usual case), the transfer function is Hermitian, and if the input is a cosine rather than a complex exponential, the output is, by linearity,

$$g[n] = \frac{1}{2}|H_\text{d}(\theta)|e^{i\varphi(\theta)}e^{i\theta n} + \frac{1}{2}|H_\text{d}(\theta)|e^{-i\varphi(\theta)}e^{-i\theta n} = |H_\text{d}(\theta)|\cos[\theta n + \varphi(\theta)].$$

The frequency of the output is the same as the frequency of the input, but the amplitude and phase are modified by the system's transfer function. A frequency selective discrete-time LTI system is commonly called a *digital filter*. The absolute value $|H_\text{d}|$ is called the *magnitude response* of the filter, and the argument φ is called the *phase response* of the filter.

Example 4.27 (Frequency responses of simple FIR systems). Returning to the two simple systems in Example 4.26, we calculate their frequency responses.

(a) The unit delay, $h = \delta[n-1]$. The transfer function is

$$H_\text{d}(\theta) = \sum_n \delta[n-1]e^{-j\theta n} = e^{-j\theta}.$$

The magnitude response is unity, and the phase response is $-\theta$, linear in frequency (compare with the shift theorem).

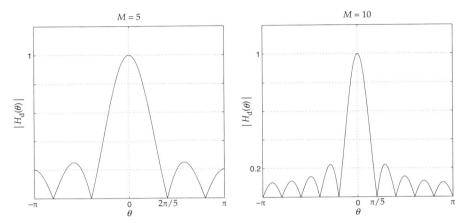

FIGURE 4.36 Magnitude response of M-point moving average filter, $M = 5$ (*left*) and $M = 10$ (*right*). The response is zero for frequencies that are integer multiples of $2\pi/M$.

(b) The moving average system, $h = \frac{1}{M}\Big(\delta[n] + \delta[n-1] + \cdots + \delta[n-(M-1)]\Big)$.
The transfer function is

$$H_d(\theta) = \frac{1}{M}\sum_{n=0}^{M-1} 1 \cdot e^{-j\theta n} = \frac{1 - e^{-jM\theta}}{1 - e^{-j\theta}} = e^{-j(M-1)\theta/2}\frac{\sin(M\theta/2)}{M\sin(\theta/2)}.$$

The magnitude response of the system is

$$|H_d(\theta)| = \left|\frac{\sin(M\theta/2)}{M\sin(\theta/2)}\right| = \frac{1}{M}\left|D_M(\theta/2\pi)\right|,$$

yet another Dirichlet kernel. Because the impulse response is normalized to a unit sum, the DC value of the magnitude response is unity (area theorem). The response goes to zero at frequencies $\theta = 2\pi m/M$ (Figure 4.36). For example, for $M = 5$, the null frequencies are $2\pi/5$ and $2 \cdot 2\pi/5$. The nature of these null frequencies is illustrated in Figure 4.37. The filter output is zero when the average of the samples in the moving window is zero. At the zero frequencies, for example, $2\pi/5$ as shown in the figure, the period is exactly the length of the window, so that the average is zero for all positions of the window. The same result is obtained if the window covers an integer number of periods, for example, two periods for $\theta = 4\pi/5$. ∎

Example 4.28 (An IIR filter). Suppose the impulse response of a digital filter is the sequence $h[n] = 0.6^n U[n]$. The transfer function is, applying Equation 4.47,

$$H_d(\theta) = \frac{1}{1 - 0.6e^{-i\theta}}.$$

This function is plotted in Figure 4.38. The magnitude response decreases with increasing frequency. A system with this characteristic is called a *lowpass filter*, because it preferentially weights low frequency Fourier components over high frequency components. ∎

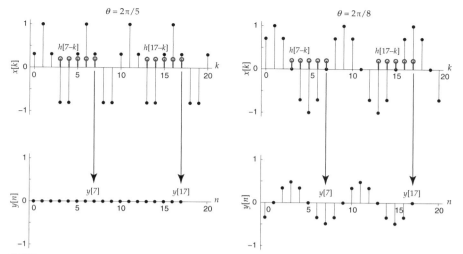

FIGURE 4.37 Time response of M-point moving average filter, $M = 5$, for sinusoids of frequency $2\pi/5$ (*left*) and $2\pi/8$ (*right*). The five samples of the averaging window (impulse response) are shown with hollow circles. The output, $y[n] = \frac{1}{5}\sum_{k=0}^{4} x[n-k]$, is zero when the averaging window covers a complete period of the input for all times n.

Example 4.29 (Upsampling with filtering). An important application of upsampling and downsampling is sample rate conversion. Given a sequence obtained by sampling a signal at one sample rate, calculate a new sequence that approximates the original signal at a different sampling rate. Suppose, for example, that samples taken at a 10 kHz rate are to be made compatible with a system that operates at a 20 kHz rate.[14] The first step is to upsample the sequence by 2, for example,

$$(\ldots\ 1\ 2\ 1\ 0\ 3\ 2\ \ldots) \overset{\uparrow 2}{\longmapsto} (\ldots\ 1\ 0\ 2\ 0\ 1\ 0\ 0\ 0\ 3\ 0\ 2\ \ldots)$$

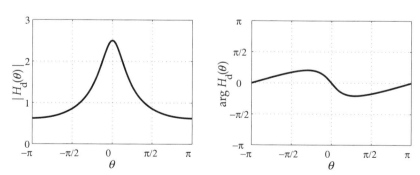

FIGURE 4.38 Transfer function of a digital filter with impulse response $h[n] = 0.6^n U[n]$. *Left:* The magnitude respose, $|H_d|$. *Right:* The phase response, $\arg H_d$.

---

[14]Good introductions to so-called multi-rate signal processing may be found in Porat (1997, pp. 461ff) and Oppenheim and Schafer (2010, pp. 179ff).

The inserted zeros are then replaced with appropriate values based on the original samples. A simple linear interpolator replaces each inserted zero with the average of the adjacent values,

$$(\ldots\ 1\ 0\ 2\ 0\ 1\ 0\ 0\ 0\ 3\ 0\ 2\ \ldots) \longmapsto \left(\ldots\ 1\ \tfrac{3}{2}\ 2\ \tfrac{3}{2}\ 1\ \tfrac{1}{2}\ 0\ \tfrac{3}{2}\ 3\ \tfrac{5}{2}\ 2\ \ldots\right)$$

That is,

$$g[n] = \begin{cases} f_{\uparrow 2}[n], & n \text{ even (an original sample)} \\ \tfrac{1}{2}f_{\uparrow 2}[n-1] + \tfrac{1}{2}f_{\uparrow 2}[n+1], & n \text{ odd (an inserted zero)} \end{cases}.$$

It turns out that a digital filter with impulse response $h[n] = \tfrac{1}{2}\delta[n-1] + \delta[n] + \tfrac{1}{2}\delta[n+1]$ will accomplish the task. (You may have noticed that this system is non-causal. A causal system that does the same calculations with a one-sample delay has impulse response $h[n] = \tfrac{1}{2}\delta[n-2] + \delta[n-1] + \tfrac{1}{2}\delta[n]$.)

The frequency response of the interpolating filter is

$$H_d(\theta) = \sum_{n=-\infty}^{\infty} h[n]e^{-in\theta} = \sum_{n=-\infty}^{\infty} \left(\tfrac{1}{2}\delta[n-1] + \delta[n] + \tfrac{1}{2}\delta[n+1]\right)e^{-in\theta}$$
$$= \tfrac{1}{2}e^{-i\theta} + 1 + \tfrac{1}{2}e^{i\theta}$$
$$= 1 + \cos\theta.$$

The Fourier transform of the interpolator's output is $G_d(\theta) = H_d(\theta)F_{\uparrow 2}(\theta)$. The effect of filtering in the frequency domain is shown in Figure 4.39. The original signal spectrum F_d is squeezed by upsampling. Ideally, if the original signal were sampled at 20 kHz, the spectrum would consist solely of the spectral bump centered at $\theta = 0$. Upsampling the 10 kHz sampled signal creates spurious replicas of the desired spectrum around $\pm\pi$. The function of the interpolation filter is to remove these undesired components. The figure shows that the simple linear interpolator partially achieves this objective. An ideal interpolating filter that completely removes the undesired

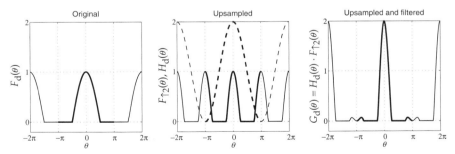

FIGURE 4.39 Sample rate conversion by upsampling and filtering, in the Fourier domain. The principal frequency domain $[-\pi, \pi)$ is highlighted with heavier lines. *Left:* The original signal spectrum, F_d. *Center:* The upsampled spectrum, $F_{\uparrow 2}$ (solid), squeezes two periods of F_d into the range $[-\pi, \pi)$. The spectral replicas centered on $\pm\pi$ are spurious, and are to be removed by the interpolating filter, H_d (dashed). *Right:* The upsampled spectrum after filtering. The spurious spectral replicas are significantly attenuated by the interpolating filter.

spectral replicas would have a rectangular, or "brick wall" frequency response, passing all frequencies less than $\frac{\pi}{4}$ and removing all higher frequencies. The design of practical filters that approximate the ideal response is an important topic in signal processing. ∎

4.9.4 Computing the DTFT

We can use the DFT to compute approximations to the DTFT.

Analysis: computing the Fourier transform
Beginning with the forward transform,

$$F_d(\theta) = \sum_{n=-\infty}^{\infty} f[n] e^{-i\theta n},$$

we sample the digital frequency at intervals $\Delta\theta = 2\pi/N$, giving

$$F_d(m\Delta\theta) = \sum_{n=-\infty}^{\infty} f[n] e^{-imn\Delta\theta} = \sum_{n=-\infty}^{\infty} f[n] e^{-i2\pi mn/N}, \quad m = 0, 1, \ldots, N-1.$$

The sum looks like a DFT, except for the limits. As a practical matter, we cannot sum over infinite limits but must truncate f to a vector $(f[0], f[1], \ldots, f[N-1])$, which we shall denote f_N. With this, the approximation to the Fourier transform, denoted $\hat{F}_d$, is

$$\hat{F}_d(m\Delta\theta) = \sum_{n=0}^{N-1} f[n] e^{-i2\pi mn/N} = \text{DFT}\{f_N\}, \quad m = 0, 1, \ldots, N-1. \quad (4.64)$$

Compare this result with Equation 4.43 for computing partial sums of a Fourier series.

Example 4.30. Consider again the decaying exponential $a^n U[n]$, $|a| < 1$, whose Fourier transform is $F_d(\theta) = \frac{1}{1-ae^{-i\theta}}$ (Equation 4.47). In Figure 4.40 we show the result of applying Equation 4.64 with two different truncations. ∎

Synthesis: inverting the Fourier transform
To invert the Fourier transform, we must compute the integral

$$f[n] = \frac{1}{2\pi} \int_{-\pi}^{\pi} F_d(\theta) e^{i\theta n} d\theta.$$

Comparing this with the Fourier series, the values $f[n]$ may be interpreted as the Fourier coefficients of the periodic function F_d. Following the same steps leading to Equation 4.41, we obtain

$$\frac{1}{N} \sum_{m=0}^{N-1} F_d\left(\frac{2\pi m}{N}\right) e^{i2\pi mn/N} = \sum_{r=-\infty}^{\infty} f[n+rN], \quad n = 0, 1, \ldots N-1. \quad (4.65)$$

The result of sampling F_d and calculating the inverse DFT is a replication of f with period N. To avoid errors caused by this replication, N must be chosen large enough

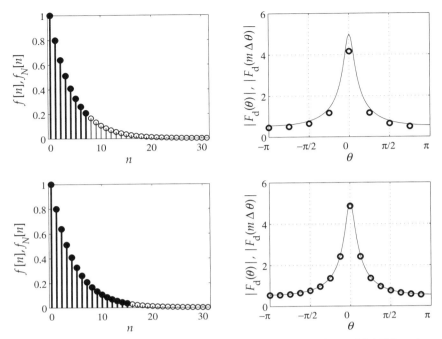

FIGURE 4.40 Discrete-time Fourier transform of the sequence $f[n] = 0.8^n U[n]$, estimated using the DFT. *Top: f* is truncated to 8 samples (filled circles, *left*) and the DFT does not agree well with the actual transform (open circles and dots, *right*). *Bottom: f* is truncated to 16 samples (*left*); the DFT is in much better agreement with the actual transform (*right*).

so that $f[n]$ for $n \geq N$ is very small compared with $f[n]$ for $N > n \geq 0$. Then, the estimate $\hat{f}$ computed from

$$\hat{f} = DFT^{-1}\left\{\left(F_d\left(\frac{2\pi m}{N}\right)\right)_{m=0}^{N-1}\right\} \qquad (4.66)$$

will be a good approximation to $f[n]$ for $n = 0, 1, \ldots, N-1$. In practice, choice of an appropriate N may require some trial and error.

Example 4.31. The calculation is illustrated in Figure 4.41 for the function $F_d(\theta) = \frac{1}{1-ae^{-i\theta}}$, whose inverse we know is $a^n U[n]$. Observe that as N increases, the errors due to the periodicity of the DFT are reduced. ∎

4.10 SUMMARY

A periodic function f that is at least absolutely integrable over one period may be expanded in a Fourier series (Equation 4.3):

$$f(x) = \sum_{n=-\infty}^{\infty} c_n \exp\left(\frac{i2\pi nx}{L}\right)$$

$$c_n = \frac{1}{L}\int_0^L f(x) \exp\left(-\frac{i2\pi nx}{L}\right) dx, \quad n \in \mathbb{Z}.$$

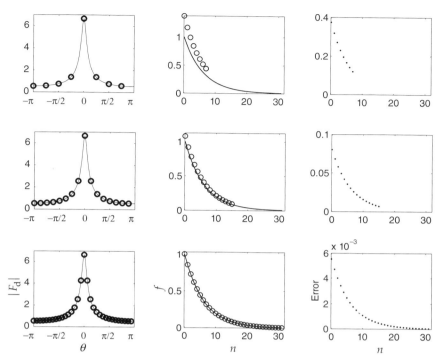

FIGURE 4.41 Inverse discrete-time Fourier transform of the function $F_d(\theta) = \frac{1}{1-0.8e^{-i\theta}}$, estimated using the DFT. *Left to right:* The function F_d and its samples (circles), the inverse transform $f[n] = 0.8^n U[n]$ and the estimated values (circles), the error $\hat{f} - f$. *Top to bottom:* $N = 8$, $N = 16$, $N = 32$. N must be chosen large enough to avoid errors due to the periodicity of the DFT.

Classical form of the Fourier series (Equation 4.5):

$$f(x) = \frac{a_0}{2} + \sum_{n=1}^{\infty} a_n \cos\left(\frac{2\pi nx}{L}\right) + \sum_{n=1}^{\infty} b_n \sin\left(\frac{2\pi nx}{L}\right)$$

$$a_n = \frac{2}{L} \int_0^L f(x) \cos\left(\frac{2\pi nx}{L}\right) dx, \quad n = 0, 1, 2, \ldots$$

$$b_n = \frac{2}{L} \int_0^L f(x) \sin\left(\frac{2\pi nx}{L}\right) dx, \quad n = 1, 2, \ldots$$

Convergence of the Fourier series improves with increasing regularity of f (Table 4.1). In particular, if f is piecewise smooth, the Fourier series converges pointwise to $\frac{1}{2}[f(x^+) + f(x^-)]$—to $f(x)$ at points of continuity, and to the average of the right and left limits at jump discontinuities. There will be overshoot near a jump. If f is continuous and at least piecewise smooth, the Fourier series converges absolutely and uniformly.

The Fourier series expansion conserves energy (preserves norm) for square-integrable functions. The Fourier line spectrum is $P_n = |c_n|^2 + |c_{-n}|^2$, $n > 0$ ($P_0 = |c_0|^2$), and is proportional to the energy in f at the frequency $\nu_n = n/L$. The

Riemann–Lebesgue lemma (Theorem 4.1) says that the Fourier spectrum rolls off to zero at high frequencies.

The DTFT (Section 4.9) is a reinterpretation of the Fourier series, exchanging time and frequency domains (Equation 4.44):

$$F_d(\theta) = \sum_{n=-\infty}^{\infty} f[n] e^{-in\theta}$$

$$f[n] = \frac{1}{2\pi} \int_{-\pi}^{\pi} F_d(\theta) e^{in\theta}\, d\theta.$$

The DTFT is chiefly used in the analysis of discrete time LTI systems (e.g., digital filters). The output of such a system is the convolution of the input with the unit-sample, or impulse, response: $g = f * h$. The frequency response of the system is contained in the transfer function H_d, which is the Fourier transform of the impulse response.

The DFT may be used to compute approximations to the Fourier series (Section 4.8) and the DTFT (Section 4.9.4).

Fourier series theorems

Theorem	Formula	Equation
Linearity	$af + bg \longleftrightarrow (af_n) + (bg_n)$	4.18
Symmetry	See Figure 4.16	
	f even $\longrightarrow b_n = 0$ (cosine series)	
	f odd $\longrightarrow a_n = 0$ (sine series)	
Riemann–Lebesgue	$\|c_n\| \to 0$ as $\|n\| \to \infty$	
Parseval	$\dfrac{1}{L}\int_0^L f(x) g^*(x)\, dx = \displaystyle\sum_{n=-\infty}^{\infty} f_n g_n^*$	4.19a
	$\dfrac{1}{L}\int_0^L \|f(x)\|^2 \, dx = \displaystyle\sum_{n=-\infty}^{\infty} \|c_n\|^2$	4.19b
Area	$c_0 = \dfrac{1}{L}\int_0^L f(x)\, dx$	4.20
	$f(0) = \displaystyle\sum_{n=-\infty}^{\infty} c_n$	
Shift	$f(x-r) \longleftrightarrow \left(e^{-i2\pi rn/L} c_n\right)$	4.21
Derivative	$f'(x) \longmapsto \left(\dfrac{i2\pi n}{L} c_n\right)$	4.22
	High frequencies boosted	
Integral	$\displaystyle\int_0^x f(\xi)\, d\xi \longmapsto (C_n) = \begin{cases} C_0 = \displaystyle\int_0^L f(\xi)\, d\xi \\ C_n = \dfrac{L}{i2\pi n} c_n, \quad n \neq 0 \end{cases}$	4.23
	($c_0 = 0$) High frequencies attenuated	
Convolution	$f \circledast h = \displaystyle\int_0^L f(\xi) h(x-\xi)\, d\xi \longleftrightarrow (Lf_n h_n)$	4.26
	$fh \longleftrightarrow (f_n) * (h_n) = \sum_{k=-\infty}^{\infty} f_k h_{n-k}$	

Discrete-time Fourier transform theorems

Theorem	Formula	Equation
Linearity	$af + bg \longleftrightarrow aF_d + bG_d$	4.48
Symmetry	See Figure 4.16	
Parseval	$\sum_{n=-\infty}^{\infty} f[n]g^*[n] = \dfrac{1}{2\pi} \int_{-\pi}^{\pi} F_d(\theta)G_d^*(\theta)\,d\theta$	4.49a
	$\sum_{n=-\infty}^{\infty} |f[n]|^2 = \dfrac{1}{2\pi} \int_{-\pi}^{\pi} |F_d(\theta)|^2\,d\theta$	4.49b
Area	$F_d(0) = \sum_{n=-\infty}^{\infty} f[n]$	4.50
	$f[0] = \dfrac{1}{2\pi} \int_{-\pi}^{\pi} F_d(\theta)\,d\theta$	
Shift	$f[n-r] \longleftrightarrow e^{-ir\theta} F_d(\theta)$	4.51
	$e^{in\beta} f[n] \longleftrightarrow F_d(\theta - \beta)$	4.52
Difference	$\Delta_1 f[n] = f[n] - f[n-1] \longleftrightarrow \left(1 - e^{-i\theta}\right) F_d(\theta)$	4.53
Cumulative sum	$\sum_{k=-\infty}^{n} f[k] \longleftrightarrow \dfrac{1}{1 - e^{-i\theta}} F_d(\theta), \quad F_d(0) = 0$	4.54
Convolution	$f * h[n] = \sum_{k=-\infty}^{\infty} f[k]h[n-k] \longleftrightarrow F_d H_d$	4.56
	$fh \longleftrightarrow F_d \circledast H_d = \dfrac{1}{2\pi} \int_{-\pi}^{\pi} F_d(\varphi)H_d(\theta - \varphi)\,d\varphi$	
Dilation	$f_{\uparrow P} \longleftrightarrow F_d(P\theta)$	4.61
	$f_{\downarrow P} \longleftrightarrow \dfrac{1}{P} \sum_{m=0}^{P-1} F_d\left(\dfrac{\theta - 2\pi m}{P}\right)$	

PROBLEMS

4.1. Let $f \in L^2[a,b]$. Show that $f \in L^1[a,b]$ as well. Hence, one can calculate Fourier coefficients for f.

4.2. Let f be bounded in the interval $[a,b]$. Show that $f \in L^1[a,b]$. Thus, one can calculate Fourier coefficients for f.

4.3. Let $f \in L^2[0,L]$. Beginning with Bessel's inequality, show that $|c_n| \to 0$ as $|n| \to \infty$. (This is a restricted version of the Riemann–Lebesgue lemma.)

4.4. Beginning with the complex Fourier series,

$$f(t) = \sum_{n=-\infty}^{\infty} c_n \exp\left(i\frac{2\pi n}{T}t\right), \quad c_n = \frac{1}{T} \int_0^T f(t) \exp\left(-i\frac{2\pi n}{T}t\right) dt,$$

derive the traditional Fourier series formulae:

$$f(t) = \frac{a_0}{2} + \sum_{n=1}^{\infty}\left[a_n \cos\left(\frac{2\pi n}{T}t\right) + b_n \sin\left(\frac{2\pi n}{T}t\right)\right]$$

$$a_n = \frac{2}{T}\int_0^T f(t)\cos\left(\frac{2\pi n}{T}t\right)dt, \qquad b_n = \frac{2}{T}\int_0^T f(t)\sin\left(\frac{2\pi n}{T}t\right)dt$$

and derive Equation 4.6.

4.5. For a real-valued function f, relate the parameters of the magnitude and phase representation, V_n and φ_n, to the classical Fourier coefficients a_n and b_n, and the complex Fourier coefficients c_n and c_{-n}.

4.6. Show that if f is real, then the coefficient c_0 is real, also.

4.7. The complex Fourier basis functions $\{e^{i2\pi nx/L}\}$ are orthogonal on any interval of length L, for example, $[0, L]$ or $[-L/2, L/2]$. Are the Fourier coefficients obtained by expanding a periodic function f on $[0, L]$ the same as those obtained by expanding on $[-L/2, L/2]$? Explain and describe the difference, if any. *Hint:* Try $f = \cos\frac{2\pi x}{L}$ first, to gain some insight into the problem, then formulate and prove a general statement.

4.8. Beginning with the sine–cosine basis (Equation 4.4), use orthogonal projections to complete the derivation of the classical Fourier series (Equation 4.5).

4.9. The Fourier coefficients for the unit amplitude square wave with period T (Equation 4.10),

$$c_n = \begin{cases} \dfrac{2}{i\pi n}, & n = \pm 1, \pm 3, \ldots \\ 0, & \text{otherwise} \end{cases}$$

are independent of T. Discuss the significance of this.

4.10. Calculate the Fourier coefficients for a square wave with arbitrary duty cycle,

$$f(x) = \begin{cases} 1, & A > x \geq 0 \\ -1, & T > x \geq A \end{cases}.$$

Check that your results reduce to Equation 4.10 when $A = \frac{T}{2}$.

4.11. *Smoothness and asymptotic behavior*
Let $f : \mathbb{R} \to \mathbb{C}$ be periodic with period L, and $\infty > p \geq 2$. Show that if $f \in C^{(p)}$, then $|c_n| < K|n|^{-p}$.

4.12. Let $f : \mathbb{R} \to \mathbb{C}$ be periodic, continuous, and piecewise smooth. Suppose also that f' is piecewise smooth. Discuss the convergence of the Fourier series for f'.

4.13. Let f have rapidly decreasing Fourier coefficients given by $c_n = a^{|n|}$, $|a| < 1$.

(a) Sum the Fourier series and show

$$f(x) = \frac{1 - a^2}{1 + a^2 - 2a\cos\frac{2\pi x}{L}}.$$

(b) Graph f for $x \in [-L, L]$ and $a = -0.9, -0.5, 0, 0.5, 0.9$.

(c) Calculate f' and graph it for $x \in [-L, L]$ and $a = -0.9, -0.5, 0, 0.5, 0.9$. Is it believable that f is infinitely continuously differentiable?

(d) Devise a way to show that $f \in C^{(\infty)}$ without appealing to Fourier theory or calculating an infinite number of derivatives.

4.14. For $f : [-\pi, \pi) \to \mathbb{R}$ given by $f(x) = |x|$,

(a) derive the complex Fourier series.

(b) at which values of x, if any, does the series fail to converge to $f(x)$?

(c) to what values does it converge at these points?

4.15. Let $f : [-\pi, \pi) \to \mathbb{R}$ be defined,

$$f(x) = \begin{cases} 0, & -\pi \leq x < 0 \\ x, & 0 \leq x < \pi \end{cases}.$$

(a) Derive the Fourier series for f.

(b) What is the sum of the Fourier series at $x = 0$ and at $x = \pi$?

(c) What is the average (DC) value of the Fourier series?

4.16. Consider the Fourier coefficients for the truncated sine wave in Example 4.2 (Equations 4.9a and 4.9b).

(a) What is the asymptotic behavior of c_n as $|n| \to \infty$? How is this consistent with the nature of the original function f?

(b) Compute and graph several partial sums of the Fourier series. What do you observe about the convergence of the series? Is this consistent with your answer to part (a)?

4.17. Derive the symmetry properties of the complex Fourier series (Theorem 4.6) when f is defined on the interval $[-L/2, L/2]$ rather than $[0, L]$.

4.18. Prove the following theorems for the complex Fourier series.

(a) Area theorem (Equation 4.20).

(b) Shift theorem (Equation 4.21).

(c) Derivative theorem (Equation 4.22).

4.19. Let f be periodic with period L and absolutely integrable over one period. Further, assume $\int_0^L f(x)dx = 0$. Show that $F(x) = \int_0^x f(u)du$ is periodic with period L, that is, $F(x) = F(x + L)$.

4.20. Prove the symmetry relationships for the classical Fourier series (corollary to Theorem 4.6).

4.21. Derive a version of Parseval's formula for the classical Fourier series. Give an expression for the power in a signal at the nth harmonic of the fundamental frequency in terms of the Fourier coefficients.

4.22. Derive versions of the following theorems for the classical Fourier series.

(a) Area theorem

(b) Shift theorem

(c) Convolution theorem

(d) Derivative theorem

(e) Integral theorem; include conditions for its validity.

4.23. For the square wave and its Fourier series (Example 4.3), verify by direct calculation that Parseval's formula (Equation 4.19b) holds. *Hint:*

$$\sum_{m=0}^{\infty} \frac{1}{(2m+1)^2} = \frac{\pi^2}{8}.$$

4.24. Calculate the complex Fourier coefficients for the following two functions.

(a) Shifted rectangle function

$$g(t) = \text{rect}\left[2\left(t - \frac{1}{4}\right)\right], \quad t \in (0, 1).$$

(b) Shifted triangle function

$$f(t) = \left(1 - 2\left|t - \frac{1}{2}\right|\right)\text{rect}\left(t - \frac{1}{2}\right) = \begin{cases} 2t, & \frac{1}{2} \geq t \geq 0 \\ 2(1-t), & 1 \geq t \geq \frac{1}{2} \\ 0, & \text{otherwise} \end{cases},$$

where rect(t), the "rectangle function," is one for $|t| < 1/2$ and 0 otherwise. Use the limits (0, 1) for both integrals. Try to express your results in terms of the sinc function,

$$\text{sinc}(x) = \frac{\sin \pi x}{\pi x}.$$

4.25. Using the shift theorem and the results in Example 4.3, derive both the complex and classical Fourier series for an even square wave,

$$f(x) = \begin{cases} -1, & -\frac{T}{4} \geq x > -\frac{T}{2} \\ 1, & \frac{T}{4} \geq x > -\frac{T}{4} \\ -1, & \frac{T}{2} \geq x > \frac{T}{4} \end{cases}.$$

4.26. A square wave signal $v(t)$ with 50% duty cycle, period T, minimum value 0, and maximum value V_0 is passed through a first-order lowpass filter described by the transfer function

$$H(\omega) = \frac{a}{a + i\omega}.$$

The steady-state output of the filter is also periodic, and using Laplace transform methods (Chapter 9), one period of the output y, from $t = kT$ to $t = (k+1)T$, is given by

$$y(t) = \begin{cases} V_0\left[1 - \dfrac{e^{-a(t-kT)}}{1 + e^{-aT/2}}\right], & (k + \frac{1}{2})T > t \geq kT \\ V_0\left[1 - \dfrac{e^{-aT/2}}{1 + e^{-aT/2}}\right]e^{-a(t-(k+\frac{1}{2})T)}, & (k+1)T > t \geq (k + \frac{1}{2})T \end{cases}.$$

(a) Calculate the complex Fourier coefficients for this periodic function. Note that, without loss of generality, you can take $k = 0$ and compute the Fourier coefficients over the interval $[0, T]$.

(b) Calculate the complex Fourier coefficients (v_n) for the square wave input v. According to linear system theory, the Fourier coefficients of the filter's output are

$$y_n = H(n/2\pi T)v_n.$$

Does this agree with your result from part (a)?

4.27. Consider the periodic function $g(x) = \exp(-ikt(x))$, where k is a constant and the function $t(x)$ is graphed below in Figure 4.42. This is a model of an optical device called a *phase grating*. The grating is made by etching a fine pattern of grooves into a piece of glass or other material. The function $t(x)$ represents the thickness of the groove profile.

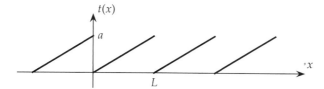

FIGURE 4.42 For Problem 4.27. Thickness function of a phase grating.

(a) What is the average power of g?

(b) Calculate the Fourier series for g.

(c) What fraction of g's average power is contained in the $n = -1$ term of the Fourier series? In the $n = +1$ term? Why are they different?

4.28. *Yet another trigonometric basis*

The complex exponential $e^{i\theta} = \cos\theta + i\sin\theta$ is sometimes abbreviated "cis θ" (cosine **i** sine). The complex Fourier basis, in this notation, is $\{\text{cis } 2\pi nx, \, n = 0, \pm1, \pm2, \dots\}$. As an alternative to the complex-valued "cis" function, a real-valued "cas" (cosine **and** sine) function is proposed, defined[15]

$$\text{cas }\theta = \cos\theta + \sin\theta.$$

(a) Show that $\{\text{cas } 2\pi nx\}$ is an orthogonal set of functions on the unit interval $[0, 1]$. Are positive and negative indices necessary to span the space, as in the complex Fourier series, or will positive indices only suffice, as in the classical Fourier series?

(b) Derive an orthogonal expansion similar to the Fourier series using the cas basis, and give expressions for the expansion coefficients.

(c) Also determine a Parseval formula for the cas expansion. Compare your results with both the classical and complex Fourier series.

(d) What is the relationship between the "cas line spectrum" and the true Fourier line spectrum?

4.29. Show that, if two functions f and g are piecewise smooth and have the same Fourier coefficients, then $f = g$.

4.30. *Full-wave rectification*

The process of converting alternating current (AC) to direct current (DC) frequently employs full-wave rectification—taking the absolute value of the AC voltage. A "110 V, 60 Hz" AC waveform is represented by the expression $v(t) = 110\sqrt{2}\,\sin(2\pi \cdot 60t)$, where t is in seconds and v is in volts. A portion of the rectified waveform is shown in Figure 4.43.

(a) In the Fourier series for this waveform, what will the fundamental frequency be?

(b) Derive an expression for the Fourier coefficients (c_n) of this waveform.

---

[15] Bracewell (1986)

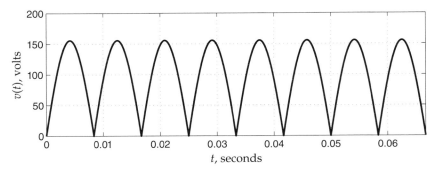

FIGURE 4.43 For Problem 4.30. Full-wave rectified 60 Hz line voltage.

(c) In a second processing step, the rectified waveform is passed through a lowpass filter, which (ideally) removes all the oscillatory components, leaving pure DC. Assuming the filter performs perfectly, what DC voltage will result?

4.31. *Half-wave rectification*
An alternative to full-wave rectification (Problem 30) is half-wave rectification, which sets the negative values of the AC voltage to zero, as shown in Figure 4.44.

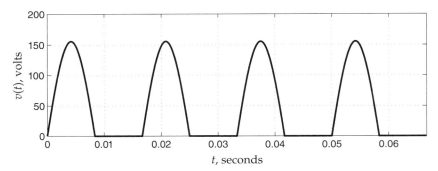

FIGURE 4.44 For Problem 4.31. Half-wave rectified 60 Hz line voltage.

(a) Calculate the Fourier coefficients (c_n) for the half-wave rectified sine wave. Compare the symmetry properties of the Fourier coefficients for the half- and full-wave rectified functions, and relate them to the properties of the functions themselves.

(b) Plot the line spectra of both the half- and full-wave rectified signals and compare them.

(c) Direct current is obtained from the rectified wave by a filtering process that removes all the oscillatory components. Based on the line spectra, which rectified signal will be easier to filter? Assuming perfect filtering, compare the DC values.

4.32. *Half-wave and full-wave rectification*
Using the shift and linearity theorems, derive the Fourier series for a full-wave rectified sine wave from the Fourier series for a half-wave rectified sine wave (see the preceding two problems).

4.33. Consider the waveform shown below in Figure 4.45, which models the output of a switching device for converting direct current to alternating current. It is a type of square wave, which you know has a Fourier spectrum consisting of a fundamental frequency and odd harmonics of the fundamental frequency. Determine the value of w that causes the amplitude of the third harmonic to be exactly zero.

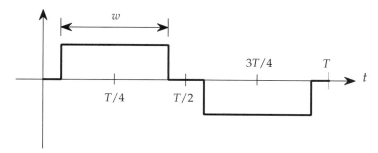

FIGURE 4.45 For Problem 4.33.

4.34. Calculate the Fourier series coefficients for the periodic function shown below in Figure 4.46. Each "pulse" is a single cycle of a sine wave of period T, and the pulses occur with period NT, where N is an integer.

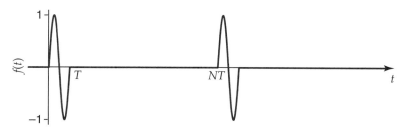

FIGURE 4.46 For Problem 4.34.

Note that when $N = 1$, this is a continuous sine wave; you can use this special case as a check on your general result.

4.35. *Harmonic distortion*

Electronic amplifiers are limited in the output voltages they can produce. When a signal exceeds this limit, it is sometimes said to "hit the rails" or "clip." Mathematically, a clipped signal may be described:

$$g(t) = \begin{cases} -A, & f < -A \\ f(t), & |f| \leq A \\ A, & f > A \end{cases}.$$

Clipping is a type of nonlinearity, and like all nonlinear systems, it causes spurious Fourier components in the output signal. One way of quantifying the severity of the nonlinearity is to measure the strength of these spurious components through Fourier analysis. With a pure sinusoidal input, the ratio of power in the spurious components to the total output power is called *total harmonic distortion*, or THD. It is usually expressed as a percentage: THD = 0% means all the output power is at the fundamental input frequency, and anything other than zero means that power is being diverted from the fundamental frequency into the harmonics.

Consider a sinusoidal input, $f(t) = V \cos 2\pi v_0 t$, applied to a clipping amplifier with limit A. Calculate the THD of the amplifier as a function of V and A. Graph your result in a meaningful way. *Hint*: You do not need to calculate the entire Fourier spectrum.

4.36. *Heat equation*
Solve the heat equation in a thin bar of length L, insulated sides, with boundary conditions $u(0, t) = u(L, t) = 0$ (ends held at zero temperature) and initial condition $u(x, 0) = u_0$ (constant).

4.37. *Heat equation*
(a) Solve the heat equation in a thin bar of length L, insulated sides, with boundary conditions $u(0, t) = 0$ (end held at zero temperature) and $\partial u/\partial x|_{(L,t)} = 0$ (other end insulated), and initial condition $u(x, 0) = u_0$ (constant).
(b) Compute and plot the solutions (e.g., using MATLAB) with $u_0 = 100$ and $L = 1$. Interpret the solution physically.

4.38. *Heat equation*
A thin bar of length L with insulated sides is held at zero temperature at one end ($x = 0$) and temperature u_0 at the other end ($x = L$).
(a) Calculate the steady-state temperature ($t \to \infty$) in the bar.
(b) Calculate a general solution for the temperature in the bar, assuming an initial distribution $u(x, t) = f(x)$.

4.39. *Vibrating string*
Consider the solution of the wave equation for the vibrating string (Equation 4.31),

$$y(x, t) = \sum_{n=1}^{\infty} b_n \sin\left(\frac{\pi n x}{L}\right) \cos\left(\frac{\pi n v t}{L}\right)$$

$$b_n = \frac{2}{L} \int_0^L f(x) \sin\frac{n\pi x}{L} dx.$$

(a) Use the trigonometric identity $\sin A \cos B = \frac{1}{2}\sin(A + B) + \frac{1}{2}\sin(A - B)$ to show that the solution may also be written

$$y(x, t) = \frac{1}{2}F(x - vt) + \frac{1}{2}F(x + vt),$$

where F is the odd periodic extension of f,

$$F(x) = \begin{cases} f(x), & L \geq x \geq 0 \\ -f(-x), & 0 > x \geq -L \\ F(x + 2L), & \text{all } x \in \mathbb{R} \end{cases}.$$

That is, the standing wave solution is the sum of two traveling waves, one moving to the left and one moving to the right.

(b) Substitute this alternate form of y into the wave equation $\frac{\partial^2 y}{\partial x^2} = \frac{1}{v^2}\frac{\partial^2 y}{\partial x^2}$ and verify that it satisfies the equation. What are the continuity requirements for f?

(c) Assuming that $f(0) = f(L) = 0$, show that the alternate form of y also satisfies the boundary conditions $y(0, t) = y(L, t) = 0$ and initial conditions $y(x, 0) = f(x)$ and $y'(x, 0) = 0$.

(d) Compute the series solution developed in Example 4.20 with $r = 0.5$, using enough terms that ripples are imperceptible. Plot the solution vs. x for $t = 0, T/4, T/2, 3T/4, T$, where T is the fundamental period of the wave. Also make a sketch of the alternative traveling wave solution for these same times, and compare.

The traveling wave form of the solution was first studied by D'Alembert.

4.40. *Vibrating string*

In the text we considered the case of a plucked string, for which the initial displacement was specified and the initial velocity was zero. Now consider the case of a struck string (e.g., a piano), for which the appropriate initial conditions are $y(x, 0) = 0$, $\left.\frac{\partial y}{\partial t}\right|_{(x,0)} = g(x)$.

If the string is struck at position $x = a$ along its length, the initial velocity may be modeled by a narrow rectangular pulse centered at $x = a$,

$$\left.\frac{\partial y}{\partial t}\right|_{(x,0)} = g(x) = \begin{cases} \mu, & a + \frac{\epsilon}{2} > x > a - \frac{\epsilon}{2} \\ 0, & \text{otherwise} \end{cases},$$

where $\epsilon \ll L$.

(a) Determine the appropriate form of the temporal part of the solution, $\psi_n(t)$, so that it can satisfy the initial conditions.

(b) The complete solution of the differential equation is

$$y(x, t) = \sum_{n=1}^{\infty} b_n \phi_n(t) \psi_n(t).$$

Calculate the Fourier coefficients $\{b_n\}$.

(c) What property does the solution have if the string is struck at its center, $a = \frac{L}{2}$?

4.41. Consider a system modeled as a string of length L on the interval $\left(-\frac{L}{2}, \frac{L}{2}\right)$, vibrating in its fundamental mode ($n = 1$) with amplitude $a_1 = A$. At a time $t = 0$, when the string's velocity is zero, the length of the string is suddenly doubled by adding a length $\frac{L}{2}$ of new string at each end (new length $L' = 2L$). Thus, the shape of the longer string at $t = 0$ is

$$y'(x, 0) = \begin{cases} y(x, 0), & x \in \left(-\frac{L}{2}, \frac{L}{2}\right) \\ 0, & \text{otherwise} \end{cases}.$$

For time $t > 0$, the longer string continues to vibrate. Calculate the amplitude a'_1 of the fundamental mode in this new configuration.

4.42. *Antennas*

Refer to Equation 4.40 for the field of a $2N + 1$ element antenna array. In this array, each emitter has the same strength. In many practical arrays, the strengths of the emitters are weighted, so that Equation 4.40 becomes

$$U_p(R, \theta, t) = \frac{A}{R} \cos(kR - \omega t) \sum_{n=-N}^{N} c_n \exp(inkd \sin \theta),$$

where the $\{c_n\}$ are the weighting factors. For example, consider the weighting function

$$c_n = \frac{1}{2}\left[1 + \cos\left(\frac{\pi n}{N}\right)\right],$$

which is one at $n = 0$ and smoothly tapers to zero at $n = \pm N$. Sum the Fourier series with these coefficients and plot the resulting radiation pattern with $d/\lambda = \frac{1}{2}$ and $N = 3$. Compare this with the unweighted result. In particular, how are the angular width of the main radiation lobe and the amplitudes of the sidelobes affected by the weighting?

4.43. *Fourier series convergence*

We saw in this chapter that the Fourier coefficients c_n of a square wave decay slowly, $O(1/n)$, and consequently the sequence of partial sums, $S_N(x)$, converges nonuniformly. The chief symptom of this nonuniform convergence is the Gibbs overshoot at the jumps in the square wave. It can be shown that the sequence of so-called Césaro sums, defined

$$K_N(x) = \frac{1}{N}\sum_{k=0}^{N-1} S_k(x),$$

does converge uniformly as $N \to \infty$.

(a) Show that the Césaro sum simplifies to the form

$$K_N(x) = \sum_{n=-(N-1)}^{N-1} \left(1 - \frac{|n|}{N}\right) c_n e^{i2\pi nx/L}.$$

That is, the Fourier coefficients are multiplied by a *convergence factor* that causes them to decay faster than $O(1/n)$.

(b) Sketch the convergence factor as a function of n for a few values of N (note that it is zero for $n \geq N$). What shape does it approach in the limit as $N \to \infty$?

(c) Compute numerically the Césaro sum $K_N(x)$ for the square wave's Fourier series for a few values of N, for example, $N = 1, 7, 15, 31, 63$. Compare with the plain sequence of partial sums $S_N(x)$.

(d) We have seen that the partial sum $S_N(x)$ may be expressed as the convolution of the original function $f(x)$ with the Dirichlet kernel $D_{2N+1}(x) = \frac{\sin(2N+1)\pi x}{\sin(\pi x)}$. Show that the corresponding expression for the Césaro sum is the convolution of $f(x)$ with the so-called *Fejér kernel*,

$$F_N(x) = \frac{1}{N}\sum_{k=0}^{N-1} D_{2k+1}(x).$$

Then show that the Fejér kernel has the simpler form,

$$F_N(x) = \frac{1}{N}\frac{\sin^2(N\pi x)}{\sin^2(\pi x)}.$$

Plot the Dirichlet kernel D_{2N+1} and Fejér kernel F_N on the same axis and compare their mainlobes and sidelobes. Discuss the appearance of the two partial sums in light of these kernels.

4.44. *Fourier series and the DFT*

(a) Calculate the DFT of the sampled square wave sequence (Figure 4.31),

$$f[n] = \begin{cases} 0, & n = 0 \\ 1, & n = 1, 2, \ldots N/2 - 1 \\ 0, & n = N/2 \\ -1, & n = N/2 + 1, N/2 + 2, \ldots N - 1 \end{cases}$$

and show that

$$\frac{1}{N}F[m] = \begin{cases} \dfrac{2}{iN} \cot \dfrac{\pi m}{N}, & m = 1, 3, \ldots N/2 - 1, N/2 + 1, \ldots N - 1 \\ 0, & m \text{ even} \end{cases}.$$

(b) Consider the limiting behavior as N becomes large. For m odd and $|m| \ll N/2$, show that

$$c_m = \frac{2}{i\pi m} \approx \begin{cases} \dfrac{1}{N}F[m], & N/2 \gg m > 0, \ m \text{ odd} \\ \dfrac{1}{N}F[m - N], & 0 > m \gg -N/2, \ m \text{ odd} \end{cases}$$

and hence that the DFT values approach the true Fourier series coefficients (Equation 4.10) as the sampling becomes progressively finer.

4.45. *Fourier series and the DFT*
Repeat Problem 4.24 using the discrete Fourier transform.

(a) Use MATLAB to compute 128 samples equally spaced in t. The sampling interval is $\Delta t = T/N$, where T is the period of the function. The samples are at $t = 0, T/N, 2T/N, \ldots, (N - 1)T/N$. Note that $t = T$ is *not* used, because this point is considered to begin the periodic extension of f, that is, $f(T) = f(0)$. So, the sampling period is $1/128$. Plot the sample sets $\{f(n\Delta t)\}$ and $\{g(n\Delta t)\}$ using MATLAB to make sure you have them right.

(b) Use the FFT function in MATLAB to compute estimates of the Fourier coefficients. Recall if $F[m]$ is the DFT of $f[n]$, then the Fourier series coefficients c_m are approximately $F[m]/N$, where N is the number of samples (128 in this case). Also compute numerical values for the exact coefficients you obtained in Problem 4.24, and compare with the DFT-derived values. Plot $\mathcal{R}e\{c_m\}$ and $\mathcal{R}e\{F[m]/N\}$ on the same graph, using different point styles (like circles and squares), and do the same for the imaginary parts on another graph. How well do the exact Fourier coefficients agree with the DFT coefficients?

(c) Repeat the calculation and comparison with coarser sampling, say $N = 8$.

4.46. *Fourier series and the DFT*
Consider the half-wave rectified cosine wave, two periods of which are drawn below in Figure 4.47.

(a) Derive the complex Fourier coefficients for $v(t)$. You may adapt your solution to Problem 4.31. Check that your coefficients have the proper symmetries.

(b) Plot the line spectrum of $v(t)$. What fraction (in percent) of the function's average power is in the second and higher harmonics? Determine an effective bandlimit B for this function.

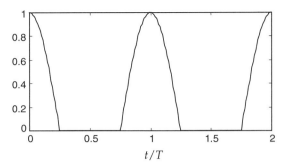

FIGURE 4.47 For Problem 4.46.

(c) Sample one period of $v(t)$ to create a discrete function $v[n]$. Note that you can get T to cancel out by writing $T = N\Delta t$. Then, you can vary the effective sampling interval by changing N. Since $v(t)$ is an even function, your sample sequence $v[n]$ must have even symmetry. Calculate (with MATLAB) the DFT $V[m]$. Verify that your DFT values have the proper symmetries.

(d) For various sampling intervals $\Delta t = T/N$, calculate and plot the DFT line spectrum. For what sampling interval does the DFT line spectrum agree "well enough" with the line spectrum you calculated in part (b)? How does this agree with the bandlimit you determined in part (b)?

4.47. *Harmonic distortion*

Consider again the problem of harmonic distortion (Problem 4.35). This time we will use the DFT to compute the amount of distortion due to clipping. Consider a sinusoidal input, $y(t) = V \cos 2\pi v_0 t$, applied to a clipping amplifier with limit A. The clipped signal may be described:

$$g(t) = \begin{cases} -A, & f < -A \\ f(t), & |f| \leq A \\ A, & f > A \end{cases}.$$

Provided that you sample fast enough, the DFT of the clipped signal will accurately represent its Fourier coefficients.

(a) The first order of business is to tell MATLAB how to create the clipped cosine wave. Here is a good way:

```
n = 0:N-1;            % Make a time vector, n = [0, 1, 2,
                      ...,N-1]
y = V*cos(2*pi*n/N);  % Samples of one period of a cosine
c = find(abs(y)>A);   % Find the points to be clipped
y(c) = sign(y(c))*A;  % Replace those points by +/- A
```

Read the online help for `find()` and `sign()` to find out what they do. For simplicity, you could use $A = 1$; then, $V \leq 1$ causes no clipping, and $V > 1$ does clip. This models the situation in an amplifier with fixed "rails."

(b) You need to determine how many samples to take. An empirical way to do this is to set V for severe clipping and calculate the DFT of y for various values of N (remember, N must be a power of two), plotting the absolute value of $Y[m]$ on a log scale so you really see how small the terms are getting. Watch the high harmonics (values of the DFT near $N/2$). When these are negligibly small, you have a high enough value of N. Why?

(c) Define the power spectrum by the formula:

$$S_y[m] = \frac{1}{N}\left(|Y[m]|^2 + |Y[N-m]|^2\right), m = 0, 1, \ldots, \frac{N}{2} - 1,$$

where Y is the DFT of y. It is conventionally plotted in decibels, $10\log_{10} S_y$. Plot power spectra of y for various degrees of clipping and observe how the harmonic structure of the signal changes. Suppose that the fundamental frequency is 500 Hz, and create a realistic frequency axis for your plots. From the power spectrum you can compute THD as a function of V and A. Graph your result in a meaningful way.

4.48. *Fourier series for discrete periodic sequences*
Consider a sequence $f: \mathbb{Z} \to \mathbb{C}$ that is periodic with period N, $f[n] = f[n+N]$. Because f is periodic, we expect it to have some kind of Fourier series expansion.

(a) Show that the appropriate basis sequences are $\phi_k = \left(\exp\left(\frac{i2\pi k n}{N}\right)\right)_{n\in\mathbb{Z}}$. Normalize them to create an orthonormal basis $\{e_k\}$. How many basis sequences are there?

(b) Construct the Fourier expansion, using the general formula $f = \sum_k \langle f, e_k\rangle e_k$, where the inner products are calculated over one period.

(c) Show that the Fourier series is actually a DFT.

4.49. *Fourier series for discrete periodic sequences*
Show that the comb sequence $\text{III}_N[n]$ has a discrete Fourier representation,

$$\text{III}_N[n] = \frac{1}{N}\sum_{m=0}^{N-1} e^{i2\pi mn/N}.$$

4.50. *Discrete-time Fourier Transform*
Consider the discrete-time signal $f[n] = a^n \cos bn\, U[n]$, $1 > a > 0$, where $U[n]$ is the unit step sequence (1 for $n \geq 0$, 0 otherwise).

(a) Derive an expression for the DTFT $F_d(\theta)$ for this signal.

(b) Plot f and F_d for $a = 0.2$ and 0.9, and for $b = \pi/10$ and $\pi/2$ (four sets in all). Describe the effects of a and b on f and on F_d.

4.51. *Discrete-time Fourier transform*
Consider the two-sided discrete-time signal $f[n] = a^{|n|}$, $|a| < 1$.

(a) Calculate the Fourier transform F_d by direct summation.

(b) Show that f may also be obtained as the even part of the one-sided exponential sequence $g[n] = a^n U[n]$.

(c) Using the facts that a real, even sequence has a real, even transform, and a real sequence with no symmetry has a Hermitian transform (see Figure 4.16), obtain the Fourier transform F_d from the Fourier transform G_d.

4.52. *Discrete-time systems*
Show that the first difference of a convolution sum obeys the relation

$$\Delta_1[f * h] = f * [\Delta_1 h] = [\Delta_1 f] * h.$$

This can be done both in the time domain, by manipulating the convolution sum, and in the frequency domain, with the difference and convolution theorems.

4.53. *Discrete-time systems*

The *step response* of a discrete-time system is the output when the input is a step sequence, $f[n] = U[n]$.

(a) Show that the impulse response is the first difference of the step response.

(b) Show that the step response is the cumulative sum of the impulse response.

4.54. *Discrete-time systems*

Calculate, by direct summation of the convolution, the step response of a system with impulse response $h[n] = a^n U[n]$, $1 > a > 0$. The step response is the output when the input is a step sequence, $f[n] = U[n]$.

4.55. *Discrete-time systems*

Determine if each of the following equations is true in general. Provide proofs of those you think are true and counterexamples for those that you think are false.

(a) $x[n] * (h[n]g[n]) = (x[n] * h[n])g[n]$.

(b) $\alpha^n x[n] * \alpha^n h[n] = \alpha^n (x[n] * h[n])$.

4.56. *Upsampling with filtering*

Consider Example 4.29. A filter is proposed that simply replaces each inserted zero with the value of the preceding sample, that is,

$$g[n] = \begin{cases} f_{\uparrow 2}[n], & n \text{ even (an original value)} \\ f_{\uparrow 2}[n-1], & n \text{ odd (an inserted zero)} \end{cases}$$

(a) Show that the impulse response of this filter is $h[n] = \delta[n] + \delta[n-1]$.

(b) Calculate the frequency response, $H_d(\theta)$.

(c) Plot $|H_d|$ and $|1 + \cos\theta|$ (the magnitude responses of the new filter and the filter in the example) on the same axes and compare them. Which one is more effective at removing undesired spectral components?

4.57. *Upsampling with filtering*

Consider Example 4.29 again, with the following interpolating filter:

$$h[n] = \sum_{k=-K}^{K} \text{sinc}\left(\frac{k}{2}\right) \delta[n-k]$$

where, recall,

$$\text{sinc } x = \frac{\sin \pi x}{\pi x}.$$

Calculate and plot h and the frequency response $H_d(\theta)$ for various values of K. Comment on the ability of the filter to remove the spurious spectral components in the vicinity of $\pm\pi$ as K increases.

CHAPTER 5

THE FOURIER TRANSFORM

So far, we have studied Fourier representations for finite sequences (the discrete Fourier transform), infinite sequences (the discrete-time Fourier transform), and functions on a finite real interval (the Fourier series).

Transform	Time domain	Frequency domain
Discrete Fourier transform	Discrete, bounded $n \in \{0, 1, \ldots, N-1\}$	Discrete, bounded $m \in \{0, 1, \ldots, N-1\}$
Fourier series	Continuous, bounded $x \in [0, L]$	Discrete, unbounded $n \in \mathbb{Z}$
Discrete-time Fourier transform	Discrete, unbounded $n \in \mathbb{Z}$	Continuous, bounded $\theta \in [-\pi, \pi)$
Fourier transform	Continuous, unbounded $x \in \mathbb{R}$	Continuous, unbounded $\nu \in \mathbb{R}$

Matching these up, there is a pattern: if one domain is bounded, the other domain is discrete, and if one domain is unbounded, the other domain is continuous. The one transform we have yet to explore should map a continuous, unbounded time domain to a continuous, unbounded frequency domain. That fourth member of the Fourier family is the (continuous-time) Fourier transform:

$$f(x) = \int_{-\infty}^{\infty} F(\nu) e^{i2\pi \nu x} d\nu,$$

$$F(\nu) = \int_{-\infty}^{\infty} f(x) e^{-i2\pi \nu x} dx.$$

This chapter begins with a heuristic development of the Fourier transform as a limiting case of the Fourier series and of the discrete-time Fourier transform. The nature of the Fourier transform—an integral over the entire real line—raises questions of what kinds of functions have Fourier transforms, what these transforms are like, and whether they can be inverted to return to the original functions. Initially many of these questions will be skipped in favor of building skill with doing calculations for well-behaved functions and cultivating intuition for the meaning of the transforms. We will develop the by-now familiar set of theorems relating a function and its

---

Fourier Transforms: Principles and Applications, First Edition. Eric W. Hansen.
© 2014 John Wiley & Sons, Inc. Published 2014 by John Wiley & Sons, Inc.

transform, many of which will be seen to have interpretations in terms of familiar physical phenomena or engineering systems. After this, we will return to questions of the existence and invertibility of the Fourier transform for broad classes of functions.

5.1 FROM FOURIER SERIES TO FOURIER TRANSFORM

Consider a Fourier series expansion of a function on the interval $[-L/2, L/2]$,

$$f(x) = \sum_{n=-\infty}^{\infty} c_n e^{i2\pi nx/L}$$

$$c_n = \frac{1}{L} \int_{-L/2}^{L/2} f(x) e^{-i2\pi nx/L} dx.$$

The nth harmonic frequency is $\nu_n = n/L$, and the spacing of the Fourier components is $\Delta \nu = 1/L$. Substituting these into the above expressions, we have

$$f(x) = \sum_{n=-\infty}^{\infty} \left[\int_{-L/2}^{L/2} f(\xi) e^{-i2\pi \nu_n \xi} d\xi \right] e^{i2\pi \nu_n x} \Delta \nu.$$

We are interested in the behavior of the Fourier series as the period of f increases to the limiting case, $L \to \infty$, where f is *aperiodic* (not periodic) or, if you will, consists of only one "period" extending from $-\infty$ to ∞. The discrete spectral lines of the Fourier series crowd closer together as L increases (Figure 5.1). The line spacing $\Delta \nu$

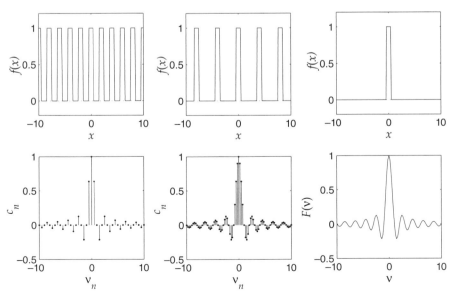

FIGURE 5.1 Visualizing the Fourier transform as the limit of a Fourier series, as $L \to \infty$. The line spectrum goes over to a continuum.

5.1 FROM FOURIER SERIES TO FOURIER TRANSFORM

becomes an infinitesimal $d\nu$, the spectral line frequency ν_n becomes a continuous frequency variable ν, and the summation becomes a Riemann integral:

$$f(x) = \int_{-\infty}^{\infty} \underbrace{\left[\int_{-\infty}^{\infty} f(\xi) e^{-i2\pi\nu\xi} d\xi\right]}_{F(\nu)} e^{i2\pi\nu x} d\nu.$$

The sequence of Fourier coefficients (c_n) has become a function $F(\nu)$. The inner integral, taking f to F, is called the forward *Fourier transform*, and the outer integral, taking F to f, is the inverse Fourier transform.

$$f(x) = \int_{-\infty}^{\infty} F(\nu) e^{i2\pi\nu x} d\nu, \tag{5.1a}$$

$$F(\nu) = \int_{-\infty}^{\infty} f(x) e^{-i2\pi\nu x} dx. \tag{5.1b}$$

As with the other three members of the Fourier family, the forward transform carries a minus sign in the complex exponential, while the inverse transform carries a plus sign. We can compactly express the forward transform by $f \longmapsto F$, $f \xrightarrow{\mathcal{F}} F$, or $F = \mathcal{F}\{f\}$, and the inverse transform by $f = \mathcal{F}^{-1}\{F\}$ or $F \xrightarrow{\mathcal{F}^{-1}} f$. The notation $f \longleftrightarrow F$ means that f and F are a Fourier transform pair: F is the forward transform of f, and f is the inverse transform of F.

We can get to the Fourier transform from the discrete-time Fourier transform by a similar heuristic procedure. This will hint at a connection between the two that will be made rigorous in the next chapter. Beginning with

$$f[n] = \frac{1}{2\pi} \int_{-\pi}^{\pi} F_{\mathrm{d}}(\theta) e^{in\theta} d\theta,$$

$$F_{\mathrm{d}}(\theta) = \sum_{n=-\infty}^{\infty} f[n] e^{-in\theta},$$

let $f[n] = f(n\Delta t)$ and $\theta = 2\pi\nu\Delta t$ (recall the discussion in Example 3.1). As Δt becomes smaller (the sampling becomes finer), the digital frequency range $[-\pi, \pi)$ maps to an ever-increasing range of analog frequency, $[-1/2\Delta t, 1/2\Delta t)$, tending to $(-\infty, \infty)$ in the limit as Δt becomes infinitesimal. A single period of the Fourier transform F_{d} becomes a spectrum over the entire real line. In the inverse transform, Δt goes over to an infinitesimal dt, the time points $n\Delta t$ crowd together into a continuous t, and the samples fuse together into a function $f(t)$:

$$f(n\Delta t) = \int_{-1/2\Delta t}^{1/2\Delta t} \left[\sum_{n=-\infty}^{\infty} f(n\Delta t) e^{-i2\pi(n\Delta t)\nu} \Delta t\right] e^{i2\pi\nu(n\Delta t)} d\nu,$$

$$\rightarrow f(t) = \int_{-\infty}^{\infty} \underbrace{\left[\int_{-\infty}^{\infty} f(t) e^{-i2\pi\nu t} dt\right]}_{F(\nu)} e^{i2\pi\nu t} d\nu.$$

Again we are at the Fourier transform pair. (You are invited to make sketches analogous to Figure 5.1 to illustrate this informal explanation.)

The Fourier transform is also found in the literature in two other forms:

$$f(t) = \frac{1}{2\pi} \int_{-\infty}^{\infty} F(\omega) e^{i\omega t} d\omega, \tag{5.2a}$$

$$F(\omega) = \int_{-\infty}^{\infty} f(t) e^{-i\omega t} dt, \tag{5.2b}$$

and

$$f(t) = \frac{1}{\sqrt{2\pi}} \int_{-\infty}^{\infty} F(\omega) e^{i\omega t} d\omega, \tag{5.3a}$$

$$F(\omega) = \frac{1}{\sqrt{2\pi}} \int_{-\infty}^{\infty} f(t) e^{-i\omega t} dt. \tag{5.3b}$$

In these, the frequency variable ω (angular frequency, radians per second) is used rather than ν whose units are cycles per second (hertz). The two are related by $\omega = 2\pi \nu$. When x is a spatial variable, the frequency ν has units of cycles per unit distance, for example, cycles/mm. One also may see the Fourier kernel written e^{ikx}, where $k = \frac{2\pi}{\lambda}$ and λ, the period of the Fourier component, is called the wavelength. The difference between Equations 5.2 and 5.3 is in the disposition of the factor $1/2\pi$, which corresponds to the normalizations that appear in the DFT ($1/N$), Fourier series ($1/L$), and discrete-time Fourier transform ($1/2\pi$).

5.2 BASIC PROPERTIES AND SOME EXAMPLES

The heuristic arguments by which we arrived at Equations 5.1 must be qualified with mathematical conditions for existence of the integrals. The various proofs are quite technical and do not contribute to a working knowledge of the transforms. The conclusion of the matter is that the Fourier transform is well defined for all functions in L^1 and L^2, though there is some subtlety in the interpretation of the integrals. For now we will consider only the simplest cases, then revisit the details later in the chapter.

Right off, we observe that the forward and inverse transforms are symmetric, differing only in the sign of the complex exponent. Thus, the forward and inverse transforms have the same properties of existence, continuity, and so on. The Fourier theorems will each have nearly identical versions for the forward and inverse transform.

The Fourier transform integral exists for all $\nu \in \mathbb{R}$ if f is absolutely integrable:

$$|F(\nu)| = \left| \int_{-\infty}^{\infty} f(x) e^{-i2\pi \nu x} dx \right| \leq \int_{-\infty}^{\infty} \left| f(x) e^{-i2\pi \nu x} \right| dx = \int_{-\infty}^{\infty} |f(x)| dx.$$

If $f \in L^1$, its Fourier transform is bounded and that bound is the L^1 norm of f— $\|F\|_\infty \leq \|f\|_1$. It can be shown that the Fourier transform is a continuous function of the frequency, ν. If f is at least piecewise continuous as well as absolutely integrable,

5.2 BASIC PROPERTIES AND SOME EXAMPLES

F inherits the continuity of the complex exponential.[1] Moreover, it can be shown that $F \to 0$ as $|\nu| \to \infty$ (Riemann–Lebesgue lemma).[2]

The fact that the Fourier transform is bounded and continuous, and even rolls off to zero at high frequencies, is not enough to guarantee that F is also absolutely integrable. Here is a classic example.

Example 5.1 (The rectangle function). The rectangle function is defined:

$$\text{rect } x = \begin{cases} 1, & |x| < \frac{1}{2} \\ \frac{1}{2}, & |x| = \frac{1}{2} \\ 0, & |x| > \frac{1}{2} \end{cases}. \tag{5.4}$$

The reason for the definition $\text{rect}(\frac{1}{2}) = \text{rect}(-\frac{1}{2}) = \frac{1}{2}$ will be explained later.

The rectangle function is useful for modeling truncation effects. In practical data analysis, for example, one never has knowledge of a signal for all time from $-\infty$ to ∞. The rectangle function models the effective acquisition window used to extract a portion of the signal for analysis. A rectangle of width X, $\text{rect}(x/X)$, is made by *dilating* or scaling x. A rectangle which turns on at $t = 0$ and off at $t = T$ is made by dilating the unit rectangle by a factor T and then shifting its center to $t = T/2$: $\text{rect}\left(\frac{t-T/2}{T}\right)$. The rectangle function (and its two-dimensional counterpart, the circle function) are also useful for modeling the spatial apertures of antennas and optical instruments. The two-dimensional rectangle, $\text{rect } x \text{ rect } y$, is used to model pixels in a digital image sensor or an image display.

The rectangle function is absolutely integrable and (trivially) piecewise continuous. The Fourier transform of the rectangle function is a simple integral:

$$\mathcal{F}\{\text{rect } x\} = \int_{-\infty}^{\infty} \text{rect } x \, e^{-i2\pi\nu x} dx = \int_{-1/2}^{1/2} e^{-i2\pi\nu x} dx$$

$$= \frac{e^{-i2\pi\nu x}}{-i2\pi\nu}\bigg|_{-1/2}^{1/2} = \frac{e^{-i\pi\nu} - e^{+i\pi\nu}}{-i2\pi\nu} = \frac{\sin \pi\nu}{\pi\nu}.$$

The function $\frac{\sin \pi\nu}{\pi\nu}$ appears often and is given the name sinc:

$$\text{sinc } \nu = \frac{\sin \pi\nu}{\pi\nu}. \tag{5.5}$$

---

[1] See Howell (2001, pp. 284–286) for a classic epsilon-delta proof of continuity for $f \in L^1$. A quicker path to the same conclusion is shown in Stade (2005, p. 300).

[2] Proofs of the Riemann–Lebesgue lemma for the Fourier transform may be found in Folland (1992, p. 217) and Stade (2005, pp. 328–329). When $F \in L^1$ as well as continuous it follows naturally, for then $\int |F| d\nu$ cannot be finite unless F rolls off at high frequencies.

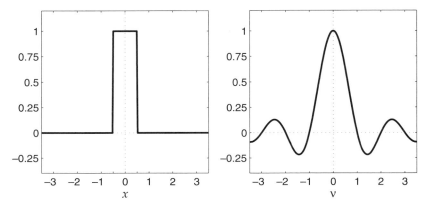

FIGURE 5.2 The rectangle function (*left*), and its Fourier transform, the sinc function (*right*).

It is the aperiodic analog of the Dirichlet kernel and arises from the same cause—truncation.³

The sinc function (Figure 5.2) is real and even, like the rectangle, and has zero crossings at integer values of ν. At $\nu = 0$, L'Hospital's rule yields the result $\text{sinc}(0) = 1$. It is bounded and continuous, as expected since the rectangle is absolutely integrable. The amplitude of the sinc function's oscillations decays with increasing frequency, as predicted by the Riemann–Lebesgue lemma. However, the rate of decay, $O(|\nu|^{-1})$, is too slow for sinc to be absolutely integrable (recall Examples 2.16 and 2.17). Consequently, the inverse Fourier transform taking sinc back to rectangle does not exist as an ordinary integral, though it will be shown later that the inverse may nevertheless be calculated, establishing the Fourier pair:

$$\text{rect } x \longleftrightarrow \text{sinc } \nu. \tag{5.6}$$

The rectangle function is a prototype for all functions with jump discontinuities—no discontinuous function can have an absolutely integrable transform. ■

Here is another common function with a jump discontinuity.

Example 5.2 (One-sided exponential function). The one-sided exponential function is

$$f(x) = e^{-x} U(x) = \begin{cases} e^{-x}, & x > 0 \\ \dfrac{1}{2}, & x = 0 \,, \\ 0, & x < 0 \end{cases} \tag{5.7}$$

---

³In fact, the Dirichlet kernel is the periodic replication of the sinc function.

where $U(x)$ is the unit step, or Heaviside function:

$$U(x) = \begin{cases} 1, & x > 0 \\ \frac{1}{2}, & x = 0 \\ 0, & x < 0 \end{cases}. \quad (5.8)$$

The step function is useful for modeling any kind of one-sided truncation or a sudden input to a dynamic system. The one-sided exponential appears as a damping function in the solutions of linear ordinary differential equations with constant coefficients. It is absolutely integrable and piecewise continuous. Calculating the Fourier transform:

$$F(\nu) = \mathcal{F}\{e^{-x}U(x)\} = \int_0^\infty e^{-x} e^{-i2\pi\nu x}\,dx = -\frac{1}{1+i2\pi\nu}\,e^{-(1+i2\pi\nu)x}\bigg|_0^\infty$$

$$= \frac{1}{1+i2\pi\nu} = \frac{1}{1+(2\pi\nu)^2} - \frac{i2\pi\nu}{1+(2\pi\nu)^2}.$$

The Fourier magnitude $|F(\nu)| = \frac{1}{\sqrt{1+(2\pi\nu)^2}}$ is bounded, $|F(\nu)| \leq |F(0)| = 1$—note that the L^1 norm of f is

$$\int_0^\infty e^{-x}\,dx = -e^{-x}\bigg|_0^\infty = 1,$$

agreeing with the bound $|F| \leq \|f\|_1$. The Fourier transform is also a continuous function of ν. Indeed, it is infinitely continuously differentiable; we will see why later, but you are invited to compare Theorem 4.4b for the Fourier series. Finally, $|F|$ decays $O(|\nu|^{-1})$ as $|\nu| \to \infty$, not fast enough for absolute integrability, because f has a jump discontinuity. Like sinc, it will be possible, eventually, to establish the inverse transform, giving the Fourier pair (Figure 5.3),

$$e^{-x}U(x) \longleftrightarrow \frac{1}{1+i2\pi\nu}. \quad (5.9)$$

∎

The next two examples show functions whose transforms are absolutely integrable.

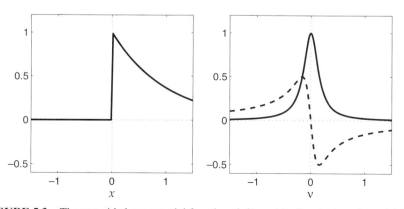

FIGURE 5.3 The one-sided exponential function (*left*), and its Fourier transform (*right*).

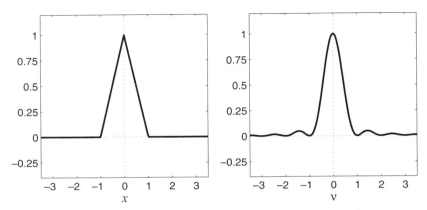

FIGURE 5.4 The triangle function (*left*), and its Fourier transform, the sinc² function (*right*).

Example 5.3 (Triangle function). The triangle function, $\Lambda(x)$, is defined

$$\Lambda(x) = \begin{cases} 1 - |x|, & |x| < 1 \\ 0, & \text{otherwise.} \end{cases} \tag{5.10}$$

It is absolutely integrable and piecewise smooth. The Fourier transform of the triangle function is

$$\mathcal{F}\{\Lambda(x)\} = \int_{-1}^{0} (1+x)e^{-i2\pi\nu x}dx + \int_{0}^{1} (1-x)e^{-i2\pi\nu x}dx$$

$$= \left[1 + \frac{1 - e^{i2\pi\nu}}{(2\pi\nu)^2}\right] + \left[-1 + \frac{1 - e^{-i2\pi\nu}}{(2\pi\nu)^2}\right] = \frac{2(1 - \cos 2\pi\nu)}{(2\pi\nu)^2} = \frac{\sin^2 \pi\nu}{(\pi\nu)^2}$$

$$= \text{sinc}^2 \nu.$$

The sinc² function is bounded and continuous (Figure 5.4), like sinc. Unlike sinc, it is absolutely integrable (Example 2.17), because it decays $O(|\nu|^{-2})$ as $|\nu| \to \infty$. The inverse transform exists as an ordinary integral (although it is not clear at this point how to actually do the calculation) and yields the transform pair:

$$\Lambda(x) \longleftrightarrow \text{sinc}^2 \nu. \tag{5.11}$$

■

Example 5.4 (Two-sided exponential). The two-sided exponential function is

$$f(x) = e^{-|x|}. \tag{5.12}$$

It is absolutely integrable, $\|f\|_1 = 2$, and piecewise smooth. We will not calculate the Fourier transform integral here. The transform will be obtained by a different method in Example 5.5. The result is

$$F(\nu) = \frac{2}{1 + (2\pi\nu)^2}.$$

Comparing this with the transform of the one-sided exponential, we see that it is, of course, bounded and continuous, but also that it decays $O(|\nu|^{-2})$ as $|\nu| \to \infty$, rapidly enough to guarantee absolute integrability. The inverse transform exists as an ordinary integral, yielding the Fourier pair (Figure 5.6):

$$e^{-|x|} \longleftrightarrow \frac{2}{1 + (2\pi\nu)^2}. \tag{5.15}$$

■

The triangle is "one step smoother" than the rectangle: piecewise smooth vs. piecewise continuous, corners at ± 1 and 0 rather than jumps at $\pm 1/2$. Likewise, the two-sided exponential is one step smoother than the one-sided exponential: piecewise smooth vs. piecewise continuous, a corner at the origin rather than a jump. In the frequency domain, the transforms of the rectangle and one-sided exponential are $O(|\nu|^{-1})$, while the transforms of the triangle and two-sided exponential are $O(|\nu|^{-2})$. Like Fourier series coefficients, the smoother a function is, the more rapidly its Fourier transform rolls off at high frequencies.

What we have observed so far about the Fourier transform and its inverse is summed up in the following theorem.

Theorem 5.1 (Fourier transform in L^1). Let $f \in L^1(\mathbb{R})$. Then its Fourier transform,

$$F(\nu) = \int_{-\infty}^{\infty} f(x) e^{-i2\pi\nu x} dx,$$

exists as an ordinary integral, is bounded and continuous for all $\nu \in \mathbb{R}$, and $|F(\nu)| \to 0$ as $|\nu| \to \infty$. If, additionally, $F \in L^1(\mathbb{R})$, then the inverse transform is an ordinary integral:

$$f(x) = \int_{-\infty}^{\infty} F(\nu) e^{i2\pi\nu x} dx,$$

and it is bounded and continuous for all $x \in \mathbb{R}$.

5.3 FOURIER TRANSFORM THEOREMS

We will now state and prove a number of Fourier transform theorems. Versions of many of them appeared earlier in connection with the DFT, Fourier series, and discrete-time Fourier transform. When you know the Fourier transform theorems and can visualize their physical meanings, you will be able to use them to simplify

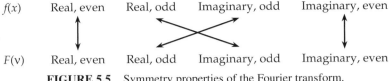

FIGURE 5.5 Symmetry properties of the Fourier transform.

calculations and to model physical systems. Very often, by creative application of the theorems, an otherwise messy calculation can easily be performed in just a few lines.

These theorems are stated for Fourier pairs, assuming that inverse transforms exist. Points where caution is advised will be noted.

Linearity
The Fourier transform is an integral, and integrals are linear operations: $\int af = a\int f$ and $\int (f+g) = \int f + \int g$. Thus, we have

Theorem 5.2 (Linearity). Let $f \longleftrightarrow F$ and $g \longleftrightarrow G$, and let $a, b \in \mathbb{C}$ be constants. Then

$$af + bg \longleftrightarrow aF + bG. \tag{5.13}$$

Symmetry
The Fourier transform obeys the same symmetries as the other transforms (Figure 5.5).

Theorem 5.3 (Symmetry). Let $f \longleftrightarrow F$. Then,

- If f is real, then F is Hermitian: $F(\nu) = F^*(-\nu)$.
- If f is even (odd), then F is even (odd).
- If f is real and even (real and odd), then F is real and even (imaginary and odd).

The converses hold, as well; for example, if F is Hermitian then f is real.

Example 5.5. The one-sided exponential (Equation 5.7) can be decomposed into the sum of an even, continuous function, and an odd, discontinuous function:

$$e^{-x}U(x) = \frac{1}{2}e^{-|x|} + \frac{1}{2}e^{-|x|}\operatorname{sgn} x,$$

where sgn is the signum function:

$$\operatorname{sgn} x = \begin{cases} 1, & x > 0 \\ \frac{1}{2}, & x = 0 \\ -1, & x < 0 \end{cases}. \tag{5.14}$$

Recall that the Fourier transform of $f(x) = e^{-x}U(x)$ is $F(\nu) = \frac{1}{1+i2\pi\nu}$ (Equation 5.9). Then, separating the even and odd parts of f and the real and imaginary parts of F,

$$e^{-x}U(x) \longmapsto \frac{1}{1+i2\pi\nu}$$

$$\frac{1}{2}e^{-|x|} + \frac{1}{2}e^{-|x|}\operatorname{sgn} x \longmapsto \frac{1}{1+(2\pi\nu)^2} - \frac{i2\pi\nu}{1+(2\pi\nu)^2}.$$

The real part of the transform is even and the imaginary part is odd. By the Fourier transform symmetries, we associate the even part of f with the real, even part of F, and the odd part of f with the imaginary, odd part of F. This yields two additional transform pairs (Figure 5.6):

$$e^{-|x|} \longleftrightarrow \frac{2}{1+(2\pi\nu)^2}, \tag{5.15}$$

$$e^{-|x|}\operatorname{sgn} x \longleftrightarrow \frac{-i4\pi\nu}{1+(2\pi\nu)^2}. \tag{5.16}$$

The one-sided exponential had a jump discontinuity and its Fourier transform was $O(|\nu|^{-1})$ as $|\nu| \to \infty$. The two-sided exponential (Equation 5.15) is free of the jump discontinuity, and this results in a more rapid decay for its Fourier transform, $O(|\nu|^{-2})$. On the other hand, $e^{-|x|}\operatorname{sgn} x$ has a jump and its Fourier transform is $O(|\nu|^{-1})$. ∎

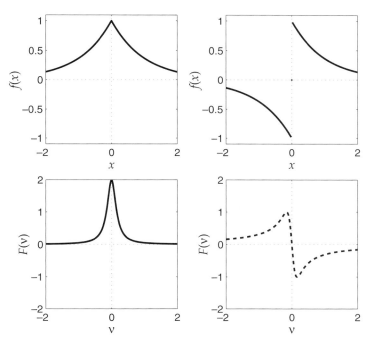

FIGURE 5.6 Two-sided exponential functions (*top*) and their Fourier transforms (*bottom*). Left: $e^{-|x|}$. Right: $e^{-|x|}\operatorname{sgn} x$. Dashed line denotes an imaginary valued function.

Other symmetry relationships concern the transforms of the reversal of a function, $f(-x)$, and the complex conjugate, $f^*(x)$.

Theorem 5.4 (Reversal). Let $f(x) \longleftrightarrow F(v)$. Then,

$$f(-x) \longleftrightarrow F(-v), \quad (5.17a)$$
$$f^*(x) \longleftrightarrow F^*(-v), \quad (5.17b)$$
$$f^*(-x) \longleftrightarrow F^*(v). \quad (5.17c)$$

Proof: Deriving the first relationship will suffice to show the flavor. By making the change of variable $\xi = -x$,

$$\int_{-\infty}^{\infty} f(-x) e^{-i2\pi v x} dx = \int_{-\infty}^{\infty} f(\xi) e^{+i2\pi v \xi} d\xi$$
$$= \int_{-\infty}^{\infty} f(\xi) e^{-i2\pi(-v)\xi} d\xi = F(-v).$$
∎

Theorem 5.5 (Repeated transform). Let $f(x) \longleftrightarrow F(v)$. Then $F(x) \to f(-v)$, or equivalently, $\mathcal{F}\{\mathcal{F}\{f(x)\}\} = f(-x)$.

Proof: Write the forward transform with $F(x)$:

$$\int_{-\infty}^{\infty} F(x) e^{-j2\pi v x} dx = \int_{-\infty}^{\infty} F(x) e^{j2\pi(-v)x} dx,$$

and this has the form of an inverse transform, with x and v exchanged, and also with a minus sign on v, that is, $f(-v)$. ∎

The repeated transform theorem doubles the number of Fourier transforms in our repertoire. Because rect $x \longmapsto$ sinc v, we also know sinc $x \longmapsto$ rect$(-v) = $ rect v.

Parseval's Formula
We know if $f \in L^1$, then F is bounded. If, in addition, $F \in L^1$, then $F \in L^2$ as well. That is, a function that is bounded and absolutely integrable is also square integrable, for then

$$\|F\|^2 = \int_{-\infty}^{\infty} |F|^2 \, dv = \int_{-\infty}^{\infty} |F| \cdot |F| \, dv \leq \sup |F| \int_{-\infty}^{\infty} |F| \, dv = \|F\|_\infty \|F\|_1 < \infty.$$

Symmetrically, if $F \in L^1$, the inverse transform is an ordinary integral and f is bounded as well as absolutely integrable, hence $f \in L^2$. We are now operating in an inner product space, and Parseval's theorem says that the Fourier transform preserves inner products and norms.

Theorem 5.6 (Parseval's formula). Let $f, g, F, G \in L^1$, where $f \longleftrightarrow F$ and $g \longleftrightarrow G$. Then

$$\int_{-\infty}^{\infty} f(x)g^*(x)dx = \int_{-\infty}^{\infty} F(v)G^*(v)dv, \tag{5.18a}$$

$$\int_{-\infty}^{\infty} |f(x)|^2 \, dx = \int_{-\infty}^{\infty} |F(v)|^2 \, dv. \tag{5.18b}$$

Proof: If $f, g, F, G \in L^1$, then they are also in L^2 and we have inner products $\langle f, g \rangle$ and $\langle F, G \rangle$. Now

$$\langle F, G \rangle = \int_{-\infty}^{\infty} F(v) G^*(v) \, dv = \int_{-\infty}^{\infty} F(v) \left(\int_{-\infty}^{\infty} g^*(x) e^{+j2\pi vx} \, dx \right) dv,$$

and by Fubini's theorem,

$$= \int_{-\infty}^{\infty} \left(\int_{-\infty}^{\infty} F(v) e^{+j2\pi vx} \, dv \right) g^*(x) \, dx = \int_{-\infty}^{\infty} f(x) g^*(x) \, dx = \langle f, g \rangle.$$

Then letting $g = f$, we also have $\|f\|_2^2 = \|F\|_2^2$. ∎

This derivation is limited to functions that are simultaneously absolutely- and square integrable, but as one would expect, Parseval's formula applies to the broader class of functions and transforms that are square integrable only (Section 5.6.2).

Shift Theorem
For functions $f : \mathbb{R} \to \mathbb{C}$, a shift $f(x - a)$ is a translation along the x-axis.

Theorem 5.7 (Shift). Let $f \longleftrightarrow F$. Let $f(x - a)$ denote a translation of f along the x-axis, and $F(v - b)$ a translation along the v-axis. Then

$$f(x - a) \longleftrightarrow e^{-i2\pi av} F(v), \tag{5.19}$$

$$e^{i2\pi bt} f(t) \longleftrightarrow F(v - b). \tag{5.20}$$

Proof: We will just derive Equation 5.19. Write down the forward transform integral:

$$\mathcal{F}\{f(x - a)\} = \int_{-\infty}^{\infty} f(x - a) e^{-i2\pi vx} dx$$

and make the change of variable $\xi = x - a$. Then,

$$\int_{-\infty}^{\infty} f(x - a) e^{-i2\pi vx} dx = \int_{-\infty}^{\infty} f(\xi) e^{-i2\pi v(\xi + a)} d\xi$$

$$= e^{-i2\pi av} \int_{-\infty}^{\infty} f(\xi) e^{-i2\pi v\xi} d\xi = e^{-i2\pi av} F(v). \blacksquare$$

Shifting does not affect the magnitude of the Fourier transform:

$$\left| e^{-i2\pi av} F(v) \right| = |F(v)|.$$

For example, consider the rectangle rect x and a shifted rectangle rect($x - a$). Their respective transforms are sinc ν and $e^{-i2\pi\nu a}$ sinc ν; their Fourier magnitudes are the same, $|\text{sinc }\nu|$. Whether the rectangle is centered at the origin or at some other position a, its spectral content is unchanged.

Example 5.6. The physical importance of the shift theorem can be illustrated with an example from audio engineering. The frequency response of an ideal audio amplifier is described by a complex-valued function $H(\nu) = |H(\nu)|e^{i\varphi(\nu)}$. The magnitude $|H|$ is called the *magnitude response* and the argument φ is called the *phase response*. These quantities express how the amplifier modifies the amplitude and phase of the Fourier components of the input signal on their way to the output, that is, for a pure tone input, $\cos 2\pi\nu_0 t$, the output is

$$|H(\nu_0)| \cos\left(2\pi\nu_0 t + \varphi(\nu_0)\right).$$

Now consider the special case of an amplifier with *linear phase*, by which we mean the phase response is linearly proportional to frequency:

$$\varphi(\nu) = -2\pi a\nu.$$

With this, a cosine input $\cos 2\pi\nu_0 t$ will be passed to the output as

$$|H(\nu_0)| \cos\left(2\pi\nu_0(t - a)\right).$$

The cosine is amplified according to the magnitude response in a frequency-dependent way and is delayed by a time a, which is *independent* of frequency. This means that all frequencies in the input signal will be delayed by the same amount of time in passing through the amplifier, and the phase relationships which existed among the Fourier components in the input will be preserved in the output. For this reason, linear phase is highly prized for high-fidelity sound reproduction. ∎

The frequency-domain version of the shift theorem (Equation 5.20) is often expressed

$$f(t)\cos(2\pi\nu_0 t) \longleftrightarrow \frac{1}{2}F(\nu - \nu_0) + \frac{1}{2}F(\nu + \nu_0). \tag{5.21}$$

This is called the *modulation theorem*.

Example 5.7 (Double sideband modulation). Audio signals are communicated over radio or optical frequency media by attaching them to a *carrier* signal through a process called *modulation*. One such process, called *double sideband*, or DSB, is a direct application of the modulation theorem. The method is diagrammed in Figure 5.7. The message signal is denoted f and, for illustrative purposes, has the triangular Fourier transform F, as shown. The carrier signal is a sinusoid of frequency ν_c. An electronic device called a *mixer* multiplies the carrier and the message, producing the time-domain signal $f(t)\cos 2\pi\nu_c t$. In the frequency domain, the modulation theorem says that the spectrum of the modulated carrier is $\frac{1}{2}F(\nu + \nu_c) + \frac{1}{2}F(\nu - \nu_c)$. The message spectrum has been shifted to a band of frequencies centered at ν_c. The scale of the drawing is exaggerated; in practice the carrier frequency is much

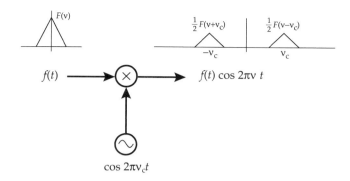

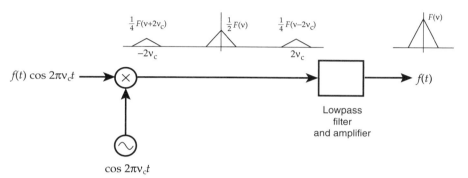

FIGURE 5.7 Double sideband modulation. (*Top*) A message signal is modulated onto a carrier signal. (*Bottom*) The message is recovered from the modulated carrier.

larger than the message bandwidth. Audio signals have bandwidth no higher than 22 kHz; radio frequency carriers are measured in MHz or GHz, and optical frequencies are orders of magnitude higher still. The message is recovered, or *demodulated*, by a nearly identical process. The modulated carrier is mixed again with a cosine of the same frequency as the carrier. (In practice, special circuitry synchronizes the receiver's local oscillator so that it has the same frequency and phase as the carrier.) In the time domain, the result is

$$f(t)\cos^2 2\pi v_c t = \frac{1}{2}f(t) + \frac{1}{2}f(t)\cos 4\pi v_c t.$$

The result contains the original message and a modulated component at twice the carrier frequency. In the frequency domain, a second application of the modulation theorem operates as follows:

$$\frac{1}{2}F(v+v_c) \longrightarrow \frac{1}{4}F(v+v_c+v_c) + \frac{1}{4}F(v+v_c-v_c) = \frac{1}{4}F(v+2v_c) + \frac{1}{4}F(v),$$

$$\frac{1}{2}F(v-v_c) \longrightarrow \frac{1}{4}F(v-v_c+v_c) + \frac{1}{4}F(v-v_c-v_c) = \frac{1}{4}F(v) + \frac{1}{4}F(v-2v_c).$$

Adding the two terms gives an output spectrum:

$$\frac{1}{4}F(\nu+2\nu_c) + \frac{1}{2}F(\nu) + \frac{1}{4}F(\nu-2\nu_c),$$

which contains the original message plus modulated components at twice the carrier frequency. These are removed by a frequency selective circuit called a *lowpass filter*, and a final amplification by a factor of two recovers the original message. ∎

Dilation theorem

Scaling the independent variable in a function, that is, $f(ax)$ or $f(x/a)$, stretches or squeezes the function along the axis. This is called *dilation*.

Theorem 5.8 (Dilation). If $f(x) \longleftrightarrow F(\nu)$ and a is a nonzero constant, then

$$f(ax) \longleftrightarrow \frac{1}{|a|}F\left(\frac{\nu}{a}\right). \tag{5.22}$$

Proof: Write down the integral, and make the change of variable $\xi = ax$. Then,

$$\int_{-\infty}^{\infty} f(ax)e^{-i2\pi\nu x}dx = \int_{-\infty}^{\infty} f(\xi)e^{-i2\pi\nu\xi/a}\frac{d\xi}{|a|} = \frac{1}{|a|}F\left(\frac{\nu}{a}\right). \qquad \blacksquare$$

In words, the dilation theorem says that if you squeeze a function in the time domain, you stretch its transform in the frequency domain, and vice versa. The spirit of the dilation theorem is the same for the discrete-time Fourier transform (Theorem 4.17), but the present form is much simpler because we are working with functions of a real variable instead of discrete sequences.

Example 5.8 (Time constant and bandwidth). Consider the function $h(t) = e^{-t/\tau}U(t) = e^{-t/\tau}U(t/\tau)$. By the dilation theorem, $H(\nu) = \frac{\tau}{1+i2\pi\tau\nu}$. The squared magnitude is $|H|^2 = \frac{\tau^2}{1+(2\pi\tau\nu)^2}$. As τ, the "time constant," increases, h decays more slowly. In the frequency domain, the squared magnitude narrows. The half-power bandwidth, the value of ν by which $|H|$ has fallen to $1/2$ of its maximum value at $\nu = 0$, is $1/2\pi\tau$, inversely proportional to τ. These effects are illustrated in Figure 5.8. ∎

You may have experience with the ideas of risetime and bandwidth of an electronic circuit. The risetime expresses the speed with which a system can respond to a step input. The bandwidth is the highest frequency that is effectively passed by the system and is a measure of the spread of the circuit's frequency response along the frequency axis. It is well known that risetime and bandwidth are inversely proportional. In order to have a small (fast) risetime, the circuit must pass high-frequency Fourier components of the step input to the output, and this requires a high bandwidth.

In a communication system, digital information is transmitted by sequences of narrow pulses. The rate at which bits can be sent is inversely proportional to the pulse

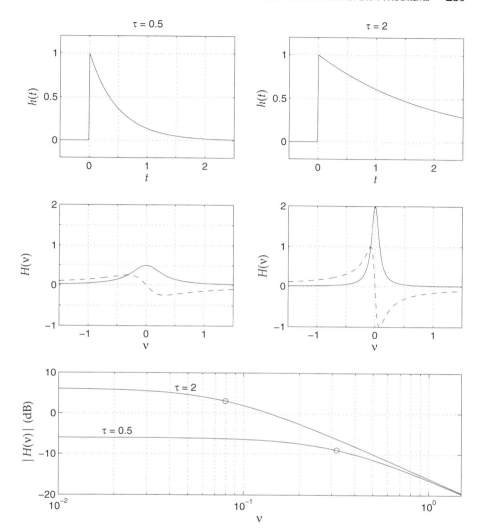

FIGURE 5.8 Dilation, illustrated by the one-sided exponential function $h(t) = e^{-t/\tau}U(t)$ and its Fourier transform, $H(\nu) = \frac{\tau}{1+i2\pi\tau\nu}$, for $\tau = 0.5$ (*left*) and $\tau = 2$ (*right*). The bottom graph shows the magnitude of the Fourier transforms, in decibels, $20\log_{10}|H(\nu)|$. There is a factor of four, or 12 decibel, difference in the magnitudes, and a factor of four difference in the half-power frequencies (circles).

width. By the dilation theorem, as a pulse is made narrower its Fourier spectrum becomes wider, and to receive it without distortion requires a higher bandwidth communication channel.

The dilation theorem is also behind the requirement that a telescope have a wide aperture (i.e., large primary mirror) in order to resolve closely spaced objects. The image of a distant point-like object, called the *point spread function*, is proportional to the (two-dimensional) spatial Fourier transform of the telescope's aperture (see Example 10.17 in a later chapter). Increasing the diameter of a telescope's primary mirror not only permits it to gather more light, but it also makes the area of the point spread function smaller so the images of distant objects become sharper.

Shift and dilation often occur together. The Fourier transform of $f(\frac{x-b}{a})$ is

$$\int_{-\infty}^{\infty} f\left(\frac{x-b}{a}\right) e^{-i2\pi v x} \, dx = \int_{-\infty}^{\infty} f(u) e^{-i2\pi v(au+b)} |a| \, du \qquad \left(\text{let } u = \frac{x-b}{a}\right)$$

$$= |a| e^{-i2\pi v b} \int_{-\infty}^{\infty} f(u) e^{-i2\pi (av)u} \, du$$

$$= |a| e^{-i2\pi v b} F(av).$$

Compactly,

$$f\left(\frac{x-b}{a}\right) \longleftrightarrow |a| e^{-i2\pi v b} F(av). \qquad (5.23)$$

One could arrive at the same result by applying the shift and dilation theorems in succession:

$$f\left(\frac{x-b}{a}\right) \longmapsto e^{-i2\pi v b} \mathcal{F}\left\{f\left(\frac{x}{a}\right)\right\} = e^{-i2\pi v b} |a| F(av).$$

This combination of theorems correctly interprets $f(\frac{x-b}{a})$ as the dilation of $f(x)$ to $f(x/a)$, which is then shifted along x by b. So, to transform the function, apply the theorems in the reverse order—shift first, then dilation. If you were to apply the dilation theorem first, you could get

$$f\left(\frac{x-b}{a}\right) \longmapsto |a| \mathcal{F}\{f(x-b)\} = |a| e^{-i2\pi v b} F(v),$$

or

$$f\left(\frac{x-b}{a}\right) \longmapsto |a| \mathcal{F}\{f(x-b)\}\Big|_{(av)} = |a| e^{-i2\pi a v b} F(av),$$

neither of which is correct.

Example 5.9. To calculate the Fourier transform of $f(t) = \text{rect}(\frac{t-T/2}{T})$, a shifted rectangle, apply Equation 5.23 with $a = T$ and $b = T/2$, yielding

$$F(v) = e^{-i\pi v T} T \, \text{sinc}(Tv).$$

∎

Differentiation and Integration

If a function is differentiable or integrable, there are convenient expressions for the transform of the derivative or integral.

Theorem 5.9 (Derivative). If f, f' are integrable (e.g., L^1 or L^2) and $f \longmapsto F$, then

$$f'(x) \longmapsto i2\pi v F(v). \qquad (5.24)$$

5.3 FOURIER TRANSFORM THEOREMS

Proof: Write $f'(x) \longmapsto \int_{-\infty}^{\infty} f'(x) e^{-i2\pi v x} \, dx$ and integrate once, by parts,

$$f'(x) \longmapsto f(x) e^{-i2\pi v x} \Big|_{-\infty}^{\infty} - \int_{-\infty}^{\infty} f(x) \cdot (-i2\pi v) e^{-i2\pi v x} \, dx.$$

Because f is integrable, $|f(x) e^{-i2\pi v x}| \to 0$ as $|x| \to \infty$, and we are left with

$$f'(x) \longmapsto i2\pi v \int_{-\infty}^{\infty} f(x) e^{-i2\pi v x} \, dx = i2\pi v F(v).$$

∎

Example 5.10. Consider the Fourier pair $\Lambda \longleftrightarrow \text{sinc}^2$. We will calculate the Fourier transform of the derivative of the triangle in three different ways to illustrate the application of different Fourier theorems.

1. The derivative of the triangle function is

$$f'(x) = \frac{d}{dx} \Lambda(x) = \begin{cases} 1, & 0 > x \geq -1 \\ -1, & 1 \geq x > 0 \\ 0, & \text{otherwise } (|x| > 1, x = 0) \end{cases}.$$

From here, we can calculate the Fourier transform by direct integration:

$$f' \longmapsto \int_{-1}^{0} e^{-i2\pi v x} \, dx + \int_{0}^{1} (-1) e^{-i2\pi v x} \, dx$$

$$= \frac{1 - e^{i2\pi v}}{-i2\pi v} - \frac{e^{-i2\pi v} - 1}{-i2\pi v} = \frac{2(1 - \cos 2\pi v)}{-i2\pi v}$$

$$= \frac{2i \sin^2 \pi v}{\pi v}.$$

2. Instead of calculating the Fourier integral, we can represent the derivative of the triangle as the sum of two shifted rectangles:

$$f' = \text{rect}\left(x + \frac{1}{2}\right) - \text{rect}\left(x - \frac{1}{2}\right).$$

Then using the shift and linearity theorems,

$$f' \longmapsto \exp\left[-i2\pi v \left(-\frac{1}{2}\right)\right] \text{sinc } v - \exp\left[-i2\pi v \left(\frac{1}{2}\right)\right] \text{sinc } v$$

$$= 2i \sin \pi v \, \text{sinc } v = \frac{2i \sin^2 \pi v}{\pi v}.$$

3. Finally, using the derivative theorem, we obtain the solution quickly:

$$f' \longmapsto i2\pi v \, \text{sinc}^2 v = \frac{2i\pi v \sin^2 \pi v}{(\pi v)^2} = \frac{2i \sin^2 \pi v}{\pi v}.$$

The functions and transforms are illustrated in Figure 5.9. ∎

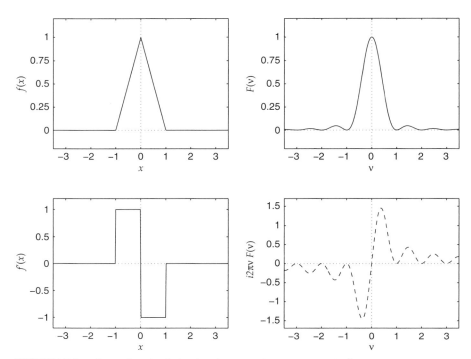

FIGURE 5.9 Illustrating the derivative theorem with the $\Lambda \longleftrightarrow \text{sinc}^2$ pair. *Top*: The triangle function and its Fourier transform. *Bottom*: The derivative of the triangle and its Fourier transform, $i2\pi\nu\,\text{sinc}^2\nu$. The derivative of the triangle, a real and even function, is a real and odd function. The Fourier transform of the triangle is real and even; the transform of the derivative is imaginary and odd.

If a function is n-times differentiable and all the derivatives are integrable, the derivative theorem may be applied n times, so that

$$f^{(n)}(x) \longmapsto (i2\pi\nu)^n F(\nu). \tag{5.25}$$

A variation on the same theme, if $x^n f$ is integrable, is

$$x^n f(x) \longmapsto \frac{F^{(n)}(\nu)}{(-i2\pi)^n}. \tag{5.26}$$

The derivation is left to the problems.

The derivative theorem underlies the application of the Fourier transform to solving differential equations. It converts derivatives in one domain into algebraic relationships in the other domain.

Example 5.11. A second-order differential equation driven by an input $f(t)$,

$$y'' + a_1 y' + a_2 y = f,$$

transforms to
$$(i2\pi\nu)^2 Y(\nu) + a_1(i2\pi\nu)Y(\nu) + a_2 Y(\nu) = F(\nu).$$

Collecting terms,
$$[(a_2 - (2\pi\nu)^2) + i2\pi a_1 \nu]Y(\nu) = F(\nu).$$

Solving for Y,
$$Y(\nu) = \frac{F(\nu)}{(a_2 - (2\pi\nu)^2) + i2\pi a_1 \nu}.$$

The solution to the differential equation is the inverse transform of Y. In linear systems analysis, the rational function
$$H(\nu) = \frac{Y(\nu)}{F(\nu)} = \frac{1}{(a_2 - (2\pi\nu)^2) + i2\pi a_1 \nu}$$

is called the *transfer function*. It often gives sufficient information about the system behavior that the actual solution $y(t)$ need not be calculated. ∎

Example 5.12 (Gaussian function). The Gaussian function, $f(x) = e^{-\pi x^2}$, has multiple applications. It is the classic "bell curve" in probability and statistics, where it is known as the normal distribution. In optics, it describes the transverse intensity profile of many laser beams. The Gaussian is also the solution of the heat equation (4.27) on an unbounded domain.

Calculating $\|f\|_1 = \int_{-\infty}^{\infty} e^{-\pi x^2} \, dx$ is not straightforward, because the Gaussian does not have an antiderivative. Instead, the norm is calculated by first calculating $\|f\|_1^2$:

$$\|f\|_1^2 = \left[\int_{-\infty}^{\infty} e^{-\pi x^2} dx\right]^2 = \left[\int_{-\infty}^{\infty} e^{-\pi x^2} dx\right]\left[\int_{-\infty}^{\infty} e^{-\pi y^2} dy\right]$$
$$= \int_{-\infty}^{\infty} \int_{-\infty}^{\infty} e^{-\pi(x^2+y^2)} \, dxdy.$$

Converting to polar coordinates, $r^2 = x^2 + y^2$, $dx\,dy = r\,dr\,d\theta$,

$$\|f\|_1^2 = \int_0^{2\pi} \int_0^{\infty} e^{-\pi r^2} r\,dr\,d\theta = 2\pi \int_0^{\infty} e^{-\pi r^2} r\,dr.$$

Now, let $u = \pi r^2$, $du = 2\pi r\,dr$; then
$$\|f\|_1^2 = \int_0^{\infty} e^{-u} du = 1 \implies \|f\|_1 = 1.$$

To calculate the Fourier transform of the Gaussian, note that
$$f'(x) = -2\pi x\,e^{-\pi x^2} = -2\pi x f(x),$$

that is, f obeys a differential equation:
$$f' + 2\pi x f = 0.$$

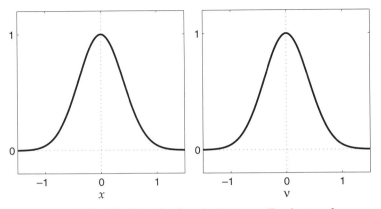

FIGURE 5.10 The Gaussian function is its own Fourier transform.

Because f and f' are absolutely integrable, we may Fourier transform the differential equation, using Equations 5.24 and 5.26:

$$i2\pi v F + 2\pi \frac{F'}{-i2\pi} = 0$$
$$\implies F' + 2\pi v F = 0.$$

This is the same differential equation, with v and F instead of x and f. The solution is a Gaussian:

$$F(v) = Ce^{-\pi v^2}.$$

To evaluate the constant C, set $v = 0$:

$$C = F(0) = \int_{-\infty}^{\infty} e^{-\pi x^2} dx = 1.$$

We have, at last, the Fourier transform pair:

$$e^{-\pi x^2} \longleftrightarrow e^{-\pi v^2}. \tag{5.27}$$

The Gaussian has the rare distinction of being an elementary function that is its own Fourier transform (Figure 5.10). ∎

Example 5.13 (Linear chirp function). The function $f(x) = e^{-\pi c x^2}$, with *complex* c, can be shown to have the same properties as the real Gaussian:

$$\int_{-\infty}^{\infty} e^{-\pi c x^2} dx = \frac{1}{\sqrt{c}}, \quad \mathcal{R}e\, c > 0, \tag{5.28}$$

$$\mathcal{F}\left\{e^{-\pi c x^2}\right\} = \frac{1}{\sqrt{c}} e^{-\pi v^2/c}. \tag{5.29}$$

The special case $c = -ib$, $f(x) = e^{+i\pi b x^2}$, is called a *complex linear chirp* function. For any function $f(x) = \exp(i\varphi(x))$, the derivative of the phase, $\varphi'(x)/2\pi$, is called the *instantaneous frequency*. The instantaneous frequency of the linear chirp function is bx; it increases with time, resembling the chirping sound of a bird. The linear chirp Fourier transform pair is

$$e^{+i\pi bx^2} \longleftrightarrow \frac{1}{\sqrt{-ib}} e^{-i\pi v^2/b}. \qquad (5.30)$$

Derivations of Equations 5.28, 5.29, and 5.30 are similar to those for the real Gaussian and are deferred to the problems. ∎

We have seen $f' \longmapsto i2\pi v F$ and $-i2\pi x f \longmapsto F'$—differentiating in one domain multiplies by x or v in the other domain. It seems that, just as we had with the Fourier series, *integrating* in one domain should *divide* by x or v in the other domain, something like

$$\int_{-\infty}^{x} f(\xi) d\xi \longmapsto \frac{F(v)}{i2\pi v}.$$

In order for this (hypothesized) transform not to blow up at the origin, F must be zero at $v = 0$. Now, as we shall soon see, if $f \longmapsto F$, then (Equation 5.35)

$$F(0) = \int_{-\infty}^{\infty} f(x) e^{-i2\pi 0 x} dx = \int_{-\infty}^{\infty} f(x) dx.$$

So in order to have $F(0) = 0$, we must require

$$\int_{-\infty}^{\infty} f(x) dx = 0;$$

f must have "zero area" or "zero DC." Now, by Leibniz' rule,

$$\frac{d}{dx} \int_{-\infty}^{x} f(\xi) d\xi = f(x)$$

and then, by the derivative theorem,

$$\frac{d}{dx} \int_{-\infty}^{x} f(\xi) d\xi \longmapsto i2\pi v \, \mathcal{F}\left\{ \int_{-\infty}^{x} f(\xi) d\xi \right\} = F(v)$$

$$\Longrightarrow \int_{-\infty}^{x} f(\xi) d\xi \longmapsto \frac{F(v)}{i2\pi v},$$

as desired. We have the following result.

Theorem 5.10 (Integral). Let f be integrable with $\int_{-\infty}^{\infty} f(x) dx = 0$, and $f \longmapsto F$. Then

$$\int_{-\infty}^{x} f(\xi) d\xi \longmapsto \frac{F(v)}{i2\pi v}. \qquad (5.31)$$

If $F(v)/i2\pi v$ is integrable, then the inverse relationship also holds.

Example 5.14 (Biquad filter). An electronic circuit having a frequency response $H(v) = \frac{1}{i2\pi v}$ is called an *integrator* and is a basic building block in designing filters

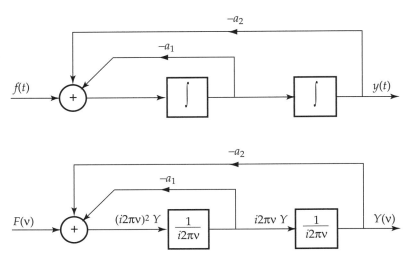

FIGURE 5.11 Biquad filter (Example 5.14). *Top:* integrators with feedback. *Bottom:* because of linearity, the integrators and signals can be replaced by their frequency-domain equivalents.

and control systems. The system shown in Figure 5.11 is made from two integrators, with *feedback* from the integrator outputs to a summing junction at the input.

Let the Fourier transform of the system output be Y. It is the integral of the input to the second integrator. Therefore, the input to that integrator is $i2\pi\nu Y$. That signal in turn is the output of the first integrator, so the input to the first integrator is $(i2\pi\nu)^2 Y$. But the input to the first integrator is also the output of the summing junction, which is a weighted sum of the input and the two integrator outputs. Equating the two, we have

$$(i2\pi\nu)^2 Y(\nu) = F(\nu) - a_1 (i2\pi\nu) Y(\nu) - a_2 Y(\nu).$$

Solving for Y,

$$Y(\nu) = \frac{F(\nu)}{\left(a_2 - (2\pi\nu)^2\right) + i2\pi a_1 \nu}.$$

The system is a filter with transfer function

$$H(\nu) = \frac{Y(\nu)}{F(\nu)} = \frac{1}{\left(a_2 - (2\pi\nu)^2\right) + i2\pi a_1 \nu}.$$ ∎

The derivative and integral theorems connect the smoothness of a function with the asymptotic behavior of its Fourier transform. As we saw with the Fourier series, differentiation makes a function less smooth and boosts the high frequencies of the Fourier transform. Integration makes a function more smooth and attenuates the high frequencies of the Fourier transform. The statement $f^{(n)}(x) \longmapsto (i2\pi\nu)^n F(\nu)$ for an n-times differentiable function can be turned around. The function $f^{(n)}$ is assumed to be L^1 so that it has a Fourier transform, which of course is bounded. Thus,

$$|F(\nu)| \leq \frac{M}{1 + |2\pi\nu|^n}.$$

The Fourier transform of f is no larger than M and decays $O(|v|^{-n})$ or better. The smoother a function is, as measured by its differentiability, the faster its Fourier transform decays as $|v| \to \infty$. With $n = 2$, for example, we have a sufficient condition for the absolute integrability of F, namely that f be twice differentiable.

Moment Theorems

The *n*-th *moment* of a function $f(x)$, denoted $\mu_f^{(n)}$, is defined

$$\mu_f^{(n)} = \int_{-\infty}^{\infty} x^n f(x)\, dx, \tag{5.32}$$

provided that the integral exists (f must decay faster than $|x|^{-n-1}$ as $|x| \to \infty$). Two moments commonly encountered in applications are the area under a function, which is $\mu_f^{(0)}$, and the *centroid*, or center-of-mass,

$$\text{Centroid} = \frac{\mu_f^{(1)}}{\mu_f^{(0)}}. \tag{5.33}$$

The *variance*, σ_f^2, measures the width, or spread, of a function. It is based on the second moment of a function about its centroid, also called the second *central moment*:

$$\sigma_f^2 = \frac{1}{\mu_f^{(0)}} \int_{-\infty}^{\infty} \left(x - \frac{\mu_f^{(1)}}{\mu_f^{(0)}}\right)^2 f(x)\, dx = \frac{\mu_f^{(2)}}{\mu_f^{(0)}} - \left(\frac{\mu_f^{(1)}}{\mu_f^{(0)}}\right)^2. \tag{5.34}$$

The moments of a function may be calculated from derivatives of its Fourier transform. The basic approach is demonstrated by the following theorem, which mirrors earlier results, Theorems 3.4 and 4.8.

Theorem 5.11 (Area). If f, F are continuous at the origin and $f \longleftrightarrow F$, then

$$\int_{-\infty}^{\infty} f(x)\,dx = F(0), \tag{5.35a}$$

$$f(0) = \int_{-\infty}^{\infty} F(v)\,dv \tag{5.35b}$$

provided that the integrals exist.

Proof: By the definition of the Fourier transform,

$$\int_{-\infty}^{\infty} f(x) e^{-i 2\pi v x}\, dx = F(v).$$

Simply set $v = 0$ on both sides. The same derivation works for $f(0)$. Continuity is required so that there is no uncertainty about the meaning of $f(0)$ and $F(0)$. ∎

Combining the area theorem with the derivative theorem, we can establish a more general pair of results. Their proofs are left to the problems.

Theorem 5.12 (Moments). If $f \longleftrightarrow F$, then

$$\mu_f^{(n)} = \int_{-\infty}^{\infty} x^n f(x)\,dx = \frac{1}{(-i2\pi)^n} F^{(n)}(0), \tag{5.36a}$$

$$\mu_F^{(n)} = \int_{-\infty}^{\infty} v^n F(v)\,dv = \frac{1}{(i2\pi)^n} f^{(n)}(0), \tag{5.36b}$$

when the indicated integrals exist and the derivatives are continuous at the origin.

Example 5.15 (Doing integrals with the moment theorems). If you know the Fourier transform of a function, you can calculate an integral by working with the transform.

1. $\displaystyle\int_{-\infty}^{\infty} \operatorname{sinc}(x)\,dx = \operatorname{rect}(0) = 1.$

2. $\displaystyle\int_{0}^{\infty} e^{-x}\,dx = \left.\frac{1}{1+i2\pi v}\right|_{v=0} = 1.$

3. $\displaystyle\int_{0}^{\infty} x e^{-ax}\,dx = \frac{1}{-i2\pi}\left[\frac{d}{dv}\frac{1/a}{1+i2\pi v/a}\right]_{v=0} = \frac{1}{-i2\pi}\left.\frac{-i2\pi/a}{(1+i2\pi v/a)^2}\right|_{v=0} = \frac{1}{a}.$ ∎

To illustrate the connection between moments and derivatives, consider the Gaussian function, $f(x) = e^{-\pi x^2/a^2}$ $(a > 0)$, which, according to the dilation theorem, has Fourier transform $F(v) = ae^{-\pi a^2 v^2}$. These functions are illustrated in Figure 5.12.

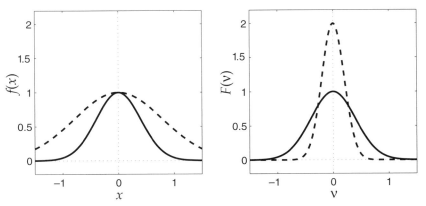

FIGURE 5.12 The Gaussian function $e^{-\pi x^2/a^2}$ (*left*), and its Fourier transform, $ae^{-\pi a^2 v^2}$ (*right*). The solid line corresponds to $a = 1$, and the dashed line to $a = 2$. The moments of f are related to the derivatives of F at the origin.

Calculating the moments of f with the moment theorem,

$$\mu_f^{(0)} = F(0) = a,$$

$$\mu_f^{(1)} = \frac{F'(0)}{-i2\pi} = \frac{1}{-i2\pi}\left[-2\pi a^3 v\, e^{-\pi a^2 v^2}\right]_{v=0} = 0,$$

$$\mu_f^{(2)} = \frac{F''(0)}{-(2\pi)^2} = \frac{1}{-i2\pi}\left[-2\pi a^3(1 - 2a^2\pi v)\, e^{-\pi a^2 v^2}\right]_{v=0} = \frac{a^3}{2\pi},$$

$$\sigma_f^2 = \frac{\mu_f^{(2)}}{\mu_f^{(0)}} - \left(\frac{\mu_f^{(1)}}{\mu_f^{(0)}}\right)^2 = \frac{a^2}{2\pi}.$$

The first moment, $\mu_f^{(1)}$, depends on the behavior of the first derivative, F', at the origin. F has zero slope at the origin, so f's first moment is zero. The second moment, $\mu_f^{(2)}$, depends on the second derivative, F'', which is a measure of curvature. Referring to Figure 5.12, the curvature of F is higher, and consequently the function is more sharply peaked, for $a = 2$ than it is for $a = 1$. Correspondingly, f is more spread out for $a = 2$ (higher variance) than it is for $a = 1$. Reversing the roles, we note that F's variance is lower for $a = 2$ than it is for $a = 1$, and f's curvature at the origin is correspondingly lower for $a = 2$ than it is for $a = 1$.

5.4 INTERPRETING THE FOURIER TRANSFORM

If, for some function f, we are able to calculate the Fourier transform F, we can informally interpret the transform as the Fourier coefficients for a set of basis functions $\{e^{i2\pi vx}\}$ indexed not by integers but by a real variable v. This interpretation does not quite work, however. The orthogonality relationship for this "continuum basis" evidently would be expressed by the inner product integral

$$\left\langle e^{i2\pi vx}, e^{i2\pi \mu x}\right\rangle = \int_{-\infty}^{\infty} e^{i2\pi(v-\mu)x} dx,$$

but the complex exponential is not absolutely integrable over the infinite limits. In particular, when $v = \mu$ the integral blows up. Moreover, in the DFT and the Fourier series, where we had a line spectrum, it was easy to relate each coefficient to the power in the signal at the corresponding frequency. This interpretation fails for the Fourier transform. If it were correct, we would be able to say, for example, that $F(v_0)$ and $F(-v_0)$ were the coefficients for a sinusoid at frequency v_0, one of many making up the original function $f(x)$. This would imply, for instance, that the Fourier transform of $\cos 2\pi v_0 x$ is

$$F(v) = \begin{cases} \frac{1}{2}, & v = \pm v_0 \\ 0, & \text{otherwise} \end{cases}.$$

But then, in the inverse Fourier transform, the integrand would be

$$F(v)e^{i2\pi v x} = \begin{cases} \dfrac{1}{2}e^{\pm i2\pi v_0 x}, & v = \pm v_0 \\ 0, & \text{otherwise} \end{cases}$$

and the integral is the area under two isolated points, which is zero—not what we want. If we try to calculate the Fourier transform integral with $f(x) = \cos 2\pi v_0 x$, we do not do any better, because cosine is not absolutely integrable or square integrable on $(-\infty, \infty)$. Ironically, sinusoids are one type of function which cannot be Fourier transformed without special handling. This is the subject of the next chapter.

So what *does* the Fourier transform say? The most physical approach is to think of it as a density function, analogous to a classical mass density, a statistical probability density function, or a quantum probability amplitude. If you have a density function $\rho(x)$ for the mass along a thin rod, you cannot say what the mass is at a particular point x_0, because a single point is massless. But you *can* calculate the mass of the entire rod by integrating the density function over the length of the rod, and you can calculate the mass of a portion of the rod by integrating the density function over that portion.

So too with the Fourier transform. We will see later that if $f \in L^2$, then $F \in L^2$ as well, and Parseval's formula (Equation 5.18b) says that the total power in f is the integral of the spectrum $|F|^2$ over all frequencies:

$$\int_{-\infty}^{\infty} |f(x)|^2 \, dx = \int_{-\infty}^{\infty} |F(v)|^2 \, dv.$$

If we interpret the squared-norm $\|f\|^2$ as power, then the squared magnitude of the Fourier transform, $|F|^2$, carries units of power/frequency or "power per unit bandwidth." The power in a band of frequencies $v_2 > |v| > v_1$ is the integral of a portion of the spectrum:

$$\int_{-v_2}^{-v_1} |F(v)|^2 \, dv + \int_{v_1}^{v_2} |F(v)|^2 \, dv.$$

The area under a point (let $v_1 \to v_2$) is zero, so we have to conclude that the fraction of power at a single frequency is zero. This poses no practical problem, because instruments for spectrum analysis have finite observation bandwidths $\Delta v = v_2 - v_1$.

5.5 CONVOLUTION

5.5.1 Definition and basic properties

The convolution of two functions f and h, denoted $f * h$, is defined:

$$f * h(x) = \int_{-\infty}^{\infty} f(\xi) h(x - \xi) \, d\xi = \int_{-\infty}^{\infty} f(x - \xi) h(\xi) \, d\xi, \tag{5.37}$$

when the integrals exist. The equivalence of the two integrals is easily verified by making a change of variable, showing that convolution is commutative, $f * h = h * f$.

In the special case that $f = 0$ for all x outside a bounded interval $[a, b]$ (the function is then said to have *bounded support*), the convolution is a finite integral:

$$f * h(x) = \int_a^b f(\xi) h(x - \xi) \, d\xi,$$

and is bounded for all x if f and h are bounded:

$$|f * h| \le (b - a) \|f\|_\infty \, \|h\|_\infty \, .$$

If f and h are one sided, equal to zero for $x < 0$, then the convolution is a finite integral:

$$f * h(x) = \int_0^x f(\xi) h(x - \xi) \, d\xi$$

but is not guaranteed to be bounded as $x \to \infty$ without additional constraints on f or h.

Let us begin, as we did with the Fourier transform, by considering what is required to have a bounded convolution:

$$\left| \int_{-\infty}^\infty f(\xi) h(x - \xi) \, d\xi \right| \le \int_{-\infty}^\infty |f(\xi) h(x - \xi)| \, d\xi. \qquad \text{(triangle inequality)}$$

The product fh is absolutely integrable if one of them is bounded, and the other is absolutely integrable:

$$\int_{-\infty}^\infty |f(\xi) h(x - \xi)| \, d\xi \le \sup |h| \int_{-\infty}^\infty |f(\xi)| \, d\xi.$$

So we have one result:

$$|f * h| \le \|f\|_1 \, \|h\|_\infty \, . \tag{5.38}$$

It can also be shown that under these same conditions, $f \in L^1$ and $h \in L^\infty$, the convolution is a *continuous* function of x, even if one or both of the functions being convolved is discontinuous.[4]

Example 5.16 (The convolution of two rectangle functions).

$$\text{rect}(t) * \text{rect}(t) = \int_{-\infty}^\infty \text{rect}(t - \tau) \text{rect}(\tau) \, d\tau = \int_{-1/2}^{1/2} \text{rect}(t - \tau) \, d\tau.$$

The rectangle $\text{rect}(t - \tau)$, as a function of τ, is time reversed and shifted to be centered at $\tau = t$. We distinguish four cases of the integral as t varies and slides the one rectangle across the other:

1. For $t < -1$, the rectangles do not overlap and the integral is zero.
2. For $0 > t \ge -1$, the integral is $\int_{-1/2}^{t+1/2} d\tau = 1 + t$.

---

[4] For a proof of the continuity of the convolution, see Gasquet and Witomski (1999, pp. 181–182).

302 CHAPTER 5 THE FOURIER TRANSFORM

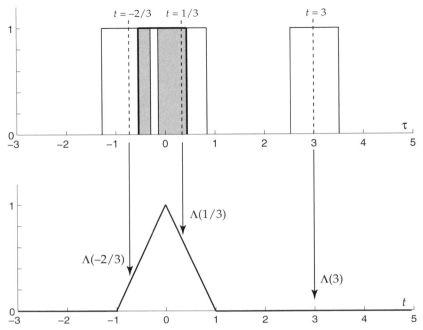

FIGURE 5.13 The convolution $\Lambda(t) = \text{rect}(t) * \text{rect}(t)$. As the time-reversed rectangle, $\text{rect}(t - \tau)$, slides across the rectangle $\text{rect}(\tau)$ (*top*), the convolution is the area under their product, indicated by the shaded area (*bottom*). The convolution is wider than the rectangles.

3. For $1 > t \geq 0$, the integral is $\int_{t-1/2}^{1/2} d\tau = 1 - t$.
4. For $t \geq 1$, the integral is zero.

Putting the pieces together, we have that $\text{rect}(t) * \text{rect}(t) = \Lambda(t)$, the triangle function (Figure 5.13). The rectangle is bounded and absolutely integrable, and the convolution is bounded and continuous. ∎

Example 5.17 (The convolution of a one-sided rectangle and a one-sided exponential).

$$g(t) = \text{rect}(t - 1/2) * e^{-t}U(t) = \int_{-\infty}^{\infty} \text{rect}(t - \tau - 1/2) e^{-\tau} U(\tau) d\tau = \int_{t-1}^{t} e^{-\tau} U(\tau) d\tau.$$

The one-sided rectangle $\text{rect}(t - \tau - 1/2)$, as a function of τ, is time reversed and shifted to the interval $\tau \in (t - 1, t)$. We distinguish three cases of the integral as t varies and slides the rectangle across the one-sided exponential:

1. For $t < 0$, the rectangle does not overlap the exponential, and the integral is zero.

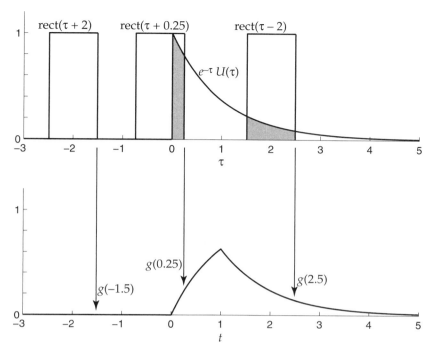

FIGURE 5.14 The convolution $g(t) = \text{rect}(t - 1/2) * e^{-t}U(t)$. As the time-reversed rectangle, $\text{rect}(t - \tau - 1/2)$, slides across the one-sided exponential function $e^{-\tau} U(\tau)$ (*top*), the convolution is the (shaded) area under their product (*bottom*). The convolution is wider than the rectangle. ∎

2. For $1 > t \geq 0$, the rectangle partially overlaps the exponential, and the integral is $\int_0^t e^{-\tau}\, d\tau = 1 - e^{-t}$.
3. For $t \geq 1$, the rectangle completely overlaps the exponential, and the integral is $\int_{t-1}^t e^{-\tau}\, d\tau = (1 - e^{-1})\, e^{-(t-1)}$.

This is the same as the charge–discharge curve for a resistor–capacitor circuit (Figure 5.14). Both functions are bounded and absolutely integrable, and the convolution is bounded and continuous. ∎

Example 5.18 (Convolution of a step and a one-sided exponential). The step function $U(t) \in L^\infty$, and the one-sided exponential $e^{-t}U(t) \in L^1$. Their convolution is

$$g(t) = U(t) * e^{-t}U(t) = \int_0^\infty e^{-(t-\tau)}U(t - \tau)\, d\tau = \left(1 - e^{-t}\right) U(t).$$

This function is bounded and continuous, but not absolutely integrable, because $g \to 1$ as $t \to \infty$. ∎

Example 5.19 (Convolution of two step functions). The convolution of two step functions is

$$g(t) = \int_0^t U(\tau)\, U(t-\tau)\, d\tau = t\, U(t).$$

The convolution is unbounded, because the step functions are not integrable. ∎

Convolution has a broadening effect on a function. In Examples 5.16 and 5.17, the result of the convolution was wider than the rectangle. If two functions f and g are supported on finite intervals $[a, b]$ and $[c, d]$, respectively, then their convolution $f * g$ is supported on the interval $[a + c, b + d]$. The width of this interval is $(b - a) + (d - c)$, the sum of the widths of the supports for f and g, respectively. For functions with either bounded or unbounded support, it is also true that centroids and variances add under convolution, that is,

$$\frac{\mu_{f*g}^{(1)}}{\mu_{f*g}^{(0)}} = \frac{\mu_f^{(1)}}{\mu_f^{(0)}} + \frac{\mu_g^{(1)}}{\mu_g^{(0)}}, \tag{5.39a}$$

$$\sigma_{f*g}^2 = \sigma_f^2 + \sigma_g^2, \tag{5.39b}$$

when the moments exist. Proofs of these width relationships are left to the problems, but here is an example.

Example 5.20 (Convolution of two rectangles, again). The rectangle rect(t) has unit area and zero centroid, by inspection. Its variance, then, is

$$\sigma_{\text{rect}}^2 = \int_{-1/2}^{1/2} t^2\, dt = \frac{1}{12}.$$

The triangle $\Lambda(x)$ has unit area and zero centroid. Its variance is

$$\sigma_\Lambda^2 = \int_{-1}^{1} t^2 (1 - |t|)\, dt = 2\int_0^1 t^2(1 - t)\, dt = \frac{1}{6}.$$

As expected, $\sigma_\Lambda^2 = \sigma_{\text{rect}}^2 + \sigma_{\text{rect}}^2$. ∎

In Example 5.18, although the convolution broadened the square edge of the step function into a gradual rise, these measures of width do not apply, because the step function has unbounded support and does not have a finite centroid or variance.

The Fourier transform of the convolution is of central importance in system theory. For that we need the convolution to be absolutely integrable. The L^1 norm of $f * h$ is

$$\|f * h\|_1 = \int_{-\infty}^{\infty} \left| \int_{-\infty}^{\infty} f(\xi) h(x - \xi)\, d\xi \right| dx$$

$$\leq \int_{-\infty}^{\infty} \int_{-\infty}^{\infty} |f(\xi)|\, |h(x - \xi)|\, d\xi\, dx = \int_{-\infty}^{\infty} |f(\xi)| \left(\int_{\infty}^{\infty} |h(x - \xi)|\, dx \right) d\xi$$

$$= \|f\|_1\, \|h\|_1 .$$

So if $f, h \in L^1$, then $f * h \in L^1$. Applying the Fourier transform to the convolution integral, we obtain the convolution theorem:

$$\begin{aligned}
\mathcal{F}\{f * h\} &= \int_{-\infty}^{\infty} \left(\int_{-\infty}^{\infty} f(\xi) h(x - \xi) d\xi \right) e^{-i2\pi v x} dx \\
&= \int_{-\infty}^{\infty} f(\xi) \left(\int_{-\infty}^{\infty} h(x - \xi) e^{-i2\pi v x} dx \right) d\xi \quad \text{(Fubini's theorem)} \\
&= \int_{-\infty}^{\infty} f(\xi) \left(e^{-i2\pi v \xi} H(v) \right) d\xi \quad \text{(shift theorem)} \\
&= \left(\int_{-\infty}^{\infty} f(\xi) e^{-i2\pi v \xi} d\xi \right) H(v) = F(v) H(v).
\end{aligned}$$

In short, if $f, h \in L^1$,

$$f * h \longmapsto FH.$$

Previously we calculated $\text{rect} \longmapsto \text{sinc}$, $\Lambda \longmapsto \text{sinc}^2$, and $\text{rect} * \text{rect} = \Lambda$. Using the above result,

$$\Lambda(x) = \text{rect}(x) * \text{rect}(x) \longmapsto \text{sinc}(v) \cdot \text{sinc}(v) = \text{sinc}^2(v).$$

To have the inverse relationship, $FH \longmapsto f * h$, we need for FH to be absolutely integrable. Since $f, h \in L^1$, we know that F and H are bounded, and in order to have $FH \in L^1$, we need for F or H to be integrable, for example,

$$\begin{aligned}
\|FH\|_1 &= \int_{-\infty}^{\infty} |FH| dv = \int_{-\infty}^{\infty} |F(v)| |H(v)| dv \\
&\leq \int_{-\infty}^{\infty} |F(v)| \sup |H| dv = \|F\|_1 \|H\|_\infty < \infty.
\end{aligned}$$

If both F and H are integrable, then we also have $F * H \in L^1$ and a symmetric *convolution theorem*:

$$f * h \longleftrightarrow FH, \tag{5.40}$$
$$fh \longleftrightarrow F * H. \tag{5.41}$$

A broad statement of the convolution theorem for functions in L^1 and L^2 will come later (Theorem 5.16).

It is not hard to show, using the convolution theorem, that in addition to being commutative, convolution is associative, $(f * g) * h = f * (g * h)$, and distributive, $(f + g) * h = f * h + g * h$. Also, the convolution of two even functions or two odd functions is even, and the convolution of an odd function and an even function is odd. And, if f is differentiable, then so is $f * h$, and $(f * h)' = f' * h$. Proofs of all these are left to the problems.

5.5.2 Convolution and Linear Systems

A *system* may abstractly be regarded as an operator $\mathcal{S}$ that maps a space of input functions to a space of output functions. For an input f and output g, we write $g = \mathcal{S}\{f\}$.

Definition 5.1 (Linear, time-invariant, causal systems).

1. A system is *linear* if $\mathcal{S}\{af_1 + bf_2\} = ag_1 + bg_2$, where $g_1 = \mathcal{S}\{f_1\}$ and $g_2 = \mathcal{S}\{f_2\}$, and a and b are constants.
2. A system is *time invariant* (or shift invariant) if $\mathcal{S}\{f(t - \tau)\} = g(t - \tau)$. That is, translating the input by an amount τ only causes the output to be translated by the same amount.
3. A system is *causal* if, for all inputs f which are zero for $t < t_0$, the outputs $g = \mathcal{S}\{f\}$ are also zero for $t < t_0$.

In this section we explore the relationship between linear, time-invariant (LTI) systems and convolution. First consider a first-order LTI system (an RC circuit, say) described by the differential equation

$$y' + \frac{1}{r}y = \frac{1}{r}x(t),$$

where r is called the *time constant* of the system. This system is LTI (any system described by a linear, ordinary differential equation with constant coefficients is LTI). Let the driving function x be a rectangular pulse of height n and width $1/n$, $x(t) = n\,\text{rect}(nt)$, and let the initial value $y(-\frac{1}{2n}) = 0$. When the input "switches on" at $t = -1/2n$, the capacitor charges according to the well-known "saturating exponential" curve:

$$y(t) = n(1 - e^{-(t+1/2n)/r}), \quad |t| < \frac{1}{2n},$$

and by $t = 1/2n$ has reached the value $y(\frac{1}{2n}) = n(1 - e^{-1/nr})$. At this point the input switches off and the capacitor begins to discharge, following the decaying exponential $e^{-(t-1/2n)/r}$. The complete output is described by the function

$$y(t) = \begin{cases} 0, & t < -\frac{1}{2n} \\ n(1 - e^{-(t+1/2n)/r}), & \frac{1}{2n} > t \geq -\frac{1}{2n} \\ n(1 - e^{-1/nr})e^{-(t-1/2n)/r}, & t \geq \frac{1}{2n} \end{cases} \quad (5.42)$$

This function is graphed in Figure 5.15. As n increases, the amplitude of the driving pulse increases but the amount of time that the capacitor is charging decreases. As n becomes very large, the pulse duration $1/n$ is very short compared with the system time constant r; the charging appears to be nearly instantaneous and the response

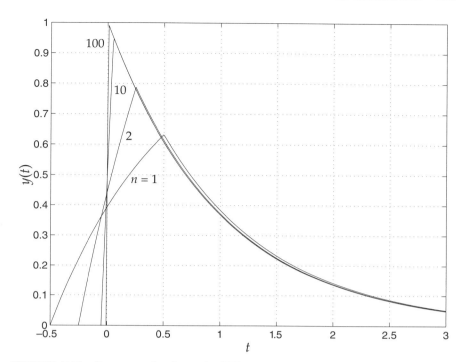

FIGURE 5.15 Response of a first-order LTI system ($r = 1$) to excitation by a unit area rectangular pulse of height n. As n increases, the pulse becomes higher and narrower. The shape of the response for large n is dominated by the decay and approaches the exponential function $\frac{1}{r}e^{-t/r}U(t)$.

appears to be all discharge. The precise width of the pulse, $1/n$, becomes negligible and we may use an idealized model created by letting $n \to \infty$, obtaining

$$y(t) = n(1 - e^{-1/nr})e^{-(t-1/2n)/]r} \to \frac{1}{r}e^{-t/r}, \quad t > 0. \quad (5.43)$$

A short, intense excitation of this sort is said to be *impulsive*. The output of a linear system in response to an impulsive input is called the *impulse response* and is typically denoted $h(t)$.

Now consider the response of an LTI system to an arbitrary input, $f(t)$. Assuming that the input is at least piecewise continuous,[5] it can be approximated arbitrarily closely by a series of rectangular pulses:

$$f(t) \approx \sum_n f(n\Delta\tau)\, \Delta\tau \, \text{rect}\left(\frac{t - n\Delta\tau}{\Delta\tau}\right).$$

The factor $\Delta\tau$ is included so that, in the limit as $\Delta\tau \to 0$, f and the approximation have the same area. The system's response g is, by linearity, approximated by a

---

[5] In fact, it is sufficient that the input be bounded and integrable ($L^1 \cap L^\infty$), but real signals are always piecewise continuous or better, and this assumption simplifies the mathematics.

superposition of its responses to the rectangular pulses, which we denote $h_{\Delta\tau}$. The pulse response $h_{\Delta\tau}$ is invariant to translation. The output of the system is

$$g(t) \approx \sum_n f(n\Delta\tau) h_{\Delta\tau}(t - n\Delta\tau) \Delta\tau.$$

Define $\tau = n\Delta\tau$. In the limit, as $\Delta\tau \to 0$, this is a classic Riemann sum which goes over to the integral:

$$g(t) = \int f(\tau) h(t - \tau) d\tau. \tag{5.44}$$

The system output is the convolution of the input f with the impulse response h.

The convolution model applies to systems in space as well as time. The image of a point source of light is the spatial impulse response, or *point spread function* of the imaging system. The image of an arbitrary ("extended") object is a superposition of point spread functions, one for each point in the ideal geometric image. The result is the convolution of the point spread function with the ideal image; a blurry image is the consequence of a broad point spread function. In like manner, the temperature distribution in a material in response to a point application of heat is also a spatial impulse response, as is the deflection of a structure in response to a point load.

A system can be linear but not time- (or space-) invariant. If an impulse is applied at time τ, the response measured at time t may depend, in general, both on τ and $t - \tau$, the time elapsed between the impulse and the measurement. A time-varying impulse response is often denoted $h(t; \tau)$ or $h(t - \tau; \tau)$. An example of a space-variant system is an camera that focuses more sharply at the center of the image field than at the edges. On the other hand, the response of a time-invariant system to an impulse at time τ depends only on the elapsed time, $t - \tau$ (Figure 5.16).

In temporal systems, causality restricts the impulse response to be one sided, $h(t) = 0, t < 0$. However, spatial systems, where the independent variable is x, are

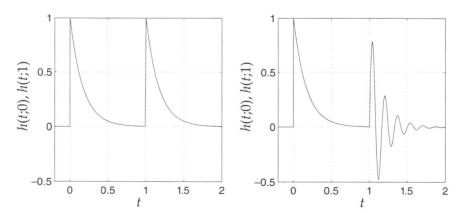

FIGURE 5.16 The responses of two systems driven by two successive impulsive inputs. The impulse response of a time-invariant system (*left*) depends only on the time elapsed since the input is applied, while the impulse response of a time-varying system (*right*) depends both on elapsed time and the absolute time when the input is applied.

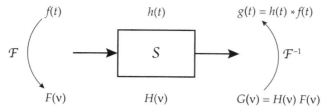

FIGURE 5.17 Two views of the input–output behavior of a linear time-invariant system. In the time domain, the output is the convolution of the input and the system's impulse response. In the frequency domain, the Fourier transform of the output is the product of the Fourier transform of the input with the system's transfer function. The transfer function is the Fourier transform of the impulse response.

not restricted by causality. The point spread function of an optical system is ideally distributed symmetrically about the geometric image point.

The response of an LTI system to a complex exponential of frequency ν, $f(t) = e^{i2\pi\nu t}$, is calculated with the convolution integral:

$$g(t) = h * f = \int_{-\infty}^{\infty} h(\tau) e^{i2\pi\nu(t-\tau)} d\tau = e^{i2\pi\nu t} \int_{-\infty}^{\infty} h(\tau) e^{-i2\pi\nu\tau} d\tau$$
$$= H(\nu) e^{i2\pi\nu t}.$$

The output is a complex exponential of the same frequency, multiplied by a complex factor $H(\nu)$, the *transfer function*. In operator notation, this can be expressed:

$$S\{e^{i2\pi\nu t}\} = H(\nu) e^{i2\pi\nu t}.$$

The result of the system operating on a complex exponential is simply a scaling by a complex number, $H(\nu)$. Complex exponentials of frequency ν are the *eigenfunctions* of linear, time-invariant systems, and the values of the transfer function, $H(\nu)$, are the corresponding *eigenvalues*.

For arbitrary inputs the convolution theorem, Equation 5.40, provides the link between the time and frequency domains for LTI systems. The output of an LTI system is the convolution of the input with the system's impulse response, $g = h * f$; the Fourier transform of the output is $G = HF$. The transfer function (frequency response) of the system, $H(\nu)$, is the Fourier transform of the impulse response. We can interpret the operation of the LTI system in the frequency domain as (1) a Fourier decomposition of the input into complex exponentials, which are the eigenfunctions of the system, (2) a weighting of the Fourier components by the transfer function, and (3) a Fourier synthesis to reconstruct the output (Figure 5.17).

5.5.3 Correlation

A close relative of convolution is correlation, defined:

$$\Gamma_{fg}(\tau) = f \star g = \int_{-\infty}^{\infty} f^*(t-\tau)g(t)dt = \int_{-\infty}^{\infty} f^*(t)g(t+\tau)dt. \quad (5.45)$$

This is sometimes also called the cross-correlation to distinguish it from the special case known as autocorrelation, $f \star f$.[6] The variable τ is called the correlation *lag*. When $\tau = 0$, the correlation has the form of an inner product of f and g, so if $f, g \in L^2$, $\Gamma_{fg}(0) = \langle f, g \rangle^*$.

The correlation can be written as a convolution:

$$f \star g = \int_{-\infty}^{\infty} f^*(t - \tau)g(t)dt = \int_{-\infty}^{\infty} f^*(-(\tau - t))g(t)dt \quad (5.46)$$
$$= f^*(-) * g,$$

where $f(-)$ denotes the time reversal of f. This leads to a Fourier representation:

Theorem 5.13 (Correlation). Let $f \longleftrightarrow F$ and $g \longleftrightarrow G$. Then,

(a) Cross-correlation

$$f \star g \longleftrightarrow F^*G \quad (5.47)$$

(b) Autocorrelation

$$f \star f \longleftrightarrow F^*F = |F|^2 \quad (5.48)$$

for functions f and g such that the correlations exist.

Proof: Begin with $f \star g = f^*(-) * g$ and use the convolution theorem with the reversal theorem (Equation 5.17), $f^*(-) \longmapsto F^*$. Then

$$f \star g = f^*(-) * g \longmapsto F^*G.$$

The autocorrelation theorem follows by setting $g = f$ and $G = F$. ∎

Properties and applications of the correlation are explored in the problems.

5.6 MORE ABOUT THE FOURIER TRANSFORM

In this section we take up further aspects of the mathematical theory of the Fourier transform:

1. Defining an inverse Fourier transform for absolutely integrable functions, and understanding its properties. The issues will be reminiscent of the convergence properties of the Fourier series.
2. Defining a Fourier transform (and inverse) for square integrable functions, which describe many physical signals.
3. Extending the definition of convolution and the convolution theorem to include square-integrable functions.

---

[6]The pentagram notation, $f \star g$, is not consistently applied in the literature. For example, Bracewell (2000) explicitly includes the complex conjugate, writing $\Gamma_{fg} = f^* \star g$, while Gray and Goodman (1995) write $f \star g$. The latter notation is used in this book.

5.6.1 Fourier inversion in L^1

When a function is absolutely integrable, it has a Fourier transform defined by an ordinary integral. That transform is bounded and continuous, and approaches zero asymptotically at high frequencies. If the function is sufficiently smooth, then its Fourier transform rolls off rapidly enough to be absolutely integrable, and the inverse Fourier transform exists as an ordinary, absolutely convergent, integral. In the previous chapter we found that certain Fourier series, for example, for the square wave, whose coefficients decayed too slowly for absolute convergence, could still have a convergent sequence of partial sums. Thus we expect for the Fourier transform that even if $F(\nu) = \mathcal{F}\{f(x)\}$ is not absolutely integrable (e.g., rect $\longmapsto$ sinc), there should still be an inverse, although it may have poor convergence properties. Ideally, we should find that $\mathcal{F}^{-1}\{\mathcal{F}\{f\}\} = f$ in some sense.

Let us assume that all we know about F is that it is the Fourier transform of an absolutely integrable function f. In the worst case, the integral $\int_{-\infty}^{\infty} F(\nu) e^{+j2\pi\nu x} d\nu$ cannot be calculated as an ordinary integral (i.e., even if we can find an antiderivative for the integrand, the integral will diverge at the infinite limits). We can make the integral converge by multiplying F by another function, called a *convergence factor*, creating a sequence of functions (F_n):

$$F_n(\nu) = K(\nu/n) F(\nu).$$

The convergence factor is designed to fall off sufficiently rapidly with ν that it tames the bad asymptotic behavior of F, but approaches one as $n \to \infty$ so that $(F_n) \to F$. Each of the F_n is integrable and is inverse transformed to a function f_n. As n increases and $(F_n) \to F$, the sequence of inverses (f_n) should also approach a limit, which we will define to be the inverse Fourier transform of F. This procedure is sometimes called *transform in the limit*.

Two common convergence factors are the Gaussian,[7]

$$K(\nu/n) = e^{-\pi(\nu/n)^2},$$

and the rectangle,

$$K(\nu/n) = \text{rect}(x/2n).$$

Each of these functions is in L^1 and approaches one as $n \to \infty$. With the Gaussian and the rectangular convergence factors, the following results can be established.

Theorem 5.14 (Transform in the limit in L^1). Let $f \in L^1$ and F be its Fourier transform.

(a) The integral

$$f_n(x) = \int_{-\infty}^{\infty} F(\nu) e^{-\pi(\nu/n)^2} e^{i2\pi\nu x} d\nu$$

---

[7] General characteristics of convergence factors are discussed in Folland (1992, pp. 208–211). Other functions, for example, the triangle and the two-sided exponential, can also be employed. The Gaussian is particularly nice, though, because it is rapidly decreasing, infinitely differentiable, and its own Fourier transform.

converges, as $n \to \infty$, to f in the L^1 norm:

$$\lim_{n \to \infty} \|f_n - f\|_1 = 0.$$

(b) If f is also piecewise continuous, then the integral

$$f_n(x) = \int_{-\infty}^{\infty} F(v) e^{-\pi(v/n)^2} e^{j2\pi vx} \, dv$$

converges to $\frac{1}{2}[f(x^-) + f(x^+)]$, as $n \to \infty$, for every x. If f is continuous, then the convergence to f is uniform.

(c) If f is also piecewise smooth, then the integral

$$f_n(x) = \int_{-n}^{n} F(v) e^{j2\pi vx} \, dv$$

converges to $\frac{1}{2}[f(x^-) + f(x^+)]$, as $n \to \infty$, for every x. If f is continuous, then the convergence to f is uniform.

Here is the gist of the argument.[8] Including the convergence factor in the inverse Fourier transform calculation, we have

$$f_n(x) = \int_{-\infty}^{\infty} F(v) K(v/n) e^{j2\pi vx} \, dv = \int_{-\infty}^{\infty} \left(\int_{-\infty}^{\infty} f(\xi) e^{-j2\pi v\xi} \, d\xi \right) K(v/n) e^{j2\pi vx} \, dv.$$

Using Fubini's theorem,

$$f_n(x) = \int_{-\infty}^{\infty} f(\xi) \left(\int_{-\infty}^{\infty} K(v/n) e^{j2\pi v(x-\xi)} dv \right) d\xi$$

then the shift and dilation theorems,

$$= \int_{-\infty}^{\infty} f(\xi) \, nk \, (n(x - \xi)) \, d\xi = f(x) * nk(nx),$$

where k is the inverse Fourier transform of K. First consider the Gaussian convergence factor, so

$$f_n(x) = \int_{-\infty}^{\infty} f(x) n e^{-\pi n^2(x-\xi)^2} d\xi.$$

The Gaussian $n e^{-\pi n^2 (x-\xi)^2}$ maintains unit area (apply the area theorem to the convergence factor) while becoming higher ($\propto n$) and narrower ($\propto n^{-1}$) as n increases. The convolution is a local smoothing, or averaging, of f, which gives greater weight to values of f around the peak at $\xi = x$. The averaging takes in progressively narrower portions of f as n increases, so that the f_n "track" f more and more closely and the norm $\|f - f_n\|_1$ converges to zero.

If f is additionally (at least) piecewise continuous, then the f_n converge pointwise to f, except at jump discontinuities in f. There, the kernel $nk(nx)$ straddles the

---

[8] Detailed proofs for Theorem 5.14 are found in Folland (1992, pp. 217–221), Dym and McKean (1972, pp. 101–104), and Gasquet and Witomski (1999, pp. 163–168).

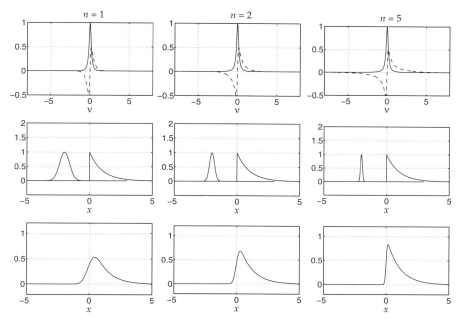

FIGURE 5.18 Fourier inversion with a Gaussian convergence factor, for the one-sided exponential, $e^{-t}U(t) \longleftrightarrow \frac{1}{1+i2\pi v}$. *Top row:* the Fourier transform multiplied by the convergence factor, $F(v)e^{-\pi(v/n)^2}$, $n = 1, 2, 5$ *(left to right)*. *Middle row:* the one-sided exponential and the Gaussian convolution kernel, divided by n for display purposes, $e^{-\pi(nx)^2}$. *Bottom row:* the inverse transform sequence $f_n(x)$. As n increases, f_n approaches the one-sided exponential, with $f_n(0) \to 1/2$, the average of the jump discontinuity.

jump; half its area is weighted by $f(x^-)$ and half by $f(x^+)$. As a result, the f_n converge at the jump to the average of $f(x^-)$ and $f(x^+)$ (we defined the rectangle function and the one-sided exponential to be $1/2$ at the jumps, to be consistent with this inversion result). Even better, if f is continuous, then the f_n converge pointwise and uniformly to f. An example of the convergence of an inverse transform with the Gaussian factor is shown in Figure 5.18.

The simple truncation in part (c) is analogous to the partial sum of a Fourier series (cf. Theorem 4.2) and gives

$$f_n(x) = f(x) * 2n \, \text{sinc}(2nx).$$

This convolution kernel, $2n \, \text{sinc}(2nx)$, also becomes higher and narrower while maintaining unit area, as n increases. Because the rectangle is discontinuous, its transform does not decay as rapidly as the Gaussian for large x. The conditions for convergence of the f_n are more stringent, and the oscillatory nature of the sinc will lead to overshoot around jump discontinuities in f. The transform in the limit procedure is illustrated in Figure 5.19.

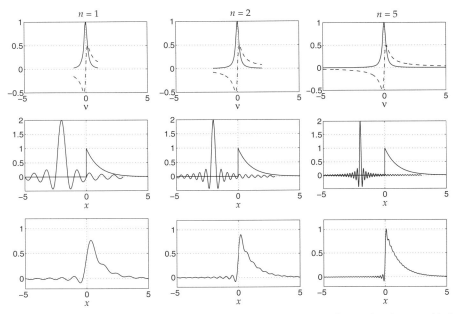

FIGURE 5.19 Fourier inversion with a rectangular convergence factor, for the one-sided exponential, $e^{-t}U(t) \longleftrightarrow \frac{1}{1+i2\pi\nu}$. *Top row:* the Fourier transform multiplied by the convergence factor, $F(\nu)\operatorname{rect}(\nu/2n)$, $n = 1, 2, 5$ (*left to right*). *Middle row:* the one-sided exponential and the sinc convolution kernel, divided by n for display purposes, $2\operatorname{sinc}(2nx)$. *Bottom row:* the inverse transform sequence $f_n(x)$. As n increases, f_n approaches the one-sided exponential, with $f_n(0) \to 1/2$, the average of the jump discontinuity. Because of the oscillations of the sinc kernel, there is overshoot around the jump which does not appear to decrease in amplitude (compare nonuniform convergence of the Fourier series of a square wave, Figure 4.7).

5.6.2 Fourier Transform in L^2

We would also like for the Fourier transform to be well defined for functions that are square integrable, since realistic physical functions have finite energy. When we were working with the Fourier series, and our functions were confined to bounded intervals like $[-\pi, \pi]$ or $[0, L]$, the function spaces were nested, $L^\infty[0, L] \subset L^2[0, L] \subset L^1[0, L]$. A square-integrable function was also absolutely integrable and guaranteed to have Fourier coefficients.

When we go to unbounded intervals, we have seen that a function that is bounded and absolutely integrable is also square integrable ($L^1(\mathbb{R}) \cap L^\infty(\mathbb{R}) \subset L^2(\mathbb{R})$). It can also be shown that a function that is continuous and absolutely integrable must also be bounded ($L^1(\mathbb{R}) \cap C(\mathbb{R}) \subset L^\infty(\mathbb{R})$), so a function that is absolutely integrable and continuous is also square integrable ($L^1(\mathbb{R}) \cap C(\mathbb{R}) \subset L^2(\mathbb{R})$). In general, however, the spaces do not simply nest. For example,

- $e^{-|x|}$ belongs to $L^1(\mathbb{R})$, $L^2(\mathbb{R})$, and $L^\infty(\mathbb{R})$.
- $|x|^{-1/2}(1+x^2)^{-1}$ belongs to $L^1(\mathbb{R})$ but not to $L^2(\mathbb{R})$ or $L^\infty(\mathbb{R})$.

- sinc x and $\frac{1}{1+i2\pi v}$ belong to $L^2(\mathbb{R})$ and $L^\infty(\mathbb{R})$ but not to $L^1(\mathbb{R})$.
- The step function $U(x)$ and constant function 1 belong to $L^\infty(\mathbb{R})$ but not to $L^1(\mathbb{R})$ or $L^2(\mathbb{R})$.

Square-integrable functions generally decay more slowly than absolutely integrable functions as $|x| \to \infty$ (compare sinc, which is in L^2, with sinc^2, which is in L^1). So, if f is square integrable but not also absolutely integrable, its Fourier transform is not defined as an ordinary integral.

The details of extending the Fourier transform from L^1 to L^2 can be found in several places.[9] The principal difference between the theorem below and the one shown previously for L^1 (Theorem 5.14) is that in L^2, *both* the forward and inverse transforms are defined as limits of sequences.

Theorem 5.15 (Fourier transform in L^2). Let $f \in L^2(\mathbb{R})$.

(a) The sequence of functions

$$F_n(v) = \int_{-\infty}^{\infty} f(x)\, e^{-\pi(x/n)^2}\, e^{-i2\pi vx}\, dx$$

converges, as $n \to \infty$, in the L^2 norm and pointwise almost everywhere, to a function $F \in L^2(\mathbb{R})$ which we define to be the Fourier transform of f. The inverse transform is similarly defined by a sequence of functions:

$$f_n(x) = \int_{-\infty}^{\infty} F(v)\, e^{-\pi(v/n)^2}\, e^{i2\pi vx}\, dv.$$

(b) If f is also piecewise continuous, then the inverse Fourier transform converges pointwise to $\frac{1}{2}[f(x^-) + f(x^+)]$, as $n \to \infty$, for every $x \in \mathbb{R}$. If f is continuous, then the convergence to f is uniform.

(c) The sequence of functions

$$F_n(v) = \int_{-n}^{n} f(x)\, e^{-i2\pi vx}\, dx$$

converges, as $n \to \infty$, in the L^2 norm (and pointwise almost everywhere), to a function $F \in L^2(\mathbb{R})$ which we define to be the Fourier transform of f. The inverse transform is similarly defined by the sequence

$$f_n(x) = \int_{-n}^{n} F(v)\, e^{i2\pi vx}\, dv.$$

The other Fourier transform theorems developed earlier continue to hold for the L^2 definition. For example, for the shift theorem,

$$\mathcal{F}\{f(x-a)\} = \lim_{n\to\infty} \mathcal{F}\{f_n(x-a)\} = \lim_{n\to\infty} e^{-j2\pi av} F_n(v) = e^{-j2\pi av} F(v).$$

---

[9] Proofs of parts (a) and (b) of Theorem 5.15 may be found in Stade (2005, pp. 321–326), Gasquet and Witomski (1999, pp. 193–196), and Dym and McKean (1972, pp. 91–101). The almost-everywhere convergence in part (c) is analogous to Carleson's theorem for the Fourier series; its lengthy and highly technical proof may be found in Grafakos (2004, pp. 796–827).

TABLE 5.1 Convolutions and products of functions. The norm of the convolution is bounded, $\|f * h\| \leq \|f\| \|h\|$, where the norms are calculated in the appropriate spaces, e.g., for $f \in L^1$ and $h \in L^2$, $\|f * h\|_2 \leq \|f\|_1 \|h\|_2$. Continuity of f and h improves the convolution (see text).

If f is	and h is	then $f * h$ is	and fh is[a]
L^1	L^1	L^1	L^1
L^1	L^2	L^2	L^1
L^1	L^∞	$L^\infty \cap C$	L^1
L^2	L^2	$L^\infty \cap C$	L^1

[a] The product relationships hold only when the functions are continuous and/or bounded. When h is bounded, the product fh inherits the norm properties of f, e.g., for $f \in L^1$, $\|fh\|_1 \leq \|f\|_1 \|h\|_\infty$.

And, of course, we have Parseval's formula, which shows that the Fourier transform is a norm-preserving map from L^2 to L^2.[10]

Both theorems (5.14 and 5.15) provide consistent, convergent definitions of the Fourier transform but not foolproof recipes for evaluating them. In practice, it may or may not be possible to evaluate a particular transform using a convergence factor. However, if the method fails, it will be for lack of an antiderivative, not because the integrand is fundamentally nonintegrable. Some other limiting procedure, like those introduced later, in Chapter 8, may do the trick.

★ 5.6.3 More about convolution

Having a Fourier transform for both L^1 and L^2, we can expand our initial statements of the convolution and product theorems (Equations 5.40 and 5.41).[11]

Theorem 5.16 (Convolution theorem for L^1 and L^2). Let $f \longleftrightarrow F$ and $h \longleftrightarrow H$.

(a) If $f, h \in L^1$, then $f * h \longleftrightarrow FH$. If, additionally, $F, H \in L^1$, then $f * h \longleftrightarrow FH$.
(b) If $f, h \in L^2$, then $f * h \longleftrightarrow FH$.
(c) If $f \in L^1$, $h \in L^2$, then $f * h \longleftrightarrow FH$ in L^2.

A more complete set of properties for convolutions and products of functions in L^1, L^2, and L^∞ are listed, without proof, in Table 5.1.[12] Proofs of the bounds are left to the problems. The bounds on convolution and products hold for functions on bounded and unbounded intervals, and also for sequences (replace the integrals by sums). The results of the previous chapter (Tables 4.2 and 4.3) are special cases.

---

[10] It is this version of Parseval's formula that is often called Plancherel's theorem. It was Plancherel who showed how to extend the Fourier transform from L^1 to L^2.

[11] Gasquet and Witomski (1999, pp. 203–204).

[12] The convolution properties are particular cases of a general result for functions in L^p spaces. See Folland (1999, pp. 240–241). For discussion and proofs of the convolution and product properties, see Gasquet and Witomski (1999, pp. 179–184, 201–206).

5.6 MORE ABOUT THE FOURIER TRANSFORM

In practice we typically have functions that are at least piecewise continuous in addition to whatever integrability properties they possess, and this improves the properties of the convolution. For example, a function that is absolutely integrable and continuous is also bounded and square integrable. The convolution of this function with another absolutely integrable function is also bounded and continuous as well as absolutely integrable. Here are some examples.

Example 5.21. The functions $f(x) = \frac{1}{x^2+1}$ and $g(x) = \frac{x}{x^2+1}$ belong to L^2 and L^1, respectively. They also belong to $L^\infty \cap C$. Using Fourier theorems and transform pairs we have already calculated, their Fourier transforms are

$$F(\nu) = \pi\, e^{-|2\pi\nu|},$$
$$G(\nu) = -i\pi\, e^{-|2\pi\nu|} \operatorname{sgn} \nu.$$

Then, using the convolution theorem,

$$f * f(x) = \mathcal{F}^{-1}\{F^2\} = \frac{2\pi}{x^2+4},$$
$$f * g(x) = \mathcal{F}^{-1}\{FG\} = \frac{\pi x}{x^2+4},$$
$$g * g(x) = \mathcal{F}^{-1}\{G^2\} = \frac{-2\pi}{x^2+4}.$$

The details are left to the problems. Observe that $f * f$ is bounded and continuous and rolls off as x^{-2}, thus it is absolutely integrable, while $f * g$ is bounded and continuous and rolls off as x^{-1}, so it is square integrable, in agreement with Table 5.1. $g * g$ is bounded and continuous, and it also happens to be absolutely integrable, although that property is not guaranteed for all convolutions of L^2 functions. ∎

Example 5.22. Again consider f and g as defined in the previous example. The step function $U(x)$ is bounded and (piecewise) continuous, but not integrable. The convolutions $f * U$ and $g * U$ are calculated:

$$f * U = \int_{-\infty}^{x} \frac{d\xi}{\xi^2+1} = \arctan(x)\Big|_{-\infty}^{x} = \frac{\pi}{2} + \arctan(x)$$

$$g * U = \int_{-\infty}^{x} \frac{\xi\, d\xi}{\xi^2+1} = \frac{1}{2}\log(x^2+1)\Big|_{-\infty}^{x}$$

In the first case, $f * U$ is bounded and continuous but not integrable, in agreement with Table 5.1. In the second case, the integral for $g * U$ does not converge; in general, we may not expect to be able to convolve a square-integrable function with a function that is merely bounded. The one-sided exponential, $e^{-x}U(x)$, is square integrable, but because of its exponential decay it exceeds the minimum requirements for square integrability. Its convolution with the step function is the saturating exponential $(1 - e^{-x})U(x)$. ∎

Example 5.23. The functions f and g in the previous two examples are bounded and continuous. Their product $fg(x) = \frac{x}{(x^2+1)^2}$ is bounded and continuous. It rolls off $O(x^3)$ for large x, so it is absolutely integrable, in agreement with Table 5.1. Using a table of integrals, the L^1 norm is

$$\|fg\|_1 = \int_{-\infty}^{\infty} \left|\frac{x}{(x^2+1)^2}\right| dx = 2\int_0^{\infty} \frac{x\,dx}{(x^2+1)^2} = 1 \leq \|f\|_1 \|g\|_\infty = \pi \cdot \frac{1}{2} = \frac{\pi}{2}.$$

∎

Example 5.24. The function $f(x) = \frac{1}{\sqrt{x}(x+1)} U(x)$ is unbounded and discontinuous. Yet, it rolls off $O(x^{3/2})$ for large x and grows slower than $1/x$ as x approaches the origin, so it is absolutely integrable, with $\|f\|_1 = \pi$. Its convolution with itself is

$$f * f(x) = \int_0^x \frac{1}{\sqrt{\xi}(\xi+1)} \frac{1}{\sqrt{x-\xi}(x-\xi+1)} d\xi = \frac{2\pi}{\sqrt{x+1}(x+2)} U(x)$$

using a table of integrals. It is discontinuous at the origin but bounded and absolutely integrable, with $\|f * f\| = \pi^2$. The product $f^2(x) = \frac{1}{x(x+1)^2} U(x)$ is unbounded and not integrable. Although it rolls off $O(x^3)$ for large x, it grows $O(x^{-1})$ as x approaches the origin. ∎

5.7 TIME–BANDWIDTH RELATIONSHIPS

The dilation theorem tells us that as a function is squeezed, its Fourier transform is stretched, and vice versa. We can quantify this relationship by defining measures of width for functions and observing their behavior in the time and frequency domains.

Equivalent Width
The *equivalent width* of a function f is simply defined to be the width of a rectangle having the same height and the same area as f:

$$W_f = \frac{\text{area}}{\text{height}} = \frac{\mu_f^{(0)}}{f(0)}. \tag{5.49}$$

The idea is illustrated in Figure 5.20. The definition requires that the maximum value of f be at the origin. If this is not the case, then the function may be shifted so that its maximum is at the origin. The equivalent width is not meaningful if the function has multiple peaks.

The equivalent width of the Fourier transform, F, may be calculated in similar fashion:

$$W_F = \frac{\mu_F^{(0)}}{F(0)}.$$

5.7 TIME–BANDWIDTH RELATIONSHIPS

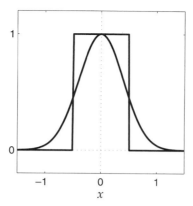

FIGURE 5.20 The equivalent width of a function f is the width of a rectangle having the same height and same area as f.

By the area theorem, $\mu_F^{(0)} = f(0)$ and $\mu_f^{(0)} = F(0)$, so we have

$$W_f W_F = 1, \tag{5.50}$$

a result which is independent of the particular form of f.

Example 5.25 (Noise equivalent bandwidth). The equivalent width is frequently used in electronics to simplify signal-to-noise calculations. Assuming that the noise at the input to a system is "white," meaning that its power per unit bandwidth N_0 is constant across all frequencies of interest, the noise power at the output of a filter having transfer function $H(\nu)$ is

$$P_N = \int_{-\infty}^{\infty} N_0 \, |H(\nu)|^2 \, d\nu,$$

that is, it is N_0 times the area under $|H(\nu)|^2$. Writing this area in terms of equivalent width,

$$P_N = N_0 \, |H(0)|^2 \, W_{|H|^2},$$

that is, N_0 times the square of the DC gain times the equivalent width. A filter (or amplifier) may have a complicated frequency response, but the output power with a white noise input is completely characterized by the DC gain and equivalent width. These are frequently included on data sheets for electronic subsystems. The equivalent width in this case is usually called the *noise equivalent bandwidth*. ∎

Mean-Square Width and the Uncertainty Principle

A second measure of width is based on variance rather than area. The *mean-square width* of a function f is defined as the variance of its squared magnitude, $\sigma^2_{|f|^2}$. The mean-square width of the Fourier transform is similarly defined as $\sigma^2_{|F|^2}$. This measure gives a meaningful result for functions that do not have sensible equivalent widths, including oscillatory wave packets. The mean-square widths of f and F have an inverse relationship, as expected, but it is more complex than the case of the equivalent width. The result is stated as a theorem. Its proof is a *tour de force* of Fourier theorems and norm calculations.[13]

Theorem 5.17 (Uncertainty Principle). Let $f, F \in L^2, f \longleftrightarrow F$, then the mean-square widths $\sigma^2_{|f|^2}$ and $\sigma^2_{|F|^2}$ obey the inequality

$$\sqrt{\sigma^2_{|f|^2} \, \sigma^2_{|F|^2}} \geq \frac{1}{4\pi}. \qquad (5.51)$$

Proof: Without loss of generality, take f and F to be centered on their centroids, so that their mean-square widths depend only on the second and zeroth moments. Then,

$$\sigma^2_{|f|^2} \, \sigma^2_{|F|^2} = \frac{\int_{-\infty}^{\infty} x^2 |f(x)|^2 dx}{\int_{-\infty}^{\infty} |f(x)|^2 dx} \frac{\int_{-\infty}^{\infty} v^2 |F(v)|^2 dv}{\int_{-\infty}^{\infty} |F(v)|^2 dv}$$

$$= \frac{\|xf\|^2 \, \|vF\|^2}{\|f\|^2 \, \|F\|^2}.$$

Now, using the derivative theorem together with Parseval's formula,

$$\|vF\|^2 = \|f'\|^2 / 4\pi^2,$$

we have

$$\sigma^2_{|f|^2} \, \sigma^2_{|F|^2} = \frac{\|xf\|^2 \, \|f'\|^2}{4\pi^2 \, \|f\|^4}.$$

Next, using the Cauchy–Schwarz inequality (Equation 2.50),

$$|\langle f, g \rangle + \langle g, f \rangle|^2 \leq 4 \|f\|^2 \, \|g\|^2,$$

we can replace the norms in the numerator, obtaining

$$\sigma^2_{|f|^2} \, \sigma^2_{|F|^2} \geq \frac{|\langle xf, f' \rangle + \langle f', xf \rangle|^2}{16\pi^2 \, \|f\|^4}.$$

Writing out the inner products in the numerator,

$$\langle xf, f' \rangle + \langle f', xf \rangle = \int_{-\infty}^{\infty} x \left(f \frac{d}{dx} f^* + f^* \frac{d}{dx} f \right) dx$$

$$= \int_{-\infty}^{\infty} x \frac{d}{dx} (ff^*) \, dx = \int_{-\infty}^{\infty} x \frac{d}{dx} |f|^2 \, dx,$$

---
[13] The proof of the uncertainty principle follows Bracewell (2000, pp. 177–178).

and integrating by parts,

$$\int_{-\infty}^{\infty} x \frac{d}{dx} |f|^2 \, dx = x|f|^2 \Big|_{-\infty}^{\infty} - \int_{-\infty}^{\infty} |f|^2 \, dx = -\|f\|^2.$$

The first term goes to zero because $|f|^2 \to 0$ faster than $1/x$ as $|x| \to \infty$. Finally, collecting everything together, we have

$$\sigma_{|f|^2}^2 \, \sigma_{|F|^2}^2 \geq \frac{\|f\|^4}{16\pi^2 \, \|f\|^4} = \frac{1}{16\pi^2}.$$

Taking square roots of both sides completes the derivation. ∎

Example 5.26 (Gaussian has minimum uncertainty). The Gaussian function is the only function known to achieve equality in the uncertainty relationship. Consider, for example, the transform pair

$$f(x) = \exp(-\pi a x^2) \longleftrightarrow F(\nu) = \frac{1}{\sqrt{a}} \exp(-\pi \nu^2 / a)$$

and calculate the mean-square widths:

$$\sigma_{|f|^2}^2 = \frac{\int_{-\infty}^{\infty} x^2 \exp(-\pi a x^2) \, dx}{\int_{-\infty}^{\infty} \exp(-\pi a x^2) \, dx} = \frac{\frac{1}{4\sqrt{2\pi a^{3/2}}}}{\frac{1}{\sqrt{2a}}} = \frac{1}{4\pi a},$$

$$\sigma_{|F|^2}^2 = \frac{\int_{-\infty}^{\infty} \nu^2 \exp(-\pi \nu^2 / a) \, d\nu}{\int_{-\infty}^{\infty} \exp(-\pi \nu^2 / a) \, d\nu} = \frac{\frac{\sqrt{a}}{4\sqrt{2\pi}}}{\frac{1}{\sqrt{2a}}} = \frac{a}{4\pi}.$$

Their product is

$$\sigma_{|f|^2}^2 \, \sigma_{|F|^2}^2 = \frac{1}{4\pi a} \frac{a}{4\pi} = \frac{1}{16\pi^2}.$$

∎

Example 5.27. At the other end of the scale, the rect–sinc transform pair exhibits maximum uncertainty. Consider, for example,

$$f(x) = \text{rect}\left(\frac{x}{a}\right) \longleftrightarrow F(\nu) = a \,\text{sinc}\,(a\nu).$$

Their mean-square widths are

$$\sigma_{|f|^2}^2 = \frac{a^2}{12}, \qquad \sigma_{|F|^2}^2 = \infty.$$

The calculations are left to the problems. ∎

5.8 COMPUTING THE FOURIER TRANSFORM

Numerical computation of the Fourier transform is appropriate when the integral cannot be done analytically or looked up in a table of transform pairs. General methods for computational integration are described in numerical analysis texts. Many algorithms are available in mathematical software packages (e.g., the `quad8` function in MATLAB). While the details vary, all numerical integration methods are based on approximating an integral $\int_a^b f(x)dx$ by a finite sum. Figure 5.21 shows an integral approximated by the sum of N rectangular areas.

The height of the nth rectangle is $f(x_n)$, and its width is $x_{n+1} - x_n$. The approximate value of the integral is

$$\int_a^b f(x)dx \approx \sum_{n=0}^{N-1} f(x_n)(x_{n+1} - x_n).$$

The approximation improves as the rectangles are made narrower and the number of rectangles is increased. More sophisticated integration methods using, for example, trapezoids rather than rectangles, and adaptive adjustment of the sample grid, achieve higher accuracy at the expense of more computation.

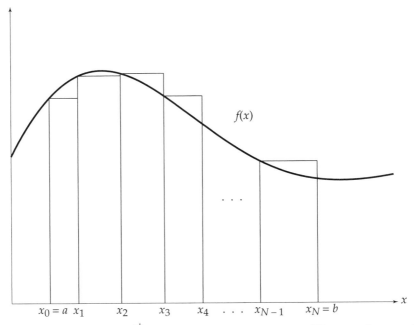

FIGURE 5.21 The integral $\int_a^b f(x)dx$ is approximated by a sum of N rectangular areas. The height of the nth rectangle is $f(x_n)$, and the width is $x_{n+1} - x_n$.

5.8 COMPUTING THE FOURIER TRANSFORM

A rectangular approximation to the Fourier transform is

$$F(v) = \int_{-\infty}^{\infty} f(x)e^{-i2\pi vx} dx$$

$$\approx \int_{-X/2}^{X/2} f(x)e^{-i2\pi vx} dx \approx \sum_{n=0}^{N-1} f(x_n)e^{-i2\pi vx_n} \Delta x,$$

where the interval $(-\frac{X}{2}, \frac{X}{2})$ is chosen such that f is acceptably close to zero for $|x| > \frac{X}{2}$, and $\Delta x = \frac{X}{N}$ is the sampling interval. Each value of v requires $O(N)$ function evaluations and multiply-adds; to compute F at M values of v with sampling interval Δv requires $O(MN)$ operations. Casting the sum into the form of a discrete Fourier transform allows it to be computed in $O(N \log N)$ operations instead, using the FFT algorithm. The focus of this section is how to do this.

Forward Transform

We begin with a rectangular approximation to the forward Fourier transform, using a uniform sampling grid, $x = n\Delta x$:

$$F(v) = \int_{-\infty}^{\infty} f(x)\, e^{-i2\pi vx} \, dx \approx \sum_{n=-\infty}^{\infty} f(n\Delta x)\, e^{-i2\pi vn\Delta x} \Delta x.$$

The sampling interval Δx is chosen sufficiently small to avoid aliasing, that is, so that F is effectively bandlimited to $|v| < \frac{1}{2\Delta x}$. The complex exponential $e^{-i2\pi vn\Delta x}$ is periodic in v with period $W = 1/\Delta x$. Consequently, the approximating sum is also periodic with period W and it is sufficient to evaluate it over a single period, $v \in [-\frac{W}{2}, \frac{W}{2})$. (The computed Fourier spectrum will give no new information about f for frequencies higher than $W/2$.)

A uniform sampling grid with M sample points is imposed on the frequency range:

$$v = m\Delta v, \quad m = -\frac{M}{2}, \ldots, 0, \ldots, \frac{M}{2} - 1,$$

$$\text{where } \Delta v = \frac{\frac{1}{2\Delta x} - \left(-\frac{1}{2\Delta x}\right)}{M} = \frac{1}{M\Delta x} = \frac{W}{M}.$$

The resolution in the frequency domain is Δv. It should be made small enough (by choosing M sufficiently large) to reveal the fine features of F. Making these substitutions, the appoximate sum is

$$F(m\Delta v) \approx \sum_{n=-\infty}^{\infty} f(n\Delta x)e^{-i2\pi mn\Delta v\Delta x} \Delta x$$

$$= \sum_{n=-\infty}^{\infty} f(n\Delta x)e^{-i2\pi mn/M} \Delta x, \quad m = -\frac{M}{2}, \ldots, \frac{M}{2} - 1.$$

Next, the sum is truncated to a finite number of terms. We shall consider two cases:

- f is one sided, for example, $f(x) = e^{-x}U(x)$.
- f is two sided, for example, $f(x) = \text{sinc } x$.

Case 1: f is one sided
Choose a maximum X sufficiently large that $f(x) \approx 0$ for $x > X$. Having already chosen Δx to avoid aliasing, this fixes the number of samples to be $N = X/\Delta x$. The truncated sum is denoted $\hat{F}$:

$$F(m\Delta v) \approx \hat{F}(m\Delta v) = \sum_{n=0}^{N-1} f(n\Delta x) e^{-i2\pi mn/M} \Delta x, \quad m = -\frac{M}{2}, \ldots, \frac{M}{2} - 1.$$

The key step in making the sum into a DFT is setting $M = N$, so the vectors $(f(n\Delta x))$ and $(\hat{F}(m\Delta v))$ have the same number of samples:

$$\hat{F}(m\Delta v) = \sum_{n=0}^{N-1} \underbrace{f(n\Delta x)}_{f[n]} e^{-i2\pi mn/N} \Delta x, \quad m = -\frac{N}{2}, \ldots, \frac{N}{2} - 1.$$

$$\underbrace{\phantom{\sum_{n=0}^{N-1} f(n\Delta x) e^{-i2\pi mn/N}}}_{\text{DFT, } F[m]}$$

For the positive frequency components, $\hat{F}(0)$ through $\hat{F}((\frac{N}{2} - 1)\Delta v)$, $\hat{F}(m\Delta v) \approx F[m]\Delta x$. The negative frequency components, $\hat{F}(-\frac{N\Delta v}{2})$ through $\hat{F}(-\Delta v)$, are obtained from the implicit periodicity of the DFT, $F[m] = F[m + N]$.

$$\hat{F}(m\Delta v) = \begin{cases} F[m + N]\Delta x, & m = -\frac{N}{2}, \ldots, -1 \\ F[m]\Delta x, & m = 0, \ldots, \frac{N}{2} - 1 \end{cases}, \quad (5.52a)$$

where $F[m] = \sum_{n=0}^{N-1} f(n\Delta x) e^{-i2\pi mn/N} = \text{DFT}\{(f(n\Delta x))\}.$ (5.52b)

That is, having computed the DFT, the positive frequency components of $\hat{F}$ are found in the first half of the DFT vector and the negative frequency components are found in the second half. A cyclic shift of the DFT by $N/2$ samples (Figure 3.9) will place the values in the proper order. In MATLAB, the cyclic shift is performed by the fftshift function (but see the problems for an alternate approach). The entire Fourier transform calculation is implemented by the code

```
Fv = fftshift( fft(fn) ) * dx,
```

where fn is the vector of samples $(f(n\Delta x))_{n=0}^{N-1}$, dx is the sampling interval Δx, and Fv is the approximate vector $(\hat{F}(m\Delta v))_{m=-N/2}^{N/2-1}$.

Case 2: f is two sided
Choose a maximum X sufficiently large that $f(x) \approx 0$ for $|x| > \frac{X}{2}$. The limits of the approximate sum range from $-\frac{N}{2}$ to $\frac{N}{2} - 1$ rather than 0 to $N - 1$:

$$F(m\Delta v) \approx \hat{F}(m\Delta v) \sum_{n=-\frac{N}{2}}^{\frac{N}{2}-1} f(n\Delta x) e^{-i2\pi mn/M} \Delta x, \quad m = -\frac{M}{2}, \ldots, \frac{M}{2} - 1.$$

Again we take $M = N$ to make the sum look like a DFT. The sum is split into two smaller sums, over the negative and nonnegative indices:

$$\hat{F}(m\Delta v) = \sum_{n=-\frac{N}{2}}^{\frac{N}{2}-1} f(n\Delta x) e^{-i2\pi mn/N} = \sum_{n=-\frac{N}{2}}^{-1} f(n\Delta x) e^{-i2\pi mn/N} + \sum_{n=0}^{\frac{N}{2}-1} f(n\Delta x) e^{-i2\pi mn/N}.$$

In the first sum, make the change of index $n' = n + N$:

$$\sum_{n=-\frac{N}{2}}^{-1} f(n\Delta x) e^{-i2\pi mn/N} = \sum_{n'=\frac{N}{2}}^{N-1} f((n'-N)\Delta x) e^{-i2\pi m(n'-N)/N}$$

$$= \sum_{n'=\frac{N}{2}}^{N-1} f((n'-N)\Delta x) e^{-i2\pi mn'/N}.$$

Then, we have

$$\hat{F}(m\Delta v) = \sum_{n=0}^{\frac{N}{2}-1} f(n\Delta x) e^{-i2\pi mn/N} + \sum_{n'=\frac{N}{2}}^{N-1} f((n'-N)\Delta x) e^{-i2\pi mn'/N}.$$

This is the DFT of the vector

$$\left(f(0), f(\Delta x), \ldots, f\left(\left(\frac{N}{2}-1\right)\Delta x\right), f\left(-\frac{N}{2}\Delta x\right), \ldots, f(-\Delta x)\right),$$

which is a cyclic shift of the vector

$$\left(f\left(-\frac{N}{2}\Delta x\right), \ldots, f(-\Delta x), f(0), f(\Delta x), \ldots, f\left(\left(\frac{N}{2}-1\right)\Delta x\right)\right).$$

The cyclic shift is necessary to place $f(0)$ in the first position of the DFT input vector. In MATLAB, the shift is performed by the `ifftshift` function (but see the problems for an alternate approach). Once the DFT is calculated, the negative frequency indices are handled in the same way as for the one-sided case. The result is

$$\hat{F}(m\Delta v) = \begin{cases} F[m+N]\Delta x, & -\dfrac{N}{2} \le m \le -1 \\ F[m]\Delta x, & 0 \le m \le \dfrac{N}{2} - 1 \end{cases}, \quad (5.53\text{a})$$

where $\quad F[m] = \sum_{n=0}^{N-1} f_c[n] e^{-i2\pi mn/N} = \text{DFT}\{f_c\} \quad (5.53\text{b})$

and $\quad f_c[n] = \begin{cases} f(n\Delta x), & n = 0, \ldots, \dfrac{N}{2} - 1 \\ f((n-N)\Delta x), & n = \dfrac{N}{2}, \ldots, N - 1 \end{cases}. \quad (5.53\text{c})$

The MATLAB code for this calculation is

```
Fv = fftshift( fft( ifftshift(fn) ) ) * dx,
```

where fn is the vector of samples $(f(n\Delta x))_{n=-N/2}^{N/2-1}$, dx is the sampling interval Δx, and Fv is the approximate vector $(\hat{F}(m\Delta v))_{m=-N/2}^{N/2-1}$. The procedure is illustrated in Figure 5.22 for the known transform pair rect $x \longleftrightarrow$ sinc v. In this particular example, the approximation for $|v| > 3$ is poor. The cause of this error is aliasing, brought on by an insufficiently small sampling interval, Δx.

Good results are obtained by proper choices of the sampling parameters Δx, Δv, X, W, and N. They are not chosen independently, but are coupled:

$$X = N\Delta x = \frac{1}{\Delta v}, \quad (5.54\text{a})$$

$$W = N\Delta v = \frac{1}{\Delta x}, \quad (5.54\text{b})$$

or, $\quad N = XW = \dfrac{1}{\Delta x \, \Delta v}. \quad (5.54\text{c})$

The quantity XW is called the *time–bandwidth product* when f is a function of time, or *space–bandwidth product* when f is a function of a spatial variable. These relationships are illustrated by Figure 5.23.

In practice, N should be a power of two to take maximum advantage of the FFT. The width X is chosen to avoid truncation errors. The correct sampling interval Δx can be found iteratively. If the approximate transform $\hat{F}$ is close to zero for high frequencies, then Δx is probably small enough. If there is doubt about this, divide Δx in half and repeat the calculation. Any appreciable differences in $\hat{F}$ are due to aliasing error and indicate that the smaller value of Δx is better. This process of halving Δx should be repeated until $\hat{F}$ is no longer changing. The frequency resolution may then be checked. If the features of the spectrum (e.g., peaks) appear to be adequately resolved, then Δv is probably small enough. If there is doubt, decrease Δv by doubling N and repeat the calculation. Appreciable differences in $\hat{F}$ are due to truncation error and indicate that the smaller value of Δv (correspondingly, a larger value of X) is better. The process of doubling N should be repeated until the appearance of $\hat{F}$ is satisfactory.

5.8 COMPUTING THE FOURIER TRANSFORM

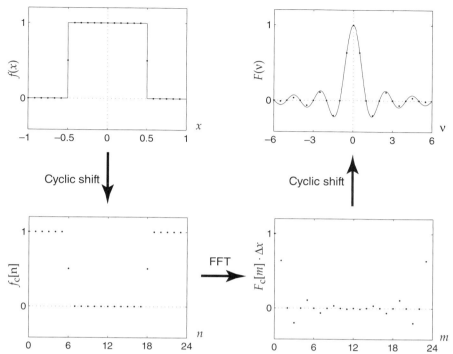

FIGURE 5.22 Using the DFT to compute the Fourier transform. Following the arrows, the process is to (1) calculate samples $f(n\Delta x)$ of the function f; (2) perform a cyclic shift of the samples so that the first element of the shifted vector f_c corresponds to $x = 0$ ($f_c[0] = f(0)$); (3) compute the FFT of f_c, and multiply by Δx; (4) cyclically shift the DFT vector $F_c \Delta x$ so that the negative frequency samples appear to the left of $v = 0$. The poor approximation for $|v| > 3$ is caused by aliasing error. The remedy is to decrease the sampling interval Δx.

Inverse Transform

The derivation of a DFT approximation for the inverse Fourier transform follows the same steps, leading up to

$$f(n\Delta x) \approx \sum_{m=-\frac{N}{2}}^{\frac{N}{2}-1} F(m\Delta v)e^{+i2\pi mn/N} \Delta v.$$

We assume that the Fourier transform F is two sided, and cyclically shift the vector $(F(m\Delta v))_{m=-N/2}^{N/2-1}$ into the vector $(F_c[m])_{m=0}^{N-1}$:

$$F_c[m] = \begin{cases} F((m+N)\Delta v), & m = -\dfrac{N}{2}, \dots, -1 \\ F(m\Delta v), & m = 0, \dots, \dfrac{N}{2} - 1 \end{cases}.$$

In many cases F is Hermitian, implying that f is real. To guarantee, therefore, that the inverse DFT f_c is real, F_c must be Hermitian. It can be shown that this requires

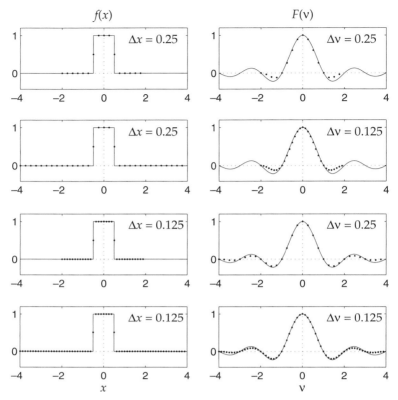

FIGURE 5.23 Time–bandwidth relationships in computing the Fourier transform. Dividing Δx by 2 to reduce aliasing error doubles the width W of the frequency interval over which samples of F are computed. Computing values of F over a wider interval requires finer sampling. Improving frequency resolution (decreasing $\Delta \nu$) requires sampling f over a wider interval.

$F_c[N/2]$ to be real. In general, however, $F(-N\Delta \nu/2)$ will be complex, so when F is Hermitian the value of $F_c[N/2]$ should be set to the real part of $F(-N\Delta \nu/2)$. This operation may also be interpreted as averaging $F(-N\Delta \nu/2)$ and $F(N\Delta \nu/2)$, which causes the imaginary parts to cancel.

With the conversion of F into F_c, the sum has the form of a DFT:

$$\sum_{m=-\frac{N}{2}}^{\frac{N}{2}-1} F(m\Delta \nu)e^{+i2\pi mn/N}\Delta \nu = \sum_{m=0}^{N-1} F_c[m]e^{+i2\pi mn/N}\Delta \nu$$

$$= \frac{1}{N}\sum_{m=0}^{N-1} F_c[m]e^{+i2\pi mn/N}\underbrace{N\Delta \nu}_{1/\Delta x} = \frac{1}{\Delta x}f_c[n].$$

5.8 COMPUTING THE FOURIER TRANSFORM

If f is known to be one sided, then $f(n\Delta x) \approx f_c[n]/\Delta x$. Otherwise, if f is two sided, a final cyclic shift is performed. The final results are

$$f(n\Delta x) \approx \hat{f}(n\Delta x),$$

where, if f is one sided,

$$\hat{f}(n\Delta x) = \frac{1}{\Delta x} f_c[n], \quad n = 0, 1, \ldots, N-1, \tag{5.55a}$$

otherwise, if f is two-sided,

$$\hat{f}(n\Delta x) = \begin{cases} \frac{1}{\Delta x} f_c[n-N], & n = -\frac{N}{2}, \ldots, -1 \\ \frac{1}{\Delta x} f_c[n], & n = 0, \ldots, \frac{N}{2}-1 \end{cases}, \tag{5.55b}$$

and where

$$f_c[n] = \frac{1}{N} \sum_{m=0}^{N-1} F_c[m] e^{+i2\pi mn/N}, \tag{5.55c}$$

$$F_c[m] = \begin{cases} F(m\Delta\nu), & m = 0, \ldots, \frac{N}{2}-1 \\ F((m-N)\Delta\nu), & m = \frac{N}{2}, \ldots, N-1 \end{cases}, \tag{5.55d}$$

$\mathcal{I}m\, F_c\left[\frac{N}{2}\right] = 0$ if F is Hermitian.

In terms of MATLAB functions,

```
fx = ifft( ifftshift(Fm) ) / dx,                    f one sided
fx = fftshift( ifft( ifftshift(Fm) ) ) / dx,        f two sided
```

where Fm contains the samples $(F(m\Delta\nu))_{m=-N/2}^{N/2-1}$, dx is the sampling interval Δx, and fx is the vector of approximate inverse transform values $(\hat{f}(n\Delta x))$.

The process is illustrated in Figure 5.24 for another known transform pair, interpreted here as an impulse response and transfer function:

$$h(t) = 2\sqrt{2\pi}B e^{-\sqrt{2}\pi Bt} \sin\sqrt{2}\pi Bt\, U(t),$$

$$H(\nu) = \frac{1}{1-(\nu/B)^2 + i\sqrt{2}(\nu/B)}. \tag{5.56}$$

The bandwidth parameter $B = 10$. The frequency sampling interval $\Delta\nu = 2$ makes the temporal width $T = 1/\Delta\nu = 0.5$ large enough to capture the entire impulse response without truncation. The temporal sampling interval is $\Delta t = \frac{1}{256} \approx 0.004$, chosen so N is a power of two (128) and W is sufficiently large (256) that H has essentially reached zero by $\nu = \pm\frac{W}{2}$. A slight error due to truncating H is still visible just before $t = 0$ in the approximated impulse response.

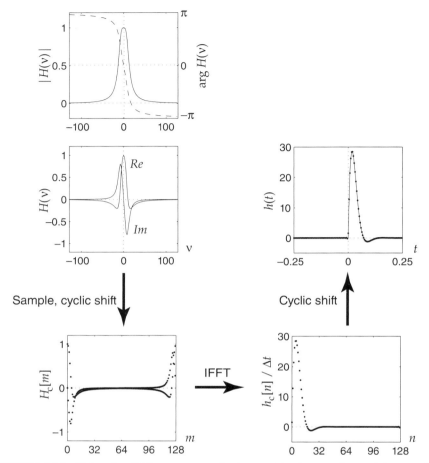

FIGURE 5.24 Using the DFT to compute the inverse Fourier transform of the transfer function $H(\nu)$ in Equation 5.56. The parameter $B = 10$. The complex values of the transfer function are shown in both polar (magnitude–phase) and cartesian (real–imaginary) forms. Sampling intervals are $\Delta x \approx 0.004$ and $\Delta \nu = 2$, and $N = 128$ points. The Hermitian transfer function has a real impulse response.

Convolution

Using the convolution theorem, $f * g \longleftrightarrow FG$, the convolution of two functions can be calculated in the frequency domain. Computationally, the procedure is to calculate the transforms F and G, multiply them together, and then calculate the inverse transform. It must not be forgotten that the DFT is being used to approximate the Fourier transform, and that the convolution theorem for the DFT is $f \circledast g \longleftrightarrow FG$—the inverse DFT of FG is the *cyclic* convolution of the vectors f and g. To demonstrate the crucial difference between cyclic and noncyclic (linear) convolution, consider

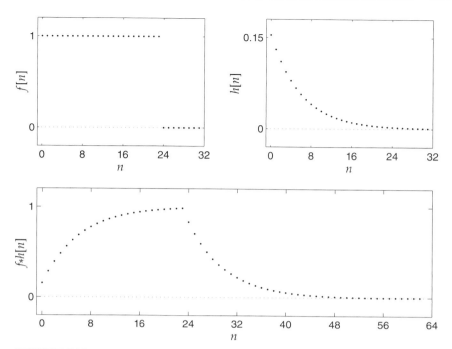

FIGURE 5.25 The linear convolution, $f * h$ (*bottom*), of two sequences f and h (*top*).

the two one-sided sequences of length $N = 32$ shown in Figure 5.25. Their linear convolution is computed by the relationship

$$f * h[n] = \sum_{k=0}^{n} f[k]h[n-k], \quad n = 0, 1, \ldots, 2N - 1.$$

The key feature to note is that the length of $f * h$ is $2N - 1 = 63$ points (compare the result for functions on $\mathbb{R}$ that the widths of the supports sum under convolution).

The cyclic convolution $f \circledast h$ of the same two sequences is shown in Figure 5.26:

$$f \circledast h[n] = \sum_{k=0}^{N-1} f[k]h[n-k], \quad n = 0, 1, \ldots, N - 1,$$

where the shift $n - k$ is interpreted modulo-N. The cyclic convolution is only $N = 32$ points long, and it is easily shown that it is the sum of periodic replicas of the linear convolution, that is,

$$f \circledast h[n] = f * h[n + N] + f * h[n] + f * h[n - N].$$

The discrepancy observed between the two convolutions in Figures 5.25 and 5.26 is caused by the overlapping replicas of $f * h$.

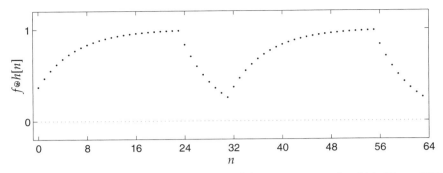

FIGURE 5.26 The cyclic convolution, $f \circledast h$, of the two sequences f and h in Figure 5.25.

Simply sampling two functions f and h, multiplying their DFTs, and computing the inverse DFT of the product will not produce the desired linear convolution $f * h$. The linear convolution is too long to fit in the space available to a cyclic convolution. The solution to the problem is to make room for the linear convolution's extra samples by appending N zeros each to the vectors f and h, that is, zero padding. The zero-padded vector f_p is defined:

$$f_p[n] = \begin{cases} f[n], & n = 0, 1, \ldots, N-1 \\ 0, & n = N, N+1, \ldots, 2N-1 \end{cases}$$

and similarly for h_p. The cyclic convolution $f_p \circledast h_p$ is shown in Figure 5.27. The overlap error has been eliminated, and the linear convolution is obtained from the cyclic convolution. This makes it possible to compute linear convolutions using the DFT, taking advantage of the FFT algorithm.

The procedure is illustrated in Figure 5.28, in which a rectangular pulse is filtered by the transfer function shown in Equation 5.56.

The sampling parameters Δt and $\Delta \nu$ and number of samples, N, are chosen to satisfy a number of criteria. The width of the pulse is 0.125, and the previous computation of the impulse response showed that its width is about 0.2. The width of the convolution will be on the order of 0.325. The input and the impulse response should be padded at least to this length, which gives the first requirement:

$$N\Delta t \geq 0.325.$$

But $\Delta \nu = 1/N\Delta t$ (Equation 5.8), thus

$$\Delta \nu \leq \frac{1}{0.325} \approx 3.$$

The transfer function's magnitude is less than 0.01 for $\nu \geq 100$, so a reasonable constraint on the frequency-domain width $W = N\Delta \nu$ is

$$N\Delta \nu \geq 200.$$

This places a constraint on the time-domain sampling interval Δt, since $\Delta t = 1/N\Delta \nu$:

$$\Delta t \leq 0.005.$$

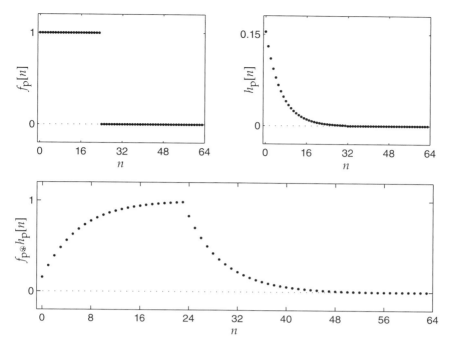

FIGURE 5.27 The cyclic convolution, $f_p \circledast h_p$ (*bottom*), of the zero padded sequences f_p and h_p (*top*). The zero padding permits the linear convolution $f * h$ to be computed using cyclic convolution.

The choice of Δt also determines how well the features in the output are resolved. For $\Delta t = 0.005$, there will be approximately $0.325/0.005 \approx 65$ sample points across the output pulse profile. Finally, we have a constraint on the number of points, N, which should be a power of two for computational efficiency. Using $\Delta t = 0.005$ and $\Delta \nu = 3$ gives $N = 1/0.015 \approx 67$. Rounding down to $N = 64$ will result in $N \Delta t = 64 \times 0.005 = 0.32$, which is less than the anticipated width of 0.325 and could consequently lead to overlap errors.

Figure 5.28 illustrates the sequence of computational steps for $\Delta t = 0.0125$, $\Delta \nu = 1$, and $N = 128$, values chosen to make visually informative graphs. Figure 5.29 compares the results for successively smaller Δt and larger N. Using the relatively large $\Delta t = 0.02$ truncates the transfer function and pulse spectra at $\nu = 25$ and causes considerable oscillatory artifact in the convolution. Decreasing Δt to the recommended value of 0.005 eliminates these errors, but $N = 64$ both truncates the output and, as expected, causes an artifact at the negative tail from insufficient zero padding. Increasing N to 128 removes the artifact and enables the positive tail of the pulse to be clearly observed. Continuing to decrease Δt and increase N does not resolve any additional features of the pulse but gives a smoother appearing graph.

The practical decision whether to compute a convolution in the time domain or in the frequency domain depends on the relative computational costs. To convolve two real-valued vectors of length N by direct summation requires N^2 multiply–add

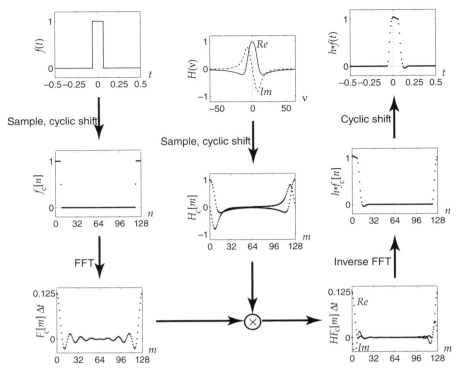

FIGURE 5.28 Using the DFT to compute a convolution by frequency-domain filtering. The transfer function (Equation 5.56) is used, with $B = 10$. The width of the rectangular pulse is 0.125. The sampling intervals are $\Delta t = 0.0125$ and $\Delta \nu = 1$, and $N = 128$. Following the arrows, (1) the pulse is sampled and cyclically shifted—all of the necessary zero padding is applied here; (2) the DFT of the pulse is calculated and multiplied by Δt to approximate the Fourier transform $F(\nu)$; (3) the transfer function is sampled and cyclically shifted—by shifting the transfer function instead of the DFT F_c, one additional shifting operation is avoided; (4) the two Fourier transforms are multiplied; (5) the inverse DFT is computed and divided by Δt to approximate the convolution $h * f$; and (6) a final cyclic shift is performed.

operations. To perform the same operation using the FFT requires that the vectors be zero padded to $2N$. Then, the DFT of each vector requires $2N \log_2 2N$ complex operations. Multiplying the DFTs requires $2N$ complex multiplications, and performing the inverse DFT adds another $2N \log_2 2N$ operations, for a total of $2N + 6N \log_2 2N$ complex operations. A complex multiply operation uses four real multiplies, which multiplies the total by four. On the other hand, some reductions are possible if the functions being convolved are real valued. If one impulse response is to be applied to a multitude of inputs, the transfer function can be precomputed and stored, saving a DFT step. Using $2N + 6N \log_2 2N = 8N + 6N \log_2 N$ as a reasonable estimate of the operation count for frequency-domain convolution, the Fourier approach is more

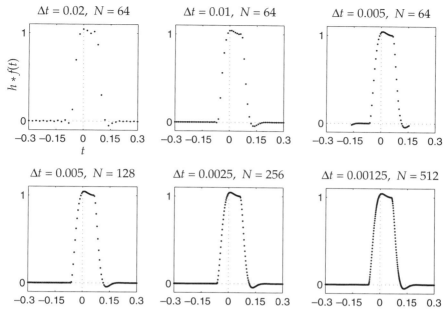

FIGURE 5.29 Using the DFT to compute a convolution by frequency-domain filtering. The input and transfer function are the same as in Figure 5.28. The plots show the results as the sampling interval Δt and number of points N are varied. With the largest sampling interval, $\Delta t = 0.02$, the output pulse shape is poorly described, and there are oscillatory artifacts characteristic of frequency-domain truncation. Decreasing Δt to 0.005 without increasing the number of points gives adequate definition of the pulse shape and eliminates the spurious oscillations, but the output is truncated and there is an artifact near $t = -0.15$ due to inadequate zero padding ($N\Delta t$ too small). Doubling the number of points to $N = 128$ displays the entire pulse with adequate resolution. Additional refinement of the sampling grid does not appear to provide additional information about the pulse.

efficient than direct summation when $8N + 6N \log_2 N < N^2$, that is, when N is a power of two and $N \geq 64$.

In many practical applications of convolution, the input f is a stream of samples being taken in real time, and the impulse response h is of fixed length. The present Fourier method cannot be applied here, but there are adaptations of frequency-domain convolution which permit f to be processed in segments.[14] Special purpose digital hardware can also rapidly compute convolutions by direct summation. Which method to use depends on the capabilities of the available hardware and a careful analysis of operation count for the particular problem.

---

[14] For methods to use the FFT to accelerate convolution when one of the sequences is much longer than the other, see Oppenheim and Schafer (2010, pp. 667–672).

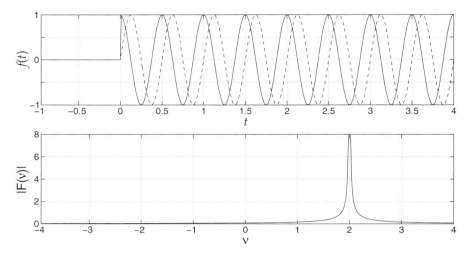

FIGURE 5.30 A sinusoidal signal, $f(t) = U(t)\exp(i2\pi bt)$, $b = 2$, that turns on at $t = 0$ (*top*) and its Fourier magnitude (*bottom*). The magnitude peaks at the frequency of the sinusoid but does not reveal the time of onset.

★ 5.9 TIME–FREQUENCY TRANSFORMS

The Fourier transform may be understood as the projection of a function f on a set of complex exponential basis functions $\{e^{i2\pi vx}\}$ that extend over the whole real axis, from $x = -\infty$ to $+\infty$. The infinite extent of the Fourier basis functions has a disadvantage, in that the Fourier transform provides no information about any changes in the frequency content of a function with time. Put another way, the Fourier transform has no ability to localize in time the frequency content of a signal. Transforms capable of time–frequency localization are important in signal processing[15] and an ongoing subject of research. In this brief section we will introduce the problem and two popular solutions: the *short-time Fourier transform* and the *wavelet transform*.

Short-Time Fourier Transform

Consider the signal $f(t) = U(t)e^{i2\pi bt}$, a sinusoid that turns on at $t = 0$. The Fourier transform of this signal is easily calculated using the modulation theorem and a result from the next chapter (Equation 6.38):

$$F(v) = \frac{1}{i2\pi(v-b)}. \tag{5.57}$$

The magnitude, plotted in Figure 5.30, correctly gives a peak (actually, a singularity) at $v = b$ but conveys no information about the time of onset. The most we can obtain, for an arbitrary onset at $t = t_0$, is a phase term $e^{-i2\pi v t_0}$, which disappears in the magnitude.

---

[15] We have already seen one such application, to image compression, in Example 2.12.

5.9 TIME–FREQUENCY TRANSFORMS

We can obtain some time localization by using a window to isolate a portion of the signal before calculating the Fourier transform. Here, we will use a rectangular window of width a and height $1/a$ (for unit area), centered at $t = u$. The transform is now a function of the window position as well as frequency:

$$F(v;u) = \int_{-\infty}^{\infty} \frac{1}{a} \text{rect}\left(\frac{t-u}{a}\right) U(t) \exp(i2\pi bt) \exp(-i2\pi vt) \, dt.$$

We identify three regimes for performing the calculation: $u < -a/2$, where the window is to the left of the step and the result is zero; $-a/2 < u < a/2$, where the window straddles the jump at $t = 0$, and $u > a/2$, where the rectangle is to the right of the jump. The integrations are straightforward. For $-a/2 < u < a/2$,

$$F(v;u) = \frac{1}{a} \int_0^{u+a/2} e^{-i2\pi(v-b)t} \, dt = \frac{u+a/2}{a} e^{-i\pi(v-b)(u+a/2)} \text{sinc}[(u+a/2)(v-b)] \quad (5.58a)$$

and for $u > a/2$,

$$F(v;u) = \frac{1}{a} \int_{u-a/2}^{u+a/2} e^{-i2\pi(v-b)t} \, dt = e^{-i2\pi(v-b)u} \text{sinc}[a(v-b)]. \quad (5.58b)$$

Taking the squared magnitudes, we have

$$F(v;u) = \begin{cases} 0, & u \leq -a/2 \\ \left(\frac{u+a/2}{a}\right)^2 \text{sinc}^2\left[(u+a/2)(v-b)\right], & -a/2 < u < a/2 \\ \text{sinc}^2[a(v-b)], & u \geq a/2 \end{cases} \quad (5.59)$$

A useful way to display this windowed Fourier transform is to show the value of the power spectrum as a brightness (or darkness) vs. frequency, v, on one axis and the location of the window, u, on the other. This is commonly known as a *spectrogram* (Figure 5.31).

Unlike the Fourier transform (5.57), the spectrogram reveals the temporal onset of the sinusoid. Before $u = -a/2$ the transform is identically zero, and after $u = a/2$ it is a sinc of width $2/a$ centered at $v = b$. The transition from full off to full on has width a. Within the transition zone, the overlap of the rectangular window with the step, over which we integrate to calculate the Fourier transform, increases from zero to a; correspondingly, the width of the sinc spectrum decreases from infinite (at $u = -a/2$) to $2/a$ (at $u = a/2$).

The window determines both the width of the spectral peak and our ability to localize the onset of the signal. By the dilation theorem and the uncertainty principle, these characteristic widths are inversely related. Making the window narrower improves the localization of onset but broadens the spectrum, which complicates the problem of resolving closely spaced frequency components. Widening the window wider sharpens the spectral peak, improving frequency resolution, but broadens the transition from off to on, making it more difficult to identify when the signal begins or to distinguish two signals with nearly simultaneous onsets.

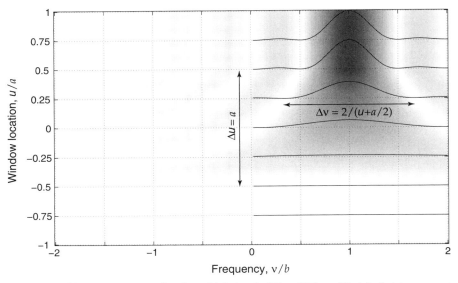

FIGURE 5.31 Spectrogram of a sinusoidal signal, $f(t) = U(t)\exp(i2\pi bt)$, that turns on at $t = 0$. The square root of the Fourier magnitude is plotted (darker is larger) to emphasize the side lobes of the sinc function that is the Fourier transform of the rectangular window. Overlaid traces show profiles of the Fourier magnitude at different window locations, from no overlap ($u < -a/2$), through a transition region surrounding the onset of the signal ($-a/2 < u < a/2$), to a steady state ($u > a/2$). The width of the transition from off to on is a, the width of the window. The frequency spread is inversely related to a.

We shall explore these ideas further with another example, the complex linear chirp signal that was introduced in Example 5.9:

$$f(x) = e^{-i\pi bx^2}.$$

The chirp has instantaneous frequency $b|x|$ and can be thought of as a sinusoid with linearly increasing frequency. Ideally, we would like for a Fourier transform to reveal this variation, for example, by a peak centered at $\nu = bx$. But the Fourier transform provides no such localization; indeed, the Fourier transform of the chirp was earlier seen to be (Equation 5.30)

$$F(\nu) = \frac{1}{\sqrt{ib}} e^{+i\pi \nu^2/b}.$$

The chirp and its transform are plotted in Figure 5.32.

As we did with the step-sinusoid, an intuitive way to obtain localized frequency information about the chirp is to view the function through an observation window w centered at $x = u$. Here, a Gaussian window[16] is analytically convenient,

---

[16] Signal modeling with Gaussian-windowed sinusoids was proposed by D. Gabor in a seminal paper (1946). It is still well-worth reading, for this and for his development of the uncertainty principle for the Fourier transform and its interpretation for communication theory.

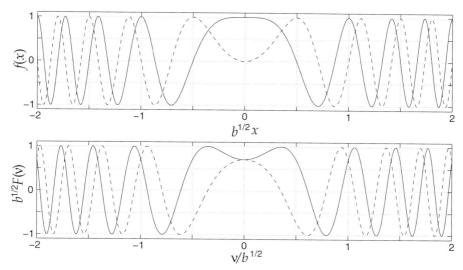

FIGURE 5.32 Real (solid) and imaginary (dashed) parts of the complex linear chirp function, $f(x) = e^{+i\pi bx^2}$ (top) and its Fourier transform, $F(v) = 1/\sqrt{-ib}\, e^{-i\pi v^2/b}$ (bottom).

$w(x-u) = \sqrt{a}\,e^{-\pi a(x-u)^2}$. The window has width (between $1/e$ points) $2/\sqrt{\pi a}$ and is normalized to have unit area. The short-time Fourier transform is

$$F(v; u) = \mathcal{F}\{e^{+i\pi bx^2}\sqrt{a}\,e^{-\pi a(x-u)^2}\}$$
$$= \sqrt{a}\,e^{-i\pi bu^2}\,\mathcal{F}\{e^{+i\pi b(x-u)^2}\,e^{-\pi a(x-u)^2}\,e^{i2\pi bux}\}$$
$$= \sqrt{a}\,e^{-i\pi bu^2}\,\mathcal{F}\{e^{+i\pi b(x-u)^2}\,e^{-\pi a(x-u)^2}\}\Big|_{v \to v-bu} \quad \text{(shift theorem)}$$
$$= \sqrt{a}\,e^{-i\pi bu^2}\,[e^{-i2\pi vu}\,\mathcal{F}\{e^{-\pi(a-ib)x^2}\}]\Big|_{v \to v-bu} \quad \text{(shift theorem)} \quad (5.60)$$
$$= \sqrt{\frac{a}{a-ib}}\,e^{-i\pi bu^2}\,e^{-i2\pi(v-bu)u}\,e^{-\pi(v-bu)^2/(a-ib)} \quad \text{(using Equation 5.30)}$$
$$= \sqrt{\frac{a}{a-ib}}\,e^{-i2\pi vu}\,e^{+i\pi bu^2}\,e^{-\pi(v-bu)^2/(a-ib)}.$$

(You can check for yourself that Equation 5.30 is recovered if a is set to zero.) As before, in order to interpret this result we will remove the phase factors by calculating the squared magnitude (power spectrum):

$$|F(v; u)|^2 = \frac{a}{\sqrt{a^2+b^2}}\,\exp\left[-\pi(v-bu)^2\left(\frac{1}{a-ib}+\frac{1}{a+ib}\right)\right]$$
$$= \frac{a}{\sqrt{a^2+b^2}}\,\exp\left[-\frac{2\pi a}{a^2+b^2}(v-bu)^2\right]. \quad (5.61)$$

This is a Gaussian, centered at $v = bu$, the instantaneous frequency of the chirp at $x = u$ (Figure 5.33).

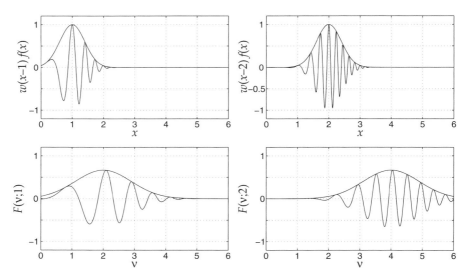

FIGURE 5.33 Short-time Fourier transform of a linear chirp, with a Gaussian window (Equation 5.60). The chirp has instantaneous frequency factor $b = 2$. The window has width parameter $a = 1$. The real parts (oscillatory) and magnitudes of $w(x - u)f(x)$ and $F(v; u)$ are shown here. *Left:* with the window centered at $u = 1$, the Fourier magnitude is a Gaussian centered at $v = 2$. *Right:* with the window centered at $u = 2$, the Fourier magnitude is a Gaussian centered at $v = 4$.

The spectrogram, $|F(v; u)|^2$, is plotted in Figure 5.34. The spectral width (measured between $1/e$ points of the Gaussian profile) is $\Delta v = \sqrt{2a/\pi}\sqrt{1 + (b/a)^2}$. Again, this is inversely related to the temporal width of the window. The spectrum is broadened by two effects: the rate of frequency variation b and the window width, proportional to $1/a$. If the window is broad (small a), so $b/a \gg 1$, then the frequency range of the chirp across the window is large, and the spectrum is broad. On the other hand, a narrow window (large a) includes a smaller range of frequencies but the window itself has a wider Fourier transform, again leading to a broader spectrum. Spreading is minimized when $b = a$, whence the width of the peak in v is $2\sqrt{a/\pi}$ and the width in u is $2/\sqrt{\pi a}$, and the reciprocal relationship between the two domains is obvious.

With these examples in hand, we now consider the short-time Fourier transform in more general terms. For a function f and window w, the transform is defined:

$$F(v; u) = \int_{-\infty}^{\infty} f(t)\, w(t - u)\, e^{-i2\pi vt}\, dt, \qquad (5.62)$$

where we shall require, in keeping with our experience with the rectangle and Gaussian windows, that w have even symmetry, be absolutely and square integrable, and possess unit norm, $\|w\|_2 = 1$.

The short-time Fourier transform $F(v; u)$ is based on a succession of partial views of the input signal f as the window is translated through locations u. It has the

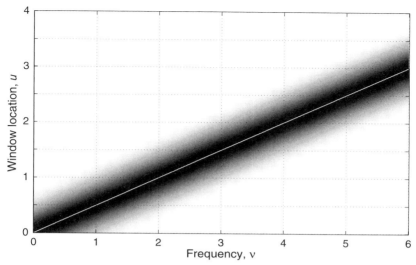

FIGURE 5.34 Spectrogram of a linear chirp, with a Gaussian window. The chirp has instantaneous frequency factor $b = 2$. The window has width parameter $a = 1$. The white line indicates the peak of the spectrogram; its slope is $1/b$, the reciprocal of the instantaneous frequency factor.

form of an inner product, though the functions $w(t - u)\, e^{-i2\pi v t}$ are not expected to constitute an orthogonal basis. Still, it seems that we ought to have enough information from these partial views to reconstruct f, that is, invert the transform. Moreover, we may ask if the transform conserves energy (i.e., if there is a Parseval formula).

If we assume that $F(v; u) \in L^1$, then its inverse Fourier transform with respect to v is simply

$$\int_{-\infty}^{\infty} F(v; u)\, e^{+i2\pi v t}\, dv = f(t)\, w(t - u).$$

Then, to get rid of the window, multiply both sides by $w(t - u)$ and integrate with respect to u:

$$\iint_{-\infty}^{\infty} F(v; u)\, w(t - u)\, e^{+i2\pi v t}\, dv\, du = \int_{-\infty}^{\infty} f(t)\, |w(t - u)|^2\, du$$

$$= f(t) \int_{-\infty}^{\infty} |w(t - u)|^2\, du = f(t),$$

because $\|w\|^2 = 1$. To check energy conservation, write

$$|F(v; u)|^2 = F(v; u) \left[\int_{-\infty}^{\infty} f(t)\, w(t - u)\, e^{-i2\pi v t}\, dt \right]^*$$

and integrate, assuming we may rearrange the order of the integrals:

$$\iint_{-\infty}^{\infty} |F(v;u)|^2 \, dv \, du = \int_{-\infty}^{\infty} f^*(t) \left[\iint_{-\infty}^{\infty} F(v;u) \, w(t-u) \, e^{+i2\pi vt} \, dv \, du \right] dt$$

$$= \int_{-\infty}^{\infty} f^*(t) f(t) \, dt = \int_{-\infty}^{\infty} |f(t)|^2 \, dt.$$

These calculations are not satisfactory proofs for $f \in L^2$. More complete proofs for the following theorem are outlined in the problems.[17]

Theorem 5.18 (Short-time Fourier transform). Let $f \in L^2$. Let $w \in L^1 \cap L^2$ and real valued, with unit norm, $\|w\|^2 = 1$, and even symmetry, $w(t) = w(-t)$. The short-time Fourier transform,

$$F(v;u) = \int_{-\infty}^{\infty} f(t) \, w(t-u) \, e^{-i2\pi vt} \, dt,$$

is invertible:

$$f(t) = \iint_{-\infty}^{\infty} F(v;u) \, w(t-u) \, e^{+i2\pi vt} \, du \, dt \qquad (5.63)$$

and norm-preserving:

$$\iint_{-\infty}^{\infty} |F(v;u)|^2 \, dv \, du = \int_{-\infty}^{\infty} |f(t)|^2 \, dt. \qquad (5.64)$$

Short-time DFT

For practical signal analysis, the short-time Fourier transform takes the form of a windowed DFT. Instead of a Gaussian window, a finite length tapered window like the Hamming window, $w[n] = 0.54 - 0.46 \cos(2\pi n/L)$, $n = 0, 1, \ldots, L$, is used. The DFT is computed over the windowed segment from $n = m$ to $n = m + L$:

$$F[k;m] = \sum_{n=m}^{m+L} f[n] \, w[n-m] \, e^{-i2\pi kn/N}, \quad k = 0, 1, \ldots, N-1. \qquad (5.65a)$$

With change of variable $r = n - m$, we have a form that looks more like a standard DFT:

$$F[k;m] = e^{-i2\pi km/N} \sum_{r=0}^{L} f[r+m] \, w[r] \, e^{-i2\pi kr/N}, \quad k = 0, 1, \ldots, N-1. \qquad (5.65b)$$

(The notation used here, $F[k;m]$, resembles that used for the DFT of a zero-padded sequence in Chapter 3, $F[m;N']$; there is no standard notation for either of these in the literature, and context should make it clear which transform is meant.) The leading phase factor in the second form drops out when the magnitude is computed and displayed. The number of frequency bins, N, is at least equal to the window length,

---

[17]Gasquet and Witomski (1999, pp. 388–392); Mallat (1999, p. 73).

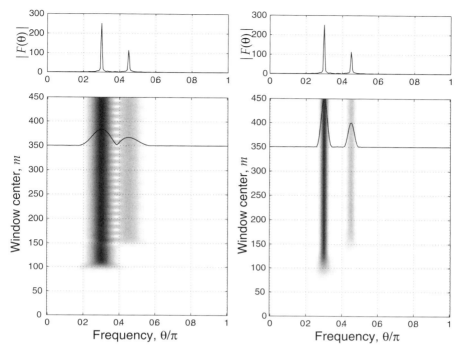

FIGURE 5.35 Spectrogram of a pair of sinusoids computed via the discrete Fourier transform. The plots on *top* show the DFT of the full 512-point signal vector, indicating spectral components at discrete frequencies $\theta = 0.3\pi$ and $\theta = 0.45\pi$. For the spectrogram on the *left*, a 32-point Hamming window was used, and the DFT was computed with zero padding to 256 points. The spectrogram on the *right* was made with a 96-point Hamming window, again with zero padding to 256 points. Traces show the spectra when the window is centered at $m = 350$. With the 32-point window, the onsets at $m = 100$ and $m = 150$ are clearly resolved, but the spectra are so broadened that they interfere. The 96-point window clearly resolves the two signals in frequency, but the temporal onsets are not as clearly determined.

$L + 1$. The windowed data segment may be zero padded to a length N greater than $L + 1$ if it is desired to interpolate the spectrum and increase the number of frequency bins. As usual, the discrete-time frequency is $\theta = 2\pi k/N$, and time is $t = n\Delta t$ or $m\Delta t$, where Δt is the sampling interval.

The application of the short-time DFT to a pair of sinusoids with different frequencies and different temporal onsets is illustrated in Figure 5.35. In Equations 5.65, the window coordinate m locates the beginning of the window, in keeping with standard nomenclature for Hamming and other windows. In the figure, this coordinate has been shifted so that m is the center of the window. Comparing windows of length 32 and 96, the narrower window more clearly resolves the onset of each signal, but results in a wider spectrum that makes it difficult to resolve closely spaced frequency components. The wider window gives good frequency resolution, but obscures the onset.

344 CHAPTER 5 THE FOURIER TRANSFORM

Like the continuous short-time Fourier transform, the short-time DFT is invertible and conserves energy. For energy conservation, consider the sum of the squared magnitude over all frequency bins:

$$\sum_{k=0}^{N-1} |F[k;m]|^2 = \sum_{r=0}^{L} \sum_{p=0}^{L} f[r+m] f^*[p+m] \, w[r] \, w^*[p] \underbrace{\left[\sum_{k=0}^{N-1} e^{-i2\pi k(r-p)/N} \right]}_{N\delta[r-p]}$$

$$= N \sum_{r=0}^{L} |f[r+m]|^2 \, |w[r]|^2 .$$

Then, summing over m,

$$\frac{1}{N} \sum_{m=0}^{M-1} \sum_{k=0}^{N-1} |F[k;m]|^2 = \sum_{r=0}^{L-1} \left[\sum_{m=0}^{M-1} |f[r+m]|^2 \right] |w[r]|^2 .$$

Because we are working with finite data records and using the DFT, all shifts are cyclic and the inner sum is the squared norm of the data vector f. Then, if the window has unit 2-norm, we recover a formula that looks like the continuous time result:

$$\frac{1}{N} \sum_{m=0}^{M-1} \sum_{k=0}^{N-1} |F[k;m]|^2 = \sum_{n=0}^{M-1} |f[n]|^2 . \qquad (5.66)$$

For inversion, a simple inverse DFT of Equation 5.65 gives

$$\frac{1}{N} \sum_{k=0}^{N-1} F[k;m] \, e^{+i2\pi kn/N} = f[n] \, w[n-m].$$

As a function of m and n, this is a sequence, or stack, of windowed data segments, each of which begins at $n = m$, ends at $n = m + L$, and is zero for all other n. Summing over m, we have

$$\frac{1}{N} \sum_{m=0}^{M-1} \sum_{k=0}^{N-1} F[k;m] \, e^{+i2\pi kn/N} = \sum_{m=0}^{M-1} f[n] \, w[n-m] = f[n] \sum_{m=0}^{M-1} w[n-m].$$

Each value of f, except for those within L of the ends of f (near 0 or $M-1$), is multiplied by the sum of all the window values. If we require the window to be normalized so that this sum is unity, $\|w\|_1 = 1$, then we have an inversion result:

$$\frac{1}{N} \sum_{m=0}^{M-1} \sum_{k=0}^{N-1} F[k;m] \, e^{+i2\pi kn/N} = f[n], \quad n = L, L+1, \ldots, M-L-1. \qquad (5.67)$$

The values at the ends are scaled by partial sums of w and can be recovered by adjusting the normalizations appropriately.[18]

Wavelet Transforms

The short-time Fourier transform analyzes a signal in terms of windowed sinusoids, for example, the Gaussian window, for which we have

$$w(t-u)e^{i2\pi vx} = \sqrt{a}e^{-\pi a(x-u)^2} e^{i2\pi vx}.$$

The width of the window, proportional to $1/a$, is fixed, even as the frequency v varies. Allowing the width of the window to vary with frequency, so that the window always encompasses the same number of periods, would enable the time localization to be sharper for more rapidly varying signals. This leads to the idea of a family of analyzing waveforms based on dilations and translations of a prototype waveform, for example,

$$\psi_{a,u}(x) = \frac{1}{\sqrt{a}}\psi\left(\frac{t-u}{a}\right), \qquad (5.68)$$

where ψ is the prototype. These analyzing waveforms, for appropriately selected ψ, are known as *wavelets*, and the expansion of a function f in terms of these, for example,

$$F(a,u) = \int_{-\infty}^{\infty} f(x)\,\psi_{a,u}(x)\,dx \qquad (5.69)$$

is called the *wavelet transform* of f. It can be shown to possess the expected properties of an orthogonal expansion, including invertibility and energy conservation. It is not a Fourier transform, in that explicit references to frequency have disappeared from view. Rather, the wavelet transform is a function of scale, a, and location, u, providing a different "look" at a signal's characteristics. Moreover, wavelets may be constructed with particular properties suitable to particular problems; one is not restricted to sinusoidal basis functions.

Wavelets are a rich field of study, with numerous applications, beyond the scope of this text.[19] We will limit our coverage to one illustrative example. The simplest

---

[18]For a properly chosen window, f can also be recovered from $F[k; m]$ for $m = 0, R, 2R, \ldots$, that is, for more widely spaced window locations. See Oppenheim and Schafer (2010, pp. 819–829). The first step in JPEG image compression, Example 2.12, is to divide an image into 8×8 pixel blocks (non-overlapping rectangular windows) and decompose each block with the discrete cosine transform. The redundant basis sets that appear in the short-time Fourier transform are also examples of frames, briefly introduced in Chapter 3 (Kovačević and Chebira, 2007b).

[19]For further reading, see example, Gasquet and Witomski (1999, pp. 395–430), Mallat (1999), Boggess and Narcowich (2001), and Percival and Walden (2000).

wavelet family is based on the Haar function:

$$\psi(x) = \begin{cases} 1, & \frac{1}{2} > x \geq 0 \\ -1, & 1 > \frac{1}{2} \geq \frac{1}{2} \\ 0, & \text{otherwise} \end{cases}.$$

The scales a and translations u are chosen to be dyadic, that is, based on powers of two, leading to the family of Haar wavelets (Figure 5.36):

$$\psi_{j,k}(x) = \frac{1}{\sqrt{2^{-j}}} \psi\left(\frac{x - 2^{-j}k}{2^{-j}}\right) = 2^{j/2} \psi(2^j x - k), \quad j = 0, 1, 2, \ldots, k = 0, \pm 1, \pm 2, \ldots \tag{5.70}$$

With this choice, the wavelets are orthonormal within one scale, $\langle \psi_{j,k}, \psi_{j,k'} \rangle = \delta[k - k']$, and also across scales, $\langle \psi_{j,k}, \psi_{j',k} \rangle = \delta[j - j']$. The expansion of a function f begins with an approximation of f by rectangles of unit width, $\phi_{0,k} = \text{rect}(x - (k + 1/2))$.[20] This approximation is denoted v_0:

$$v_0(x) = \sum_{k=-\infty}^{\infty} \langle f, \phi_{0,k} \rangle \phi_{0,k}(x).$$

Next, projecting f onto the Haar wavelets at scale $j = 0$ yields a function w_0:

$$w_0(x) = \sum_{k=-\infty}^{\infty} \langle f, \psi_{0,k} \rangle \psi_{0,k}(x).$$

The $\{\psi_{0,k}\}$ are orthogonal to the $\{\phi_{0,k}\}$, so w_0 is orthogonal to v_0, and the sum $v_0 + w_0$ approximates f at scale $j = 1$. This approximation is denoted v_1. Projecting f onto the wavelets at scale $j = 1$ gives a function w_1 that is orthogonal to both v_0 and w_0, and hence v_1, because of the orthogonality of the wavelets. Combining w_1 with v_1 gives a finer approximation to f, $v_2 = v_0 + w_0 + w_1$. Continuing at higher scales gives increasingly fine approximations:

$$v_J(x) = v_0(x) + \sum_{j=0}^{J} w_j(x),$$

---

[20] In the language of wavelet theory, the $\phi_{0,k}$ are called *scaling functions*. Every wavelet family has a companion set of scaling functions which may also be dilated and translated. Quaintly, the prototypes ϕ and ψ are also called the *father wavelet* and *mother wavelet*, respectively.

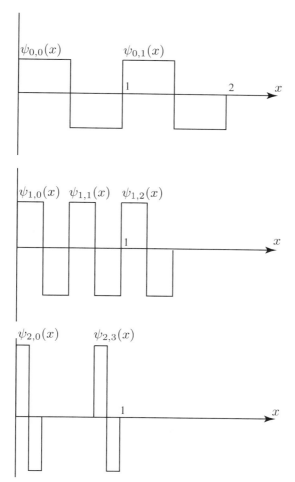

FIGURE 5.36 Representative Haar wavelets $\psi_{j,k}(x) = 2^{j/2}\psi(2^j x - k)$. They are orthonormal both within and across scales, $\langle \psi_{j,k}, \psi_{j,k'} \rangle = \delta[k - k']$ and $\langle \psi_{j,k}, \psi_{j',k} \rangle = \delta[j - j']$.

where

$$v_0(x) = \sum_k \langle f, \phi_{0,k} \rangle \, \phi_{0,k}(x),$$
$$w_j(x) = \sum_k \langle f, \psi_{j,k} \rangle \, \psi_{j,k}(x).$$
(5.71)

The v_0 and w_j are mutually orthogonal views of f at the various scales. The w_j may be thought of as the details of f at scale j; the function v_0 is the remainder after all the scale information has been extracted by the wavelets. It can be shown that $\|f - v_J\|_2 \to 0$ as $J \to \infty$, that is, the set $\{\phi_{0,k}, \psi_{0,k}, \psi_{1,k}, \ldots\}$ constitutes a basis for L^2.

The set of all functions w_j constructed from the $\{\psi_{j,k}\}$ comprises a subspace of L^2, which we denote W_j. The $\{\psi_{j,k}\}$ are an orthonormal basis for W_j. Likewise, the set of all v_0 constructed from the $\phi_{0,k}$ is a subspace of L^2, denoted V_0, having an orthonormal basis $\{\phi_{0,k}\}$. These subspaces, $V_0, W_1, W_2, \ldots$ are mutually orthogonal. The wavelet expansion thus produces a partitioning of L^2 into the direct sum of mutually orthogonal subspaces:

$$L^2 = V_0 \oplus W_1 \oplus W_2 \oplus \cdots$$

That is, any function in L^2 may be written as a sum of functions at each of the scales, with successive functions providing more and more detail.

The wavelet expansion is illustrated in Figure 5.37. The function f consists of 128 samples of a sinusoid whose frequency changes abruptly, upon which has

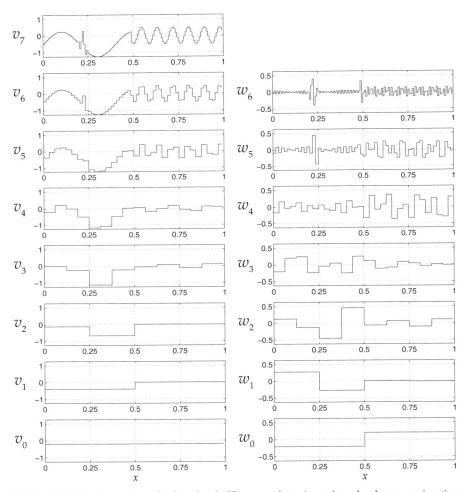

FIGURE 5.37 Expansion of a function in Haar wavelets. At each scale, the approximation $v_j = v_{j-1} + w_{j-1}$. The pulse near $x = 0.2$, the jump at $x = 0.5$, and the higher frequency portion of the function, $x > 0.5$, project strongly onto the finer scales and appear mostly in the detail functions w_4 through w_6. The lower frequency sine for $x < 0.5$ projects strongly onto the coarser scales and appears mostly in the detail functions w_1 through w_3.

been superimposed a narrow sinc-shaped pulse. The approximations v_j and detail functions w_j are shown for scales $j = 0$ through $j = 7$.

There are two abrupt features in f, the pulse near $x = 0.2$ and the jump at $x = 0.5$. Both of these project strongly onto the narrower, finer scale, wavelets $\phi_{4,k}, \ldots, \phi_{6,k}$ and appear in the corresponding detail functions $v_4, \ldots, v_6$ and approximations $w_4, \ldots, w_6$. At coarser scales, these features disappear. The higher frequency portion of the sinusoid, for $x > 0.5$, also projects strongly onto the finer scale wavelets and appears in the finer scale details and approximations. At coarser scales, this sinusoid is nearly invisible. On the other hand, the lower frequency portion, for $x < 0.5$, projects strongly onto the coarser wavelets and appears in the coarser details and approximations.

The detail functions are analogous to the terms of a Fourier series, and the approximations are like the partial sums. There is a correspondence between finer scales and higher frequencies. But because the wavelet basis functions are localized, unlike the Fourier basis functions which extend over the entire real line, the wavelet expansion more easily captures and localizes temporal changes in a function. For these reasons, wavelets have found wide application in a variety of fields. The interested reader is referred to the aforementioned references for further study.

5.10 SUMMARY

The Fourier Transform

$$F(v) = \int_{-\infty}^{\infty} f(x) e^{-i2\pi v x} \, dx,$$

$$f(x) = \int_{-\infty}^{\infty} F(v) e^{i2\pi v x} \, dv, \tag{5.1}$$

where the integrals are interpreted appropriately for L^1 or L^2 (Theorems 5.14, 5.15).

Fourier Transform Pairs Derived in this Chapter

$$\text{rect } x \longleftrightarrow \text{sinc } v \tag{5.6}$$

$$\Lambda(x) \longleftrightarrow \text{sinc}^2 v \tag{5.11}$$

$$e^{-x} U(x) \longleftrightarrow \frac{1}{1 + i2\pi v} \tag{5.9}$$

$$e^{-|x|} \longleftrightarrow \frac{2}{1 + (2\pi v)^2} \tag{5.15}$$

$$e^{-|x|} \text{sgn } x \longleftrightarrow \frac{-i4\pi v}{1 + (2\pi v)^2} \tag{5.16}$$

$$e^{-\pi x^2} \longleftrightarrow e^{-\pi v^2} \tag{5.27}$$

$$e^{+i\pi b x^2} \longleftrightarrow \frac{1}{\sqrt{-ib}} e^{-i\pi v^2 / b}. \tag{5.30}$$

Fourier Transform Theorems

Theorem	Formula	Equation
Parseval	$\int_{-\infty}^{\infty} |f(x)|^2 \, dx = \int_{-\infty}^{\infty} |F(\nu)|^2 \, d\nu$	5.18b
	$\int_{-\infty}^{\infty} f(x) g^*(x) dx = \int_{-\infty}^{\infty} F(\nu) G^*(\nu) d\nu$	5.18a
Area	$\int_{-\infty}^{\infty} f(x) dx = F(0)$	5.35
Moment	$\mu_f^{(n)} = \int_{-\infty}^{\infty} x^n f(x) dx = \dfrac{F^{(n)}(0)}{(-i2\pi)^n}$	5.36
Equivalent width	$W_f = \dfrac{\mu_f^{(0)}}{f(0)}, \; W_f W_F = 1$	5.49, 5.50
Mean-square width	$\sigma_{|f|^2}^2 = \dfrac{\int_{-\infty}^{\infty} x^2 |f(x)|^2 dx}{\int_{-\infty}^{\infty} |f(x)|^2 dx}$	5.34
Uncertainty principle	$\sigma_{|f|^2}^2 \, \sigma_{|F|^2}^2 \geq \dfrac{1}{4\pi}$	5.51
Linearity	$af + bg \longleftrightarrow aF + bG$	5.13
Symmetry	See Figure 5.5 $f(-x) \longleftrightarrow F(-\nu)$, etc.	5.17
Shift	$f(x - a) \longleftrightarrow e^{-i2\pi a\nu} F(\nu)$	5.23
	$e^{i2\pi bx} f(x) \longleftrightarrow F(\nu - b)$	5.20
Modulation	$f(x) \cos 2\pi \nu_0 x \longleftrightarrow \dfrac{1}{2} F(\nu - \nu_0) + \dfrac{1}{2} F(\nu + \nu_0)$	5.21
Dilation	$f(ax) \longleftrightarrow \dfrac{1}{|a|} F\left(\dfrac{\nu}{a}\right)$	5.22
Shift-Dilate	$f\left(\dfrac{x - b}{a}\right) \longleftrightarrow |a| e^{-i2\pi\nu b} F(a\nu)$	5.23
Derivative	$f'(x) \longleftrightarrow i2\pi\nu F(\nu)$	5.24
	$f^{(n)}(x) \longleftrightarrow (i2\pi\nu)^n F(\nu)$	5.25
	$x^n f(x) \longleftrightarrow \dfrac{F^{(n)}(\nu)}{(-i2\pi)^n}$	5.26
Integral	$\int_{-\infty}^{x} f(\xi) \, d\xi \longleftrightarrow (i2\pi\nu)^{-1} F(\nu)$	5.31
Convolution	$f * g \longleftrightarrow FG$	5.40
Product	$fg \longleftrightarrow F * G$	5.41
Correlation	$f \star g \longleftrightarrow F^* G$	5.47
	$f \star f \longleftrightarrow F^* F = |F|^2$	5.48

PROBLEMS

5.1. Using the Fourier theorems, prove the following transform pairs. Sketch (or plot with MATLAB) the functions. Use solid lines for real parts, dashed lines for imaginary parts.
   (a) $\text{rect}(x - \frac{1}{2}) \longmapsto \exp(-i\pi v) \text{sinc}(v)$.
   (b) $\exp(-2x^2) \longmapsto \sqrt{\frac{\pi}{2}} \exp t(-\frac{\pi^2 v^2}{2})$.
   (c) $\Lambda(x - 1) - \Lambda(x + 1) \longmapsto -2i \sin 2\pi v \, \text{sinc}^2 v$.
   (d) $\text{rect}(\frac{x-2}{4}) \longmapsto 4\exp(-i4\pi v) \text{sinc}(4v)$.

5.2. Using Fourier theorems, prove the following transforms:
   (a) $x \exp(-\pi x^2) \longmapsto -iv \exp(-\pi v^2)$.
   (b) $x \, \text{rect} \, x \longmapsto \frac{i}{2\pi^2 v^2}(\pi v \cos \pi v - \sin \pi v)$.

5.3. Calculate and sketch accurately the Fourier transforms of the following functions:
   (a) The Gabor function, $f(x) = \exp(-\pi a^2 x^2) \cos 2\pi bx$.
   (b) $f(t) = te^{-t}U(t)$.

5.4. The Gaussian, or normal, distribution, used in statistics, has the functional form:

$$f(x) = \frac{1}{\sqrt{2\pi}\sigma} \exp\left[-\frac{(x-\mu)^2}{2\sigma^2}\right].$$

Use the Fourier theorems to calculate the Fourier transform of f. In statistics, this is called the *characteristic function* or *moment generating function*. Then, use the moment theorem to show that the first and second moments of the Gaussian are

$$\mu^{(1)} = \mu,$$
$$\mu^{(2)} = \sigma^2 + \mu^2.$$

5.5. Using Fourier theorems, calculate the transforms of the following functions:
   (a) $f(x) = \text{rect}(\frac{x}{3} + 1)$.
   (b) $f(x) = \text{rect}(x - 1)e^{i\pi x}$.
   (c) $f(x) = \exp(-x^2)$.
   (d) $f(x) = \exp(-2\pi(x+1)^2)$.

5.6. Suppose we have two functions, f and g, and their respective Fourier transforms, F and G. If $F = G$, show that $f = g$ (a.e.). Thus, the Fourier transform of a function is unique.

5.7. *Repeated transforms*
   (a) Show that applying the forward Fourier transform twice in succession to $f(x)$ gives $f(-x)$, that is, $\mathcal{F}\{\mathcal{F}\{f(x)\}\} = f(-x)$. This result is important because wave diffraction obeys a forward Fourier transform, and in solving wave propagation problems, you can only use forward transforms.
   (b) Using this result, show that if $f(x) \longmapsto F(v)$, then $F(x) \longmapsto f(-v)$. This is simply a matter of keeping your x's and v's straight!

5.8. Devise and prove versions of the reversal symmetries (Theorem 5.4) for the DFT, Fourier series, and discrete-time Fourier transform.

5.9. Consider a signal composed of cosines, truncated by a rectangular window:

$$f(t) = [\cos 2\pi v_1 t + \cos 2\pi v_2 t] \operatorname{rect}\left(\frac{t}{T}\right).$$

The frequencies of the cosines, v_1 and v_2, are closely spaced. Calculate the Fourier transform of this signal, which you know will consist of two pairs of sinc functions. If the frequencies are too close, the sinc functions will partially overlap and sum, and it will be difficult to identify separate peaks in the Fourier transform. Explore the relationship between the resolvable frequency separation, $\Delta v = |v_1 - v_2|$ and the width of the window function, T. Derive an approximate expression connecting Δv and T.

5.10. The transfer function of a first-order lowpass filter is $H(v) = \frac{1}{1+i2\pi v\tau}$, where τ is the time constant.

(a) The cutoff frequency v_c is defined by the relationship $|H(v_c)|^2 = \frac{1}{2}|H(0)|^2$. The cutoff frequency is sometimes called the half-power frequency, because the input power is reduced by a factor of two at this frequency. Calculate the cutoff frequency of this filter.

(b) Calculate the noise equivalent bandwidth of this filter. Compare with the cutoff frequency.

5.11. The *ideal bandpass filter* (Figure 5.38) is defined by the transfer function

$$H(v) = \operatorname{rect}\left(\frac{v + v_0}{B}\right) + \operatorname{rect}\left(\frac{v - v_0}{B}\right).$$

Using Fourier transform theorems, find the impulse response $h(t)$ of this filter (i.e., the inverse Fourier transform of the transfer function). Pick $v_0 = 100$ and $B = 10$, and plot $h(t)$ with MATLAB (it is easier than sketching!).

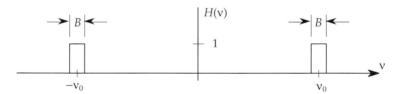

FIGURE 5.38 For Problem 5.11 Frequency response of the ideal bandpass filter.

5.12. Calculate the Fourier transform of the Haar wavelet function (Equation 5.70, Figure 5.36).

5.13. Beginning with the transform pairs

$$e^{-|x|} \longmapsto \frac{2}{1 + (2\pi v)^2}$$

$$e^{-|x|} \operatorname{sgn} x \longmapsto \frac{-i4\pi v}{1 + (2\pi v)^2},$$

use Fourier theorems to show

$$\frac{1}{1 + x^2} \longmapsto \pi e^{-|2\pi v|}$$

$$\frac{x}{1 + x^2} \longmapsto -i\pi e^{-|2\pi v|} \operatorname{sgn} v.$$

5.14. The *Hermite–Gaussian wavefunctions* are important in optics and atomic physics. They are defined:

$$\psi_n(x) = H_n(\sqrt{2\pi}x)\exp(-\pi x^2), \quad n = 0, 1, 2, \ldots,$$

where the $\{H_n\}$ are Hermite polynomials defined in turn via the Rodrigues formula:[21]

$$H_n(x) = (-1)^n \exp(+x^2)\left[\frac{d^n}{dx^n}\exp(-x^2)\right].$$

(a) Show that $H_0(x) = 1$, $H_1(x) = 2x$, and $H_2(x) = 4x^2 - 2$. Plot the wavefunctions $\psi_0(x)$, $\psi_1(x)$, and $\psi_2(x)$.

(b) Calculate the Fourier transforms $\Psi_0(\nu)$, $\Psi_1(\nu)$, and $\Psi_2(\nu)$. What can you conjecture about the Fourier transform of the general wavefunction $\psi_n(x)$?

(c) Prove the conjecture you made in (b), that is, derive a general expression for the Fourier transform $\Psi_n(\nu)$.

5.15. *Chirp functions*

(a) Following the area calculation in Example 5.12, derive Equation 5.28. Why is it necessary that $\mathcal{R}e\, c > 0$?

(b) Following the Fourier transform calculation in 5.12, derive Equation 5.29.

(c) Finally, let $c = a - ib$ and show that, as $a \to 0$, the linear chirp Fourier transform pair in Equation 5.30 is obtained.

5.16. The derivative of a function f is defined:

$$f'(t) = \lim_{\Delta t \to 0}\frac{f(t+\Delta t) - f(t)}{\Delta t},$$

which suggests the finite-difference approximation:

$$f'(t) \approx g(t) = \frac{f(t) - f(t - \Delta t)}{\Delta t},$$

where Δt is now a finite time delay. A linear, time-invariant system which might be used to implement this operation is shown in Figure 5.39.

(a) Calculate the transfer function of this system, $H(\nu) = \frac{G(\nu)}{F(\nu)}$.

(b) According to the derivative theorem, an ideal differentiation filter has transfer function $H(\nu) = i2\pi\nu$. How well does the transfer function you calculated above agree with the ideal? What can be done to improve the approximation?

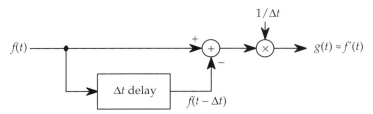

FIGURE 5.39 For Problem 5.16. Finite difference approximation to the derivative.

---

[21] Abramowitz and Stegun (1972, 22.11).

5.17. Generalize the proof of the area theorem (Equation 5.35) to prove the moment theorem (Equation 5.36).

5.18. Derive an expression for the variance of a function 5.34, in terms its Fourier transforms and their derivatives, evaluated at the origin.

5.19. *Expanding the Fourier theorems*
All the Fourier theorems so far have been stated as forward transforms, for example, $f * g$ transforms to FG, and $f(x - a)$ transforms to $e^{-i2\pi a v} F(v)$. There are "inverse" versions of the theorems, too. Find the inverse Fourier transforms of the following, and compare the results with their "forward" counterparts.
 (a) $F(v) * G(v)$.
 (b) $F(v - a)$.
 (c) $\frac{d}{dv} F(v)$.

5.20. The following integral, which has the form of an inner product, looks difficult.
$$\int_{-\infty}^{\infty} \text{sinc}\left[2B\left(t - \frac{n}{2B}\right)\right] \text{sinc}\left[2B\left(t - \frac{m}{2B}\right)\right] dt.$$
However, by applying Fourier transform theorems (such as dilation, shift, and Parseval), the problem can be moved into the frequency domain, where it is much easier.
 (a) Carefully apply the dilation and shift theorems to find the Fourier transform of $\text{sinc}[2B(t - \frac{n}{2B})]$.
 (b) Use Parseval's theorem to change the integral into an easier form, and carry out the calculation for all integers m, n. *Hint:* Make a sketch of the integrand. What does the result say about the signals $\text{sinc}[2B(t - \frac{n}{2B})]$ for integer n? What happens if the signals have the more general form $\text{sinc}[2B(t - nT)]$, where T is not necessarily equal to $1/2B$?

5.21. Let $f, g \in L^1$, and let F, G be their Fourier transforms. Show that
$$\int_{-\infty}^{\infty} F(v) g(v) dv = \int_{-\infty}^{\infty} f(x) G(x) dx.$$

5.22. Using Fourier theorems, evaluate the integral $\int_{-\infty}^{\infty} \text{sinc}^2(2x) dx$.

5.23. The function shown in Figure 5.40 could be the response of an underdamped second-order system to a sudden excitation. It is described by the equation $f(t) = \exp(-t/3) \cos 4\pi t\, U(t)$.

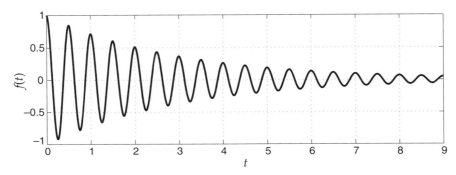

FIGURE 5.40 For Problem 5.23.

(a) Calculate and sketch the Fourier transform $F(v)$, labeling important features of your graph.

(b) Describe what happens to the graph of F if the exponential is changed from $\exp(-t/3)$ to $\exp(-2t)$.

5.24. A popular filter in image processing has impulse response $h(x) = \frac{d^2}{dx^2} a \exp(-\pi a^2 x^2)$, $a > 0$, which is the second derivative of a Gaussian. Calculate and sketch the transfer function of this filter, $H(v)$, labeling your graph to show the effect of the parameter a on its width and height.

5.25. *Frequency-division multiplexing*

Through the use of modulation, large numbers of voice messages can be packed, or *multiplexed*, together for efficient transmission on long distance radio or optical frequency communication links. A simple example of this is diagrammed in Figure 5.41.

(a) Using $B = 4$ kHz, $v_1 = 72$ kHz, and $v_2 = 80$ kHz, sketch the output spectrum $G(v)$.

(b) Design a method for recovering the original messages from the multiplexed signal.

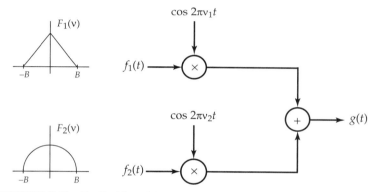

FIGURE 5.41 For Problem 5.25: System for frequency-division multiplexing.

5.26. A modulation and filtering system is shown in Figure 5.42. The input $x(t)$ is multiplied by a unit-amplitude 20 kHz cosine wave and applied to a system (called a *highpass filter*) which allows frequencies above 20 kHz to pass unattenuated but rejects frequencies below 20 kHz. The output of the highpass filter is multiplied by unit-amplitude 25 kHz

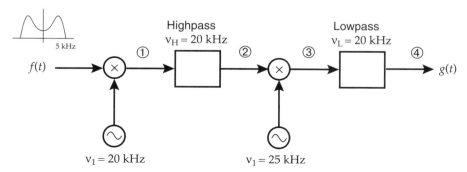

FIGURE 5.42 For Problem 5.26. A modulation and filtering system.

cosine wave and applied to a system (a *lowpass filter*) which allows frequencies below 20 kHz to pass unattenuated but rejects frequencies above 20 kHz.

Assume the input $x(t)$ has the Fourier spectrum shown, which is limited to frequencies below 5 kHz.

(a) Sketch accurately the signal spectrum at each of the numbered points in the system. Compare the output spectrum $Y(v)$ with the input spectrum $X(v)$.

(b) Draw a block diagram for a system which will take $y(t)$ as its input and recover $x(t)$ at its output.

5.27. *Narrowband filters*

The frequency response of a narrowband filter can be described by the transfer function

$$H(\omega) = \frac{Q}{1 - iQ\frac{1-\omega^2}{\omega}},$$

where $\omega = 2\pi v$ is angular frequency (rads/s as opposed to Hz). The parameter Q determines the narrowness of the frequency response. The center frequency ω_0 is that positive frequency which maximizes $|H(\omega)|^2$. The bandwidth of the filter is defined by the two positive half-power frequencies ω_1 and ω_2, for which $|H(\omega_{1,2})|^2 = \frac{1}{2}|H(\omega_0)|^2$. The bandwidth, B, is the difference of the half-power frequencies, $B = |\omega_2 - \omega_1|$.

(a) Derive the following expressions for the center frequency, half-power frequencies, and bandwidth, assuming that Q is large:

$$\omega_0 = 1,$$

$$\omega_{1,2} = 1 \pm \frac{1}{2Q},$$

$$B = \frac{1}{Q}.$$

(b) Show that a narrowband filter with arbitrary center frequency ω_0 is obtained by substituting ω/ω_0 for ω in the expression for $H(\omega)$. What is the bandwidth of this filter?

5.28. *Spectrum analysis*

The basic idea behind analog spectrum analyzers is narrowband filtering of the signal spectrum. With the filter centered at $v = v_0$ denoted by the transfer function $H(v; v_0)$, the estimated spectral density at v_0 is proportional to the energy passed by the filter:

$$|F(v_0)|^2 = \int_{-\infty}^{\infty} |H(v; v_0)|^2 \; |F(v)|^2 \; dv. \tag{5.72}$$

The resolution of a spectrum analyzer is defined in terms of *adjacent channel separation*. If a signal consists of two frequency components at v_1 and v_2, and the analyzer filter is centered on $v_0 = v_1$, then any energy from the component at v_2 which "leaks" into the passband of the filter contributes error to the spectral estimate. To minimize this error, a very narrow filter is needed. The half-power bandwidth (see the previous problem) is a standard measure of resolution.

(a) Figure 5.43 is a time-domain block diagram of a spectrum analyzer using a narrowband filter with a variable center frequency. Verify that this system produces the operation described by Equation 5.72, above. *Hint:* Consider $T \gg \frac{1}{v_0}$. What is the resolution of this system?

FIGURE 5.43 For Problem 5.28. Analog spectrum analyzer using a variable narrowband filter.

(b) Consider now the system shown in Figure 5.44, which uses a modulator followed by a filter with fixed center frequency v'_0. Assume $v'_0 \gg v_0$. Verify that the output of this system is also the spectrum defined by Equation 5.72. What is the resolution of this system? Compare with your result from Part (a). What advantage does the second approach have?

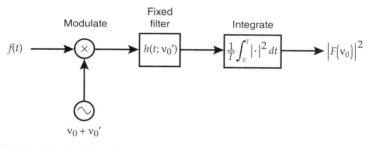

FIGURE 5.44 For Problem 5.28. Analog spectrum analyzer using a fixed narrowband filter with modulation.

5.29. *Antenna beamwidth.*

The electric field amplitude in the far-field of an antenna is given by the Fourier transform relationship:

$$P(\sin\theta) \propto \int_{-\infty}^{\infty} E\left(\frac{x}{\lambda}\right) \exp\left(-i2\pi\frac{x}{\lambda}\sin\theta\right) d\left(\frac{x}{\lambda}\right), \quad (5.73)$$

where $E(x/\lambda)$ is the electric field in the antenna aperture (distance measured in units of wavelength), and θ is the azimuth angle from the center of the aperture to the far-field observation point. A rectangular aperture of width A has a field pattern that goes like $\text{sinc}(A \sin\theta)$. This pattern is plotted as shown in Figure 5.45.

A measure of the quality of an antenna is how directionally selective it is. This can be measured by the full width at half-maximum (FWHM) of the radiation pattern $|P(\theta)|^2$.

(a) Calculate and plot the FWHM (in degrees) of the pattern $|P(\theta)|^2 \propto \text{sinc}^2(A \sin\theta)$, as a function of the aperture width A (measured in wavelengths).

(b) Suppose it is desired to have an antenna pattern whose FWHM corresponds to an area 18 miles wide at a distance of 100 miles. At an operating frequency of 430 MHz, how large is this antenna? If the operating frequency of this antenna is changed to 10 GHz, what is the beamwidth of the antenna?

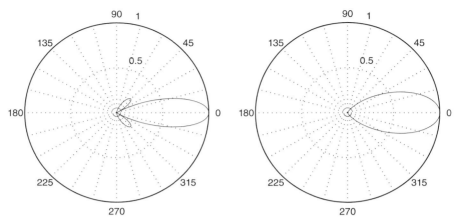

FIGURE 5.45 For Problems 5.29 and 5.30. *Left*: array pattern of unapodized antenna. *Right*: antenna apodized with triangular taper (Bartlett window).

5.30. *Antenna sidelobe suppression.*

Consider further the far-field radiation pattern of an antenna, Equation 5.73 and Figure 5.45. The radiation pattern consists of a strong central lobe called the *mainlobe* and weaker, off-center lobes called *sidelobes*. The mainlobe of the antenna pattern is directed toward $\theta = 0°$, as expected, but the *sidelobes* point toward $\pm 45°$. If this were a radar receiver, then an airplane at a 45° bearing could be mistaken for one coming in at 0°. Sidelobe suppression is one of the principal practical issues in antenna design.

Apodization ("removing the feet") is a method for lowering the sidelobe levels in an antenna pattern by tapering the aperture field distribution to zero at the edges. For example, if the field amplitude tapers linearly from the center to the edges (a triangular apodization), then the other pattern shown in Figure 5.45 results. The sidelobes are indeed reduced, but the mainlobe is considerably wider (FWHM is about 50° vs. 35° for the unapodized antenna). This is a classic tradeoff.

Your task in this problem is to compare the effects of various apodizations. This will help develop your skill with Fourier transform manipulations and also give you insight into a useful Fourier application area. A side benefit, if you are interested in signal processing, is that the antenna apodization problem is analogous to a method for designing digital filters.

Here are the apodizations you will consider. All of them are based on a maximum aperture width of A.

Rectangular (unapodized):	$\text{rect}(x/A\lambda)$
Bartlett (triangular):	$\Lambda(2x/A\lambda)$
Hanning (raised cosine):	$0.5[1 + \cos(2\pi x/A\lambda)]\text{rect}(x/A\lambda)$
Hamming (another kind of raised cosine):	$[0.54 + 0.46\cos(2\pi x/A\lambda)]\text{rect}(x/A\lambda)$.

Calculate the far-field pattern $P(\sin\theta)$ for each of these aperture distributions, and plot them *together* on a decibel scale, that is, $20\log_{10}|P(\sin\theta)|$ vs. $\sin\theta$ over the range $(-1, 1)$, which corresponds to the angular range $(-\pi/2, \pi/2)$. (Use a standard Cartesian format, not a polar plot.) Use a sufficiently small value of A that you get three or four sidelobes to either side of the mainlobe—this will take some experimentation. If necessary, normalize the patterns so that the values at $\theta = 0$ are all unity (0 dB) and be careful of zeros causing the logarithm to blow up.

Measure from your graph the mainlobe widths and maximum sidelobe levels of the four patterns. What do you observe? Which apodization do you think is best?

5.31. *Convolution*

(a) Calculate, by direct integration, the convolution $\text{rect}(\frac{x}{a}) * \text{rect}(\frac{x}{b})$, where $b > a$, and sketch accurately the result.

(b) From the sketch, express the result in terms of the triangle function, and calculate its Fourier transform.

(c) Calculate the Fourier transform of $\text{rect}(\frac{x}{a}) * \text{rect}(\frac{x}{b})$ using the convolution theorem, and compare with your result from part (b). Show that they are equal.

5.32. *Convolution*

Using the convolution theorem (and other theorems, as needed), calculate the following and sketch the results:

(a) $\text{sinc } 2x * \text{sinc}(x - 1)$.

(b) $\text{sinc}(x + 1) * \text{sinc } 3x$.

5.33. *Convolution*

Let the functions f and g be supported on the intervals $[a, b]$ and $[c, d]$, respectively. Show that their convolution $f * g$ is supported on the interval $[a + c, b + d]$, and the widths of the supports sum under convolution.

5.34. *Convolution*

Using the convolution and moment theorems,

(a) Show that centroid of the convolution $f * g$ is the sum of the centroids of f and g.

(b) Show that the variance of the convolution $f * g$ is the sum of the variances of f and g.

You will need to calculate the moments $\mu_{f*g}^{(0)}$, $\mu_{f*g}^{(1)}$, and $\mu_{f*g}^{(2)}$ in terms of the moments for f and g alone, then combine them with Equations 5.33 and 5.34.

5.35. *Convolution*

Use the convolution theorem (and other theorems, as appropriate) to perform the following calculations:

(a) Calculate the convolution $\text{sinc } ax * \text{sinc } bx$.

(b) Show that

$$\exp(-ax^2) * \exp(-bx^2) = \sqrt{\frac{\pi}{a+b}} \exp\left[-\frac{ab}{a+b}x^2\right].$$

Verify that the convolution is wider (more spread out) than either of the two original functions.

5.36. *Convolution*

Let $f \in L^1(\mathbb{R})$ and $h \in L^2(\mathbb{R})$, and show that the 2-norm of the convolution is bounded, $\|f * h\|_2 \le \|f\|_1 \|h\|_2$ (Table 5.1). *Hint:* Begin by bounding the convolution with the triangle inequality:

$$|f * h| \le \int_{-\infty}^{\infty} |f(\xi) h(x - \xi)| \, d\xi.$$

Then rewrite the integrand:

$$|f(\xi)h(x-\xi)| = |f(\xi)|^{1/2}\left(|f(\xi)|^{1/2}\,|h(x-\xi)|\right)$$

and apply the Cauchy–Schwarz inequality to obtain a bound on $|f * h|^2$.

5.37. *Convolution*
Let $f \in L^1(\mathbb{R})$ and $h \in L^\infty(\mathbb{R})$. Show that the convolution is bounded, $\|f * h\|_\infty \leq \|f\|_1 \|h\|_\infty$ (Table 5.1). *Hint:* Begin by bounding the convolution with the triangle inequality:

$$|f * h| \leq \int_{-\infty}^{\infty} |f(\xi)h(x-\xi)|\,d\xi.$$

5.38. *Convolution*
Let $f, h \in L^2(\mathbb{R})$. Show that the convolution is bounded, $\|f * h\|_\infty \leq \|f\|_2 \|h\|_2$ (Table 5.1). *Hint:* Use the Cauchy–Schwarz inequality.

5.39. *Convolution*
Using the convolution theorem (and other theorems, as appropriate), complete the calculations in Example 5.21:

$$\frac{1}{x^2+1} * \frac{1}{x^2+1} = \frac{2\pi}{x^2+4}.$$

$$\frac{1}{x^2+1} * \frac{x}{x^2+1} = \frac{\pi x}{x^2+4}.$$

$$\frac{x}{x^2+1} * \frac{x}{x^2+1} = \frac{-2\pi}{x^2+4}.$$

5.40. *Convolution*
For the functions in the previous problem, calculate norms and verify the bounds in Table 5.1.

5.41. *Convolution*
Using the convolution theorem (and other theorems, as appropriate) prove the following results:
(a) $\frac{d}{dx}(f * g) = f' * g = f * g'$, where it is assumed that f and g are differentiable.
(b) The convolution of two odd functions is even. (The same approach will work to show that the convolution of two even functions is even, and the convolution of an even function and an odd function is odd.)
(c) Convolution is associative: $f * (g * h) = (f * g) * h$.
(d) Convolution is distributive over addition: $f * (g + h) = f * g + f * h$.
(e) $|f * g| \leq \int_{-\infty}^{\infty} |FG|\,dv$. (This is a useful bound on the output of a filter.)

5.42. *Convolution*

Determine if each of the following statements is true in general. Provide proofs of those you think are true and counterexamples for those that you think are false:

(a) If $g(x) = f(x) * h(x)$, then $g(2x) = 2f(2x) * h(2x)$.

(b) If $g(x) = f(x) * h(x)$, then $\text{Even}\{g(x)\} = f(x) * \text{Even}\{h(x)\} + \text{Even}\{f(x)\} * h(x)$.

5.43. *Heat and diffusion*

The one-dimensional heat equation, or diffusion equation, is

$$\frac{\partial^2 u}{\partial x^2} = \frac{1}{k}\frac{\partial u}{\partial t}.$$

Given an initial spatial distribution $u(x, 0)$, this equation describes how that distribution spreads spatially (diffuses) in time. In Section 4.5, the Fourier series was used to solve the heat equation on a finite interval. In this problem the Fourier transform is used to solve the same equation on the entire real line.

(a) Let the *spatial* Fourier transform of u be $U(v, t)$. Fourier transform both sides of the equation to get a first-order ordinary differential equation (in the time variable). The solution of this ODE is straightforward. Derive the result:

$$U(v, t) = U(v, 0)\exp\left[-(2\pi v)^2 kt\right],$$

which, viewed as a function of v, is a Gaussian. Compare with the Fourier series solution of the heat equation on a bounded domain (Section 4.5).

(b) To get back to $u(x, t)$ inverse transform the expression for $U(v, t)$ with respect to v. Show that this results in a convolution in the space domain:

$$u(x, t) = u(x, 0) * h(x, t)$$

$$h(x, t) = \frac{1}{2\sqrt{\pi kt}}\exp\left(-\frac{x^2}{4kt}\right).$$

(c) The *kernel* of the convolution, $h(x, t)$, is a Gaussian in the space domain whose variance (spread) is proportional to kt. Accurately sketch (or plot, using the computer) the kernel for the following values of t: $0, 1/k, 1/4k$, and $1/16k$. What is the area $\int h(x, t)dx$? Interpret your results physically—using diffusion or heat.

5.44. Use the correlation theorem (and other theorems, as appropriate). For a real-valued function f:

(a) Show that the autocorrelation of f is always even.

(b) Show that the autocorrelation function has its maximum value at zero lag, that is, $|\Gamma(\tau)| \leq |\Gamma(0)|$.

5.45. Show that the autocorrelation of a complex-valued function f is always Hermitian.

5.46. *Convolution and correlation*

(a) Sometimes autocorrelation $(f \star f)$ and autoconvolution $(f * f)$ give the same result, but usually they do not. What condition(s) must f satisfy in order for the two results to be the same? Give an example of a function that does, and one that does not, satisfy your conditions, and calculate the autocorrelation and autoconvolution for each, to illustrate your point. You can do the calculations numerically with Matlab.

(b) Using the convolution theorem, $f * g \longmapsto FG$, show that convolution commutes, that is, $f * g = g * f$, but correlation does not. Express Γ_{gf} in terms of Γ_{fg}.

5.47. In several important applications of cross-correlation, f is an unknown signal and g is a known, fixed template. The cross-correlation measures the resemblance of the unknown signal to the template. Show that the cross-correlation can be obtained as the output of a linear filter, that is, $\Gamma_{fg} = f * h$, and determine what the impulse response, h, must be.

5.48. Two functions f and g, which are assumed to be real, positive, and identical except for a translation ($g(x) = f(x-a)$), can be registered (aligned) by finding the lag τ that maximizes their correlation function $\Gamma_{fg}(\tau)$. An alternative is to find the value of τ which *minimizes* the integral of the squared difference $f(x-\tau) - g(x)$, that is, the squared norm $\|f_\tau - g\|^2$, where $f_\tau(x) = f(x-\tau)$. Show that this is equivalent to maximizing Γ_{fg}, that is, that $\|f_\tau - g\|^2$ is minimized when Γ_{fg} is maximized, and vice versa.

5.49. *Uncertainty*
Carry out the mean-square width calculations for the transform pair $\text{rect}(\frac{x}{a}) \longleftrightarrow a \,\text{sinc}\, av$ (Example 5.27).

5.50. *Quantum mechanics*
In quantum mechanics, the position and momentum of a particle are described by wavefunctions $\psi(x)$ and $\phi(p)$, respectively, also called probability amplitudes. The probability of finding a particle in the interval $(x, x+dx)$ is $|\psi(x)|^2 dx$. The probability that the particle is in the interval (a, b) is

$$P = \int_a^b |\psi(x)|^2 dx,$$

and because the particle must be somewhere, $\int_{-\infty}^{\infty} |\psi(x)|^2 dx = 1$. Likewise, the probability that the particle's momentum is in the range $(p, p+dp)$ is $|\phi(p)|^2 dp$, and like the position, $\int_{-\infty}^{\infty} |\phi(p)|^2 dp = 1$. (Note that the wavefunctions are square integrable; the theory of the Hilbert space L^2 plays an important role in quantum theory.)

The *expected* (average) values of the particle's position and momentum are given by the first moment integrals:

$$\langle x \rangle = \int_{-\infty}^{\infty} x|\psi(x)|^2 dx = \int_{-\infty}^{\infty} \psi^*(x) x \psi(x) dx,$$

$$\langle p \rangle = \int_{-\infty}^{\infty} p|\phi(p)|^2 dp = \int_{-\infty}^{\infty} \phi^*(p) p \phi(p) dp.$$

The average momentum may also be calculated from the position wavefunction, like this:

$$\langle p \rangle = \int_{-\infty}^{\infty} \psi^*(x) \frac{-ih}{2\pi} \frac{d}{dx} \psi(x) dx,$$

where h is *Planck's constant*. In the language of quantum mechanics, $\frac{-ih}{2\pi}\frac{d}{dx}$ is called the "momentum operator in position space."

It is an interesting fact that the two wavefunctions are connected by a Fourier transform:

$$\phi(p) = \frac{1}{\sqrt{h}} \int_{-\infty}^{\infty} \psi(x) e^{-i2\pi px/h} dx,$$

$$\psi(x) = \frac{1}{\sqrt{h}} \int_{-\infty}^{\infty} \phi(p) e^{+i2\pi px/h} dp.$$

That is, if Ψ is the Fourier transform of ψ, then $\phi(p) = \frac{1}{\sqrt{h}}\Psi(p/h)$. For these two functions, the Fourier uncertainty relationship $\Delta x \Delta v \geq 1/4\pi$ becomes $\Delta x \Delta p \geq h/4\pi$. This is the celebrated Heisenberg uncertainty principle.

Your task in this problem is to use the Fourier transform to show that the two expressions for average momentum are equivalent:

$$\langle p \rangle = \int_{-\infty}^{\infty} \phi^*(p) p \phi(p) dp = \int_{-\infty}^{\infty} \psi^*(x) \frac{-ih}{2\pi} \frac{d}{dx} \psi(x) dx.$$

After all that introduction, this is really just a Fourier transform problem. Some of the theorems you may need are the derivative and Parseval.

5.51. *Infrared spectroscopy*

Infrared (IR) spectroscopy is an important tool in organic chemistry. Molecules have modes of vibration (as though the atoms were masses connected by springs) with frequencies in the range of infrared light. Probing a substance with infrared light and observing the resonant frequencies gives insight into the chemical composition and molecular structure. IR spectroscopy is done in two ways. The older method is to use a variable-wavelength source to illuminate the specimen, noting the strengths of the resonant responses as the input wavelength is adjusted. This is analogous to using a signal generator and an oscilloscope to measure the frequency response of an electronic circuit. The more modern technique uses a setup like the one in Figure 5.46.

The light from the source, a wave oscillating with frequency ω, is modeled by a complex exponential $e^{i\omega t}$. It is divided by a beam splitter, reflected off a pair of mirrors, and recombined by the beam splitter. One mirror is stationary, and the other is movable, so that the light wave following one path is delayed by an adjustable time $\Delta t = 2x/c$ relative to the light wave following the other path. The light reaching the specimen is proportional to the sum

$$e^{i\omega t} + e^{i\omega(t+2x/c)} = e^{ikct}(1 + e^{i2kx}),$$

where $\omega = kc$ and $k = 2\pi/\lambda$, λ being the wavelength of the light.

The specimen's absorption response to the incident radiation is $G(k)$. In chemistry labs, spectra are not plotted as functions of frequency (ν or ω) or wavelength λ, but of

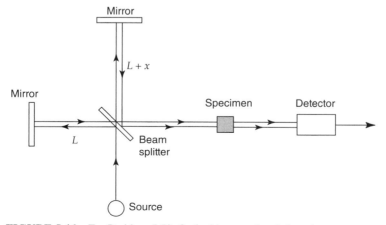

FIGURE 5.46 For Problem 5.52. Optical layout of an infrared spectrometer.

inverse wavelength $1/\lambda$, which is called *wavenumber* and measured in units of "reciprocal centimeters," cm^{-1}. The usual range of wavenumber is 400 to 4000 cm^{-1}. After passing through the specimen, the complex amplitude of the light reaching the detector is

$$G(k)(1 + e^{i2kx})e^{ikct}.$$

The detector responds to the intensity of the light and integrates out the rapid temporal fluctuations, leaving $|G(k)(1 + e^{i2kx})|^2$. This is the result for a single wavelength; the total detector response is the integral of this over k. The final result is a signal at the output of the detector that depends on the mirror displacement, x:

$$f(x) = \int_{-\infty}^{\infty} |G(k)(1 + e^{i2kx})|^2 \, dk.$$

(a) Derive an expression for the spectrum $|G(k)|^2$ in terms of the measured function $f(x)$. (You will then see why this technique is called Fourier Transform Infrared Spectroscopy, or FTIR.)

(b) Simple FTIR instruments used in instructional chemistry labs have a range of mirror travel (x) of about 0.5 cm. In a research-grade instrument, this range can be 1 m or more. Based on your understanding of Fourier theory, how does this make the research instrument superior to the instructional version?

5.52. *Radar*

In radar, pulses of radio frequency energy are transmitted at a target, which reflects a portion of the energy back to the receiver. The reflected pulse is detected and used to obtain information about the target, such as range (distance from the antenna) and range rate (speed along a line between antenna and target). The radar signal is a pulsed waveform:

$$g(t) = a(t) \cos 2\pi v_0 t,$$

where v_0 is the carrier frequency and $a(t)$ describes the pulse shape (or, envelope). If the target is a range R from the transmitter, the pulse undergoes a round-trip delay $t_R = 2R/c$, where c is the speed of light. Further, if the target is moving at a range rate s, then the pulse will experience a Doppler frequency shift given by $v_D = 2v_0 s/c$. The Doppler shift is very small compared to the carrier frequency, $v_D \ll v_0$, and the pulse duration T is many cycles of the carrier, $T \gg 1/v_0$. Hence, to a very good approximation, the Doppler effects on the pulse shape $a(t)$ are negligible. The resulting Doppler- and time-shifted radar return is

$$g_R(t) = a(t - t_R) \cos[2\pi(v_0 + v_D)(t - t_R)].$$

(a) Assuming a rectangular envelope $a(t) = \text{rect}(\frac{t}{T})$, $T \gg 1/v_0$, calculate and sketch the Fourier transforms of the transmitted pulse $g(t)$ and the received pulse $g_R(t)$.

When the return pulse reaches the receiver, it must be detected, usually in the presence of noise. The goal is to locate the return pulse in time, which gives the range of the target. It can be shown that the optimum linear time-invariant filter for pulse detection (in the absence of Doppler effects) has impulse response

$$h(t) = g(-t),$$

that is, it is a time-reversed version of the signal. This system is called a *matched filter*. The output of the matched filter is passed through an *envelope detector*, which removes the carrier and takes the absolute value of what is left (Figure 5.47).

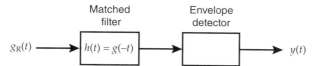

FIGURE 5.47 For Problem 5.53. Matched filter for radar pulse detection.

(b) Assuming that the target's range rate is zero (no Doppler), calculate and sketch the output $y(t)$ of the system. How may the range information be extracted from y?

(c) Reasoning in the frequency domain, explain what happens to this system when the received signal has Doppler shift as well as time delay.

Finally, consider the system shown in Figure 5.48.

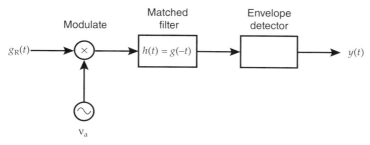

FIGURE 5.48 For Problem 5.53. Matched filter system for radar pulse detection with Doppler.

The return signal is assumed to be both time- and Doppler shifted. The matched filter is augmented by a modulator whose frequency v_a can be adjusted.

(d) Describe the output $y(t, v_a)$ of this system as v_a is varied. Show how both range and Doppler information are obtainable.

5.53. *Computing the Fourier transform*

(a) When using the DFT to approximate the Fourier transform of a function $f(x)$, $x > 0$, we calculate samples $f[n] = f(n\Delta x)$ and then compute the DFT:

$$F[m] = \sum_{n=0}^{N-1} f[n] e^{-i2\pi mn/N}.$$

The DFT vector $(F[m])_{m=0}^{N-1}$ contains the positive frequency components followed by the negative frequency components:

$$F\Delta x = \left(\hat{F}(0), \ldots, \hat{F}\left(\left(\tfrac{N}{2} - 1\right)\Delta v\right), \hat{F}\left(-\tfrac{N}{2}\Delta v\right), \ldots, \hat{F}(-\Delta v) \right).$$

Applying the MATLAB command fftshift to the DFT vector rotates it so that the negative frequency components come first, followed by the positive frequency components. Show that the same result is obtained without using fftshift by calculating the DFT of $(-1)^n f[n]$. (Use the shift theorem.)

(b) When using the DFT to approximate the Fourier transform of a two-sided function $f(x)$, we calculate samples $f(n\Delta x)$, $n = -\tfrac{N}{2}, \ldots, 0, 1, \ldots, \tfrac{N}{2} - 1$. The MATLAB ifftshift command is applied to rotate this vector so that the first element of the vector

$(f[n])_{n=0}^{N-1}$ is $f(0)$. Show that the same effect is obtained, without using ifftshift, by calculating the DFT of f and then multiplying the DFT by $(-1)^m$. (Use the shift theorem.)

5.54. *Discrete-time Fourier transform, again*
Make an appropriate definition of equivalent width for a discrete-time signal $(f[n])_{n\in\mathbb{Z}}$ and derive a result analogous to Equation 5.50.

5.55. *Discrete-time Fourier transform, again*
Derive a version of the moment theorem for the discrete-time Fourier transform.

5.56. *Short-time Fourier transform*
Show that the short-time Fourier transform conserves energy (Theorem 5.18).

(a) First, note that the short-time Fourier transform is an inner product, and use Parseval's formula to show

$$F(v;u) = \int_{-\infty}^{\infty} F(\xi + v) W(\xi) e^{+i2\pi\xi u} d\xi,$$

that is, the Fourier transform of $F(v;u)$ with respect to u is $F(\xi + v) W(\xi)$.

(b) Next, apply Parseval's formula again to show

$$\iint_{-\infty}^{\infty} |F(v;u)|^2 \, dv \, du = \iint_{-\infty}^{\infty} |F(\xi + v) W(\xi)|^2 \, dv \, d\xi$$
$$= \|f\|^2 \, \|w\|^2 = \|f\|^2.$$

5.57. *Short-time Fourier transform*
Derive the inverse formula for the short-time Fourier transform (Theorem 5.18).

(a) From the previous problem, we know that $F(v;u) = \mathcal{F}^{-1}\{F(\xi + v) W(\xi)\}$. Use this with Parseval's formula to show

$$\iint_{-\infty}^{\infty} F(v;u) \, w(t-u) \, e^{+i2\pi vt} \, du \, dt = \iint_{-\infty}^{\infty} F(v+\xi) \, |W(\xi)|^2 \, e^{i2\pi(v+\xi)t} \, dv \, d\xi.$$

(b) Use Fubini's theorem to write

$$\iint_{-\infty}^{\infty} F(v+\xi) \, |W(\xi)|^2 \, e^{i2\pi(v+\xi)t} \, dv \, d\xi = \int_{-\infty}^{\infty} |W(\xi)|^2 \left[\int_{-\infty}^{\infty} F(v+\xi) \, e^{i2\pi(v+\xi)t} \, dv \right] d\xi.$$

and then show that the right-hand side is $f(t)$.

CHAPTER 6
GENERALIZED FUNCTIONS

The Fourier transform developed in the previous chapter applies to functions which are absolutely or square integrable on the real line (members of $L^1(\mathbb{R})$ or $L^2(\mathbb{R})$). These classes are large and encompass many functions of importance in engineering and physics, but there are some surprising omissions—sine, cosine, the step and signum functions, and powers of x, to name a few. These are brought into the picture by the introduction of *generalized functions*, the subject of this chapter.

The best-known generalized function is the delta function $\delta(x)$, which models impulsive phenomena like sudden shocks and point charge or mass distributions. The chapter begins with some physical situations that naturally lead to the introduction of the delta function. Some operational rules for manipulating delta functions and using them for practical calculations are then developed. This leads into a broader discussion of generalized function theory. It is shown that all the ordinary functions we have worked with so far, and more, are in fact generalized functions, as well as objects like the delta function with behaviors that defy the traditional definition of "function." All generalized functions possess derivatives of all orders—continuity is no longer a restriction. All generalized functions possess Fourier transforms which are themselves generalized functions—integrability is no longer the barrier it was. The chapter concludes by revisiting sampling and the Fourier series in light of generalized functions. This will unify the Fourier analysis of periodic and aperiodic functions, finite and infinite sequences—everything we have done so far.

6.1 IMPULSIVE SIGNALS AND SPECTRA

We begin by returning to the subject of linear, time-invariant systems. In Section 5.5.2, we saw that the response of the first-order differential equation

$$y' + \frac{1}{\tau}y = \frac{1}{\tau}x$$

to a rectangular pulse, $n \operatorname{rect}(nt)$, was

$$y(t) = \begin{cases} 0, & t < -1/2n \\ n(1 - e^{-(t+1/2n)\tau}), & 1/2n > t \geq -1/2n \\ n(1 - e^{-1/n\tau})e^{-(t-1/2n)/\tau}, & t \geq 1/2n \end{cases}.$$

Fourier Transforms: Principles and Applications, First Edition. Eric W. Hansen.
© 2014 John Wiley & Sons, Inc. Published 2014 by John Wiley & Sons, Inc.

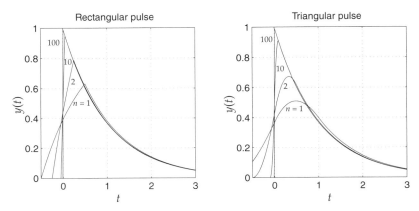

FIGURE 6.1 Response of a first-order system ($\tau = 1$) to excitation by a unit area pulse (height n) with rectangular (*left*) and triangular (*right*) profile. As n increases so that the pulsewidth is much smaller than the time constant ($1/n \ll \tau$), the pulse shape has a negligible effect on the overall response. Both responses converge to $y(t) = \frac{1}{\tau} e^{-t/\tau}$.

As n became very large, so that the pulse duration $1/n$ was very short compared with the system time constant τ, the response was seen to approach the limit:

$$y(t) = \frac{1}{\tau} e^{-t/\tau}, \quad t > 0. \tag{6.1}$$

Interpreted as an RC circuit, this corresponds to an instantaneous charging of the capacitor followed by an exponential discharge.

Now, let us do the problem again, but this time using a triangular input pulse, $x(t) = n\Lambda(nt)$. The solution of the differential equation is

$$y(t) = \begin{cases} 0, & t < -1/n \\ n^2(t + 1/n) - n^2\tau(1 - e^{-(t+1/n)/\tau}), & 0 > t \geq -1/n \\ -n^2(t - 1/n) + n^2\tau(1 - e^{-t/\tau}) - n^2\tau(1 - e^{-1/n\tau})e^{-t/\tau}, & 1/n > t \geq 0 \\ n^2\tau(1 - e^{-1/n\tau})^2 e^{-(t-1/n)/\tau}, & t \geq 1/n \end{cases}.$$

and it is plotted in Figure 6.1. When the triangular pulse width is comparable to the time constant of the system, the risetime of the response is comparable to the decay time and the shape of the pulse has a noticeable effect on the response. However, as n increases to where $1/n \ll \tau$, the risetime becomes negligible compared with the decay time. As $n \to \infty$, the limit of the response to the triangular pulse input is $e^{-t/\tau}/\tau$, just as with the rectangular pulse. Evidently, the precise shape of the pulse does not matter once it has become sufficiently narrow.

Informally for now, we will call a pulse with infinitesimal width and unit area a *unit impulse*, and write $\delta(t)$ to indicate a unit impulse applied at $t = 0$. The physical effect of the impulse is to instantaneously change the state of the system (the capacitor voltage, say). The area of the impulse determines how high the response rises before beginning to decay. Just prior to $t = 0$, the output is $y(0^-) = 0$. The impulse is applied at $t = 0$, and the output just after this is $y(0^+) = 1/\tau$. There is a jump discontinuity in the output in response to the impulse.

6.1 IMPULSIVE SIGNALS AND SPECTRA

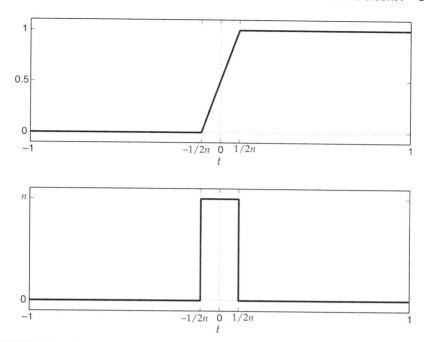

FIGURE 6.2 Modeling the step function, $U(t)$. Instead of a jump at $t = 0$, the step is modeled by a steep ramp (*top*). Its derivative, U', is a rectangular pulse $n \operatorname{rect} nt$ (*bottom*).

A similar situation occurs when an inductive circuit is driven with a step input. The voltage across the inductor in a series RL circuit obeys the differential equation

$$v' + \frac{R}{L}v = v'_{\text{in}}.$$

Let $v_{\text{in}} = U(t)$ and write $\tau = L/R$, so

$$v' + \frac{1}{\tau}v = U'(t).$$

The step function is not differentiable at $t = 0$. We get around this by approximating the step by a sequence of ramps (Figure 6.2). The derivative of one such ramp, approximating the derivative U' of the step function, is $n \operatorname{rect} nt$. The differential equation is now, approximately,

$$v' + \frac{1}{\tau}v = n \operatorname{rect} nt.$$

Within a factor of τ, this is the differential equation solved earlier. The solution, as $n \to \infty$, will be

$$v(t) = e^{-t/\tau}, \quad t > 0.$$

The driving function $U'(t)$ instantaneously changes the inductor voltage $v(t)$. We expect this on physical grounds; the inductor voltage changes instantaneously to

370 CHAPTER 6 GENERALIZED FUNCTIONS

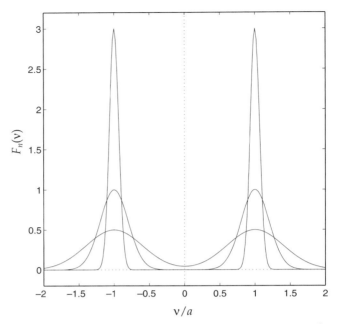

FIGURE 6.3 The sequence of Fourier transforms of $f_n(x) = \exp(-\pi(x/n)^2) \cos(2\pi ax)$, a cosine function multiplied by a Gaussian convergence factor.

oppose an instantaneous change in its current. Mathematically, it appears that the derivative of the step function has the same effect as an impulse, and instead of regarding the step as undifferentiable, we might say $U'(t) = \delta(t)$.[1]

For a third example, we consider the Fourier transform of $f(x) = \cos 2\pi ax$. This function is not integrable on $\mathbb{R}$, but with a Gaussian convergence factor the functions

$$f_n(x) = e^{-\pi(x/n)^2} \cos 2\pi ax, \quad n > 0$$

are integrable, and as $n \to \infty$ the sequence converges to f. Using the modulation and dilation theorems, we can easily calculate a sequence of transforms:

$$F_n(\nu) = \frac{n}{2} \exp(-\pi n^2 (\nu - a)^2) + \frac{n}{2} \exp(-\pi n^2 (\nu + a)^2).$$

This sequence is plotted in Figure 6.3. There are two Gaussian pulses, centered at $\nu = \pm a$, each with height $n/2$, width proportional to $1/n$, and area $1/2$.

According to the definition of the Fourier transform in L^2, the limit of this sequence is expected to be the Fourier transform F. However, while each of the

---

[1] A problem similar to this, driving a transmission line with rectangular telegraph pulses, motivated Heaviside's development, in the 1890s, of "operational methods," which we know today as the Laplace transform. Heaviside noted that the derivative of a step is usefully treated as impulsive. See Heaviside (1950, p. 133).

individual functions f_n is square integrable, the limit $f(x) = \cos 2\pi ax$ is not. Correspondingly, although the functions F_n are each in $L^2(\mathbb{R})$ (by Parseval's formula), they blow up as $n \to \infty$ and fail to converge to an ordinary function.

On the other hand, the Fourier transforms F_n are concentrated around $\nu = \pm a$. If we took an integer number of periods of $\cos 2\pi ax$ and calculated the Fourier series, the result would be a pair of spectral lines at $\nu = \pm a$. The tendency of the Gaussian pulses to become very high and thin, resembling a line spectrum as the sequence f_n approaches the "pure" cosine, seems to be correct. Taking a cue from the two previous examples, we might say that the Gaussian pulses become a pair of impulses as $n \to \infty$ and (provisionally) write

$$\cos 2\pi ax \longleftrightarrow \frac{1}{2}\delta(\nu - a) + \frac{1}{2}\delta(\nu + a)$$

to denote the cosine spectrum.

To show that this is not just wishful thinking, we will calculate the inverse Fourier transform, using the rectangular sequence $n \operatorname{rect} n\nu$ to represent the impulses (you would rightfully be suspicious of using the Gaussians again). Then the Fourier transform of the cosine is, approximately,

$$F_n(\nu) = \frac{1}{2} n \operatorname{rect}(n(\nu - a)) + \frac{1}{2} n \operatorname{rect}(n(\nu + a)).$$

The inverse transform, using the shift and dilation theorems, is

$$f_n(x) = \frac{1}{2} e^{i2\pi ax} \operatorname{sinc}\left(\frac{x}{n}\right) + \frac{1}{2} e^{-i2\pi ax} \operatorname{sinc}\left(\frac{x}{n}\right)$$
$$= \cos(2\pi ax) \operatorname{sinc}\left(\frac{x}{n}\right).$$

In the limit as $n \to \infty$, the sinc gets wider and approaches $\operatorname{sinc}(0) = 1$ everywhere, leaving just $\cos(2\pi ax)$, as desired.

In the real world, infinitely high and narrow pulses do not exist. But there are numerous physical examples of highly localized quantities for which impulses provide useful idealizations, such as sudden shocks, voltage spikes, spectral lines, point masses, point charges, dipole moments, and point radiators. Using a sequence of pulses representation, the next section develops an operational definition for impulses and a set of rules for mathematically manipulating them.

6.2 THE DELTA FUNCTION IN A NUTSHELL

The preceding examples showed three different pulse sequences that exhibit impulsive behavior: rectangle, $(n \operatorname{rect}(nx))_{n=1}^\infty$, triangle, $(n\Lambda(nx))_{n=1}^\infty$, and Gaussian, $(ne^{-\pi n^2 x^2})_{n=1}^\infty$. The pulses all have height n, width proportional to $1/n$, and unit area. By themselves, they grow without bound and cannot be regarded as converging to any reasonable function in the limit. But in the examples of the RC and RL circuits, physical damping effects smeared (integrated) the pulses and produced final results that were both physically and mathematically reasonable. In the Fourier transform

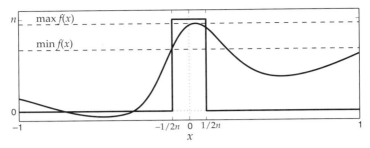

FIGURE 6.4 Integrating a continuous function against a rectangular pulse sequence, $(n \operatorname{rect} nx)_{n=1}^{\infty}$. The dashed lines indicate the maximum and minimum of f on the interval $\left(-\frac{1}{2n}, \frac{1}{2n}\right)$. As $n \to \infty$, the maximum and minimum values both approach $f(0)$.

example, the pulse sequences modeling the impulsive spectra likewise were "tamed" by passing through the inverse transform integral.

Definition by Pulse Sequences

Let us, then, consider the simplest calculation involving an impulsive pulse sequence, a rectangular pulse $n \operatorname{rect}(nx)$, integrated against a continuous function $f(x)$ (Figure 6.4):

$$\int_{-\infty}^{\infty} n \operatorname{rect}(nx) f(x) \, dx = \int_{-1/2n}^{1/2n} nf(x) \, dx.$$

The integrand $nf(x)$ is bounded above and below by its maximum and minimum values on the interval $\left(-\frac{1}{2n}, \frac{1}{2n}\right)$. The integral is likewise trapped between two values,

$$\int_{-1/2n}^{1/2n} n \max f \, dx = \max f \quad \text{and} \quad \int_{-1/2n}^{1/2n} n \min f \, dx = \min f,$$

and because f is continuous, as $n \to \infty$ and the interval $\left(-\frac{1}{2n}, \frac{1}{2n}\right)$ shrinks to $(0^-, 0^+)$, the maximum and minimum values both approach $f(0)$. Therefore,

$$\lim_{n \to \infty} \int_{-\infty}^{\infty} f(x) \, n \operatorname{rect}(nx) \, dx = f(0).$$

As a shorthand for this limit of the sequence of integrals, we write

$$\int \delta(x) f(x) \, dx = \lim_{n \to \infty} \int_{-\infty}^{\infty} n \operatorname{rect}(nx) f(x) \, dx = f(0). \tag{6.2}$$

The symbol $\delta(x)$ is called the *unit impulse* or *delta function*. We understand that $\delta(x)$ is not to be interpreted as the pointwise limit of the pulse sequence, for we have already seen that the sequence grows without bound. Rather, the integral $\int \delta(x) f(x) \, dx$ is just a *symbol*, a convenient notational device, for the limiting value of the sequence of integrals in Equation 6.2.

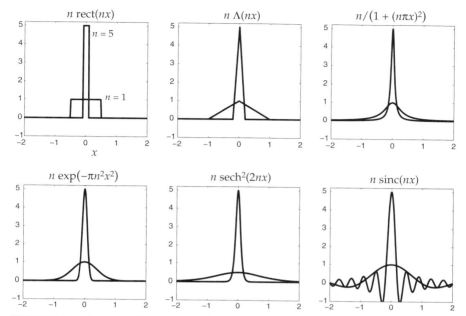

FIGURE 6.5 Several pulse sequences that can be used to model an impulse. In each graph, the cases $n = 1$ and $n = 5$ are shown.

Equation 6.2 is not the only possible definition for the impulse. The derivation would be essentially the same if we used triangular pulses, $n\Lambda(nx)$. And by a different argument[2] it is shown that Gaussian pulses have the same property:

$$\lim_{n \to \infty} \int_{-\infty}^{\infty} n e^{-\pi n^2 x^2} f(x)\, dx = f(0).$$

In fact, any "peaky" unit area function can be made into an impulse sequence (Figure 6.5).[3]

You may have heard the delta function defined this way: it has infinite height, zero width, and unit area. This is problematic, because height × width = area but the statement "$\infty \times 0 = 1$" makes no sense. It is closer to the truth to say that $\delta(x)$ is the limit of some unit area pulse sequence $n\psi(nx)$ as $n \to \infty$, but we have to qualify this: the sequence of pulses does not have a limit function, but the sequence of integrals $\int_{-\infty}^{\infty} n\psi(nx) f(x)\, dx$ does have a limiting value, and in the sense of this limit of integrals one can say that that $n\psi(nx) \to \delta(x)$. This is not the normal behavior one expects from a function. The delta function is, in fact, *not* a function in the ordinary

---

[2] For proofs that the Gaussian pulse sequence defines a delta function, see Howell (2001, pp. 427–429) and Lighthill (1958, p. 17).

[3] Such functions are called "identity sequences" or "approximate identities." See Howell (2001, pp. 415–430).

sense of the term, but the name is entrenched in physics and engineering and we shall continue to use it here.

In most applications we may bypass the pulse sequence and simply define the impulse in terms of its fundamental *sifting property*:

$$\int \delta(x) f(x)\, dx = f(0). \quad (6.3)$$

The impulse extracts, or "sifts out," the value of f at the impulse location $x = 0$. Letting $f = 1$ reveals that δ has unit area:

$$\int_{-\infty}^{\infty} \delta(x)\, dx = 1. \quad (6.4)$$

Scaling
Setting $f = a$, a constant function, the impulse extracts $f(0) = a$:

$$\int_{-\infty}^{\infty} a\,\delta(x)\, dx = a. \quad (6.5)$$

Scaling a unit impulse by a constant a changes its area to a.

Shifting
Performing the basic integral with a shifted pulse sequence leads to the definition for a shifted impulse:

$$\int_{-\infty}^{\infty} \delta(x-b) f(x)\, dx = \lim_{n\to\infty} \int_{-\infty}^{\infty} n\,\text{rect}(n(x-b)) f(x)\, dx$$

$$= \lim_{n\to\infty} \int_{-\infty}^{\infty} n\,\text{rect}(n\xi) f(\xi + b)\, d\xi$$

$$= \int_{-\infty}^{\infty} \delta(\xi) f(\xi + b)\, d\xi = f(b). \quad (6.6)$$

(Evidently we would have been justified in pretending that δ is a function and performing the change of variable directly on the integral $\int \delta(x-b) f(x)\, dx$ rather than going through the pulse sequence—more about this later.) The shifted impulse sifts out the value of f at the impulse location $x = b$. When $f = 1$, we have $\int_{-\infty}^{\infty} \delta(x-b)\, dx = 1$; the shifted unit impulse also has unit area.

The unit impulse $\delta(x)$ is portrayed graphically by an arrow of unit height, as shown in Figure 6.6. The scaled impulse $a\delta(x)$ with area a is drawn as an arrow of height a. A shifted unit impulse $\delta(x-b)$ is drawn as an unit height arrow at $x = b$.

Dilation
We may define a dilated impulse $\delta(ax)$ by a dilated pulse sequence, say $n\,\text{rect}(nax)$. The pulses are still centered at $x = 0$ and will still behave impulsively as $n \to \infty$, but their areas are $\frac{1}{|a|}$ rather than unity. They no longer represent a unit impulse but rather

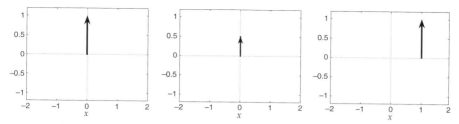

FIGURE 6.6 Impulse functions. *Left to right*: the unit impulse $\delta(x)$; the scaled impulse $\frac{1}{2}\delta(x)$ has area $\frac{1}{2}$; the shifted unit impulse $\delta(x-1)$.

a scaled impulse $\frac{1}{|a|}\delta(x)$. Formally,

$$\int_{-\infty}^{\infty} \delta(ax)f(x)\,dx = \lim_{n\to\infty} \int_{-\infty}^{\infty} n\,\mathrm{rect}(nax)f(x)\,dx$$

$$= \lim_{n\to\infty} \int_{-\infty}^{\infty} n\,\mathrm{rect}(n\xi)f\left(\frac{\xi}{a}\right)\frac{d\xi}{|a|}$$

$$= \int_{-\infty}^{\infty} \delta(\xi)\frac{1}{|a|}f\left(\frac{\xi}{a}\right)d\xi = \frac{1}{|a|}f(0)$$

$$\therefore \delta(ax) = \frac{1}{|a|}\delta(x). \tag{6.7}$$

Example 6.1 (Simple operations with delta functions).

1. $\int_{-\infty}^{\infty} \delta(x-1)\cos\pi x\,dx = \cos\pi = -1$

2. $\int_{-\infty}^{\infty} \delta(2x)\,\mathrm{rect}\,x\,dx = \int_{-\infty}^{\infty} \frac{1}{2}\delta(x)\,\mathrm{rect}\,x\,dx = \frac{1}{2}$

3. $\int_{-\infty}^{\infty} \delta(x+1)e^{-x}U(x)\,dx = e^{-(-1)}U(-1) = 0$

4. $\int_{-\infty}^{\infty} \delta(3x-2)x^2\,dx = \int_{-\infty}^{\infty} \delta\left(3\left(x-\frac{2}{3}\right)\right)x^2\,dx = \int_{-\infty}^{\infty} \frac{1}{3}\delta\left(x-\frac{2}{3}\right)x^2\,dx = \frac{4}{27}$

5. $\int_{0}^{\infty} \delta(x-1)x^2\,dx = \int_{-\infty}^{\infty} \delta(x-1)x^2 U(x)\,dx = (1)^2 U(1) = 1$ ∎

Nonlinear Dilation

Let $h(x)$ be a continuous function with a zero crossing at a point $x = c$: $h(c) = 0$, $h'(c) \neq 0$. We may consider a more general dilation of the delta function, $\delta(h(x))$, of which $\delta(ax+b)$ is a special case. We know that the delta function is localized to the point where its argument is zero; in this case, at $x = c$. In the vicinity of $x = c$, h is approximated to first order by $h(x) = h(c) + h'(c)(x-c) = h'(c)(x-c)$. Then $\delta(h(x)) = \delta(h'(c)(x-c))$, and using Equation 6.7,

$$\delta(h(x)) = \frac{1}{|h'(c)|}\delta(x-c).$$

If h has multiple zero crossings c_k, then $\delta(h(x))$ produces an impulse at each root.

$$\delta(h(x)) = \sum_k \frac{1}{|h'(c_k)|} \delta(x - c_k), \qquad (6.8)$$

where $h(c_k) = 0$, $h'(c_k) \neq 0$

Example 6.2. In the delta function $\delta(x^2 - 1)$, $h(x) = x^2 - 1$ has roots at $x = \pm 1$. The slope is $h'(x) = 2x$. Thus,

$$\delta(x^2 - 1) = \frac{1}{2}\delta(x + 1) + \frac{1}{2}\delta(x - 1).$$

We may obtain some additional insight by representing the delta function as a sequence of rectangles:

$$\delta_n(x^2 - 1) = n \operatorname{rect}(n(x^2 - 1)) = \begin{cases} n, & |x^2 - 1| < 1/2n \\ n/2, & |x^2 - 1| = 1/2n \\ 0, & \text{otherwise} \end{cases}.$$

One of the rectangles extends from $x = \sqrt{1 - 1/2n}$ to $\sqrt{1 + 1/2n}$, and the other from $-\sqrt{1 + 1/2n}$ to $-\sqrt{1 - 1/2n}$. Each has height n and width $\sqrt{1 + 1/2n} - \sqrt{1 - 1/2n}$. As n becomes large, the area is

$$n \cdot [(1 + 1/4n) - (1 - 1/4n) + O(1/n^2)] = 1/2 + O(1/n) \to 1/2.$$

That is, as each rectangle approaches a delta function, it does so with an area approaching $1/2$. ∎

Example 6.3. Another interesting example is $\delta(\sin(2\pi x))$. There are zeros leading to delta functions at $x = k$. The areas are

$$\frac{1}{|2\pi \cos(2\pi k)|} = \frac{1}{2\pi}.$$

Thus,

$$\delta(\sin(2\pi x)) = \sum_{k=-\infty}^{\infty} \frac{1}{2\pi} \delta(x - k),$$

an infinite train of impulses of area $1/2\pi$, located at the integers. This is known as a *comb function* and will be studied in detail later. ∎

The Impulse and the Step

What about the statement that $U'(x) = \delta(x)$? Integrating both sides would imply

$$U(x) = \int_{-\infty}^{x} \delta(\xi) \, d\xi. \qquad (6.9)$$

If, for simplicity, the impulse is modeled by a rectangular pulse sequence ($n \operatorname{rect} nx$), it is easy to develop a sense of what Equation 6.9 is saying. Integrating, we obtain

(look at Figure 6.2 again)

$$\int_{-\infty}^{x} n \operatorname{rect} n\xi \, d\xi = \begin{cases} 0, & x < -\frac{1}{2n} \\ \frac{1}{2} + nx, & \frac{1}{2n} \geq x \geq -\frac{1}{2n} \\ 1, & x > \frac{1}{2n} \end{cases}.$$

As $n \to \infty$ and the rectangles become narrower and higher, the ramp between $x = -\frac{1}{2n}$ and $x = \frac{1}{2n}$ becomes steeper (slope n) and narrower (width $1/n$). The integral for $x > \frac{1}{2n}$ is always unity, because the pulses have unit area. In the limit, the result is a step function. When we integrate across a delta function, there is no contribution until $x = 0^-$. As the delta function is crossed we collect all of its area at once, and there is no further contribution for $x > 0^+$.

The relationship between the step and the impulse enables us to define the derivative of a function that has jump discontinuities.

Example 6.4 (Derivative of the rectangle function). A rectangle can be written as the difference of two steps:

$$\frac{d}{dx} \operatorname{rect} x = \frac{d}{dx}\left[U\left(x + \tfrac{1}{2}\right) - U\left(x - \tfrac{1}{2}\right)\right]$$

$$= \delta\left(x + \tfrac{1}{2}\right) - \delta\left(x - \tfrac{1}{2}\right).$$

The derivative of the positive jump at $x = -\frac{1}{2}$ is a positive-going impulse, and the derivative of the negative jump at $x = \frac{1}{2}$ is a negative-going impulse (Figure 6.7). ∎

Example 6.5. Let $f(x) = \cos 2\pi x \operatorname{sgn} x$ (Figure 6.8). For $x < 0, f = -\cos 2\pi x$ and $f' = 2\pi \sin 2\pi x$. For $x > 0, f = \cos 2\pi x$ and $f' = -2\pi \sin 2\pi x$. Right at $x = 0$, there is a jump of height 2, from -1 to 1. Ordinarily, the function would be undifferentiable here, but we have just seen that the derivative of a unit step is a unit impulse. By linearity, the derivative of a jump of height 2 is an impulse with area 2. Hence, we have

$$f'(x) = \begin{cases} 2\pi \sin 2\pi x, & x < 0 \\ 2\delta(x), & x = 0 \\ -2\pi \sin 2\pi x, & x > 0 \end{cases}$$

$$= -2\pi \sin 2\pi x \operatorname{sgn} x + 2\delta(x)$$

In general, any piecewise continuous function can be expressed as the sum of a piecewise smooth function and a piecewise constant function consisting solely of jumps. The derivative of this function will be the sum of the ordinary derivatives of the continuous parts and a set of impulses representing the derivatives of the jumps. ∎

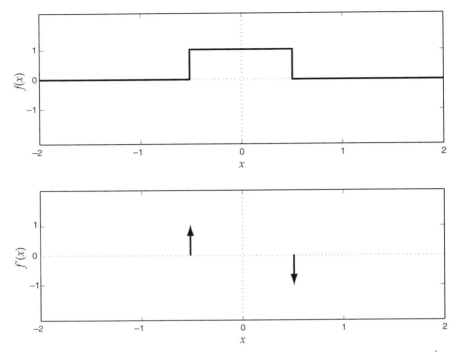

FIGURE 6.7 The rectangle function $f(x) = \operatorname{rect} x$ (*top*) and its derivative $f'(x) = \delta(x + \frac{1}{2}) - \delta(x - \frac{1}{2})$ (*bottom*). The positive and negative impulses at $x = \mp\frac{1}{2}$ correspond to the positive and negative jumps in the rectangle.

Example 6.6. The derivative of the one-sided exponential function $f(x) = \frac{1}{2}e^{-3x}U(x)$ is

$$f'(x) = \begin{cases} 0, & x < 0 \\ (f(0^+) - f(0^-))\,\delta(x), & x = 0 \text{ (the jump)} \\ -\frac{3}{2}e^{-3x}, & x > 0 \text{ (the continuous part)} \end{cases}$$

$$= \frac{1}{2}\delta(x) - \frac{3}{2}e^{-3x}U(x).$$ ∎

Derivative of the Delta Function

Suppose we represent the delta function by a rectangle sequence, $n\operatorname{rect}(nx)$. The derivative of this rectangular pulse is $n\delta(x + 1/2n) - n\delta(x - 1/2n)$—a positive impulse at the leading edge of the rectangle and a negative impulse at the trailing edge. Does this sequence have a meaningful limit? To find out, we integrate it against a continuous function:

$$\int_{-\infty}^{\infty} [n\delta(x + 1/2n) - n\delta(x - 1/2n)] f(x)\,dx = nf(-1/2n) - nf(1/2n)$$

$$= -\frac{f(1/2n) - f(-1/2n)}{1/n}.$$

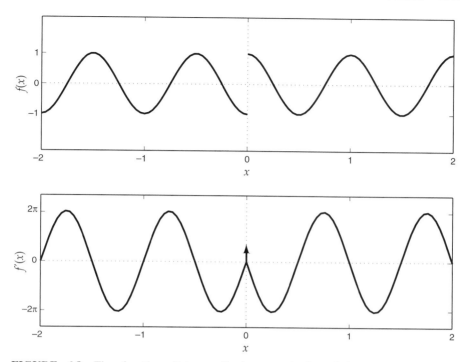

FIGURE 6.8 The function $f(x) = \cos(2\pi x)\,\text{sgn}\,x$ (*top*) and its derivative $f'(x) = -2\pi \sin(2\pi x)\,\text{sgn}\,x + 2\delta(x)$ (*bottom*).

As $n \to \infty$, it becomes a derivative:

$$-\frac{f(1/2n) - f(-1/2n)}{1/n} \to -f'(0).$$

We define the derivative of the delta function, $\delta'(x)$, by the derivative of the pulse sequence for $\delta(x)$ and assign the operational rule

$$\int_{-\infty}^{\infty} \delta'(x) f(x)\, dx = -f'(0). \tag{6.10}$$

Fourier Transforms

Let us apply the sifting property (6.3) to the inverse Fourier transform. The Fourier kernel $e^{+i2\pi vx}$ is continuous, so we may insert an impulse into the inverse Fourier transform, obtaining

$$\mathcal{F}^{-1}\{\delta(v)\} = \int_{-\infty}^{\infty} \delta(v)\, e^{+i2\pi vx}\, dv = e^{+i2\pi 0 x} = 1.$$

We cannot perform the forward transform as an ordinary integral because $f(x) = 1$ is not absolutely integrable. Instead, model f as the limit of a sequence of functions,

$f_n(x) = e^{-\pi(x/n)^2}$; as $n \to \infty$, $f_n \to 1$. The Fourier transform of f_n is, using the dilation theorem,

$$F_n(\nu) = n e^{-\pi n^2 \nu^2},$$

a sequence that defines the impulse (Figure 6.5).

By an identical calculation, the inverse Fourier transform of the shifted impulse $\delta(\nu - a)$ is $e^{i2\pi ax}$. By the linearity of the Fourier transform,

$$\mathcal{F}^{-1}\left\{\frac{1}{2}\delta(\nu - a) + \frac{1}{2}\delta(\nu + a)\right\} = \frac{1}{2}e^{+i2\pi ax} + \frac{1}{2}e^{-i2\pi ax} = \cos 2\pi ax,$$

and also,

$$\mathcal{F}^{-1}\left\{\frac{1}{2i}\delta(\nu - a) - \frac{1}{2i}\delta(\nu + a)\right\} = \frac{1}{2i}e^{+i2\pi ax} - \frac{1}{2i}e^{-i2\pi ax} = \sin 2\pi ax.$$

These transforms are graphed in Figure 6.9.

We can even calculate the Fourier transform of δ':

$$\mathcal{F}\{\delta'(x)\} = \int_{-\infty}^{\infty} \delta'(x) e^{-i2\pi \nu x}\, dx = -\frac{d}{dx} e^{-i2\pi \nu x}\bigg|_{x=0} = i2\pi \nu\, e^{-i2\pi \nu 0} = i2\pi \nu.$$

In light of the derivative theorem for the Fourier transform, this result should not be surprising. The repeated transform theorem, Equation 5.5, then gives the result $x \longmapsto -\frac{1}{i2\pi}\delta'(\nu)$. How interesting—$x$ is not integrable, but it has a Fourier transform!

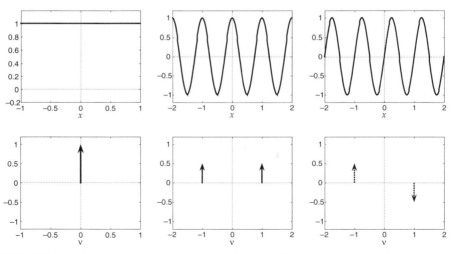

FIGURE 6.9 *Left to right*: the unit function 1 and its Fourier transform, $\delta(\nu)$; the cosine function $\cos 2\pi x$ and its Fourier transform, $\frac{1}{2}\delta(\nu - 1) + \frac{1}{2}\delta(\nu + 1)$; the sine function $\sin 2\pi x$ and its Fourier transform, $\frac{1}{2i}\delta(\nu - 1) - \frac{1}{2i}\delta(\nu + 1)$.

Collecting our results, we have several new transform pairs:

$$1 \longleftrightarrow \delta(\nu) \tag{6.11}$$
$$e^{i2\pi x} \longleftrightarrow \delta(\nu - 1) \tag{6.12}$$
$$\cos 2\pi x \longleftrightarrow \frac{1}{2}\delta(\nu - 1) + \frac{1}{2}\delta(\nu + 1) \tag{6.13}$$
$$\sin 2\pi x \longleftrightarrow \frac{1}{2i}\delta(\nu - 1) - \frac{1}{2i}\delta(\nu + 1) \tag{6.14}$$
$$x \longleftrightarrow -\frac{1}{i2\pi}\delta'(\nu) \tag{6.15}$$

Example 6.7 (Mixed transforms). Sums of impulses and ordinary functions may appear in the same Fourier transform. For example, using the linearity of the Fourier transform,

$$1 + \Lambda(x) \longmapsto \delta(\nu) + \text{sinc}^2(\nu).$$
$$2\sin(\pi x) + e^{-2\pi x^2} \longmapsto -i\delta(\nu - 1/2) + i\delta(\nu + 1/2) + \frac{1}{\sqrt{2}}e^{-\pi \nu^2/2}.$$

∎

By linearity, we expect the Fourier transform of a function f defined by a Fourier series (i.e., a periodic function) to be

$$F(\nu) = \mathcal{F}\left\{\sum_{n=-\infty}^{\infty} c_n e^{i2\pi nx/L}\right\} = \sum_{n=-\infty}^{\infty} c_n \mathcal{F}\{e^{i2\pi nx/L}\}.$$

For the moment, we will assume that the infinite series can actually be Fourier-transformed term-by-term. To calculate the Fourier transform of the complex exponential, we use the dilation theorem together with Equations 6.12 and 6.7:

$$\mathcal{F}\{e^{i2\pi nx/L}\} = \mathcal{F}\{e^{i2\pi (n/L)x}\} = \frac{1}{|n/L|}\delta\left(\frac{\nu}{n/L} - 1\right)$$
$$= \frac{L}{|n|}\delta\left[\frac{L}{n}\left(\nu - \frac{n}{L}\right)\right] = \delta\left(\nu - \frac{n}{L}\right).$$

Finally, we have

$$F(\nu) = \sum_{n=-\infty}^{\infty} c_n \delta\left(\nu - \frac{n}{L}\right).$$

The Fourier spectrum is a set of discrete spectral lines, represented by impulses located at the frequencies of the harmonics.

Convolution

A convolution result also follows from the sifting property. Consider

$$f(x) * \delta(x-a) = \int_{-\infty}^{\infty} f(\xi)\,\delta[(x-\xi) - a]\,d\xi = \int_{-\infty}^{\infty} f(\xi)\,\frac{1}{|-1|}\delta[\xi - (x-a)]\,d\xi$$
$$= f(x-a), \tag{6.16}$$

if f is everywhere continuous (later we will see that the continuity requirement can be removed). The effect of convolution with the delta function is to shift f by a along the x axis. One way to remember the distinction between product and convolution, $f\delta$ and $f * \delta$, is

"Multiplication sifts, convolution shifts."

Assuming that the convolution theorem applies when a delta function is involved (it does), we can see in the frequency domain

$$\mathcal{F}\{f(x) * \delta(x-a)\} = F(\nu)e^{-i2\pi\nu a} = \mathcal{F}\{f(x-a)\},$$

in agreement with the shift theorem.

Example 6.8 (Modulation as convolution). The modulation theorem (Equation 5.21)

$$f(t)\cos(2\pi\nu_0 t) \longmapsto \frac{1}{2}F(\nu-\nu_0) + \frac{1}{2}F(\nu+\nu_0)$$

may be usefully interpreted as a convolution in the frequency domain:

$$f(t)\cos(2\pi\nu_0 t) \longmapsto F(\nu) * \mathcal{F}\{\cos(2\pi\nu_0 t)\}$$
$$= F(\nu) * \left[\frac{1}{2}\delta(\nu-\nu_0) + \frac{1}{2}\delta(\nu+\nu_0)\right]$$
$$= \frac{1}{2}F(\nu-\nu_0) + \frac{1}{2}F(\nu+\nu_0).$$

Convolution provides an efficient route to the Fourier transform of the Gaussian-tapered cosine in Figure 6.3:

$$e^{-\pi(x/b)^2}\cos 2\pi ax \longmapsto be^{-\pi(b\nu)^2} * \left[\frac{1}{2}\delta(\nu-a) + \frac{1}{2}\delta(\nu+a)\right]$$
$$= \frac{b}{2}e^{-\pi n^2(\nu-a)^2} + \frac{b}{2}e^{-\pi n^2(\nu+a)^2}.$$

Convolution with the Gaussian causes the delta functions to "melt". As b increases the impulses sharpen. ∎

This completes the basic set of operational rules for manipulating the delta function. They are summarized in Table 6.1 and again at the end of the chapter.

6.3 GENERALIZED FUNCTIONS

The delta function and related mathematical objects have been intuitively understood and used in physics and engineering since the nineteenth century. Kirchoff employed a sequence of Gaussian pulses to model the singularity at the origin of a diverging spherical wave. Heaviside used the step function and its derivative in his analyses of telegraph cables. Dirac, in his development of quantum mechanics, used the delta function to express the orthogonality of families of functions with continuous,

TABLE 6.1 Basic operational rules for the delta function.

Rule	Formula
Unit area	$\int_{-\infty}^{\infty} \delta(x)\,dx = 1$
Sifting	$\int_{-\infty}^{\infty} f(x)\delta(x-a)\,dx = f(a)$
with δ'	$\int_{-\infty}^{\infty} f(x)\delta'(x-a)\,dx = -f'(a)$
Convolution	$f(x) * \delta(x-a) = f(x-a)$
vs. sifting	$f(x)\delta(x-a) = f(a)\delta(x-a)$
Step function	$U(x) = \int_{-\infty}^{x} \delta(\xi)\,d\xi,\ U'(x) = \delta(x)$
Dilation	$\delta(ax) = \frac{1}{|a|}\delta(x)$
nonlinear	$\delta(h(x)) = \frac{1}{|h'(c)|}\delta(x-c),\ h(c) = 0$
Fourier transforms	$1 \longleftrightarrow \delta(\nu)$
	$e^{i2\pi x} \longleftrightarrow \delta(\nu - 1)$
	$\cos(2\pi x) \longleftrightarrow \frac{1}{2}\delta(\nu - 1) + \frac{1}{2}\delta(\nu + 1)$
	$\sin(2\pi x) \longleftrightarrow \frac{1}{2i}\delta(\nu - 1) - \frac{1}{2i}\delta(\nu + 1)$
	$x \longleftrightarrow -\frac{1}{i2\pi}\delta'(\nu)$

rather than discrete, indices. Recall, for the Fourier series we wrote $\langle e^{i2\pi nx}, e^{i2\pi mx}\rangle = \delta[n-m]$, the unit sample or Kronecker delta. The equivalent statement for the Fourier transform, as we shall see, is $\langle e^{i2\pi \nu x}, e^{i2\pi \mu x}\rangle = \delta(\nu - \mu)$, using the "Dirac delta."

These early applications were based largely on physical intuition. The calculations were justified by the physical correctness of the results. In the mid-twentieth century Schwartz developed a rigorous mathematical structure for what he called *distributions*. Lighthill followed a decade later with his theory of *generalized functions*, based on a careful development of the physically appealing sequence of functions approach we followed in the preceding section.[4]

Our earlier observation that $U' = \delta$ suggests that ordinary functions like the step and impulsive objects like the delta function are not really that far apart. Our objective in this section is to present a common framework for the two via the theory of generalized functions. The delta function is the best known example of a generalized function, and the sequence-of-pulses approach of the preceding section is sufficient to obtain its most important properties, including its Fourier transform. We will see that ordinary functions, including nonintegrable functions like the step and signum as well as all the functions in L^2, are also generalized functions.

---

[4]See Schwartz (1951). For an accessible introduction to Schwartz' approach, in English, see Strichartz (1994), especially the first four chapters. Also, see Folland (1992, Chapter 9). The sequence of functions approach to generalized functions is beautifully and succinctly presented in Lighthill (1958). For a more comprehensive treatment see Jones (1982).

6.3.1 Functions and Generalized Functions

We know that a function f is a mapping from one set of numbers, for example, $\mathbb{R}$ or $[a, b]$, to another set of numbers, for example, $\mathbb{R}$ or $\mathbb{C}$. In elementary calculus, two functions f and g are equal if they agree pointwise, $f(x) = g(x)$, for all x in their common domain. Equivalently, $f = g$ if their difference is zero, $(f - g)(x) = 0$ for all x. When we began to look at normed spaces, we augmented this strict pointwise equality with the idea of equivalence in norm: two functions in a normed space may differ at some points (comprising a set of measure zero) but be considered equivalent if $\|f - g\| = 0$.

When we met the delta function in the previous section, we lost these touchpoints. The sequence of functions $\delta_n(x) = n\,\text{rect}(nx)$, by which we defined the delta function, is not Cauchy (try it), and while it converges to zero for $x \neq 0$, it diverges at $x = 0$. There is no well-defined function that is the limit of $\delta_n(x)$ either pointwise or in norm. Yet we did find that the sequence of numbers produced by the integrals $\int \delta_n(x) f(x)\, dx$ was convergent, and that is how we defined the limit of the sequence (δ_n) to be δ. We also saw that more than one pulse sequence could be used for δ_n (e.g., $n\,\text{rect}(nx)$, $n\Lambda(nx)$) with the same results, so in that sense they were equivalent, even though individual members of the sequences were certainly not equal: $n\,\text{rect}(nx) \neq n\Lambda(nx)$, either pointwise or in norm, by our familiar criteria (e.g., calculate $\|n\,\text{rect}(nx) - n\Lambda(nx)\|_1$).

Yet another way to think about equivalent functions is expressed by the following theorem for inner product spaces.

Theorem 6.1. Let V be an inner product space. Two vectors $f, g \in V$ are equivalent if and only if $\langle f, \varphi \rangle = \langle g, \varphi \rangle$ for all vectors $\varphi \in V$.

Proof: Suppose $f = g$. Then $\|f - g\| = 0$. Consider the inner product $\langle f - g, \varphi \rangle$. By Cauchy–Schwarz,

$$|\langle f - g, \varphi \rangle| \leq \|f - g\|\,\|\varphi\| = 0.$$

Since the magnitude of the inner product is bounded above by zero, it is zero, and by the linearity of the inner product, $\langle f - g, \varphi \rangle = \langle f, \varphi \rangle - \langle g, \varphi \rangle = 0$, so $\langle f, \varphi \rangle = \langle g, \varphi \rangle$.

Now suppose that $\langle f, \varphi \rangle = \langle g, \varphi \rangle$ for all $\varphi \in V$. Then

$$\langle f, \varphi \rangle - \langle g, \varphi \rangle = \langle f - g, \varphi \rangle = 0,$$

that is, the function $f - g$ is orthogonal to *every* vector φ in the space. The only element of V that has this distinction is the zero vector (and, for functions, those that are equal to zero almost everywhere). So $f - g = 0$, therefore $f = g$. ∎

The theorem implies that a function f may be characterized not by its values (a table of x vs. $f(x)$), but by its inner products with all the functions in some space (a table of φ vs. $\langle f, \varphi \rangle$). We can think of f as specifying a mapping not from $\mathbb{R}$ to $\mathbb{C}$, but from V to $\mathbb{C}$. The functions $\varphi \in V$ "test" f and produce a characteristic set of numbers according to the inner product $\langle f, \varphi \rangle$. If another function g makes the same

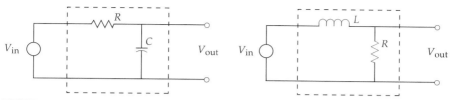

FIGURE 6.10 On the basis of sinusoidal test inputs these two circuits are indistinguishable. The frequency response of the RC circuit (*left*) is $V_{\text{out}}(\nu)/V_{\text{in}}(\nu) = 1/(1 + i2\pi\nu RC)$, and the frequency response of the RL circuit (*right*) is $V_{\text{out}}(\nu)/V_{\text{in}}(\nu) = 1/(1 + i2\pi\nu R/L)$.

inner products as f when it is applied to the same set of "testing functions" φ, then g is considered to be equivalent to f.

This notion of equivalence has an analogy in system theory. If you have a "black box" containing an unknown assembly of components, you can characterize it by applying test inputs, for example, with a sine wave generator and measuring the magnitude and phase of the outputs. On the basis of your measurements, you might hypothesize that the box contains an RC circuit with a one-second time constant $RC = 1$. But it could also be an RL circuit with time constant $L/R = 1$ (Figure 6.10). There are other ways to tell the circuits apart, but if they respond identically to the set of sinusoidal test inputs, they must be considered equivalent with respect to those test inputs.

This new approach—characterizing a mapping by its action on a set of testing functions—provides common ground for ordinary functions and the delta function. We can diagram the relationship as follows:

Ordinary functions	Delta function
Convergent sequence, $f_n \to f$	Divergent sequence, δ_n
Convergent sequence, $\int_{-\infty}^{\infty} f_n(x)\,\varphi(x)\,dx$	Convergent sequence, $\int_{-\infty}^{\infty} \delta_n(x)\,\varphi(x)\,dx$
Limit is ordinary integral, $\int_{-\infty}^{\infty} f(x)\,\varphi(x)\,dx$	Limit is symbolic, $\int \delta(x)\,\varphi(x)\,dx$

When we write, for an ordinary function,

$$\underbrace{\int_{-\infty}^{\infty} f(x)\,\varphi(x)\,dx}_{\text{ordinary integral}} = \lim_{n \to \infty} \underbrace{\int_{-\infty}^{\infty} f_n(x)\,\varphi(x)\,dx}_{\text{ordinary integral}}$$

we have actual integrals of ordinary functions on both sides of the expression.[5] When we write $\int \delta(x)\,\varphi(x)\,dx$ (with or without limits of integration), it is not an actual

---
[5] Actually, one must prove that one can take the limit under the integral sign. More about this later.

integral, but a convenient symbol for the limit of a sequence of integrals:

$$\underbrace{\int \delta(x)\,\varphi(x)\,dx}_{\text{symbol}} = \lim_{n \to \infty} \underbrace{\int_{-\infty}^{\infty} \delta_n(x)\,\varphi(x)\,dx}_{\text{ordinary integral}} = \varphi(0).$$

In both cases—ordinary function and delta function—the underlying model is a sequence of ordinary functions that produce a convergent sequence of integrals, a mapping not from numbers to numbers, but from testing functions to numbers.

A mapping from functions to numbers, a "function of a function," is called a *functional*.[6] We may write $T[\varphi]$ to denote the number obtained by operating on an input function φ with a functional T. What we call the delta function is actually the delta functional, defined $T_\delta[\varphi] = \varphi(0)$. An important category of functionals consists of those that are generated by inner products:

$$T[\varphi] = \int_{-\infty}^{\infty} f(x)\,\varphi(x)\,dx.$$

Such functionals are linear, in that $T[a\varphi_1 + b\varphi_2] = aT[\varphi_1] + bT[\varphi_2]$, for input functions φ_1, φ_2 and constants a, b. They are also bounded (by Hölder's inequality) and continuous, meaning that if (φ_k) is a sequence of functions converging to zero, then the sequence $(T[\varphi_k])$ also converges to zero.[7] One way to define a generalized function is as a linear continuous functional on a space of testing functions. This is a little abstract for our purposes, so we shall follow a different path similar to our development of the delta function in the preceding section.

6.3.2 Generalized Functions as Sequences of Functions

Specifying a set of testing functions is the first step in building a theory of generalized functions. One function space in particular is best suited for developing generalized functions for Fourier analysis. They are called *Schwartz functions* or (quaintly) *good functions*. These and three other important function classes are defined next. Table 6.2 gives some common examples of these classes.

Definition 6.1 (Rapid descent, slow growth, fairly good, and good functions). Let $f : \mathbb{R} \to \mathbb{C}$ be a function.

(a) f has *rapid descent*, or is *rapidly decreasing*, if it and all its derivatives decay faster than $(1 + x^2)^{-N}$ as $|x| \to \infty$, for all N.

(b) f is of *slow growth* or *slowly increasing* if it can be made integrable (L^1) by dividing it by a polynomial, for example,

$$\int_{-\infty}^{\infty} \frac{|f(x)|\,dx}{(1 + x^2)^N} < \infty, \quad N > 0.$$

---

[6]Champeney, 1987, pp. 119–122; Folland, 1992, pp. 304–308.

[7]Properly, the *Riesz representation theorem* states that any linear continuous functional has the form of an inner product, not the other way around as we have stated here. Boundedness and continuity can be shown to go together; a linear functional that is bounded is also continuous, and vice versa.

TABLE 6.2 Examples of slow growth, fairly good, and good functions. The classes are nested: good ⊂ fairly good ⊂ continuous and slow growth ⊂ slow growth. x^{-1} has polynomial decay but a nonintegrable singularity at the origin. It may be treated as a generalized function. The last entries, e^x, sinh x, and cosh x, exhibit rapid growth.

Function	$\mathcal{K}$ Slow growth	$\mathcal{C} \cap \mathcal{K}$ Continuous, SG	Fairly good	$\mathcal{S}$ Good
$e^{-\pi x^2}$	•	•	•	•
$e^{-1/(1-x^2)}\mathrm{rect}(x/2)$	•	•	•	•
polynomials $P(x)$	•	•	•	
$\frac{P(x)}{Q(x)}$, Q has no real roots	•	•	•	
e^{ix}, $\sin x$, $\cos x$	•	•	•	
$x^n e^{ix}$	•	•	•	
sinc x	•	•	•	
Bandlimited functions	•	•	•	
$e^{i\pi x^2}$	•	•	•	
$e^{-|x|}$	•	•		
$\Lambda(x)$	•	•		
$x \log |x|$	•	•		
$\log |x|$	•			
$|x|^{-1/2}$	•			
$U(x)$, sgn x, rect x	•			
x^{-1}				
e^x, sinh x, cosh x				

(c) f is a *fairly good function* if it is infinitely continuously differentiable, and if it and all its derivatives are slowly growing.

(d) f is a *good function* if it is infinitely continuously differentiable and rapidly decreasing. Good functions are also called *Schwartz functions*.

Various properties of these function classes are readily established. The point of these relationships is that routine operations performed on good functions do not affect their "goodness," and that "less good" functions are improved by operating on them with good functions. Proofs are left to the problems.

- The set of all slow growth functions is a vector space (denoted $\mathcal{K}$).
- The set of all good functions is a vector space (denoted $\mathcal{S}$).
- The derivative of a good function is good.
- The sum, product, or convolution of two good functions is good.
- The product of a fairly good function and a good function is good.
- The product of a slow growth function and a good function is bounded, absolutely integrable, and goes to zero as $|x| \to \infty$.
- The convolution of a rapidly decreasing function and a good function is good.
- The convolution of a slow growth function and a good function is infinitely continuously differentiable.

It should also be apparent that good functions belong to L^1 and L^2 (in fact, $S \subset L^p$ for all $p \geq 1$), and that $L^1 \subset \mathcal{K}$.

We shall choose the good functions S for our testing functions.[8] They are so smooth and decay so rapidly that we can shift, dilate, differentiate, so on, with impunity. Moreover, the Fourier transform of a good function is good. To see why, recall the intimate connection between a function's smoothness and the asymptotic behavior of its Fourier transform. If a function is r-times differentiable, its Fourier transform decays $O(|\nu|^{-(r+1)})$. (Think: rect $\longmapsto$ sinc, $\Lambda \longmapsto$ sinc^2, and Gaussian $\longmapsto$ Gaussian.) If a function is infinitely continuously differentiable ($C^{(\infty)}$), its transform will decay faster than any polynomial, that is, be rapidly decreasing. Conversely, the Fourier transform of a rapidly decreasing function is $C^{(\infty)}$. A good function possesses both of these complementary attributes, and so does its Fourier transform.

Generalized functions are defined using good functions, starting with the following.

Definition 6.2 (Generalized function). A sequence $(g_n)_{n=0}^{\infty}$ of good functions is *regular* if the sequence of numbers (b_n), where

$$b_n = \int_{-\infty}^{\infty} g_n(x)\, \varphi(x)\, dx,$$

converges for all good testing functions φ. A *generalized function* g is a regular sequence of good functions (g_n). The symbol $\int g(x)\, \varphi(x)\, dx$ is defined to be the limit of the sequence of integrals:

$$\int g(x)\, \varphi(x)\, dx = \lim_{n \to \infty} \int_{-\infty}^{\infty} g_n(x)\, \varphi(x)\, dx.$$

The integrals $\int_{-\infty}^{\infty} g_n(x)\, \varphi(x)\, dx$ will exist because the g_n are good, so the product $g_n \varphi$ is good and therefore integrable. The question of regularity concerns the convergence of the sequence of values that come out of the integrals. Not every such sequence is convergent. For example, the sequence of Gaussian pulses $(n \exp(-\pi n^2 x^2))$ have unit area, and the sequence of integrals with testing functions converges to $\varphi(0)$. It is a regular sequence and defines the generalized function $\delta(x)$. On the other hand, the functions in the sequence $(n^2 \exp(-\pi n^2 x^2))$ have areas proportional to n. Although they are also good functions, their integrals with testing functions, equal to $n\varphi(0)$, grow without bound as $n \to \infty$. The sequence does not define a generalized function.

The sequence (g_n) may also converge to an ordinary function, pointwise or in norm. If it does, we say it defines a *regular* generalized function. If it does not (e.g., the delta function), we say the generalized function is *singular*. It can be shown that all ordinary functions $f \in L^p$, $\infty > p \geq 1$, and all slowly growing functions are

---

[8] Generalized functions based on good testing functions are also called *tempered distributions, temperate distributions*, or *distributions of slow growth*. The set of all tempered distributions is called S' and can be shown to be a vector space. In this text when we say "generalized function" we mean only "tempered distribution." Other classes of generalized function using different testing functions are described in several places, including Zemanian (1987), Folland (1992), Gasquet and Witomski (1999), and Champeney (1987).

regular generalized functions.[9] Thus, generalized functions encompass the ordinary functions we worked with in earlier chapters, particularly those in L^1 and L^2, together with slowly growing functions and the impulses.

The notion of convergence in Definition 6.2 can be applied to more than sequences of good functions.

Definition 6.3 (Weak convergence[10]). A sequence of generalized functions (f_n) is said to be *weakly convergent* if the sequence of numbers (b_n), where

$$b_n = \int f_n(x)\,\varphi(x)\,dx,$$

is convergent for all $\varphi \in S$. The sequence (f_n) is said to converge weakly to a generalized function f if

$$\int f(x)\,\varphi(x)\,dx = \lim_{n \to \infty} \int f_n(x)\,\varphi(x)\,dx.$$

For example, both δ and δ' can be represented by good (Gaussian-based) sequences, but we have also seen that a delta function can be represented by a weakly convergent sequence of rectangles, and the derivative of a delta function can be represented by a weakly convergent sequence of delta functions, $\delta'_n(x) = n\delta(x + 1/2n) - n\delta(x - 1/2n)$.

6.3.3 Calculus of Generalized Functions

The expansion from ordinary functions to generalized functions necessitates a rethinking of how to do calculus. This section shows how the familiar ways of manipulating functions are redefined for generalized functions.

Equivalence

Because a generalized function is defined by its action on testing functions, two generalized functions that act the same way for all testing functions are considered to be equivalent.

Definition 6.4 (Equivalence of generalized functions). Let f and g be generalized functions. If

$$\int f(x)\,\varphi(x)\,dx = \int g(x)\,\varphi(x)\,dx,$$

for all testing functions $\varphi \in S$, they are *equivalent*, and we write $f = g$.

---

[9] See Lighthill (1958, pp. 22–23) and Strichartz (1994, pp. 92–95) for a way to construct an approximating sequence for an ordinary function. The basic idea is to smear out the function by convolution with a unit area good function $n\rho(nx)$, then multiply the result by a second good function $\psi(x/n)$, where $\psi(0) = 1$. The convolution gives infinite continuous differentiability, and the product gives rapid descent. As $n \to \infty$, $n\rho(nx)$ becomes sharper, $\psi(x/n)$ flattens out, and the sequence converges to the original function.

[10] What we are calling weak convergence is more properly called *temperate convergence*, because our generalized functions are tempered distributions. The idea of weak convergence is applicable to generalized functions defined over other spaces of testing functions besides S. See Folland (1992, pp. 314, 334).

Example 6.9. Let f and g be generalized functions defined by the sequences $f_n(x) = n\exp(-\pi n^2 x^2)$ and $g_n(x) = \frac{n}{2}\operatorname{sech}^2(nx)$. Both sequences have delta function behavior:

$$\int_{-\infty}^{\infty} n\exp\left(-\pi n^2 x^2\right) \varphi(x)\,dx \to \varphi(0)$$

$$\int_{-\infty}^{\infty} \frac{n}{2}\operatorname{sech}^2(nx)\,\varphi(x)\,dx \to \varphi(0)$$

so we say $f = g = \delta$. ∎

For ordinary functions f and g, in addition to equivalence we have scaling (af), addition ($f + g$), multiplication (fg), shifting ($f(x - b)$), dilation ($f(ax)$), even and odd symmetry, real and imaginary parts, and differentiation (f'). What we need to do now is see how these ideas extend to generalized functions. An operation on a generalized function is always consistent with the operation on an ordinary function, via the generalized definition of equivalence.

Scaling, Adding, Shifting, Dilating

Generalized functions can be scaled, added, shifted, and dilated. First, consider scaling. If f is an ordinary function, then af is defined by $(af)(x) = a \cdot f(x)$—just take the values of $f(x)$ and multiply them by a. We cannot do this for a generalized function, because it may not have nice values (think: $\delta(0) = ?$). Instead, we develop a consistent definition by replacing the generalized function by a sequence of good functions, working with the sequence of ordinary integrals, and then taking a limit to get back to a generalized function:

$$\int af(x)\,\varphi(x)\,dx = \lim_{n\to\infty}\int_{-\infty}^{\infty} af_n(x)\,\varphi(x)\,dx.$$

Because f_n is an ordinary function, we can transfer the scaling factor from f_n to φ:

$$\int_{-\infty}^{\infty} f_n(x)\,a\varphi(x)\,dx.$$

The scaled testing function $a\varphi$ is still a good function, so we know the sequence of integrals will converge to

$$\int f(x)\,a\varphi(x)\,dx.$$

Thus we have the definition of a scaled generalized function:

$$\int af(x)\,\varphi(x)\,dx = \int f(x)\,a\varphi(x)\,dx, \tag{6.17}$$

that is, af is the generalized function which operates on a testing function φ to produce the same result as would be obtained by operating on a scaled testing function $a\varphi$ with the generalized function f.

6.3 GENERALIZED FUNCTIONS

The method used may be diagrammed like this:

$$\text{Generalized} \quad \int af(x)\,\varphi(x)\,dx = \int f(x)\,a\varphi(x)\,dx$$

$$\downarrow \text{sequence} \qquad\qquad \uparrow n\to\infty$$

$$\text{Ordinary} \quad \int_{-\infty}^{\infty} af_n(x)\,\varphi(x)\,dx = \int_{-\infty}^{\infty} f_n(x)\,a\varphi(x)\,dx$$

The result for generalized functions is obtained by passing through the realm of ordinary functions. Applying the same method to addition, shifting, and dilation yields

$$\int (f+g)(x)\,\varphi(x)\,dx = \int f(x)\,\varphi(x)\,dx + \int g(x)\,\varphi(x)\,dx, \tag{6.18}$$

$$\int f(x-b)\,\varphi(x)\,dx = \int f(x)\,\varphi(x+b)\,dx, \tag{6.19}$$

$$\int f(ax)\,\varphi(x)\,dx = \int f(x)\,\frac{1}{|a|}\varphi\left(\frac{x}{a}\right)dx. \tag{6.20}$$

Each of these seems intuitive—what works for generalized functions must also work for ordinary functions, and conversely, a good clue to what works for generalized functions is to consider what works for ordinary functions. In many cases, we may manipulate generalized functions as though they were ordinary functions, without explicitly invoking an underlying sequence of ordinary functions. When in doubt, though, one may always resort to a sequence to be sure.

Example 6.10. Suppose that we want to know what a scaled delta function does. We cannot talk about values of $a\delta(x)$, because δ does not have values in the ordinary sense. So we use the definition (Equation 6.17):

$$\int a\delta(x)\,\varphi(x)\,dx = \int \delta(x)\,a\varphi(x)\,dx.$$

The definition pushes the factor a off the δ and onto the testing function φ, an ordinary function that we know how to scale. The result is

$$\int a\delta(x)\,\varphi(x)\,dx = \int \delta(x)\,a\varphi(x)\,dx = a\varphi(0).$$

That is, $a\delta$ is the generalized function that maps the testing function φ to $a\varphi(0)$. ∎

Example 6.11. Show that $2\delta(x) + \delta(x) = 3\delta(x)$.

$$\int (2\delta(x) + \delta(x))\varphi(x)\,dx = \int 2\delta(x)\,\varphi(x)\,dx + \int \delta(x)\,\varphi(x)\,dx$$
$$= 2\varphi(0) + \varphi(0) = 3\varphi(0).$$

This is the same result as $\int 3\delta(x)\,\varphi(x)\,dx$, so by the definition of equivalence, $2\delta(x) + \delta(x) = 3\delta(x)$. ∎

Example 6.12 (Shifted delta function).

$$\int \delta(x-1)\,\varphi(x)\,dx = \int \delta(x)\,\varphi(x+1)\,dx = \varphi(1).$$

$\delta(x-1)$ is the generalized function that maps φ to $\varphi(1)$. ∎

Example 6.13 (Dilated delta function). Earlier we argued that $\delta(ax) = \frac{1}{|a|}\delta(x)$ (Equation 6.7). Now we will get it from the definition (6.20):

$$\int \delta(ax)\,\varphi(x)\,dx = \int \delta(x)\,\frac{1}{|a|}\varphi\left(\frac{x}{a}\right)\,dx = \frac{1}{|a|}\varphi(0).$$

This is the same result as $\int \frac{1}{|a|}\delta(x)\,\varphi(x)\,dx$, so we conclude $\delta(ax) = \frac{1}{|a|}\delta(x)$. ∎

Even and Odd Generalized Functions

Like ordinary functions, a generalized function is even if $f(x) = f(-x)$ and odd if $f(x) = -f(-x)$. To understand what $f(-x)$ means for a generalized function, use the definition of dilation with $a = -1$:

$$\int f(-x)\,\varphi(x)\,dx = \int f(x)\,\varphi(-x)\,dx.$$

This leads to generalized definitions for even and odd in terms of actions on testing functions:

Even $$\int f(x)\,\varphi(x)\,dx = \int f(x)\,\varphi(-x)\,dx \qquad (6.21\text{a})$$

Odd $$\int f(x)\,\varphi(x)\,dx = -\int f(x)\,\varphi(-x)\,dx \qquad (6.21\text{b})$$

That is, f is even if it acts identically on a testing function and a reversed version of the testing function. It is odd if acts identically except for a sign flip.

Example 6.14. The Fourier transforms for $\sin 2\pi x$ and $\cos 2\pi x$ are

$$\sin 2\pi x \longmapsto \frac{1}{2i}[\delta(\nu-1) - \delta(\nu+1)],$$

$$\cos 2\pi x \longmapsto \frac{1}{2}[\delta(\nu-1) + \delta(\nu+1)].$$

The transform of the cosine is even, because

$$\int \frac{1}{2}[\delta(\nu-1) + \delta(\nu+1)]\,\varphi(\nu)\,d\nu = \frac{1}{2}[\varphi(1) + \varphi(-1)]$$

$$\int \frac{1}{2}[\delta(\nu-1) + \delta(\nu+1)]\,\varphi(-\nu)\,d\nu = \frac{1}{2}[\varphi(-1) + \varphi(1)].$$

Verification that the transform of the sine is odd is left to the reader. The symmetry can also be seen in the graphs in Figure 6.9. ∎

Real and Imaginary Generalized Functions

Generalized functions can also be real, imaginary, or complex. A simple example of an imaginary generalized function is $\frac{1}{2i}\delta(v+1) - \frac{1}{2i}\delta(v-1)$, the Fourier transform of $\sin 2\pi x$. In terms of action on a testing function, a generalized function f is real if

$$\int f(x)\,\varphi(x)\,dx = \left[\int f(x)\,\varphi^*(x)\,dx\right]^*, \qquad (6.22a)$$

and imaginary if

$$\int f(x)\,\varphi(x)\,dx = -\left[\int f(x)\,\varphi^*(x)\,dx\right]^*. \qquad (6.22b)$$

The derivation is left to the problems.

Example 6.15. Using the definition, show that $\frac{1}{2i}\delta(v+1) - \frac{1}{2i}\delta(v-1)$ is imaginary (this is obvious by inspection, but it will do for demonstrating how the definition works).

$$\int \left[\frac{1}{2i}\delta(v+1) - \frac{1}{2i}\delta(v-1)\right]\varphi(x)\,dx = \frac{1}{2i}(\varphi(-1) - \varphi(1))$$

$$-\left[\int \left(\frac{1}{2i}\delta(v+1) - \frac{1}{2i}\delta(v-1)\right)\varphi^*(x)\,dx\right]^* = -\left[\frac{1}{2i}(\varphi^*(-1) - \varphi^*(1))\right]^*$$

$$= \frac{1}{2i}(\varphi(-1) - \varphi(1)) \qquad \blacksquare$$

Differentiation

Let f be a generalized function defined by a regular sequence of good functions (f_n). The derivatives f_n' are also good, and we will define the generalized derivative by means of this sequence. Integrate f_n' against a testing function:

$$\int_{-\infty}^{\infty} f_n'(x)\,\varphi(x)\,dx.$$

This is an ordinary integral and may be integrated by parts:

$$\int_{-\infty}^{\infty} f_n'(x)\,\varphi(x)\,dx = f_n\,\varphi\Big|_{-\infty}^{\infty} - \int_{-\infty}^{\infty} f_n(x)\,\varphi'(x)\,dx.$$

Since f_n and φ are both good, so is $f_n\varphi$, therefore (rapidly decreasing) $f_n\varphi \to 0$ as $|x| \to \infty$, leaving just the second term on the right. The derivative φ', being a good function, is just another testing function. The sequence of integrals on the right is, therefore, convergent, showing that the sequence of integrals on the left is convergent. Taking limits on both sides, we have a definition for the generalized derivative f':

$$\int f'(x)\,\varphi(x)\,dx = -\int f(x)\,\varphi'(x)\,dx. \qquad (6.23)$$

Equation 6.23 indicates that the derivative of a generalized function is another generalized function, which can always be differentiated again to yield another generalized function, for example,

$$\int f''(x)\,\varphi(x)\,dx = -\int f'(x)\,\varphi'(x)\,dx = \int f(x)\,\varphi''(x)\,dx.$$

Any number of derivatives of f are simply pushed off to derivatives of the testing function. But the testing function, being a good function, is infinitely continuously differentiable, so the integral will always be well defined. This shows that *all generalized functions possess well-defined generalized derivatives of all orders.* We are no longer blocked by discontinuities. Moreover, it can be shown that every generalized function, regular or singular, is a finite sum of derivatives of continuous, slowly growing, functions, and vice versa.[11]

Example 6.16 (Step and impulse). The function $xU(x)$, a one-sided ramp, is continuous and slowly growing, hence it is a regular generalized function. Its derivative is the unit step $U(x)$, an ordinary function and regular generalized function defined by the integral

$$\int U(x)\varphi(x)\,dx = \int_0^\infty \varphi(x)\,dx.$$

Then, using Equation 6.23,

$$\int U'(x)\varphi(x)\,dx = -\int U(x)\varphi'(x)\,dx$$

$$= -\int_0^\infty \varphi'(x)\,dx = -\varphi(x)\Big|_0^\infty = \varphi(0).$$

Because U' has the same action as a delta function, $\int \delta(x)\,\varphi(x)\,dx = \varphi(0)$, we conclude (again) that $\delta(x) = U'(x)$ and see, as well, that $\delta(x) = \frac{d^2}{dx^2} xU(x)$. ∎

Example 6.17 (Derivatives of the delta function). Earlier we calculated the derivative of the delta function by differentiating a rectangular pulse sequence. There is now a more direct way, using Equation 6.23:

$$\int \delta'(x)\,\varphi(x)\,dx = -\int \delta(x)\,\varphi'(x)\,dx = -\varphi'(0). \tag{6.24}$$

While δ sifts out the value of φ at the origin, δ' sifts out the value of the derivative φ' at the origin. The definition of generalized derivative can be applied any number of times, that is,

$$\int \delta^{(r)}(x)\varphi(x)\,dx = -\int \delta^{(r-1)}(x)\varphi'(x)\,dx$$

$$\vdots$$

$$= (-1)^r \int \delta(x)\varphi^{(r)}(x)\,dx$$

$$= (-1)^r \varphi^{(r)}(0), \tag{6.25}$$

and we see that $\delta^{(r)}$ sifts out the value of the rth derivative of φ at the origin. ∎

---

[11] Strichartz (1994, pp. 77–80).

6.3 GENERALIZED FUNCTIONS 395

Example 6.18 (Derivative of $|x|$). Because $|x|$ has a corner at $x = 0$, its derivative at $x = 0$ is undefined by ordinary calculus. Treating $|x|$ as a generalized function removes this restriction:

$$\int |x|' \varphi(x)\,dx = -\int_{-\infty}^{\infty} |x|\, \varphi'(x)\,dx$$

$$= -\int_{-\infty}^{0} (-x)\, \varphi'(x)\,dx - \int_{0}^{\infty} x\, \varphi'(x)\,dx.$$

Now integrate by parts:

$$= x\varphi(x)\Big|_{-\infty}^{0} - \int_{-\infty}^{0} \varphi(x)\,dx - x\varphi(x)\Big|_{0}^{\infty} + \int_{0}^{\infty} \varphi(x),$$

and because $x\varphi(x)$ is rapidly decreasing,

$$= -\int_{-\infty}^{0} \varphi(x)\,dx + \int_{0}^{\infty} \varphi(x) = \int_{-\infty}^{\infty} \operatorname{sgn} x\, \varphi(x)\,dx.$$

Thus $|x|' = \operatorname{sgn}$, and because $\operatorname{sgn}(0) = 0$, we may reasonably define the derivative of $|x|$ to be zero at the origin. ∎

Example 6.19 (x^{-1} as a generalized function). The function $f(x) = 1/x$ is singular at the origin and not integrable in the ordinary sense, except over bounded intervals $[a, b]$ not including the origin. It is the derivative of $\log |x|$, and $\log |x|$ can be shown to be an ordinary function of slow growth (see the problems), which is itself the derivative of the continuous, slowly growing function $x \log |x| - x$. Let us therefore try to define $1/x$ as the generalized derivative of $\log |x|$. Apply the definition of generalized derivative (Equation 6.23):

$$\int \left(\frac{d}{dx} \log |x|\right) \varphi(x)\,dx = -\int \log |x|\, \varphi'(x)\,dx$$

$$= -\lim_{\epsilon \to 0} \left[\int_{-\infty}^{-\epsilon} \log |x|\, \varphi'(x)\,dx + \int_{\epsilon}^{\infty} \log |x|\, \varphi'(x)\,dx\right],$$

since $\log |x|$ is an ordinary function of slow growth. Integrate the first term by parts,

$$\int_{-\infty}^{-\epsilon} \log |x|\, \varphi'(x) = \log |x|\, \varphi(x)\Big|_{-\infty}^{-\epsilon} - \int_{-\infty}^{-\epsilon} \frac{\varphi(x)}{x}\,dx$$

$$= \log(\epsilon)\varphi(-\epsilon) - \int_{-\infty}^{-\epsilon} \frac{\varphi(x)}{x}\,dx.$$

Performing the same manipulation of the second term and combining,

$$\int \log |x|\, \varphi'(x)\,dx = \lim_{\epsilon \to 0} \left[(\varphi(-\epsilon) - \varphi(\epsilon))\log(\epsilon) - \int_{-\infty}^{-\epsilon} \frac{\varphi(x)}{x}\,dx - \int_{\epsilon}^{\infty} \frac{\varphi(x)}{x}\,dx\right].$$

The testing function possesses continuous derivatives of all orders, and for small ϵ can be approximated $\varphi(\epsilon) = \varphi(0) + \varphi'(0)\epsilon + O(\epsilon^2)$. Thus $\varphi(\epsilon) - \varphi(-\epsilon) = 2\varphi'(0)\epsilon +$

$O(\epsilon^2)$, which dies faster than $\log(\epsilon)$ grows as $\epsilon \to 0$, so $(\varphi(-\epsilon) - \varphi(\epsilon))\log(\epsilon) \to 0$. We are left with

$$\int \left(\frac{d}{dx}\log|x|\right)\varphi(x)\,dx = \lim_{\epsilon \to 0}\left[\int_{-\infty}^{-\epsilon}\frac{\varphi(x)}{x}\,dx + \int_{\epsilon}^{\infty}\frac{\varphi(x)}{x}\,dx\right] = \mathcal{P}\int_{-\infty}^{\infty}\frac{\varphi(x)}{x}\,dx.$$

This Cauchy principal value integral is always defined; write

$$\int_{-\infty}^{-\epsilon}\frac{\varphi(x)}{x}\,dx + \int_{\epsilon}^{\infty}\frac{\varphi(x)}{x}\,dx = \int_{\epsilon}^{\infty}\frac{\varphi(x) - \varphi(-x)}{x}\,dx.$$

For large x, the integrand decays rapidly and for small x, the integrand approaches $2\varphi'(0)$.

We will therefore define the generalized function $\mathcal{P}(1/x) = (\log|x|)'$ via the Cauchy principal value integral:

$$\int \mathcal{P}(1/x)\varphi(x)\,dx = \mathcal{P}\int_{-\infty}^{\infty}\frac{\varphi(x)}{x}\,dx. \tag{6.26}$$

Powers of x^{-1} may also be defined by repeated differentiation of $\log|x|$.[12] ∎

The preceding calculations were conveniently done without considering any underlying sequences of functions. However, using a particular sequence can be helpful in cultivating intuition about the action of a generalized function.

Example 6.20. We can visualize δ' using the sequence of derivatives of the Gaussian sequence, $\delta_n(x) = ne^{-\pi n^2 x^2}$:

$$\delta'_n(x) = \frac{d}{dx}ne^{-\pi n^2 x^2} = -2\pi n^3 x\, e^{-\pi n^2 x^2}.$$

The sequence of functions is shown in Figure 6.11. The peaks are located at $x = \pm\frac{1}{\sqrt{2\pi n}}$, their heights are $\mp n^2\sqrt{2\pi}e^{-1/2}$, and their areas are $\mp n$, respectively. For large n you can imagine that the narrow peak at $x = -\frac{1}{\sqrt{2\pi n}}$ will pick off $n\varphi\left(-\frac{1}{\sqrt{2\pi n}}\right)$, and the peak at $x = \frac{1}{\sqrt{2\pi n}}$ will pick off $n\varphi\left(\frac{1}{\sqrt{2\pi n}}\right)$. Then,

$$\int_{-\infty}^{\infty}\delta'_n(x)\varphi(x)\,dx \approx n\varphi\left(-\frac{1}{\sqrt{2\pi n}}\right) - n\varphi\left(\frac{1}{\sqrt{2\pi n}}\right) = -\frac{\varphi\left(\frac{1}{\sqrt{2\pi n}}\right) - \varphi\left(-\frac{1}{\sqrt{2\pi n}}\right)}{1/n},$$

and this tends to $-\varphi'(0)$ as $n \to \infty$. ∎

Products of Generalized Functions

Multiplication by a delta function, $f(t)\delta(t - t_0)$, is used later in the chapter to model sampling the value of a function f at a time t_0. The key to properly handling these and other problems is to know what kinds of functions may be multiplied by generalized functions.

---

[12] See Lighthill (1958, pp. 35–40), and Gasquet and Witomski (1999, pp. 258–261) for a discussion of generalized functions based on powers x^{-p}.

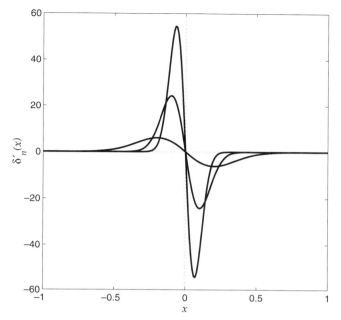

FIGURE 6.11 A sequence of pulses, derivatives of the Gaussian pulse, which serve to define the generalized derivative $\delta'(x)$.

Let f and g be generalized functions, form the product fg, and write the usual integral

$$\int fg(x)\,\varphi(x)\,dx.$$

If the product $g\varphi$ is a good function, we may push the g from fg to $g\varphi$ and have a valid integral of f against the testing function $g\varphi$:

$$\int f(x)\,g\varphi(x)\,dx.$$

(It may be easier to see this if you represent f by a sequence (f_n).) It can be shown that, for arbitrary f, this will work if the function g is fairly good (so that the product fg is good).[13]

Theorem 6.2 (Products of ordinary and generalized functions). Let $f \in S'$ be a generalized function, g be a fairly good function, and $\varphi \in S$ be a testing function. The product fg is a generalized function defined by

$$\int f(x)g(x)\,\varphi(x)\,dx = \int f(x)\,g(x)\varphi(x)\,dx. \qquad (6.27)$$

[13] Jones, 1982, pp. 164–166.

Example 6.21. Show that

$$x\delta'(x) = -\delta(x). \tag{6.28}$$

Integrate the left side against a testing function and push the factor of x from δ' to φ:

$$\int x\delta'(x)\,\varphi(x)\,dx = \int \delta'(x)\,x\varphi(x)\,dx.$$

We are justified in making this step because x is a fairly good function, so $x\varphi(x)$ is a good function. Now, apply the result for the derivative of the delta function (Equation 6.24):

$$\int \delta'(x)\,x\varphi(x)\,dx = -\int \delta(x)\,\frac{d}{dx}(x\varphi(x))dx$$
$$= -\int \delta(x)\,(\varphi(x) + x\varphi'(x))dx$$
$$= -\varphi(0) - 0 \cdot \varphi'(0) = -\varphi(0).$$

This is the same result as $\int -\delta(x)\,\varphi(x)\,dx$, proving $x\delta' = -\delta$.

We can also use the Gaussian-derived sequence $(x\delta'_n) = (-2\pi n^3 x^2\, e^{-\pi n^2 x^2})$ to define $x\delta'(x)$. This sequence is plotted in Figure 6.12. The peaks are located at $x = \pm\frac{1}{\sqrt{\pi n}}$, have height $-2ne^{-1}$, and area $-\frac{1}{2}$. In the integral $\int x\delta'_n(x)\varphi(x)\,dx$ for

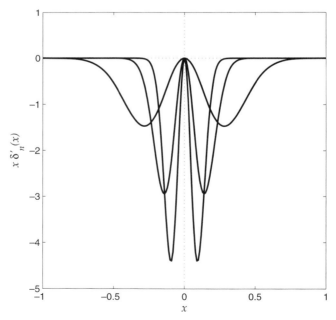

FIGURE 6.12 A sequence of pulses that serves to define the generalized function $x\delta'(x)$, which is equivalent to $-\delta(x)$.

large n, the narrow peaks will pick off $\varphi(\pm\frac{1}{\sqrt{\pi n}})$ times their areas $(-\frac{1}{2})$, giving, approximately,

$$\int_{-\infty}^{\infty} x\delta'_n(x)\varphi(x)\,dx \approx -\frac{1}{2}\varphi\left(-\frac{1}{\sqrt{\pi n}}\right) - \frac{1}{2}\varphi\left(\frac{1}{\sqrt{\pi n}}\right).$$

In the limit as $n \to \infty$, $\varphi(-\frac{1}{\sqrt{\pi n}}) = \varphi(\frac{1}{\sqrt{\pi n}}) = \varphi(0)$, and we have the result

$$\lim_{n\to\infty}\int_{-\infty}^{\infty} x\delta'_n(x)\varphi(x)\,dx = -\varphi(0),$$

the same as $\int -\delta(x)\varphi(x)\,dx$. So we see, graphically, how $x\delta'(x) = -\delta(x)$. The graphical result seems counterintuitive—a sequence of functions that are all zero at the origin is equivalent to a generalized function that is zero everywhere except the origin. ∎

Some functions are close to fairly good, for example, step and signum are slowly growing but discontinuous. We may still be able to form well-defined products with these functions by resorting to sequences.[14] A slowly growing function with a jump discontinuity can be made infinitely continuously differentiable by convolving it with a unit area good function $n\rho(nx)$; as $n \to \infty$, $g_n(x) = g(x) * n\rho(nx) \to g(x)$. Or, we can happen to know a continuous approximation, for example, $\tanh(nx)$ is fairly good and $\lim_{n\to\infty} \tanh(nx) = \text{sgn}(x)$. Then the product $g_n\varphi$ is good and the product fg is defined:

$$\int f(x)g(x)\,\varphi(x)\,dx = \lim_{n\to\infty}\int f(x)\,g_n(x)\varphi(x)\,dx,$$

when the limit exists.

Example 6.22. If f is continuous at $x = a$, then

$$f(x)\delta(x-a) = f(a)\delta(x-a). \tag{6.29}$$

This is the sifting property, without an integral to smooth out the impulse. To prove it, set up the integral

$$\int f(x)\delta(x-a)\,\varphi(x)\,dx,$$

and push f over to φ:

$$\int f(x)\delta(x-a)\,\varphi(x)\,dx = \int \delta(x-a)\,f(x)\varphi(x)\,dx.$$

If f is everywhere continuous and slowly growing, then $f\varphi$ is good and we have immediately

$$\int \delta(x-a)\,f(x)\varphi(x)\,dx = f(a)\varphi(a).$$

[14] Champeney, 1987, p. 142.

This is the same result obtained from $\int f(a)\delta(x-a)\varphi(x)\,dx$. If f is continuous at $x = a$ but has jumps elsewhere, a sequence of fairly good functions could be constructed, $f_n(x) = f(x) * n\rho(nx)$. Then

$$\int \delta(x-a) f_n(x) \varphi(x)\,dx = f_n(a)\varphi(a) \to f(a)\varphi(a),$$

but this really is more than needed, because the delta function acts only at the one point $x = a$ and ignores what the function is doing away from $x = a$.

Sample calculations using this rule are $x\delta(x) = 0$ and $\sin(\pi x)\,\delta\!\left(x - \tfrac{1}{4}\right) = \sin\!\left(\tfrac{\pi}{4}\right)\delta\!\left(x - \tfrac{1}{4}\right) = \tfrac{1}{\sqrt{2}}\delta\!\left(x - \tfrac{1}{4}\right)$.

Equation 6.29 can be extended. If f is continuously differentiable at $x = a$, then

$$f(x)\,\delta'(x-a) = f(a)\delta'(x-a) - f'(a)\delta(x-a). \tag{6.30}$$

Proof (and further generalization) is left to the problems. ∎

Example 6.23. Show that $x\mathcal{P}(1/x)$ is a generalized function equivalent to 1.

Write down the integral, and perform the usual maneuver of pushing x over to φ:

$$\int x\mathcal{P}(1/x)\,\varphi(x)\,dx = \int \mathcal{P}(1/x)\,x\varphi(x)\,dx.$$

Now apply the definition of $\mathcal{P}(1/x)$ (Equation 6.26):

$$\int \mathcal{P}(1/x)\,x\varphi(x)\,dx = \mathcal{P}\int_{-\infty}^{\infty} \frac{x\varphi(x)}{x}\,dx.$$

We can cancel the xs top and bottom because the principal value integration excludes the origin until the final limit is taken. Thus,

$$\mathcal{P}\int_{-\infty}^{\infty} \frac{x\varphi(x)}{x}\,dx = \mathcal{P}\int_{-\infty}^{\infty} \varphi(x)\,dx = \int_{-\infty}^{\infty} \varphi(x)\,dx.$$

This is the same result as $\int 1 \cdot \varphi(x)\,dx$, so we conclude that $x\mathcal{P}(1/x) = 1$. ∎

Not every product can be defined, even with sequences. Here is a notable failure.

Example 6.24 (δ^2 is undefined). If we attempt to define $\delta^2 = \delta \cdot \delta$ and apply it to a testing function, we obtain

$$\int (\delta(x) \cdot \delta(x))\,\varphi(x)\,dx = \int \delta(x)\,(\delta(x)\varphi(x))\,dx.$$

But $\delta(x)\varphi(x) = \varphi(0)\delta(x)$ is not a good function, so the attempt fails. The first impulse wants to sift out the value of the second impulse at the origin, but this makes no sense.

To see this in a different way, let the delta function be defined by the sequence of functions $\delta_n(x) = ne^{-\pi n^2 x^2}$. The functions $\delta_n^2(x) = n^2 e^{-2\pi n^2 x^2}$ have area $\frac{n}{\sqrt{2}}$, which grows without bound. The sequence of integrals

$$\int_{-\infty}^{\infty} \delta_n^2(x) \varphi(x) \, dx$$

does not converge. ∎

Convolution of Generalized Functions

Convolution is important, of course, because of the need to model the passage of signals, including sines and cosines, through linear time-invariant systems. To arrive at a consistent definition for the convolution of two generalized functions, consider the integral with a testing function:

$$\int f * g(x) \, \varphi(x) \, dx = \int \left[\int f(\xi) g(x - \xi) \, d\xi \right] \varphi(x) \, dx,$$

and formally change the order of integration:

$$= \int f(\xi) \left[\int g(x - \xi) \, \varphi(x) \, dx \right] d\xi.$$

This is a meaningful expression if the inner integral works out to a good function in ξ. The integral is the convolution $g(-) * \varphi$ of a reversed version of g with φ. So, g must be a generalized function which, when convolved with a good function, results in a good function.[15]

Theorem 6.3 (Convolution of generalized functions). Let f, g be generalized functions. If, for one of these (say g), $g * \varphi \in \mathcal{S}$, then the convolution $f * g$ is defined as a generalized function such that

$$\int f * g(x) \, \varphi(x) \, dx = \int f(x) \left[\int g(\xi - x) \varphi(\xi) \, d\xi \right] dx$$

$$= \int f(x) \left[\int g(\xi) \varphi(\xi + x) \, d\xi \right] dx. \qquad (6.31)$$

It is readily shown that convolution commutes, $f * g = g * f$. It can also be shown that convolution obeys associativity, $(f * g) * h = f * (g * h)$, under certain conditions.[16] The existence of $f * g$ as a generalized function guarantees the existence of a generalized derivative that follows the usual formula, $(f * g)' = f' * g = f * g'$, and may be repeated to calculate derivatives of all orders. Moreover, it can be shown that the convolution of a generalized function with a good function is infinitely continuously differentiable (see the problems). So again, we see that convolution is

---

[15] If $g * \varphi$ is good, then so is $g(-) * \varphi$. Regarding convolution of generalized functions, see Champeney (1987, pp. 139–144) and Gasquet and Witomski (1999, pp. 297ff).

[16] See Champeney (1987, p. 143) and Gasquet and Witomski (1999, pp. 306–307). Also, see the problems for some examples.

a smoothing operation, and good functions are capable of a remarkable degree of repair work on ill-behaved functions.

Example 6.25 (Convolution with a delta function). Consider Equation 6.31 with $f = \delta(x - a)$. The inner integral is

$$\int f(\xi)\, \varphi(\xi + x)\, d\xi = \int \delta(\xi - a)\, \varphi(\xi + x)\, d\xi = \varphi(x + a),$$

which is a good function. The outer integral then gives

$$\int g(x)\, \varphi(x + a)\, dx = \int g(x - a)\, \varphi(x)\, dx.$$

This shows that, for *any* generalized function g, regular or singular, $\delta(x - a) * g(x) = g(x - a)$. We are not restricted to continuous functions, as in the earlier derivation (Equation 6.16). ∎

Example 6.26 (Some more generalized convolutions). Verification of most of these relationships is left to the problems.

1. Two delta functions can be convolved:

$$\int (\delta(x - a) * \delta(x - b))\, \varphi(x)\, dx = \int \delta(x - a)\,(\delta(-x - b) * \varphi(x))\, dx$$

$$= \int \delta(x - a)\,(\delta(x + b) * \varphi(x))\, dx$$

$$= \int \delta(x - a)\, \varphi(x + b)\, dx = \varphi(a + b).$$

This result would also be obtained from $\int \delta(x - (a + b))\, \varphi(x)\, dx$, therefore

$$\delta(x - a) * \delta(x - b) = \delta(x - (a + b)). \tag{6.32}$$

Convolution with a delta function shifts. In the convolution $\delta(x - a) * \delta(x - b)$, the first delta function shifts the second one by a, from $x = b$ to $x = a + b$.

2. Convolution with the derivative of a delta function:

$$\delta' * f = \delta * f' = f', \tag{6.33a}$$
$$\delta^{(r)} * f = f^{(r)}. \tag{6.33b}$$

3. A function of slow growth can be convolved with a rapidly decreasing function, for example, $\cos 2\pi x * e^{-x} U(x)$ (this is the model of a sine wave passing through a first-order LTI system).

4. A function of slow growth can be convolved with a bounded function of bounded support, for example, $\cos 2\pi x * \operatorname{rect} x$, $x^2 * \Lambda(x)$.

5. Two polynomials (including constant functions) cannot be convolved, two sinusoids cannot be convolved, a polynomial cannot be convolved with a sinusoid, so on. ∎

One-sided ordinary functions are important because of their connection with causal linear systems. For example, consider the convolution of two step functions, $U(x) * U(x)$. A straightforward calculation yields

$$U * U = \int_{-\infty}^{\infty} U(\xi) U(x - \xi) \, d\xi = \begin{cases} 0, & x < 0 \\ \int_0^x d\xi = x, & x \geq 0 \end{cases},$$

$$= xU(x),$$

and by inspection this is slowly growing, hence it is a generalized function. However, this example falls outside the view of Theorem 6.3, for

$$U * \varphi = \int_{-\infty}^{x} \varphi(\xi) \, d\xi$$

is not guaranteed to be a good function of x (while it has sufficient differentiability, it may not be rapidly decreasing as $x \to \infty$). We can get around this problem by writing U as the limit of a rapidly decreasing sequence of functions, for example, $e^{-x/n}U(x)$. The convolution $e^{-x/n}U(x) * \varphi(x)$ is good, and Theorem 6.3 says that the convolution

$$U(x) * e^{-x/n}U(x) = n(1 - e^{-x/n})U(x)$$

is a generalized function. The sequence converges weakly to $xU(x)$, as desired.[17]

$$\lim_{n \to \infty} \int_{-\infty}^{\infty} n\left(1 - e^{-x/n}\right) U(x) \varphi(x) \, dx = \int_{-\infty}^{\infty} \lim_{n \to \infty} n\left(1 - e^{-x/n}\right) U(x) \varphi(x) \, dx$$

$$= \int_{-\infty}^{\infty} xU(x) \varphi(x) \, dx.$$

In the more general case of two one-sided slowly growing functions, let them both be approximated by sequences f_n and g_n of rapidly decreasing one-sided functions, for example, by

$$g_n(x) = g(x) e^{-x/n} U(x).$$

Then their convolution $f_n * g_n$ will be a generalized function, according to Theorem 6.3. The weak limit of the sequence, if it exists, is a generalized function and defined to be the convolution[18]

$$f * g = \lim_{n \to \infty} f_n * g_n.$$

---

[17] One must justify taking the limit under the integral sign. If a sequence of ordinary functions (f_n) converges pointwise to an ordinary function f and $|f_n(x)| \leq |g(x)|$, where g is a function of slow growth (we say that f_n is *dominated* by g), then

$$\lim_{n \to \infty} \int_{-\infty}^{\infty} f_n(x) \varphi(x) \, dx = \int_{-\infty}^{\infty} \lim_{n \to \infty} f_n(x) \varphi(x) \, dx = \int_{-\infty}^{\infty} f(x) \varphi(x) \, dx.$$

In this case $n(1 - e^{-x/n})U(x)$ converges pointwise to $xU(x)$ and is dominated by x. This is an application of the *dominated convergence theorem* of real analysis. For elementary discussions, see Folland (1992, p. 83) and Champeney (1987, pp. 25–26).

[18] Jones, 1982, pp. 200–201.

By this maneuver, two one-sided polynomials, one-sided sinusoids, so on, may be convolved, if they are both right sided or both left sided, for example, $U(x) * xU(x)$ or $U(x-1) * \cos 2\pi x\, U(x)$, but not $U(x) * U(-x)$.

6.4 GENERALIZED FOURIER TRANSFORM

6.4.1 Definition

The Fourier transform is defined, for ordinary functions, as an integral:

$$F(v) = \int_{-\infty}^{\infty} f(x)\, e^{-i2\pi vx}\, dx,$$

with appropriate interpretations depending on the function spaces involved. We have seen that it also gives a sensible result if f is a delta function or the derivative of a delta function. We seek a definition of the Fourier transform that works for generalized functions in general. Consider the generalized function's underlying regular sequence of good functions, (f_n). The Fourier transform of a good function is a good function (this is why we picked the good functions to build our generalized functions), so we Fourier transform each of the f_n to obtain a sequence (F_n). By Parseval's theorem (because good functions are square integrable),

$$\int_{-\infty}^{\infty} F_n(v)\, \varphi(v)\, dv = \int_{-\infty}^{\infty} f_n(x)\, \Phi(x)\, dx, \tag{6.34}$$

where φ and $\Phi = \mathcal{F}^{-1}\{\varphi\}$ are both testing functions. The sequence of integrals on the right is convergent because (f_n) is regular, therefore the sequence of integrals on the left is also convergent and (F_n) defines a generalized function F which we define to be the Fourier transform of f. Moreover, in the limit, Equation 6.34 becomes

$$\int F(v)\, \varphi(v)\, dv = \int f(x)\, \Phi(x)\, dx.$$

This leads to the definition of the Fourier transform for generalized functions.

Theorem 6.4 (Generalized Fourier transform). Let f and F be generalized functions defined by regular sequences of good functions (f_n) and (F_n), respectively. Then the following statements are equivalent:

(a) F is the Fourier transform of f if $f_n \longleftrightarrow F_n$.
(b) F is the Fourier transform of f if and only if

$$\int F(v)\, \varphi(v)\, dv = \int f(x)\, \Phi(x)\, dx \tag{6.35}$$

for all Fourier pairs $\varphi, \Phi \in \mathcal{S}$.

Let us see how this plays out for some of the generalized Fourier transforms we have already calculated.

Example 6.27.

1. Previously, we calculated $\delta(x) \longleftrightarrow 1$ by the sifting property of the delta function and also using sequences. Using Equation 6.35 instead,

$$\int \mathcal{F}\{\delta(x)\}\,\varphi(v)\,dv = \int \delta(x)\,\Phi(x)\,dx = \Phi(0).$$

But by the area theorem,

$$\Phi(0) = \int_{-\infty}^{\infty} \varphi(v)\,dv = \int_{-\infty}^{\infty} 1 \cdot \varphi(v)\,dv.$$

Therefore we identify $\mathcal{F}\{\delta(x)\} = 1$.

2. Previously, we calculated $\sin 2\pi x \longleftrightarrow \frac{1}{2i}\delta(v-1) - \frac{1}{2i}\delta(v+1)$ using sequences. Using Equation 6.35 instead,

$$\int \mathcal{F}\{\sin 2\pi x\}\,\varphi(v)\,dv = \int_{-\infty}^{\infty} \sin 2\pi x\,\Phi(x)\,dx$$

$$= \int_{-\infty}^{\infty} \Phi(x)\frac{1}{2i}e^{i2\pi x}\,dx - \int_{-\infty}^{\infty} \Phi(x)\frac{1}{2i}e^{-i2\pi x}\,dx$$

$$= \frac{1}{2i}\varphi(1) - \frac{1}{2i}\varphi(-1).$$

We recognize that δ is the generalized function that sifts out isolated values of a testing function, in particular:

$$\frac{1}{2i}\varphi(1) - \frac{1}{2i}\varphi(-1) = \int \left(\frac{1}{2i}\delta(v-1) - \frac{1}{2i}\delta(v+1)\right)\varphi(v)\,dv.$$

Therefore we identify

$$\mathcal{F}\{\sin 2\pi x\} = \frac{1}{2i}\delta(v-1) - \frac{1}{2i}\delta(v+1).$$

∎

Transform calculations are often facilitated by representing generalized functions as sequences. Weakly convergent sequences of generalized functions have the following very useful property. The proof is deferred to the problems.

Theorem 6.5 (Generalized transform in the limit). Let (f_n) be a sequence of generalized functions, converging weakly to a generalized function f. Let F_n be the generalized Fourier transform of f_n. Then the sequence of transforms (F_n) converges weakly to a generalized function F, the Fourier transform of f.

Example 6.28 (Fourier transforms of sgn x and x^{-1}). The signum function

$$\operatorname{sgn} x = \begin{cases} 1, & x > 0 \\ 0, & x = 0 \\ -1, & x < 0 \end{cases}$$

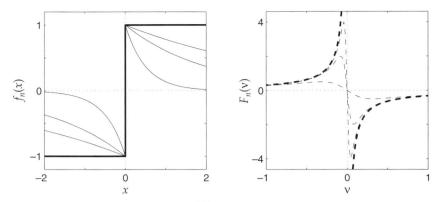

FIGURE 6.13 The sequence $f_n(x) = e^{-|x|/n}\operatorname{sgn} x$, converging to $\operatorname{sgn} x$ (*left*), and its Fourier transform (*right*). Heavy lines indicate the Fourier transform pair $\operatorname{sgn} x \longleftrightarrow 1/i\pi v$.

is discontinuous and not integrable. We multiply it by the convergence factor $e^{-|x|/n}$, creating a sequence of rapidly-decreasing functions:

$$f_n(x) = e^{-|x|/n}\operatorname{sgn} x.$$

Now, the Fourier transform of $f_n(x)$ is

$$\begin{aligned} F_n(v) &= \int_{-\infty}^{\infty} e^{-|x|/n}\operatorname{sgn}(x)e^{-i2\pi vx}\,dx \\ &= -\int_{-\infty}^{0} e^{x/n}e^{-i2\pi vx}\,dx + \int_{0}^{\infty} e^{-x/n}e^{-i2\pi vx}\,dx \\ &= \frac{1}{-1/n + i2\pi v} + \frac{1}{1/n + i2\pi v} = \frac{-i4\pi v}{(1/n)^2 + (2\pi v)^2}. \end{aligned}$$

The F_n are dominated by $\frac{1}{|\pi v|}$ away from the origin and converge pointwise to $\frac{1}{i\pi v}$ as $n \to \infty$ (Figure 6.13). The sequence of integrals

$$\int_{-\infty}^{\infty} \frac{-i4\pi v}{(1/n)^2 + (2\pi v)^2}\,\varphi(v)\,dv$$

converges weakly to the integral

$$\int_{-\infty}^{\infty} \frac{1}{i\pi v}\,\varphi(v)\,dv,$$

which we interpret as a Cauchy principal value integral. We define the Fourier transform $\operatorname{sgn} x \longleftrightarrow 1/i\pi v$.

The Fourier transform of $f(x) = x^{-1}$ can be calculated by carrying out the above steps in reverse, leading to two generalized Fourier transforms:

$$\operatorname{sgn} x \longleftrightarrow \frac{1}{i\pi v}, \tag{6.36a}$$

$$\frac{1}{x} \longleftrightarrow -i\pi \operatorname{sgn} v, \tag{6.36b}$$

where x^{-1} and $(i\pi v)^{-1}$ are interpreted using the Cauchy principal value. (Note, by the way, that these generalized functions obey the "real and odd ⟷ imaginary and odd" symmetry rule.) ∎

Example 6.29. The convolution of two ordinary functions is wider than either of the original functions—the rect-triangle example being typical. But the generalized Fourier pair $x^{-1} \longleftrightarrow -i\pi \operatorname{sgn} v$ violates this. The convolution of $\frac{1}{i\pi x}$ with itself, taken to the frequency domain, is

$$\frac{1}{i\pi x} * \frac{1}{i\pi x} \longmapsto (-\operatorname{sgn}(v))^2 = 1,$$

from which we conclude that

$$\frac{1}{i\pi x} * \frac{1}{i\pi x} = \delta(x). \tag{6.37}$$

∎

6.4.2 Fourier Theorems

Many of the Fourier theorems for ordinary functions work for the Fourier transforms of generalized functions. Theorems that require a function to be evaluated at a particular point, such as the area and moment theorems, do not carry over to generalized functions ($\int 1\, dx = \delta(0)$?). Neither does Parseval's theorem, which holds only for functions in L^2. In these derivations, the second definition of the Fourier transform (Equation 6.35) provides simpler paths to the results than the first using sequences.

Linearity
The rules for scaling and adding generalized functions immediately show that the generalized Fourier transform is linear:

$$af + bg \longleftrightarrow aF + bG.$$

Example 6.30 (Fourier transform of the step function). The step function, $U(x)$, can be expressed in terms of the signum function,

$$U(x) = \frac{1}{2}(1 + \operatorname{sgn}(x)).$$

By the linearity of the Fourier transform,

$$U(x) \longmapsto \frac{1}{2}\mathcal{F}\{1\} + \frac{1}{2}\mathcal{F}\{\operatorname{sgn}(x)\}$$

$$= \frac{1}{2}\delta(v) + \frac{1}{i2\pi v}.$$

Therefore (Figure 6.14),

$$U(x) \longleftrightarrow \frac{1}{2}\delta(v) + \frac{1}{i2\pi v}. \tag{6.38}$$

∎

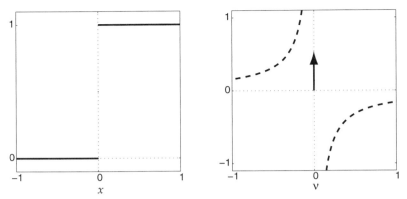

FIGURE 6.14 The step function $U(x)$ (*left*) and its Fourier transform, $\frac{1}{2}\delta(\nu) - \frac{i}{2\pi\nu}$ (*right*).

Shift Theorem

Begin by writing down the integral $\int f(x-a)\,\Phi(x)\,dx$ and make a change of variable to push the shift onto the testing function (Equation 6.19):

$$\int f(x-a)\,\Phi(x)\,dx = \int f(x)\,\Phi(x+a)\,dx.$$

Now invoke Theorem 6.4:

$$\int f(x)\,\Phi(x+a)\,dx = \int F(\nu)\,\mathcal{F}^{-1}\{\Phi(x+a)\}\,d\nu$$
$$= \int F(\nu)\,[e^{-i2\pi\nu a}\varphi(\nu)]\,d\nu,$$

then using Theorem 6.2,

$$= \int [e^{-i2\pi\nu a}F(\nu)]\,\varphi(\nu)\,d\nu.$$

Thus, $f(x-a) \longleftrightarrow e^{-i2\pi\nu a}F(\nu)$. An identical calculation proves the converse, $e^{i2\pi x b}f(x) \longleftrightarrow F(\nu - b)$.

Dilation Theorem

Using the same method, it can be shown that $f(ax) \longleftrightarrow \frac{1}{|a|}F\left(\frac{\nu}{a}\right)$. The verification is left to the problems.

Symmetries

We shall show that if f is a real and even generalized function, then F is also real and even. Proof of the rest of the symmetry relationships follows the same logic. Because f is real and even, we know (Equations 6.21 and 6.22)

$$\int f(x)\,\Phi(x)\,dx = \left[\int f(x)\,\Phi^*(x)\,dx\right]^* = \int f(x)\,\Phi(-x)\,dx,$$

6.4 GENERALIZED FOURIER TRANSFORM

and we want to show that F is real and even:

$$\int F(v)\varphi(v)\,dv = \left[\int F(v)\varphi^*(v)\,dv\right]^* = \int F(v)\varphi(-v)\,dv.$$

This is just a matter of applying the definitions and being careful about the Fourier transforms of the testing functions. First, show that F is even:

$$\int F(v)\varphi(v)\,dv = \int f(x)\Phi(x)\,dx$$
$$= \int f(x)\Phi(-x)\,dx \qquad (f \text{ is even})$$
$$= \int F(v)\mathcal{F}^{-1}\{\Phi(-x)\}\,dv = \int F(v)\varphi(-v)\,dv. \quad \text{(Equation 5.17)}$$

Then show that F is real:

$$\int F(v)\varphi(v)\,dv = \int f(x)\Phi(x)\,dx$$
$$= \left[\int f(x)\Phi^*(x)\,dx\right]^* \qquad (f \text{ is real})$$
$$= \left[\int F(v)\mathcal{F}^{-1}\{\Phi^*(x)\}\,dv\right]^*$$
$$= \left[\int F(v)\varphi^*(-v)\,dv\right]^* \qquad \text{(Equation 5.17)}$$
$$= \left[\int F(v)\varphi^*(v)\,dv\right]^*. \qquad (F \text{ is even})$$

Derivative

To justify the derivative theorem, $f'(x) \longmapsto i2\pi v F(v)$, for generalized Fourier transforms,

$$\int f'(x)\Phi(x)\,dx = -\int f(x)\Phi'(x)\,dx \qquad \text{(Generalized derivative)}$$
$$= -\int F(v)\mathcal{F}^{-1}\{\Phi'(x)\}\,dv.$$

Now use the ordinary derivative theorem on Φ':

$$= -\int F(v)[-i2\pi v\,\varphi(v)]\,dv.$$

This integral is well defined because $v\varphi(v)$ is good. Finally, push the factor of $i2\pi v$ over to F (Theorem 6.2):

$$= \int i2\pi v\,F(v)\,\varphi(v)\,dv.$$

A particularly nice application of the derivative theorem is given by the following example.

Example 6.31 (Fourier transform of a polynomial). Polynomials, because they grow without bound as $|x| \to \infty$, are not absolutely- or square integrable and are not Fourier transformable as ordinary functions. But they grow slowly and are well defined as generalized functions. We will derive a generalized Fourier transform for the polynomial $\sum_{n=0}^{N} a_n x^n$. Apply the derivative theorem repeatedly:

$$x^n f(x) \longleftrightarrow \frac{F^{(n)}(\nu)}{(-i2\pi)^n},$$

with $f = 1$ and $F = \delta$, obtaining

$$x^n \longleftrightarrow \frac{\delta^{(n)}(\nu)}{(-i2\pi)^n}, \tag{6.39}$$

and insert this into the polynomial:

$$\sum_{n=0}^{N} a_n x^n \longleftrightarrow \sum_{n=0}^{N} \frac{a_n}{(-i2\pi)^n} \delta^{(n)}(\nu). \tag{6.40}$$

The Fourier transform of a polynomial comes out in terms of derivatives of the delta function. ∎

Example 6.32 (Signum function, again). Here is another way to calculate the Fourier transform of the signum. Signum has a jump of $+2$ at the origin and is constant everywhere else, so its derivative is $\text{sgn}' = 2\delta$. Apply the derivative theorem:

$$\mathcal{F}\{\text{sgn}' x\} = \mathcal{F}\{2\delta(x)\} = 2$$
$$= i2\pi\nu \, \mathcal{F}\{\text{sgn } x\}$$
$$\Rightarrow \mathcal{F}\{\text{sgn } x\} = \frac{2}{i2\pi\nu} = \frac{1}{i\pi\nu}.$$

A similar approach yields a simple calculation of the Fourier transforms of x^{-1} and $\log|x|$ (see the problems). ∎

Example 6.33 (Fourier transform of $|x|$). Here is a harder problem:

$$\int \mathcal{F}\{|x|\} \, \varphi(\nu) \, d\nu = \int |x| \, \Phi(x) \, dx = \int x \, \text{sgn } x \, \Phi(x) \, dx$$
$$= \int \text{sgn } x \, (x\Phi(x)) \, dx.$$

Use $\text{sgn } x \longmapsto 1/i\pi\nu$ and apply the derivative theorem to $x\Phi(x)$:

$$= \int \frac{1}{i\pi\nu} \frac{\varphi'(\nu)}{i2\pi} \, d\nu.$$

Then use the definition of generalized derivative (Equation 6.23):

$$= -\int \left(\frac{-1}{2(i\pi)^2} \frac{1}{\nu}\right) \varphi'(\nu) \, d\nu = \int \frac{1}{2(i\pi)^2} \frac{1}{\nu^2} \varphi(\nu) \, d\nu$$

$$\Rightarrow |x| \longmapsto \frac{2}{(i2\pi\nu)^2}.$$

6.4 GENERALIZED FOURIER TRANSFORM

TABLE 6.3 The relationship of smoothness and asymptotic behavior of the Fourier transform continues to hold for generalized functions. Each function f in the table is the derivative of the function above it. Differentiation makes a function less smooth and causes its transform F to decay more slowly. If a function has jumps, its derivatives are singular and the Fourier transforms are slowly growing rather than decreasing.

Smoothness		$f(x)$	$F(\nu)$	Asymptotic
$C^{(\infty)}$ $\vdots$		$\exp(-\pi x^2)$	$\exp(-\pi \nu^2)$	Rapidly decreasing $\vdots$
$C^{(1)}$		$\frac{1}{2}x^2 \operatorname{sgn} x$	$-\dfrac{1}{i4\pi^3 \nu^3}$	Decreasing
$C^{(0)}$		$\lvert x \rvert$	$-\dfrac{1}{2\pi^2 \nu^2}$	Decreasing
PC		$\operatorname{sgn} x$	$\dfrac{1}{i\pi \nu}$	Decreasing
Singular		$2\delta(x)$	2	Slowly growing
Singular $\vdots$		$2\delta'(x)$	$i4\pi\nu$	Slowly growing $\vdots$

By linearity, we can also calculate the Fourier transform of $xU(x)$:

$$xU(x) = \frac{1}{2}(\lvert x \rvert + x) \longmapsto \frac{1}{(i2\pi\nu)^2} - \frac{1}{i4\pi}\delta'(\nu) = -\frac{1}{(2\pi\nu)^2} + \frac{i}{4\pi}\delta'(\nu).$$

■

Beginning with the Fourier series and continuing through our study of Fourier transforms in the Chapter 5, we have observed a relationship between the smoothness of a function and the asymptotic behavior of its Fourier transform as $\nu \to \infty$. The smoother a function is, the more rapidly its transform decays at high frequencies. Infinitely continuously differentiable functions have the most rapidly decaying transforms, epitomized by Gaussian⟵⟶Gaussian. At the other extreme are piecewise smooth and piecewise continuous functions like the triangle and rectangle, whose transforms decay $O(\nu^{-2})$ and $O(\nu^{-1})$, respectively. The story ended with functions having jumps, because they were undifferentiable.

The apparatus of generalized functions removes this barrier. We can keep differentiating, and keep Fourier transforming, and watch what happens (Table 6.3). The signum function's Fourier transform is $O(\nu^{-1})$, as expected, because it has a jump. Differentiating signum produces an impulse, whose Fourier transform is constant. Differentiating the impulse produces δ', whose Fourier transform is actually increasing with frequency, $O(\nu)$. As smoothness runs from $C^{(\infty)}$ to piecewise continuous to singular, the behavior of the Fourier transform runs, without interruption, from rapidly decreasing to slowly decaying to slowly growing.

The same trends hold for the relationship of a function's asymptotic behavior and the smoothness of its Fourier transform. A rapidly decreasing function has an infinitely smooth transform. The slowly decaying functions $\operatorname{sinc} x$ and x^{-1} have piecewise constant transforms, $\operatorname{rect} x$ and $i\pi \operatorname{sgn} \nu$, respectively. If we push past the slow decay to a constant function, the transform is impulsive, and going further to polynomials, which are slowly growing, we find transforms that contain derivatives of impulses.

Products and Convolutions

Earlier, we determined sufficient conditions for the existence of the product and convolution of two generalized functions:

- The product fg exists as a generalized function if the product of one of the functions with a good function is a good function.
- The convolution $f * g$ exists as a generalized function if the convolution of one of the functions with a good function is a good function.

We now seek conditions under which we have a convolution theorem, $f * g \longleftrightarrow FG$. The derivation of a generalized convolution theorem is a straightforward exercise in applying the definitions, watching along the way for the conditions such that the steps are valid.

Apply the definition of the generalized Fourier transform to the convolution:

$$\int \mathcal{F}\{f * g\}\, \varphi(\nu)\, d\nu = \int f * g(x)\, \Phi(x)\, dx,$$

and then apply the definition for generalized convolution (Equation 6.31):

$$= \int f(x)\, (g(-x) * \Phi(x))\, dx.$$

If $g * \Phi$ is good, we may apply the Fourier transform definition again:

$$= \int F(\nu)\, \mathcal{F}^{-1}\{g(-x) * \Phi(x)\}\, d\nu$$

$$= \int F(\nu) G(\nu)\, \varphi(\nu)\, d\nu$$

$$\therefore \mathcal{F}\{f * g\} = FG.$$

The key condition is that the convolution $g * \Phi$ must be a good function (this is just the condition for $f * g$ to exist in the first place). Because g and G are a Fourier pair, a condition imposed on one is reflected in the other. If g is sufficiently well behaved that $g * \Phi(x)$ is good, then G will likewise be such that $G\varphi(\nu)$ is good.

We have the following version of the convolution theorem.[19]

Theorem 6.6 (Generalized convolution theorem). Let f and g be generalized functions with Fourier transforms F and G, and let $\varphi, \Phi \in \mathcal{S}$ be a Fourier pair. If $g * \Phi$ is a good function (equivalently, $G\varphi$ is a good function), then $f * g$ and FG are generalized functions and $f * g \longleftrightarrow FG$, that is, $\int f * g(x)\, \Phi(x)\, dx = \int FG(\nu)\, \varphi(\nu)\, d\nu$.

To illustrate, we will revisit the earlier examples (Example 6.26).

---

[19] See Strichartz (1994, pp. 52–53). More statements about generalized convolutions and the Fourier transform may be found in Champeney (1987, pp. 139–144).

Example 6.34 (Generalized convolutions in the frequency domain).

1. Two delta functions, $\delta(x-a) * \delta(x-b) = \delta(x-(a+b))$.
   - $\delta(-x-b) * \Phi(x) = \delta(x+b) * \Phi(x) = \Phi(x+b)$, and $\Phi(x+b) \in \mathcal{S}$.
   - $\mathcal{F}\{\delta(x-b)\} = e^{-i2\pi bv}$, and $e^{-i2\pi bv}\varphi(v) \in \mathcal{S}$.
   - $\mathcal{F}\{\delta(x-a) * \delta(x-b)\} = e^{-i2\pi av}e^{-i2\pi bv} = e^{-i2\pi(a+b)v} = \mathcal{F}\{\delta(x-(a+b))\}$.

2. Convolution with the derivative of a delta function, $\delta' * f = f'$.
   - $\delta'(-x) * \Phi(x) = -\delta'(x) * \Phi(x) = -\Phi'(x)$, and $\Phi' \in \mathcal{S}$.
   - $\mathcal{F}\{\delta'(x)\} = i2\pi v$ is slowly growing, and $i2\pi v \varphi(v) \in \mathcal{S}$.
   - $\mathcal{F}\{\delta' * f\} = i2\pi v F(v) = \mathcal{F}\{f'(x)\}$.

3. A function of slow growth can be convolved with a rapidly decreasing function. Consider $x * e^{-x}U(x)$:
   - $e^{-x}U(x) * \varphi \in \mathcal{S}$.
   - $e^{-x}U(x) \longmapsto \dfrac{1}{1+i2\pi v}$ is fairly good, and $\dfrac{\varphi(v)}{1+i2\pi v} \in \mathcal{S}$.
   - By the convolution theorem,

$$x * e^{-x}U(x) = \mathcal{F}^{-1}\left\{\frac{1}{-i2\pi}\delta'(v) \cdot \frac{1}{1+i2\pi v}\right\}$$

$$= \int_{-\infty}^{\infty} -\frac{1}{i2\pi}\delta'(v) \cdot \frac{1}{1+i2\pi v} e^{i2\pi vx}\, dv$$

$$= -\frac{1}{i2\pi} \cdot \left[-\frac{d}{dv}\frac{e^{i2\pi vx}}{1+i2\pi v}\right]_{v=0}$$

$$= \frac{1}{i2\pi} \left.\frac{i2\pi x(1+i2\pi v) - i2\pi e^{i2\pi vx}}{(1+i2\pi v)^2}\right|_{v=0}$$

$$= x - 1.$$

   Surprised? Work it out in the time domain (see the problems).

4. A function of slow growth can be convolved with a bounded function of bounded support, for example, $\cos 2\pi x * \text{rect } 2x$ (this models a cosine smoothed by a filter with a rectangular impulse response).
   - $\text{rect} * \Phi \in \mathcal{S}$.
   - $\text{rect } 2x \longmapsto \frac{1}{2}\text{sinc}(\frac{v}{2})$ is fairly good, and $\text{sinc}(\frac{v}{2})\varphi(v) \in \mathcal{S}$.
   - Using the convolution theorem,

$$\cos 2\pi x * \text{rect } 2x = \mathcal{F}^{-1}\left\{\left[\frac{1}{2}\delta(v+1) + \frac{1}{2}\delta(v-1)\right] \cdot \frac{1}{2}\text{sinc}\left(\frac{v}{2}\right)\right\}$$

$$= \mathcal{F}^{-1}\left\{\frac{1}{4}\left[\text{sinc}\left(-\frac{1}{2}\right)\delta(v+1) + \text{sinc}\left(\frac{1}{2}\right)\delta(v-1)\right]\right\}$$

$$= \mathcal{F}^{-1}\left\{\frac{1}{2\pi}\delta(v+1) + \frac{1}{2\pi}\delta(v-1)\right\}$$

$$= \frac{1}{\pi}\cos 2\pi x.$$

   This convolution can also be checked in the time domain (see the problems).

5. Two polynomials cannot be convolved. If they could, then in the frequency domain we would have, for example, $\mathcal{F}\{x * x^2\} = \frac{1}{-i2\pi}\delta'(\nu) \cdot \frac{1}{(-i2\pi)^2}\delta''(\nu)$, but $\delta' \cdot \delta''$ is undefined. ∎

The convolution theorem may work for sequences if it does not for the functions themselves. Earlier we saw that two one-sided functions of slow growth may be convolved, for example, $U(x) * U(x) = xU(x)$. The Fourier transform of $xU(x)$ may be calculated by the derivative theorem in the frequency domain:

$$xU(x) \longmapsto \frac{1}{-i2\pi}\frac{d}{d\nu}\left[\frac{1}{2}\delta(\nu) + \frac{1}{i2\pi\nu}\right] = -\frac{1}{(2\pi\nu)^2} + \frac{i}{4\pi}\delta'(\nu).$$

(Note, by the way, that the Fourier transform is Hermitian.) However, these functions are problematic for the convolution theorem:

- $U * \Phi$ is not a good function.
- $\left(\frac{1}{2}\delta(\nu) + \frac{1}{i\pi\nu}\right)\varphi(\nu)$ is not a good function.
- $\left(\frac{1}{2}\delta(\nu) + \frac{1}{i\pi\nu}\right)$ cannot be squared.

However, as we did with the definition of generalized convolution, we can try to use sequences and take limits. The method is diagrammed below:

$$\begin{array}{ccc} f * g & & FG \\ \downarrow \text{sequence} & & \uparrow n\to\infty \\ f_n * g_n & \longleftrightarrow & F_n G_n \end{array}$$

Represent the step functions by the rapidly decreasing sequence $e^{-x/n}U(x)$, which has Fourier transform $F_n(\nu) = \frac{n}{1+i2\pi n\nu}$. Now $e^{-x/n}U(x) * \Phi(x)$ is good, because of the rapid decrease of the exponential, and $\frac{n}{1+i2\pi n\nu}\varphi(\nu)$ is also good, because $\frac{n}{1+i2\pi n\nu}$ is infinitely continuously differentiable. Thus the generalized convolution theorem yields

$$f_n * g_n(x) = xe^{-x/n}U(x)$$

$$F_n G_n(\nu) = \left(\frac{n}{1+i2\pi n\nu}\right)^2$$

and both $f_n * g_n$ and $F_n G_n$ are generalized functions. Verification that these sequences converge to $xU(x)$ and its Fourier transform is left to the problems.

6.5 SAMPLING THEORY AND FOURIER SERIES

Sampled functions were introduced in Chapter 3. A sequence $(f[n])$ is created by measuring the values of a continuous function f at regular intervals, $f[n] = f(n\Delta x)$, where Δx is the sampling interval. Because $f(x)\delta(x - a) = f(a)\delta(x - a)$, a sampled function may be modeled as a train of weighted impulses, the product of $f(x)$ with a train of unit impulses. Denoting the impulse-sampled function by $f_s(x)$,

$$f_s(x) = f(x) \sum_{n=-\infty}^{\infty} \delta(x - n\Delta x) = \sum_{n=-\infty}^{\infty} f(n\Delta x)\delta(x - n\Delta x)$$

$$= \sum_{n=-\infty}^{\infty} f[n]\delta(x - n\Delta x).$$

A periodic function $f(x)$ is the replication of a function $f_0(x)$ at regular intervals $x = nL$. A single replica of f_0 at $x = a$ is just a shift, which can be modeled by convolution with an impulse, $f_0(x - a) = f_0(x) * \delta(x - a)$. Infinite replication of f_0 into a periodic function can then be modeled by the convolution of f_0 with a train of impulses spaced by the period L:

$$f(x) = f_0(x) * \sum_{n=-\infty}^{\infty} \delta(x - nL) = \sum_{n=-\infty}^{\infty} f_0(x) * \delta(x - nL)$$

$$= \sum_{n=-\infty}^{\infty} f_0(x - nL).$$

The infinite impulse train, which we shall call a *comb function*, is central to both sampling and replication and connects them both through the Fourier transform. The key result of this section is that the Fourier transform of a train of weighted impulses (a sampled function) is periodic, and conversely, the Fourier transform of a periodic function is a train of impulses weighted by Fourier series coefficients. This leads to a more complete picture of Fourier series, sampling, bandlimited functions, aliasing, and the Nyquist sampling rate. It also permits a "grand unification" of the various Fourier representations—DFT, Fourier series, discrete-time Fourier transform, and (continuous-time) Fourier transform.

We begin with a connection between the Fourier transform and the Fourier series for ordinary functions.

6.5.1 Fourier Series, Again

In Chapter 4 it was shown that a periodic function f that is integrable on a finite interval $[0, L]$ may possess a Fourier series representation:

$$f(x) = \sum_{n=-\infty}^{\infty} c_n e^{i2\pi nx/L},$$

where L is the period and the Fourier coefficients are

$$c_n = \frac{1}{L} \int_0^L f(x) e^{-i2\pi nx/L} \, dx.$$

The integral for the Fourier coefficients can be made to look like a Fourier transform if a unit period of f is defined:

$$f_0(x) = \begin{cases} f(x), & 0 \leq x < L \\ 0, & \text{otherwise.} \end{cases}$$

In terms of f_0 and its Fourier transform F_0, the Fourier coefficients are

$$c_n = \frac{1}{L}\int_{-\infty}^{\infty} f_0(x) e^{-i2\pi nx/L}\, dx = \frac{1}{L}F_0\left(\frac{n}{L}\right).$$

The following theorem expresses this idea in a more general form.

Theorem 6.7. Let f be a continuous and bounded function that decays faster than $O(|x|^{-1})$ as $|x| \to \infty$. Let $F = \mathcal{F}\{f\}$, and assume that it is bounded and decays faster than $O(|v|^{-1})$ as $|v| \to \infty$. Form a periodic function $\tilde{f}$ by replicating f with period L:

$$\tilde{f} = \sum_{k=-\infty}^{\infty} f(x + kL).$$

Then $\tilde{f}$ has a Fourier series:

$$\tilde{f}(x) = \sum_{k=-\infty}^{\infty} f(x+kL) = \sum_{n=-\infty}^{\infty} \frac{1}{L}F\left(\frac{n}{L}\right) e^{i2\pi nx/L}. \tag{6.41}$$

These series converge absolutely and uniformly.

Proof: Proof that $\tilde{f}$ is integrable on $[0, L]$, and periodic, $\tilde{f}(x) = \tilde{f}(x+L)$, is taken up in the problems. The Fourier coefficients are calculated from

$$c_n = \frac{1}{L}\int_0^L \left[\sum_{k=-\infty}^{\infty} f(x+kL)\right] e^{-i2\pi nx/L}\, dx.$$

Begin with a finite sum, then exchange summation and integration and change variables $\xi = x + kL$:

$$\frac{1}{L}\int_0^L \left[\sum_{k=-N}^{N} f(x+kL)\right] e^{-i2\pi nx/L}\, dx = \sum_{k=-N}^{N}\frac{1}{L}\int_{kL}^{(k+1)L} f(\xi) e^{-i2\pi n(\xi-kL)/L}\, d\xi$$

$$= \sum_{k=-N}^{N}\frac{1}{L}\int_{kL}^{(k+1)L} f(\xi) e^{-i2\pi n\xi/L}\, d\xi = \frac{1}{L}\int_{-NL}^{(N+1)L} f(\xi) e^{-i2\pi n\xi/L}\, d\xi.$$

Now, as $N \to \infty$, this becomes a Fourier transform integral, and so

$$c_n = \frac{1}{L}\int_{-\infty}^{\infty} f(\xi) e^{-i2\pi\xi(n/L)}\, d\xi = \frac{1}{L}F\left(\frac{n}{L}\right).$$

The Fourier coefficients will exist merely if $f \in L^1(\mathbb{R})$, but by specifying the rates of decay for f and F, we obtain absolute and uniform convergence for the periodic replication of f and for the Fourier series. ∎

Example 6.35 (Square wave). The replication of a rectangle function $\mathrm{rect}\left(\frac{x}{A}\right)$ with period $L > 2A$ results in a square wave with duty cycle $\frac{A}{L}$. The Fourier transform of the rectangle is $A \operatorname{sinc} A\nu$, and using Equation 6.41, the Fourier coefficients for the square wave are obtained by sampling the Fourier transform:

$$c_n = \frac{A}{L}\operatorname{sinc}\left(\frac{nA}{L}\right).$$

When $A = L/2$, we recover a familiar result (compare Example 4.3):

$$c_n = \frac{1}{2}\operatorname{sinc}\left(\frac{n}{2}\right) = \begin{cases} \frac{1}{2}, & n = 0 \\ 0, & n \text{ even} \\ \frac{(-1)^{(n-1)/2}}{\pi n}, & n \text{ odd} \end{cases}$$

The lack of continuity of the rectangle and the consequent slow decay of the sinc function prevent the Fourier series from converging uniformly. ∎

A remarkable corollary of Theorem 6.7 is the *Poisson sum formula*, obtained by setting $x = 0$ in Equation 6.41:

$$\sum_{k=-\infty}^{\infty} f(kL) = \sum_{n=-\infty}^{\infty} \frac{1}{L} F\left(\frac{n}{L}\right). \qquad (6.42)$$

Example 6.36. The Poisson sum formula can provide a way to sum infinite series. One such series that appears in a later chapter is $\sum_{n=-\infty}^{\infty} \frac{1}{n^2+b^2}$. If there is a function related by Fourier transformation to the terms being summed, the Poisson sum formula can be applied. Here, the transform pair $e^{-|x|} \longleftrightarrow \frac{2}{1+(2\pi\nu)^2}$ is the key. By the dilation theorem, $\frac{1}{\nu^2+b^2} \longleftrightarrow \frac{\pi}{b}e^{-2\pi b|x|}$ ($b > 0$). The Poisson sum formula then says

$$\sum_{n=-\infty}^{\infty} \frac{1}{n^2+b^2} = \sum_{k=-\infty}^{\infty} \frac{\pi}{b} e^{-2\pi b|k|}.$$

The sum on the right is a pair of geometric series:

$$\sum_{k=-\infty}^{\infty} \frac{\pi}{b} e^{-2\pi b|k|} = -\frac{\pi}{b} + \sum_{k=-\infty}^{0} \frac{\pi}{b}(e^{+2\pi b})^k + \sum_{k=0}^{\infty} \frac{\pi}{b}(e^{-2\pi b})^k$$

$$= -\frac{\pi}{b} + \frac{2\pi}{b}\sum_{k=0}^{\infty}(e^{-2\pi b})^k = -\frac{\pi}{b} + \frac{2\pi}{b}\frac{1}{1-e^{-2\pi b}}$$

$$= \frac{\pi}{b}\frac{1+e^{-2\pi b}}{1-e^{-2\pi b}} = \frac{\pi}{b}\coth \pi b.$$

Thus, we have the sum of the infinite series:

$$\sum_{n=-\infty}^{\infty} \frac{1}{n^2+b^2} = \frac{\pi}{b}\coth \pi b.$$

Even if a closed form sum is not possible, the Poisson sum formula may lead to a better numerical approximation. In the present example, $\frac{1}{n^2+b^2}$ decays $O(n^{-2})$, but $e^{-2\pi b|k|}$ is rapidly decreasing. With $b = 1$, the exponential series converges to an answer within 10^{-14} (MATLAB "long" precision) of the true value of $\pi \coth \pi$ in 11 terms. The other is only within 10^{-2} after 401 terms! ∎

6.5.2 Periodic Generalized Functions

A function is periodic if $f(x + L) = f(x)$. A generalized function is periodic if

$$\int f(x)\,\varphi(x)\,dx = \int f(x)\,\varphi(x-L)\,dx. \tag{6.43}$$

One important example of a periodic generalized function is the *comb function*:

$$\mathrm{III}(x) = \sum_{n=-\infty}^{\infty} \delta(x-n). \tag{6.44}$$

It is a shorthand notation for an infinite impulse train. Other periodic generalized functions can be obtained by differentiating periodic functions; for example, think about what happens if you differentiate a square wave, or take the second derivative of a triangle wave.

The Fourier series of a periodic generalized function has the same form as an ordinary Fourier series:

$$\sum_{n=-\infty}^{\infty} c_n e^{i2\pi nx/L}.$$

Consider the partial sum $S_N(x) = \sum_{n=-N}^{N} c_n e^{i2\pi nx/L}$, which is a regular generalized function. The sequence of partial sums converges weakly to a generalized function f if the sequence of integrals

$$\int_{-\infty}^{\infty} \left[\sum_{n=-N}^{N} c_n e^{i2\pi nx/L} \right] \Phi(x)\,dx$$

converges as $N \to \infty$. As we did in the derivation of Equation 6.41, for fixed N, we may take the integral inside the summation, and obtain

$$\sum_{n=-N}^{N} c_n \int_{-\infty}^{\infty} e^{i2\pi nx/L} \Phi(x)\,dx = \sum_{n=-N}^{N} c_n \varphi\left(\frac{n}{L}\right),$$

where $\varphi = \mathcal{F}^{-1}\{\Phi\}$. Because φ is a good function, the doubly infinite sequence of values $(\varphi(n/L))$ is rapidly decreasing as $|n| \to \infty$. The sequence of Fourier coefficients (c_n) can be slowly growing as $|n| \to \infty$ and the sum $\sum_{n=-N}^{N} c_n \varphi(n/L)$ will still converge absolutely as $N \to \infty$.

Suppose we have an ordinary periodic function f whose Fourier series is absolutely and uniformly convergent (i.e., f is continuous and piecewise smooth on $[0, L]$). As a regular generalized function it can be repeatedly differentiated term-by-term, each time multiplying the Fourier coefficients by $i2\pi n/L$. After k derivatives, the coefficient sequence is $((i2\pi n/L)^k c_n)$ and is never worse than slowly growing in n.

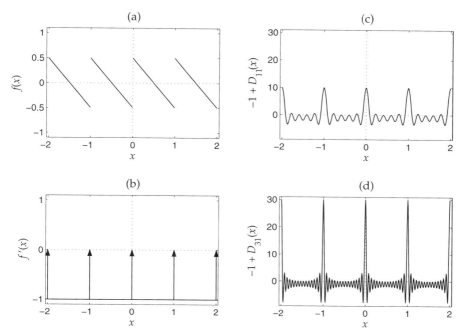

FIGURE 6.15 (a) A sawtooth function. (b) The derivative of the sawtooth has impulses at the jumps, $f'(x) = -1 + \text{III}(x)$. The Fourier series representation of f', obtained by differentiating the Fourier series of the sawtooth, is $-1 + \sum_{n=-\infty}^{\infty} 1 \cdot e^{i2\pi nx}$. It is weakly convergent to a generalized function. (c) The fifth partial sum of the Fourier series, $-1 + \sum_{n=-5}^{5} e^{i2\pi nx} = -1 + D_{11}(x)$. (d) The 15th partial sum, $-1 + D_{31}(x)$. The sequence of partial sums converges weakly to a comb function, $\text{III}(x)$.

Any such derived series will therefore be weakly convergent. We conclude that all periodic generalized functions have weakly convergent Fourier series.[20]

A periodic generalized function, written as a weakly convergent Fourier series, can also be Fourier transformed term-by-term. The result is

$$\tilde{F}(\nu) = \sum_{n=-\infty}^{\infty} c_n \mathcal{F}\{e^{i2\pi nx/L}\} = \sum_{n=-\infty}^{\infty} c_n \delta\left(\nu - \frac{n}{L}\right). \quad (6.45)$$

Example 6.37 (Sawtooth wave and its derivative). Consider the sawtooth wave shown in Figure 6.15. By inspection, $c_0 = 0$. The rest of the Fourier coefficients are

$$c_n = \int_0^1 \left(\frac{1}{2} - x\right) e^{-i2\pi nx} \, dx = \frac{1}{i2\pi n}.$$

---

[20] For detailed discussions of periodic generalized functions and their Fourier series expansions, see Gasquet and Witomski (1999, pp. 335–342), Folland (1992, pp. 320–323), Champeney (1987, pp. 170–176), and Lighthill (1958, pp. 58–75).

The coefficients are decaying, but only $O(|n|^{-1})$, because of the jumps in the sawtooth. Though it converges nonuniformly as an ordinary Fourier series (we expect overshoot at the jumps), it converges weakly and that is all we need to treat f as a generalized function. Differentiating the Fourier series term-by-term, the Fourier coefficients become $i2\pi n\, c_n = 1, n \neq 0$. We can look at the derived series in two ways.

First, we know the series will converge weakly to the (generalized) derivative of f. The derivative of each linear part is -1, and the derivative of each jump is a delta function. Therefore,

$$f'(x) = -1 + \sum_{k=-\infty}^{\infty} \delta(x-k),$$

a constant plus an infinite train of impulses (a comb function). The comb function $\sum_{n=-\infty}^{\infty} \delta(x-n)$ is weakly convergent:

$$\int \left[\sum_{n=-N}^{N} \delta(x-n)\right] \varphi(x)\, dx = \sum_{n=-N}^{N} \int \delta(x-n)\varphi(x)\, dx = \sum_{n=-N}^{N} \varphi(n),$$

and this sum converges as $N \to \infty$ because φ is good. So f' is a well-defined generalized function.

Second, look at the sequence of partial sums of the Fourier series:

$$S_N = \sum_{N \geq |n| > 0} 1 \cdot e^{i2\pi nx} = -1 + \sum_{n=-N}^{N} e^{i2\pi nx} = -1 + D_{2N+1}(x),$$

a constant plus a Dirichlet kernel. Some of these partial sums are shown in Figure 6.15. Over one period, say $x \in [-\frac{1}{2}, \frac{1}{2}]$, D_{2N+1} has height $2N+1$, zero crossings spaced by $\frac{1}{2N+1}$, and unit area. It is plain that the peaks of the Dirichlet kernel are converging toward delta functions. We may consider the sequence of Dirichlet kernels $(D_{2n+1}(x))_{n=1}^{\infty}$ to be a representation for the comb function. ∎

The Comb Function

We saw in the previous example that the Fourier series for the comb function is

$$\text{III}(x) = \sum_{n=-\infty}^{\infty} 1 \cdot e^{i2\pi nx}.$$

The series is weakly convergent and may be Fourier transformed term-by-term to give a very nice result

$$\text{III}(x) \longleftrightarrow \text{III}(\nu). \tag{6.46}$$

This can also be derived without passing through the Fourier series (see the problems).

A dilated comb function, $\text{III}(x/\Delta x)$, is

$$\text{III}\left(\frac{x}{\Delta x}\right) = \sum_{n=-\infty}^{\infty} \delta\left(\frac{x}{\Delta x} - n\right) = \sum_{n=-\infty}^{\infty} \Delta x\, \delta(x - n\Delta x).$$

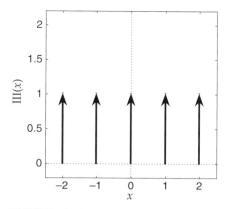
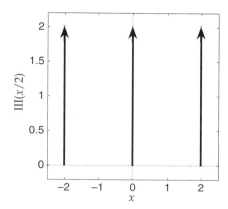

FIGURE 6.16 The comb function, III(x), is an infinite train of unit impulses (*left*). The dilated comb, III$\left(\frac{x}{\Delta x}\right)$, is a train of impulses with spacing Δx and height Δx (*right*).

As the impulses spread out (increasing Δx), their heights (areas) increase (Figure 6.16). A train of unit impulses with period Δx is

$$\sum_{n=-\infty}^{\infty} \delta(x - n\Delta x) = \frac{1}{\Delta x} \text{III}\left(\frac{x}{\Delta x}\right). \tag{6.47}$$

By the dilation theorem, the Fourier transform of this impulse train is

$$\mathcal{F}\left\{\sum_{n=-\infty}^{\infty} \delta(x - n\Delta x)\right\} = \mathcal{F}\left\{\frac{1}{\Delta x}\text{III}\left(\frac{x}{\Delta x}\right)\right\} = \text{III}(\Delta x\, \nu)$$

$$= \sum_{n=-\infty}^{\infty} \delta(\Delta x\, \nu - n) = \sum_{n=-\infty}^{\infty} \Delta x\, \delta\left(\nu - \frac{n}{\Delta x}\right). \tag{6.48}$$

The comb function transform pair obeys the same reciprocal spreading principle as other Fourier transforms: the spacing of the impulses in the frequency domain is the inverse of the spacing in the time domain.

6.5.3 The Sampling Theorem

Sampling and Replication
A sampled function $f_s(x)$ is represented by the product of $f(x)$ with the dilated comb function (Figure 6.17):

$$f_s(x) = f(x) \cdot \frac{1}{\Delta x}\text{III}\left(\frac{x}{\Delta x}\right). \tag{6.49}$$

Convolving a function with a comb produces a periodic replication (Figure 6.18):

$$f(x) * \frac{1}{\Delta x}\text{III}\left(\frac{x}{\Delta x}\right) = \sum_{n=-\infty}^{\infty} f(x - n\Delta x). \tag{6.50}$$

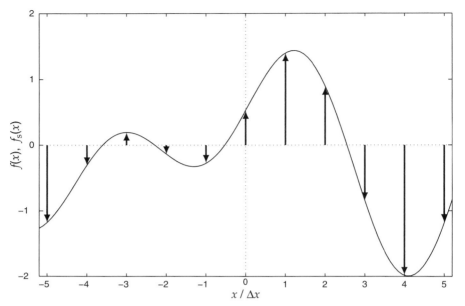

FIGURE 6.17 A function $f(x)$ sampled at points $x = k\Delta x$ can be modeled as a weighted impulse train, $f_s(x) = f(x) \cdot \frac{1}{\Delta x}\text{III}\left(\frac{x}{\Delta x}\right)$.

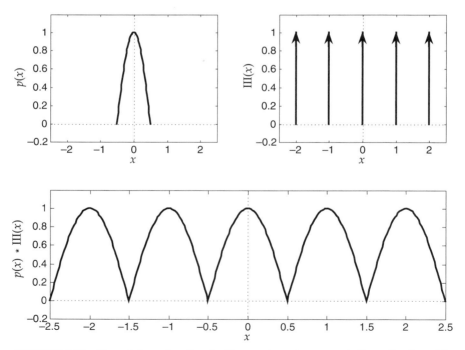

FIGURE 6.18 Convolving a function f with a comb produces a periodic replication of f.

The Fourier transform of the impulse-sampled function is

$$f_s(x) \longmapsto \mathcal{F}\left\{f(x) \cdot \frac{1}{\Delta x} \mathrm{III}\left(\frac{x}{\Delta x}\right)\right\} = F(\nu) * \mathrm{III}(\Delta x \, \nu) \tag{6.51a}$$

and the Fourier transform of a periodic replication is

$$\sum_{n=-\infty}^{\infty} f(x - n\Delta x) \longmapsto \mathcal{F}\left\{f(x) * \frac{1}{\Delta x} \mathrm{III}\left(\frac{x}{\Delta x}\right)\right\} = F(\nu) \, \mathrm{III}(\Delta x \, \nu). \tag{6.51b}$$

The Fourier transform of a sampled function is a periodic replication of the transform of the unsampled function. The Fourier transform of a periodically replicated function is an impulse sampling of the transform of the base function. *Sampling and replication are a transform pair.* We have seen one of these relationships applied to the Fourier series (Theorem 6.7). We will now use the other to derive a classic result.

Bandlimited Functions
A function $f(x)$ is *bandlimited* if its Fourier transform $F(\nu)$ is zero for frequencies ν higher than a finite value B. If f is bandlimited and in L^2, then $F \in L^2(-B, B)$, which means it is also in $L^1(-B, B)$ (recall $L^2(a, b) \subset L^1(a, b)$ for bounded intervals (a, b), Figure 4.5). Then, because $F \in L^1(-B, B)$, f is also bounded and continuous (in fact, we shall see in Chapter 8 that bandlimited functions are infinitely continuously differentiable, e.g., sinc, whose Fourier transform is rect).[21]

Sampling Theorem
Because square-integrable bandlimited functions are bounded and continuous, their samples are well defined. As the following theorem shows, under appropriate conditions a bandlimited function can be reconstructed from its samples.[22]

Theorem 6.8 (Sampling theorem). Let f be a bandlimited function ($F(\nu) = 0$ for $|\nu| > B > 0$) and have finite energy, $f \in L^2(\mathbb{R})$. If $\Delta x < 1/2B$, then f may be reconstructed from samples $f(n\Delta x)$, according to the formula

$$f(x) = \sum_{n=-\infty}^{\infty} f(n\Delta x) \, \mathrm{sinc}\left(\frac{x - n\Delta x}{\Delta x}\right). \tag{6.52}$$

Proof: Because f is bandlimited and has finite energy, it is continuous and bounded; the samples $f(n\Delta x)$ are well defined. The Fourier transform of the impulse-sampled function is (Equation 6.51)

$$F_s(\nu) = F(\nu) * \mathrm{III}(\Delta x \, \nu) = \sum_{n=-\infty}^{\infty} \frac{1}{\Delta x} F\left(\nu - \frac{n}{\Delta x}\right).$$

The Fourier transform is a periodic replication of F, the Fourier transform of f. If $\Delta x < 1/2B$, then the replicas, or "spectral islands", will not overlap; otherwise they will overlap, causing aliasing (Figure 6.19).

---
[21] Papoulis (1977, pp. 184–191) discusses several additional properties of bandlimited functions.
[22] Brief historical notes on the development of the sampling theorem can be found in Marks (1991, pp. 1–4).

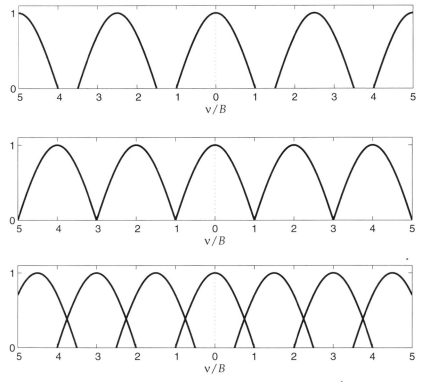

FIGURE 6.19 A function with bandlimit B is sampled at rate $v_s = \frac{1}{\Delta x}$. The spectrum is replicated by sampling. *Top*: The case of oversampling, $v_s > 2B$. The spectral replicas are well separated. *Middle*: The case of critical sampling, $v_s = 2B$. *Bottom*: The case of undersampling, $v_s < 2B$. The spectral islands overlap (aliasing).

Assuming the spectral islands are well separated, the central ($n = 0$) island, which is the Fourier transform of f, can be isolated from the others by an ideal lowpass filter, $\Delta x \, \text{rect}(\Delta x \, v)$. Applying the lowpass filter gives

$$F(v) = F_s(v) \, \Delta x \, \text{rect}(\Delta x v),$$

and inverting the Fourier transform yields the desired result,

$$f(x) = f_s(x) * \text{sinc}\left(\frac{x}{\Delta x}\right) = \sum_{n=-\infty}^{\infty} f(n\Delta x) \, \delta(x - n\Delta x) * \text{sinc}\left(\frac{x}{\Delta x}\right)$$
$$= \sum_{n=-\infty}^{\infty} f(n\Delta x) \, \text{sinc}\left(\frac{x - n\Delta x}{\Delta x}\right).$$

∎

The sampling and reconstruction process is illustrated in Figure 6.20. The sinc functions interpolate between the sample values to supply the rest of the original function.

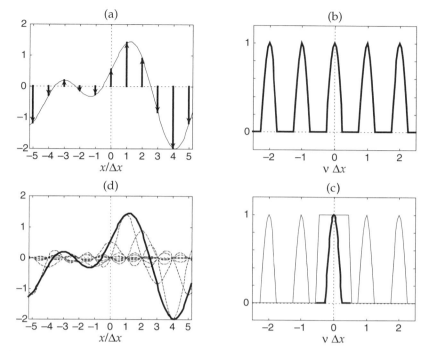

FIGURE 6.20 The sampling theorem. (a) A bandlimited function f is sampled at $\Delta x > 1/2B$. (b) The spectrum of the sampled function is the Fourier transform F, replicated at $\nu = n/\Delta x$. (c) Because f is bandlimited, the spectral replicas do not overlap, and the central ($n = 0$) term can be isolated by an ideal lowpass filter. (d) The function f is recovered by interpolating its sample values with sinc functions (dashed lines).

⋆ *A Vector Space Interpretation*

The sampling theorem may be interpreted as an orthogonal expansion in L^2.[23] Let $\psi_n(x) = \mathrm{sinc}(x - n)$ be a shifted sinc function. The dilated and shifted sinc functions,

$$\psi_n\left(\frac{x}{\Delta x}\right) = \mathrm{sinc}\left(\frac{x}{\Delta x} - n\right) = \mathrm{sinc}\left(\frac{x - n\Delta x}{\Delta x}\right),$$

are orthogonal:

$$\left\langle \psi_n\left(\frac{x}{\Delta x}\right), \psi_m\left(\frac{x}{\Delta x}\right) \right\rangle = \left\langle \mathrm{sinc}\left(\frac{x - n\Delta x}{\Delta x}\right), \mathrm{sinc}\left(\frac{x - m\Delta x}{\Delta x}\right) \right\rangle$$

$$= \left\langle e^{-i2\pi n\Delta x \nu} \Delta x \, \mathrm{rect}(\Delta x \, \nu), e^{-i2\pi m\Delta x \nu} \Delta x \, \mathrm{rect}(\Delta x \, \nu) \right\rangle \quad \text{(Parseval)}$$

$$= (\Delta x)^2 \int_{-1/2\Delta x}^{1/2\Delta x} e^{-i2\pi(n-m)\Delta x \nu} \, d\nu = \Delta x \, \delta[n - m].$$

---

[23] The orthogonal expansion interpretation of the sampling theorem is outlined in Gasquet and Witomski (1999, pp. 357–359) and in the comprehensive paper by Unser (2000).

Let V_B denote the set of square-integrable functions with bandlimit B. It can be shown that it is a subspace of L^2. The $\psi_n(x/\Delta x)$ are individually bandlimited to B and are in L^2. A linear combination of them is also bandlimited; indeed, by the sampling theorem, any bandlimited L^2 function can be expressed as a linear combination of the ψ_n. Therefore, the set $\{\psi_n(x/\Delta x)/\sqrt{\Delta x}\}_{n\in\mathbb{Z}}$ is an orthonormal basis for the subspace V_B.

For any function $f \in L^2$, the inner product $\langle f, \psi_n(x/\Delta x)/\sqrt{\Delta x}\rangle$ is the projection of f onto the nth basis vector of V_B, and the orthogonal projection of f onto V_B, which is the best bandlimited approximation to f, is the sum:

$$\hat{f}(x) = \sum_n \left\langle f, \frac{1}{\sqrt{\Delta x}} \psi_n\left(\frac{x}{\Delta x}\right) \right\rangle \frac{1}{\sqrt{\Delta x}} \psi_n\left(\frac{x}{\Delta x}\right) = \sum_n \left\langle f, \frac{1}{\Delta x} \psi_n\left(\frac{x}{\Delta x}\right) \right\rangle \psi_n\left(\frac{x}{\Delta x}\right).$$

The inner products are, using Parseval's theorem,

$$\left\langle f, \frac{1}{\Delta x} \psi_n\left(\frac{x}{\Delta x}\right) \right\rangle = \left\langle F, e^{-i2\pi n \Delta x \nu} \text{rect}(\Delta x\, \nu) \right\rangle$$
$$= \int_{-\infty}^{\infty} [F(\nu) \text{rect}(\Delta x\, \nu)] e^{i2\pi n \Delta x \nu}\, d\nu$$
$$= \mathcal{F}^{-1}\{F(\nu) \text{rect}(\Delta x\, \nu)\}|_{x=n\Delta x}.$$

The product $F(\nu) \text{rect}(\Delta x\, \nu)$ represents the application of an ideal lowpass filter to f, resulting in a function which is bandlimited to $|\nu| < 1/2\Delta x$. The expansion coefficients $\langle f, \psi_n(x/\Delta x)/\sqrt{\Delta x}\rangle$ are samples of the filter's output at $x = n\Delta x$. If f is already bandlimited ($f \in V_B$), the coefficients are precisely the samples of f, $f[n] = f(n\Delta x)$. Otherwise, f is bandlimited before sampling to prevent aliasing. The necessary antialiasing filter is implicit in the orthogonal projections onto V_B.

Sampled Sinusoidal Functions
The Fourier transform of a sinusoidal function $\cos 2\pi \nu_0 x$ is a pair of impulses at $\pm\nu_0$. This function may also be considered bandlimited to any $B > \nu_0$, but it is not square integrable. The requirement for square integrability in the sampling theorem does two things: it guarantees that the function is bounded and continuous and (see the problems) provides a Parseval relationship between the energy in the signal and the squares of the samples. A sinusoid, even a finite sum of sinusoids, is also bounded and continuous and will have well-defined samples. It can be shown that the sinc-function interpolation can recover a sinusoidal function from its samples.[24] One must be careful, however, not to sample a sinusoid right at the Nyquist rate, $\Delta x = 1/2\nu_0$, because this is precisely the spacing of zero crossings, and there is a chance that the samples could fall at or near the zero crossings!

6.5.4 Discrete-time Fourier Transform

The duality of the discrete-time Fourier transform with the Fourier series enables the properties of a generalized discrete-time Fourier transform to be obtained easily. If a

---
[24] See Gasquet and Witomski (1999, p. 356) for a proof.

sequence $(f[n])_{n\in\mathbb{Z}}$ is in ℓ^1, it possesses a Fourier transform $F_d(\theta)$ that is bounded, continuous, and periodic with period 2π. The extension to generalized functions is simple: the Fourier transform of a slowly growing sequence ($|f[n]| < C(1+|n|)^N$ for some finite C and N) is a periodic generalized function. Here are some examples.

Example 6.38 (Fourier transforms of slowly growing sequences).

1. *Constant function.* The Fourier transform of a constant is obtained using a convergence factor. A rectangle is simple:

$$1 \longmapsto \lim_{N\to\infty} \sum_{n=-N}^{N} e^{-in\theta} = \lim_{N\to\infty} D_N\left(\frac{\theta}{2\pi}\right)$$

$$= \sum_{k=-\infty}^{\infty} 2\pi\delta(\theta - 2\pi k) = \text{III}\left(\frac{\theta}{2\pi}\right).$$

On the principal interval $[-\pi, \pi]$, $1 \longmapsto 2\pi\delta(\theta)$. We will usually just write it this way, and interpret the digital frequencies modulo-2π.

2. *Sine and cosine.* Using the shift theorem,

$$e^{i\alpha n} \longmapsto 2\pi\delta(\theta - \alpha).$$

Then, by linearity,

$$\cos\alpha n \longmapsto \pi\delta(\theta - \alpha) + \pi\delta(\theta + \alpha),$$
$$\sin\alpha n \longmapsto -i\pi\delta(\theta - \alpha) + i\pi\delta(\theta + \alpha).$$

3. *Signum function.* The discrete signum function is

$$\text{sgn}[n] = \begin{cases} 1, & n > 0 \\ 0, & n = 0 \\ -1, & n < 0 \end{cases}.$$

The Fourier transform is obtained via a convergence factor, $a^{|n|}$, $1 > a > 0$:

$$a^{|n|}\text{sgn}[n] \longmapsto \sum_{n=-\infty}^{-1}(-1)\cdot a^{-n}e^{-in\theta} + \sum_{n=1}^{\infty} 1\cdot a^n e^{-in\theta}$$

$$= -\sum_{n=0}^{\infty} a^n e^{in\theta} + 1 + \sum_{n=0}^{\infty} a^n e^{-in\theta} - 1$$

$$= -\frac{1}{1-ae^{i\theta}} + \frac{1}{1-ae^{-i\theta}} = -\frac{i2a\sin\theta}{(1+a^2) - 2a\cos\theta},$$

and letting $a \to 1$,

$$\text{sgn}[n] \longmapsto -\frac{i\sin\theta}{1-\cos\theta} = \frac{1}{2i}\cot\left(\frac{\theta}{2}\right). \tag{6.53}$$

4. *Step function.* The discrete step function is usually defined

$$U[n] = \begin{cases} 1, & n \geq 0 \\ 0, & n < 0 \end{cases}.$$

In continuous time, the step function is defined to be equal to $\frac{1}{2}$ at $x = 0$, and $U(x) = \frac{1}{2}(1 + \operatorname{sgn} x)$, If this is carried through to discrete time, then the step function is $U[n] = \frac{1}{2}\delta[n] + \frac{1}{2}(1 + \operatorname{sgn}[n])$. In continuous time, the value at a single point makes no difference to the Fourier transform, but not in discrete time. By linearity, the Fourier transform of $\frac{1}{2}(1 + \operatorname{sgn}[n])$ is

$$\frac{1}{2}(1 + \operatorname{sgn}[n]) \longmapsto \pi\delta(\theta) + \frac{1}{2i}\cot\left(\frac{\theta}{2}\right)$$

and then, applying linearity again,

$$U[n] \longmapsto \frac{1}{2} + \pi\delta(\theta) + \frac{1}{2i}\cot\left(\frac{\theta}{2}\right). \tag{6.54}$$

A straightforward derivation using a convergence factor leads to the result

$$U[n] \longmapsto \sum_{n=0}^{\infty} a^n e^{in\theta} = \frac{1}{1 - ae^{-i\theta}}.$$

So how does this become Equation 6.54? First, separate the real and imaginary parts by rationalizing the denominator:

$$\frac{1}{1 - ae^{-i\theta}} = \frac{1 - a\cos\theta}{(1 + a^2) - 2a\cos\theta} - \frac{ia\sin\theta}{(1 + a^2) - 2a\cos\theta}.$$

The second term becomes $\frac{1}{2i}\cot\left(\frac{\theta}{2}\right)$ as $a \to 1$. The first term is an "improper" fraction in $\cos\theta$ and is reduced by division to

$$\frac{1 - a\cos\theta}{(1 + a^2) - 2a\cos\theta} = \frac{1}{2} + \frac{1}{2}\frac{1 - a^2}{(1 + a^2) - 2a\cos\theta}.$$

It remains to verify that the second term of this expression becomes $\pi\delta(\theta)$ as $a \to 1$. It can be shown that this term has constant area π, it goes to zero as $a \to 1$ except at $\theta = 0$, and has height $\frac{1+a}{2(1-a)}$ at $\theta = 0$. This means that the term is a sequence of functions converging to a delta function $\pi\delta(\theta)$. ∎

If a periodic function $f(x)$ has Fourier coefficients (c_n), then the Fourier coefficients of its derivative $f'(x)$ are $\frac{i2\pi n}{L} c_n$. The converse, interpreted for the discrete-time Fourier transform ($L = 2\pi$), is that if a sequence $f[n]$ has Fourier transform $F_d(\theta)$, then the Fourier transform of $nf[n]$ is the (generalized) derivative, $iF'_d(\theta)$.

Example 6.39 (Ramp sequences).

1. The Fourier transform of $f[n] = n$ is obtained from the derivative of the transform of the unit sequence:

$$1 \longmapsto 2\pi\delta(\theta)$$
$$n \longmapsto i2\pi\delta'(\theta). \tag{6.55}$$

2. The Fourier transform of a one-sided ramp, $f[n] = nU[n]$, is obtained from the derivative of the transform of the step function,

$$U[n] \longmapsto \frac{1}{2} + \pi\delta(\theta) + \frac{1}{2i}\cot\left(\frac{\theta}{2}\right),$$
$$nU[n] \longmapsto i\pi\delta'(\theta) - \frac{1}{2}\csc^2\left(\frac{\theta}{2}\right). \quad (6.56)$$

■

For a final example, we consider the Fourier transform of the discrete comb sequence,

$$\text{III}_N[n] = \sum_{k=-\infty}^{\infty} \delta[n - kN].$$

Example 6.40 (Discrete comb sequence). The transform of the discrete comb is

$$\text{III}_N[n] \longmapsto \sum_{n=-\infty}^{\infty}\left(\sum_{k=-\infty}^{\infty} \delta[n - kN]\right)e^{-in\theta} = \sum_{k=-\infty}^{\infty} e^{-ikN\theta}.$$

The partial sum $\sum_{k=-K}^{K} e^{-ikN\theta}$ is a Dirichlet kernel, $D_{2K+1}\left(\frac{N\theta}{2\pi}\right)$. There are zero crossings at $\theta = \frac{2\pi}{N(2K+1)}$ and peaks of height $2K+1$ at $\theta = \frac{2\pi}{N}$. Each period has constant area 2π. Each peak becomes a delta function as $K \to \infty$, and the transform of $\text{III}_N[n]$ is another comb:

$$\text{III}_N[n] \longmapsto \text{III}\left(\frac{N\theta}{2\pi}\right). \quad (6.57)$$

The period of the discrete comb is N, and the period of its transform is $2\pi/N$. The limiting cases $N \to 1$ and $N \to \infty$ are considered in the problems. ■

6.6 UNIFYING THE FOURIER FAMILY

6.6.1 Basis Functions and Orthogonality Relationships

The DFT and Fourier series were developed conceptually as orthogonal expansions on trigonometric basis functions. The basis vectors for the DFT are $\{\phi_m\}_{m=0}^{N-1}$, where

$$\phi_m = (e^{i2\pi mn/N})_{n=0}^{N-1}.$$

The orthogonality relationship is

$$\langle \phi_k, \phi_m \rangle = \sum_{n=0}^{N-1} \phi_k[n]\, \phi_m^*[n] = N\delta[k - m].$$

The Fourier series basis functions are $\{\phi_n\}_{n=-\infty}^{\infty}$, where

$$\phi_n(x) = e^{i2\pi nx/L},$$

and the orthogonality relationship is

$$\langle \phi_k, \phi_m \rangle = \int_0^L \phi_k(x)\,\phi_m(x)\,dx = L\delta[k-m].$$

Using the delta function, we can now interpret the Fourier transform and the discrete-time Fourier transforms as expansions on orthogonal bases. Evidently the basis functions for the Fourier transform are the complex exponentials $e^{i2\pi vx}$, where v is continuous rather than a discrete index. The orthogonality condition expressed as the integral

$$\int_{-\infty}^{\infty} e^{i2\pi vx} e^{-i2\pi \mu x}\,dx$$

does not exist in the ordinary sense. But, applying the transform pair $1 \longleftrightarrow \delta(v)$ (Equation 6.12), we have a generalized orthogonality relationship for the Fourier transform basis set $\{e^{i2\pi vx}\}_{v \in \mathbb{R}}$:

$$\int_{-\infty}^{\infty} e^{i2\pi vx} e^{-i2\pi \mu x}\,dx = \delta(v - \mu).$$

For the discrete-time Fourier transform, the basis functions are $\{e^{in\theta}\}_{\theta \in [-\pi,\pi)}$. Their orthogonality relationship is

$$\sum_{n=-\infty}^{\infty} e^{in\theta} e^{-in\phi} = \lim_{N \to \infty} D_{2N+1}\left(\frac{\theta - \phi}{2\pi}\right) = 2\pi\delta(\theta - \phi).$$

6.6.2 Sampling and Replication

The fact that the comb function is its own transform, with the result that $f \cdot \text{III} \longleftrightarrow F * \text{III}$, shows that sampling in one domain always results in periodicity in the other domain. Let f be a continuous aperiodic function, and let F be its Fourier transform. This is the only Fourier transform where both domains are continuous. The Fourier series is discrete in the frequency domain and periodic in the time domain. The discrete-time Fourier transform is discrete in the time domain and periodic in the frequency domain. The DFT is discrete and periodic in both domains. These three transforms are obtained from the continuous time Fourier pair $f \longleftrightarrow F$ by sampling and replication.

1. *Sampled* $\longmapsto$ *replicated*. The Fourier transform of a sampled function, $f_s(x) = f(x) \cdot \frac{1}{\Delta x}\text{III}\left(\frac{x}{\Delta x}\right)$, is a periodic replication of F:

$$f(x) \cdot \frac{1}{\Delta x}\text{III}\left(\frac{x}{\Delta x}\right) \longmapsto F(v) * \text{III}(\Delta x\, v).$$

The impulse sampled function is also written

$$f(x) \cdot \frac{1}{\Delta x}\text{III}\left(\frac{x}{\Delta x}\right) = \sum_{n=-\infty}^{\infty} f[n]\,\delta(x - n\Delta x),$$

where $f[n] = f(n\Delta x)$. The Fourier transform, by a direct calculation, is $\sum_{n=-\infty}^{\infty} f[n] e^{-i2\pi n \Delta x \nu}$. Compare this expression with the discrete-time Fourier transform of $(f[n])$,

$$F_d(\theta) = \sum_{n=-\infty}^{\infty} f[n] e^{-in\theta}.$$

If we set $\theta = 2\pi \nu \Delta x$, the usual mapping from analog frequency to digital frequency, then we see that F_d is the same as $F * \text{III}$, that is, F_d is the periodic replication of F:

$$F(\nu) * \text{III}(\Delta x \, \nu) = \sum_k \frac{1}{\Delta x} F\left(\nu + \frac{k}{\Delta x}\right)$$

$$= \sum_k \frac{1}{\Delta x} F\left(\frac{\theta + 2\pi k}{2\pi \Delta x}\right) = F_d(\theta)\bigg|_{\theta = 2\pi \nu \Delta x}.$$

This may be diagrammed as follows:

$$\begin{array}{ccc} f(x) & \xrightarrow{\mathcal{F}} & F(\nu) \\ \cdot \text{III} \downarrow & & *\text{III} \downarrow \\ f[n] & \xrightarrow{\mathcal{F}} & F_d(\theta) \end{array}$$

If f is bandlimited, then $F_d|_{\theta \in [-\pi, \pi)}$ is identically $F/\Delta x$, and $f(x)$ may be recovered from its samples according to the sampling theorem.

2. *Replicated* $\longmapsto$ *sampled*. The Fourier transform of the periodically replicated function, $\tilde{f}(x) = f(x) * \frac{1}{L} \text{III}\left(\frac{x}{L}\right)$, is

$$f(x) * \frac{1}{L} \text{III}\left(\frac{x}{L}\right) \longmapsto F(\nu) \cdot \text{III}(L\nu)$$

$$= \sum_{n=-\infty}^{\infty} \frac{1}{L} F\left(\frac{n}{L}\right) \delta\left(\nu - \frac{n}{L}\right).$$

A periodic function has a line spectrum. The samples of F are the Fourier coefficients of $\tilde{f}$, $c_n = \frac{1}{L} F\left(\frac{n}{L}\right)$.

$$\begin{array}{ccc} f(x) & \xrightarrow{\mathcal{F}} & F(\nu) \\ *\text{III} \downarrow & & \cdot \text{III} \downarrow \\ \tilde{f}(x) & \xrightarrow{\mathcal{F}} & c_n \end{array}$$

If f has bounded support smaller than L, then one period of $\tilde{f}$ is identically f, and F may be recovered from the Fourier coefficients by an interpolation analogous to the sampling theorem.

3. *Replicated, sampled* $\longmapsto$ *Sampled, replicated*. We restrict attention to the case $L = N\Delta x$, so that the sequence $(\tilde{f}[n]) = (\tilde{f}(n\Delta x))$ will also be periodic. We know $\tilde{f} \longmapsto \sum_m c_m \delta(\nu - m/L)$. Sampling $\tilde{f}$ then replicates the transform, with period $1/\Delta x = N/L$; that is, the sequence of Fourier coefficients (c_m) is replicated with

period N. At the mth frequency $\nu = m/L = m/N\Delta x$, the Fourier coefficient is a superposition of coefficients:

$$\tilde{c}_m = \cdots c_{m-N} + c_m + c_{m+N} + \cdots.$$

This new sequence is periodic, as expected. Only N coefficients are unique, say $\tilde{c}_0$ through $\tilde{c}_{N-1}$. We previously derived (Equation 4.41)

$$\tilde{F}[m] = N \sum_{r=-\infty}^{\infty} c_{m+rN}, \quad m = 0, 1, \ldots, N-1,$$

where $\tilde{F}[m]$ is the discrete Fourier transform of the single period, $(\tilde{f}[n])_{n=0}^{N-1}$. Thus,

$$\tilde{c}_m = \frac{1}{N}\tilde{F}[m].$$

The Fourier transform of a sampled, replicated function is the DFT:

$$\begin{array}{ccc} \tilde{f}(x) & \xrightarrow{\mathcal{F}} & c_m \\ \cdot\text{III}\downarrow & & *\text{III}\downarrow \\ \tilde{f}[n] & \xrightarrow{\mathcal{F}} & \tilde{F}[m] \end{array}$$

If f is bandlimited, then the values of the DFT are identically the Fourier coefficients and $\tilde{f}$ may be recovered from $\tilde{f}[n]$ by the sampling theorem. However, it is impossible for f simultaneously to be bandlimited and of bounded support, so the unreplicated $f(x)$ cannot be recovered from the replicated samples $\tilde{f}[n]$.

4. *Sampled, replicated $\longmapsto$ replicated, sampled*. We know that $(f[n]) \longmapsto F_d(\theta)$. Replicating the sample sequence with period $L = N\Delta x$ then results in sampling the Fourier transform with sampling interval $\Delta\nu = 1/N\Delta x$. Using the continuous-to-discrete frequency mapping, this is equivalent to sampling $F_d(\theta)$ with $\Delta\theta = 2\pi\Delta\nu\Delta x = 2\pi/N$:

$$F_d(\theta) \longrightarrow \sum_m F_d(2\pi m/N)\,\delta(\theta - 2\pi m/N).$$

The sample sequence $(F_d(2\pi m/N))$ is periodic, as expected, $F_d(2\pi(m+N)/N) = F_d(2\pi m/N)$. Only N samples are required to describe the function completely, say $F_d(0)$ through $F_d(2\pi(N-1)/N)$. These values are identically the DFT of $(\tilde{f}[n])$:

$$F_d(2\pi m/N) = \sum_{k=-\infty}^{\infty} f[k]\,e^{-i(2\pi m/N)k}.$$

The complex exponential is periodic in k with period N, so we may write $k = n + rN$ and rewrite the sum:

$$F_d(2\pi m/N) = \sum_{r=-\infty}^{\infty} \sum_{n=0}^{N-1} f[n+rN]\, e^{-i(2\pi m/N)(n+rN)}$$

$$= \sum_{n=0}^{N-1} \left(\sum_{r=-\infty}^{\infty} f[n+rN] \right) e^{-i2\pi mn/N}$$

$$= \sum_{n=0}^{N-1} \tilde{f}[n]\, e^{-i2\pi mn/N} = \tilde{F}[m].$$

The Fourier transform of a replicated, sampled function is also the DFT:

$$\begin{array}{ccc} f[n] & \xrightarrow{\mathcal{F}} & F_d \\ {\scriptstyle *\mathrm{III}} \downarrow & & \downarrow {\scriptstyle \cdot\mathrm{III}} \\ \tilde{f}[n] & \xrightarrow{\mathcal{F}} & \tilde{F}[m] \end{array}$$

These suggest that the sampling and replication may be done in either order, that is,

$$\left(f(x) * \frac{1}{N\Delta x}\mathrm{III}\left(\frac{x}{N\Delta x}\right)\right) \cdot \frac{1}{\Delta x}\mathrm{III}\left(\frac{x}{\Delta x}\right) = \left(f(x) \cdot \frac{1}{\Delta x}\mathrm{III}\left(\frac{x}{\Delta x}\right)\right) * \frac{1}{N\Delta x}\mathrm{III}\left(\frac{x}{N\Delta x}\right)$$

Direct proof of this is left to the problems.

The connections among the four Fourier transforms may be combined into a single diagram (Figure 6.21).[25]

6.7 SUMMARY

Operational Rules for Generalized Functions

Definition	$\int f\varphi = \lim_{n\to\infty} \int_{-\infty}^{\infty} f_n(x)\varphi(x)\, dx$ for all $\varphi \in S$	Definition 6.2
Equivalence	$f = g$ if $\int f\varphi = \int g\varphi$ for all $\varphi \in S$	Definition 6.4
Derivative	$\int f'\varphi = -\int f\varphi'$	Equation 6.23
Fourier transform	If $f_n \to f$ and $f_n \longleftrightarrow F_n$, then $F_n \to F$ $\int f\Phi = \int F\varphi$, where $\varphi \longleftrightarrow \Phi$	Equation 6.35

[25] Also see Kammler (2000, pp. 31–37) for a similar diagram and development.

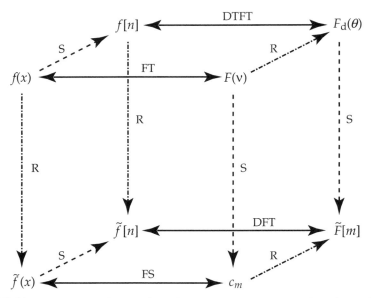

FIGURE 6.21 The Fourier family of transforms related by sampling (S) and periodic replication (R). Bidirectional arrows indicate invertible Fourier transforms. The sampling and replication operations are unidirectional, except: a function can be reconstructed from its sample sequence by interpolation if its Fourier transform has bounded support and the sampling interval is sufficiently small (sampling theorem). Consult the text for mathematical details of each mapping.

Operational Rules for the Delta Function

Unit area	$\int_{-\infty}^{\infty} \delta(x)dx = 1$	Equation (6.4)
Sifting	$\int_{-\infty}^{\infty} f(x)\delta(x-a)dx = f(a)$	
	$f(x)\delta(x-a) = f(a)\delta(x-a)$	Equation (6.3)
Convolution	$f(x) * \delta(x-a) = f(x-a)$	Equation (6.16)
Step function	$U(x) = \int_{-\infty}^{x} \delta(\xi)\,d\xi,\; U'(x) = \delta(x)$	Equation (6.9)
Dilation	$\delta(ax) = \frac{1}{|a|}\delta(x)$	Equation (6.7)
Derivative	$\int_{-\infty}^{\infty} f(x)\delta'(x-a)dx = -f'(a)$	Equation (6.10)
	$f(x) * \delta'(x-a) = f'(x-a)$	

Fourier Transform Pairs Derived in this Chapter

$$1 \longleftrightarrow \delta(\nu) \qquad (6.11)$$

$$e^{i2\pi x} \longleftrightarrow \delta(\nu - 1) \qquad (6.12)$$

$$\cos(2\pi x) \longleftrightarrow \frac{1}{2}\delta(\nu - 1) + \frac{1}{2}\delta(\nu + 1) \qquad (6.13)$$

$$\sin(2\pi x) \longleftrightarrow \frac{1}{2i}\delta(\nu - 1) - \frac{1}{2i}\delta(\nu + 1) \qquad (6.14)$$

$$\operatorname{sgn} x \longleftrightarrow \frac{1}{i\pi\nu} \qquad (6.36a)$$

$$\frac{1}{x} \longleftrightarrow -i\pi \operatorname{sgn} \nu \qquad (6.36b)$$

$$U(x) \longleftrightarrow \frac{1}{2}\delta(\nu) + \frac{1}{i2\pi\nu} \qquad (6.38)$$

$$x^n \longleftrightarrow \frac{\delta^{(n)}(\nu)}{(-i2\pi)^n} \qquad (6.39)$$

$$\delta^{(n)}(x) \longleftrightarrow (i2\pi x)^n$$

$$\operatorname{III}(x) \longleftrightarrow \operatorname{III}(\nu) \qquad (6.46)$$

Discrete-Time Fourier Transform Pairs Derived in this Chapter

$$1 \longleftrightarrow 2\pi\delta(\theta) \qquad \text{(Example 6.38)}$$

$$e^{i\alpha n} \longleftrightarrow 2\pi\delta(\theta - \alpha)$$

$$\cos(\alpha n) \longleftrightarrow \pi\delta(\theta - \alpha) + \pi\delta(\theta + \alpha)$$

$$\sin(\alpha n) \longleftrightarrow -i\pi\delta(\theta - \alpha) + i\pi\delta(\theta - \alpha)$$

$$\operatorname{sgn}[n] \longleftrightarrow \frac{1}{2i}\cot\left(\frac{\theta}{2}\right)$$

$$U[n] \longleftrightarrow \frac{1}{2} + \pi\delta(\theta) + \frac{1}{2i}\cot\left(\frac{\theta}{2}\right)$$

$$n \longleftrightarrow i2\pi\delta'(\theta) \qquad \text{(Equation 6.55)}$$

$$nU[n] \longleftrightarrow i\pi\delta'(\theta) - \frac{1}{2}\csc^2\left(\frac{\theta}{2}\right) \qquad (6.56)$$

$$\operatorname{III}_N[n] \longleftrightarrow \operatorname{III}\left(\frac{N\theta}{2\pi}\right) \qquad (6.57)$$

Mappings Between Transforms (Section 6.6.2)
Periodization (Fourier series)

$$\tilde{f}(x) = \sum_k f(x + kL), \quad x \in [0, L]$$

$$c_n = \frac{1}{L}F\left(\frac{n}{L}\right), \quad n \in \mathbb{Z}.$$

Sampling (Discrete-time Fourier transform)

$$f[n] = f(n\Delta x), \quad n \in \mathbb{Z}$$

$$F_d(\theta) = \sum_k \frac{1}{\Delta x}F(\nu + k/\Delta x)\Big|_{\nu=\theta/2\pi\Delta x} = \sum_k \frac{1}{\Delta x}F\left(\frac{\theta + 2\pi k}{2\pi\Delta x}\right), \quad \theta \in [-\pi, \pi)$$

Sampled and periodic (Discrete Fourier transform)

$$L = N\Delta x$$
$$\tilde{f}[n] = \tilde{f}(n\Delta x)$$
$$= \sum_k f((n+kN)\Delta x) = \sum_k f[n+kN], \quad n = 0, 1, \ldots N-1$$
$$\tilde{F}[m] = N \sum_k c_{m+kN} = F_d\left(\frac{2\pi m}{N}\right), \quad m = 0, 1, \ldots, N-1$$

PROBLEMS

6.1. *Tempered distributions*
  (a) Verify that the set $\mathcal{S}$ of good functions is a vector space and, in fact, is a subspace of L^2.
  (b) Show that the dual space $\mathcal{S}'$ of tempered distributions is itself a vector space.

6.2. Derive the dilation formula (Equation 6.20):
$$\int f(ax)\,\varphi(x)\,dx = \int f(x)\,\frac{1}{|a|}\varphi\left(\frac{x}{a}\right)dx.$$

6.3. In Example 6.2 it was shown that $\delta(x^2 - 1) = \frac{1}{2}\delta(x+1) + \frac{1}{2}\delta(x-1)$. Follow the sequence of pulses method in that example to show that $\delta(x^2)$ is undefined.

6.4. Perform the following calculations:
  (a) $\displaystyle\int_{-\infty}^{\infty} \delta(x-1)\cos\pi x\,dx.$
  (b) $\displaystyle\int_0^{\infty} \delta(x+2)\sin(4x)\,dx.$
  (c) $\displaystyle\int_{-\infty}^{\infty} \delta(2x-1)e^{-x}\,dx.$

6.5. A certain generalized function f has the property that when integrated against a testing function φ, the following result is obtained:
$$\int_{-\infty}^{\infty} f(x)\varphi(x)\,dx = \frac{1}{2}\varphi(1) - \frac{1}{2}\varphi(-1).$$
Give an expression for f.

6.6. Derive the symmetry relationships for generalized functions (Equations 6.21).

6.7. Verify the definitions for real and imaginary generalized functions (Equations 6.22). *Hint:* Begin by deriving a definition for the complex conjugate f^* of a generalized function. Then recall that the imaginary part of a real function is zero, and the real part of an imaginary function is zero.

6.8. Derive a definition for a Hermitian generalized function that is analogous to the definitions for real–imaginary and even–odd generalized functions.

6.9. Use the definition of generalized derivative to show that $\operatorname{sgn}' = 2\delta$.

6.10. In the text, $1/x$ as a generalized function is defined as the generalized derivative of $\log |x|$. Show that $\log |x|$ is slowly growing, that is, that

$$\int_{-\infty}^{\infty} \frac{|\log |x||\, dx}{1+x^2} < \infty.$$

6.11. In the text the identity $x\delta' = -\delta$ was derived. This problem extends the definition.

(a) Express $x^2\delta'$ in terms of linear combinations of δ and its derivatives. Generalize to $x^n\delta'$, $n \geq 1$.

(b) Express $x\delta''$ and $x^2\delta''$ in terms of linear combinations of δ and its derivatives.

(c) Derive the general formula

$$x^n \delta^{(m)}(x) = \begin{cases} 0, & m < n \\ (-1)^n \frac{m!}{(m-n)!} \delta^{(m-n)}(x), & m \geq n \end{cases}. \quad (6.58)$$

6.12. Let f and g be generalized functions and assume that their convolution $f * g$ is defined.

(a) Show that convolution commutes, $f * g = g * f$.

(b) Show that the convolution has a generalized derivative, and that $(f * g)' = f' * g = f * g'$.

6.13. Derive the generalized convolution relationships for $\delta^{(r)}$ (Equation 6.33).

6.14. Under what circumstances is multiplication associative, $(fg)h = f(gh)$? Try this example:

(a) $(xx^{-1})\delta(x)$.

(b) $x^{-1}(x\delta)$.

(c) $x(x^{-1}\delta(x))$.

6.15. *Associativity of convolution*

(a) Assume that $f * g$ is well defined as a generalized function. Under what conditions is $(f * g) * h$ well defined as a generalized function?

(b) With the conditions you assumed in (a), show that $(f * g) * h = f * (g * h)$.

6.16. In Chapter 5, we saw that for square integrable f and g, the product of their Fourier transforms is square integrable and so $\mathcal{F}^{-1}\{FG\} = f * g \in L^2$. But the convolution of two square-integrable functions is only guaranteed to be bounded and continuous, so the forward transform $f * g \longmapsto FG$ could not be justified. Using the fact that a square-integrable function is also a regular generalized function, can you now resolve this paradox?

6.17. *Derivative of the delta function*

(a) What is the "area" under δ'? Is it an even function or an odd function?

(b) Derive Equation 6.30:

$$f(x)\delta'(x) = f(0)\delta'(x) - f'(0)\delta(x).$$

(c) Generalize part (b) to an expression for $f(x)\delta^{(r)}(x)$.

6.18. Calculate, by carrying out the convolution integral, $x * e^{-x}U(x)$. Compare with the result in Example 6.34(c).

6.19. Calculate, by carrying out the convolution integral, $\cos 2\pi x * \text{rect}(2x)$. Compare with the result in Example 6.34(d).

6.20. *Cauchy principal value*
Show that the generalized Fourier transform of $\mathcal{P}(1/x)$ may be calculated by the integral
$$\int_{-\infty}^{\infty} (-i2\pi v) \operatorname{sinc}(2vx)\, dx$$
and from here, show that $\mathcal{P}(1/x) \longmapsto -i\pi \operatorname{sgn} v$.

6.21. Verify the dilation theorem for the generalized Fourier transform. (Follow the plan of the proof of the shift theorem.)

6.22. Calculate the Fourier transform of $|x| = x \operatorname{sgn} x$ by convolution.

6.23. The following two sequences, derived in the text, are claimed to converge to the transform pair $U(x) * U(x) = xU(x) \longleftrightarrow -\frac{1}{(2\pi v)^2} + \frac{i}{4\pi}\delta'(v)$:
$$f_n * g_n(x) = xe^{-x/n}U(x),$$
$$F_n G_n(v) = \left(\frac{n}{1 + i2\pi nv}\right)^2.$$

(a) Show that $xe^{-x/n}U(x) \to xU(x)$ as $n \to \infty$ (this is the easy part).

(b) Separate the sequence for $F_n G_n$ into real and imaginary parts, and show that the real part becomes $-\frac{1}{(2\pi v)^2}$ as $n \to \infty$.

(c) The members of the sequence $\frac{n}{1+(\pi nv)^2}$ are unit area peaks, equivalent to a delta function, $\delta(v)$. Use this fact to show that the imaginary part of $F_n G_n$ becomes $\frac{1}{4\pi}\delta'(v)$ as $n \to \infty$.

6.24. *Weak convergence*
Let (f_n) be a sequence of generalized functions, weakly convergent to a generalized function f. Show that the sequence of derivatives (f_n') converges weakly to f'.

6.25. *Weak convergence*
Prove Theorem 6.5.

6.26. *Derivative theorem*
In Example 6.32, the derivative theorem was used to calculate the Fourier transform of $\operatorname{sgn} x$.

(a) Use a similar approach to calculate the Fourier transform of x^{-1}.

(b) Calculate the Fourier transform of $\log |x|$.

6.27. *Derivative theorem*
By applying the derivative theorem to the transform pair $\Lambda(x) \longleftrightarrow \operatorname{sinc}^2 v$, calculate the second derivative of $\Lambda(x)$. Also explain in the space domain.

6.28. Use the generalized Fourier theorems to calculate the transforms of the following functions:

(a) $\operatorname{rect}(x) \operatorname{sgn}(x)$.

(b) $e^{-t}U(t) \cos 2\pi at$.

(c) $\sin 2\pi bt\, U(t)$.

(d) e^{ix}.

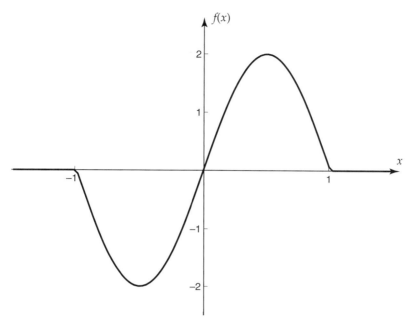

FIGURE 6.22 For Problem 6.30.

6.29. Calculate and sketch accurately the Fourier transform of the following functions:
  (a) $\frac{1}{2}\cos(2\pi x) + 3\sin(3\pi x)$.
  (b) $2 - \sin(\pi x/2)$.
  (c) $1 + \cos^2(2\pi x)$.

6.30. Calculate the Fourier transform of the function shown in Figure 6.22.

6.31. Use the convolution theorem to prove the following transform pairs:
  (a) $\text{rect}(x)\cos(\pi x) \longleftrightarrow \frac{1}{2}\text{sinc}\left(v + \frac{1}{2}\right) + \frac{1}{2}\text{sinc}\left(v - \frac{1}{2}\right)$.
  (b) $x\,\text{rect}(x) \longleftrightarrow \frac{i}{2\pi}\frac{d}{dv}\text{sinc}\,v$.
  (c) $\Lambda(x) * \text{sgn}(x) \longleftrightarrow -i\frac{\text{sinc}^2 v}{\pi v}$.

6.32. The *ideal bandpass filter* (Figure 6.23) is defined by the transfer function

$$H(v) = \text{rect}\left(\frac{v + v_0}{B}\right) + \text{rect}\left(\frac{v - v_0}{B}\right).$$

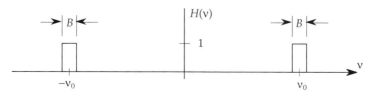

FIGURE 6.23 For Problem 6.32. Frequency response of the ideal bandpass filter.

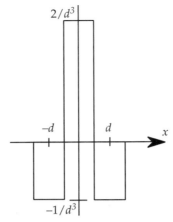

FIGURE 6.24 For Problem 6.33.

Express H as a convolution and use the convolution theorem to calculate the Fourier transform.

6.33. Consider the function h shown in Figure 6.24.

(a) Express this function as a convolution of a single rectangle with three delta functions.

(b) Calculate the Fourier transform H of this function.

(c) Consider the convolution $g = h * \varphi$ of h with a good function φ. Discuss the behavior of g as $d \to 0$. What operation does "$h *$" approximate?

6.34. *Signals with finite average power*

Pure sinusoidal functions are not square integrable and Parseval's formula is not applicable. For such signals, we may work instead with the long-time averaged power, defined:

$$P_{\text{avg}} = \lim_{T \to \infty} \frac{1}{2T} \int_{-T}^{T} |f(t)|^2 \, dt.$$

(a) Using Fourier theorems, show that

$$P_{\text{avg}} = \lim_{T \to \infty} \int_{-\infty}^{\infty} F \star F(\xi) \operatorname{sinc}(2T\xi) \, d\xi,$$

where F is the Fourier transform of f and $\star$ denotes correlation.

(b) Now let $f(t) = A \cos(2\pi \nu_0 t)$, and show that the expression you derived in (a) gives $P_{\text{avg}} = A^2/2$, which we know is the correct average power for f.

6.35. *Image deblurring*

A certain camera records an image while in constant linear motion in the x direction, as shown in Figure 6.25. We shall assume the imaging is ideal in the y direction and concentrate on the x direction. The recorded image (blurred) is related to the ideal image by a filtering operation.

(a) Let T be the exposure time and v the velocity of the camera. For simplicity, assume the camera shutter is open from $t = -T/2$ to $t = T/2$ (rather than from 0 to T).

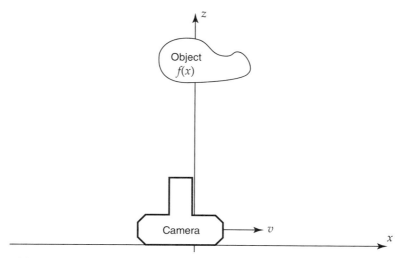

FIGURE 6.25 For Problem 6.35. Recording an image with a moving camera.

Show that the blurred image $g(x)$ may be expressed in terms of the ideal image $f(x)$ using the convolution relationship:

$$g(x) = Tf(x) * \frac{1}{vT}\text{rect}\left(\frac{x}{vT}\right).$$

(b) What is the transfer function $H(\nu)$ of this imaging system (ν is spatial frequency)?

(c) It is proposed to attempt deblurring of $g(x)$ by filtering with a transfer function $K(\nu) = \frac{1}{H(\nu)}$, called the *inverse filter*. The idea is expressed by

$$\hat{F}(\nu) = K(\nu)G(\nu) = \frac{1}{H(\nu)}H(\nu)F(\nu) = F(\nu).$$

Can we recover $F(\nu)$ perfectly? *Hint:* Consider the effect of noise, due to recording, which is added to $g(x)$ after the blurring. If perfect recovery is not possible, how well could we do?

(d) It is claimed that, in the absence of noise, the ideal image $f(x)$ can be recovered from the recorded image $g(x)$ by differentiation. Analyze this proposal. If the ideal image has a finite width X, what relationship must hold between X, v, and T if deblurring is to be possible in this way? What would the effects of noise be on this system?

6.36. *Modulation*

Modulation is the process of impressing information (speech, video, data, etc.) on a "carrier" signal for the purpose of transmitting that information over a communication channel (radio waves in air, electric current in a telephone wire, lightwave in a fiberoptic cable, etc.). The simplest form of carrier modulation is amplitude modulation, or AM.

The general equation describing AM is

$$y(t) = [1 + mx(t)]\cos(2\pi\nu_c t),$$

where $x(t)$ is the information, or message, ν_c is the frequency of the carrier signal, and m is a system parameter called the modulation depth (normally, $m \le 1$). Simple electronic circuits are capable of performing this operation. Assume that $x(t)$ is a pure tone (not very interesting, but easy to analyze), $x(t) = \cos(2\pi v_m t)$, where $v_m \ll v_c$. (In AM radio, v_c is on the order of 1000 kHz, while v_m is less than 5 kHz.)

(a) Sketch the modulated signal $y(t)$, labeling accurately the salient features of the graph. Because of the large difference between v_c and v_m, you will have to be creative in designing your sketch.

(b) Calculate and plot the Fourier spectrum $Y(v)$, labeling accurately the salient features of the graph.

(c) How do these plots change as the message frequency and modulation depth are varied?

6.37. Let $g = f * h$, for example, g is the output of an LTI system with impulse response h. Prove the following relationships:

(a) Autocorrelation $g \star g = (f \star f) * (h \star h)$.

(b) Cross-correlation $f \star g = (f \star f) * h$.

(c) A *pseudo-random binary sequence* is a deterministic (non-random) sequence f with the property that $f \star f \approx \delta$. Suppose h is unknown and f is a pseudo-random binary sequence. Use the above results to design a method for "identifying" the impulse response from a suitably designed experiment.

6.38. *Periodic replication*

Let $f \in L^1(\mathbb{R})$.

(a) Show that the replication,

$$\tilde{f} = \sum_{k=-\infty}^{\infty} f(x + kL),$$

is absolutely integrable on $[0, L]$. Begin with a partial sum, $\sum_{k=-N}^{N} f(x + kL)$, and show

$$\int_0^L \left| \sum_{k=-N}^{N} f(x + kL) \right| dx \leq \|f\|_1$$

for all N.

(b) Then show that $\tilde{f}(x) = \tilde{f}(x + L)$.

6.39. *Fourier transform of the comb function*

The Fourier transform of the comb function can be calculated without passing through the Fourier series. Fourier transform the comb function term-by-term:

$$\mathcal{F}\{\mathrm{III}(x)\} = \sum_n \mathcal{F}\{\delta(x - n)\} = \sum_n e^{-i2\pi v n}.$$

Now integrate this result against a testing function and show that

$$\int \mathcal{F}\{\mathrm{III}\}\, \varphi(v)\, dv = \sum_{n=-\infty}^{\infty} \Phi(n) = \sum_{n=-\infty}^{\infty} \varphi(n).$$

From here, show that $\mathcal{F}\{\mathrm{III}(x)\}$ and $\mathrm{III}(v)$ are equivalent generalized functions.

6.40. (a) Sketch the function $f(x) = \Lambda(x/4)\, \mathrm{III}(2x)$, then calculate the Fourier transform $F(v)$.

(b) We know $\delta(x) * \delta(x) = \delta(x)$. What is $\delta(x) * \mathrm{III}(x)$? What about $\mathrm{III}(x) * \mathrm{III}(x)$?

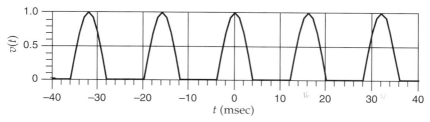

FIGURE 6.26 For Problem 6.42. A half-wave rectified sinusoid.

6.41. (a) Using comb and other functions, as needed, give a concise expression for the rectified cosine function, $f(x) = |\cos 2\pi ax|$.

(b) Using the Fourier transform, calculate the Fourier series coefficients of $f(x)$.

6.42. For the half-wave rectified sinusoid (Figure 6.26),

(a) Develop a mathematical model in the form of a convolution of a comb with the product of a cosine and rect.

(b) Calculate the Fourier transform of the unit period and from this derive the Fourier coefficients.

6.43. Consider the periodic waveforms that are produced by periodically replicating a half-cosine, $\cos(2\pi t/T) \operatorname{rect}(2t/T)$. Replicating with period T produces a half-wave rectified sinusoid, and replicating with period $T/2$ produces a full-wave rectified sinusoid (see the previous problems). Here, consider replication of the half-cosine with an arbitrary period T/N, $N = 1, 2, \ldots$. Derive concise expressions for the periodic waveform and its Fourier transform, make graphs for a few values of N, and compare them.

6.44. Show that the Dirichlet kernel is the periodic replication of the sinc function. Begin with the definition

$$D_{2N+1}(x) = \sum_{n=-N}^{N} e^{i2\pi nx}$$

and show that its Fourier transform is

$$\mathcal{F}\{D_{2N+1}(x)\} = \operatorname{rect}\left(\frac{\nu}{2N+1}\right) \cdot \operatorname{III}(\nu),$$

then calculate the inverse transform. Plot a few replicas of the sinc to see how this works.

6.45. Amplitude-modulated radio signals (Problem 6.36) are usually demodulated by envelope detection. An envelope detector consists of a half-wave rectifier followed by a lowpass filter. The half-wave rectifier blocks the negative half cycles of the modulated carrier, as shown in Figure 6.27.

(a) Derive a mathematical model for the rectified AM signal, and calculate its Fourier transform, assuming that the message spectrum is bandlimited to 5 kHz and the carrier frequency is 1000 kHz.

(b) A lowpass filter follows the half-wave rectifier to complete the envelope detector. Use your model to explain the function of this filter.

6.46. *Bandlimited functions*
Let $f \in L^2$ be bandlimited, $F(\nu) = 0$ for $|\nu| > B > 0$. Show that the set V_B of all such functions is a linear subspace of L^2.

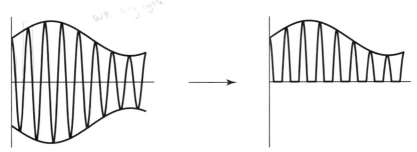

FIGURE 6.27 For Problem 6.45. Half-wave rectification of an amplitude-modulated waveform.

6.47. *Sampling theorem*

The sampling theorem is derived in the text for bandlimited, finite energy functions. Another case of interest is bandlimited periodic functions. A function of the form

$$f(x) = \sum_{n=-N}^{N} c_n e^{i2\pi nx/L}$$

is bandlimited, but does not have finite energy. Nevertheless, show that it can be reconstructed from uniformly spaced samples $f(k\Delta x)$ according to Equation 6.52.

6.48. *Sampling theorem, alternate derivation*

Let f be bandlimited and have finite energy. The Fourier transform F has bounded support and can be written as a Fourier series:

$$F(\nu) = \sum_{n=-\infty}^{\infty} c_n e^{-i2\pi n\nu/2B}, \quad |\nu| < B.$$

(a) Show that the Fourier coefficients are $c_n = \frac{1}{2B} f\left(\frac{n}{2B}\right)$.

(b) Perform an inverse Fourier transform of F and show that

$$f(x) = \sum_{n=-\infty}^{\infty} f\left(\frac{n}{2B}\right) \operatorname{sinc}\left[2B\left(x - \frac{n}{2B}\right)\right].$$

6.49. The system shown in Figure 6.28 models a device called a *chopper*, used in laser experiments. A function f is multiplied by a unit-amplitude square wave of period T and 50% duty cycle (time on = time off), producing a "chopped" output g.

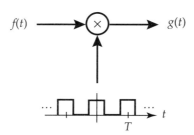

FIGURE 6.28 For Problem 6.49. Square-wave modulation system.

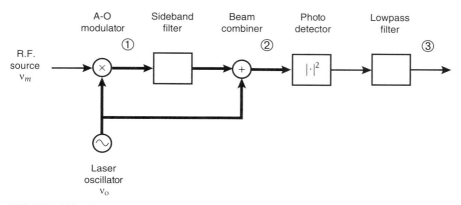

FIGURE 6.29 For Problem 6.50. Acousto-optic spectrum analyzer. Heavy lines indicate an optical beam. Lighter lines indicate an electrical signal path.

(a) Derive an expression for the Fourier transform $G(\nu)$ of the output.

(b) Assuming that f is bandlimited with bandwidth less than $\frac{1}{2T}$, make an accurate sketch of $G(\nu)$. You may assume that $F(\nu)$ has a convenient shape (e.g., a triangle).

6.50. *Spectrum analysis*

Devices called acousto-optic modulators enable very high frequency modulation of a laser beam. A block diagram for a spectrum analyzer using an acousto-optic modulator is shown in Figure 6.29.

Assume that the laser and the radio frequency (RF) source are sinusoids, with frequencies $\nu_o = 5 \times 10^{14}$ Hz and $\nu_m = 150$ MHz, respectively. The acousto-optic modulator behaves much like a diffraction grating (or, if you will, an AM modulator), producing an undiffracted beam and two diffracted orders, one to either side of the undiffracted beam. The diffraction angle increases with the modulating frequency ν_m. The "sideband filter" is a barrier which completely blocks the undiffracted light and the lower-frequency diffracted order. The remaining diffracted order is recombined with the laser beam, and the sum is incident on a photodetector, modeled as a square-law device. The output of the detector is an electrical signal. This signal is finally passed through a lowpass filter which blocks all frequencies above 400 MHz.

(a) Find and plot the spectra at the indicated points in the diagram.

(b) Suppose the laser actually oscillates in two frequencies, separated by $\Delta \nu = 414$ MHz. Describe the effect this has on the operation of the spectrum analyzer. In particular, what is the practical range of input frequency ν_m which may reliably be used in this system?

6.51. *Reverberation*

Here is a simple model of reverberation in a room (or, intentional reverb in an audio system). Consider a device (called a *delay line*) which introduces a pure time delay from input to output: $\text{Out}(t) = \text{In}(t - T)$. The output of the delay line is fed back and added to the input, with an intermediate gain factor, a (Figure 6.30).

(a) Show that the input–output behavior of this system is described by the *difference* equation

$$g(t) - ag(t - T) = f(t).$$

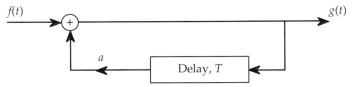

FIGURE 6.30 For Problem 6.51. Model of a reverberant environment.

In an auditorium, the time T would be the round-trip delay time between the stage and the back wall of the room. Then it hits the wall at the front of the auditorium and goes round again. The factor a models whether the room is "lively" or "dead." This factor will be between zero and one, unless there is a PA system—then, a fraction of the sound could be picked up by the microphone, amplified, and sent back out again.

(b) Find the impulse response of this system by injecting an impulse, $f(t) = \delta(t)$, and tracing its fate through the equation.

(c) Apply the Fourier transform to the difference equation and determine the transfer function of the system. Note that if $a > 1$, you cannot get a transform (why?). Sketch the magnitude of the transfer function, labeling salient features (like the effects of a and T).

6.52. *Analog to digital converters*

Many A/D converters include a "sample and hold" circuit at the input, which consists of an analog switch and a capacitor. The analog switch closes for a short time, sometimes called the "aperture time," enabling the input voltage to charge the capacitor. The capacitor voltage is then converted to digital form by the rest of the electronics. Charging a capacitor is basically an integration operation, which means that the A/D does not convert the instantaneous input voltage, but an integral of the input voltage. Let $v_{sh}(t)$ be the capacitor voltage, and $v_{in}(t)$ be the input voltage. The two are approximately related by the integral

$$v_{sh}(t) = \frac{1}{T_a} \int_{t-T_a}^{t} v_{in}(\tau)\, d\tau,$$

where T_a is the aperture time, and t is the time when a sample is to be taken.

(a) Let $v_s(t)$ be the signal obtained by impulse sampling $v_{sh}(t)$ at a sampling interval $\Delta t > T_a$. Derive an equation for v_s in terms of v_{in}, comb, and rect.

(b) Calculate the Fourier transform of v_s. What is the effect of the finite aperture time on the spectrum?

6.53. *Digital-to-analog converters*

According to the sampling theorem, an ideal lowpass filter can (theoretically) reconstruct a bandlimited signal perfectly from its samples. In the time domain, the impulse response of the filter, which is a sinc function, interpolates between the samples. The usual digital-to-analog converter (D/A) does not work this way. Instead, it connects the samples by an operation called a zero-order hold (ZOH) — that is, the output of the converter is a constant equal to the sample value until the next sample comes along. This results in the characteristic "stairstep" appearance of D/A outputs (Figure 6.31).

(a) Model a general sampled signal by a train of weighted impulses (multiplication by comb). Use this to derive a concise model for the ZOH and D/A output. This

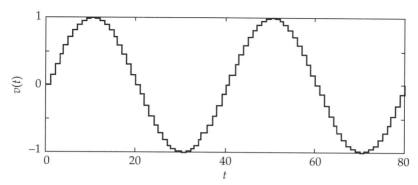

FIGURE 6.31 For Problem 6.53. The stairstep output characteristic of a digital-to-analog converter.

same model will work for images composed of square pixels, for example, computer displays and digital television.

(b) Using the model you developed in (a), find the Fourier transform of the D/A output and compare it with the transform of the unreconstructed sampled signal (make a sketch). What does the Fourier transform tell you about the quality of the reconstruction?

(c) Why is it advisable to follow a D/A converter with a lowpass filter? Why is it that your eye does not see the individual pixels on a computer screen (provided, of course, that the dot pitch of the monitor is fine enough)?

(d) A "first-order hold" (if such a thing could be built) would link the samples by line segments, so the reconstructed signal is piecewise linear instead of piecewise constant. This, too, can be modeled as a filter having a particular impulse response. What is the impulse response, and how does the transform of the reconstructed signal compare to the ZOH result? Which is better?

6.54. *Digital cameras*

In a digital camera, the film is replaced by a two-dimensional array of tiny photosensors. For example, a camera might have 16 megapixels (4096 × 4096), each of which is 2 μm square. Assume, for purposes of this problem, that the sensors are packed closely (with no gaps). Also consider only one row of the chip—a 4096-element array—to avoid the complications of doing two-dimensional transforms (the subject of a later chapter).

(a) Model a single photosensor as a rectangle function, R, of appropriate width. The output of a single photosensor is the integral of the incident intensity distribution, $f(x)$, over the area of the sensor. Let g be the output of the sensor array. Derive a model for g of the general form $g = (f * R) \, \text{III} \, R$, where the inner rectangle models a single sensor, and the (wider) outer rectangle models the finite width of the entire array.

(b) Calculate and sketch the Fourier transform of g. Interpret your model in the frequency domain.

6.55. *Imaging systems*

Image manipulation applications typically enable the user to specify the scale of an image, for example, in "pixels per inch." For example, a 128 × 128 image printed at 10 pixels per inch will be 12.8 inches on a side. You can imagine that with each pixel

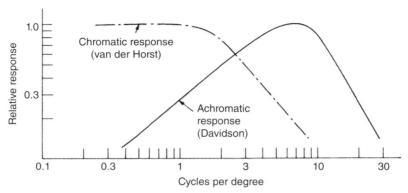

FIGURE 6.32 For Problem 6.55. Spatial frequency response of the human visual system. From Pratt (2007, p. 42, used by permission).

being 0.1 inch square, the image will appear "blocky." Printing the image at a much smaller scale, say 100 pixels per inch, will result in a smoother visual appearance, but the image will be much smaller, only 1.28 inch square. So it is interesting to ask, what is the largest scale at which the individual pixels appear to blend together satisfactorily, giving a smooth, non-blocky appearance?

You could approach this by trial and error, printing the image several times at different scales and picking the best one, but a good solution is possible using the Fourier transform. And, as it turns out, the theoretical prediction works well in practice.

The solution requires two things—a model of the "pixellated" image and a model of the human visual system.

(a) The digital image is obtained by sampling a continuous image $f(x)$ at an interval Δx (working in one dimension for convenience). Each sample is represented in the printed output by a pixel of width w, which can be modeled as a rectangle function. The scale of the image is $1/w$ pixels/inch. Derive a mathematical model for the pixellated image, using the functions f, comb, and rect.

The spatial frequency response of the human visual system is shown in Figure 6.32. The frequency variable is "cycles per degree," which permits the viewing distance to be taken into account. For example, consider a sinusoidal test image of period $L = 1$ cm. At the "standard reading distance" of 18 in $\approx$ 46 cm, one period of the sinusoid subtends an angle of 1/46 radian. The spatial frequency is 46 cycles/radian $\approx$ 0.8 cycles/degree. At a distance of 1 m, the test image appears to be smaller, and the spatial frequency higher. The angular subtense of one period is now $1/100 = 0.01$ radians, for a spatial frequency of 100 cycles/radian $\approx$ 1.75 cycles/degree. In general, the spatial frequency in cycles/degree for a spatial sinusoid of period L viewed at a distance D is

$$\nu = \frac{\pi}{180} \times \frac{D}{L}.$$

The eye uses two photoreceptors—the rod cells are achromatic (responding to shades of gray) and are responsible for vision in reduced light. The cone cells (there are three types for three different spectral bands) take care of color vision, but do not have the low light sensitivity of the rods. The cone cells are better at lower spatial frequencies than the rod cells, but the rod cells have a higher spatial bandwidth than the cones.

The response is partly optical (depending on the diameter of the pupil of the eye and the condition of the cornea and lens) and partly neural (signal processing occurring

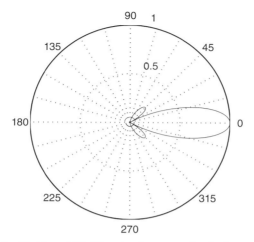

FIGURE 6.33 For Problem 6.56. Directional pattern of a single rectangular aperture.

at the retina). The relative response curve is measured with sinusoidal test targets at a particular input intensity level. A relative response of 0.5 at some frequency means that a sinusoidal target at that frequency requires twice the input contrast in order to be visible than a target at a frequency which has a response of 1.0. At normal light levels, where both rods and cones are engaged, the response to 0.8 cycles/degree is 1.0. At 1.75 cycles/degree this decreases to about 0.8.

You can regard this relative response curve as the transfer function of a lowpass filter, applied to the spatial frequency spectrum of the pixellated image. The key to removing pixellation effects is to decrease the pixel size to a point where the eye response filters out the spurious harmonic frequency components created by pixellation.

(b) Based on the Fourier transform of your model developed in part (a), and the eye response function, determine an image scale in pixels/inch where you believe the image will stop looking blocky and start looking "natural." Your answer should be somewhere between 30 and 100 pixels/inch.

6.56. *Antenna arrays*

The far-field radiation pattern of an antenna is given by the Fourier transform relationship:

$$P(\sin\theta) \propto \int_{-\infty}^{\infty} E\left(\frac{x}{\lambda}\right) \exp\left(-i2\pi \frac{x}{\lambda} \sin\theta\right) d\left(\frac{x}{\lambda}\right),$$

where $E(x/\lambda)$ is the electric field in the antenna aperture (distance measured in units of wavelength) and θ is the azimuth angle from the center of the aperture to the far-field observation point. A rectangular aperture of width A has a radiation pattern that goes like $\text{sinc}(A \sin\theta)$ (Figure 6.33).

There is an inverse relationship between the aperture size and the beamwidth (dilation theorem); it is desirable to have a wide antenna. This can be achieved either by making a wide antenna (such as the dish at the radio observatory at Arecibo, Puerto Rico) or by making an array of smaller antennas (such as the Very Large Array in New Mexico). The array approach has several advantages, as we shall see below.

(a) Assume first that the basic radiator has a very small aperture, such that it can be regarded as a point source (delta function). The array shall consist of N such

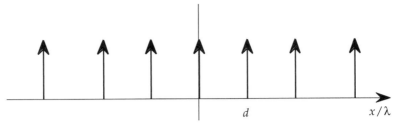

FIGURE 6.34 For Problem 6.56. An array consisting of N point radiators.

radiators arranged in a line, with spacing d, and $Nd = A$ (Figure 6.34; N is odd in this figure.)

You can model this array as the product of a comb function and a rect function of appropriate width. Calculate the radiation pattern of this antenna array. You should obtain

$$A(s) = \sum_{n=-\infty}^{\infty} \frac{1}{d} \operatorname{sinc}\left[\frac{Nd}{\lambda}\left(s - \frac{n\lambda}{d}\right)\right],$$

where $s = \sin\theta$. Even though this appears to be different from the result of the earlier Fourier series analysis

$$A(s) = \frac{\sin\left(\frac{N\pi d}{\lambda} s\right)}{\sin\left(\frac{\pi d}{\lambda} s\right)}$$

they are the same.

Sketch the magnitude of the pattern for θ between $-\pi/2$ and $\pi/2$. Compare this with the pattern from the plain rectangular aperture of width A. Note the similarities and differences. If you wish, you can explore the result with MATLAB, choosing $a = 1$, and a range of values for N. The neat polar plots are obtained with the `polar()` command.

(b) Something we could not do conveniently with the Fourier series analysis was account for finite-sized array elements. That is, each element of the array, rather than being a point source (modeled as a delta function), is, say, a rectangular source of width $D < d$. (For example, an ultrasound imaging array is composed of square piezo-electric elements.) Modify the point array model (part (a)) to include finite sources and calculate the radiation pattern. What is the effect of the finite sources (make a sketch)?

6.57. *Reflection seismology*

In reflection seismology, the layers of the earth are mapped by bouncing acoustic waves off the interfaces between the layers. A simple two-layer earth model is shown in Figure 6.35.

An acoustic wavelet $w(t)$ is sent downward into the earth. At each interface, part of the downward propagating wavelet is reflected back toward the surface, and the rest is transmitted into the next layer. The reflection coefficients are R_1 and R_2, and the corresponding transmission coefficients are T_1 and T_2. (In reality, the transmission coefficients depend on whether you are going up or down through the interface, but that need not concern us here.) In principle, there may be multiple (secondary) reflections

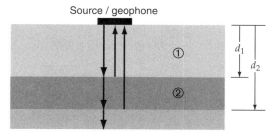

FIGURE 6.35 For Problem 6.57. Two-layer model of the earth for reflection seismology.

between layers, but in practice the reflections are often weak enough ($R_1, R_2 \ll 1$) that a "primaries only" model is valid. The acoustic velocities in the layers are v_1 and v_2, and the depths of the interfaces are d_1 and d_2. The signal $y(t)$ received back at the geophone is a linear superposition of reflected wavelets.

(a) Calculate an expression for $y(t)$, the received signal, in terms of $w(t)$ and the various parameters of the model.

(b) Show that y is equivalent mathematically to the output of a system with impulse response w which is excited by a string of impulse functions, where each impulse corresponds to one reflection. Thus, the seismic deconvolution problem consists of computing an input reflectivity sequence (R_k) from knowledge of the received signal y and the wavelet w.

6.58. *Discrete-time Fourier transform*

(a) Calculate the (continuous-time) Fourier transform, $F(\nu)$, of the damped cosine function, $f(t) = e^{-at} \cos 2\pi bt$.

(b) Sample this function to obtain a discrete-time function $f[n] = \Delta t f(n \Delta t)$ and calculate the discrete-time Fourier transform, $F_d(\theta)$.

(c) Truncate $f[n]$ to N samples ($n = 0, 1, \ldots, N-1$) and compute the DFT, $F[m]$. Plot all three transforms on an appropriate common scale, for $a = 1$ and $b = 2$. What are acceptable values of Δt and $N (= 2^p)$ in order to have good agreement among $F(\nu)$, $F_d(\theta)$, and $F[m]$?

6.59. *Sampling theory*

Let the energy in a signal $f(x)$ be defined

$$E = \int_{-\infty}^{\infty} |f(x)|^2 dx.$$

(Naturally, we assume $f \in L^2$.) Show that if f is bandlimited with maximum frequency W, then

$$E = \frac{1}{2W} \sum_{n=-\infty}^{\infty} \left| f\left(\frac{n}{2W}\right) \right|^2.$$

6.60. *Sampling theory*

A *square-law device* is a nonlinear system whose output g is the square of its input f, $g = f^2$. Suppose it is known that f is bandlimited to W Hz. The output g is to be sampled and later reconstructed from its samples. At what rate must g be sampled?

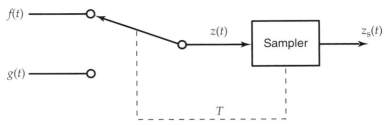

FIGURE 6.36 For Problem 6.61. Interlaced sampling.

6.61. *Sampling theory*
Two signals, $f(t)$ and $g(t)$, each bandlimited to W Hz, are multiplexed and sampled as shown in Figure 6.36.

At times $t = nT$, the switch flips from one input to the other, and the ideal sampler following the switch produces impulse samples. Thus the sample values alternate, $f(nT)$, $g((n+1)T), f((n+2)T), g((n+3)T), \ldots$.

(a) Calculate the Fourier transform $Z_s(v)$ in terms of the signal spectra $F(v)$ and $G(v)$.

(b) What must the sampling rate be (in terms of W) to avoid aliasing?

6.62. *Digital filtering*
We know that convolution models the input–output behavior of an LTI system: $g = h * f$, where f is the input, g is the output, and h is the impulse response.

(a) Show that, if f and h are bandlimited to $|v| < B$, then g is also bandlimited to $|v| < B$.

(b) Because f and h are bandlimited, they can be written using sinc function interpolation, for example,

$$f(t) = \sum_{m=-\infty}^{\infty} f[m] \operatorname{sinc}\left(\frac{t - m\Delta t}{\Delta t}\right), \quad f[m] = f(m\Delta t).$$

Hence, the convolution of $f(t)$ and $h(t)$ is

$$g(t) = \left[\sum_{m=-\infty}^{\infty} f[m] \operatorname{sinc}\left(\frac{t - m\Delta t}{\Delta t}\right)\right] * \left[\sum_{k=-\infty}^{\infty} h[k] \operatorname{sinc}\left(\frac{t - k\Delta t}{\Delta t}\right)\right].$$

Derive the result

$$g(t) = \sum_{m=-\infty}^{\infty} \sum_{k=-\infty}^{\infty} f[m] h[k] \operatorname{sinc}\left(\frac{t - (m+k)\Delta t}{\Delta t}\right) \Delta t$$

and then, by applying the sampling theorem, obtain

$$g(t) = \sum_{n=-\infty}^{\infty} g[n] \operatorname{sinc}\left(\frac{t - n\Delta t}{\Delta t}\right),$$

$$g[n] = g(n\Delta t) = \sum_{k=-\infty}^{\infty} f[k] h[n-k] \Delta t.$$

That is, as long as f and h are bandlimited, $g(t)$ can be reconstructed from the discrete-time convolution of the samples of f and h. (Note the presence of the factor Δt.) This inspires the use of digital computing to implement filtering operations on analog signals: sample the input f (with an A/D converter), perform the discrete-time convolution with the sampled impulse response, and reconstruct the output (with a D/A converter). This is called *digital filtering*.

6.63. *Discrete comb sequence*

The Fourier transform of the discrete comb sequence $\text{III}_N[n]$ is another comb function, $\text{III}\left(\frac{N\theta}{2\pi}\right)$.

(a) Show that, as $N \to 1$, the transform pair becomes $1 \to 2\pi\delta(\theta)$.

(b) Show that, as $N \to \infty$, the transform pair becomes $\delta[n] \longleftrightarrow 1$. *Hint:* The key is to show that $\text{III}\left(\frac{N\theta}{2\pi}\right) \to 1$ as $N \to \infty$. Consider the integral of one period of the comb with a testing function,

$$\int_{-\pi}^{\pi} \text{III}\left(\frac{N\theta}{2\pi}\right) \Phi(\theta)\, d\theta,$$

and show that this is a sum which, becomes, as $N \to \infty$, the Riemann integral $\int_{-\pi}^{\pi} \Phi(\theta)\, d\theta$.

6.64. *Discrete-time Fourier transform*

The Fourier transform of the signum function is $\text{sgn}\, x \longmapsto \frac{1}{i\pi\nu}$. The discrete-time Fourier transform of the signum sequence is $\text{sgn}[n] \longmapsto \frac{1}{2i}\cot\left(\frac{\theta}{2}\right)$. By the principle of sampling and replication, we would expect the transform of the signum sequence to be the periodic replication of the transform of the signum function. Plot both functions on comparable axes, that is, using the mapping $\theta = 2\pi\nu\Delta x$, and see if this assertion makes sense.

6.65. *Simultaneous sampling and replication*

Let f be bounded and continuous. Show that both sides of the expression below are well-defined generalized functions, and that they are, in fact, equivalent:

$$\left(f(x) * \frac{1}{N\Delta x}\text{III}\left(\frac{x}{N\Delta x}\right)\right) \cdot \frac{1}{\Delta x}\text{III}\left(\frac{x}{\Delta x}\right)$$
$$= \left(f(x) \cdot \frac{1}{\Delta x}\text{III}\left(\frac{x}{\Delta x}\right)\right) * \frac{1}{N\Delta x}\text{III}\left(\frac{x}{N\Delta x}\right).$$

CHAPTER 7

COMPLEX FUNCTION THEORY

Up to this point we have worked exclusively with real- or complex-valued functions of real or integer variables, for example,

- $f(t) = e^{-t}U(t)$, a real-valued function of the real variable t;
- $F(v) = \dfrac{1}{1 + i2\pi v}$, a complex-valued function of the real variable v; and
- $f[n] = e^{i\theta n}$, a complex-valued function of the integer variable n.

A complex-valued function may be separated into its real and imaginary parts, $F(v) = F_r(v) + iF_i(v)$. Once this is done, it can be manipulated according to the methods of ordinary real calculus, applied individually to the real and imaginary parts, for example,

$$\frac{dF}{dv} = \frac{dF_r}{dv} + i\frac{dF_i}{dv}$$

$$\int F(v)dv = \int F_r(v)dv + i \int F_i(v)dv.$$

The subject of the next two chapters is functions of a *complex* variable. The motivation for this study is twofold. First, there are insights and methods from complex analysis applicable to solving Fourier transform problems. Second, there are other useful transforms, related to the Fourier transform, whose use requires some facility with complex variable theory. These are the Laplace transform,

$$F_L(s) = \int_0^\infty f(t)e^{-st}dt,$$

and its discrete-time counterpart, the Z transform,

$$F_Z(z) = \sum_{n=0}^\infty f[n]z^{-n}.$$

The transform variables s and z are complex, $s = \sigma + i\omega$ and $z = re^{i\theta}$.

This chapter explains the basic properties of complex functions, including continuity, differentiability, and the all-important idea of an analytic function. The

---

Fourier Transforms: Principles and Applications, First Edition. Eric W. Hansen.
© 2014 John Wiley & Sons, Inc. Published 2014 by John Wiley & Sons, Inc.

following two chapters are devoted to complex integration and applications to transforms.

7.1 COMPLEX FUNCTIONS AND THEIR VISUALIZATION

A complex function is a mapping from complex numbers to complex numbers. A complex function $f(z)$ has a real and imaginary part, each of which is a function of the real and imaginary parts of z:

$$w = f(z) = f(x + iy) = u(x, y) + iv(x, y),$$

or of the modulus and argument of z:

$$w = f(z) = f(re^{i\theta}) = u(r, \theta) + iv(r, \theta).$$

As in real analysis, a graphical representation of a complex function can provide useful information about the function's behavior. It is easy to represent a scalar function of a single real variable, say $y = f(x)$, by a graph showing the mapping from x to y. Graphing a complex function, $w = f(z)$, is more difficult, because there are four dimensions (x, y, u, v) rather than two (x, y). One way is to plot $u(x, y)$ and $v(x, y)$ as surfaces above the complex Z-plane. Another is to show how points, curves, and regions in the complex Z-plane map to corresponding points, curves, and regions in the complex W-plane.

Example 7.1. The following curves occur frequently in the study of complex functions (Figure 7.1).

Horizontal line:

$$z = x + iy_0.$$

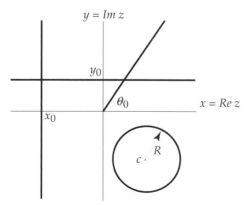

FIGURE 7.1 Four basic curves in the complex plane: horizontal line $z = x + iy_0$, vertical line $z = x_0 + iy$, ray $z = re^{i\theta_0}$, and circle, $z = c + Re^{i\theta}$.

Vertical line:
$$z = x_0 + iy.$$

Ray at angle θ_0:
$$z = re^{i\theta_0}, \quad r \geq 0.$$

Circle of radius R, centered at $z = c$:
$$|z - c| = R$$
$$z = c + Re^{i\theta}, \quad \pi \geq \theta > -\pi.$$

■

Here are three simple examples of complex mappings.

Example 7.2 ($f(z) = a + bz$). When $b = 1$, the mapping is a simple translation, $w = z + a$. A circle centered at the origin in the Z-plane, $z = Re^{i\theta}$, maps to $w = z + a = a + Re^{i\theta}$. It becomes a circle centered at $w = a$ in the W-plane.

When $a = 0$ and b is real, the mapping is a dilation, for example, the unit circle $z = e^{i\theta}$ maps to $w = bz = be^{i\theta}$, a circle of radius $|b|$ in the W-plane.

When b is complex, write $b = \beta e^{i\varphi}$. With z also expressed in polar form, $z = re^{i\theta}$, we have $w = bz = \beta r e^{i(\theta + \varphi)}$. A curve in the Z-plane is dilated by β and rotated by φ in the W-plane.

■

Example 7.3 ($f(z) = z^2$). With $z = re^{i\theta}$, the result in the W-plane is $w = r^2 e^{i2\theta}$. The magnitude of z is squared and the argument is doubled. When $z = x$, on the real axis, the result is familiar, $w = x^2$; the two points $z = \pm a$ both map to $w = a^2$. On the complex plane, we understand this via the doubling of the argument. A negative real number has $\arg z = \pi$, for example, $-2 = 2e^{i\pi}$. The result of squaring this number is $w = 2^2 e^{i2\pi} = 4$. The negative real axis in Z is mapped to the positive real axis in W. In the more general case, consider what happens to the unit disk, $|z| \leq 1$. Owing to the doubling of the argument, the upper half of the disk, $\pi \geq \theta > 0$, maps to $2\pi \geq \phi > 0$, the entire unit disk in the W-plane. The lower half of the disk, $0 \geq \theta > -\pi$, maps to $0 \geq \phi > -2\pi$, which is also the entire unit disk. Two distinct points in Z map to the same point in W (Figure 7.2).

■

Example 7.4 ($f(z) = z^{-1}$). Surface plots are shown in Figure 7.3. As the origin $z = 0$ is approached, the real and imaginary parts, and the modulus, grow rapidly. By analogy with the real function x^{-1}, we expect $f(z)$ to "blow up," grow without bound, as $z \to 0$. The implications of this singular behavior will be discussed in greater detail later.

Some region maps for the function are shown in Figure 7.4. We see that the real and imaginary axes in the Z-plane map to the real and imaginary axes in the W-plane. However, lines that are displaced from the axes, either horizontally or vertically, appear to map to circles in the W-plane. We also observe that circular curves centered at the origin of the Z-plane appear to map to circular curves centered at the origin of the W-plane. But note two things about the image circles. The curve in the Z-plane

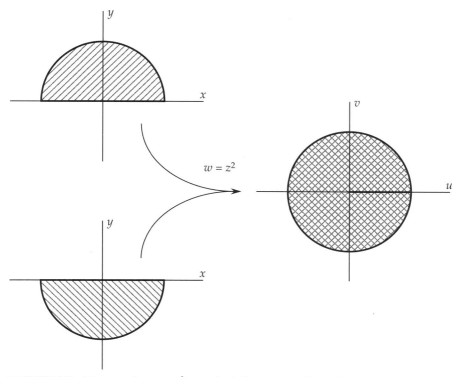

FIGURE 7.2 The mapping $w = z^2$ maps both the upper and lower halves of the of the unit disk to the full unit disk. The real axis is mapped to the positive real axis.

with the largest radius maps to the curve in the W-plane having the smallest radius, and conversely. Also, while all the curves begin on the positive real axis and end in the fourth quadrant, their images in the W-plane terminate in the first quadrant. The directions of the curves are reversed between the two planes, from counterclockwise to clockwise. The surface plot does a better job of displaying singularities than the region plot; on the other hand, the region plot better displays the geometry of the mapping. ∎

Our graphical observations about the mapping $f = z^{-1}$ can be studied in more detail using algebra and analytic geometry. To begin, let $z = x + iy$. Then,

$$w = \frac{1}{x+iy} = \frac{x-iy}{x^2+y^2}$$

$$= \frac{x}{x^2+y^2} + i\frac{-y}{x^2+y^2} = u(x,y) + iv(x,y).$$

Or, in polar coordinates,

$$w = f(z) = \frac{1}{z} = \frac{1}{re^{i\theta}} = \frac{1}{r}e^{-i\theta} = \frac{1}{r}\cos\theta - i\frac{1}{r}\sin\theta.$$

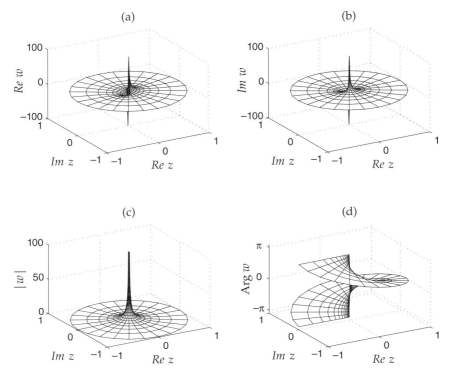

FIGURE 7.3 Surface plots of $w = z^{-1}$. (a,b) Real and imaginary parts. Notice the change of sign on opposite sides of the singular point $z = 0$. (c,d) Modulus and argument. The modulus blows up at $z = 0$. The argument is discontinuous (also see Figure 7.7).

Both the rectangular and polar forms of the mapping show that the function does indeed blow up as $x, y \to 0$ or $r \to 0$. On opposite sides of the singular point, the angle θ changes by π, causing a sign change in both the cosine and the sine, and hence in $\mathcal{R}e\, z$ and $\mathcal{I}m\, z$, as observed in Figure 7.3.

We next consider the mappings of particular curves. The x and y axes are simple: substituting $z = x$ and $z = iy$, respectively, we obtain $w = x^{-1}$ and $w = -iy^{-1}$. The image of the real axis is purely real, but the origin is mapped to infinity and the point at infinity is mapped to the origin. Likewise, the imaginary z-axis maps to the imaginary w-axis, with the origin mapping to infinity and the point at infinity mapping to the origin.[1] Furthermore, going from negative to positive along the y-axis produces v values that run from positive to negative. The displaced lines in Figure 7.4

---

[1] The origin is a single point that may be approached from any direction in the plane, for example, from above and below on the imaginary axis, and from left and right on the real axis. Its image under the mapping $f(z) = z^{-1}$, which is approached by moving away from the origin in any direction, for example, along the real and imaginary axes, is also regarded as a single point "at infinity."

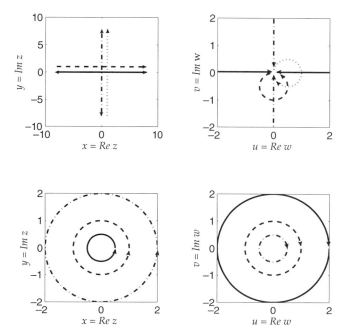

FIGURE 7.4 Curves in the Z-plane mapped to curves in the W-plane by the function $w = z^{-1}$. (*Top*) x and y axes map to the u and v axes, respectively, but the point at infinity maps to the origin, and vice versa. Lines displaced from the axes appear to map to circular curves. (*Bottom*) Circular curves map to circular curves, but the radii and sense of direction are inverted.

are $z = x + i$ and $z = 1 + iy$. We will just consider the first one, leaving the second to the problems:

$$\frac{1}{x+i} = \frac{x}{x^2+1} + i\frac{-1}{x^2+1} = u + iv.$$

According to Figure 7.4, this function should describe a circle, centered at a point below the origin (dashed curve). To verify this, we derive one equation relating u and v. Solving for x in terms of u and v,

$$\frac{u}{v} = \frac{x/(x^2+1)}{-1/(x^2+1)} = -x,$$

we can eliminate it, obtaining

$$v = \frac{-1}{\left(\frac{u}{v}\right)^2 + 1},$$

which simplifies to

$$u^2 + v^2 + v = 0.$$

Finally, completing the square in v,

$$u^2 + \left(v + \frac{1}{2}\right)^2 = \frac{1}{4},$$

which is indeed the equation of a circle of radius $\frac{1}{2}$, centered at $(u, v) = (0, -\frac{1}{2})$, or $w = 0 - i\frac{1}{2}$.

The function $z = re^{i\theta}$, where $r = $ constant and θ ranges from 0 to 2π, describes a circle of radius r, traversed in a counterclockwise (positive) direction. Applying the mapping $w = z^{-1}$ to the circle yields

$$w = \frac{1}{r}e^{-i\theta},$$

which is also a circle, of radius $\frac{1}{r}$. A circle of radius $\frac{1}{2}$, for example, maps to a circle of radius 2. Furthermore, arg w runs from 0 to -2π in the clockwise (negative) direction. Finally, any point inside the original circle, because it has a radius smaller than r, will map to a point with radius greater than $\frac{1}{r}$, which is outside the image circle in the W-plane. The interior and exterior of the circles are swapped. If you imagine yourself walking around the circle in the Z-plane in the positive direction, the interior of the circle is on your left. The image of this interior region is also on your left as you walk in the negative direction around the circle in the W-plane.

7.2 DIFFERENTIATION

For a function of a real variable f, at those points x on the real line where the function $f(x)$ is finite, single valued, and continuous, the derivative is defined

$$f'(x) = \lim_{\Delta x \to 0} \frac{f(x + \Delta x) - f(x)}{\Delta x}, \tag{7.1}$$

provided the limit exists and is the same whether 0 is approached from the right ($\Delta x \to 0^+$) or from the left ($\Delta x \to 0^-$). The importance of these conditions is illustrated by the following examples.

Example 7.5.

- $f(x) = x^{-1}$ has an infinite discontinuity at $x = 0$, and hence $f'(0)$ does not exist.
- The square root, $f(x) = x^{1/2}$, is double valued except at $x = 0$ (Figure 7.5); it can be differentiated if we first specify whether we are considering the positive ($+\sqrt{x}$) or negative ($-\sqrt{x}$) portion, or *branch*, of the mapping. On the positive branch, $f'(x) = \frac{1}{2\sqrt{x}}$, while on the negative branch, $f'(x) = -\frac{1}{2\sqrt{x}}$.
- The step function, $U(x)$, is discontinuous at $x = 0$, and barring recourse to generalized functions, is not differentiable there. Likewise, piecewise smooth functions like $|x|$ and $\Lambda(x)$ are not ordinarily differentiable at their "corners" because the limits (Equation 7.1) from the left and from the right are different.

∎

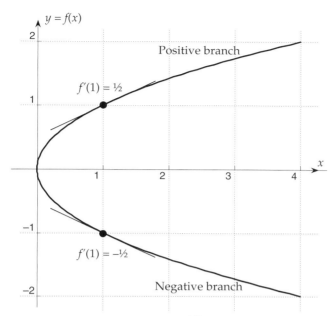

FIGURE 7.5 The square root mapping, $f(x) = x^{1/2}$, has two branches corresponding to the positive and negative roots. On the positive branch, the derivative is positive, $f'(x) = \frac{1}{2\sqrt{x}}$, and on the negative branch, the derivative is negative, $f'(x) = -\frac{1}{2\sqrt{x}}$.

The ideas of existence, continuity, and differentiability carry over in a natural way from real to complex analysis. A function $f(z)$ is defined at $z = z_0$ if and only if $f(z_0)$ is a single, finite number. $f(z) = z^{-1}$ is undefined at the origin, because $|f(0)|$ is not finite. Even if z^{-1} is redefined to be zero at the origin, it still has an infinite discontinuity there. The square root mapping, $f(z) = z^{1/2}$, will be seen to be double valued, like its real variable counterpart; it must be restricted in some way in order to be a function.

A function $f(z)$ is continuous at $z = z_0$ if and only if $\lim_{z \to z_0} f(z) = f(z_0)$, independent of the direction of approach to z_0. For functions defined on the real line, the limit can be approached only from the left or from the right, but on the complex plane the point z_0 can be approached along a path from any direction, and the limit must be the same regardless of which path is taken. To illustrate, consider first $f(z) = z + 1$, and take the limit as $z \to 1$, approaching from above ($z = 1 + i\delta$) and from the left ($z = 1 - \delta$), as shown in Figure 7.6.

If we approach this point from above,

$$\lim_{z \to 1} z + 1 = \lim_{\delta \to 0^+} 2 + i\delta = 2 = f(1).$$

Approaching from the left,

$$\lim_{z \to 1} z + 1 = \lim_{\delta \to 1^-} 2 - \delta = 2 = f(1).$$

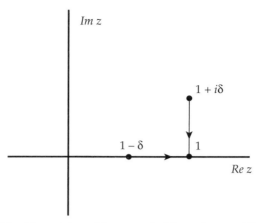

FIGURE 7.6 The point $z = 1$ is approached from above and from the left.

In general, for an approach from any direction,

$$|z - z_0| = |z - 1| = |\delta|$$

$$|f(z) - f(z_0)| = |z + 1 - 2| = |z - 1| = |\delta|, \text{ also,}$$

and as $\delta \to 0$ from any direction, bringing z and z_0 together, the distance $|\delta|$ between $f(z)$ and $f(z_0)$ also goes to zero. So, f is continuous at $z = 1$. The classic example of a discontinuous function is Arg z, the principal value of the argument, which jumps from $-\pi$ to π in passing from the third quadrant to the second quadrant across the negative real axis (Figure 7.7).

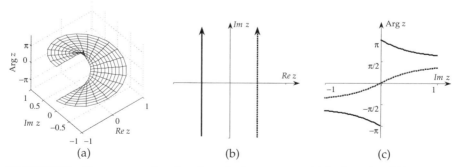

FIGURE 7.7 The argument function, Arg z, is discontinuous. (a) Surface plot. (b) Two paths in the complex Z-plane. The solid path goes from the third to second quadrant, crossing the negative real axis. The dashed path goes from the fourth to first quadrant, crossing the positive real axis. (c) Arg z as the paths in (b) are traversed. The argument is discontinuous between the second and third quadrants (solid line) and continuous between the first and fourth quadrants (dashed line).

The derivative for a function of a complex variable is defined like the real derivative.

Definition 7.1. Let $f : \mathbb{C} \to \mathbb{C}$ be a function. If the limit

$$\lim_{\Delta z \to 0} \frac{f(z_0 + \Delta z) - f(z_0)}{\Delta z} \tag{7.2}$$

exists and is independent of the path taken by Δz in approaching zero, then it is defined to be $f'(z_0)$, the derivative of f at z_0.

Example 7.6 ($f(z) = z^2$). Suppose we are interested in the derivative of this function at $z = 1$. We can approach this point from above, along the path $z = 1 + i\delta$. In this case, $\Delta z = i\delta$.

$$\frac{f(z + \Delta z) - f(z)}{\Delta z} = \frac{(1 + i\delta)^2 - 1^2}{i\delta} = \frac{1 + 2i\delta - \delta^2 - 1}{i\delta} = 2 + i\delta$$

and this goes to 2 as $\delta \to 0$. We can also approach it from the left, along the path $z = 1 - \delta$: $\Delta z = -\delta$:

$$\frac{f(z + \Delta z) - f(z)}{\Delta z} = \frac{(1 - \delta)^2 - 1^2}{-\delta} = \frac{1 - 2\delta + \delta^2 - 1}{-\delta} = 2 - \delta.$$

This also goes to 2 as $\delta \to 0$. Of course, these are only two of the infinitely many paths which Δz could take. In general,

$$\frac{f(z + \Delta z) - f(z)}{\Delta z} = \frac{(z + \Delta z)^2 - z^2}{\Delta z} = 2z + \Delta z.$$

As $\Delta z \to 0$ along any path, the result is $2z$, the same as in real analysis, and at $z = 1$, $f'(1) = 2$. ∎

Example 7.7 (The squared modulus, $g(z) = |z|^2$). We begin again with the particular case, attempting to calculate $g'(1)$ from above and from the left. From above,

$$\frac{g(z + \Delta z) - g(z)}{\Delta z} = \frac{|1 + i\delta|^2 - 1^2}{i\delta} = \frac{1 + \delta^2 - 1}{i\delta} = -i\delta,$$

which goes to zero as $\delta \to 0$. From the left, we have

$$\frac{g(z + \Delta z) - g(z)}{\Delta z} = \frac{|1 - \delta|^2 - 1^2}{-\delta} = \frac{1 - 2\delta + \delta^2 - 1}{-\delta} = 2 - \delta,$$

and this goes to 2 as $\delta \to 0$. Something is wrong here—the two limits are different, indicating that the derivative does not exist at $z = 1$. In general,

$$\frac{g(z + \Delta z) - g(z)}{\Delta z} = \frac{|z + \Delta z|^2 - |z|^2}{\Delta z} = \frac{z^* \Delta z + z \Delta z^* + \Delta z \Delta z^*}{\Delta z} = z^* + z\frac{\Delta z^*}{\Delta z} + \Delta z^*.$$

Now, taking the limit from above, $\Delta z = i\delta$,

$$\frac{g(z + \Delta z) - g(z)}{\Delta z} = z^* - z - i\delta = -2i\,\mathcal{I}m\,z - i\delta,$$

which goes to $-2i\mathcal{I}m\,z$ as $\delta \to 0$. From the left, $\Delta z = -\delta$, and

$$\frac{g(z+\Delta z) - g(z)}{\Delta z} = z^* + z - \delta = 2\,\mathcal{R}e\,z - \delta.$$

This goes to $2\,\mathcal{R}e\,z$ as $\delta \to 0$. The *only* place where these two limits are the same is the origin, where $\mathcal{R}e\,z = -\mathcal{I}m\,z$. Here, we find

$$\frac{g(0+\Delta z) - g(0)}{\Delta z} = \frac{|\Delta z|^2 - 0}{\Delta z} = \Delta z^*,$$

which goes to zero as $\Delta z \to 0$ from any direction. We conclude that $|z|^2$ is differentiable only at the origin, and there, the derivative is zero. ∎

The Cauchy–Riemann Equations

We will now develop a general test for differentiability of a function of a complex variable. Let $f(z) = u(x,y) + iv(x,y)$, where $z = x + iy$. Also define $\Delta z = \Delta x + i\Delta y$. Evaluate Equation 7.2 again, taking limits horizontally and vertically (denoted f'_x and f'_y, respectively):

$$f'_x(z) = \lim_{\Delta x \to 0} \frac{u(x+\Delta x, y) - u(x,y)}{\Delta x} + \lim_{\Delta x \to 0} i\,\frac{v(x+\Delta x, y) - v(x,y)}{\Delta x},$$

$$f'_y(z) = \lim_{\Delta y \to 0} \frac{u(x, y+\Delta y) - u(x,y)}{i\Delta y} + \lim_{\Delta y \to 0} i\,\frac{v(x, y+\Delta y) - v(x,y)}{i\Delta y}.$$

Each of the limits becomes a partial derivative, for example,

$$\lim_{\Delta x \to 0} \frac{u(x+\Delta x, y) - u(x,y)}{\Delta x} = \frac{\partial u}{\partial x}.$$

Hence,

$$f'_x(z) = \frac{\partial u}{\partial x} + i\frac{\partial v}{\partial x},$$

$$f'_y(z) = -i\frac{\partial u}{\partial y} + \frac{\partial v}{\partial y},$$

and since the derivative must be independent of direction, $f' = f'_x = f'_y$, and

$$f' = \frac{\partial u}{\partial x} + i\frac{\partial v}{\partial x} = -i\frac{\partial u}{\partial y} + \frac{\partial v}{\partial y}.$$

Separating the real and imaginary parts, we obtain a pair of equations,

$$\frac{\partial u}{\partial x} = \frac{\partial v}{\partial y} \quad \text{and} \quad \frac{\partial v}{\partial x} = -\frac{\partial u}{\partial y}. \tag{7.3}$$

These are known as the *Cauchy–Riemann equations*.

If a function f is differentiable at a point $z = z_0$, it will satisfy the Cauchy–Riemann equations there, making them a necessary condition for differentiability. It is possible, however, to find functions which satisfy the Cauchy–Riemann equations at some point but are not differentiable there. With an added continuity requirement, the Cauchy–Riemann equations can be shown to be necessary *and* sufficient for the existence of the derivative. The resulting theorem is proved in many complex variable books:[2]

Theorem 7.1 (Differentiability of complex functions). Let $f : \mathbb{C} \to \mathbb{C}$ be a function, mapping $z = x + iy$ to $f(z) = u(x, y) + iv(x, y)$. f is differentiable at $z = z_0$ if and only if the partial derivatives $\frac{\partial u}{\partial x}, \frac{\partial u}{\partial y}, \frac{\partial v}{\partial x},$ and $\frac{\partial v}{\partial y}$ are all continuous and satisfy the Cauchy–Riemann equations at z_0.

Example 7.8. Consider again the functions $f(z) = z^2$ and $g(z) = |z|^2$. Express them in terms of their real and imaginary parts:

$$z^2 = (x + iy)^2 = (x^2 - y^2) + i2xy,$$
$$|z|^2 = |x + iy|^2 = x^2 + y^2 + i0.$$

For z^2, $u = x^2 - y^2$ and $v = 2xy$. Computing the partial derivatives,

$$\frac{\partial u}{\partial x} = 2x \quad \frac{\partial u}{\partial y} = -2y$$
$$\frac{\partial v}{\partial x} = 2y \quad \frac{\partial v}{\partial y} = 2x.$$

It is apparent that the partial derivatives are continuous and the Cauchy–Riemann equations are satisfied, for all x and y; hence, z^2 is differentiable everywhere. The derivative $f'(z)$ can be computed either as $f'_x(z) = \frac{\partial u}{\partial x} + i\frac{\partial v}{\partial x} = 2x + i2y = 2z$ or $f'_y(z) = -i\frac{\partial u}{\partial y} + \frac{\partial v}{\partial y} = i2y + 2x = 2z$, although the difference quotient (7.2) is quicker.

It is a different story with $|z|^2$, for which $u = x^2 + y^2$ and $v = 0$:

$$\frac{\partial u}{\partial x} = 2x \quad \frac{\partial u}{\partial y} = 2y$$
$$\frac{\partial v}{\partial x} = 0 \quad \frac{\partial v}{\partial y} = 0.$$

The partial derivatives are continuous, but the Cauchy–Riemann equations are satisfied only when $2x = 0$ and $2y = 0$—the origin, $z = 0$. The derivative there is zero. ∎

Example 7.9 ($f(z) = z^{-1}$). To apply the Cauchy–Riemann conditions, let $z = x + iy$ and express f in terms of its real and imaginary parts:

$$f(z) = \frac{1}{x + iy} = \frac{x}{x^2 + y^2} + i\frac{-y}{x^2 + y^2} = u(x, y) + iv(x, y).$$

Then, calculating the partial derivatives,

$$\frac{\partial u}{\partial x} = -\frac{x^2 - y^2}{(x^2 + y^2)^2} \quad \frac{\partial u}{\partial y} = -\frac{2xy}{(x^2 + y^2)^2}$$
$$\frac{\partial v}{\partial x} = \frac{2xy}{(x^2 + y^2)^2} \quad \frac{\partial v}{\partial y} = -\frac{x^2 - y^2}{(x^2 + y^2)^2}.$$

[2] See, for example, Flanagan (1983), pp. 117–118.

The partial derivatives exist and are continuous everywhere except $z = 0$, and excepting this point, the Cauchy–Riemann equations are satisfied. The derivative is

$$f'(z) = f'_x(z) = \frac{\partial u}{\partial x} + i\frac{\partial v}{\partial x} = -\frac{x^2 - y^2 - 2ixy}{(x^2+y^2)^2}$$

$$= -\frac{(z^*)^2}{|z|^4} = -\frac{(z^*)^2}{(zz^*)^2} = -\frac{1}{z^2}, \qquad (7.4)$$

which is algebraically the same as the result from real analysis for $f(x) = x^{-1}$. The difference quotient (7.2) gives the same result:

$$\frac{d}{dz}\frac{1}{z} = \lim_{\Delta z \to 0} \frac{\frac{1}{z+\Delta z} - \frac{1}{z}}{\Delta z} = \lim_{\Delta z \to 0} \frac{\frac{z-(z+\Delta z)}{z(z+\Delta z)}}{\Delta z} = \lim_{\Delta z \to 0} -\frac{1}{z^2 + z\Delta z} = -\frac{1}{z^2}. \qquad \blacksquare$$

The Cauchy–Riemann equations can also be written in polar form. A function $f(z) = u(r,\theta) + iv(r,\theta)$ is differentiable if and only if the partial derivatives of u and v with respect to r and θ are continuous, and

$$\frac{\partial u}{\partial r} = \frac{1}{r}\frac{\partial v}{\partial \theta}, \qquad \frac{\partial v}{\partial r} = -\frac{1}{r}\frac{\partial u}{\partial \theta}. \qquad (7.5a)$$

In terms of these partial derivatives, the derivative of $f = z^{-1} = e^{-i\theta}/r$ is

$$f' = e^{-i\theta}\left(\frac{\partial u}{\partial r} + i\frac{\partial v}{\partial r}\right) = \frac{e^{-i\theta}}{r}\left(\frac{\partial v}{\partial \theta} - i\frac{\partial u}{\partial \theta}\right). \qquad (7.5b)$$

The derivation of these equations is left to the problems.

7.3 ANALYTIC FUNCTIONS

Virtually everything important in complex analysis hinges on the idea of an *analytic function*. As a preliminary step, some terms are defined—open set, connected set, and domain—which enable precise description of regions in the complex plane. Some of these are illustrated in Figure 7.8.

Definition 7.2 (Sets in the complex plane). A *neighborhood* of a point $z = z_0$ is a set of points $\{z : |z - z_0| < \delta\}$, $\delta > 0$. A *deleted neighborhood* of a point $z = z_0$ is just the neighborhood with z_0 removed, that is, the set $\{z : 0 < |z - z_0| < \delta\}$. An *open set* is a set of points in which every point can be surrounded by a neighborhood which also belongs to that set. A *closed set* is one which is not open. Every point in a *connected set* can be joined to every other point in the set by a polygonal path made of points in the set. A *domain* in the complex plane is an open, connected set.

The interior of the unit circle, $\{z : |z| < 1\}$, is an open set. The unit circle together with its interior, $\{z : |z| \leq 1\}$, is not open, because any neighborhood of a point on the circle itself will include points which are outside the circle, and hence outside the set. An open or closed set can be connected or not—they are independent

7.3 ANALYTIC FUNCTIONS   467

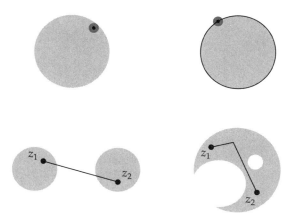

FIGURE 7.8 *Top, left*: An open set does not include a boundary. For any point in the set, a neighborhood can be set up which is contained in the set. *Top, right*: A closed set includes a boundary. If a point is on the boundary of a closed set, any neighborhood, no matter how small, will include points outside the set. *Bottom, left*: A disconnected set. A line connecting the points z_1 and z_2 contains points not in the set. *Bottom, right*: A connected set. The points z_1 and z_2 are connected by a polygonal path in the set. An open, connected set is a domain.

properties. The interior of the unit circle is a connected set. On the other hand, the union of sets $\{z : |z| < 1\} \cup \{z : |z - 4| < 1\}$, which consists of the interiors of two disjoint circles, is not connected. A line joining a point in the first set with a point in the second set will include points not in either set. A set with a "hole" in it, like the unit disk with the origin removed, $\{z : 0 < |z| < 1\}$, is still connected, because you can connect two points on opposite sides of the origin with a polygonal path (for example, a pair of line segments). The interior of the unit circle is open and connected, so it is a domain.

Definition 7.3 (Analytic function).   A function $f : \mathbb{C} \to \mathbb{C}$ is *analytic* at a point $z = z_0$ if the derivative f' exists at z_0 and in a neighborhood of z_0. A function is *analytic in a domain* if it is analytic at every point in the domain. The largest domain where a function f is analytic is called its *domain of analyticity*. If a function's domain of analyticity is the entire complex plane, that is, if f is analytic everywhere, it is called an *entire* function.

The important thing about being analytic is that *the derivative cannot exist just at z_0; it must also exist in an open disk of some radius $\delta > 0$ centered at z_0.*

Originally, f' was required to be continuous in the neighborhood as well, but it was later shown by Goursat that a complex derivative is always continuous, making the mere existence of the derivative necessary and sufficient for analyticity. If the partial derivatives are continuous and the Cauchy–Riemann equations are satisfied everywhere in a neighborhood of z_0 (including z_0), then f is analytic at z_0.

Points where a function fails to be analytic are also important.

Definition 7.4 (Isolated singularities). If a function f is not analytic at some point $z = z_0$, but is analytic everywhere else in a neighborhood of z_0 (i.e., in a deleted neighborhood of z_0), the point z_0 is called an *isolated singularity* of f. If, for some integer $\infty > n > 0$, the limit $\lim_{z \to z_0}(z - z_0)^n f(z)$ is finite and nonzero, then the isolated singularity is called a *pole* of order n. When $n = 1$, the pole is called *simple*; a double pole has $n = 2$, so on. If the limit fails to exist for all finite n, then z_0 is called an *essential singularity*.

Example 7.10. Let us revisit the functions we studied in the preceding section.

$f(z) = z^2$ was found to be differentiable everywhere. It is an entire function, because its domain of analyticity is the entire complex plane.

$f(z) = z^{-1}$ is differentiable everywhere except at the isolated singularity $z = 0$, which is a simple pole. For any $z \neq 0$, no matter how close to the origin, $f'(z)$ will exist, and you can establish a neighborhood around z that excludes the origin. The domain of analyticity for f is the complex plane with $z = 0$ excluded, sometimes called the *punctured* complex plane.

$f(z) = |z|^2$ is differentiable nowhere except $z = 0$. Although the derivative exists right at the origin, you cannot establish a neighborhood of the origin, however small, in which f is differentiable. Hence, f is not analytic anywhere. Its domain of analyticity is the empty set.

$f(z) = \text{Arg } z$ is discontinuous along the negative real axis, but continuous everywhere else. It cannot be analytic along the negative real axis because of the discontinuity. (It can also be shown that, despite continuity everywhere else in the complex plane, Arg z fails to satisfy the Cauchy–Riemann conditions anywhere, a far more serious deficiency.) ∎

Certain combinations of analytic functions are also analytic.

Theorem 7.2. Let f and g be functions which are analytic on some common domain. On this domain,

$$f + g \text{ and } f - g \text{ are analytic,} \tag{7.6a}$$
$$fg \text{ is analytic,} \tag{7.6b}$$
$$f/g \text{ is analytic except where } g = 0, \tag{7.6c}$$
$$f \circ g = f(g) \text{ is analytic.} \tag{7.6d}$$

Proof: These combination rules are proven using the Cauchy–Riemann equations. For illustration, we prove the sum rule (Equation 7.6a). Let $f(z) = u_f(x, y) + iv_f(x, y)$ and $g(z) = u_g(x, y) + iv_g(x, y)$. Then,

$$f + g = (u_f + iv_f) + (u_g + iv_g) = (u_f + u_g) + i(v_f + v_g).$$

The partial derivatives are $\frac{\partial}{\partial x}(u_f + u_g) = \frac{\partial u_f}{\partial x} + \frac{\partial u_g}{\partial x}$, so on. And because f and g are analytic, their associated partial derivatives are continuous and satisfy the

Cauchy–Riemann equations. We may therefore substitute $\frac{\partial u_f}{\partial x} = \frac{\partial v_f}{\partial y}$, so on, yielding

$$\frac{\partial}{\partial x}(v_f + v_g) = -\frac{\partial u_f}{\partial y} - \frac{\partial u_g}{\partial y} = -\frac{\partial}{\partial y}(u_f + u_g),$$

$$\frac{\partial}{\partial y}(v_f + v_g) = \frac{\partial u_f}{\partial x} + \frac{\partial u_g}{\partial x} = \frac{\partial}{\partial x}(u_f + u_g),$$

and this is what was to be proved. ∎

The sum, product, quotient, and chain rules from real analysis also extend to the complex plane:

$$(f \pm g)' = f' \pm g', \tag{7.7a}$$
$$(fg)' = f'g + fg', \tag{7.7b}$$
$$(f/g)' = \frac{f'g - fg'}{g^2}, \tag{7.7c}$$
$$(f \circ g)' = (f' \circ g) g'. \tag{7.7d}$$

It is easy to show that $f(z) = z$ is an entire function. It then follows from the product rule (Equation 7.6b) that $z^2 = z \cdot z$ is entire, and (by induction) that all positive integer powers of z are entire. Furthermore, by the sum rule (Equation 7.6a) all polynomials are entire functions. The ratio of two polynomials, $f(z) = \frac{b(z)}{a(z)}$, is called a *rational function*. If the numerator and denominator polynomials $b(z)$ and $a(z)$ are coprime (have no common roots), then it follows from the quotient rule (Equation 7.6c) that the rational function f is analytic everywhere except at the roots of $a(z)$. These roots are the poles of f. The case of common roots is discussed in the problems. If you are familiar with transfer functions from linear system theory, you know that they are rational functions of the complex Laplace transform variable s, $H(s) = \frac{b(s)}{a(s)}$. Now you also know that they are analytic everywhere in the complex plane except at their poles.

What about the complex conjugate function, $f(z) = z^*$? It is continuous and single valued. By definition, $z^* = x - iy$, so $u = x$ and $v = -y$. Applying the Cauchy–Riemann equations,

$$\frac{\partial u}{\partial x} = 1 \quad \frac{\partial u}{\partial y} = 0,$$
$$\frac{\partial v}{\partial x} = 0 \quad \frac{\partial v}{\partial y} = -1.$$

Because $\frac{\partial u}{\partial x} \neq \frac{\partial v}{\partial y}$, for all x and y, we conclude that z^* is not analytic anywhere in the complex plane. It follows that combinations involving z^*, such as $\mathcal{R}e\, z = \frac{z+z^*}{2}$, $\mathcal{I}m\, z = \frac{z-z^*}{2i}$, and $|z|^2 = zz^*$, are not analytic.

Next we explore what happens to other common functions when they are applied to complex numbers. These divide into two classes: those based on the exponential function, which are relatively easy to deal with, and their inverses, which are based on the logarithm and are more difficult.

7.4 exp z AND FUNCTIONS DERIVED FROM IT

The exponential function, exp z, is an entire function of z. To see this, write

$$\exp z = \exp(x + iy) = e^x e^{iy} = e^x \cos y + i e^x \sin y$$

and apply the Cauchy–Riemann equations, with $u = e^x \cos y$ and $v = e^x \sin y$:

$$\frac{\partial u}{\partial x} = e^x \cos y \quad \frac{\partial u}{\partial y} = -e^x \sin y$$
$$\frac{\partial v}{\partial x} = e^x \sin y \quad \frac{\partial v}{\partial y} = e^x \cos y.$$

The partial derivatives are clearly continuous and the Cauchy–Riemann equations are satisfied everywhere.

Surface plots reveal something of this function's character. Referring to Figure 7.9, observe that the real and imaginary parts are exponential in the direction of the real axis, and oscillatory in the imaginary direction. The fact that the real part oscillates like $\cos y$ and the imaginary part like $\sin y$ is also evident. The modulus and argument are $|w| = \exp x$ and y, respectively. The plot shows the principal value of the argument, which varies between $-\pi$ and π.

Region plots (Figure 7.10) display several additional features.

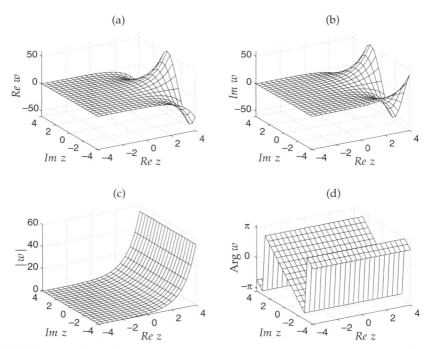

FIGURE 7.9 Surface plots of the exponential function, $w = \exp z$. (a,b) The real and imaginary parts behave exponentially in the direction of the real axis and oscillate in the direction of the imaginary axis. (c) The modulus, $|w| = \exp x$. (d) Principal value of $\arg w$.

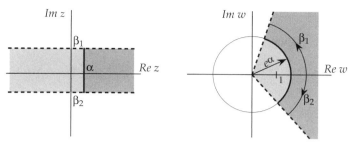

FIGURE 7.10 The exponential function $w = \exp z$ maps horizontal lines $z = i\beta$ to rays at an angle β (dashed). The vertical line segment $z = \alpha + iy$, $y \in (\beta_2, \beta_1)$ is mapped to an arc of radius e^α, with $\arg w \in (\beta_2, \beta_1)$ (solid). When $\beta_1 - \beta_2 = 2k\pi$, the arc is a circle. The region $\mathcal{R}e\, z < \alpha$ maps to the interior of the circle, and $\mathcal{R}e\, z > \alpha$ maps to the exterior of the circle. When $\alpha = 0$, the circle becomes the unit circle.

Horizontal lines in the Z-plane, $z = x + i\beta$, map to rays extending from the origin in the W-plane at angle β, $w = e^x e^{i\beta}$. The negative extreme of the line, $x \to -\infty$, maps to $w = 0$. Vertical lines in the Z-plane, $z = \alpha + iy$, map to circles in the W-plane of radius e^α, $w = e^\alpha e^{iy}$. As shown in the figure, a vertical line segment maps to an arc. The imaginary axis, $z = iy$, maps to the unit circle, since $e^\alpha = e^0 = 1$. Points in the left-half Z-plane ($x < 0$) map inside the unit circle ($e^x < 1$), and points in the right-half Z-plane map outside the unit circle. (What is the image in the W-plane of a rectangular area in the Z-plane?)

The circular and hyperbolic functions are defined in terms of the exponential in the usual way:

$$\sin z = \frac{e^{iz} - e^{-iz}}{2i}, \tag{7.8a}$$

$$\cos z = \frac{e^{iz} + e^{-iz}}{2}, \tag{7.8b}$$

$$\tan z = \frac{\sin z}{\cos z}, \tag{7.8c}$$

$$\sinh z = \frac{e^z - e^{-z}}{2}, \tag{7.8d}$$

$$\cosh z = \frac{e^z + e^{-z}}{2}, \tag{7.8e}$$

$$\tanh z = \frac{\sinh z}{\cosh z}. \tag{7.8f}$$

Algebraically, the complex definitions are identical to those for real variables. This means that all the trigonometric identities apply to the complex case, for example,

$$\cos(z_1 + z_2) = \cos z_1 \cos z_2 - \sin z_1 \sin z_2.$$

The trigonometric functions are sums and ratios of entire functions, so they are also analytic everywhere, except for $\tan z$ and $\tanh z$, which fail to be analytic where $\cos z = 0$ and $\cosh z = 0$, respectively. Their derivatives are identical in form to their real counterparts: $\dfrac{d \sin z}{dz} = \cos z$, so on.

Expanding $\sin z$ and $\cos z$ into their real and imaginary parts,

$$\sin z = \frac{e^{i(x+iy)} - e^{-i(x+iy)}}{2i} = \frac{e^{-y}(\cos x + i \sin x) - e^{y}(\cos x - i \sin x)}{2i}$$
$$= \left(\frac{e^y + e^{-y}}{2}\right)\sin x + i\left(\frac{e^y - e^{-y}}{2}\right)\cos x$$
$$= \sin x \cosh y + i \cos x \sinh y \tag{7.9a}$$

and

$$\cos z = \cos x \cosh y - i \sin x \sinh y. \tag{7.9b}$$

In the complex plane, $\sin z$ and $\cos z$ behave as combinations of circular and hyperbolic functions. If z is purely real, we get the circular functions; if z is purely imaginary, their behavior is hyperbolic. Replacing z by iz produces general relationships between circular and hyperbolic functions on the complex plane:

$$\sin iz = \frac{e^{-z} - e^{+z}}{2i} = i \sinh z, \tag{7.10a}$$

$$\cos iz = \frac{e^{-z} + e^{+z}}{2} = \cosh z, \tag{7.10b}$$

$$\tan iz = \frac{i \sinh z}{\cosh z} = i \tanh z, \tag{7.10c}$$

$$\sinh iz = \frac{e^{iz} - e^{-iz}}{2} = i \sin z, \tag{7.10d}$$

$$\cosh iz = \frac{e^{iz} + e^{-iz}}{2} = \cos z, \tag{7.10e}$$

$$\tanh iz = \frac{i \sin z}{\cos z} = i \tan z. \tag{7.10f}$$

These relationships say, for example, that the cosine function can take on values outside the real interval $[-1, 1]$: $\cos 2i = \cosh 2 = 3.7622$, and $\cos(\pi + i2) = -\cosh 2 = -3.7622$.

Complex versions of other functions exist, too: the gamma function $\Gamma(z)$ and the Bessel functions $J_n(z)$, to name two. Their properties are beyond the scope of this text, but helpful summaries can be found in mathematical handbooks.[3]

7.5 LOG z AND FUNCTIONS DERIVED FROM IT

All the transcendental functions of the previous section, and the polynomial functions (of degree greater than 1) as well, are many-to-one mappings, for example, $z^3 = 1$ for $z = 1, \frac{-1+i\sqrt{3}}{2}, \frac{-1-i\sqrt{3}}{2}$, and $\sin z = 0$ for $z = 0, \pm\pi, \pm 2\pi, \ldots$. This means that if we need to solve an equation like $z^n = 4$ or $\cos z = 3i$ for z, we will be inverting a many-to-one mapping and should expect the solutions, $z = 4^{1/n}$ and $z = \cos^{-1}(3i)$, to be multivalued. We need a way to ensure that we obtain all the possible solutions and

---

[3] See, for example, Abramowitz and Stegun, 1972; Jahnke *et al.*, 1960.

face an additional problem if we want to differentiate these multivalued mappings, because a function must be single valued in order to be differentiated.

The logarithm is the foundation for these inverse functions. Using the logarithm, we can define the nth root $z^{1/n}$ of a complex number, rational powers $z^{m/n}$, arbitrary irrational, even complex, powers, z^c, and inverse trigonometric functions.

7.5.1 The Logarithm Function

The logarithm is defined via the exponential. For a complex z, $\log z$ (recall that we use "log" rather than "ln" to denote the natural logarithm) is any complex number for which $z = \exp(\log z)$. The unit circle provides some convenient examples.

Example 7.11 (Some complex logarithms). We know from real analysis that $\log 1 = 0$: $e^0 = 1$. In the complex plane, $e^{i2\pi}$ is also equal to 1, so $i2\pi$ is a complex logarithm of 1. In fact, $e^{i2\pi k} = 1$ for all integer k, so 1 has an infinite number of complex logarithms, $\log 1 = i2\pi k$. On the other side of the unit circle, $e^{i\pi} = -1$; moreover, $e^{i\pi + i2\pi k} = -1$, so -1 has an infinite number of logarithms, $\log(-1) = i\pi + i2\pi k$. Also, $e^{i\pi/2 + i2\pi k} = i$, and $\log i = i\frac{\pi}{2} + i2\pi k$. ∎

To calculate the logarithm of an arbitrary z, write $\log z = u + iv$ and $z = re^{i\theta}$. Then,

$$z = re^{i\theta} = \exp(u + iv) = e^u e^{iv},$$

and we identify $u = \log r = \log |z|$ and $v = \theta = \arg z$. Hence, we have the basic definition of the logarithm:

$$\log z = \log r + i\theta = \log |z| + i \arg z. \tag{7.11}$$

We want to know if the logarithm is an analytic function. We will examine the usual things:

- Is $\log z$ finite and single valued?
- Is $\log z$ continuous?
- Does $\log z$ satisfy the Cauchy–Riemann conditions?

The real logarithm, $\log x$, blows up at $x = 0$. For the complex logarithm, $|\log z| = \sqrt{(\log r)^2 + \theta^2}$; this blows up when $r = 0$. The complex logarithm, therefore, has a singularity at $z = 0$, and any domain of analyticity for $\log z$ will necessarily exclude the origin.

What about being single valued? The argument is multivalued (Figure 7.11), so the logarithm is, as well. The principal value of the logarithm, denoted Log z, is defined using the principal value of the argument, $\pi \geq \text{Arg } z > -\pi$ (Equation 1.11). All other values are obtained by adding integer multiples of $2\pi i$:

$$\text{Log } z = \text{Log } |z| + i\text{Arg } z,$$
$$\log z = \text{Log } z + i2\pi k = \text{Log } |z| + i\text{Arg } z + i2\pi k, \quad k = 0, \pm 1, \ldots \tag{7.12}$$

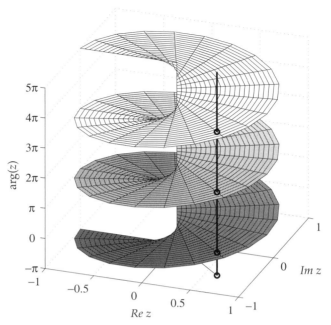

FIGURE 7.11 The argument arg z is multivalued. The same point in the Z-plane is repeatedly addressed by taking laps around the origin. Each lap changes the argument by 2π.

Choosing a value of k identifies a particular 2π range of argument over which the logarithm is single valued. There may be a practical reason to select a particular value of k in the same way that physical considerations often dictate the choice of a positive or negative square root. Lacking any such constraints at this point, we will leave the choice of k to be arbitrary.

Another way of visualizing the multivalued nature of the logarithm is via a region plot (Figure 7.12). The complex Z-plane which is addressed by arguments between $-\pi$ and π is mapped by the logarithm to a horizontal strip in the W-plane, with $\mathcal{I}m\, w \in (-\pi, \pi]$. Other ranges of argument map to different strips. Consider a point z in the Z-plane whose argument is defined to be between $-\pi$ and π, as shown in the figure. Its image under the logarithm mapping will appear in the strip with $\mathcal{I}m\, w \in (-\pi, \pi]$. Now travel once around the origin of the Z-plane in the positive direction, returning to z. You are at the same point in the Z-plane, as defined by the real and imaginary parts, but the argument has increased by 2π, and because the argument is different, the logarithm is different. *This* point maps to the next higher strip in the W-plane. Had you moved in the negative direction around the origin, you would have decreased the argument of z by 2π, and the corresponding logarithm would have been in a lower strip in the W-plane. Making repeated trips around $z = 0$ moves you to successive strips in the W-plane. This is how the multiple values of the logarithm are generated. You can, of course, verify that all the points so generated

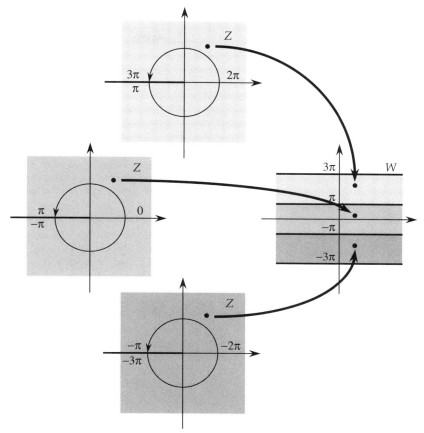

FIGURE 7.12 The logarithm is multivalued. The same point in the Z-plane is mapped to an infinite number of points in the W-plane, separated by $i2\pi$. The Z-plane is mapped into strips of width 2π in the W-plane. Each 2π range of argument maps to a different strip.

are valid logarithms of z by applying the exponential function to them and observe that because $\exp(i2\pi) = 1$, they all map to the same value.

As it stands, the logarithm is not analytic, because it is multivalued. This deficiency is remedied by restricting the range of the logarithm to one of the strips in Figure 7.12. This, in turn, restricts the argument of z to one of the 2π intervals $(-\pi + i2\pi k, \pi + i2\pi k]$. Choosing a particular k, we now have a domain $\{z : |z| > 0, \arg z \in (-\pi + 2\pi k, \pi + 2\pi k]\}$ where the logarithm is finite and single valued.

Next we take up continuity. For all points z_0 in the eventual domain of analyticity, we require $\lim_{z \to z_0} \log z = \log z_0$ for all z in a neighborhood of z_0, independent of the path z takes to get to z_0. As we have defined it, the argument is discontinuous on the negative real axis. Any neighborhood of a point on the negative real axis, no matter how small, will contain points above and below the axis. However, any point above or below the axis, no matter how close to the axis it is, possesses a neighborhood

FIGURE 7.13 Discontinuity of the argument illustrated for the principal value. *Left*: A neighborhood of a point on the negative real axis will always contain points whose arguments are close to π and others whose arguments are close to $-\pi$. *Right*: On the other hand, for a point above or below the axis, a neighborhood can always be found in which the argument is continuous.

of points in which the argument is continuous, because the axis can be excluded by making the neighborhood small enough (Figure 7.13).

Discontinuity in the argument makes the logarithm discontinuous as well. The discontinuity is avoided by further restricting the domain to exclude the negative real axis. On this new domain, $\{z : |z| > 0, \arg z \in (-\pi + 2\pi k, \pi + 2\pi k)\}$, the logarithm is finite, single valued, and continuous. (Values of the logarithm can still be calculated on the negative real axis using arg $z = \pi$, but it cannot be differentiated there because it is discontinuous.)

Finally, we consider whether the logarithm function satisfies the Cauchy–Riemann conditions on this domain. Applying the Cauchy–Riemann equations in polar form (Equation 7.5), with $u(r, \theta) = \log r$ and $v(r, \theta) = \theta$, the partial derivatives are continuous and

$$\frac{\partial u}{\partial r} = \frac{1}{r}\frac{\partial v}{\partial \theta} = \frac{1}{r}$$
$$\frac{\partial v}{\partial r} = -\frac{1}{r}\frac{\partial u}{\partial \theta} = 0$$

everywhere in the domain.

Definition 7.5 (Branch). Let $f : \mathbb{C} \to \mathbb{C}$ be a mapping. The restriction of f to a domain $D \subseteq \mathbb{C}$ on which f is single valued and continuous is called a *branch* of f.

The result of these successive restrictions of the domain is this: a branch of $\log z$ is analytic in the domain created by selecting a 2π range of argument (fixing k) and cutting the nonpositive real axis out of the complex plane. The set of points $\{z : \arg z = \pi\}$ which were excluded from the domain to define the branch is called, appropriately enough, a *branch cut*.

The branch cut is a boundary between the domains that specify different branches of the logarithm. Let z_1 be a point in the second quadrant of the Z-plane, and z_2 be a point in the third quadrant, and consider a path connecting z_1 to z_2 (Figure 7.14). You have two options: a path like A that detours around the branch cut through the first and fourth quadrants on its way to the third, or a path like B that crosses the branch cut in going directly from the second quadrant to the third. The image of z_2, its logarithm, depends on the path taken. Along path A, the argument of a point on the path always stays between $-\pi$ and π; its image under the logarithm

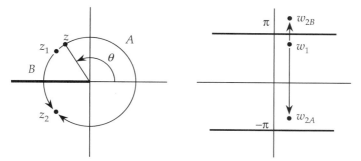

FIGURE 7.14 Crossing a branch cut moves you to a different branch of the logarithm.

function is always on the strip with $\mathcal{I}m\, w \in (-\pi, \pi)$. The image points w_1 and w_{2A} are on the same strip, or the same branch of the logarithm. However, moving continuously from z_1 to z_2 along path B, as though the branch cut were not there, causes the argument of z to pass beyond π to a value greater than π. The image of this path crosses from one strip in the W-plane to the next one above. The image points w_1 and w_{2B} are on different strips or different branches of the logarithm. If you want to differentiate or integrate a branch of any multivalued f along a path in the complex plane, you must stay in that branch's domain and not choose a path that crosses the branch cut to a different domain.

But what if the path you are interested in must cross a branch cut? With reference again to Figure 7.14, suppose you really need for the logarithm to be continuous between the second and third quadrants. There is a way out. Suppose we were to define $\arg z$ to live between 0 and 2π rather than between $-\pi$ and π. This constitutes a valid domain, and results in a valid branch of the logarithm, one which is analytic in the cut plane with the nonnegative real axis excluded. Figure 7.15 revisits the situation depicted in Figure 7.14, with this new positioning of the branch cut.

Now it is path A that crosses the branch cut, placing w_1 and w_{2A} on different branches of the logarithm. Path B, on the other hand, stays within a single domain, and the image points w_1 and w_{2B} are on the same branch of the logarithm.

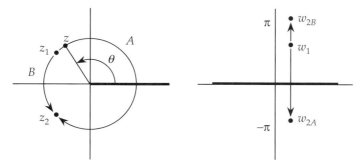

FIGURE 7.15 Moving the branch cut from the nonpositive real axis to the nonnegative real axis creates a domain of analyticity which includes the negative numbers but excludes the positive numbers.

The choice of branch cut is not limited to the real axis. If you need to, you can place the branch cut at an arbitrary angle in the complex plane. For example, if you want $\log z$ to be analytic on the whole real axis (excepting $z = 0$, of course), you can place the branch cut along the negative imaginary axis, $\theta = -\pi/2$.

All the possible branch cuts for $\log z$ share one point, $z = 0$, in common. This point is called a *branch point* of the function. The logarithm has one branch point. Some functions have more than one and more complicated sets of branch cuts. A multivalued mapping is not single valued in a neighborhood of a branch point. If you make a full lap around a branch point, you will be sure to cross a branch cut and hop between branches—the beginning and the end of the path will not map to the same value. This provides a way to test a point to see if it is a branch point. Define a neighborhood of the point and a closed path in that neighborhood which encircles the point. If you completely traverse the path and the image of the starting point is different than the image of the ending point, then you have found a branch point.

Example 7.12 (Branch point for logarithm). To check that $z_0 = 0$ is the branch point of $\log z$, go around the circle $z = \epsilon e^{i\theta}$ from $\theta = 0$ to $\theta = 2\pi$ (Figure 7.16). The image of the starting point is $w = \text{Log } \epsilon + i0$, but the image of the ending point is $w = \text{Log } \epsilon + i2\pi$. These are different points in the W-plane, verifying that $z = 0$ is a branch point.

On the other hand, consider the point $z_0 = 2$ and the path $z = 2 + \epsilon e^{i\theta}$ which encircles it. The magnitude of $z = 2 + \epsilon e^{i0}$ is $2 + \epsilon$, and its argument is zero. Its logarithm, therefore, is $w = \text{Log }(2 + \epsilon) + i0 = \text{Log }(2 + \epsilon)$. At the other end of the path, $z = 2 + \epsilon e^{i2\pi}$, the magnitude $|z| = 2 + \epsilon$, and the argument is still zero. The logarithm is also $\text{Log }(2 + \epsilon) + i0 = \text{Log }(2 + \epsilon)$, which proves that $z = 2$ is not a branch point of $\log z$. ■

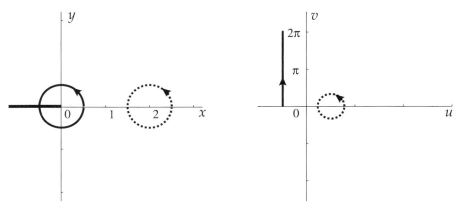

FIGURE 7.16 Two circular paths in the Z-plane (*left*) and their images under $w = \log z$ (*right*). Encircling the branch point at $z = 0$ (solid curve), the images of $z = \epsilon e^{i0}$ and $z = \epsilon e^{i2\pi}$ are different points. When the branch point is not encircled, (dashed curve), the points $z = 2 + \epsilon e^{i0}$ and $z = 2 + \epsilon e^{i2\pi}$ map to the same point.

A branch point is not an isolated singularity. The function z^{-1} is singular at the origin, but it is analytic in a deleted neighborhood of the origin, proving that $z = 0$ is an isolated singularity of the function. On the other hand, although $\log z$ is also singular at the origin, there is a branch cut extending from $z = 0$ so that there is no deleted neighborhood of the origin in which $\log z$ is analytic.

7.5.2 The Square Root, Revisited

The square root function, $y = x^{1/2}$, is the solution of the quadratic equation $y^2 - x = 0$; a quadratic equation has two roots, hence there are two square roots for x. For real x, these two square roots are commonly denoted $\pm\sqrt{x}$ for $x > 0$, and $\pm i\sqrt{|x|}$ for $x < 0$. Now consider the complex square root, $z^{1/2}$. With $z = \exp(\log z)$,

$$z^{1/2} = \exp\left(\frac{1}{2}\log z\right) = \exp\left(\frac{\text{Log }|z|}{2}\right) \exp\left(\frac{i \arg z}{2}\right)$$
$$= \sqrt{|z|} \exp\left(\frac{i \arg z}{2}\right),$$

where $\sqrt{|z|}$ denotes the positive square root of $|z|$. The argument is multivalued, so we write

$$z^{1/2} = \sqrt{|z|} \exp\left(\frac{i(\text{Arg } z + 2\pi k)}{2}\right)$$
$$= \exp(i\pi k)\sqrt{|z|} \exp\left(\frac{i\text{Arg } z}{2}\right), \quad k = 0, \pm 1, \ldots.$$

Now consider the result as k takes on various values.

- $k = 0$:

$$z^{1/2} = \exp(i0)\sqrt{|z|} \exp\left(\frac{i\text{Arg } z}{2}\right) = \sqrt{|z|} \exp\left(\frac{i\text{Arg } z}{2}\right).$$

- $k = 1$:

$$z^{1/2} = \exp(i\pi)\sqrt{|z|} \exp\left(\frac{i\text{Arg } z}{2}\right) = -\sqrt{|z|} \exp\left(\frac{i\text{Arg } z}{2}\right).$$

- $k = 2$:

$$z^{1/2} = \exp(i2\pi)\sqrt{|z|} \exp\left(\frac{i\text{Arg } z}{2}\right) = +\sqrt{|z|} \exp\left(\frac{i\text{Arg } z}{2}\right),$$

which is the same as the $z = 0$ case.
- $k = 3$:

$$z^{1/2} = \exp(i3\pi)\sqrt{|z|} \exp\left(\frac{i\text{Arg } z}{2}\right) = -\sqrt{|z|} \exp\left(\frac{i\text{Arg } z}{2}\right),$$

the same as the $z = 1$ case.

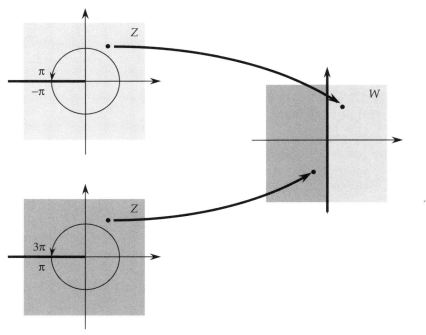

FIGURE 7.17 The square root has two branches: $k = 0, \pm 2, \ldots$ and $k = \pm 1, \pm 3, \ldots$. The branch cut divides two domains of analyticity.

The pattern is clear: only $k = 0$ and $k = 1$ yield unique values for $z^{1/2}$, and one is the negative of the other. This is the complex origin of the two square roots:

$$z^{1/2} = \pm\sqrt{|z|} \exp\left(\frac{i\text{Arg } z}{2}\right). \tag{7.13}$$

It is easy to see how this more general function reduces to the real cases. When z is a positive real number, Arg $z = 0$, and we just get $z^{1/2} = \pm\sqrt{x}$. When z is a negative real number, Arg $z = \pi$, and $z^{1/2} = \pm e^{i\pi/2}\sqrt{|x|} = \pm i\sqrt{|x|}$.

The structure of the square root mapping is illustrated in Figure 7.17. When arg $z \in (-\pi, \pi]$, the argument of the square root, arg w, is between $-\pi/2$ and $\pi/2$—the right-half W-plane. When arg $z \in (\pi, 3\pi]$, we have arg $w \in (\pi/2, 3\pi/2]$, the left-half W-plane. All other domains, for example, arg $z \in (3\pi, 5\pi]$, map to one of these two branches. Just as with the logarithm function, a path that crosses the branch cut moves you from one branch of the square root to the other. But unlike the logarithm, crossing the branch cut a second time in the same direction (making two laps around $z = 0$) puts you back on the first branch, rather than on a third, different, branch. Multiple laps around the origin in the Z-plane just move you back and forth between the two branches of the square root function.

Moving the branch cut from the negative real axis to the positive real axis yields a square root function which is analytic for negative real numbers (Figure 7.18).

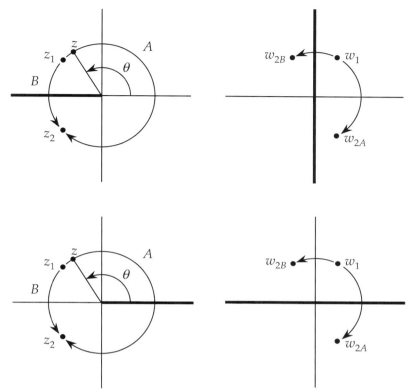

FIGURE 7.18 Branches of the square root function. *Top*: The branch cut placed along the negative real axis. w_1 and w_{2A} are on the same branch; w_{2B} is on the other branch. *Bottom*: The branch cut placed along the positive real axis. Now w_1 and w_{2B} are on the same branch, while w_{2A} is on the other branch.

The generalization to the nth root is straightforward:

$$\begin{aligned} z^{1/n} &= \sqrt[n]{|z|} \exp\left[i\frac{1}{n}(\text{Arg } z + 2\pi k)\right] \\ &= e^{i2\pi k/n} \sqrt[n]{|z|} \exp\left(\frac{i\text{Arg } z}{n}\right), \quad k = 0, 1, \ldots, n-1. \end{aligned} \quad (7.14)$$

For example, the "Nth roots of unity" are

$$1^{1/N} = e^{i2\pi k/N}, \quad k = 0, 1, \ldots, N-1,$$

which, you may recall, appear in the discrete Fourier transform. These are complex numbers, spaced uniformly around the unit circle at angles $\theta_k = \dfrac{2\pi k}{N}$.

The function $f(z) = z^{1/n}$ is analytic in the cut plane $\{z : |z| > 0, \text{ arg } z \in (-\pi + 2\pi k, \pi + 2\pi k)\}$. Each value of $k = 0, 1, \ldots, n-1$ corresponds to a different branch

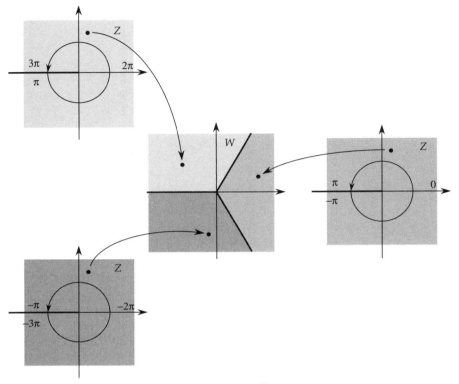

FIGURE 7.19 The cube root, $z^{1/3}$, has three branches.

of the function. The square root has two branches, the cube root has three branches, and so on (Figure 7.19).

7.5.3 Rational powers, $z^{m/n}$

A rational power of a complex number, $z^{m/n}$, is the combination of the mth power, which is single valued, and the nth root, which is n valued. Using the complex logarithm,

$$z^{m/n} = \exp\left(\frac{m}{n} \log z\right) = \exp\left[\frac{m}{n} \left(\text{Log}\, |z| + i\text{Arg}\, z + i2\pi k\right)\right]$$
$$= |z|^{m/n} \exp\left(i\frac{m}{n} \text{Arg}\, z\right) \exp\left(\frac{i2\pi km}{n}\right), \quad k = 0, 1, \ldots, n-1,$$

where $|z|^{m/n} = \sqrt[n]{|z|^m} = \left(\sqrt[n]{|z|}\right)^m$. You need to be careful that the fraction m/n is in simplest terms. If m and n share a common factor, then there will actually be fewer than n unique values of $z^{m/n}$. For example, if $m = 2$ and $n = 4$, and you neglect to reduce m/n to $1/2$, you will generate:

$$z^{2/4} = \exp\left(ik2\pi\frac{2}{4}\right) \sqrt[4]{|z|^2} \exp\left(i\frac{2}{4}\text{Arg } z\right), \quad k = 0, 1, 2, 3,$$

$$= \exp(ik\pi)\sqrt{|z|} \exp\left(i\frac{\text{Arg } z}{2}\right), \quad k = 0, 1, 2, 3,$$

$$= +\sqrt{|z|} \exp\left(i\frac{\text{Arg } z}{2}\right), \quad -\sqrt{|z|} \exp\left(i\frac{\text{Arg } z}{2}\right),$$

$$+ \sqrt{|z|} \exp\left(i\frac{\text{Arg } z}{2}\right), \quad -\sqrt{|z|} \exp\left(i\frac{\text{Arg } z}{2}\right).$$

Two of the values are redundant.

The function $z^{m/n}$ is analytic in the cut plane, just like $\log z$ and $z^{1/n}$, but the branch structure is more complicated than the square root and cube root. To illustrate, take $m = 2$, $n = 3$ (Figure 7.20). The domain defined by $\arg z \in (-\pi, \pi)$ maps to a sector in the W-plane defined by $\arg w \in (-2\pi/3, 2\pi/3)$. The next domain, $\arg z \in (\pi, 3\pi)$, maps to $\arg w \in (2\pi/3, 2\pi)$. The next, $\arg z \in (3\pi, 5\pi)$, maps to $\arg w \in (2\pi, 4\pi/3)$. Because $\exp(i4\pi/3) = \exp(-i2\pi/3)$, we see that the three sectors in the W-plane do adjoin and cover the plane *twice*. Three full laps around the Z-plane result in two full laps around the W-plane.

In general with $z^{m/n}$, n laps around the Z-plane result in m laps around the W-plane. As with $\log z$ and $z^{1/n}$, the branch cut may extend from the branch point at $z = 0$ in any direction in the Z-plane, with a corresponding definition of $\arg z$.

7.5.4 Irrational and Complex Powers of z

The function $f(z) = z^c$, where c is an arbitrary real or complex number, is evaluated using the logarithm function, as follows. Again beginning with

$$z^c = \exp(c \log z),$$

let $c = a + ib$, then

$$z^c = \exp[(a + ib)(\text{Log } |z| + i\text{Arg } z + i2\pi k)]$$
$$= \exp[a \text{Log } |z| - b\text{Arg } z - 2\pi bk] \exp[i(a\text{Arg } z + b \text{Log } |z| + 2\pi ak)],$$
$$k = 0, \pm 1, \ldots \quad (7.15)$$

In general, there will be an infinite number of values for a particular choice of z and c. It looks messy, but it is the most general possible power of z, and there is no power you can't calculate using it. For example, in the unlikely event that you ever need π^π, substitute $z = \pi$, $a = \pi$, and $b = 0$ into Equation 7.15:

$$\pi^\pi = e^{\pi \text{Log } \pi} e^{i2\pi^2 k}, \quad k = 0, \pm 1, \ldots.$$

The principal value ($k = 0$) is the one your calculator gives, $e^{\pi \text{Log } \pi} = 36.4621596072$. But there is also a set of complex values for $k \neq 0$.

Like all the others, z^c is analytic in the cut plane we have been seeing throughout this section. When c is real but irrational, a single 2π domain in the Z-plane maps to a $2\pi c$ sector in the W-plane. Unlike the rational (m/n) case, though, the boundaries

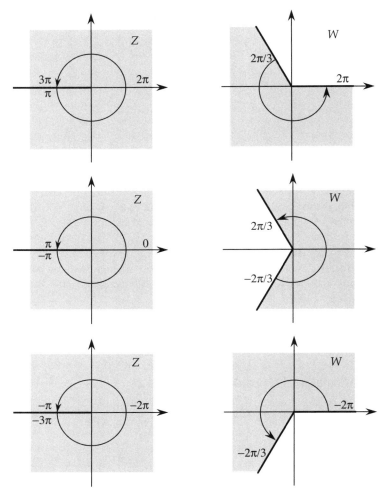

FIGURE 7.20 $w = z^{2/3}$ has three branches. Each one maps a 2π domain in the Z-plane to a $4\pi/3$ range in the W-plane.

of these sectors will never meet up again as k runs through the integers. Like the logarithm, you have an infinite number of branches. The branch structure when c is complex is very complicated and will not be discussed further here.

7.5.5 The Square Root of a Polynomial

We shall require, on occasion, the square root of a polynomial such as $f(z) = (z^2 - 1)^{1/2}$. It is simple enough to program this function with, say, MATLAB and compute values. It is not so simple to determine domains of analyticity and the structure of the branches, since f is a combination of a many-to-one mapping (z^2) and a one-to-many mapping (square root).

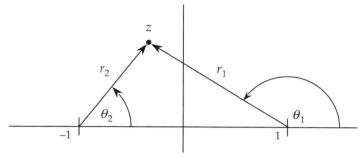

FIGURE 7.21 Setup for calculating $f(z) = (z^2 - 1)^{1/2} = (z+1)^{1/2}(z-1)^{1/2}$.

Factoring $z^2 - 1 = (z+1)(z-1)$, f may be written $f(z) = (z+1)^{1/2}(z-1)^{1/2}$, the product of two simpler square roots. We know how to deal with these. Let (Figure 7.21)

$$z - 1 = r_1 e^{i\theta_1},$$
$$z + 1 = r_2 e^{i\theta_2}.$$

Then,

$$f(z) = (r_1 r_2)^{1/2} \exp\left(i\frac{\theta_1 + \theta_2}{2}\right).$$

We have two arguments to define, $\theta_1 = \arg(z-1)$ and $\theta_2 = \arg(z+1)$, hence two branch cuts to position. Let us begin with the usual principal branch of the argument, so $\theta_1, \theta_2 \in (-\pi, \pi)$ (Figure 7.22).

We are interested in the continuity of f as the real axis is approached from above and below. So, consider points just above and below the real axis at $z = a$, $z = 0$, and $z = -a$, $a > 1$ (Figure 7.23).

There is no branch cut to the right of $z = 1$. As $z = a$ is approached from above and below, the angles θ_1 and θ_2 both approach 0 continuously. The argument of the square root, $(\theta_1 + \theta_2)/2$, is also continuous, approaching 0 at $z = a$. We anticipate that f will be analytic to the right of $z = 1$.

As $z = 0$ is approached from above and below, it is a different story. From above, θ_1 approaches π and θ_2 approaches 0. From below, θ_1 approaches $-\pi$, and θ_2 still approaches 0. So from above, the argument of the square root is approaching $\pi/2$, while from below it approaches $-\pi/2$. Now, $\exp(i\pi/2) = i$, while $\exp(-i\pi/2) = -i$, so f is discontinuous across the real axis here.

Finally, consider $z = -a$. As $z = -a$ is approached from above, both θ_1 and θ_2 tend to π, while from below they both tend to $-\pi$, since each angle has to stay between $-\pi$ and π to avoid its branch cut. The argument of the square root is π just above the axis and $-\pi$ just below the axis. This looks like another discontinuity. But $\exp(-i\pi) = \exp(i\pi) = -1$, so in fact, f is continuous at $z = -a$! The branch cuts appear to cancel, but what actually happens is that you simultaneously cross from one branch to the other and then back again. The resulting domain of analyticity for f is the complex plane cut just along the real axis between $z = -1$ and $z = 1$.

Another way of visualizing the same thing is shown in Figure 7.24. Looking at the argument, the discontinuities at the branch cuts are apparent. There

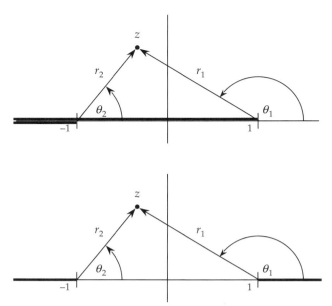

FIGURE 7.22 Two of the possible choices of branch cut for calculating $f(z) = (z+1)^{1/2}(z-1)^{1/2}$. *Top*: $\theta_1, \theta_2 \in (-\pi, \pi)$. *Bottom*: $\theta_1 \in (0, 2\pi)$, $\theta_2 \in (-\pi, \pi)$.

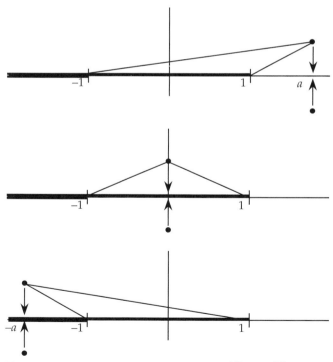

FIGURE 7.23 Exploring the continuity of $f(z) = (z+1)^{1/2}(z-1)^{1/2}$ at the real axis. *Top*: To the right of both branch cuts. *Middle*: At one branch cut. *Bottom*: At both branch cuts.

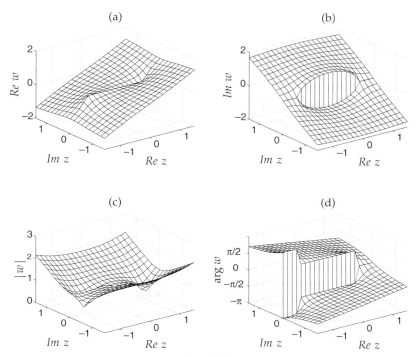

FIGURE 7.24 Surface plots of $f(z) = (z^2 - 1)^{1/2}$. (a) real part; (b) imaginary part; (c) modulus; and (d) argument. The imaginary part is discontinous across the real axis between $z = -1$ and $z = 1$, but continuous everywhere else. The real part is continuous, but not smooth, between $z = -1$ and $z = 1$. The apparent discontinuity in the argument to the left of $z = -1$ is a jump of 2π, which does not affect the value of the function.

is a jump of π between $z = -1$ and $z = 1$ due to the one branch cut, then a jump of 2π to the left of $z = -1$ due to the combined effects of both branch cuts. But a jump of 2π in argument does not affect the value of the function, since $\exp(i2\pi) = 1$. The jump in the imaginary part is caused by the π jump in the argument. The imaginary part is proportional to $\sin \arg w$; above the jump, $\sin(\arg w) = \sin \frac{\pi}{2} = 1$, while below the jump, $\sin(\arg w) = \sin \frac{-\pi}{2} = -1$. The real part does not have a jump, because it is proportional to $\cos(\arg w)$ and $\cos \frac{\pi}{2} = \cos \frac{-\pi}{2} = 0$. Instead, the real part is zero on the line between $z = -1$ and $z = 1$, but its slope is discontinuous across the line. The modulus is continuous (why?).

As long as we take paths in the complex plane that go around branch cuts rather than across them, we will be assured of staying on one branch of the function. The consequence of ignoring the branch cut is shown in Figure 7.25. The figure shows three paths in the Z-plane connecting two points, $z = 2i$ and $z = -2i$. One path goes to the right, avoiding both branch cuts of f. The second path goes straight from $2i$ to $-2i$, crossing over one branch cut. The third path goes to the left, crossing both branch cuts. The images of the three paths under f are also shown. Observe that the

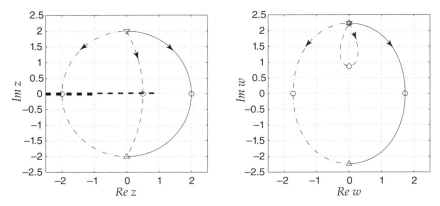

FIGURE 7.25 The difference between crossing a branch cut and going around it, for $f(z) = (z^2 - 1)^{1/2}$. *Left*: Paths in the Z-plane connecting $z = 2i$ with $z = -2i$. The starting and ending points are marked with downward and upward directed triangles, respectively, and the points where the paths cross the real axis are marked with circles. One path goes around both branch cuts of f, one crosses one of them, and the third crosses both. *Right*: The mappings under f of these paths, in the W-plane. The first and third paths take different routes to the same destination, but the second path goes somewhere else—to the other branch of the function.

images of the first and third paths, crossing zero and two branch cuts, end at the same point, $-i\sqrt{5}$. The image of the second path, which crosses one branch cut, ends at a different point, $+i\sqrt{5}$, which is, in fact, the value of the function on its other branch.

If you require a branch of $(z^2 - 1)^{1/2}$ to be continuous across the real axis between -1 and 1, then you can move one of the branch cuts, defining $\theta_1 \in (0, 2\pi)$ (Figure 7.22). Now the domain of analyticity is the complex plane, cut along the real axis for $\mathcal{R}e\, z > 1$ and $\mathcal{R}e\, z < -1$. As with the logarithm, if you want a branch whose domain includes the whole real axis (excepting $z = \pm 1$), you can hang the branch cuts under the axis and define the arguments θ_1 and θ_2 accordingly.

7.5.6 Inverse Trigonometric Functions

We use $\arccos z$ as an example of how the inverse trigonometric functions are calculated. We seek the solution of the equation $z = \cos w$. Begin with the definition of the cosine in terms of the exponential:

$$\cos w = \frac{e^{iw} + e^{-iw}}{2},$$

and let $\xi = e^{iw}$, so

$$z = \frac{\xi + \xi^{-1}}{2}.$$

Collect the terms into a polynomial in ξ:
$$\xi^2 - 2z\xi + 1 = 0.$$

Using the quadratic formula,
$$\xi = z + (z^2 - 1)^{1/2} = z + i(1 - z^2)^{1/2}.$$

(The two roots are implicitly indicated by writing the square root as $(\cdot)^{1/2}$.) Finally, substituting $\xi = e^{iw}$ and solving for w gives
$$w = \arccos z = -i \log \left[z + i(1 - z^2)^{1/2} \right].$$

The function will have a "double infinity" of values, owing to the infinite number of logarithm values and the two square roots.

Here are all the inverse trig functions, collected together:

$$\arcsin z = -i \log \left[iz + (1 - z^2)^{1/2} \right], \qquad (7.16a)$$
$$\arccos z = -i \log \left[z + i(1 - z^2)^{1/2} \right], \qquad (7.16b)$$
$$\arctan z = \frac{i}{2} \log \left(\frac{i+z}{i-z} \right), \qquad (7.16c)$$
$$\operatorname{arcsinh} z = \log \left[z + (z^2 + 1)^{1/2} \right], \qquad (7.16d)$$
$$\operatorname{arccosh} z = \log \left[z + (z^2 - 1)^{1/2} \right], \qquad (7.16e)$$
$$\operatorname{arctanh} z = \frac{1}{2} \log \left(\frac{1+z}{1-z} \right). \qquad (7.16f)$$

The domains of analyticity of these functions are complicated and will not be discussed further here.

7.6 SUMMARY

1. A function $f : \mathbb{C} \to \mathbb{C}, f(z) = u(x, y) + iv(x, y)$ is analytic at a point z_0 if the partial derivatives $\frac{\partial u}{\partial x}, \frac{\partial u}{\partial y}, \frac{\partial v}{\partial x}, \frac{\partial v}{\partial y}$ are continuous and satisfy the Cauchy–Riemann equations
$$\frac{\partial u}{\partial x} = \frac{\partial v}{\partial y}, \quad \frac{\partial u}{\partial y} = -\frac{\partial v}{\partial x}$$
at z_0 and in a neighborhood of z_0.

2. An isolated singularity is a point z_0 at which a function fails to be analytic, although the function is analytic in a deleted neighborhood of z_0, for example, $\epsilon > |z - z_0| > 0$.

3. A domain in the complex plane is an open, connected set.

4. A function is analytic in a domain D if it is analytic for every point $z \in D$. A function's domain of analyticity is the largest such domain.

5. A function is entire if it is analytic for all $z \in \mathbb{C}$. That is, its domain of analyticity is the entire complex plane.

6. A mapping $f : \mathbb{C} \to \mathbb{C}$ may in general be multivalued (one-to-many). The restriction of a multivalued f to a domain D in which f is single valued and analytic is called a branch of f. A line which is excluded from $\mathbb{C}$ to create such a domain is called a branch cut. If a function is evaluated along a path in the complex plane which does not cross a branch cut, all the values will come from the same branch. A branch point is one end of a branch cut. Encircling a branch point will always cross a branch cut.

7. Allowing for isolated singularities and branch cuts, the common algebraic and transcendental functions have well-defined generalizations to the complex plane.

PROBLEMS

7.1. Consider the *bilinear transform* $w = \dfrac{1+z}{1-z}$, which is important in signal processing.

  (a) Use MATLAB to calculate w for $z = iy$, and plot the resulting locus of points in the complex plane (`plot(real(w),imag(w))`). Be sure to use `axis equal` so that the proportions are correct.

  (b) Once you see what the curve looks like from the graph, deduce a mathematical formula for it. (*Hint*: What is $|w|$?)

  (c) Where does the left half of the Z-plane map under this transformation?

7.2. Consider the complex function $z = \exp(sT)$, where $s = \sigma + i\omega$ may be thought of as the traditional complex Laplace transform variable. Where do the following regions of the s plane map in the z plane? Explain both analytically and graphically.

  (a) The imaginary axis, positive and negative.

  (b) The real axis, positive and negative.

  (c) The left and right half-planes.

  (d) The origin ($s = 0$) and extreme left half-plane ($s = -\infty$).

  (e) The general point $s = a + ib$. Discuss the effect of the parameter T on where this point maps.

  (f) A line of "constant damping," $\omega = -\sigma$, for $\sigma < 0$.

  This mapping is very important in digital signal processing and in the design of computerized feedback control systems.

7.3. Derive the polar form of the Cauchy–Riemann equations and express the derivative $f'(z)$ in terms of the partial derivatives of $u(r, \theta)$ and $v(r, \theta)$ (Equation 7.5),

7.4. Use the polar form of the Cauchy–Riemann equations to examine the analyticity of $f(z) = 1/z$.

7.5. Use the polar form of the Cauchy–Riemann conditions to show that Arg z is not analytic.

7.6. Using surface and region plots, explore the properties of z^2 and $|z|^2$. Carefully design your graphs to reveal important features of the functions. You should be able to observe that one function is analytic and the other is not.

7.7. Prove the product rule (Equation 7.6b).

7.8. Prove the quotient rule (Equation 7.6c).

7.9. Prove the composition rule (Equation 7.6d).

7.10. Consider the rational function $f(z) = \dfrac{b(z)}{a(z)}$, where the numerator and denominator polynomials are not coprime. Let z_0 be a root of multiplicity one for both polynomials: $a(z) = (z - z_0)\alpha(z)$ and $b(z) = (z - z_0)\beta(z)$. Show that z_0 is not a pole of f. In linear system theory, this is called "pole-zero cancellation."

7.11. Identify all the singular points of the function cosech $z = \dfrac{1}{\sinh z}$.

7.12. Use the MATLAB mesh function to plot the real and imaginary parts of $\cos z$ as surfaces above the $x - y$ plane. You should observe sinusoidal behavior in the x direction, and exponential (cosh, sinh) behavior in the y direction.

7.13. Determine the singular points and domains of analyticity for the following complex functions. It is unnecessary to use the Cauchy–Riemann equations.

(a) $f(z) = \dfrac{z}{z^2 + 3z + 2}$

(b) $f(z) = \dfrac{z+1}{z^2 - 1}$

(c) $f(z) = \text{sinc } z$

(d) $f(z) = \dfrac{1}{\sin z}$

(e) $f(z) = \exp i\pi z^2$

(f) $f(z) = \dfrac{\log z}{z - 1}$

(g) $f(z) = \dfrac{\sin \pi z}{e^{\pi z^2} - 1}$

7.14. Where in the complex plane are the following functions analytic? Which are entire functions? If the derivative exists in a domain, find an expression for $f'(z)$ in terms of z. The Cauchy–Riemann equations are not necessary.

(a) $f(z) = 2z^2 + 3$

(b) $f(z) = \exp(z^2)$

(c) $f(z) = z + 1/z$

(d) $f(z) = \dfrac{z+1}{z^2 + z + 2}$

(e) $f(z) = \dfrac{1}{\cos z}$

(f) $f(z) = \sin(i \tan z)$

7.15. *Complex functions and field theory*

There are certain relationships in complex variable theory which connect it to the theory of two-dimensional vector fields.

(a) Show that the real and imaginary parts of an analytic function $f = u + iv$ each satisfy Laplace's equation: $\nabla^2 u(x, y) = 0$ and $\nabla^2 v(x, y) = 0$. A function which satisfies Laplace's equation is called *harmonic* (nothing to do with harmonic in the Fourier sense).

(b) If a function is analytic, its real and imaginary parts satisfy the Cauchy–Riemann conditions. Given a real harmonic function $u(x, y)$, show how to calculate a function $v(x, y)$ such that $u + iv$ is analytic. This function v is called the *harmonic conjugate* of u. Demonstrate your method with $u(x, y) = x^2 - y^2$. What is the complex function $f(z)$? *Answer*: $f(z) = z^2$. Show that $\nabla^2 u(x, y) = 0$ and $\nabla^2 v(x, y) = 0$ in this particular case.

(c) The real and imaginary parts u and v so described have a nice interpretation as a potential function and stream function. That is, if $u(x, y)$ is an electrostatic or thermal potential function, then the gradient ∇u is proportional to the electric field or heat flow field. Show that the field vector is tangent to the stream function $v(x, y)$, in general and for the particular u and v you used in Parts (a) and (b).

(d) Hence, show that the equipotential lines $u(x, y) = $ const are orthogonal to the streamlines $v(x, y) = $ const, in general and in the particular case. Plot a few equipotential lines and streamlines to visualize this. MATLAB's `contour` function can be used to make the plots.

7.16. Is the function $g(z) = \text{rect}\,|z|$ analytic? Why or why not?

7.17. Locate all the singularities of the function $f(z) = 1/\cos(\pi z^2/2)$.

7.18. Prove that the *n*th root function, $f(z) = z^{1/n}$, is analytic in the cut plane $\{z : |z| > 0, \arg z \in (-\pi + 2\pi k, \pi + 2\pi k)\}$.

7.19. Show that the general form of z^c, Equation 7.15, reduces to the particular forms for z^n, $z^{1/n}$, and $z^{m/n}$, when $c = n$, $1/n$, and m/n, respectively.

7.20. *Multivalued functions*
Consider the function $f(z) = z^{1/2}$ with the branch cut placed along the negative real axis. Calculate and plot, using MATLAB, the real and imaginary parts of $f(z)$ along four paths:

(a) Along $z = x + i\epsilon$ and $z = x - i\epsilon$, $\epsilon > 0$ (horizontal, just above and below the real axis). In MATLAB, `eps` can be used to make a very small number.

(b) Along $z = 1 + iy$ and $z = -1 + iy$ (vertical, left and right of the imaginary axis).

In each part, graph the real parts together in one plot and the imaginary parts together in another plot to facilitate comparisons (four plots in all). Discuss your results, especially the effect of the branch cut.

7.21. *Multivalued functions*
Consider the function $f(z) = (z^2 - 1)^{1/2}$. This function has two branch points and therefore two branch cuts to place.

(a) Place both branch cuts so that $\arg(z \pm 1) \in (-\pi, \pi]$. Calculate and plot, using MATLAB, the real and imaginary parts of $f(z)$ along two paths:
- $z = x + i\epsilon$, $\epsilon > 0$ (just above the real axis).
- $z = x - i\epsilon$, $\epsilon > 0$ (just below the real axis).

(b) Repeat, with the branch cuts placed so that $\arg(z + 1) \in (-\pi, \pi]$ and $\arg(z - 1) \in (0, 2\pi]$.

In each part, graph the real parts together in one plot and the imaginary parts together in another plot to facilitate comparisons (four plots in all). Discuss your results, especially the effect of the branch cut.

7.22. Repeat the discussion of Section 7.5.5 for the function $f(z) = (z^3 - z)^{1/2}$. Consider at least two configurations of the branch cuts.

7.23. Determine all the roots of the following equations, and identify the principal values.
   (a) $z^3 + 2 = 0$
   (b) $e^{1+iz} - 2 = 0$

7.24. In addition to deriving an expression for each of the following, calculate the principal value of each solution. Compare with the value that MATLAB or your pocket calculator computes, for example, for (a), what you get by typing `(1+i)^(1/3)`.
   (a) Find all solutions of the equation $z^3 = 1 + i$.
   (b) Find all solutions of the equation $e^{z+1} = i\pi$.
   (c) Find all values of $i^{1/5}$.
   (d) Find all values of $\log(1 - i\sqrt{3})$.

7.25. (a) Find all possible solutions to the equation $\sin w = 1 + i$.
   (b) Consider the branch of $z^{1/2}$ defined by a branch cut along the negative real axis. On this branch, let $1^{1/2} = -1$ (i.e., the "negative" square root). Give a general formula for this branch of $z^{1/2}$. What is the value of $i^{1/2}$ on this branch? Do the equations $z^{1/2} - 3 = 0$ and $z^{1/2} + 3 = 0$ have solutions within the domain of analyticity of the branch? Explain why or why not, and give any solutions that you find.
   (c) Devise a (simple) set of branch cuts for the function $\arccos z$ so that it is analytic along almost the entire real axis and in the upper half-plane. On this branch, let $\arccos 0 = \pi/2$. What are the values of $\arccos -2$, $\arccos i$, and $\arccos 2$?

7.26. (a) The MATLAB function `angle(z)` computes the argument of the complex number z. Verify experimentally that `angle(z)`, in fact, computes the principal value of $\arg z$. One way to do this is to compute `angle(z)` for values of z between -2π and 2π. Where is the branch cut?
   (b) Write your own function which computes the correct value of $\arg(z)$ when the branch cut is along the positive real axis ($2\pi > \arg z \geq 0$) and demonstrate that it works. *Hint*: Some of the values returned by `angle(z)` fall into the appropriate range, and some do not. Think about how to modify the ones that do not.

CHAPTER 8

COMPLEX INTEGRATION

In this chapter we will develop techniques and applications of integration for complex functions. The complex integral $\int_A^B f(z)\,dz$, with $z = x + iy$, is an integral in the complex plane with respect to the two real variables x and y along a path connecting $z = A$ and $z = B$. The path need not be a straight line, but can be an arbitrary curve. In this way complex integration is very much like the line integrals one encounters in physics, and this is where the chapter begins. From here the unique properties of complex integrals are developed, with some methods for their evaluation. Finally, we shall see how real definite integrals can often be evaluated using complex integration. The methods are applied to the Fourier transform in this chapter and to related transforms in the next chapter.

8.1 LINE INTEGRALS IN THE PLANE

Review of Real Line Integrals
Line integration is introduced in physics via the definition of work:

$$W = \int_A^B \mathbf{f} \cdot d\mathbf{s},$$

where the vector $\mathbf{f}$ is a force field and A and B are the endpoints of the path along which the force acts. The path of integration is specified in *parametric* form, as a vector-valued function of a single variable. In two dimensions,

$$\mathbf{f} = (f_x, f_y),$$

$$\mathbf{s}(t) = (x(t), y(t)),$$

$$d\mathbf{s} = (x'(t), y'(t))\,dt,$$

and the integral becomes

$$\int_A^B \mathbf{f} \cdot d\mathbf{s} = \int_{t_A}^{t_B} (f_x, f_y) \cdot (x'(t), y'(t))\,dt$$

$$= \int_{t_A}^{t_B} f_x(x(t), y(t))\,x'(t)\,dt + \int_{t_A}^{t_B} f_y(x(t), y(t))\,y'(t)\,dt,$$

Fourier Transforms: Principles and Applications, First Edition. Eric W. Hansen.
© 2014 John Wiley & Sons, Inc. Published 2014 by John Wiley & Sons, Inc.

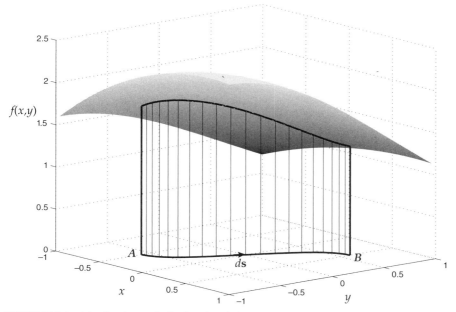

FIGURE 8.1 The line integral of a function f along a path from A to B is the area of a curved surface extending vertically from the path in the (x, y) plane to its intersection with the surface $z = f(x, y)$.

with $A = (x(t_A), y(t_A))$ and $B = (x(t_B), y(t_B))$. Scalar fields $f(x, y)$ are integrated with respect to arc length (ds taken along the path, Figure 8.1). The integral can be written either as

$$\int_{(x_1,y_1)}^{(x_2,y_2)} f(x, y)\, ds = \int_{x_1}^{x_2} f(x, y(x))\, dx$$

or, with the path specified parametrically:

$$\int_{(x_1,y_1)}^{(x_2,y_2)} f(x, y)\, ds = \int_{t_1}^{t_2} f(x(t), y(t))\, \|s'(t)\|\, dt.$$

The path must be at least piecewise smooth. It can be broken into subpaths, each with its own parametrization:

$$\int_A^B f(x, y)\, ds = \int_A^Q f(x, y)\, ds + \int_Q^B f(x, y)\, ds,$$

where Q is on the path between A and B (Figure 8.2). As with ordinary integrals, traversing a path in the reverse direction changes the sign of the integral, $\int_B^A = -\int_A^B$.

In general, the value of the integral depends on the path taken from A to B. An important exception occurs in physics, where the work done in a conservative force field is path independent.

FIGURE 8.2 A path of integration can be broken into subpaths. The integral from A to B is the sum of the integral from A to Q and the integral from Q to B.

Complex Line Integrals

Complex line integrals are a straightforward adaptation of vector field integrals. A complex function $f(z) = u(x, y) + iv(x, y)$, with its real and imaginary parts, is similar to a two-dimensional vector field. The integral of f along a path Γ in the complex plane is

$$\int_\Gamma f(z)dz = \int_\Gamma (u + iv)\,(dx + i\,dy)$$
$$= \int_\Gamma (udx - vdy) + i \int_\Gamma (vdx + udy). \tag{8.1}$$

The most common paths of integration used in practice are combinations of straight line segments and arcs:

$$\begin{array}{lll} \text{Horizontal line:} & z = x + iy_0, & dz = dx \\ \text{Vertical line:} & z = x_0 + iy, & dz = idy \\ \text{Ray at angle } \theta_0: & z = re^{i\theta_0}, & dz = e^{i\theta_0} dr \\ \text{Arc of radius } R: & z = Re^{i\theta}, & dz = iRe^{i\theta} d\theta. \end{array}$$

For example, the integral of a complex function along the vertical line $z = 1 + iy$ is

$$\int_\Gamma f(z)\,dz = i \int_{-\infty}^{\infty} f(1 + iy)\,dy,$$

and the integral along a semicircular arc of radius R in the upper half-plane (UHP) is

$$\int_\Gamma f(z)\,dz = iR \int_0^\pi f(Re^{i\theta})\,e^{i\theta}\,d\theta.$$

The ML Inequality

In one-dimensional real analysis, an integral is bounded above by the area of a rectangle whose height is the maximum absolute value of the function, and whose length is the length of the interval of integration (Figure 8.3).

The *ML* inequality is the extension of this idea to complex integration. If (a) $f(z)$ is continuous on a path Γ, (b) $|f(z)| \leq M$ for all $z \in \Gamma$, and (c) $L =$ the length of Γ, then

$$\left| \int_\Gamma f(z)dz \right| \leq \int_\Gamma |f(z)|\,dz \leq ML. \tag{8.2}$$

We shall repeatedly use this bound in calculations.

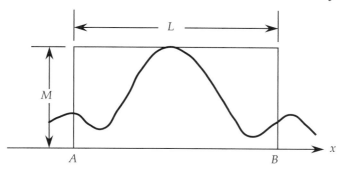

FIGURE 8.3 The integral of a function is bounded above by the product of the maximum absolute value, M, and the length of the interval, L.

8.2 THE BASIC COMPLEX INTEGRAL: $\oint_\Gamma z^n dz$

When $n \geq 0$, z^n is an analytic function, but when $n \leq -1$, z^n has a pole of order n at $z = 0$. We consider what happens when this function is integrated on three different closed paths (Figure 8.4).

On a circle of radius R centered at $z = 0$, we have $z = Re^{i\theta}$ and $dz = iRe^{i\theta} d\theta$. The integral becomes

$$\oint_\Gamma z^n dz = \int_0^{2\pi} iR^{n+1} e^{i(n+1)\theta} d\theta$$

$$= \begin{cases} \dfrac{R^{n+1}}{n+1} e^{i(n+1)\theta} \Big|_0^{2\pi} = 0, & n \neq -1 \\ i\theta \Big|_0^{2\pi} = 2\pi i, & n = -1 \end{cases} \quad (8.3)$$

Several of these integrals are illustrated in Figures 8.5, 8.6, 8.7, and 8.8.

The same result is obtained if the contour is square, although it is a bit more work. We will go through it in detail. The integral is evaluated in four segments (Figure 8.4):

$$\oint_\Gamma z^n dz = \int_{\Gamma_1} + \int_{\Gamma_2} + \int_{\Gamma_3} + \int_{\Gamma_4}.$$

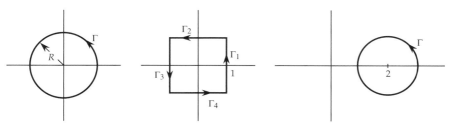

FIGURE 8.4 Three contours of integration in the Z-plane. *Left:* a circular contour of radius R, centered at the origin. *Center:* a unit square contour, centered at the origin. *Right:* a unit circle contour, centered at $z = 2$.

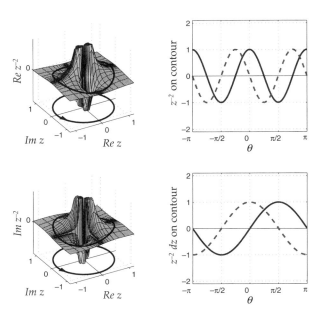

FIGURE 8.5 Integrating $f(z) = z^{-2}$ on the unit circle. *Left*: surface plots of $\mathcal{R}ef$ and $\mathcal{I}mf$, with the complex plane shown underneath. The contour of integration is shown both in the complex plane and overlaid on the surface. As the contour is traversed, the value of f, shown as the elevation of the surface plots, rises and falls (*top, right*). The area under the graph of $f(z)\,dz$ (*bottom, right*) is the integral. The integrals of both the real and imaginary parts are zero, by inspection.

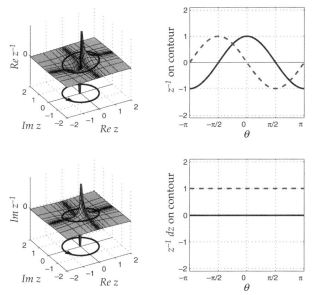

FIGURE 8.6 Integrating $f(z) = z^{-1}$ on the unit circle. *Left*: surface plots of $\mathcal{R}ef$ and $\mathcal{I}mf$, with the complex plane shown underneath. The contour of integration is shown both in the complex plane and overlaid on the surface. As the contour is traversed, the value of f, shown as the elevation of the surface plots, rises and falls (*top, right*). The area under the graph of $f(z)\,dz$ (*bottom, right*) is the integral. The real part is zero. By inspection, the area under the imaginary part is 2π.

8.2 THE BASIC COMPLEX INTEGRAL: $\oint_\Gamma z^n dz$

First we consider the $n \neq -1$ case. On Γ_1, $z = 1 + iy$, $dz = idy$. The integral is

$$\int_{\Gamma_1} z^n dz = i \int_{-1}^{1} (1 + iy)^n dy$$

$$= \frac{-(1 + iy)^{n+1}}{n+1} \bigg|_{-1}^{1} = \frac{(1+i)^{n+1} - (1-i)^{n+1}}{n+1}.$$

On Γ_2, $z = x + i$, $dz = dx$. Note that the path goes right to left, so we integrate from $x = 1$ to $x = -1$:

$$\int_{\Gamma_2} z^n dz = \int_{1}^{-1} (x+i)^n dx$$

$$= \frac{(x+i)^{n+1}}{n+1} \bigg|_{1}^{-1} = \frac{-(1+i)^{n+1} + (-1+i)^{n+1}}{n+1}.$$

On Γ_3, $z = -1 + iy$, $dz = idy$, and because the path is directed downward, we integrate from $y = 1$ to $y = -1$:

$$\int_{\Gamma_3} z^n dz = i \int_{1}^{-1} (-1 + iy)^n dy$$

$$= \frac{-(-1+iy)^{n+1}}{n+1} \bigg|_{1}^{-1} = \frac{(-1+i)^{n+1} - (-1-i)^{n+1}}{n+1}.$$

On Γ_4, $z = x - i$, $dz = dx$, and we integrate left to right, from $x = -1$ to $x = 1$:

$$\int_{\Gamma_4} z^n dz = \int_{-1}^{1} (x-i)^n dx$$

$$= \frac{(x-i)^{n+1}}{n+1} \bigg|_{-1}^{1} = \frac{(1-i)^{n+1} + (-1-i)^{n+1}}{n+1}.$$

The sum of the four terms is seen to be zero.

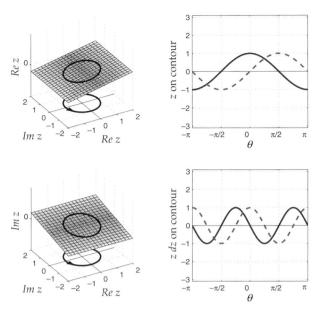

FIGURE 8.7 Integrating $f(z) = z$ on the unit circle. *Left*: surface plots of $\mathcal{R}ef$ and $\mathcal{I}mf$, with the complex plane shown underneath. The contour of integration is shown both in the complex plane and overlaid on the surface. As the contour is traversed, the value of f, shown as the elevation of the surface plots, rises and falls (*top, right*). The area under the graph of $f(z)\,dz$ (*bottom, right*) is the integral. The integrals of both real and imaginary parts are zero.

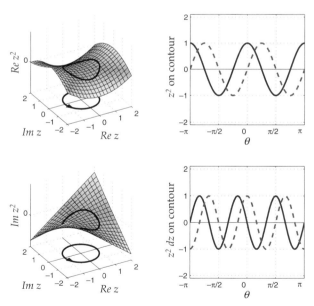

FIGURE 8.8 Integrating $f(z) = z^2$ on the unit circle. *Left*: surface plots of $\mathcal{R}ef$ and $\mathcal{I}mf$, with the complex plane shown underneath. The contour of integration is shown both in the complex plane and overlaid on the surface. As the contour is traversed, the value of f, shown as the elevation of the surface plots, rises and falls (*top, right*). The area under the graph of $f(z)\,dz$ (*bottom, right*) is the integral. The integrals of both real and imaginary parts are zero.

When $n = -1$, we have

$$\int_{\Gamma_1} \frac{dz}{z} = \int_{-1}^{1} \frac{i\,dy}{1+iy}$$

$$= \int_{-1}^{1} \frac{i(1-iy)\,dy}{1+y^2} = \int_{-1}^{1} \frac{i\,dy}{1+y^2} + \int_{-1}^{1} \frac{y\,dy}{1+y^2}$$

$$= i\arctan y \Big|_{-1}^{1} + \frac{1}{2}\log(1+y^2)\Big|_{-1}^{1} = \frac{i\pi}{2} + 0 = \frac{i\pi}{2},$$

$$\int_{\Gamma_2} \frac{dz}{z} = \int_{1}^{-1} \frac{dx}{x+i}$$

$$= \int_{1}^{-1} \frac{(x-i)\,dx}{x^2+1} = -\int_{-1}^{1} \frac{x\,dx}{x^2+1} + \int_{-1}^{1} \frac{i\,dx}{x^2+1}$$

$$= -\frac{1}{2}\log(x^2+1)\Big|_{-1}^{1} + i\arctan y \Big|_{-1}^{1} = 0 + \frac{i\pi}{2} = \frac{i\pi}{2},$$

$$\int_{\Gamma_3} \frac{dz}{z} = \int_{1}^{-1} \frac{i\,dy}{-1+iy} = \int_{-1}^{1} \frac{i(-d\xi)}{-1+i(-\xi)} = \int_{-1}^{1} \frac{i\,d\xi}{1+i\xi}$$

$$= \int_{\Gamma_1} = \frac{i\pi}{2},$$

$$\int_{\Gamma_4} \frac{dz}{z} = \int_{-1}^{1} \frac{dx}{x-i} = \int_{1}^{-1} \frac{-d\xi}{(-\xi)-i} = \int_{1}^{-1} \frac{d\xi}{\xi+i}$$

$$= \int_{\Gamma_2} = \frac{i\pi}{2}.$$

Summing the pieces, we obtain $\oint_\Gamma \frac{dz}{z} = i2\pi$. The integration of $f(z) = z^{-1}$ on a square contour is shown in Figure 8.9.

On the other hand, if the contour does not enclose the origin, the result is zero. Consider a circular contour of unit radius centered at $z = z_0$, where $|z_0| > 1$ so that it does not enclose the singularity at the origin. On this contour, $z = z_0 + e^{i\theta}$, $dz = ie^{i\theta}\,d\theta$. Then,

$$\oint_\Gamma z^n\,dz = \begin{cases} \int_0^{2\pi} i(z_0 + e^{i\theta})^n e^{i\theta}\,d\theta = \frac{(z_0 + e^{i\theta})^{n+1}}{n+1}\Big|_0^{2\pi} = 0, & n \neq -1 \\ \int_0^{2\pi} \frac{ie^{i\theta}\,d\theta}{z_0 + e^{i\theta}} = \log(z_0 + e^{i\theta})\Big|_0^{2\pi} = 0, & n = -1 \end{cases}.$$

The integration of $f(z) = z^{-1}$ on a circular contour centered at $z_0 = 1 - i$ is illustrated in Figure 8.10.

502 CHAPTER 8 COMPLEX INTEGRATION

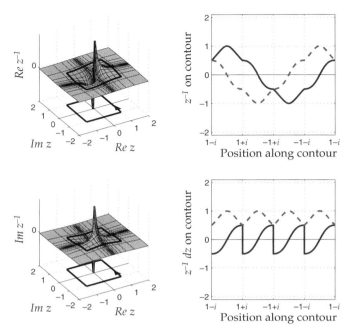

FIGURE 8.9 Integrating $f(z) = z^{-1}$ on a square contour enclosing the singularity at the origin. *Left*: surface plots of $\mathcal{R}e f$ and $\mathcal{I}m f$, with the complex plane shown underneath. The contour of integration is shown both in the complex plane and overlaid on the surface. As the contour is traversed, the value of f, shown as the elevation of the surface plots, rises and falls (*top, right*). The area under the graph of $f(z)\,dz$ (*bottom, right*) is the integral. The integral of the real part is zero, by inspection, but the area under the imaginary part is nonzero. In fact, it is 2π. Compare Figure 8.6.

Here are four observations about these examples:

- When the contour of integration encloses the singularity z^{-1}, the integral is $i2\pi$, whether the contour is square or circular.
- When the contour of integration does not enclose the singularity z^{-1}, the integral is zero.
- The integral of a nonnegative power of z on a closed contour is zero.
- The integral of a higher-order singularity, for example, z^{-2}, is zero, whether it is enclosed by the contour of integration or not.

We can sort these out with the aid of two important results: Cauchy's integral theorem and Cauchy's integral formula.

8.3 CAUCHY'S INTEGRAL THEOREM

The first important general result in complex integration is Cauchy's integral theorem. First, we need to define some terms.

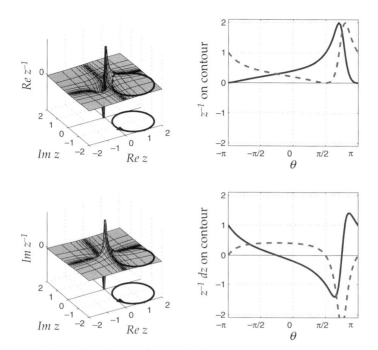

FIGURE 8.10 Integrating $f(z) = z^{-1}$ on a unit circular contour centered at $z_0 = 1 - i$, which does not enclose the singularity at the origin. *Left*: surface plots of $\mathcal{R}ef$ and $\mathcal{I}mf$, with the complex plane shown underneath. The contour of integration is shown both in the complex plane and overlaid on the surface. As the contour is traversed, the value of f, shown as the elevation of the surface plots, rises and falls (*top, right*). The area under the graph of $f(z)\,dz$ (*bottom, right*) is the integral. The integrals of both the real and imaginary parts are zero, by inspection.

Definition 8.1 (Connected domains). A *domain* is an open, connected set. A domain is *simply connected* if any closed curve in that domain encloses only points in the domain.

A domain is not simply connected if it has a "hole" in it (Figure 8.11). A domain with one hole is called doubly connected. Two holes make a triply connected domain, etc.

An annular domain like $2 > z > 1$ is not simply connected. Nor is the domain of analyticity for z^{-1}, created by excluding the singular point $z = 0$ (puncturing the complex plane).

Definition 8.2 (Simple, closed contour). A *simple contour* (path) does not cross over itself. A *simple closed contour* divides the complex plane into two domains; one is the interior of the contour, and the other is the exterior. The counter-clockwise direction around a closed contour is defined to be the positive direction.

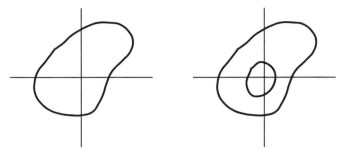

FIGURE 8.11 *Left:* a simply connected domain. *Right:* a doubly connected domain has a "hole" in it.

A simple closed contour cannot cross over itself (e.g., make a figure-eight), because that would create more than two domains (Figure 8.12).

In the previous section we integrated z^n around circular contours. When the contour did not enclose a singularity, that is, the function was analytic on and inside the contour, the integral was zero, regardless of the value of n or the radius of the circle. Let us now ask what happens if we integrate an arbitrary analytic function around an arbitrary simple closed contour, $\oint_\Gamma f(z)\,dz$.

We write the contour integral in the form 8.1:

$$\oint_\Gamma f(z)\,dz = \oint_\Gamma (u\,dx - v\,dy) + i\oint_\Gamma (v\,dx + u\,dy).$$

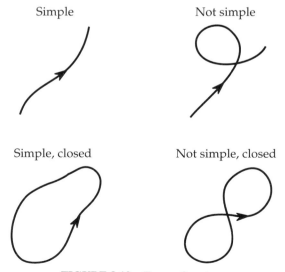

FIGURE 8.12 Types of contour.

To simplify it, we will use Green's theorem from multivariable calculus, which is usually written

$$\oint_\Gamma (P\,dx + Q\,dy) = \iint_R \left(\frac{\partial Q}{\partial x} - \frac{\partial P}{\partial y} \right) dx\,dy \tag{8.4}$$

and expresses a line integral around a simple closed loop Γ as an integral over the region R enclosed by Γ. (If P and Q are the x and y components of a force field, then the loop integral is the work done in moving around Γ. The difference of partial derivatives, $\frac{\partial Q}{\partial x} - \frac{\partial P}{\partial y}$, is the curl of the field.) The first partial derivatives must be continuous. With this assumption, we can write

$$\oint_\Gamma (u\,dx - v\,dy) = \iint_R -\left(\frac{\partial v}{\partial x} + \frac{\partial u}{\partial y} \right) dx\,dy$$

$$\oint_\Gamma (v\,dx + u\,dy) = \iint_R \left(\frac{\partial u}{\partial x} - \frac{\partial v}{\partial y} \right) dx\,dy.$$

Because f is analytic, the Cauchy–Riemann equations (7.3) give

$$\frac{\partial v}{\partial x} + \frac{\partial u}{\partial y} = 0 \quad \text{and} \quad \frac{\partial u}{\partial x} - \frac{\partial v}{\partial y} = 0,$$

so the integrands of both integrals are zero, and consequently $\oint_\Gamma f(z)\,dz = 0$.

A more technical derivation avoids Green's theorem and removes the requirements of a simple closed contour and continuous partial derivatives.[1] The result is expressed in the following theorem.

Theorem 8.1 (Cauchy's integral theorem). Let f be analytic in a simply connected domain D. For any closed contour Γ inside D,

$$\oint_\Gamma f(z)\,dz = 0. \tag{8.5}$$

Cauchy's integral theorem explains some of the results of the previous section. When $n \geq 0$, z^n is analytic everywhere in the plane, hence any integral over any closed contour will be zero (not just the example we performed). Furthermore, when $n \leq -1$, z^n is singular at $z = 0$, but analytic everywhere else in the complex plane. Hence, an integral over a contour which does not include the origin will be an integral of an analytic function, and again, Cauchy's integral theorem says the result will be zero. What Cauchy's integral theorem does not explain is what happens when the contour encloses a singularity. That will come later.

Cauchy's integral theorem has two other useful consequences.

---

[1] Marsden and Hoffman (1998, pp. 123–142). This stronger version is also known as the *Cauchy–Goursat theorem*. In this approach, Corollaries 8.2 and 8.3 and Theorem 8.4 are first established by other methods and then used in the proof of Cauchy's integral theorem. An even stronger version relaxes the requirement that the domain be simply connected: if f is analytic in D and the closed path Γ can be smoothly deformed (shrunk) to a point in D, then the integral around the path is zero. Such a path will not enclose a "hole" in the region, else it cannot be smoothly deformed to a point.

Corollary 8.2 (Deformation of contour). Consider two simple closed contours Γ_1 and Γ_2 and a function $f(z)$ which is analytic on each contour. If one contour can be smoothly deformed into the other without crossing any singularity of f, then

$$\oint_{\Gamma_1} f(z)dz = \oint_{\Gamma_2} f(z)dz.$$

Proof: Consider Figure 8.13. The two contours are joined in two places by line segments, creating two closed paths enclosing singularity-free regions: Γ_{abcd} and Γ_{efgh}. By Cauchy's integral theorem,

$$\oint_{\Gamma_{abcd}} = \int_a + \int_b + \int_c + \int_d = 0$$

$$\oint_{\Gamma_{efgh}} = \int_e + \int_f + \int_g + \int_h = 0.$$

Furthermore, because paths b and h are identical but oppositely directed, as are paths d and f:

$$\int_b = -\int_h \quad \text{and} \quad \int_d = -\int_f.$$

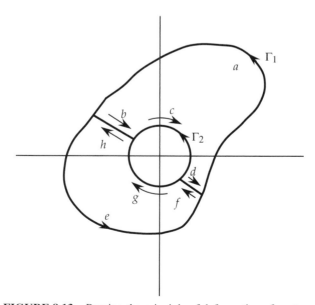

FIGURE 8.13 Proving the principle of deformation of contour.

Making these substitutions,

$$\oint_{\Gamma_{abcd}} + \oint_{\Gamma_{efgh}} = \int_a - \int_h + \int_c - \int_f + \int_e + \int_f + \int_g + \int_h$$

$$= \int_a + \int_c + \int_e + \int_g = 0.$$

But

$$\oint_{\Gamma_1} = \int_a + \int_e \quad \text{and} \quad \oint_{\Gamma_2} = -\left(\int_c + \int_g\right).$$

Therefore,

$$\oint_{\Gamma_1} - \oint_{\Gamma_2} = 0$$

which is what we needed to prove. ∎

The principle of deformation of contour enables an arbitrary (and inconveniently shaped) contour to be replaced by one made up of convenient sections (lines and arcs) (Figure 8.14).

It explains why, in the earlier discussion of $\oint_\Gamma \frac{dz}{z}$, the radius of the circular contour did not matter, nor did it matter whether the contour was square or circular. All of these contours could be smoothly deformed into each other without crossing the singularity at $z = 0$.

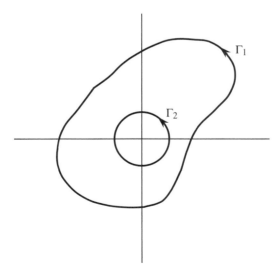

FIGURE 8.14 An arbitrary contour can be deformed into a convenient contour, without changing the value of the integral, if there are no singularities between them.

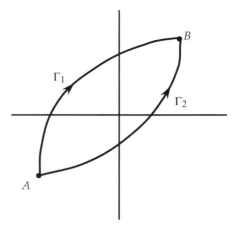

FIGURE 8.15 Path independence. The integral from A to B is independent of the path taken.

Corollary 8.3 (Path independence). Let f be analytic in a domain D, and let two points A and B belong to D. The integral $\int_A^B f(z)dz$ along a path in D is independent of the path taken from A to B.

Proof: The proof follows directly from Cauchy's integral theorem. With reference to Figure 8.15, the integral around the closed loop is zero, because f is analytic on and inside the loop. Going around the loop in the positive direction, path Γ_1 is traversed backward, so $\int_{\Gamma_2} - \int_{\Gamma_1} = 0$. Therefore, $\int_{\Gamma_1} = \int_{\Gamma_2}$. ∎

Here are some examples of the application of Cauchy's integral theorem to the evaluation of integrals.

Example 8.1. Integrate $f(z) = \frac{2}{z-1}$ around the contour shown in Figure 8.16a.

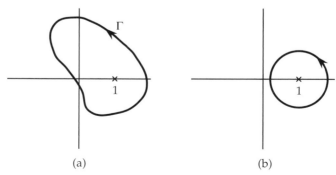

FIGURE 8.16 Integrating around a single pole. (a) The given contour. (b) The contour is deformed into a convenient shape.

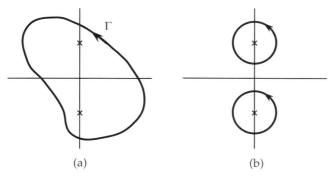

FIGURE 8.17 Integrating around a pair of poles. (a) The given contour. (b) The final pair of contours.

The contour may be deformed into a circle centered at $z = 1$ (Figure 8.16b). On this contour, $z = 1 + Re^{i\theta}$ and $dz = iRe^{i\theta}\,d\theta$. Making these substitutions,

$$\oint_\Gamma = \int_0^{2\pi} \frac{2iRe^{i\theta}\,d\theta}{(1+Re^{i\theta})-1} = \int_0^{2\pi} 2i\,d\theta = i4\pi.$$

■

Example 8.2. Integrate $f(z) = \frac{z}{z^2+1}$ around the contour shown in Figure 8.17a. We have a pair of poles to integrate. We begin by rewriting the integral with a partial fraction expansion:

$$\oint_\Gamma \frac{z\,dz}{z^2+1} = \oint_\Gamma \left(\frac{1/2}{z+i} + \frac{1/2}{z-i}\right) dz = \frac{1}{2}\oint_\Gamma \frac{dz}{z+i} + \frac{1}{2}\oint_\Gamma \frac{dz}{z-i}.$$

We now have two integrals, each around a single pole. We can deform the contour to a convenient shape around each pole (Figure 8.17b) and proceed as in the previous example. Around $z = i$, the contour is $z = i + Re^{i\theta}$, $dz = iRe^{i\theta}\,d\theta$; around $z = -i$, we have $z = -i + Re^{i\theta}$, $dz = iRe^{i\theta}\,d\theta$. Hence,

$$\oint_\Gamma \frac{z\,dz}{z^2+1} = \frac{1}{2}\int_0^{2\pi} \frac{iRe^{i\theta}\,d\theta}{(i+Re^{i\theta})-i} + \frac{1}{2}\int_0^{2\pi} \frac{iRe^{i\theta}\,d\theta}{(-i+Re^{i\theta})+i}$$

$$= \frac{1}{2}\int_0^{2\pi} i\,d\theta + \frac{1}{2}\int_0^{2\pi} i\,d\theta = i\pi + i\pi = i2\pi.$$

■

Example 8.3. Integrate the function $f(z) = \frac{1}{z^2+2}$ around the semicircular contour shown in Figure 8.18a.

We begin again by writing f in a partial fraction expansion:

$$f(z) = \frac{1}{z^2+2} = \frac{i/2\sqrt{2}}{z+i\sqrt{2}} + \frac{-i/2\sqrt{2}}{z-i\sqrt{2}},$$

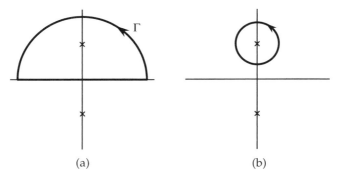

FIGURE 8.18 When you integrate a complex function, sometimes the contour does not include all the singularities. (a) The given contour. (b) The final contour.

which again yields two integrals:

$$\oint_\Gamma \frac{dz}{z^2+2} = \frac{i}{2\sqrt{2}} \oint_\Gamma \frac{dz}{z+i\sqrt{2}} - \frac{i}{2\sqrt{2}} \oint_\Gamma \frac{dz}{z-i\sqrt{2}}.$$

In the first integral, the contour does not encircle the pole at $z = -i\sqrt{2}$. The integrand is analytic on and inside the contour, and Cauchy's integral theorem gives zero. The second integral is attacked using deformation of contour. The semicircular contour is deformed into a convenient circular contour (Figure 8.18b). On this contour, $z = i\sqrt{2} + Re^{i\theta}$ and $dz = iRe^{i\theta}\, d\theta$, $R < 2\sqrt{2}$. Hence,

$$\oint_\Gamma \frac{dz}{z^2+2} = -\frac{i}{2\sqrt{2}} \int_0^{2\pi} \frac{iRe^{i\theta}\, d\theta}{(i\sqrt{2}+Re^{i\theta}) - i\sqrt{2}}$$

$$= -\frac{i}{2\sqrt{2}} \int_0^{2\pi} i\, d\theta = -\frac{i}{2\sqrt{2}} \times i2\pi = \frac{\pi}{\sqrt{2}}. \quad \blacksquare$$

The following result makes an interesting connection between complex integrals and real integrals.[2]

Theorem 8.4 (Fundamental theorem of calculus). Let f be analytic in a simply connected domain D. For every $z, z_0 \in D$, the integral

$$F(z) = \int_{z_0}^z f(\zeta)\, d\zeta$$

along any path in D from z_0 to z is a function of z which is analytic in D, and whose derivative is

$$F'(z) = f(z).$$

---

[2]LePage, 1980, pp. 100–102.

8.3 CAUCHY'S INTEGRAL THEOREM

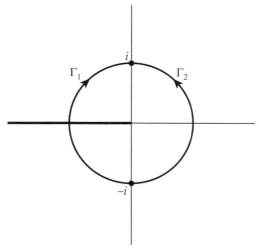

FIGURE 8.19 Two different paths from $-i$ to i. One crosses the branch cut for $\log z$.

Thus, every analytic function is integrable. Moreover, if an analytic function F is the antiderivative of an analytic function f, then

$$\int_{z_1}^{z_2} f(z)\,dz = F(z_2) - F(z_1). \tag{8.6}$$

Example 8.4. Consider the integral of $f(z) = z$ along a quarter circle Γ from $z = 1$ to $z = i$. On this path, $z = e^{i\theta}$, $\frac{\pi}{2} \geq \theta \geq 0$, and $dz = ie^{i\theta}\,d\theta$. We have

$$\int_\Gamma z\,dz = \int_0^{\pi/2} ie^{i2\theta}\,d\theta = \frac{1}{2}e^{i2\theta}\Big|_0^{\pi/2} = -1.$$

Or, using Equation 8.6,

$$\int_1^i z\,dz = \frac{z^2}{2}\Big|_1^i = \frac{-1-1}{2} = -1. \qquad\blacksquare$$

Example 8.5. One must be careful about branch cuts. We know that the derivative of $\log z$ is z^{-1}, but compare the integrals $\int_{\Gamma_1} \frac{dz}{z}$ and $\int_{\Gamma_2} \frac{dz}{z}$, where both paths run from $z = -i$ to $z = i$, but in different ways (Figure 8.19).

On Γ_1, $z = e^{i\theta}$ with $-\frac{\pi}{2} \geq \theta \geq -\frac{3\pi}{2}$, while on Γ_2, $\frac{\pi}{2} \geq \theta \geq -\frac{\pi}{2}$. The two integrals are

$$\int_{\Gamma_1} \frac{dz}{z} = \log z\Big|_{e^{-i\pi/2}}^{e^{-i3\pi/2}} = -\frac{i3\pi}{2} - \left(-\frac{i\pi}{2}\right) = -i\pi$$

$$\int_{\Gamma_2} \frac{dz}{z} = \log z\Big|_{e^{-i\pi/2}}^{e^{i\pi/2}} = \frac{i\pi}{2} - \left(-\frac{i\pi}{2}\right) = i\pi.$$

The value of the integral is *not* path independent because the antiderivative, log z, has a branch cut crossed by Γ_1.

This gives another interpretation of the result $\oint_\Gamma \frac{dz}{z} = i2\pi$ when Γ encircles the origin. At the beginning of the path, arg $z = \theta_0$, and at the end, arg $z = \theta_0 + 2\pi$. The path crosses the branch cut, and consequently,

$$\oint_\Gamma \frac{dz}{z} = \log z \Big|_{e^{i\theta_0}}^{e^{i(\theta_0+2\pi)}} = i(\theta_0 + 2\pi) - i\theta_0 = i2\pi.$$

For any other integer power of z, the antiderivative is an integer power of z. There is no branch cut, and the integral around the path evaluates to zero. ∎

The converse of Cauchy's integral theorem gives a sufficient condition for a function f to be analytic.[3]

Theorem 8.5 (Morera's theorem). Let f be a continuous function on a domain D, and let Γ be a simple closed contour in D whose interior is in D. If, for all such contours Γ,

$$\oint_\Gamma f(z)\, dz = 0,$$

then f is analytic in D.

8.4 CAUCHY'S INTEGRAL FORMULA

This companion to the Cauchy integral theorem is the key to all complex integrals over closed contours.

Theorem 8.6 (Cauchy's integral formula). Let f be analytic on and inside a simple closed contour Γ. Let z_0 be a point in the interior of Γ. Then

$$f(z_0) = \frac{1}{2\pi i} \oint_\Gamma \frac{f(z)\,dz}{z - z_0}. \tag{8.7}$$

Proof: By the principle of deformation of contours, we can replace the arbitrary contour Γ by a more convenient one, namely a circle of radius r centered at z_0 (Figure 8.20).

Next, rewrite the integral using the obvious fact that $f(z) = f(z) - f(z_0) + f(z_0)$, obtaining

$$\oint_\Gamma \frac{f(z)\,dz}{z - z_0} = \oint_{|z-z_0|=r} \frac{f(z_0)\,dz}{z - z_0} + \oint_{|z-z_0|=r} \frac{(f(z) - f(z_0))\,dz}{z - z_0}.$$

[3] Flanagan, 1983, pp. 189–190; Hahn and Epstein, 1996, p. 127.

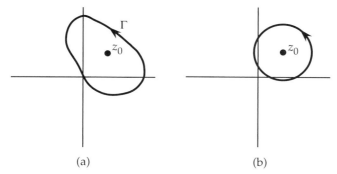

FIGURE 8.20 For proof of Cauchy's integral formula. (a) An arbitrary contour around $z = z_0$. (b) A convenient contour centered at $z = z_0$.

The first integral on the right is

$$\oint_{|z-z_0|=r} \frac{f(z_0)dz}{z - z_0} = f(z_0) \oint_{|z-z_0|=r} \frac{dz}{z - z_0} = 2\pi i f(z_0).$$

It remains to show that the second integral is equal to zero. By the principle of deformation of contour, the value of the integral is independent of the contour—in particular, it is independent of the radius r. We can, as a result, take r to be arbitrarily small. Now, using the *ML* inequality, we can place an upper bound on its value. The absolute value of the integrand on the contour is

$$\frac{|f(z) - f(z_0)|}{|z - z_0|} = \frac{|f(z) - f(z_0)|}{r}.$$

Furthermore, because f is analytic, it is also continuous, so we know that $|f(z) - f(z_0)| < \epsilon$ for z such that $|z - z_0| < \delta$, with ϵ becoming smaller as δ decreases. (This is the classic "epsilon-delta" definition of continuity, which was also developed by Cauchy.) Hence, choosing $r < \delta$, we can bound the integrand above by ϵ/r, which is the M part of the ML inequality. The L part is the circumference of the contour, $2\pi r$. The product ML, therefore, is $\epsilon/r \times 2\pi r = 2\pi\epsilon$, so

$$\left| \oint_{|z-z_0|=r} \frac{(f(z) - f(z_0))dz}{z - z_0} \right| \leq 2\pi\epsilon.$$

We can take r as small as we want without changing the value of the integral. Hence, $\delta > r$ can be made arbitrarily small, and ϵ follows suit because of the continuity of f. The upper bound, $2\pi\epsilon$, becomes smaller and smaller as r approaches zero. It must be, therefore, that the integral is actually equal to zero. ∎

Cauchy's integral formula says two things. As written (Equation 8.7), it says that the value of f at any point z_0 inside a closed region is determined by the values of f on the boundary of that region. This can be used to solve certain boundary-value problems in physics, such as finding the potential inside a conducting cylinder when the walls of the cylinder have a particular potential distribution.[4] However, if we turn it around, we obtain this form:

$$\oint_\Gamma \frac{f(z)dz}{z-z_0} = 2\pi i f(z_0), \tag{8.8}$$

which says that the integral of a function of the form $g(z) = \frac{f(z)}{z-z_0}$, where f is analytic, is given by the value of f at the pole, $z = z_0$. We can use this form to evaluate integrals.

Example 8.6. Integrate $g(z) = \frac{1}{z-a}$ on any contour encircling $z = a$. This, of course, is just the problem of integrating a single pole, which we have found on several occasions to give $i2\pi$ as the result. Cauchy's integral formula yields this directly. The function g is the ratio of two entire functions, 1 and $z - a$, and is therefore analytic except where the denominator goes to zero, $z = a$. We apply Cauchy's integral formula with $f(z) = 1$ and $z_0 = a$:

$$\oint_\Gamma \frac{dz}{z-a} = 2\pi i f(a) = 2\pi i. \qquad\blacksquare$$

Example 8.7. Integrate $g(z) = \frac{\cos z}{z-\pi/4}$ on the unit circle centered at the origin. The given function g is the ratio of two entire functions, $\cos z$ and $z - \pi/4$. It is therefore analytic except where the denominator is zero, $z = \pi/4$. This singularity is inside the contour of integration, because $\pi/4 < 1$. We apply Cauchy's integral formula with $f(z) = \cos z$ and $z_0 = \pi/4$. The result is

$$\oint_{|z|=1} \frac{\cos z\, dz}{z-\pi/4} = 2\pi i \cos(\pi/4) = 2\pi i \times \frac{1}{\sqrt{2}} = i\sqrt{2}\pi. \qquad\blacksquare$$

Example 8.8. Integrate $g(z) = \frac{\sin z}{z^2-1}$ on a contour enclosing the points $z = \pm 1$. Factoring the denominator,

$$g(z) = \frac{\sin z}{(z+1)(z-1)},$$

we see there are two singularities, at $z = -1$ and $z = 1$. By deformation of contour, this integral can be decomposed into the sum of two integrals:

$$\oint_\Gamma \frac{\sin z\, dz}{z^2-1} = \oint_{\Gamma_1} \frac{\sin z\, dz}{z^2-1} + \oint_{\Gamma_2} \frac{\sin z\, dz}{z^2-1},$$

where Γ_1 encloses just the pole at $z = -1$ and Γ_2 encloses just the pole at $z = 1$ (in a manner similar to Example 8.2). Inside Γ_1, $\frac{\sin z}{z-1}$ is analytic, and inside Γ_2, $\frac{\sin z}{z+1}$ is analytic. We may therefore rewrite the integrals:

$$\oint_\Gamma \frac{\sin z\, dz}{z^2-1} = \oint_{\Gamma_1} \frac{\frac{\sin z}{z-1}\, dz}{z+1} + \oint_{\Gamma_2} \frac{\frac{\sin z}{z+1}\, dz}{z-1}.$$

---
[4]Wunsch, 1994, pp. 200–205.

In this form, each integral can be evaluated using Cauchy's integral formula, yielding

$$\oint_\Gamma \frac{\sin z\, dz}{z^2 - 1} = i2\pi \left.\frac{\sin z}{z - 1}\right|_{z=-1} + i2\pi \left.\frac{\sin z}{z + 1}\right|_{z=1}$$

$$= i2\pi \frac{\sin(-1)}{-2} + i2\pi \frac{\sin(1)}{2} = i2\pi \sin(1).$$ ∎

Extended Cauchy Formula

A more general version of Cauchy's integral formula can be established. Returning to Equation 8.7,

$$f(\zeta) = \frac{1}{2\pi i} \oint_\Gamma \frac{f(z)dz}{z - \zeta}.$$

If we differentiate both sides with respect to ζ and take the derivative under the integral sign (in the same way we would do for a real integral, see Chapter 1), we find

$$f'(\zeta) = \frac{1}{2\pi i} \frac{d}{d\zeta} \oint_\Gamma \frac{f(z)dz}{z - \zeta} = \frac{1}{2\pi i} \oint_\Gamma \frac{d}{d\zeta} \frac{f(z)dz}{z - \zeta}$$

$$= \frac{1}{2\pi i} \oint_\Gamma \frac{f(z)dz}{(z - \zeta)^2}.$$

The singularity $z = \zeta$ is inside the closed path Γ, and f is analytic on Γ, therefore the integrand is finite on Γ. The path has finite length, so by the *ML* inequality (Equation 8.2), we know the integral is finite.

This result says that the derivative of f, which we know exists because f is analytic, can actually be calculated by *integrating f*! Furthermore, f' is a continuous function of ζ by the continuity of $(z - \zeta)^2$ and in fact, it can be shown that f' is analytic for all ζ inside Γ. We may therefore repeat the calculation, obtaining

$$f''(\zeta) = 2 \cdot \frac{1}{2\pi i} \oint_\Gamma \frac{f(z)dz}{(z - \zeta)^3}$$

$$f^{(3)}(\zeta) = 6 \cdot \frac{1}{2\pi i} \oint_\Gamma \frac{f(z)dz}{(z - \zeta)^4}$$

$$\vdots$$

$$f^{(n)}(\zeta) = \frac{n!}{2\pi i} \oint_\Gamma \frac{f(z)dz}{(z - \zeta)^{n+1}}.$$

These results are collected in an extended version of the Cauchy formula, due to Goursat.[5]

Theorem 8.7 (Extended Cauchy integral formula). If a function f is analytic within a domain, then it possesses derivatives of all orders in that domain, which are

---

[5] For a complete proof, see Hahn and Epstein (1996, p. 122).

themselves analytic functions in that domain. Furthermore, if f is analytic on and inside a simple closed contour Γ and if z_0 is inside Γ, then

$$f^{(n)}(z_0) = \frac{n!}{2\pi i} \oint_\Gamma \frac{f(z)}{(z-z_0)^{n+1}} dz. \tag{8.9}$$

Rearranging (Equation 8.9) slightly gives a formula for calculating more general complex integrals:

$$\oint_\Gamma \frac{f(z)}{(z-z_0)^{n+1}} dz = 2\pi i \frac{f^{(n)}(z_0)}{n!}. \tag{8.10}$$

Example 8.9 (The basic integral $\oint_\Gamma (z-a)^k dz$, revisited). Earlier (Figures 8.5, 8.6, 8.7, and 8.8), we observed that the integral $\oint_\Gamma z^k dz$ evaluates to $2\pi i$ when $k = -1$ and 0 otherwise. Cauchy's integral theorem explained why the integral is zero for $k \geq 0$, namely, the integrand is analytic. For $k \leq -1$, the more general integrand $(z-a)^k$ has the form $\frac{f(z)}{(z-z_0)^{n+1}}$ with $z_0 = a$, $n = -(k+1)$, and $f(z) = 1$, a constant. When $k = -1$, Cauchy's integral formula directly gives $2\pi i f(0) = 2\pi i$. For $k < -1$ (double poles and higher), the extended Cauchy formula requires that we differentiate f one or more times, but because $f = 1$ all the derivatives are zero and so the integrals are zero. ∎

The Cauchy integral formula, in its extended version, enables us to calculate the integrals of functions $g(z)$ which can be written as the ratio of an analytic function $f(z)$ and a polynomial $a(z)$. Let $a(z)$ have roots $z_1, z_2, \ldots, z_p$, with multiplicities $n_1, n_2, \ldots, n_p$. Now consider the integral of $g(z)$ on a simple closed contour. We can use the principle of deformation of contour to reduce this integral to the sum of integrals around individual singularities (roots of $a(z)$), as shown in Figure 8.21.

The original contour is smoothly deformed into a "shrink-wrapped" contour that nearly encircles each singularity. The integral is the sum of contributions from each nearly circular path around the singularities, together with the contributions

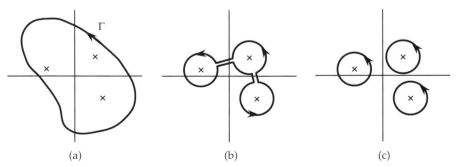

FIGURE 8.21 An integral on a contour encircling several isolated singularities (a) simplified by deformation of contour (b) to a sum of integrals, each one taken around each singularity (c).

from the line segments connecting the circular pieces. Now consider the limit as each pair of line segments connecting two circular sections becomes infinitesimally close. The integrals along these line segments will be equal and opposite, since they are the same path traversed in opposite directions. They will, therefore, cancel, leaving only the integrals around the circular contours.

The integral becomes a sum of integrals on the separate contours:

$$\oint_\Gamma g(z)dz = \sum_{z_k \text{ inside } \Gamma} \oint_{|z-z_k|=\epsilon} g(z)\,dz.$$

On the kth such circular contour, write

$$g(z) = \frac{(z-z_k)^{n_k} g(z)}{(z-z_k)^{n_k}},$$

which is valid because $z \neq z_k$ on the contour. The denominator of this expression is a pole of order n_k. The numerator is analytic on and inside the contour, because

$$(z-z_k)^{n_k} g(z) = \frac{(z-z_k)^{n_k} f(z)}{a(z)}$$

$$= \frac{(z-z_k)^{n_k} f(z)}{(z-z_1)^{n_1} \cdots (z-z_k)^{n_k} \cdots (z-z_p)^{n_p}},$$

and the pole at $z = z_k$ is cancelled by the factor $(z-z_k)^{n_k}$ in the numerator. Now the integral on the kth contour can be written using the extended Cauchy formula (Equation 8.10):

$$\oint_{|z-z_k|=\epsilon} g(z)\,dz = \oint_{|z-z_k|=\epsilon} \frac{(z-z_k)^{n_k} g(z)}{(z-z_k)^{n_k}}\,dz$$

$$= \frac{2\pi i}{(n_k-1)!} \left[\frac{d^{n_k-1}}{dz^{n_k-1}}(z-z_k)^{n_k} g(z)\right]_{z=z_k}.$$

The quantity

$$\frac{1}{(n_k-1)!} \lim_{z \to z_k} \left[\frac{d^{n_k-1}}{dz^{n_k-1}}(z-z_k)^{n_k} g(z)\right]$$

is called the *residue* of g at z_k, denoted $\text{Res}[g, z_k]$. The reason for this name will be made clear in the following section. The integral is therefore a sum of residues:

$$\oint_\Gamma g(z)dz = 2\pi i \sum_{z_k \text{ inside } \Gamma} \text{Res}[g, z_k]. \qquad (8.11)$$

Example 8.10. Integrate $g(z) = \frac{\sin z}{(z-1)(z-i)^2}$ on a circle $|z| = 2$. There is a single pole at $z = 1$ and a double pole at $z = i$. Both are inside the contour, which has radius 2. The contour is "shrink-wrapped" to two circles centered at 1 and i. Hence,

$$\oint_{|z|=2} \frac{\sin z\,dz}{(z-1)(z-i)^2} = \oint_{|z-1|=\epsilon} \frac{\sin z\,dz}{(z-1)(z-i)^2} + \oint_{|z-i|=\epsilon} \frac{\sin z\,dz}{(z-1)(z-i)^2}.$$

The first contour includes the single pole at $z = 1$. The other factor in the denominator, $(z - i)^2$, is analytic on and inside this contour. For the purpose of applying Cauchy's integral formula, we group this factor with the $\sin z$ in the numerator. Likewise, the second contour includes the double pole at $z = i$ but not the pole at $z = 1$, which may be lumped together with $\sin z$. We then have

$$\oint_{|z|=2} \frac{\sin z \, dz}{(z-1)(z-i)^2} = \oint_{|z-1|=\epsilon} \frac{\frac{\sin z}{(z-i)^2} \, dz}{(z-1)} + \oint_{|z-i|=\epsilon} \frac{\frac{\sin z}{z-1} \, dz}{(z-i)^2}.$$

The first integral is evaluated using Cauchy's integral formula and yields

$$\oint_{|z-1|=\epsilon} \frac{\frac{\sin z}{(z-i)^2} \, dz}{(z-1)} = 2\pi i \left. \frac{\sin z}{(z-i)^2} \right|_{z=1} = \frac{2\pi i \sin(1)}{(1-i)^2} = -\pi \sin(1),$$

where we have used the fact that $(1 - i)^2 = -2i$. The second integral involves a double pole and must be evaluated using the extended formula:

$$\oint_{|z-i|=\epsilon} \frac{\frac{\sin z}{z-1} \, dz}{(z-i)^2} = 2\pi i \frac{1}{(2-1)!} \lim_{z \to i} \frac{d}{dz} \frac{\sin z}{z-1}$$

$$= 2\pi i \lim_{z \to i} \frac{(z-1)\cos z - \sin z}{(z-1)^2} = 2\pi i \frac{-(1-i)\cos(i) - \sin(i)}{(1-i)^2}$$

$$= 2\pi i \frac{-(1-i)\cosh(1) - i\sinh(1)}{(1-i)^2}$$

$$= -\pi \left(-(1-i)\cosh(1) - i\sinh(1) \right).$$

Finally, combining the two integrals,

$$\oint_{|z|=2} \frac{\sin z \, dz}{(z-1)(z-i)^2} = -\pi \sin(1) - \pi \left(-(1-i)\cosh(1) - i\sinh(1) \right)$$

$$= \pi \left(\cosh(1) - \sin(1) \right) - i\pi \left(\cosh(1) - \sinh(1) \right)$$

$$= \pi \left(\cosh(1) - \sin(1) \right) - i\pi/e.$$

(Not every answer is pretty.) ∎

More Properties of Analytic Functions

The infinite differentiability of analytic functions follows from the extended Cauchy integral formula. Several other interesting and useful consequences of the Cauchy integral formula are collected here.[6]

---

[6] Proofs may be found in Wunsch (1994, pp. 190ff), among other sources.

8.4 CAUCHY'S INTEGRAL FORMULA

Theorem 8.8 (Gauss' mean value theorem). Let f be analytic on a simply connected domain, D, and let C be a circle of radius R, centered at z_0, contained in this domain. Then $f(z_0)$ is the average of f on the circle:

$$f(z_0) = \frac{1}{2\pi i} \oint_C \frac{f(z)}{z - z_0} dz \quad \text{with } z = z_0 + Re^{i\theta}$$

$$= \frac{1}{2\pi} \int_0^{2\pi} f\left(z_0 + Re^{i\theta}\right) d\theta.$$

Theorem 8.9 (Maximum modulus theorem). Let f be a nonconstant function which is analytic in a bounded domain D and continuous on the boundary Γ. Then f attains its maximum modulus (i.e., $|f|$ is maximized) only at certain points on Γ.

Theorem 8.10 (Minimum modulus theorem). Let f be a nonzero function which is analytic in a bounded domain D and continuous on the boundary Γ. Then f attains its minimum modulus ($|f|$ is minimized) only at certain points on Γ.

Theorem 8.11 (Liouville's theorem). Let f be a bounded entire function. Then f is constant.

Example 8.11. We can illustrate these with the function $f(z) = z^{-1}$. Let C be a unit circle centered at $z = 2$, and D be the interior of C. This center was chosen so that the circle does not intersect or include the pole at the origin. The value $f(2) = \frac{1}{2}$. To apply the mean value theorem, write

$$\frac{1}{2\pi} \int_0^{2\pi} f(2 + e^{i\theta}) d\theta = \frac{1}{2\pi} \int_0^{2\pi} \frac{d\theta}{2 + e^{i\theta}}.$$

Let $u = e^{i\theta}$, then $d\theta = -iu^{-1} du$, and the integral is

$$\int \frac{-i du}{u(2+u)} = -i \int \left[\frac{1/2}{u} + \frac{-1/2}{2+u}\right] du$$

$$= \frac{-i}{2} \log u + \frac{i}{2} \log(2+u) = \frac{i}{2} \log\left(1 + \frac{2}{u}\right).$$

Changing variables back to θ,

$$\frac{1}{2\pi} \int_0^{2\pi} \frac{d\theta}{2 + e^{i\theta}} = \frac{i}{4\pi} \log\left(1 + 2e^{-i\theta}\right)_0^{2\pi}.$$

As $\theta \to 2\pi$, $1 + 2e^{-i\theta}$ approaches a complex number with modulus 3 and argument -2π. Its logarithm is $\log 3 - i2\pi$. As $\theta \to 0$, $1 + 2e^{-i\theta}$ approaches a complex number with modulus 3 and argument 0. Its logarithm is $\log 3$. Thus, the integral is

$$\frac{1}{2\pi} \int_0^{2\pi} \frac{d\theta}{2 + e^{i\theta}} = \frac{i}{4\pi} (\log 3 - i2\pi - \log 3) = \frac{1}{2},$$

as expected.

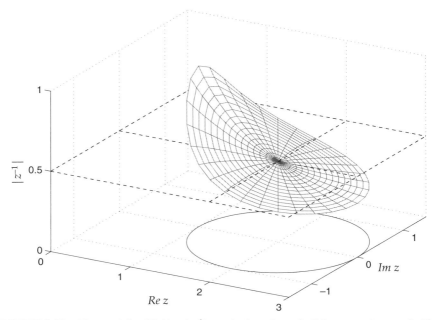

FIGURE 8.22 The modulus $|f(z)| = |z^{-1}|$ graphed on the unit disk centered at $z = 2$. The maximum and minimum modulus values occur on the boundary of the disk.

Now we check the maximum and minimum modulus theorems. In Figure 8.22, $|f|$ is graphed for z on and inside the unit circle centered at $z = 2$. The maximum modulus is observed to occur on the circle at $z = 1$ and the minimum occurs on the circle at $z = 3$. Evaluating the function on this circle, $z = 2 + e^{i\theta}$,

$$\left| \frac{1}{2 + e^{i\theta}} \right| = \frac{1}{\sqrt{5 + 4\cos\theta}},$$

which is maximized for $\theta = \pi$ ($z = 1$) and minimized for $\theta = 0$ ($z = 3$). If C enclosed the pole at the origin, the maximum modulus (in this case, infinite) would occur inside C rather than on the boundary. Thus we see why analyticity is a condition of the maximum modulus theorem.

Liouville's theorem does not apply to $f(z) = z^{-1}$ because f is not entire. The function $f(z) = z^2$ is entire and is unbounded as $|z| \to \infty$. So is the entire function $\exp z$. The function $f(z) = \sin(\mathcal{R}e\, z)$ is bounded, but it is not analytic, because $\mathcal{R}e\, z$ is not analytic. The constant function $f(z) = 1$ is both bounded and entire (all its derivatives are zero, for all z). ∎

8.5 LAURENT SERIES AND RESIDUES

Cauchy's integral formula permits us to evaluate integrals of functions that can be written as ratios of analytic functions and polynomials. We now extend this to functions g in which the denominator, $a(z)$, is not a polynomial, for example, $g(z) = \frac{1}{\cos z}$. To begin, we introduce series expansions of complex functions.

8.5.1 Laurent series

In real analysis, a function which is $n + 1$ times differentiable on an open interval (a, b) can be expanded, for $x, x_0 \in (a, b)$, in a Taylor series:

$$f(x) = f(x_0) + \frac{f'(x_0)}{1!}(x - x_0) + \frac{f''(x_0)}{2!}(x - x_0)^2 + \cdots$$
$$+ \frac{f^{(n)}(x_0)}{n!}(x - x_0)^n + R_n(x, x_0),$$

where $c \in (a, b)$. The last term is called the remainder. If $f \in C^{(\infty)}(a, b)$, then the series is infinite (a power series) and there is no remainder. The series does not necessarily converge for all $x, x_0 \in \mathbb{R}$. At worst, it will converge only if $x = x_0$; better, it may converge for $|x - x_0| < r$, where r is called the *radius of convergence* of the series. The series diverges for $|x - x_0| > r$.

In the complex plane, a function f which is analytic in a domain D possesses derivatives of all orders in D and can be represented by a Taylor series without remainder:

$$f(z) = c_0 + c_1(z - z_0) + c_2(z - z_0)^2 + \cdots$$
$$c_n = \frac{f^{(n)}(z_0)}{n!}.$$

The expansion is valid in a disk centered at z_0, with radius r extending to the nearest singularity; making the disk any larger encloses the singularity and violates the analyticity requirement. Again, r is the radius of convergence of the series (Figure 8.23).

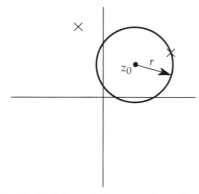

FIGURE 8.23 A function f which is analytic at a point z_0 has a Taylor series expansion about z_0. The series converges in a disk centered at z_0 and extending to the nearest singularity of f.

On this domain, integrating f results in zero (Cauchy's integral theorem), and consequently the series integrates term-by-term to zero.

Example 8.12. The Taylor series for $f(z) = \frac{1}{z+1}$ about the point $z = 0$ is obtained in the usual way:

$$f(z) = f(0) + f^{(1)}(0)z + \frac{f^{(2)}(0)}{2!}z^2 + \cdots$$

$$= 1 - (z+1)^{-2}\Big|_{z=0} z + 2(z+1)^{-3}\Big|_{z=0} \frac{z^2}{2!} + \cdots$$

$$= 1 - z + z^2 - z^3 + \cdots = \sum_{n=0}^{\infty} (-z)^n.$$

Applying the ratio test,

$$\left| \frac{(-z)^{n+1}}{(-z)^n} \right| = |z| < 1,$$

we see that the series is absolutely convergent for $|z| < 1$, that is, inside the unit circle. The radius of convergence is one. As expected, the region of convergence does not extend beyond the pole at $z = -1$. ∎

Example 8.13. The function $f(z) = \frac{\cos \pi z}{z}$ has a singularity at $z = 0$, but is analytic away from the origin. The Taylor series expansion for this function about the point $z_0 = 1$ works out to

$$f(z) \approx (2-z) \sum_{n=0}^{\infty} c_{2n}(z-1)^{2n},$$

where $c_0 = -1$

$$c_{2n} = c_{2(n-1)} - \frac{(-\pi^2)^n}{(2n)!}.$$

Because $(2n)!$ grows faster than π^{2n}, the coefficients eventually settle down to a constant value, and for large N the ratio test gives $|z-1|^2 < 1$, which says that the series converges in a circle of radius 1 centered at $z_0 = 1$. As expected, again, the region of convergence does not extend beyond the pole at the origin.

Approximations based on N-term partial sums of this series are shown in Figure 8.24 for $N = 5$ (11th order approximation) and $N = 9$ (18th order approximation) for real z. Observe that the higher order approximation comes closer to the singularity at the origin, but diverges more sharply beyond $x = 2$. Increasing the order of approximation only results in worse behavior outside the region of convergence. ∎

These examples show that singular behavior cannot be captured by a Taylor series. The remedy for isolated singularities like the poles in these examples is to include *negative* powers of z in the series expansion.

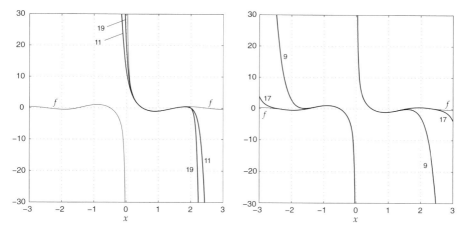

FIGURE 8.24 Two series expansions of $f(z) = \frac{\cos \pi z}{z}$, graphed along the real axis, $z = x$. *Left:* the Taylor series expansion around $z = 1$ diverges for $|z - 1| > 1$. As the order of the truncated series increases from 11 to 19, it approaches the singularity at the origin more closely, but diverges more strongly beyond $x = 2$. *Right:* Laurent series expansion around $z = 0$. The series is convergent for all $x \neq 0$. Although the truncated series, being a polynomial, blows up as $|x|$ increases, it approximates f more closely over a larger range of x as the order increases from 9 to 17.

Example 8.14. In the vicinity of $z = 0$, we write, for $f(z) = \frac{\cos \pi z}{z}$, the ratio of the power series for $\cos \pi z$ and z,

$$\frac{\cos \pi z}{z} = \frac{1 - \frac{(\pi z)^2}{2!} + \frac{(\pi z)^4}{4!} - \frac{(\pi z)^6}{6!} + \cdots}{z} = \frac{1}{z} - \frac{\pi^2 z}{2!} + \frac{\pi^4 z^3}{4!} - \frac{\pi^6 z^5}{6!} + \cdots.$$

Again applying the ratio test, we have $\left| \frac{\pi^{2n} z^{2n-1}/(2n)!}{\pi^{2(n-1)} z^{2n-3}/(2(n-1))!} \right| = \left| \frac{\pi^2 z^2}{2n(2n-1)} \right| < 1$. That is,

$$|z^2| < \frac{2n(2n-1)}{\pi^2} \to \infty \text{ as } n \to \infty,$$

which says that the series converges for all z away from the origin.

Partial sums of this series for $N = 5$ (9th order) and $N = 9$ (17th order) for real z are shown in Figure 8.24. Because of the z^{-1} term the series matches the singular behavior of f at the origin, for both positive and negative x. The partial sums, which are polynomials, must eventually blow up as $|x|$ increases toward infinity. However, unlike the Taylor series, as N increases the region of convergence grows and the correspondence with f becomes steadily better. ∎

The series expansion in this example is called a *Laurent series*. It can be shown[7] that any function f analytic in an *annular* domain $r_2 > |z - z_0| > r_1$ has a

---
[7]Hahn and Epstein, 1996, pp. 139–141; Wunsch, 1994, pp. 264–268.

Laurent expansion about z_0 given by

$$f(z) = c_{-N}(z-z_0)^{-N} + \cdots + c_{-1}(z-z_0)^{-1} + c_0 + c_1(z-z_0) + c_2(z-z_0)^2 + \cdots$$
$$= \sum_{n=-N}^{\infty} c_n(z-z_0)^n \tag{8.12}$$

with coefficients given by

$$c_n = \frac{1}{2\pi i} \oint_{\Gamma_0} \frac{f(z)\,dz}{(z-z_0)^{n+1}}, \tag{8.13}$$

where Γ_0 is a simple closed contour in the annular domain enclosing the inner boundary $|z - z_0| = r_1$. The Laurent series converges absolutely in the annular domain. The inner and outer radii are typically marked by singularities (Figure 8.25).

If z_0 is a singularity, then $r_1 = 0$ and the annulus is a punctured disk. If f has other singularities, then the one closest to z_0 fixes the outer boundary of the region of convergence. If the function has only the one singular point at z_0, the outer radius is infinite and the annulus becomes the punctured plane.

The coefficients of the positive powers are identical to Taylor series coefficients (compare Equations 8.10 and 8.13), and the corresponding portion of the Laurent series is analytic in the region of convergence. It is also known as the *regular part* of the expansion. The portion consisting of the negative powers, which captures the singular behavior, is called the *principal part*.

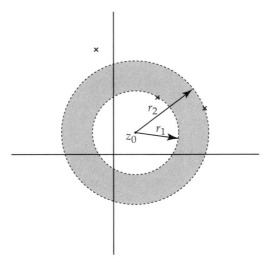

FIGURE 8.25 A function f which is analytic in an annulus has a Laurent series expansion about z_0. The series converges in the annulus, which is bounded by singularities.

Example 8.15 ($\cos(\pi z)/z$, revisited). Earlier we calculated the Laurent series for $f(z) = \frac{\cos \pi z}{z}$ by dividing the power series for $\cos \pi z$ by z. Here we will calculate a few terms using Equation 8.13:

$$c_{-2} = \frac{1}{2\pi i} \oint_{\Gamma_0} \frac{\cos(\pi z)/z \, dz}{z^{-2+1}} = \frac{1}{2\pi i} \oint_{\Gamma_0} \cos(\pi z) \, dz = 0, \quad \text{(Cauchy theorem)}$$

$$c_{-1} = \frac{1}{2\pi i} \oint_{\Gamma_0} \frac{\cos(\pi z) \, dz}{z} = \cos(\pi 0) = 1, \quad \text{(Cauchy formula)}$$

$$c_0 = \frac{1}{2\pi i} \oint_{\Gamma_0} \frac{\cos(\pi z) \, dz}{z^2} = -\pi \sin(\pi 0) = 0, \quad \text{(Ext. Cauchy formula)}$$

$$c_1 = \frac{1}{2\pi i} \oint_{\Gamma_0} \frac{\cos(\pi z) \, dz}{z^3} = -\frac{\pi^2}{2} \cos(\pi 0) = -\frac{\pi^2}{2}. \quad \text{(Ext. Cauchy formula)}$$

∎

The Laurent expansion can be performed about any z_0, but the most interesting case for our purposes is when z_0 is itself an isolated singularity of f. The isolated singularities of a function may be classified according to the Laurent expansion. Let z_0 be an isolated singularity of f and expand f in a Laurent series (Equation 8.12). We identify three cases:

- If the Laurent series has no negative powers, then it is really a Taylor series, and f is analytic. If f looks like it ought to have a singularity (e.g., sinc $z = (\sin \pi z)/\pi z$) but $\lim_{z \to z_0} f(z)$ is finite, then z_0 is a *removable singularity*. Writing sinc z as a series, for example,

$$\text{sinc } z = \frac{\sin \pi z}{\pi z} = \frac{\pi z - \frac{(\pi z)^3}{3!} + \cdots}{\pi z} = 1 - \frac{(\pi z)^2}{3!} + \cdots,$$

we see that the series has no principal part, indicating that sinc z is analytic. Or, calculating c_{-1} directly,

$$c_{-1} = \frac{1}{2\pi i} \oint_{\Gamma_0} \frac{\sin(\pi z)/z \, dz}{z^{-1+1}} = \frac{1}{2\pi i} \oint_{\Gamma_0} \frac{\sin(\pi z) \, dz}{z} = \sin(\pi 0) = 0.$$

- If the Laurent series has negative powers down to $-N$, then z_0 is a *pole of order* N. For example,

$$\frac{\cos z}{z^2} = \frac{1 - \frac{z^2}{2!} + \frac{z^4}{4!} - \cdots}{z^2} = \frac{1}{z^2} - \frac{1}{2} + \frac{z^2}{4!} - \cdots.$$

The principal part goes down to z^{-2}, so there is a double pole at $z = 0$.

- If the Laurent series has negative powers forever ($n \to -\infty$), then z_0 is an *essential singularity* (sometimes called, informally, an "infinite order pole"). The classic example is $\exp(1/z)$. Using the Taylor series for $\exp z$, we write

$$\exp(1/z) = 1 + \frac{1}{z} + \frac{(1/z)^2}{2!} + \frac{(1/z)^3}{3!} + \cdots$$

$$= \cdots + \frac{1}{3!z^3} + \frac{1}{2z^2} + \frac{1}{z} + 1,$$

and readily observe that the principal part does not terminate.

8.5.2 Residues and Integration

As we saw earlier, the integral of a function on a contour enclosing several isolated singularities can be performed as the sum of integrals around the individual singularities. We have just seen that the function can be expanded (in principle) in a Laurent series around each of these singularities. In the region of convergence of each of these series, the function is analytic and the series converges absolutely. Integrating it term-by-term along a simple closed contour around $z = z_0$ in the region of convergence, we obtain

$$\oint_\Gamma f(z)dz = \oint_\Gamma \left[\cdots + c_{-2}(z-z_0)^{-2} \right] dz + \oint_\Gamma c_{-1}(z-z_0)^{-1} dz$$
$$+ \oint_\Gamma \left[c_0 + c_1(z-z_0) + \cdots \right] dz \qquad (8.14)$$
$$= 0 + 2\pi i c_{-1} + 0 = 2\pi i c_{-1}.$$

Only the c_{-1} term contributes to the integral; all the other terms integrate to zero. (This is also the result obtained by setting $n = -1$ in Equation 8.13.) The coefficient c_{-1} is called the *residue* of f at z_0, denoted $\text{Res}[f, z_0]$. For each of the singularities, the integral will be given by a residue. This leads again to the result (Equation 8.11):

$$\oint_\Gamma f(z)dz = 2\pi i \sum_{z_k \text{ inside } \Gamma} \text{Res}\left[f, z_k\right],$$

but this time we have derived it without the restriction that f be the ratio of an analytic function and a polynomial. We only require that it be possible to calculate the residues of f at the singularities. What we need next is a way to do this, preferably without having to derive the series expansions.

Residue at a Simple Pole

Consider a function with a simple (first order) pole at z_0. The Laurent expansion is

$$f(z) = c_{-1}(z-z_0)^{-1} + c_0 + c_1(z-z_0) + \cdots.$$

At $z = z_0$, all the terms go to zero except one, which blows up. If we multiply $f(z)$ by $(z-z_0)$, we get

$$(z-z_0)f(z) = c_{-1} + c_0(z-z_0) + c_1(z-z_0)^2 + \cdots,$$

and in the limit as $z \to z_0$, only c_{-1}, the residue, remains. This is very convenient. For a function f with a simple pole at $z = z_0$,

$$\text{Res}[f, z_0] = \lim_{z \to z_0} (z - z_0) f(z). \tag{8.15}$$

When f is the ratio of two functions, $b(z)/a(z)$, there is an even simpler result, which follows from L'Hospital's rule. We assume that $b(z_0) \neq 0$, so that the point $z = z_0$ is not a removable singularity. Because $a(z_0) = 0$, the limit

$$\lim_{z \to z_0} (z - z_0) f(z) = \lim_{z \to z_0} \frac{(z - z_0) b(z)}{a(z)}$$

is indeterminate of the form $0/0$. Applying L'Hospital's rule,

$$\lim_{z \to z_0} \frac{(z - z_0) b(z)}{a(z)} = \lim_{z \to z_0} \frac{(z - z_0) b'(z) + b(z)}{a'(z)} = \frac{b(z_0)}{a'(z_0)}.$$

Therefore, when z_0 is a simple pole and $b(z_0) \neq 0$,

$$\text{Res}\left[\frac{b(z)}{a(z)}, z_0\right] = \frac{b(z_0)}{a'(z_0)}. \tag{8.16}$$

Example 8.16. The function $\tan z$ is singular at $z = \pi/2$. It so happens that the singularity is a simple pole. Using Equation 8.16 with the fact that $\tan z = \frac{\sin z}{\cos z}$,

$$\text{Res}\left[\tan z, \frac{\pi}{2}\right] = \frac{\sin\left(\frac{\pi}{2}\right)}{-\sin\left(\frac{\pi}{2}\right)} = -1.$$

∎

Residue at a Multiple Pole

Suppose f has a double pole at z_0. Its Laurent expansion around z_0 is

$$f(z) = c_{-2}(z - z_0)^{-2} + c_{-1}(z - z_0)^{-1} + c_0 + c_1(z - z_0) + \cdots.$$

If we take $\lim_{z \to z_0} (z - z_0) f(z)$ as before, we get

$$\lim_{z \to z_0} \left[c_{-2}(z - z_0)^{-1} + c_{-1} + c_0(z - z_0) + c_1(z - z_0)^2 + \cdots \right],$$

which blows up because of the term $c_{-2}(z - z_0)^{-1}$. If instead we multiply by $(z - z_0)^2$, we get

$$(z - z_0)^2 f(z) = c_{-2} + c_{-1}(z - z_0) + c_0(z - z_0)^2 + c_1(z - z_0)^3 + \cdots,$$

and if we take the limit as $z \to z_0$ of this, we will get c_{-2} rather than c_{-1}. However, differentiating once,

$$\frac{d}{dz}(z - z_0)^2 f(z) = c_{-1} + 2c_0(z - z_0) + 3c_1(z - z_0)^2 + \cdots,$$

and now the limit will give c_{-1}, as desired. The process is repeated for higher order poles: multiply f by $(z - z_0)^n$ to clear out the singularities, differentiate $n - 1$ times to remove the terms below c_{-1}, then take the limit to make all the terms above c_{-1} disappear. A general residue formula results, which looks very much like the extended Cauchy integral formula.

For a pole of order N at $z = z_0$, the residue is

$$\text{Res}[f, z_0] = \frac{1}{(N-1)!} \lim_{z \to z_0} \frac{d^{N-1}}{dz^{N-1}} \left[(z-z_0)^N f(z) \right]. \tag{8.17}$$

This method breaks down for essential singularities, because $N \to \infty$. The only thing you can do then is calculate, by some means, a few terms of the Laurent expansion. Sometimes, this is not too hard. For example, using the Taylor series for $\exp(z)$,

$$\exp(1/z) = 1 + (1/z) + \frac{(1/z)^2}{2!} + \cdots,$$

and we see that $c_{-1} = 1$.

Even when a singularity is not essential, it can sometimes be simpler to expand a few terms of the Laurent series rather than use the general formula (Equation 8.17) with its several derivatives.

Example 8.17. Using the residue formula (Equation 8.17), the residue of $f = \frac{z-1}{\sin^2 z}$ at $z = 0$ is (it so happens that $N = 2$)

$$\lim_{z \to 0} \frac{d}{dz} \frac{z^2(z-1)}{\sin^2 z} = \lim_{z \to 0} \frac{(3z^2 - 2z)\sin^2 z - 2z^2(z-1)\sin z \cos z}{\sin^4 z} = 1.$$

The last limit requires L'Hospital's rule. But we could also expand f in a Laurent series by dividing $z - 1$ by the Taylor series for $\sin^2 z$:

$$\frac{z-1}{\sin^2 z} = \frac{z-1}{(z - z^3/3! + O(z^5))^2} = \frac{z-1}{z^2 - z^4/3 + O(z^6)}$$

$$= -\frac{1}{z^2} + \frac{1}{z} - \frac{1}{3} + \cdots,$$

and read off $c_{-1} = 1$. This required no calculus, just some algebra and knowledge of the Taylor series for $\sin z$. ∎

Determining the Order of a Pole

The general formula (Equation 8.17) assumes that you know the order N of the pole. Suppose you guess wrong. If you guess too low, then you will not multiply by a high enough power of $(z - z_0)$ to clear out the singular terms, and when you take the limit, it will blow up. On the other hand, if you guess too high, you will multiply by too high a power of $(z - z_0)$, and you could extract the wrong value. For example, if you guess $N = 3$ when the pole is really second order, you will get

$$\lim_{z \to z_0} \frac{d^2}{dz^2} \left[c_{-2}(z - z_0) + c_{-1}(z - z_0)^2 + c_0(z - z_0)^3 + O((z - z_0)^4) \right]$$
$$= \lim_{z \to z_0} \left[2c_{-1} + 6c_0(z - z_0) + O((z - z_0)^2) \right]$$
$$= 2c_{-1}.$$

To derive a method for correctly determining the order of a pole, consider the product $(z - z_0)^M f(z)$, where f is represented by its Laurent expansion and M is a nonnegative integer. Then

$$(z - z_0)^M f(z) = c_{-N}(z - z_0)^{M-N} + \cdots + c_{-1}(z - z_0)^{M-1} + c_0(z - z_0)^M + \cdots.$$

We distinguish three cases:

1. $M > N$—If we multiply by too high a power of $(z - z_0)$, the result will be a polynomial in $(z - z_0)$, which will evaluate to zero for $z = z_0$.
2. $M < N$—If we multiply by too low a power, there will still be negative powers of $(z - z_0)$ present, which will blow up as $z \to z_0$.
3. $M = N$—If we hit it just right, we will get

$$c_{-N} + \cdots + c_{-1}(z - z_0)^{N-1} + c_0(z - z_0)^N + c_1(z - z_0)^{N+1} + \cdots,$$

and this will be finite ($= c_{-N}$) as $z \to z_0$.

Thus, to determine the order of a pole, guess a value of M and calculate

$$\lim_{z \to z_0} (z - z_0)^M f(z).$$

Then,

1. If the limit is infinite, M is too low.
2. If the limit is zero, M is too high.
3. If the limit is finite and nonzero, M is the order of the pole.
   (a) If $M = 1$, then the limit calculated is c_{-1} and you have the residue.
   (b) If $M > 1$, carry out the residue calculation for an Mth order pole.

Example 8.18. The function $f(z) = \frac{1}{\cos z}$ has singularities everywhere that $\cos z = 0$: $z = \pm \frac{\pi}{2}, \pm \frac{3\pi}{2}, \ldots$. We want to know what their orders are. The singular points may be written $z_k = \pm(2k + 1)\frac{\pi}{2}, k = 0, 1, \ldots$. Let us try a first-order pole:

$$\lim_{z \to z_k} \frac{z - z_k}{\cos z} = \lim_{z \to z_k} \frac{1}{-\sin z} \qquad \text{(L'Hospital's rule)}$$
$$= -\frac{1}{\pm \sin(2k + 1)\frac{\pi}{2}} = \pm 1.$$

The result is finite, so the poles are first order. Had we guessed $M = 2$, we would have obtained

$$\lim_{z \to z_k} \frac{(z - z_k)^2}{\cos z} = \lim_{z \to z_k} \frac{2(z - z_k)}{-\sin z} = 0.$$

∎

Example 8.19. Returning to an earlier example, it was claimed that $\tan z$ has a simple pole at $z = \frac{\pi}{2}$. We just showed, in the previous example, that all the poles of $\frac{1}{\cos z}$ are first order. Hence, $\tan z = \frac{\sin z}{\cos z}$ has first-order poles at $z = \pm(2k+1)\frac{\pi}{2}, k = 0, 1, \ldots$, and in particular, at $z = \frac{\pi}{2}$. ∎

Example 8.20. Determine the locations and orders of the singularities of $f(z) = \frac{z}{\sin^2 z}$. We know that $\sin z = 0$ for $z = k\pi, k = 0, \pm 1, \ldots$. The numerator is zero for $z = 0$ and will cancel, in whole or in part, the singularity at $z = 0$. We must consider this case separately from the others.

At $z = 0$, try a first-order pole:

$$\lim_{z \to 0} z f(z) = \lim_{z \to 0} \frac{z^2}{\sin^2 z} = \lim_{z \to 0} \frac{2z}{2 \sin z \cos z}.$$

This limit is indeterminate (0/0), so apply L'Hospital's rule again:

$$\lim_{z \to 0} z f(z) = \lim_{z \to 0} \frac{2}{2 \cos^2 z - 2 \sin^2 z} = -2.$$

The limit is finite; the pole at $z = 0$ is first order.

At other values, $z = k\pi$,

$$\lim_{z \to k\pi} (z - k\pi) f(z) = \lim_{z \to k\pi} \frac{(z - k\pi)z}{\sin^2 z} = \lim_{z \to k\pi} \frac{2z - k\pi}{2 \sin z \cos z} = \frac{k\pi}{0}.$$

This limit is not finite, so the poles must be higher than first order. Try second order. This is actually a reasonable guess, because if we expand the numerator and denominator in Taylor series around $z = k\pi$, we get

$$\frac{z}{\sin^2 z} = \frac{k\pi + (z - k\pi)}{(z - k\pi)^2 + O((z - k\pi)^6)} = \frac{k\pi}{(z - k\pi)^2} + \frac{1}{(z - k\pi)} + \cdots.$$

The leading term indicates that the poles are second order. To confirm this,

$$\lim_{z \to k\pi} (z - k\pi)^2 f(z) = \lim_{z \to k\pi} \frac{(z - k\pi)^2 z}{\sin^2 z} = \lim_{z \to k\pi} \frac{3z^2 - 4k\pi z + k^2/\pi^2}{2 \sin z \cos z}.$$

This limit is indeterminate (0/0), so apply L'Hospital's rule again:

$$\lim_{z \to k\pi} (z - k\pi)^2 f(z) = \lim_{z \to k\pi} \frac{6z - 4k\pi}{2 \cos^2 z - 2 \sin^2 z} = k\pi.$$

The limit is finite. All the other poles are second order. ∎

8.6 USING CONTOUR INTEGRATION TO CALCULATE INTEGRALS OF REAL FUNCTIONS

Complex integration is a useful tool for integrating functions of a real variable. The basic idea is this: certain integrals of functions of a *real* variable can be performed by embedding them within complex integrals, which are then evaluated using residues and the *ML* inequality.

8.6.1 Trigonometric Integrals

The first method transforms integrals of the form

$$I = \int_0^{2\pi} f(\cos\theta, \sin\theta)\, d\theta, \tag{8.18}$$

where f is a rational function of $\cos\theta$ and $\sin\theta$, into complex contour integrals. By the Euler formulae, $\cos\theta = \frac{e^{i\theta}+e^{-i\theta}}{2}$ and $\sin\theta = \frac{e^{i\theta}-e^{-i\theta}}{2i}$, which are equal to $\frac{z+z^{-1}}{2}$ and $\frac{z-z^{-1}}{2i}$, respectively, when $z = e^{i\theta}$ (i.e., on the unit circle). Hence, if we make the changes of variable

$$\cos\theta = \frac{z+z^{-1}}{2} \quad \text{and} \quad \sin\theta = \frac{z-z^{-1}}{2i}, \tag{8.19}$$

the function f becomes a ratio of polynomials in z. The path of integration is the unit circle, $z = e^{i\theta}$, and $dz = ie^{i\theta}\,d\theta$, or $d\theta = -iz^{-1}dz$. The original integral (Equation 8.18) becomes

$$I = \oint_{|z|=1} f\left(\frac{z+z^{-1}}{2}, \frac{z-z^{-1}}{2i}\right)(-iz^{-1})\,dz. \tag{8.20}$$

Example 8.21.

$$I = \int_0^{2\pi} \frac{d\theta}{2+\sin\theta}$$

Making the substitutions,

$$I = \int_0^{2\pi} \frac{d\theta}{2+\sin\theta} = -i\oint_{|z|=1} \frac{1}{2+\frac{z-z^{-1}}{2i}} z^{-1}\,dz$$

$$= \oint_{|z|=1} \frac{2}{z^2+4iz-1}\,dz.$$

The integrand has two poles, at $z = i(-2 \pm \sqrt{3}) \approx -0.27i, -3.73i$. One of these is inside the unit circle and contributes its residue to the integral (Figure 8.26).
Using Equation (8.16),

$$\text{Res}\left[\frac{2}{z^2+4iz-1}, i(-2+\sqrt{3})\right] = \frac{2}{2z+4i}\bigg|_{z=i(-2+\sqrt{3})} = \frac{1}{i\sqrt{3}},$$

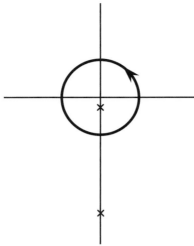

FIGURE 8.26 Contour for integrating $\oint_{|z|=1} \frac{2}{z^2+4iz-1} dz$.

and the integral is

$$I = 2\pi i \, \text{Res}\left[\frac{2}{z^2+4iz-1}, i(-2+\sqrt{3})\right] = \frac{2\pi}{\sqrt{3}}.$$

∎

Example 8.22 (Discrete-time Fourier transform). We will calculate the inverse discrete-time Fourier transform of $F_d(\theta) = \frac{1}{1-ae^{-i\theta}}$, given by the integral (Equation (4.44)b):

$$f[n] = \frac{1}{2\pi}\int_{-\pi}^{\pi} \frac{e^{in\theta}\, d\theta}{1-ae^{-i\theta}}.$$

Make the substitution $z = e^{i\theta}$, $d\theta = -iz^{-1}dz$, and integrate around the unit circle:

$$f[n] = \frac{1}{2\pi i}\oint_{|z|=1} \frac{z^{n-1}\, dz}{1-az^{-1}} = \frac{1}{2\pi i}\oint_{|z|=1} \frac{z^n\, dz}{z-a}.$$

When $n \geq 0$, there is only a simple pole at $z = a$, and its residue is a^n. The integral is $\frac{1}{2\pi i} 2\pi i\, a^n = a^n$. When $n < 0$, then z^n is really an $|n|$th order pole at $z = 0$. The residue at $z = a$ is still a^n. To calculate the residue at $z = 0$, let $m = -n$. Then, according to the residue formula (Equation 8.17),

$$\text{Res}\left[\frac{1}{z^m(z-a)}, 0\right] = \frac{1}{(m-1)!}\lim_{z\to 0}\frac{d^{m-1}}{dz^{m-1}}\frac{1}{z-a}$$

$$= \frac{1}{(m-1)!}\lim_{z\to 0}(-1)^{m-1}(m-1)!\,(z-a)^{-m} = (-1)^{m-1}(-a)^{-m}$$

$$= -a^n.$$

8.6 USING CONTOUR INTEGRATION TO CALCULATE INTEGRALS OF REAL FUNCTIONS

The integral is $2\pi i$ times the sum of the residues, which is $a^n + (-a^n) = 0$. Therefore, the integral is zero for $n < 0$, and we have the final result:

$$f[n] = \begin{cases} a^n, & n \geq 0 \\ 0, & n < 0 \end{cases}.$$

This sequence converges as $n \to \infty$ for $|a| < 1$, and the result agrees with Equation 4.47. This method is generally applicable to those $F_d(\theta)$ which can be expressed as rational functions of $\sin\theta$ and $\cos\theta$, a class of functions which includes the transfer functions of discrete LTI systems. ∎

8.6.2 Improper Integrals

Next we show how contour integration is used to evaluate real integrals with infinite limits, $\int_{-\infty}^{\infty} f(x)dx$ and $\int_0^{\infty} f(x)dx$.

Integrals of the Form $\int_{-\infty}^{\infty} f(x)dx$

Consider the complex integral $\oint_\Gamma f(z)\,dz$ where the contour Γ is the combination of a segment along the real axis and a semicircle in the upper half-plane (UHP) (Figure 8.27).

On the flat part of the contour, $z = x$. On the semicircular arc, $z = Re^{i\theta}$, $\theta \in [0, \pi]$. So we have

$$\oint_\Gamma f(z)dz = \int_{-R}^{R} f(x)dx + \oint_{\Gamma_R} f(z)dz.$$

In the limit as $R \to \infty$, the first term, $\int_{-R}^{R} f(x)dx$, becomes the Cauchy principal value of a real integral with infinite limits (Equation 1.30):

$$\lim_{R \to \infty} \int_{-R}^{R} f(x)dx = \mathcal{P}\int_{-\infty}^{\infty} f(x)dx.$$

All such improper integrals will be taken as Cauchy principal values.

Here is the strategy for evaluating real integrals with infinite limits by embedding them in contour integrals. If $f(z)$ has singularities in the UHP, they will be

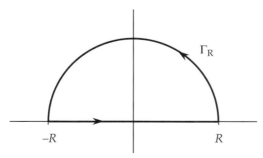

FIGURE 8.27 Semicircular contour for integrals of the form $\int_{-\infty}^{\infty} f(x)dx$.

enclosed by the contour Γ for some sufficiently large R. Using residues, evaluate the contour integral. Next, as $R \to \infty$, the integral along the real axis becomes the principal value of the integral we seek, $\int_{-\infty}^{\infty} f(x)dx$. It remains to determine what the integral on the semicircle, Γ_R, becomes in the limit as $R \to \infty$. For this we appeal to the *ML* inequality and endeavor to show that the integral on Γ_R vanishes in the limit.

Informally, here is how the *ML* inequality works for us. Each pole of f behaves approximately as $1/R$ when you are far away from it. If f has three poles, say, it decays like $1/R^3$ sufficiently far from the origin. So the "M" part of the *ML* inequality is on the order of $1/R^3$. The "L" part is the length of the contour, which is πR for the semicircle. Hence, the integral is bounded above by $ML = 1/R^3 \times \pi R = \pi/R^2$, and this upper bound goes to zero as $R \to \infty$. So even though the contour is getting longer, the function is dying faster, and the net result is a vanishing integral on the semicircular contour.

Example 8.23.

$$I = \int_{-\infty}^{\infty} \frac{dx}{1+x^2}.$$

Consider the integral on a semicircular contour in the UHP:

$$\oint_\Gamma \frac{dz}{1+z^2} = \int_{-R}^{R} \frac{dx}{1+x^2} + \oint_{\Gamma_R} \frac{dz}{1+z^2}.$$

You can check for yourself that a semicircle in the lower half-plane (LHP) would work too, but the advantage of the UHP contour is that we get to traverse the contour in the positive direction.

There are two first-order poles, at $z = \pm i$. Only one of these ($z = i$) is inside the contour, and its residue is

$$\text{Res}[f, i] = \frac{1}{2z}\bigg|_{z=i} = \frac{1}{2i}.$$

(Although the other pole at $z = -i$ is not encircled by the contour, it still influences the integral through its presence in the residue calculation at $z = i$). The contour integral is $2\pi i \cdot \frac{1}{2i} = \pi$. As $R \to \infty$, the integral along the x-axis becomes I, the integral we seek. Finally, we attack the integral on Γ_R with the *ML* inequality. On Γ_R, $z = Re^{i\theta}$, and

$$M = \max \left| \frac{1}{1+R^2 e^{i2\theta}} \right| = \frac{1}{\min |1+R^2 e^{i2\theta}|} = \frac{1}{\min |1+R^2 \cos 2\theta + iR^2 \sin 2\theta|}$$

$$= \frac{1}{\min \sqrt{1+R^4+2R^2 \cos 2\theta}}.$$

For $\theta \in [0, \pi]$, the denominator is minimized when $\theta = \frac{\pi}{2}$, giving

$$M = \frac{1}{\sqrt{1+-2R^2+R^4}} = \frac{1}{R^2-1}.$$

8.6 USING CONTOUR INTEGRATION TO CALCULATE INTEGRALS OF REAL FUNCTIONS

As we would expect, with two poles and no zeroes the integrand behaves like $1/R^2$ for large R. The length of the semicircle is πR, so the ML bound is

$$\left| \oint_{\Gamma_R} \frac{dz}{1+z^2} \right| \leq \frac{\pi R}{R^2 - 1},$$

which goes to 0 as $R \to \infty$. Putting the pieces together, we get the final result, $I = \pi$. This is, by the way, the result you would get if you looked it up in a table of integrals: $(1+x^2)^{-1}$ is the derivative of $\arctan x$, and

$$\int_{-\infty}^{\infty} \frac{dx}{1+x^2} = \arctan x \Big|_{-\infty}^{\infty} = \frac{\pi}{2} - \left(-\frac{\pi}{2}\right) = \pi.$$ ∎

Integrals of the Form $\int_0^\infty f(x)dx$

Next we consider integrals of the form $\int_0^\infty f(x)dx$. If f is an even function, we simply rewrite the integral

$$\int_0^\infty f(x)dx = \frac{1}{2} \int_{-\infty}^\infty f(x)dx$$

and proceed as above. On the other hand, if f is odd we have to work a bit harder. Consider, for example, the integral

$$I = \int_0^\infty \frac{x\,dx}{x^4 + 1}.$$

We can embed this integral in the contour integral $\oint_\Gamma \frac{z\,dz}{z^4+1}$, provided we can find an appropriate contour. We know we need one piece of Γ to overlap the positive real axis, since that is where we will get I. We also want an arc of radius R, so that we can use the ML inequality. It appears that we must close the contour with a segment from the end of the arc back to the origin, so that Γ is a pie-shaped sector rather than a semicircle (Figure 8.28).

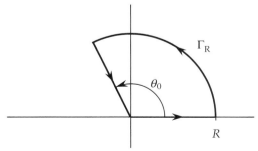

FIGURE 8.28 Pie-shaped contour for integrals of the form $\int_0^\infty f(x)dx$.

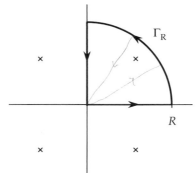

FIGURE 8.29 Quarter-circle contour for $\oint_\Gamma \frac{z\,dz}{z^4+1}$.

This segment will be parametrized by $z = re^{i\theta_0}$ and $dz = e^{i\theta_0}dr$, integrating from $r = R$ back to $r = 0$. The question is what to choose for θ_0. For this particular example, the integral along the return segment is

$$\int_R^0 \frac{re^{i2\theta_0}dr}{r^4 e^{i4\theta_0}+1}.$$

If we choose $\theta_0 = \pi/2$, we will have $e^{i2\theta_0} = e^{i\pi} = -1$, and the integral becomes

$$\int_R^0 \frac{re^{i2\theta_0}dr}{r^4 e^{i4\theta_0}+1} = \int_0^R \frac{r\,dr}{r^4+1},$$

which is identical to I, the integral we seek, as $R \to \infty$. We have a way to proceed.

Using a quarter-circle contour (Figure 8.29), the problem becomes

$$\oint_\Gamma \frac{z\,dz}{z^4+1} = 2\int_0^R \frac{x\,dx}{x^4+1} + \oint_{\Gamma_R} \frac{z\,dz}{z^4+1}.$$

There are four simple poles, at $z = (-1)^{1/4}$. One of these, $z = e^{i\pi/4}$, is inside Γ, and its residue is

$$\operatorname{Res}\left[\frac{z}{z^4+1}, e^{i\pi/4}\right] = \left.\frac{z}{4z^3}\right|_{z=e^{i\pi/4}} = \frac{1}{4i}.$$

Next, on the arc, the *ML* inequality gives

$$\left|\oint_{\Gamma_R} \frac{z\,dz}{z^4+1}\right| \leq \frac{1}{R^3} \cdot \frac{\pi R}{2} = \frac{\pi}{2R^2},$$

and this goes to zero as $R \to \infty$. Therefore, we have

$$\oint_\Gamma \frac{z\,dz}{z^4+1} = 2\pi i \cdot \frac{1}{4i} = \frac{\pi}{2} = 2I,$$

yielding the final result, $I = \pi/4$.

8.6 USING CONTOUR INTEGRATION TO CALCULATE INTEGRALS OF REAL FUNCTIONS

This method readily generalizes to integrals of the form $\int_0^\infty \frac{x^m \, dx}{x^n+1}$, $m < n$. The trick is choose a pie-shaped contour where the returning segment is at an angle θ_0 such that $e^{in\theta_0} = 1$.

8.6.3 Singular Integrals

When the integrand $f(x)$ becomes infinite at some point $x = a$ between the limits of integration, the integral is singular and must be evaluated as a Cauchy principal value:

$$P \int_A^B f(x)dx = \lim_{\epsilon \to 0} \left(\int_A^{a-\epsilon} + \int_{a+\epsilon}^B \right) f(x)dx.$$

When the limits are infinite,

$$P \int_{-\infty}^\infty f(x)dx = \lim_{R \to \infty} \lim_{\epsilon \to 0} \left(\int_{-R}^{a-\epsilon} + \int_{a+\epsilon}^R \right) f(x)dx.$$

The principal value calculation approaches the singularity and the infinite limits symmetrically, in the hope that the contributions on opposite sides of the singularity will sum to a finite value. When this happens it is because the integrand changes sign in passing through $x = a$, so that the areas to the left and right of a have opposite sign and partially cancel.

We illustrate the method with $I = \int_{-\infty}^\infty \frac{dx}{x^3+1}$. The integrand has a singularity at $x = -1$. The Cauchy principal value of the integral is

$$P \int_{-\infty}^\infty \frac{dx}{x^3+1} = \lim_{R \to \infty} \lim_{\epsilon \to 0} \left(\int_{-R}^{-1-\epsilon} + \int_{-1+\epsilon}^R \right) \frac{dx}{x^3+1}.$$

We seek to embed this in a contour integral of the form $\oint_\Gamma \frac{dz}{z^3+1}$. The complex function $f(z)$ has three poles, at $z = -1, e^{i\pi/3}, e^{-i\pi/3}$. The contour will include a segment along the real axis from $x = -R$ to $x = -1 - \epsilon$, and another from $x = -1 + \epsilon$ to $x = R$. We connect $x = R$ back to $x = -R$ with a semicircle of radius R and can anticipate that the contour integral on this arc will go to zero as $R \to \infty$, using the ML inequality. This leaves the gap between $x = -1 - \epsilon$ and $x = -1 + \epsilon$. We cannot integrate through the pole at $z = -1$, so we have to go around it. We will bump over it with a semicircle of radius ϵ and then see what happens in the limit as $\epsilon \to 0$ (Figure 8.30).

This closes the contour, and we proceed, with

$$\oint_\Gamma \frac{dz}{z^3+1} = \int_{-R}^{-1-\epsilon} + \int_{\Gamma_\epsilon} + \int_{-1+\epsilon}^R + \int_{\Gamma_R}.$$

The pole at $z = e^{i\pi/3}$ is inside the contour. Its residue is

$$\text{Res}\left[\frac{1}{z^3+1}, e^{i\pi/3}\right] = \frac{1}{3z^2}\bigg|_{z=e^{i\pi/3}} = \frac{1}{3e^{i2\pi/3}},$$

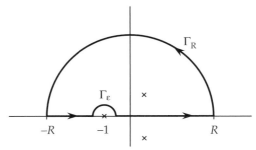

FIGURE 8.30 Contour for calculating $\int_{-\infty}^{\infty} \frac{dx}{x^3+1}$. The path along the real axis is indented above the pole at $z = -1$.

giving, for the contour integral, $i\frac{2\pi}{3}e^{-i2\pi/3}$. The two segments along the real axis, in the limit as $R \to \infty$ and $\epsilon \to 0$, yield the integral we seek. On the large semicircle, Γ_R, the *ML* inequality gives

$$\left| \oint_{\Gamma_R} \frac{dz}{z^3+1} \right| \leq \frac{1}{R^3} \times \pi R = \frac{\pi}{R^2} \to 0 \text{ as } R \to \infty.$$

This leaves the integral on the indentation, $\oint_{\Gamma_\epsilon}$. There, $z = -1 + \epsilon e^{i\theta}$, $dz = i\epsilon e^{i\theta} d\theta$, and we integrate from $\theta = \pi$ to $\theta = 0$. Near the pole, the function has its maximum modulus for $\theta = 0$:

$$\left| \frac{1}{z^3+1} \right|_{z \in \Gamma_\epsilon} \leq \frac{1}{3\epsilon - 3\epsilon^2 + \epsilon^3}.$$

The length of the contour is $\pi\epsilon$, giving an *ML* upper bound on the integral:

$$\left| \oint_{\Gamma_\epsilon} \frac{dz}{z^3+1} \right| \leq \frac{\pi\epsilon}{3\epsilon - 3\epsilon^2 + \epsilon^3} = \frac{\pi}{3 - 3\epsilon + \epsilon^2}.$$

This bound is finite as $\epsilon \to 0$. Even though the function is blowing up $O(\epsilon^{-1})$, the path is getting shorter $O(\epsilon)$. The result is a finite value for the integral, given by

$$\lim_{\epsilon \to 0} \int_0^\pi \frac{-id\theta}{3 - 3\epsilon e^{i\theta} + \epsilon^2 e^{i2\theta}}.$$

The bounded integrand and finite path length permit the limiting and integration operations to be exchanged:

$$\lim_{\epsilon \to 0} \int_0^\pi \frac{-id\theta}{3 - 3\epsilon e^{i\theta} + \epsilon^2 e^{i2\theta}} = \int_0^\pi \lim_{\epsilon \to 0} \frac{-id\theta}{3 - 3\epsilon e^{i\theta} + \epsilon^2 e^{i2\theta}}$$

$$= \int_0^\pi \frac{-id\theta}{3} = -\frac{i\pi}{3}.$$

8.6 USING CONTOUR INTEGRATION TO CALCULATE INTEGRALS OF REAL FUNCTIONS

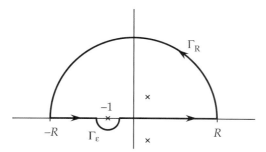

FIGURE 8.31 The contour may also be indented below the pole.

Finally, assemble the pieces:

$$\lim_{R\to\infty}\lim_{\epsilon\to 0} \oint_\Gamma \frac{dz}{z^3+1} = \frac{2\pi i e^{-i2\pi/3}}{3}$$

$$= I + \lim_{\epsilon\to 0}\oint_{\Gamma_\epsilon} + \lim_{R\to\infty}\oint_{\Gamma_R} = I - \frac{i\pi}{3} + 0,$$

and solve for I:

$$I = \frac{2\pi i e^{-i2\pi/3}}{3} + \frac{i\pi}{3} = \frac{2\pi i}{3}\frac{-1-i\sqrt{3}}{2} + i\frac{\pi}{3}$$

$$= \frac{\pi}{\sqrt{3}}.$$

Some of the constituent terms had imaginary parts, but when everything was added up the imaginary parts cancelled out. This should be expected, since we are evaluating a real integral and must get a real-valued result. A complex value for a real integral indicates that an error has been made somewhere in the calculation.

You might be wondering why we indented the contour above the pole rather than below. This could be done (Figure 8.31), but it would be a bit more work. Consider: With two poles inside the contour, we calculate two residues for the contour integral. Then, we evaluate the contribution around the indentation ($\epsilon \to 0$). When all these terms are combined at the end, we will in fact get the same answer.

Here is a general result for indented contours. When you integrate part of the way around a *simple* pole on an arc $\Gamma_\epsilon(\alpha)$ of radius ϵ and angle α (Figure 8.32), the result is

$$\lim_{\epsilon\to 0}\oint_{\Gamma_\epsilon(\alpha)} = i\alpha \, \text{Res}[f, z_0], \tag{8.21}$$

where α is positive if the arc is traversed in the positive direction, and negative otherwise ($|\alpha| = 2\pi$ for a full encirclement). To show this, consider the Laurent expansion of a function with a simple pole (we may, without loss of generality, take the pole to be at the origin):

$$f(z) = c_{-1}z^{-1} + c_0 + c_1 z + \cdots.$$

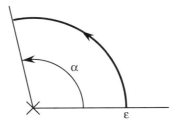

FIGURE 8.32 When a single pole is partially encircled, it contributes a fraction of its residue to the integral.

On the arc, $z = \epsilon e^{i\theta}$, $dz = i\epsilon e^{i\theta} d\theta$, and

$$\oint_{\Gamma_\alpha} f(z)dz = \lim_{\epsilon \to 0} \int_{\theta_0}^{\theta_0 + \alpha} i\epsilon e^{i\theta} \left[c_{-1}(\epsilon e^{i\theta})^{-1} + c_0 + c_1 \epsilon e^{i\theta} + \cdots \right] d\theta$$

$$= \lim_{\epsilon \to 0} \int_{\theta_0}^{\theta_0 + \alpha} i\left[c_{-1} + c_0 \epsilon e^{i\theta} + \cdots \right] d\theta$$

$$= \int_{\theta_0}^{\theta_0 + \alpha} ic_{-1} d\theta = i\alpha c_{-1}.$$

8.6.4 Integrals with Multivalued Functions

Integrals like $\int_0^\infty \frac{\sqrt{x}}{x^2+1}$, $\int_{-\infty}^\infty \frac{\text{Log}|x|}{x^2+4}$, and $\int_{-\infty}^\infty \frac{dx}{\sqrt{x^2+1}}$ involve multivalued functions. Embedding the integral in a contour integral leads to contour integrals in planes with a branch cut. We illustrate the issues with the integral

$$I = \int_0^\infty \frac{dx}{\sqrt{x}(x+2)}.$$

Consider the complex integral $\oint_\Gamma \frac{z^{-1/2} dz}{z+2}$ with an appropriate contour. The integral we seek is taken over the limits $x \in (0, \infty)$, so we know that one segment of the contour must be on the positive real axis. If we try to use a pie-shaped contour as before, with an arc of included angle θ_0, the return path, $re^{i\theta_0}$, must be such that $z + 2 = re^{i\theta_0} + 2$ is identical to $x + 2$. This will only occur if $\theta_0 = 2\pi$, that is, the arc portion of the contour is a full circle! However, we also have a branch cut which cannot be crossed by the contour. The only way to satisfy all these constraints is to place the branch cut along the positive real axis. The contour will go from 0 to R above the branch cut, around a circle of radius R to the bottom of the branch cut, and back in to the origin under the branch cut. A final circle of radius ϵ avoids the branch point (Figure 8.33).

8.6 USING CONTOUR INTEGRATION TO CALCULATE INTEGRALS OF REAL FUNCTIONS 541

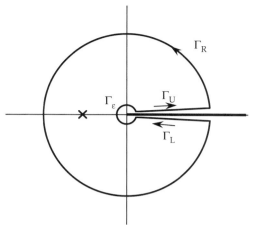

FIGURE 8.33 Contour for integrating $\oint_\Gamma \frac{z^{-1/2} dz}{z+2}$.

We will take limits as $R \to \infty$ and $\epsilon \to 0$. The integral over this contour is, therefore,

$$\oint_\Gamma \frac{z^{-1/2} dz}{z+2} = \oint_{\Gamma_U} + \oint_{\Gamma_R} + \oint_{\Gamma_L} + \oint_{\Gamma_\epsilon}.$$

Inside the contour there is one first-order pole, at $z = -2$. The residue is evaluated using the positive branch of $z^{1/2}$ because in the original integral, $\sqrt{x}$ refers to the positive branch of $x^{1/2}$:

$$\text{Res}\left[\frac{z^{-1/2}}{z+2}, -2\right] = (-2)^{-1/2} = \frac{1}{i\sqrt{2}}.$$

The contour integral is $\oint_\Gamma = 2\pi i \cdot 1/i\sqrt{2} = \pi\sqrt{2}$. Now consider the four path segments in turn.

Above the branch cut, we take the upper contour Γ_U to be a ray $z = re^{i\varphi}$ at an infinitesimal angle φ, and extending from $r = \epsilon$ to $r = R$. The argument of the square root will be $\varphi/2$. As $\varphi \to 0$, Γ_U moves down toward the real axis, and we obtain

$$\int_{\Gamma_U} \to \int_0^\infty \frac{|x|^{-1/2} e^{-i0} dx}{x+2} = \int_0^\infty \frac{dx}{\sqrt{x}(x+2)} = I,$$

the integral we seek. Below the branch cut, we take the lower contour Γ_L to be a ray $z = re^{i(2\pi-\varphi)}$, where φ is again an infinitesimal angle, and r comes in from R to ϵ. The argument of the square root below the branch cut will be $\pi - \varphi/2$. As $\varphi \to 0$, Γ_L moves up toward the real axis, and we obtain

$$\int_{\Gamma_L} \to \int_\infty^0 \frac{|x|^{-1/2} e^{-i(\pi-0)} dx}{x+2} = -\int_0^\infty \frac{-|x|^{-1/2} dx}{x+2} = \int_0^\infty \frac{dx}{\sqrt{x}(x+2)} = I$$

as well. On the large circle, Γ_R, we apply the *ML* inequality:

$$\left|\oint_{\Gamma_R}\right| \leq 2\pi R \times \max \left|\frac{e^{-i\theta/2}}{\sqrt{R}(Re^{i\theta}+2)}\right| = \frac{2\pi\sqrt{R}}{R+2} \to 0 \text{ as } R \to \infty.$$

On the small circle around the branch point, Γ_ϵ, we also apply the *ML* inequality:

$$\left|\oint_{\Gamma_\epsilon}\right| \leq 2\pi\epsilon \times \max \left|\frac{e^{-i\theta/2}}{\sqrt{\epsilon}(\epsilon e^{i\theta}+2)}\right| = \frac{2\pi\sqrt{\epsilon}}{\epsilon+2} \to 0 \text{ as } \epsilon \to 0.$$

Assembling the pieces, we have

$$\oint_\Gamma \frac{z^{-1/2}dz}{z+2} = \pi\sqrt{2} = \int_{\Gamma_U} + \oint_{\Gamma_R} + \int_{\Gamma_L} + \oint_{\Gamma_\epsilon} = 2I,$$

from which we obtain, finally, $I = \pi/\sqrt{2}$.

More complicated integrations with branch cuts can be imagined, but the principles remain the same:

- The contour of integration must not cross a branch cut.
- When the contour runs along a branch cut, the integrand must be evaluated using the argument of z specified by the branch cut.
- Indent around branch points and use the *ML* inequality.

If a pole is on a branch cut, the contour is indented around the pole without crossing the branch cut (Figure 8.34). The contribution of the pole is obtained in the usual way, being careful to get the argument of z right, according to the branch cut.

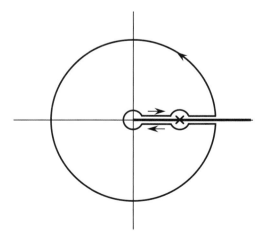

FIGURE 8.34 When a pole is on a branch cut, the contour is indented on both the upper and lower paths.

8.7 COMPLEX INTEGRATION AND THE FOURIER TRANSFORM

In earlier chapters it was shown how to calculate Fourier transforms by direct evaluation of the Fourier integral and by using Fourier transform theorems to express a desired transform in terms of known transform pairs. Up to this point, direct integration has relied on finding the antiderivative of $f(x)e^{-i2\pi v x}$ or $F(v)e^{i2\pi v x}$ for a particular f or F. If an antiderivative could not be found, the integral could not be done. This blockage can sometimes be removed by using complex integration.

The general approach is introduced by the following example. In an earlier chapter we calculated the Fourier transform of $f(t) = e^{-at}U(t)$, which was $F(v) = \frac{1}{a+i2\pi v}$. Let us now consider how to do the inverse transform:

$$I = \int_{-\infty}^{\infty} \frac{e^{i2\pi vt}}{a+i2\pi v} dv.$$

The transform F is not absolutely integrable, so we will try to perform the inversion as a transform in the limit with a rectangular convergence factor (Theorem 5.14),

$$I = \lim_{R\to\infty} \int_{-R}^{R} \frac{e^{i2\pi vt}}{a+i2\pi v} dv,$$

and use complex integration. Consider

$$\oint_\Gamma \frac{e^{i2\pi zt} dz}{a+i2\pi z}$$

on the usual semicircular contour. There is a simple pole at $z = ia/2\pi$. The residue is

$$\text{Res}\left[\frac{e^{i2\pi zt}}{a+i2\pi z}, \frac{ia}{2\pi}\right] = \frac{e^{-at}}{2\pi i},$$

and the contour integral is e^{-at}. This looks promising.

Next we apply the ML inequality on the semicircular arc, $z = Re^{i\theta}$.

$$\left|\oint_{\Gamma_R} \frac{e^{i2\pi zt}dz}{a+i2\pi z}\right| \leq \pi R \times \max \left|\frac{\exp(-2\pi Rt \sin\theta)}{\sqrt{a^2+(2\pi R)^2 - 4\pi aR\sin\theta}}\right|,$$

where θ ranges from 0 to π. As $R \to \infty$, it appears that the denominator will cancel the R out front. As for the exponential, $\exp(-2\pi Rt\sin\theta)$, because $\theta \in [0,\pi]$ we will have $\sin\theta \geq 0$. We know that $R > 0$, as well. But what about t? If t is positive, the exponential will decay as $R \to \infty$, except at the endpoints, $\theta = 0$ and $\theta = \pi$. As either of the endpoints is approached, $\sin\theta$ approaches 0, and $|\exp(-2\pi Rt\sin\theta)|$ approaches 1 as its upper bound. The ML bound, then, would be $\frac{1}{2}$ as $R \to \infty$. But 0 and π are isolated points; for all other θ the exponential decays and experience suggest that the integral will still converge to 0 as desired. Something much worse happens if t is negative: the exponential grows rather than decays, and the integral is unbounded.

The following result provides the tool we need to resolve these issues.

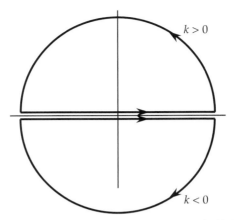

FIGURE 8.35 Jordan's lemma holds for $k > 0$ in the upper half-plane, and $k < 0$ in the lower half-plane. In the lower half-plane, the contour is traversed in the negative direction.

Theorem 8.12 (Jordan's lemma). Let f be a function which is analytic (a) in the UHP domain $D_U = \{z = Re^{i\theta} : R > R_0 > 0, \theta \in [0, \pi]\}$ or (b) in the LHP domain $D_L = \{z = Re^{i\theta} : R > R_0 > 0, \theta \in [-\pi, 0]\}$. That is, f is analytic in the U(L)HP except for a finite number of isolated singularities which are within R_0 of the origin. Further, assume that $\lim_{R \to \infty} |f(Re^{i\theta})| = 0$ uniformly with respect to θ. Then, on a semicircular contour Γ_U (Γ_L) of radius R in the U(L)HP, centered at the origin (Figure 8.35),

$$\lim_{R \to \infty} \oint_{\Gamma_U} f(z)e^{ikz}dz = 0 \quad k > 0$$
$$\lim_{R \to \infty} \oint_{\Gamma_L} f(z)e^{ikz}dz = 0 \quad k < 0. \tag{8.22}$$

Proof: We will prove Jordan's lemma for $k > 0$. The proof for $k < 0$ is identical. On the contour, $z = Re^{i\theta}$, $\theta \in [0, \pi]$ and $dz = iRe^{i\theta}d\theta$. Making these substitutions,

$$\left| \oint_{\Gamma_U} f(z)e^{ikz}dz \right| = \left| \int_0^\pi f(Re^{i\theta}) e^{-kR\sin\theta} e^{ikR\cos\theta} iRe^{i\theta} d\theta \right|$$
$$\leq \int_0^\pi |f(Re^{i\theta})| e^{-kR\sin\theta} R d\theta.$$

The uniform convergence of f to zero as $R \to \infty$ enables us to bound it above by ϵ, independent of θ, and pull it outside the integral, giving

$$\left| \oint_{\Gamma_U} f(z)e^{ikz}dz \right| \leq \epsilon R \int_0^\pi e^{-kR\sin\theta} d\theta = 2\epsilon R \int_0^{\pi/2} e^{-kR\sin\theta} d\theta,$$

where in the last step we used the fact that the integrand is symmetric about $\theta = \pi/2$.

8.7 COMPLEX INTEGRATION AND THE FOURIER TRANSFORM

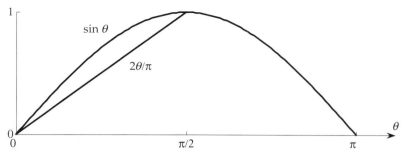

FIGURE 8.36 On the interval $(0, \pi/2)$, $\sin\theta > 2\theta/\pi$. Consequently, $e^{-\sin\theta} < e^{-2\theta/\pi}$.

We need to show that this upper bound goes to zero. The strategy is to replace $e^{-R\sin\theta}$ by something larger (i.e., bound it above) and show that the new integral goes to zero. On the interval $(0, \pi/2)$, $\sin\theta > 2\theta/\pi$ (Figure 8.36), so $e^{-kR\sin\theta} < e^{-2kR\theta/\pi}$, and we have

$$\left| \oint_{\Gamma_R} f(z)e^{ikz} dz \right| \leq 2\epsilon R \int_0^{\pi/2} e^{-kR\sin\theta} d\theta$$

$$< 2\epsilon R \int_0^{\pi/2} e^{-2kR\theta/\pi} d\theta = \frac{\pi\epsilon}{k}\left(1 - e^{-kR}\right)$$

$$< \frac{\pi\epsilon}{k}.$$

Finally, $\epsilon \to 0$ as $R \to \infty$, completing the proof. ∎

Corollary 8.13. If a function fulfills the conditions for Jordan's lemma in the UHP and LHP,

$$\int_{-\infty}^{\infty} f(x)e^{ikx} dx = \begin{cases} 2\pi i \sum_{z_m \in \text{UHP}} \text{Res}\left[f(z)e^{ikz}, z_m\right], & k > 0 \\ -2\pi i \sum_{z_m \in \text{LHP}} \text{Res}\left[f(z)e^{ikz}, z_m\right], & k < 0 \end{cases}. \quad (8.23)$$

The minus sign in the $k < 0$ case occurs because the contour is traversed in the negative direction (Figure 8.35).

When applying Jordan's lemma to the Fourier transform, the parameter k will correspond to a time or frequency variable, for example, $k = 2\pi t$. Frequency and time can be both positive and negative, so it will be necessary to perform two integrals, one for $k > 0$ and one for $k < 0$, closing the contour in the UHP and LHP, respectively.

Example 8.24. We will work through the previous example completely, using Jordan's lemma. We seek to calculate the inverse Fourier transform of $F(\nu) = \frac{1}{a+i2\pi\nu}$. Take $F(z) = \frac{1}{a+i2\pi z}$ and $k = 2\pi t$. We must first consider whether the conditions are

546   CHAPTER 8   COMPLEX INTEGRATION

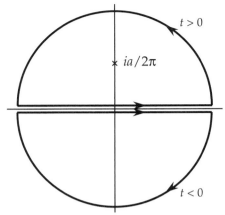

FIGURE 8.37 Contours for calculating the Fourier transform $\int_{-\infty}^{\infty} \frac{e^{i2\pi vt} dv}{a+i2\pi v}$. The contour in the upper half-plane is for $t > 0$, and the one in the lower half-plane is for $t < 0$.

met. It is easy to see (Figure 8.37) that F is analytic in the UHP for $|z| > a/2\pi$ and in the entire LHP. We must next check that $F(z) \to 0$ uniformly on the semicircular arcs as $R \to \infty$. With $z = Re^{i\theta}$,

$$\left|\frac{1}{a+i2\pi z}\right| = \left|\frac{1}{a+i2\pi Re^{i\theta}}\right| = \frac{1}{\sqrt{(2\pi R)^2 - 4\pi aR \sin\theta + a^2}}$$

In the UHP, the denominator has a minimum at $\theta = \pi/2$ for all R, so

$$\sqrt{(2\pi R)^2 - 4\pi aR \sin\theta + a^2} \geq \sqrt{(2\pi R - a)^2} = 2\pi R - a,$$

and so

$$\left|\frac{1}{a+i2\pi z}\right| \leq \frac{1}{2\pi R - a},$$

for all θ in the UHP. In the LHP, the denominator has minima at $\theta = -\pi$ and $\theta = 0$, and

$$\left|\frac{1}{a+i2\pi z}\right| \leq \frac{1}{\sqrt{(2\pi R)^2 + a^2}},$$

for all θ in the LHP. Both of these bounds go to zero as $R \to \infty$, independently of θ, verifying the uniformity of the convergence. Hence, Jordan's lemma is applicable.

Taking $k = 2\pi t$, we close the contour in the UHP for $t > 0$ and in the LHP for $t < 0$ (Figure 8.37). One pole is encircled in the UHP, and the complex integral is e^{-at}, as previously calculated. Now we may assert that the integral on the semicircular arc is zero for $t > 0$, by Jordan's lemma. Closing the contour in the LHP, there are no poles within the contour, so the complex integral is zero. The integral on the semicircular arc is also zero for $t < 0$, by Jordan's lemma.

8.7 COMPLEX INTEGRATION AND THE FOURIER TRANSFORM

When $t = 0$, the Fourier kernel drops out, and we have to calculate the integral $\int_{-\infty}^{\infty} \frac{dv}{a+i2\pi v}$. The easiest way here is to rationalize the denominator, obtaining

$$\int_{-\infty}^{\infty} \frac{dv}{a + i2\pi v} = \int_{-\infty}^{\infty} \frac{a\,dv}{a^2 + (2\pi v)^2} - \int_{-\infty}^{\infty} \frac{i2\pi v\,dv}{a^2 + (2\pi v)^2}.$$

The second integrand is an odd function, so we know the integral will be zero. The other integral has a known antiderivative (or adapt Example 8.23), leading to the result

$$\int_{-\infty}^{\infty} \frac{a\,dv}{a^2 + (2\pi v)^2} = \frac{1}{2\pi} \int_{-\infty}^{\infty} \frac{2\pi/a}{1 + (2\pi v/a)^2} dv = \frac{1}{2\pi} \arctan\left(\frac{2\pi v}{a}\right)\bigg|_{-\infty}^{\infty} = \frac{1}{2}.$$

Putting all the pieces together, we finally have

$$\mathcal{F}^{-1}\left\{\frac{1}{1+i2\pi v}\right\} = \int_{-\infty}^{\infty} \frac{e^{i2\pi vt}\,dv}{1 + i2\pi v} = \begin{cases} e^{-at}, & t > 0 \\ \frac{1}{2}, & t = 0 \\ 0, & t < 0 \end{cases}$$

$$= e^{-at} U(t).$$

The value of $\frac{1}{2}$ at $t = 0$ is what we expect—at a jump discontinuity, the transform takes on the average value of the right and left limits. ∎

Example 8.25. Calculate the inverse Fourier transform of $1/i\pi v$ which we know should turn out to be sgn x. We will need to take the integral as a Cauchy principal value because of the singularity at $v = 0$:

$$\mathcal{F}^{-1}\left\{\frac{1}{i\pi v}\right\} = \mathcal{P}\int_{-\infty}^{\infty} \frac{1}{i\pi v} e^{i2\pi vx}\,dv = \lim_{R\to\infty}\lim_{\epsilon\to 0}\int_{-R}^{-\epsilon} + \int_{\epsilon}^{R} \frac{1}{i\pi v} e^{i2\pi vx}\,dv.$$

We will use the complex integral

$$\oint_\Gamma \frac{1}{i\pi z} e^{i2\pi xz}\,dz$$

on indented contours, avoiding the pole at the origin (Figure 8.38).

To use Jordan's lemma, we observe that $|1/i\pi z| = 1/\pi R \to 0$ as $R \to \infty$, so we are assured that the integral along Γ_{R+} in the UHP will go to zero as $R \to \infty$, for $2\pi x > 0$. The integral along Γ_{R-} in the LHP will also go to zero for $2\pi x < 0$. There are no other poles inside either contour, so $\oint_{\Gamma_{R+}} = \oint_{\Gamma_{R-}} = 0$. However, there will be contributions from the indentations. For $x > 0$,

$$\oint_{\Gamma_{\epsilon+}} = -i\pi\,\text{Res}\left[\frac{e^{i2\pi xz}}{i\pi z}, 0\right] = -1$$

(the path is traversed in the negative direction). The integrals along the real axis become I, the integral we seek, as $R \to \infty$ and $\epsilon \to 0$. Thus, we have

$$\oint_{\text{UHP}} \to I + \oint_{\Gamma_{R+}} + \oint_{\Gamma_{\epsilon+}} = I + 0 - 1 = 0,$$

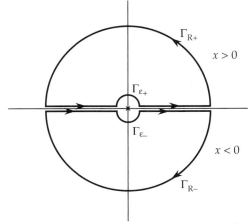

FIGURE 8.38 Contours for calculating the Fourier transform $\int_{-\infty}^{\infty} \frac{e^{i2\pi vx} dv}{i\pi v}$. The contour in the upper half-plane is for $x > 0$, and the one in the lower half-plane is for $x < 0$. Both contours are indented around the pole at the origin.

so $I = 1$ for $x > 0$. For $x < 0$, the contribution from the indented contour (traversed in the positive direction) is

$$\oint_{\Gamma_{\varepsilon-}} = +i\pi \operatorname{Res}\left[\frac{e^{i2\pi xz}}{i\pi z}, 0\right] = 1.$$

Thus,

$$\oint_{\text{LHP}} \to I + \oint_{\Gamma_{R-}} + \oint_{\Gamma_{\varepsilon-}} = I + 0 + 1 = 0,$$

so $I = -1$ for $x < 0$. Finally, when $x = 0$, the integral becomes

$$\mathcal{P}\int_{-\infty}^{\infty} \frac{dv}{i\pi v} = 0,$$

because $1/i\pi v$ is an odd function. Putting it all together,

$$\mathcal{F}^{-1}\left\{\frac{1}{i\pi v}\right\} = \mathcal{P}\int_{-\infty}^{\infty} \frac{e^{i2\pi vx}}{i\pi v} dv = \begin{cases} 1, & x > 0 \\ 0, & x = 0 \\ -1, & x < 0 \end{cases}$$

$$= \operatorname{sgn} x. \qquad \blacksquare$$

Once this result is in hand, it is a simple matter to calculate the inverse Fourier transform of sinc v (see the problems).

Similar methods apply to integrals of the form

$$\int_{-\infty}^{\infty} f(t)\cos\omega t\, dt \quad \text{and} \quad \int_{-\infty}^{\infty} f(t)\sin\omega t\, dt.$$

8.7 COMPLEX INTEGRATION AND THE FOURIER TRANSFORM 549

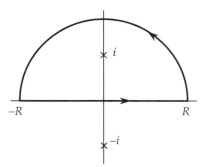

FIGURE 8.39 Semicircular contour for integrating $\int_{-\infty}^{\infty} \frac{\cos x\, dx}{x^2+1}$.

Because $e^{i\omega t} = \cos \omega t + i \sin \omega t$, either of these integrals can be performed (for real f) by calculating the Fourier transform $\int_{-\infty}^{\infty} f(t)e^{i\omega t} dt$ and extracting either the real or imaginary part. Indeed, we *must* do them in this way, as we shall now show.

Let us try to integrate $\int_{-\infty}^{\infty} \frac{\cos x\, dx}{x^2+1}$. Proceeding naively, consider the contour integral $\oint_\Gamma \frac{\cos z\, dz}{z^2+1}$ on the usual semicircular contour (Figure 8.39).

There is one pole inside the contour, and its residue is $\operatorname{Res}\left[\frac{\cos z}{z^2+1}, i\right] = \frac{\cos i}{2i}$, so the contour integral evaluates to $\pi \cos i = \pi \cosh 1$. The portion along the real axis is the integral we seek. Next we apply the *ML* inequality to the integral on the semicircular arc:

$$\left|\int_{\Gamma_R} \frac{\cos z\, dz}{z^2+1}\right| \leq \pi R \times \max \left|\frac{\cos R e^{i\theta}}{R^2 e^{i2\theta}+1}\right|$$

$$= \pi R \times \max \left[\frac{\cos R e^{i\theta} \cos R e^{-i\theta}}{(R^2 e^{i2\theta}+1)(R^2 e^{-i2\theta}+1)}\right]^{1/2}$$

$$= \pi R \times \max \left[\frac{1}{2} \frac{\cosh(2R \sin \theta) + \cos(2R \cos \theta)}{R^4 + 2R^2 \cos 2\theta + 1}\right]^{1/2}.$$

Without going any farther, we can see we are in trouble. The presence of $\cosh(2R \sin \theta)$ dooms this upper bound to blow up as $R \to \infty$. The *ML* inequality fails, and so consequently our attempt at calculating the integral fails.

The problem is that the cosine is unbounded on the semicircular arc. The sine function presents the same difficulty:

$$\left|\sin R e^{i\theta}\right| = \left[\sin R e^{i\theta} \sin R e^{-i\theta}\right]^{1/2}$$

$$= \left[\frac{1}{2}\left(\cosh(2R \sin \theta) - \cos(2R \cos \theta)\right)\right]^{1/2}.$$

However, if we combine the two into a complex exponential,

$$\cos(R e^{i\theta}) + i \sin(R e^{i\theta}) = \exp\left(i R e^{i\theta}\right) = \exp\left(i R (\cos \theta + i \sin \theta)\right)$$

$$= e^{-R \sin \theta} e^{iR \cos \theta},$$

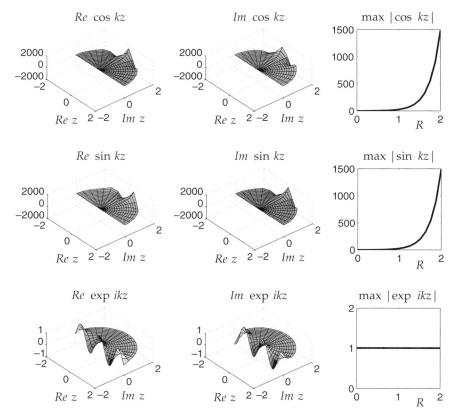

FIGURE 8.40 *Left to right*: Real part, imaginary part, and maximum modulus of $\cos kz$, $\sin kz$, and $\exp(ikz)$. The functions $\cos kz$ (*top*) and $\sin kz$ (*middle*) grow exponentially in the upper half-plane for $k > 0$, but $\exp(iz)$ (*bottom*) is bounded.

their bad features will cancel each other. On an arc in the UHP, $\theta \in [0, \pi]$, $\sin \theta$ is nonnegative and

$$\left|\exp\left(iRe^{i\theta}\right)\right| = e^{-R\sin\theta} \leq 1.$$

If the rest of the integrand dies away as $R \to \infty$, we win (Figure 8.40).

Example 8.26. To calculate $\int_{-\infty}^{\infty} \frac{\cos x \, dx}{x^2+1}$, consider the integral

$$\int_{-\infty}^{\infty} \frac{e^{ix} dx}{x^2 + 1} = \int_{-\infty}^{\infty} \frac{(\cos x + i \sin x) dx}{x^2 + 1}$$

and take the real part of the result. Proceeding with the complex integral $\oint_\Gamma \frac{e^{iz} dz}{z^2+1}$, we want to use Jordan's lemma with $k = 1$, so we will seek to close the contour in the UHP. Take $f(z) = \frac{1}{z^2+1}$ and observe that f has only one isolated singularity in the

UHP, at $z = i$, and f is analytic for $|z| > 1$. Next show that f goes to zero uniformly as $|z| \to \infty$. In the UHP, $\frac{1}{z^2+1} = \frac{1}{R^2 e^{i2\theta}+1}$, with $\theta \in [0, \pi]$. Now,

$$\left|\frac{1}{R^2 e^{i2\theta} + 1}\right| = \frac{1}{\sqrt{R^4 + 2R^2 \cos 2\theta + 1}}.$$

The denominator has a minimum at $\theta = \pi/2$, so we may write

$$\sqrt{R^4 + 2R^2 \cos 2\theta + 1} \le \sqrt{R^4 - 2R^2 + 1} = R^2 - 1,$$

and $\left|\frac{1}{z^2+1}\right| \le \frac{1}{R^2-1}$. As $R \to \infty$, $\left|\frac{1}{z^2+1}\right| \to 0$ uniformly. Jordan's lemma applies. Using Equation 8.23 with $k = 1$,

$$\int_{-\infty}^{\infty} \frac{e^{ix} \, dx}{x^2 + 1} = 2\pi i \operatorname{Res}\left[\frac{e^{iz}}{z^2 + 1}, i\right] = \pi e^{-1}.$$

Thus,

$$\int_{-\infty}^{\infty} \frac{\cos x \, dx}{x^2 + 1} = \operatorname{Re} \pi e^{-1} = \pi e^{-1}.$$

The value of the companion integral,

$$\int_{-\infty}^{\infty} \frac{\sin x \, dx}{x^2 + 1} = \operatorname{Im} \pi e^{-1} = 0,$$

could have been anticipated from the odd symmetry of the integrand. ∎

Example 8.27. Calculate the area under the sinc function, $\int_{-\infty}^{\infty} \frac{\sin \pi x}{\pi x} \, dx$. Using the area theorem, if you know the Fourier transform of sinc, you have the answer easily. But let us attack it directly. Because $\sin \pi x$ is the imaginary part of $e^{i\pi x}$, we can calculate the integral

$$\mathcal{P} \int_{-\infty}^{\infty} \frac{e^{i\pi x}}{\pi x} \, dx$$

and take the imaginary part of the result. Consider the complex integral

$$\oint_{\Gamma} \frac{e^{i\pi z}}{\pi z} \, dz$$

on a half-circle contour in the UHP ($k = \pi > 0$), indented at the origin (Figure 8.38, again). All the criteria for Jordan's lemma are obviously met, and the integral we seek reduces to

$$\mathcal{P} \int_{-\infty}^{\infty} \frac{e^{i\pi x}}{\pi x} \, dx = -\oint_{\Gamma_\epsilon} \frac{e^{i\pi z} \, dz}{\pi z}$$

$$= i\pi \operatorname{Res}\left[\frac{e^{i\pi z}}{\pi z}, 0\right] = i.$$

Therefore, $\int_{-\infty}^{\infty} \frac{\sin \pi x}{\pi x} \, dx = \operatorname{Im} i = 1.$ ∎

Example 8.28. Calculate the area under the sinc² function. Knowing the Fourier transform of either sinc or sinc², this area calculation is easy, using Parseval's theorem or the area theorem, respectively. But again, it is instructive to calculate the integral directly.

In the previous example we replaced $\sin \pi x$ by $e^{i\pi x}$ and took the imaginary part of the integral. One might think that here, we would replace $\sin^2 \pi x$ by $(e^{i\pi x})^2 = e^{i2\pi x}$. But this is wrong, because $\mathcal{I}m \, e^{i2\pi x}$ is $\sin 2\pi x$, not $\sin^2 \pi x$. We have to go a different way. Using a trigonometric identity, we can write $\sin^2 \pi x = \frac{1}{2}(1 - \cos 2\pi x)$, which gives the integral

$$\int_{-\infty}^{\infty} \text{sinc}^2 x \, dx = \int_{-\infty}^{\infty} \frac{1 - \cos 2\pi x}{2\pi^2 x^2} \, dx.$$

This integral cannot be broken into the sum of two integrals, because each integral will be infinite (do you see why?). Rather, they must be considered as one unit. We write $1 - \cos 2\pi x = \mathcal{R}e \left(1 - e^{i2\pi x}\right)$ and proceed as above with a complex integral, that is,

$$\oint_\Gamma \frac{1 - e^{i2\pi z}}{2\pi^2 z^2} \, dz.$$

This integrand appears to have a double pole at the origin, but one of them is removable:

$$\lim_{z \to 0} z \cdot \frac{1 - e^{i2\pi z}}{2\pi^2 z^2} = \lim_{z \to 0} \frac{1 - e^{i2\pi z}}{2\pi^2 z}$$
$$= \lim_{z \to 0} \frac{-i2\pi e^{i2\pi z}}{2\pi^2} = \frac{1}{i\pi},$$

demonstrating that the pole is actually first order.

The complex integral is taken around the usual semicircular contour, with an indentation above the pole at $z = 0$ (cf. Figure 8.38). There being no singularities within the contour, $\oint_\Gamma = 0$. The contribution from the pole is due to the indentation:

$$\lim_{\epsilon \to 0} \oint_{\Gamma_\epsilon} = -\frac{1}{2} \, 2\pi i \, \text{Res} \left[\frac{1 - e^{i2\pi z}}{2\pi^2 z^2}, 0 \right] = -1.$$

Next, we have to show that the complex integral goes to zero on the large semicircle Γ_R as $R \to \infty$. If we break the integral into two pieces, each piece has to go to zero by itself. The first integrand, $\frac{1}{2\pi^2 z^2}$, clearly presents no problems. The second integrand,

8.7 COMPLEX INTEGRATION AND THE FOURIER TRANSFORM

$\frac{e^{i2\pi z}}{2\pi^2 z^2}$, satisfies the conditions for Jordan's lemma (with $k = 2\pi$). Hence, $\oint_{\Gamma_R} \to 0$ as $R \to \infty$. We therefore have

$$\lim_{R \to \infty} \lim_{\epsilon \to 0} \oint_{\Gamma} = 0$$

$$= \int_{-\infty}^{\infty} \operatorname{sinc}^2 x \, dx + \lim_{R \to \infty} \oint_{\Gamma_R} + \lim_{\epsilon \to 0} \oint_{\Gamma_\epsilon}$$

$$= \int_{-\infty}^{\infty} \operatorname{sinc}^2 x \, dx - 1,$$

so

$$\int_{-\infty}^{\infty} \operatorname{sinc}^2 x \, dx = 1.$$

Here is yet another solution. Integrating $\operatorname{sinc}^2 x$ once by parts,

$$\int_{-\infty}^{\infty} \frac{\sin^2 \pi x}{\pi^2 x^2} \, dx = -\frac{\sin^2 \pi x}{\pi^2 x}\bigg|_{-\infty}^{\infty} + \int_{-\infty}^{\infty} \frac{2\pi \sin \pi x \cos \pi x}{\pi^2 x} \, dx$$

$$= 0 + \int_{-\infty}^{\infty} \frac{\sin 2\pi x}{\pi x} \, dx,$$

and making a change of variable $\xi = 2x$,

$$\int_{-\infty}^{\infty} \operatorname{sinc}^2 x \, dx = \int_{-\infty}^{\infty} \frac{\sin \pi \xi}{\pi \xi} \, d\xi = 1$$

(using a previous result). ∎

This example shows that there is often more than one way to solve a problem. The more tools you have in your kit, the more problems you can solve, and with experience, you will get better at picking the right tool for a particular job. As much as possible, you want to avoid the situation where, figuratively, you only have a hammer so you have to make every problem look like a nail!

Finally, here is a nice example of Fourier transformation using complex integration on a different sort of contour.

Example 8.29. Calculate the Fourier transform of the Gaussian, $f(x) = e^{-\pi x^2}$. The approach here, calculating the Fourier integral, is more direct than the differential equation method in Section 5.2. Working first with the Fourier integral,

$$\int_{-\infty}^{\infty} e^{-\pi x^2} e^{-i2\pi vx} dx = \int_{-\infty}^{\infty} e^{-\pi(x^2 + i2vx)} dx.$$

Complete the square in the exponent,

$$x^2 + i2vx = x^2 + i2vx - v^2 + v^2 = (x + iv)^2 + v^2,$$

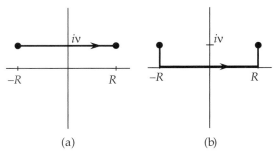

FIGURE 8.41 Contour for calculating the Fourier transform of the Gaussian, $e^{-i\pi x^2}$. (a) The original path of integration. (b) An equivalent, and more convenient, path.

obtaining a similar integral,

$$\int_{-\infty}^{\infty} e^{-\pi x^2} e^{-i2\pi v x} dx = \int_{-\infty}^{\infty} e^{-\pi v^2} e^{-\pi (x+iv)^2} dx.$$

This new integral can be written as the integral of the complex function $e^{-\pi v^2} e^{-\pi z^2}$ on a path $z = x + iv$, where v is constant and x ranges from $-\infty$ to ∞ (Figure 8.41a):

$$\int_{-\infty}^{\infty} e^{-\pi v^2} e^{-\pi (x+iv)^2} dx = \lim_{R \to \infty} \int_{-R+iv}^{R+iv} e^{-\pi v^2} e^{-\pi z^2} dz.$$

The function $e^{-\pi z^2}$ is analytic in the entire complex plane. Therefore, the integral from $-R + iv$ to $R + iv$ is independent of the path taken. We can deform the path so that it runs along the real axis (Figure 8.41b):

$$\int_{-R+iv}^{R+iv} e^{-\pi v^2} e^{-\pi z^2} dz = \int_{-R+iv}^{-R+i0} + \int_{-R+i0}^{R+i0} + \int_{R+i0}^{R+iv}.$$

As $R \to \infty$, the integral along the real axis is just

$$\int_{-\infty}^{\infty} e^{-\pi v^2} e^{-\pi x^2} dx = e^{-\pi v^2} \int_{-\infty}^{\infty} e^{-\pi x^2} dx = e^{-\pi v^2}.$$

Along one vertical segment of the contour, we have

$$\int_{-R+iv}^{-R+i0} = -\int_0^v e^{-\pi v^2} e^{-\pi (-R+iy)^2} i \, dy$$

$$= ie^{-\pi R^2} e^{-\pi v^2} \int_0^v e^{+\pi y^2} e^{i2\pi Ry} dy.$$

Apply the *ML* inequality,

$$\left| ie^{-\pi R^2} e^{-\pi v^2} \int_0^V e^{+\pi y^2} e^{i2\pi Ry} dy \right| \leq e^{-\pi R^2} e^{-\pi v^2} \max \left| e^{+\pi y^2} e^{i2\pi Ry} \right| \times V$$

$$= e^{-\pi R^2} e^{-\pi v^2} e^{+\pi v^2} \times V$$

$$= V e^{-\pi R^2}.$$

This goes to zero as $R \to \infty$, for all values of v. The same fate befalls the integral on the other vertical segment. Putting the pieces together, we have the transform:

$$\mathcal{F}\left\{ e^{-\pi x^2} \right\} = 0 + e^{-\pi v^2} + 0 = e^{-\pi v^2}. \qquad \blacksquare$$

★Special Properties of Bandlimited Functions

As a further illustration of the significant interplay between the Fourier transform and complex analysis, we shall explore some special properties of bandlimited functions. Recall that a function $f(t)$ is bandlimited if its Fourier transform $F(v)$ is identically zero for $|v| > B$. We have already seen (Theorem 6.8) that a bandlimited function can be, in principle, exactly reconstructed from samples taken at a spacing Δt no greater than $1/2B$. Here, as there, we will restrict attention to bandlimited functions that are also square integrable, that is, they have finite energy $E = \int_{-\infty}^{\infty} |f(t)|^2 \, dt = \int_{-B}^{B} |F(v)|^2 \, dv$.

Because F is in $L^2(-B, B)$, it is also in $L^1(-B, B)$ (Figure 4.5; also see the problems). Then, because $F \in L^1(-B, B)$, f is bounded and continuous as well as square integrable. By the derivative theorem,

$$f'(t) = \int_{-B}^{B} i2\pi v \, F(v) \, e^{i2\pi vt} \, dv.$$

That is, f' is also bandlimited. Moreover, its Fourier transform $i2\pi v F$ is also square integrable (see the problems), and by the same argument, is absolutely integrable, so f' is also bounded, square integrable, and continuous. Repeated application of these steps leads to the conclusion that f is infinitely continuously differentiable in addition to being bounded and square integrable. This is an intuitively satisfying result, for we know that the smoother a function is, the more rapidly its Fourier transform goes to zero as $|v| \to \infty$, and vice versa. The Fourier transform of a bandlimited function goes exactly to zero at $|v| = B$, which is an extremely rapid decay, so the bandlimited function ought to be extremely smooth. (Consider, for example, the sinc $\longleftrightarrow$ rect Fourier pair.)

Now, moving to complex analysis, let t be a complex variable. On the real line, all of f's properties still hold, but what happens when we move off the real line into the complex plane? We may well expect

$$f(t) = \int_{-B}^{B} F(v) \, e^{i2\pi vt} \, dv$$

to be analytic everywhere, inheriting the analytic properties of $e^{i2\pi vt}$ because F is rather well behaved. Indeed, this turns out to be the case.[8]

---

[8] Apply the Cauchy–Riemann conditions, assuming that one may differentiate under the integral sign. For a proof that does not require this assumption see Papoulis (1977, p. 186).

Furthermore, it can be shown that in the complex plane a bandlimited function f can grow no faster than a simple exponential as $|t| \to \infty$, $|f(t)| \le \sqrt{2BE}\, e^{2\pi B|t|}$ (see the problems). A function bounded in this way is said to be of *exponential type*.[9]

So we have that a bandlimited function $f(t)$, continued into the complex plane by letting t be complex, is entire (even better than infinitely continuously differentiable) and of exponential type (asymptotically bounded as $|t| \to \infty$). The remarkable fact is that *the converse is also true*. The complete result may be stated as follows. The proof is formidable and not included here.[10]

Theorem 8.14 (Paley–Wiener). Let $f \in L^2$ on the real axis. f is bandlimited if and only if, in the complex plane, it is an entire function of exponential type.

A fundamental consequence of the Paley–Wiener theorem is the fact that a function cannot simultaneously be bandlimited and time limited. The archetypal transform pair is rect $\longleftrightarrow$ sinc. Any bandlimited function f has a Fourier transform of the form $F = G \cdot$ rect, and the convolution theorem then gives $f = g * $ sinc. It certainly seems unlikely that such a convolution can be time limited. The conclusive demonstration comes through complex analysis.

Suppose that f is bandlimited. By the Paley–Wiener theorem, it is an entire function in the complex plane. Suppose f is also time limited. Then it is zero for $|t|$ greater than some T. On any interval where f is constant, all of its derivatives are zero, and it has a Taylor series expansion with only one term, the constant (here, zero). Now, because f is entire in the complex plane, there are no singularities and the radius of convergence of this (trivial) series is infinite.[11] So, f is zero not only for $|t| > T$ on the real axis, but throughout the complex plane, including the whole real axis. That is, the only function which is simultaneously bandlimited and time limited is trivial, $f = 0$.[12]

8.8 SUMMARY

1. *ML* inequality (Equation 8.2):

$$\left| \int_\Gamma f(z)\,dz \right| \le \underbrace{\max_{z \in \Gamma} |f(z)|}_{\text{``}M\text{''}} \times \underbrace{\text{Length of } \Gamma}_{\text{``}L\text{''}}$$

for f continuous on Γ.

---

[9] Entire functions of exponential type are treated comprehensively in Boas (1954).

[10] Papoulis (1977, pp. 258–260), referring to Boas (1954). For the original development, see Paley and Wiener (1934, pp. 1–13). For an extension to generalized functions, see Strichartz (1994, pp. 112–117).

[11] Extending a real Taylor series to represent a complex function is an application of the principle of *analytic continuation*. This topic is covered in many complex variable texts, for example, the previously cited works of Wunsch, LePage, and Hahn and Epstein. A particularly nice introduction is Saff and Snider (1976, pp. 224–232).

[12] The reader who wishes to pursue further the deep connections between complex analysis and Fourier transforms is referred to Dym and McKean (1972) and Paley and Wiener (1934).

2. Cauchy's integral theorem (Equation 8.5):

$$\oint_\Gamma f(z)dz = 0$$

for f analytic on and inside a simple closed contour Γ.

3. Two important consequences of Cauchy's integral theorem, deformation of contour (Corollary 8.2) and path independence (Corollary 8.3), enable arbitrary contours to be replaced by convenient ones.

4. Cauchy's integral formula (Equations 8.8 and 8.10).

$$\oint_\Gamma \frac{f(z)dz}{z - z_0} = 2\pi i f(z_0),$$

$$\oint_\Gamma \frac{f(z)}{(z - z_0)^{n+1}} dz = 2\pi i \frac{f^{(n)}(z_0)}{n!},$$

where f is analytic on and inside a simple closed contour Γ, and $z = z_0$ is inside (not on) Γ.

5. Laurent series. If f has a pole of order N at $z = z_0$, then on a punctured disk $r > |z - z_0| > 0$,

$$f(z) = \underbrace{c_{-N}(z - z_0)^{-N} + \cdots + c_{-1}(z - z_0)^{-1}}_{\text{principal part}} + \underbrace{c_0 + c_1(z - z_0) + \cdots}_{\text{regular part}}$$

$$c_{-1} = \text{Residue of } f \text{ at } z_0$$

6. Residue formula (Equations 8.11, 8.16, and 8.17):

$$\oint_\Gamma g(z)dz = 2\pi i \sum_{z_k \text{ inside } \Gamma} \text{Res}\left[g, z_k\right],$$

where, for a general Nth order pole,

$$\text{Res}[g, z_k] = \frac{1}{(N-1)!} \lim_{z \to z_k} \frac{d^{N-1}}{dz^{N-1}} \left[(z - z_k)^N g(z)\right]$$

and

$$\text{Res}\left[\frac{b(z)}{a(z)}, z_k\right] = \frac{b(z_k)}{a'(z_k)},$$

when $g(z) = b(z)/a(z)$ and z_k is a simple pole.

7. Several types of real integral, including the Fourier transform, can be performed by including the real axis as part of a closed contour in the complex plane and evaluating a complex integral.

PROBLEMS

8.1. Calculate the integral $I = \oint_\Gamma \frac{dz}{z}$ on the contour shown in Figure 8.42, by direct integration (not using Cauchy's theorem).

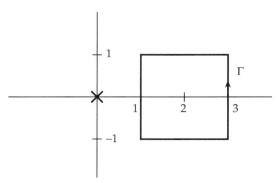

FIGURE 8.42 Contour of integration for Problem 8.1.

8.2. Using the Cauchy integral theorem and Cauchy integral formula, evaluate the following integrals:

(a) $\oint_\Gamma \frac{z-1}{z^2+3} \, dz$, where Γ is a circle of radius 1 centered at $z = i$.
Answer: $-\frac{\pi}{\sqrt{3}} + i\pi$

(b) $\oint_\Gamma \frac{\cos(z-1)}{z^2-z-2} \, dz$, where Γ is a circle of radius 3 centered at the origin.
Answer: $\frac{2\pi i}{3}[\cos(1) - \cos(2)]$

(c) $\oint \frac{\exp z}{(z-1)^2(z+1)} \, dz$, around the square formed by $x = \pm 4, y = \pm 4$.
Answer: $i\pi \cosh(1)$

(d) $\oint_\Gamma \frac{e^z \, dz}{z^2 - \frac{5}{2}z+1}$, where Γ is the unit circle.

8.3. Generalize the derivation of Equation 8.17 to obtain a formula for the other coefficients c_{-n} in the principal part of a function's Laurent expansion.

8.4. For each of the functions below, identify and classify all the isolated singularities, and calculate their residues. Where indicated, also calculate an integral.

(a) $f(z) = \frac{z}{z^2-z-2}$.

(b) $f(z) = \frac{z-1}{z^2+3z-4}$; calculate $\oint_{|z|=2} f(z) dz$.

(c) $f(z) = \frac{z}{\cos \pi z - 1}$; calculate $\oint_{|z|=\frac{3}{2}} f(z) dz$.

(d) $f(z) = \frac{\cos \pi z}{z^3-3z-2}$.

(e) $f(z) = \frac{z+1}{e^{i\pi z}+1}$.

(f) $f(z) = z \sin\left(\frac{1}{z}\right)$. This result may be counterintuitive.

8.5. Calculate the integral $\int_{-\pi}^{\pi} \frac{d\theta}{a+b\cos\theta} = \frac{2\pi}{\sqrt{a^2-b^2}}$ for $a > b \geq 0, a$ and b real.

8.6. Calculate the Fourier series coefficients, $\{c_n\}$, for the function $f(\theta) = \frac{2}{2+\cos\theta}$. *Hint*: Show that, because of symmetry, it is sufficient to calculate the coefficients only for $n \geq 0$ or $n \leq 0$.

8.7. Calculate, by contour integration,

(a) $\int_{-\infty}^{\infty} \frac{dx}{x^2+x+1} = \frac{2\pi}{\sqrt{3}}$.

(b) $\int_{-\infty}^{\infty} \frac{x^2+1}{x^4+1} dx = \pi\sqrt{2}$.

(c) $\int_{-\infty}^{\infty} \frac{x}{x^3+1} dx$.

8.8. Using contour integration, calculate the L^2 norm of the function $f(x) = x/(1+x^2)$.

8.9. Show that $\int_0^{\infty} \frac{x^m}{x^n+1} dx = \frac{\pi}{n \sin\left(\frac{\pi(m+1)}{n}\right)}$, where n and m are nonnegative integers and $n - m \geq 2$. *Hint*: Use the pie-shaped contour.

8.10. Calculate the integral $f[n] = \frac{1}{2\pi i} \oint_\Gamma F(z) z^{n-1} dz$ (which happens to be the inverse Z transform, see Chapter 9) for the function $F(z) = \frac{1}{1-0.5z^{-1}}$, where Γ is a closed contour encircling all singularities. Consider the $n \geq 0$ case (easy) separately from the $n < 0$ case (harder).

8.11. Calculate the integral $g(t) = \int_{-\infty}^{\infty} \frac{dx}{(t-x)(1+x^2)}$, where the integral is interpreted as a Cauchy principal value. This type of integral is called a *Hilbert transform* and is important in physics and in communication theory (also see Chapter 9).

8.12. Consider the functions $f(x) = 1/(x^2+1)$ and $g(x) = x/(x^2+1)$. Calculate, by direct integration (no Fourier transforms), the following convolutions, which were calculated in Chapter 5 using the convolution theorem.

(a) $f * f(x) = \frac{2\pi}{x^2+4}$.

(b) $f * g(x) = \frac{\pi x}{x^2+4}$.

(c) $g * g(x) = \frac{-2\pi}{x^2+4}$.

8.13. *Integrals with branch cuts*

(a) Calculate the integral $\int_{-\infty}^{\infty} \frac{\log|x| dx}{x^2+4}$. (Consider the complex integral $\oint_\Gamma \frac{\log z \, dz}{z^2+4}$ on a closed semicircular contour in the UHP.)

(b) Calculate the integral $\int_{-\infty}^{\infty} \frac{\log|x| dx}{x^2+1}$. Use the same contour as part (a).

8.14. The noise equivalent bandwidth of an electronic filter is defined to be the equivalent width of the squared magnitude of its frequency response:

$$\text{NEB} = \frac{\int_{-\infty}^{\infty} |H(\omega)|^2 d\omega}{|H(0)|^2}.$$

The NEB is a convenient single-number measure of a real filter's noise removing ability. If white (spectrally flat) noise with power density N_0 watts/unit bandwidth is put through the filter, the output noise power is just N_0 NEB.

The family of *Butterworth* filters is defined by a frequency response of the form

$$|H(\omega)|^2 = \frac{1}{(\omega/a)^{2n}+1},$$

where a is the cutoff (half-power) frequency, $|H(a)|^2 = 1/2$, and n is the order of the filter (number of poles in the transfer function $H(s)$).

(a) Using contour integration, calculate the integral $\int_{-\infty}^{\infty} |H(\omega)|^2 d\omega$ and show that
$$\text{NEB} = \frac{\pi a}{n \sin\left(\frac{\pi}{2n}\right)}.$$
Use a pie-shaped contour and use the fact that the integrand is even.

(b) Plot NEB vs. n. Determine an asymptotic value for large n. Interpret the graph and the asymptote physically.

8.15. Using contour integration, show that $\int_0^\infty \frac{x^\beta \, dx}{x^2+a^2} = \frac{\pi a^\beta}{2a \cos \frac{\beta \pi}{2}}$, where $a > 0$, $x^\beta \geq 0$, and $-1 < \beta < 1$. Use the contour in Figure 8.33.

8.16. Calculate, using complex integration, the areas under the sinc function and the Dirichlet kernel:

(a) $\int_{-\infty}^{\infty} \operatorname{sinc} x \, dx$

(b) $\int_{-\pi}^{\pi} \frac{\sin(2N+1)\theta}{\sin \theta} d\theta$

Comments:

- If you substitute $\sin(2N+1)\theta = \frac{e^{i(2N+1)\theta} - e^{-i(2N+1)\theta}}{2i}$, and then let $z = e^{i\theta}$, as usual, you will end up having to do two contour integrals, one of which is pretty tedious (try this and see where it leads, but do not actually do the integrals). Instead, observe that $\int_{-\pi}^{\pi} \frac{\sin(2N+1)\theta}{\sin \theta} d\theta = \mathcal{I}m \int_{-\pi}^{\pi} \frac{e^{i(2N+1)\theta}}{\sin \theta} d\theta$, and you only have to do one complex integration (then take the imaginary part of the result).

- While the original integrand, $\frac{\sin(2N+1)\theta}{\sin \theta}$, has removable singularities at $\theta = k\pi$, the new integrand, $\frac{e^{i(2N+1)\theta}}{\sin \theta}$, actually has poles at $\theta = k\pi$. They happen to sit on the path of integration, but you know what to do about that.

8.17. Calculate the inverse Fourier transform of sinc ν. Begin by writing $\sin \pi \nu = \frac{e^{i\pi\nu} - e^{-i\pi\nu}}{2i}$, then use the calculation in the text for $\frac{1}{i\pi\nu}$.

8.18. Calculate the inverse Fourier transform of $\operatorname{sinc}^2(\nu)$. One way, of course, is to use the result of the previous problem together with the convolution theorem. But it is instructive to calculate the integral directly.

8.19. Show that, for an n-times differentiable function f,
$$\int_a^b \frac{f(x)}{x^{n+1}} dx = \frac{1}{n!} \int_a^b \frac{f^{(n)}(x)}{x} dx - \sum_{k=0}^{n-1} \frac{(n-k-1)!}{n!} \left. \frac{f^{(k)}(x)}{x^{n-k}} \right|_a^b.$$

This result, which is based on integration by parts, can be a useful first step in simplifying integrals with higher order poles. Note that all integrals may be taken as Cauchy principal values if required.

8.20. Calculate the integral $\int_{-\infty}^{\infty} \operatorname{sinc}^3(x) \, dx$ by two methods:

(a) Direct integration.

(b) Using Fourier theorems and known results for sinc and sinc^2.

8.21. Use contour integration to calculate $\mathcal{F}^{-1}\left\{\frac{2}{1+(2\pi\nu)^2}\right\} = e^{-|x|}$.

8.22. Use complex integration to show that the Fourier transform of sech πx is sech $\pi \nu$. Here are some steps you should follow:

(a) First, sech $\pi x = \frac{1}{\cosh \pi x}$. Show that sech πx has an infinite number of first-order poles, at the locations $i/2 + ik$, where k is an integer.

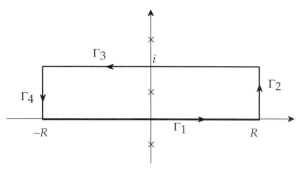

FIGURE 8.43 For Problem 8.22. Contour for calculating the Fourier transform of sech πx.

(b) Using the usual semicircular contour will result in an infinite series of residues as the radius of the semicircle, R, goes to infinity. This is clearly an inconvenient approach. In problems of this type, a rectangular contour, like the one shown in Figure 8.43, is often useful.

So consider the complex integral $\oint_\Gamma \frac{e^{-i2\pi vz} dz}{\cosh \pi z}$. As usual, the integral along the real axis (Γ_1) is the one we seek. You should find that the integral along the upper horizontal segment (Γ_3) is closely related to the integral along Γ_1.

(c) Then, use the *ML* inequality to show that the integrals along Γ_2 and Γ_4 go to zero as $R \to \infty$.

(d) Calculating the residue at the pole and putting all the pieces together, you should obtain sech πv.

8.23. Using the partial fraction expansion of $\frac{1}{\sin \pi x}$ (see Example 9.30 in the next chapter), show that

$$D_N(x) = \frac{\sin(2N+1)\pi x}{\sin \pi x} = (2N+1) \sum_{k=-\infty}^{\infty} \text{sinc}[(2N+1)(x-k)].$$

That is, the Dirichlet kernel is the periodic replication of the sinc.

8.24. *Bandlimited functions*

(a) Let $f \in L^2$ be bandlimited, with Fourier transform $F \in L^2(-B, B)$. The total energy is $\|f\|_2^2 = \|F\|_2^2 = E$. Using the Cauchy–Schwarz inequality, show that $F \in L^1(-B, B)$ as well, and $\|F\|_1 \leq \sqrt{2BE}$. *Hint*: Write $F(v)$ as a product, $F(v)$ rect$(v/2B)$.

(b) Show that $f^{(n)}$ is also bandlimited, and its Fourier transform $(i2\pi v)^n F(v)$ also has finite energy:

$$\int_{-\infty}^{\infty} |(i2\pi v)^n F(v)|^2 \, dv = \int_{-B}^{B} |(i2\pi v)^n F(v)|^2 \, dv \leq M < \infty,$$

and calculate the upper bound M. *Hint*: Use the Cauchy–Schwarz inequality again.

(c) Assuming that one may differentiate under the integral sign,[13] apply the Cauchy–Riemann conditions to

$$f(t) = \int_{-B}^{B} F(v) e^{i2\pi vt} \, dv$$

with complex t, and show that f is an entire function in the complex plane.

(d) Verify that f is of exponential type, with $|f(t)| \leq \sqrt{2BE} \, e^{2\pi B|t|}$. *Hint:* In the Fourier integral,

$$f(t) = \int_{-B}^{B} F(v) e^{i2\pi vt} \, dv,$$

write $t = |t| (\cos \varphi + i \sin \varphi)$, where $\varphi = \text{Arg}(t)$. Then show

$$\left| e^{i2\pi vt} \right| = e^{-2\pi v |t| \sin \varphi} \leq e^{2\pi B |t|}$$

for $|v| \leq B$.

---

[13] F must be bounded to guaranteed the validity of this move. See Folland (2002, p. 192).

CHAPTER 9

LAPLACE, Z, AND HILBERT TRANSFORMS

In this chapter we look at three other important transforms that are closely related to the Fourier transform. The venerable Laplace transform,

$$F_L(s) = \int_0^\infty f(t)\, e^{-st}\, dt,$$

is widely used in the analysis and design of dynamic systems. In addition to its "operational" properties, we will explore some of its connections with complex analysis and the Fourier transform. The Z transform,

$$F_Z(z) = \sum_{n=0}^\infty f[n] z^{-n},$$

fulfills the same role for discrete-time systems that the Laplace transform does for continuous-time systems and has a similar relationship to the discrete-time Fourier transform. Finally, we consider some important properties of one-sided functions, such as the impulse responses of causal LTI systems. These relationships, which exist for all four Fourier transforms, are collectively known as Hilbert transforms.

9.1 THE LAPLACE TRANSFORM

We have earlier seen how to define and calculate the Fourier transform for functions which are absolutely integrable (L^1), square integrable (L^2), or slowly growing (e.g., polynomials), and for generalized functions (e.g., $\delta(x)$). The Laplace transform extends the capability of the Fourier transform to certain important functions of *rapid growth* which are not Fourier transformable.

9.1.1 Definition, Basic Properties

The Laplace transform is typically applied in situations where the functions of interest are one sided, that is, zero for $t < 0$. So to begin, consider a one-sided function

---

Fourier Transforms: Principles and Applications, First Edition. Eric W. Hansen.
© 2014 John Wiley & Sons, Inc. Published 2014 by John Wiley & Sons, Inc.

$f : [0, \infty) \to \mathbb{C}$ which is not absolutely integrable. If f can be "rescued" by a convergence factor—in particular, if $|f(t)e^{-\sigma t}|$ is integrable for some real σ—then a Fourier transform can be calculated:

$$\mathcal{F}\{f(t)e^{-\sigma t}\} = \int_0^\infty f(t)e^{-\sigma t}e^{-i2\pi \nu t}dt.$$

Earlier, this was a strategy for calculating transforms of functions like $U(t)$. After calculating the Fourier transform of $f(t)e^{-\sigma t}$, the limit as $\sigma \to 0$, if it existed, was defined to be the Fourier transform of f. But if the limit does not exist, or if we choose not to take the limit, we can define a Fourier-like transform that depends on the convergence parameter σ as well as the frequency $\omega = 2\pi \nu$.

Definition 9.1 (Laplace transform). Let $f : [0, \infty) \to \mathbb{C}$. If $f(t)e^{-\sigma t} \in L^1$ for some real σ, then the Laplace transform of f is defined:

$$F_L(s) = \int_0^\infty f(t)e^{-st}dt, \qquad (9.1)$$

where $s = \sigma + i\omega = \sigma + i2\pi \nu$. The operator notation for the Laplace transform is $F_L = \mathcal{L}\{f\}$.

The Laplace transform was advanced in the early 1900s as a method for solving differential equations in electrical circuit theory[1] and remains a valuable tool for solving linear, constant-coefficient ODEs. Its popularity is due to three features.

1. Like the Fourier transform, it converts linear, constant-coefficient ODEs (i.e., from LTI systems) to algebraic equations, which are easily solved.
2. The Laplace transform allows initial conditions and driving functions, and hence homogeneous and particular solutions, to be handled straightforwardly in one step rather than separately.
3. Solutions obtained by Laplace transform can include unstable (growing) behavior as well as stable (decaying) behavior. In many cases, sufficient knowledge about the dynamic behavior of an LTI system can be inferred from the Laplace transform without returning to a time-domain expression for the solution.

Example 9.1. Consider the following initial value problem:

$$y' + 2y = U(t), \quad y(0) = -1.$$

In the classical method, one first solves the homogeneous equation, $y' + 2y = 0$, obtaining $y_h(t) = Ae^{-2t}$. Then, by educated guesswork, one obtains the particular solution; in this case, the simple result is $y_p(t) = 1$. The complete solution is $y(t) = Ae^{-2t} + 1$. The initial condition (just prior to switching the input on) requires $y(0) = A + 1 = -1$, so $A = -2$. Thus, $y(t) = 1 - 2e^{-2t}$, $t > 0$. The difficulty of the method increases with higher order equations and more complicated driving functions.

---

[1] See the historical notes in Carslaw and Jaeger (1941, pp. viii–xvi).

The transform approach begins by taking the Laplace transform of both sides of the equation, using identities which will be proven later,

$$\mathcal{L}\{y'\} = sY_L(s) - y(0) \quad \text{and} \quad \mathcal{L}\{U(t)\} = \frac{1}{s}$$

The differential equation is transformed into an algebraic equation:

$$sY_L(s) + 1 + 2Y_L(s) = (s+2)Y_L(s) + 1 = \frac{1}{s}.$$

Solving for Y_L,

$$Y_L(s) = \underbrace{\frac{-1}{s+2}}_{\text{from initial condition}} + \underbrace{\frac{1}{s(s+2)}}_{\text{from driving function}}.$$

Inverting the transforms (these will also be proven later),

$$y(t) = \underbrace{-e^{-2t}}_{\text{homogeneous}} + \underbrace{1 - e^{-2t}}_{\text{particular}} = 1 - 2e^{-2t}, \quad t > 0. \quad \blacksquare$$

Extensive tables of Laplace transform pairs have been developed,[2] and with one of these and a few simple rules, many practical problems in linear system theory can be solved. Here we will endeavor to develop a deeper understanding of the Laplace transform beyond the simple rules, making connections with the Fourier transform, generalized functions, and complex analysis.

Existence of the Laplace Transform—Region of Convergence

The Laplace transform integral converges absolutely for those values of s whose real parts, σ, are sufficiently large to overcome any growing tendencies in f:

$$\left| \int_0^\infty f(t) e^{-st} \, dt \right| \le \int_0^\infty \left| f(t) e^{-(\sigma + i\omega)t} \right| dt = \int_0^\infty |f(t)| \, e^{-\sigma t} \, dt < \infty. \quad (9.2)$$

This condition may hold for all σ or some σ or may fail to hold for any σ. If the integral converges for some value $\sigma = \sigma_0$, then it may be shown to converge for all $\sigma > \sigma_0$ (i.e., more aggressive convergence factors). The smallest such σ_0 is called σ_a, the *abcissa of convergence*, and the half-plane $\mathcal{R}e\ s > \sigma_a$ is called the *region of convergence*.

Example 9.2. Let $f(t) = e^{at}U(t)$. When $a < 0$ this is the one-sided decaying exponential, whose Fourier transform is known. When $a = 0$ this is the step function, whose Fourier transform exists as a generalized function. When $a > 0$ this is a growing exponential, which is not Fourier transformable. But for any $\sigma > a$, the product $f(t)e^{-\sigma t} = e^{-(\sigma - a)}U(t)$ decays exponentially and is therefore absolutely integrable. The Laplace transform of f may be calculated:

$$F_L(s) = \int_0^\infty e^{at} e^{-st} dt = -\frac{e^{-(s-a)t}}{s-a} \bigg|_0^\infty = \frac{1}{s-a} - \lim_{R \to \infty} \frac{e^{-(s-a)R}}{s-a}.$$

[2] Cannon (1967, pp. 731–755) and Abramowitz and Stegun (1972).

Now, with $s = \sigma + i\omega$, $e^{-(s-a)R} = e^{-(\sigma-a)R} e^{-i\omega R}$. The second factor oscillates with unit amplitude. The first factor decays exponentially as $R \to \infty$ if $\sigma - a > 0$, giving a convergent integral. Thus, we have the result:

$$F_L(s) = \mathcal{L}\{e^{at}U(t)\} = \frac{1}{s-a}, \quad \mathcal{R}e\, s > a \tag{9.3}$$

The abcissa of convergence is a, and the region of convergence is the half-plane to the right of $\sigma = a$.

For a decaying exponential, for example, e^{-2t}, the Laplace transform exists for any $\sigma > -2$ and is $\frac{1}{s+2}$. The Laplace transform of the unit step function ($a = 0$) exists for any $\sigma > 0$ and is $\frac{1}{s}$. Even a growing exponential like e^{2t} can be transformed— here is where the Laplace transform has an edge over the Fourier transform. For $\sigma > 2$, the convergence factor $e^{-\sigma t}$ dies faster than e^{2t} grows. The Laplace transform of $e^{2t}U(t)$ exists and is $\frac{1}{s-2}$.

The Laplace transform of $e^{at}U(t)$ has a pole at $s = -a$. When $a < 0$ (decaying exponential), this pole is in the left half of the S-plane, and when $a > 0$ (growing exponential), the pole is in the right half-plane. The more rapid the decay, the farther to the left is the pole, and the more rapid the growth, the farther to the right is the pole. When $a = 0$ (step function), the pole is at $s = 0$, on the imaginary axis. ∎

Example 9.3 (Laplace transform of a cosine). Let $f(t) = \cos bt\, U(t)$. The Laplace transform integral is

$$\mathcal{L}\{\cos bt\} = \int_0^\infty \frac{1}{2}(e^{ibt} + e^{-ibt})e^{-st}\, dt$$

$$= \int_0^\infty \frac{1}{2} e^{-(s-ib)t}\, dt + \int_0^\infty \frac{1}{2} e^{-(s+ib)t}\, dt$$

$$= \frac{1}{2} \frac{1}{-(s-ib)} e^{-(s-ib)t}\Big|_0^\infty + \frac{1}{2} \frac{1}{-(s+ib)} e^{-(s+ib)t}\Big|_0^\infty.$$

The exponentials, $e^{-\sigma t}e^{-i(\omega \pm b)}$, will go to zero at the upper limit for any $\sigma > 0$. This establishes the region of convergence for this transform, $\mathcal{R}e\, s > 0$. At the lower limit, both exponentials are 1, and we obtain

$$\mathcal{L}\{\cos bt\} = \frac{1}{2}\frac{1}{s-ib} + \frac{1}{2}\frac{1}{s+ib} = \frac{s}{s^2+b^2}, \quad \mathcal{R}e\, s > 0. \tag{9.4}$$

The Laplace transform has two poles, at $s = \pm ib$, and a zero at $s = 0$. ∎

For some functions, the Laplace transform integral converges uniformly as well as absolutely. Recall that an infinite series converges uniformly if $\sup_x |f(x) - S_N(x)| \to 0$ as $N \to \infty$—the maximum error gets smaller as more terms are included in the series (Definition 4.2). Analogously, an integral $\int_0^\infty f(x, y)\,dy$ converges uniformly if

$$\sup_x \left| \int_0^\infty f(x, y)\,dy - \int_0^A f(x, y)\,dy \right| = \sup_x \left| \int_A^\infty f(x, y)\,dy \right| \to 0 \text{ as } A \to \infty.$$

The partial integral $\int_0^A$ fulfills the same role as the partial sum S_N of an infinite series.[3] If an integral converges absolutely and uniformly, then limits may be taken under the integral sign, $\lim_{x \to b} \int_0^\infty f(x, y)\,dy = \int_0^\infty \lim_{x \to b} f(x, y)\,dy$, and if f is differentiable in x, $\frac{d}{dx} \int_0^\infty f(x, y)\,dy = \int_0^\infty \frac{\partial}{\partial x} f(x, y)\,dy$.

The *functions of exponential order* are a large class of useful functions, for which it is relatively easy to prove absolute and uniform convergence as well as other properties of the Laplace transform.

Definition 9.2 (Exponential order). A function $f \colon [0, \infty) \to \mathbb{C}$ is of exponential order[4] if there are real constants α, $C > 0$, and $T > 0$, such that f is bounded above by an exponential:

$$|f(t)| < Ce^{\alpha t} \tag{9.5}$$

for $t > T$.

For a function of exponential order, then, $|e^{-\alpha t} f(t)| < C$ for for sufficiently large t.

Example 9.4 (Some functions of exponential order and their Laplace transforms).

1. $e^{2t} U(t)$ is of exponential order, because $e^{2t} e^{-\alpha t}$ will decay to zero for any $\alpha \geq 2$. We have already seen that $F_L(s) = \frac{1}{s-2}$, $\mathcal{R}e\, s \geq 2$.

2. Any finite sum of simple exponentials, $\sum_{k=1}^N A_k e^{b_k t}$, is of exponential order, with $\alpha \geq \max\{b_k\}$. Generalizing the previous result, we have

$$\mathcal{L}\left\{ \sum_{k=1}^N A_k e^{b_k t} \right\} = \sum_k \frac{A_k}{s - b_k}, \quad \mathcal{R}e\, s > \max\{b_k\}. \tag{9.6}$$

3. $\cos bt\, U(t)$ is of exponential order. $|\cos bt\, U(t)| \leq 1 < Ce^{\alpha t}$ for any $\alpha \geq 0$ and $C > 1$. And, as we saw earlier, the Laplace transform is

$$\mathcal{L}\{\cos bt\, U(t)\} = \frac{s}{s^2 + b^2}, \quad \mathcal{R}e\, s > 0.$$

---
[3] Folland (2002, pp. 336 ff).
[4] Exponential order is not the same as exponential type (e.g., Theorem 8.14). Exponential type is a property of entire functions in the complex plane. Exponential order is a property of functions on the real line.

4. $t^n U(t)$ is of exponential order, $\alpha > 0$, for all n. (No polynomial can grow faster than an exponential.) Integrating the Laplace integral once by parts,

$$\int_0^\infty t^n e^{-st}\, dt = \left.\frac{t^n e^{-st}}{-s}\right|_0^\infty - \int_0^\infty n t^{n-1} \frac{e^{-st}}{-s}\, dt$$

$$= \frac{n}{s} \int_0^\infty t^{n-1} e^{-st}\, dt, \quad \mathcal{R}e\, s > 0.$$

We can repeat this step $n-1$ more times, obtaining at last

$$\mathcal{L}\{t^n U(t)\} = \int_0^\infty t^n e^{-st}\, dt = \frac{n!}{s^n} \int_0^\infty e^{-st}\, dt$$

$$= \frac{n!}{s^{n+1}}, \quad \mathcal{R}e\, s > 0. \tag{9.7}$$

5. The functions $t^n e^{at} e^{ibt} U(t)$ which comprise the solutions of linear, constant-coefficient, ordinary differential equations are of exponential order ($\alpha > a$). Using the above result for t^n and a change of variable, we have

$$\mathcal{L}\{t^n e^{at} e^{ibt} U(t)\} = \int_0^\infty t^n e^{at} e^{ibt} e^{-st}\, dt = \int_0^\infty t^n e^{-(s-(a+ib))t}\, dt$$

$$= \frac{n!}{(s-(a+ib))^{n+1}}, \quad \mathcal{R}e\, s > a. \tag{9.8}$$

6. The square pulse $\operatorname{rect}\left(\frac{t-T/2}{T}\right)$ is of exponential order. It is identically zero for $t > T$, so the exponential can be growing or decaying ($\alpha > -\infty$). The Laplace transform is

$$\mathcal{L}\left\{\operatorname{rect}\left(\frac{t-T/2}{T}\right)\right\} = \int_0^T e^{-st}\, dt = \left.\frac{e^{-st}}{-s}\right|_0^T$$

$$= \frac{1 - e^{-sT}}{s}, \quad \mathcal{R}e\, s > -\infty. \tag{9.9}$$

There is a removable singularity at $s = 0$. This function is entire, and the region of convergence is the whole complex plane, $\mathcal{R}e\, s > -\infty$. This is true in general for time-limited functions (compare the Paley–Wiener theorem, 8.14).

7. $\operatorname{sinc}(t)\, U(t)$ is of exponential order. There is a removable singularity at $t = 0$. The function decays like $1/t$, slower than an exponential, but multiplication by any decaying exponential ($\alpha > 0$) speeds up the decay satisfactorily. The Laplace transform is

$$\mathcal{L}\{\operatorname{sinc}(t)\, U(t)\} = \frac{1}{\pi} \arctan \frac{\pi}{s}, \quad \mathcal{R}e\, s > 0. \tag{9.10}$$

The details of the calculation are deferred to the problems.

8. e^{+t^2} is not of exponential order. Because t^2 grows faster than t as $t \to \infty$, there is no way that $e^{-\alpha t}$ can overcome e^{t^2}, for any choice of α. (In contrast, e^{-t^2} decays rapidly and is of exponential order for all choices of α, positive and negative.) ∎

In addition to absolute and uniform convergence, it can be shown that the Laplace transform of a function of exponential order goes to zero as $\sigma \to \infty$ in the region of convergence (consider the above examples). These properties are also more generally true,[5] but most functions of practical interest are of exponential order and we shall confine attention to them.

The Laplace transform is bounded and analytic in its region of convergence. We have boundedness as a consequence of absolute convergence. We will not attempt a proof of analyticity, but simply observe that convergence implies an absence of singularities, which in turn implies analyticity.[6]

★ More about the Region of Convergence—The Two-Sided Transform

In the simple example $e^{at}U(t) \longmapsto \frac{1}{s-a}$ worked earlier, the Laplace transform integral was convergent only for $\mathcal{R}e\, s > a$. The complex function $\frac{1}{s-a}$ is analytic everywhere except at the pole, $s = a$, but it only represents a Laplace transform for $\mathcal{R}e\, s > a$, where the transform integral converges.

One might ask if $\frac{1}{s-a}$ is the Laplace transform of some other function in the opposite half-plane, $\mathcal{R}e\, s < a$. Convergence of the Laplace integral for $e^{at}U(t)$ hinges on the behavior of the complex exponential $e^{-(s-a)t}$. If $\mathcal{R}e\, s < a$, the exponential blows up as $t \to \infty$ and decays as $t \to -\infty$. Imagine then, the function $e^{at}U(-t)$ which is zero for $t > 0$, and consider how the Laplace transform for this function should be defined. If we simply change the limits of integration in Equation 9.1 to accommodate the change in domain, we have

$$F_L(s) = \int_{-\infty}^{0} e^{at} e^{-st} dt = -\frac{e^{-(s-a)t}}{s-a}\bigg|_{-\infty}^{0}$$

$$= -\frac{1}{s-a} + \lim_{T \to -\infty} \frac{e^{-(s-a)T}}{s-a}.$$

The limit will exist if $e^{-(s-a)T}$ decays for *negative* T; that is, $\mathcal{R}e\,(s-a)$ must be less than zero, or $\mathcal{R}e\, s < a$. So, we will have

$$\mathcal{L}\{e^{at}U(-t)\} = -\frac{1}{s-a}, \quad \mathcal{R}e\, s < a.$$

Comparing this with the transform of $e^{at}U(t)$, we see that the pole is at the same location, but the region of convergence is flipped (Figure 9.1).

The Laplace transforms of the functions $e^{at}U(t)$ and $e^{at}U(-t)$ are superficially the same (they are, after all, just two halves of the same function). They are distinguished by the region of convergence.

---

[5] See LePage (1980, pp. 289–298) for a discussion of functions not of exponential order; also Papoulis (1977, pp. 224–225). It can be shown that the Laplace transform $F(s)$ will go to zero as $s \to \infty$ along any ray in the region of convergence. And, similar to the Riemann–Lebesgue lemma for the Fourier transform, $F(\sigma + i\omega) \to 0$ as $|\omega| \to \infty$ for every σ in the region of convergence.

[6] A proof may be found in LePage (1980, pp. 297–299).

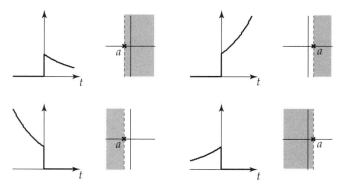

FIGURE 9.1 Regions of convergence for the Laplace transforms of two functions. *Top, left*: $f(t) = e^{at}U(t)$, $a < 0$. *Bottom, left*: $f(t) = e^{at}U(-t)$, $a < 0$. *Top, right*: $f(t) = e^{at}U(t)$, $a < 0$. *Bottom, right*: $f(t) = e^{at}U(-t)$, $a < 0$.

We will say that a function f which is zero for $t \geq 0$ is *left sided* and define the Laplace transform of a left-sided function:

$$F_L(s) = \int_{-\infty}^{0} f(t)e^{-st}dt. \tag{9.11}$$

The left-sided and right-sided transforms may be combined into a general two-sided transform:

$$F_L(s) = \int_{-\infty}^{\infty} f(t)e^{-st}dt. \tag{9.12}$$

It can be evaluated as the sum of two integrals:

$$\begin{aligned} F_L(s) &= \int_{-\infty}^{0} f(t)e^{-st}dt + \int_{0}^{\infty} f(t)e^{-st}dt \\ &= \int_{0}^{\infty} f(-t)e^{st}dt + \int_{0}^{\infty} f(t)e^{-st}dt. \end{aligned} \tag{9.13}$$

One usually uses the Laplace transform to solve initial value problems, or to analyze real linear systems, which must be causal. Noncausal systems occasionally turn up in certain applications of system theory and may be approximated in practice by digital filters.

Example 9.5. Consider the function $f(t) = e^{a|t|}$, the two-sided exponential, which decays in both directions when $a < 0$. The left-sided piece of the transform is

$$\int_{-\infty}^{0} f(t)e^{-st}dt = \int_{-\infty}^{0} e^{-at}e^{-st}dt = \frac{1}{s+a}, \quad \mathcal{R}e\, s < -a$$

and the right-sided piece is

$$\int_{0}^{\infty} f(t)e^{-st}dt = \int_{0}^{\infty} e^{at}e^{-st}dt = \frac{1}{s-a}, \quad \mathcal{R}e\, s > a.$$

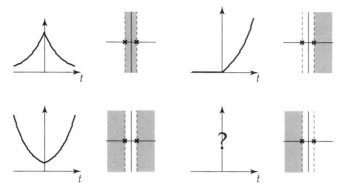

FIGURE 9.2 Regions of convergence for the Laplace transforms of four functions. All four transforms have poles at $s = \pm a$. Top, left: $f(t) = e^{a|t|}$, $a < 0$. Bottom, left: $f(t) = e^{a|t|}$, $a > 0$. Top, right: $f(t) = 2\sinh at\, U(t)$. Bottom, right: what is $f(t)$ with this region of convergence?

The two-sided transform will exist in their common region of convergence. If $a < 0$, so f is decaying in both directions, then the pole at $s = a$ is in the left half-plane and the pole at $s = -a$ is in the right half-plane. The regions of convergence, $\mathcal{R}e\, s < -a$ for the left-sided piece, and $\mathcal{R}e\, s > a$ for the right-sided piece, overlap in a strip between the two poles, $-a > \mathcal{R}e\, s > a$. So we say

$$\mathcal{L}\{e^{a|t|}\} = \frac{1}{s+a} - \frac{1}{s-a} = \frac{2a}{s^2 - a^2}, \quad -a > \mathcal{R}e\, s > a.$$

On the other hand, if $a > 0$, so f grows in both directions, the two regions of convergence are disjoint. The pole for the left-sided piece is in the left half-plane at $s = -a$, and the region of convergence extends to the left, $\mathcal{R}e\, s < -a$. The pole for the right-sided piece is in the right half-plane at $s = a$, and the region of convergence extends to the right, $\mathcal{R}e\, s > a$. There being no common region of convergence, the Laplace transform fails to exist anywhere (Figure 9.2).

Now let us consider a different function, $f(t) = 2\sinh at\, U(t)$, which is one sided. Its Laplace transform is

$$F_L(s) = \int_0^\infty 2\sinh at\, e^{-st}\, dt = \int_0^\infty e^{at} e^{-st}\, dt - \int_0^\infty e^{-at} e^{-st}\, dt$$

$$= \frac{1}{s-a} - \frac{1}{s+a} = \frac{2a}{s^2 - a^2}, \quad \mathcal{R}e\, s > |a|.$$

Except for the different region of convergence, this transform is the same as the one for $e^{a|t|}$ (Figure 9.2). There is yet another function whose transform is $\frac{2a}{s^2-a^2}$, but with region of convergence $\mathcal{R}e\, s < -a$. Can you figure out what it is? ∎

The region of convergence is a critical part of the specification of a two-sided transform. In the usual system analysis application of the Laplace transform, where all functions are right sided, the region of convergence is always the half-plane to the right of the rightmost singularity, so it is not necessary to specify it explicitly. In a two-sided world, however, if you want to determine the time-domain function $f(t)$ that corresponds to the function $F_L(s)$, you must specify the region of convergence.

You may recall from system theory that a right half-plane pole indicates that a system is unstable, that is, the time response grows without bound as t increases. A left half-plane pole corresponds to a stable system, one whose time response decays as t increases. The systems under question are, of course, causal; their time responses are right sided. If we suspend disbelief for a moment and consider an *anticausal* system, one with a left-sided time response, we find the opposite. A stable response decays as t decreases (toward $-\infty$), and an unstable response grows without bound as $t \to -\infty$. A pole in the right half-plane may belong to an unstable, causal system or a stable, anticausal system, and a pole in the left half-plane may belong to a stable, causal system or an unstable, anticausal system.

Relationship of the Laplace and Fourier Transforms

A stable, causal system has its poles in the left half-plane. The region of convergence is the half-plane to the right of the rightmost pole and includes the imaginary axis. A stable, anticausal system has its poles in the right half-plane. The region of convergence is the half-plane to the left of the leftmost pole, including the imaginary axis. The region of convergence of the Laplace transform for a stable, two-sided system response is a strip that also includes the imaginary axis.

If a system response is stable, whether it is right sided, left sided, or two sided, the regions of convergence of the Laplace transform include the imaginary axis. The Laplace transform may then be evaluated along the imaginary axis, $s = i\omega$:

$$F_L(i\omega) = \int_{-\infty}^{\infty} f(t) e^{-i\omega t} dt,$$

which we recognize as the Fourier transform of $f(t)$, that is,

$$F(\nu) = F_L(i2\pi\nu). \tag{9.14}$$

Example 9.6. In a previous chapter we calculated the Fourier transform pair, $e^{-t}U(t) \longmapsto \frac{1}{1+i2\pi\nu}$ (Equation 5.9). The Laplace transform of the same function is $\frac{1}{s+1}$ (Equation 9.3), and substituting $s = i2\pi\nu$, we have $\frac{1}{1+i2\pi\nu}$. ∎

This relationship between the Laplace and Fourier transforms is the reason, in circuit theory, that one evaluates the transfer function $H_L(s)$, which is a Laplace transform, along the $i\omega$ axis in order to obtain the frequency response, a Fourier transform (Figure 9.3). The figure shows the complex magnitude of the transfer function of a classic second-order Butterworth filter, $H_L(s) = \frac{1}{s^2+\sqrt{2}s+1}$. There are two poles, at $s = \frac{-1\pm i}{\sqrt{2}}$. One may visualize the magnitude as a rubber sheet that is "lifted" by the two poles. As an observer moves up the imaginary axis toward the poles, the magnitude of the frequency response $|H_L(i\omega)|$ increases. The response is maximized in the near vicinity of the poles then falls off as the observation point moves away from the poles to higher values of ω.

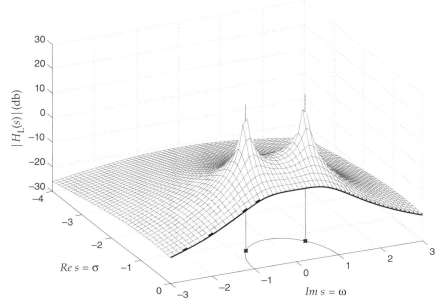

FIGURE 9.3 The Laplace transform, evaluated along the imaginary axis, gives the Fourier transform. Shown here is the complex magnitude of the transfer function $H_L(s) = \frac{1}{s^2 + \sqrt{2}s + 1}$, a second-order Butterworth filter with $\omega_c = 1$. The magnitude response, $|H_L(i\omega)|$, is the section along the imaginary axis. The pole locations are shown in the complex plane below the surface. The poles "lift" the magnitude surface in such a way that the profile along the imaginary axis is the characteristic Butterworth frequency response.

Here is another example that illustrates the relationship between the Laplace and Fourier transforms.

Example 9.7 (Fourier and Laplace transforms of the step function). The Fourier transform of the unit step function $f(t) = U(t)$ was calculated in Chapter 6. The Laplace transform is obtained from the transform of $e^{at}U(t)$ (Equation 9.3) by setting $a = 0$. Comparing the two results,

$$F(\nu) = \frac{1}{2}\delta(\nu) + \frac{1}{i2\pi\nu}$$
$$F_L(s) = \frac{1}{s}, \quad \mathcal{R}e\, s > 0. \tag{9.15}$$

The region of convergence for the Laplace transform does not include the imaginary axis. Attempting to obtain the Fourier transform by evaluating $F_L(i2\pi\nu)$ gives $\frac{1}{i2\pi\nu}$ and misses the delta function. However, observe what happens if we take the limit of F_L as $s = \sigma + i2\pi\nu$ approaches the imaginary axis through the region of convergence ($\sigma \to 0^+$). This is, in effect, the same thing we did in Chapter 6 when we calculated the Fourier transform of the signum and step functions by using exponential convergence factors. Writing

$$F_L(\sigma + i2\pi\nu) = \frac{1}{\sigma + i2\pi\nu} = \frac{\sigma - i2\pi\nu}{\sigma^2 + (2\pi\nu)^2}$$

and taking the limit,

$$\lim_{\sigma \to 0^+} F_L(\sigma + i2\pi\nu) = \lim_{\sigma \to 0^+} \frac{\sigma}{\sigma^2 + (2\pi\nu)^2} + i \lim_{\sigma \to 0^+} \frac{-2\pi\nu}{\sigma^2 + (2\pi\nu)^2}.$$

The limit of the imaginary part is easily seen to be $-\frac{1}{2\pi\nu}$. For the real part, note

$$\frac{\sigma}{\sigma^2 + (2\pi\nu)^2} = \frac{\sigma^{-1}}{1 + (2\pi\nu/\sigma)^2}.$$

The function $(1 + (\pi x)^2)^{-1}$ has unit area (compare Example 8.23), and the sequence of functions $\left(\frac{n}{1+(n\pi x)^2}\right)_{n>0}$ converges, in the generalized sense, to the delta function $\delta(x)$ as $n \to \infty$. Comparing σ^{-1} with n, the family of functions $\left(\frac{\sigma^{-1}}{1+(2\pi\nu/\sigma)^2}\right)_{\sigma \geq 0}$ is seen to converge to $\delta(2\nu) = \frac{1}{2}\delta(\nu)$ as $\sigma \to 0$. Thus, we may say

$$\lim_{\sigma \to 0^+} F_L(\sigma + i2\pi\nu) = \frac{1}{2}\delta(\nu) + \frac{1}{i2\pi\nu} = F(\nu). \qquad \blacksquare$$

9.1.2 Laplace Transforms of Generalized Functions

Generalized functions were introduced in Chapter 6 to accommodate important functions that failed to meet the normal existence criteria for the Fourier transform—in particular, functions of slow growth (step, signum, sine, cosine, polynomials) and singularities (the delta function and its derivatives). The Laplace transform takes care of one-sided functions of slow growth without any special handling, because all of these functions are of exponential order, for example, we have seen $U(t) \longmapsto 1/s$, $tU(t) \longmapsto 1/s^2$, $\cos bt\, U(t) \longmapsto s/(s^2 + b^2)$. These transforms converge absolutely and uniformly and are analytic in the right half-plane, and they all are observed to go to zero as $\mathcal{R}e\, s \to \infty$, reminiscent of the Riemann–Lebesgue lemma for Fourier transforms. In this section we will develop the Laplace transform for the delta function and its derivatives; a full treatment of the Laplace transform for generalized functions is beyond the scope of this text.[7]

Recall that a generalized function g is defined by its action on a set of testing functions φ, denoted by an integral symbol $\int g(t)\,\varphi(t)\,dt$. When g happens to be a regular function, this is an actual integral, but when it is a singular function, we interpret the symbol as the limit of a sequence of actual integrals, for example, for the delta function,

$$\int \delta(t)\,\varphi(t)\,dt = \lim_{n \to \infty} \int_{-\infty}^{\infty} n e^{-\pi n^2 t^2} \varphi(t)\,dt = \varphi(0).$$

[7] Zemanian (1987, Chapter 8); Doetsch (1974), Chapters 12–14.

The Laplace transform is made to fit this formalism by defining it as the action of a right-sided generalized function on a right-sided testing function $e^{-st}U(t)$.[8]

But we run into a problem when we attempt to calculate the Laplace transform of $\delta(t)$. In the expression

$$F_L(s) = \int_0^\infty \delta(t) e^{-st}\, dt,$$

it seems like we ought to be able to use the sifting property and obtain $F_L(s) = 1$; indeed, that is the result you may remember from an earlier course in system theory. But representing the delta function as a sequence of pulses like $ne^{-\pi n^2 t^2}$, which are symmetric about the origin, we find that the lower limit of the integral, $t = 0$, is in the center of the pulse. As $n \to \infty$, half of the pulse will always be excluded from the integral.

The solution is to redefine (slightly) the Laplace transform integral so that the lower limit is $t = 0^-$, that is,

$$F_L(s) = \int_{0^-}^\infty f(t) e^{-st}\, dt = \lim_{\epsilon \to 0} \int_{0-\epsilon}^\infty f(t) e^{-st}\, dt. \qquad (9.16)$$

The lower limit of the integral is approached from below, through negative values of t. Now, in the sequence of integrals

$$\int_{0-\epsilon}^\infty ne^{-\pi n^2 t^2} e^{-st}\, dt,$$

as n increases and the pulses become more and more localized around 0, the integral includes more and more of each pulse, and in the limit we have the expected result $\delta(t) \longmapsto 1$. If the lower limit is instead taken to be 0^+, the integral includes less and less of each pulse as $n \to \infty$, and in the limit the result will be zero.

The admission of singularities forces a distinction between 0^- and 0^+, that is, 0 approached from below and 0 approached from above. Integrating from 0^- to ∞ includes the actions of singularities at the origin, while integrating from 0^+ to ∞ excludes them. This distinction becomes important when we use the Laplace transform to solve differential equations with driving functions and initial conditions. For regular functions, the same results are obtained whether the lower limit is 0^- or 0^+. Therefore, we shall almost always use the 0^- definition of the Laplace transform (Equation 9.16), but when it is necessary to distinguish, we shall denote them $\mathcal{L}_-$ and $\mathcal{L}_+$:

$$\mathcal{L}\{\delta(t)\} = \mathcal{L}_-\{\delta(t)\} = 1. \qquad (9.17)$$

---

[8] The transform kernel $e^{-st}U(t)$ is not a "good function"; although it is infinitely continuously differentiable for $t > 0$, it is not rapidly decreasing for all s nor is it continuously differentiable at $t = 0$. This technical difficulty is resolved in various ways. See Zemanian (1987) and Doetsch (1974); also Beerends, et al. (2003, Chapter 13).

We can also calculate transforms for the derivatives of the delta function. We know (Equation 6.25) that

$$\int \delta^{(n)}(t)\, \varphi(t)\, dt = (-1)^n \varphi^{(n)}(0).$$

The transform kernel e^{-st} is infinitely continuously differentiable, so this result applies to the Laplace transform of $\delta^{(n)}$ for all n:

$$\mathcal{L}\{\delta^{(n)}(t)\} = \int_{0-}^{\infty} \delta^{(n)}(t)\, e^{-st}\, dt = (-1)^n \frac{d^n}{dt^n} e^{-st} \bigg|_0$$

$$= (-1)^n (-s)^n e^{-st} \bigg|_0 = s^n. \tag{9.18}$$

The Laplace transforms of the delta function and its derivatives are analytic functions. Indeed, they are entire functions, not having any singularities in the complex plane. Unlike the Laplace transforms of regular functions, however, they do not go to zero as $\mathcal{R}e\, s \to \infty$; rather, they are constant (when $n = 0$) or blow up (when $n > 0$).

9.1.3 Laplace Transform Theorems

There are several important theorems for the Laplace transform. Some are nearly identical to Fourier transform theorems, and they are stated here without proof. Others are unique to the Laplace transform. Except as noted, they apply to generalized functions.

Linearity

Theorem 9.1 (Linearity). If f has Laplace transform F_L, $\mathcal{R}e\, s > \sigma_f$, and g has Laplace transform G_L, $\mathcal{R}e\, s > \sigma_g$, then

$$\mathcal{L}\{f + g\} = F_L + G_L, \quad \mathcal{R}e\, s > \max\{\sigma_f, \sigma_g\}. \tag{9.19}$$

The region of convergence of $F_L + G_L$ is the intersection of their respective regions of convergence.

Example 9.8 (Laplace transform of $(1 - e^{at})U(t)$). The "saturating exponential" function $(1 - e^{at})U(t)$ is the difference of two functions with known Laplace transforms:

$$U(t) \longmapsto \frac{1}{s}, \quad \mathcal{R}e\, s > 0,$$

$$e^{at} U(t) \longmapsto \frac{1}{s-a}, \quad \mathcal{R}e\, s > \mathcal{R}e\, a.$$

Using linearity, then,

$$(1 - e^{at})U(t) \longmapsto \frac{1}{s} - \frac{1}{s-a} = \frac{-a}{s(s-a)}. \tag{9.20}$$

The intersection of the regions of convergence is $\mathcal{R}e\, s > \max(0, \mathcal{R}e\, a)$. ■

Example 9.9 (Laplace transform of sin $bt\, U(t)$). The sine function is the sum of two complex exponentials, $\sin bt = \frac{1}{2i} e^{ibt} - \frac{1}{2i} e^{-ibt}$. Using Equation 9.3 with $a = \pm ib$,

$$\mathcal{L}\{\sin bt\, U(t)\} = \mathcal{L}\left\{\frac{1}{2i}e^{ibt}\right\} + \mathcal{L}\left\{-\frac{1}{2i}e^{-ibt}\right\}$$
$$= \frac{1}{2i}\frac{1}{s-ib} - \frac{1}{2i}\frac{1}{s+ib} = \frac{b}{s^2+b^2}, \quad \mathcal{R}e\, s > 0. \tag{9.21}$$

Symmetries

The Laplace transform has symmetry properties similar to those of the Fourier transform.

Theorem 9.2 (Symmetry). If f has Laplace transform $F_L(s)$, $\mathcal{R}e\, s > \sigma_f$, then

$$f^*(t) \longleftrightarrow F_L^*(s^*), \tag{9.22}$$

and if f is real, so $f = f^*$,

$$F_L^*(s) = F_L(s^*). \tag{9.23}$$

When $s = i\omega$, the latter relationship becomes $F_L^*(i\omega) = F_L(-i\omega)$, that is, $F_L(i\omega)$ is Hermitian. The region of convergence of F_L^* is the same as the region of convergence of F_L.

Shift and Modulation

Theorem 9.3 (Shift theorem). If f has Laplace transform F_L, $\mathcal{R}e\, s > \sigma_a$, then the Laplace transform of $f(t-T)U(t-T)$, $T > 0$ (shifted to the right), is

$$\mathcal{L}\{f(t-T)U(t-T)\} = e^{-sT}F_L(s), \quad \mathcal{R}e\, s > \sigma_a. \tag{9.24}$$

Also, for any α, real or complex,

$$\mathcal{L}\{e^{-\alpha t}f(t)\} = F_L(s+\alpha), \quad \mathcal{R}e\, s > \sigma_a - \mathcal{R}e\, \alpha. \tag{9.25}$$

The step function, $U(t-T)$, is here to remind us that f is assumed right sided, because we are using the right-sided definition of the transform. The Laplace transform of a left-shifted function, $T < 0$, is not covered by this theorem. Because the lower limit of the integral cannot go below $t = 0$, a left shift causes f to be truncated, which can turn f into a different function with a very different transform. For example, a misapplication of the theorem would give $\delta(t+1) \longmapsto e^{+s}$, while in fact $\delta(t+1) \longmapsto 0$. (However, e^{+s} is the two-sided Laplace transform of $\delta(t+1)$.)

To illustrate Equation 9.25, consider the earlier example, with $f(t) = e^{-2t}U(t)$. The Laplace transform is $F_L(s) = \frac{1}{s+2}$, $\mathcal{R}e\, s > -2$. If we multiply $f(t)$ by e^{-t}, then according to the theorem, the Laplace transform of $e^{-t}f(t)$ is $F_L(s+1) = \frac{1}{(s+1)+2} = \frac{1}{s+3}$, $\mathcal{R}e\, s > -2 - 1 = -3$. We see directly that this is so, for $e^{-t}f(t) = e^{-3t}U(t)$, which has Laplace transform $\frac{1}{s+3}$, $\mathcal{R}e\, s > -3$. Multiplying the function by e^{-t} causes it to decay faster; the more rapid decay shows up in the S-plane by the pole of the transform moving to the left. If we multiply $f(t)$ by e^{-ibt}, the transform becomes $\frac{1}{s+(2+ib)}$. The pole shifts up or down in the S-plane, and the region of convergence is still $\mathcal{R}e\, s > -2$.

Dilation

Theorem 9.4 (Dilation). If f has Laplace transform F_L, $\mathcal{R}e\, s > \sigma_f$, then the Laplace transform of $f(at)$, $a > 0$, is

$$\mathcal{L}\{f(at)\} = \frac{1}{a} F_L\left(\frac{s}{a}\right), \quad \mathcal{R}e\, s > \frac{\sigma_f}{a}. \tag{9.26}$$

For example, changing $e^{-t}U(t)$ to $e^{-2t}U(2t) = e^{-2t}U(t)$ changes the transform from $\frac{1}{s+1}$ to $\frac{1}{2}\frac{1}{s/2+1} = \frac{1}{s+2}$. The decay rate of the exponential is doubled, and the pole is pushed farther into the left half-plane by a factor of two.

Differentiation

Theorem 9.5 (Derivative theorem). If f and its derivative f' are Laplace transformable and $f \longmapsto F_L$, then

$$f' \longmapsto sF_L \tag{9.27}$$

Proof: Following the usual procedure for the derivative of a generalized function, Equation 6.23,

$$\int_{0^-}^{\infty} f'(t)\, e^{-st}\, dt = -\int_{0^-}^{\infty} f(t)\, \frac{d}{dt} e^{-st}\, dt = \int_{0^-}^{\infty} f(t)\, s e^{-st}\, dt = sF_L(s).$$

∎

If f is a regular function, it is sufficient that f be piecewise smooth and of exponential order. In addition, steps and impulses in f are differentiable and Laplace transformable as generalized functions.

Example 9.10. Derivatives of one-sided functions are often impulsive. For example,

$$\frac{d}{dt} e^{-at} U(t) = -ae^{-at} U(t) + e^{-at} \underbrace{U'(t)}_{=\delta(t)} = \delta(t) - ae^{-at} U(t).$$

The delta function results from the jump at $t = 0$ and may be thought of as the derivative between $t = 0^-$ and $t = 0^+$. The other term, $-ae^{-at}U(t)$, is the conventional derivative for $t > 0^+$. Both terms contribute to the Laplace transform:

$$\mathcal{L}\left\{\frac{d}{dt} e^{-at} U(t)\right\} = \int_{0^-}^{\infty} \left[\delta(t) - ae^{-at} U(t)\right] e^{-st}\, dt = 1 - \frac{a}{s+a} = \frac{s}{s+a}.$$

And according to the derivative theorem, the transform is $s \cdot \mathcal{L}\{e^{-at}U(t)\} = \frac{s}{s+a}$. ∎

9.1 THE LAPLACE TRANSFORM

Example 9.11. The Laplace transforms of $\sin bt\, U(t)$ and $\cos bt\, U(t)$ are $\frac{b}{s^2+b^2}$ and $\frac{s}{s^2+b^2}$, respectively. Now, $\frac{d}{dt}\sin bt\, U(t) = b\cos bt\, U(t) + \sin bt\, \delta(t) = b\cos bt\, U(t)$, and

$$\mathcal{L}\left\{\frac{d}{dt}\sin bt\, U(t)\right\} = \frac{bs}{s^2+b^2},$$

in agreement with the derivative theorem. For the derivative of cosine, the theorem says

$$\mathcal{L}\left\{\frac{d}{dt}\cos bt\, U(t)\right\} = \frac{s^2}{s^2+b^2},$$

which does not look like the Laplace transform of $\sin bt\, U(t)$, although $\frac{d}{dt}\cos bt = -b\sin bt$. The difference is due to the right-sidedness of these functions. There is a jump in the cosine at the origin, which produces an impulse in the derivative:

$$\frac{d}{dt}\cos bt\, U(t) = -b\sin bt\, U(t) + \cos bt\, \delta(t) = \delta(t) - b\sin bt\, U(t)$$

and $\mathcal{L}\left\{\frac{d}{dt}\cos bt\, U(t)\right\} = \mathcal{L}\left\{\delta(t) - b\sin bt\right\} = 1 - \frac{b^2}{s^2+b^2} = \frac{s^2}{s^2+b^2}.$ ∎

When there is a jump in f at the origin, the derivative has two terms. One is an impulse at the origin caused by differentiating the jump, and the other is the derivative of f away from the jump ($t > 0^+$). The size of the jump is $f(0^+)$, so we may write this relationship

$$f'(t)\Big|_{t>0^-} = f(0^+)\,\delta(t) + f'(t)\Big|_{t>0^+}.$$

Then, Laplace transforming both sides, we have

$$\underbrace{\int_{0^-}^{\infty} f'(t)\,e^{-st}\,dt}_{=sF_L(s)} = f(0^+) + \int_{0^+}^{\infty} f'(t)\,e^{-st}\,dt,$$

which leads to another common version of the derivative theorem,

$$\mathcal{L}_+\{f'(t)\} = s\mathcal{L}_-\{f\} - f(0^+). \tag{9.28}$$

Example 9.12 (Step and impulse). The impulse is the derivative of the step function, $\delta(t) = U'(t)$. The 0^- and 0^+ transforms of the impulse are

$$\mathcal{L}_-\{\delta(t)\} = 1 \quad \text{and} \quad \mathcal{L}_+\{\delta(t)\} = 0,$$

depending on whether the impulse is included in the Laplace integral or not. The derivative theorem gives, for the transform of $U'(t)$,

$$\mathcal{L}_-\{U'(t)\} = s\mathcal{L}_-\{U(t)\} = s\cdot\frac{1}{s} = 1,$$

$$\mathcal{L}_+\{U'(t)\} = s\mathcal{L}_-\{U(t)\} - U(0^+) = s\cdot\frac{1}{s} - 1 = 0,$$

which agrees with the results for the impulse. ∎

The derivative theorem can be iterated to give the Laplace transform of higher-order derivatives, for example, for a twice-differentiable function,

$$\mathcal{L}_-\{f''\} = s^2 \mathcal{L}_-\{f\}$$
$$\mathcal{L}_+\{f''\} = s\mathcal{L}_-\{f'\} - f'(0^+)$$
$$= s[s\mathcal{L}_-\{f\} - f(0^+)] - f'(0^+) = s^2 \mathcal{L}_+\{f\} - sf(0^+) - f'(0^+),$$

and for an n-times differentiable function,

$$\mathcal{L}_-\{f^{(n)}\} = s^n \mathcal{L}_-\{f\}$$

$$\mathcal{L}_+\{f^{(n)}\} = s^n \mathcal{L}_+\{f\} - \sum_{k=1}^{n} s^{n-k} f^{(k-1)}(0^+). \tag{9.29}$$

Also, as we had with the Fourier transform, there is a relationship for derivatives of the Laplace transform. For a function f of exponential order (or any other f such that the Laplace integral converges uniformly), we may differentiate under the integral sign:

$$\frac{d}{ds} \int_0^\infty f(t) e^{-st} dt = \int_0^\infty f(t) \frac{d}{ds} e^{-st} dt = \int_0^\infty -tf(t) e^{-st} dt.$$

Repeating n times, we have the following result:

Theorem 9.6 (S-domain derivative theorem). If the Laplace transform integral of f converges uniformly, then

$$\mathcal{L}\{t^n f(t)\} = (-1)^n \frac{d^n F_L}{ds^n}. \tag{9.30}$$

Example 9.13 (Laplace transform of a ramp). For the ramp function $tU(t)$, the theorem gives

$$tU(t) \longmapsto -\frac{d}{ds} \mathcal{L}\{U(t)\} = -\frac{d}{ds}\frac{1}{s} = s^2. \tag{9.31}$$
∎

Integration

Theorem 9.7 (Integral theorem). Let $g(t) = \int_0^t f(\tau) d\tau$. Then

$$G_L(s) = \mathcal{L}\left\{\int_0^t f(\tau) d\tau\right\} = \frac{F_L(s)}{s} \tag{9.32}$$

Proof: The integral $g(t)$ is continuous and differentiable, so $g' = f$ and by the derivative theorem,

$$g' \longmapsto sG_L(s) = F_L.$$
∎

The integral $g(t)$ is also a right-sided function, so the region of convergence of its Laplace transform is the half-plane to the right of all singularities of $F_L(s)/s$.

Convolution and Product of Functions

Theorem 9.8 (Convolution theorem). Let f and g have Laplace transforms F_L and G_L, with abcissae of convergence σ_f and σ_g, respectively. The Laplace transform of the convolution of f and g,

$$f * g = \int_0^t f(\tau) g(t - \tau) \, d\tau,$$

is

$$\mathcal{L}\{f * g\} = F_L G_L, \quad \mathcal{R}e \, s > \max(\sigma_f, \sigma_g). \tag{9.33}$$

When we studied the Fourier transform we saw that the convolution theorem is the link between time- and frequency-domain descriptions of linear, time-invariant systems. The convolution theorem with the Laplace transform fulfills the same role for causal LTI systems.

Example 9.14 (A causal LTI system). Revisiting an earlier example (Example 9.1), but with zero initial conditions,

$$y' + 2y = f(t), \quad y(0) = 0,$$

after Laplace transforming both sides and collecting terms,

$$Y_L(s) = \frac{1}{s+2} F_L(s) = H_L(s) F_L(s),$$

where $H_L(s) = \frac{1}{s+2}$ is the system's transfer function. Applying the convolution theorem,

$$y(t) = h(t) * f(t).$$

When the input f is an impulse, $y = h$ is the impulse response. Using the known pair $e^{at} U(t) \longleftrightarrow \frac{1}{s-a}$, we have $h(t) = e^{-2t} U(t)$.

With a step function input, we can calculate the output directly in the time domain (recall Example 5.18):

$$y(t) = h * U(t) = e^{-2t} U(t) * U(t) = \frac{1}{2}(1 - e^{-2t}) U(t).$$

We can also calculate the output using the convolution theorem. The Laplace transforms of $e^{-2t} U(t)$ and $U(t)$ are $1/(s+2)$ and $1/s$, respectively. Then the transform of the output is

$$Y_L(s) = \frac{1}{s+2} \frac{1}{s} = \frac{-1/2}{s+2} + \frac{1/2}{s},$$

and the inverse transform is, using linearity,

$$y(t) = -\frac{1}{2} e^{-2t} U(t) + \frac{1}{2} U(t) = \frac{1}{2}(1 - e^{-2t}) U(t).$$

When initial conditions are zero, the output of a causal LTI system is the convolution of the impulse response and the driving function. When the initial conditions are not zero, the $h * f$ term is joined by additional terms representing the system's response to the initial conditions. ∎

Theorem 9.9 (Product theorem). Let f and g have Laplace transforms F_L and G_L, with abcissae of convergence σ_f and σ_g, respectively. The Laplace transform of the product fg is

$$\mathcal{L}\{fg\} = \frac{1}{2\pi i} \int_{c-i\infty}^{c+i\infty} F_L(z) G_L(s-z)\, dz = \frac{1}{2\pi i} \int_{c-i\infty}^{c+i\infty} F_L(s-z) G_L(z)\, dz,$$
$$\mathcal{R}e\, s > \sigma_f + \sigma_g, \tag{9.34}$$

when the integral exists. The path of integration is a vertical line in the complex Z-plane, in the common region of convergence of $F_L(z)$ and $G_L(s-z)$, or of $F_L(s-z)$ and $G_L(z)$.

Proof: The derivation of this result requires the inverse Laplace transform integral, to be discussed later. The inverse transform of $F_L(s)$ is

$$f(t) = \frac{1}{2\pi i} \int_{c-i\infty}^{c+i\infty} F_L(z)\, e^{+zt}\, dz,$$

where c is a real number in the region of convergence of F_L (and, as we shall see, also in the region of convergence of $G_L(s-z)$). Substitute this integral for f in the Laplace integral for $\mathcal{L}\{fg\}$:

$$\mathcal{L}\{fg\} = \int_0^\infty \left[\frac{1}{2\pi i} \int_{c-i\infty}^{c+i\infty} F_L(z)\, e^{+zt}\, dz\right] g(t)\, e^{-st}\, dt.$$

Assuming that the order of the integrations may be reversed,[9] we have the double integral:

$$\mathcal{L}\{fg\} = \frac{1}{2\pi i} \int_{c-i\infty}^{c+i\infty} F_L(z) \left[\int_0^\infty g(t)\, e^{-(s-z)t}\, dt\right] dz.$$

The inner integral is $G_L(s-z)$. The path of integration must be located in the common region of convergence of $F_L(z)$ and $G_L(s-z)$ in order to guarantee convergence. The second form in Equation 9.34 is obtained by writing $g(t)$ as an inverse transform and substituting it into the integral for $\mathcal{L}\{fg\}$. ∎

Example 9.15 (Laplace transform of $t \sin bt\, U(t)$.). We begin with $f(t) = tU(t) \longmapsto \frac{1}{s^2}$, $\mathcal{R}e\, s > 0$, and $g(t) = \sin bt\, U(t) \longmapsto \frac{b}{s^2+b^2}$, $\mathcal{R}e\, s > 0$. Form the integral,

$$\frac{1}{2\pi i} \int_{c-i\infty}^{c+i\infty} F_L(s-z) G_L(z)\, dz = \frac{1}{2\pi i} \int_{c-i\infty}^{c+i\infty} \frac{1}{(s-z)^2} \frac{b}{z^2+b^2}\, dz.$$

[9] We may anticipate that this is allowed based on experience with the Fourier transform. See LePage (1980), pp. 343–347) for a discussion.

The path of integration is located in the intersection of the half-planes $\operatorname{Re}(s-z) > 0$ and $\operatorname{Re} z > 0$, that is, the strip $\operatorname{Re} s > \operatorname{Re} z > 0$. There are poles at $z = s$ and at $z = \pm ib$. The integrand falls off as $|z|^{-4}$ in both the right and left half-planes, so the contour may be closed either to the left or to the right. Closing to the right encompasses the double pole at $z = s$. The integral is

$$\frac{1}{2\pi i} \int_{c-i\infty}^{c+i\infty} \frac{1}{(s-z)^2} \frac{b}{z^2+b^2}\, dz = -\operatorname{Res}\left[\frac{1}{(s-z)^2} \frac{b}{z^2+b^2}, z=s\right]$$

$$= -\frac{d}{dz}\frac{b}{z^2+b^2}\bigg|_{z=s} = \frac{2bs}{(s^2+b^2)^2}, \quad \operatorname{Re} s > 0.$$

Closing the contour to the left gives the same result; the calculation is left as an exercise for the reader. ∎

Parseval's Formula

Theorem 9.10 (Parseval). Let f and g have Laplace transforms F_L and G_L, respectively. Then,

$$\int_0^\infty f(t) g^*(t)\, dt = \frac{1}{2\pi i} \int_{c-i\infty}^{c+i\infty} F_L(s) G_L^*(-s^*)\, ds. \tag{9.35}$$

The path of integration is a vertical line in the complex S-plane, in the common region of convergence of $F_L(s)$ and $G_L^*(-s)$. Proof follows from the product theorem.

Initial and Final Values

The asymptotic behaviors of the Laplace transform as $s \to 0$ and as $s \to \infty$ are closely related to the behavior of the time function as $t \to \infty$ and $t \to 0$, respectively.

Theorem 9.11 (Initial value theorem). If f is differentiable and has Laplace transform F_L, then

$$\lim_{\sigma \to \infty} \sigma F_L(\sigma) = f(0^+), \tag{9.36}$$

when the limit exists. The limit is taken along the real axis, $s = \sigma + i0 = \sigma$.

Proof: $sF_L(s)$ and $f(0^+)$ are connected by the derivative theorem (9.28), $\mathcal{L}_+\{f'(t)\} = sF_L - f(0^+)$. In its region of convergence, the Laplace transform goes to zero as $\sigma \to \infty$. Thus,

$$\lim_{\sigma \to \infty} \mathcal{L}_+\{f'(t)\} = 0$$

$$= \lim_{\sigma \to \infty} \sigma F_L(\sigma) - f(0^+). \quad ∎$$

We may identify the following cases of interest:

- If the Laplace transform falls off asymptotically like $1/s$, then the limit will be finite. This case results if F_L is rational, $F_L(s) = \frac{\prod_m (s-b_m)}{\prod_n (s-a_n)}$, and the number of poles is one more than the number of zeros, for example, $\frac{s}{s^2+b^2} \longmapsto \cos bt\, U(t)$, which $\to 1$ as $t \to 0^+$.

- If the Laplace transform falls off faster than $1/s$, for example, $1/s^2$, then the limit will be zero. This is the case if the number of poles is more than one greater than the number of zeros, for example, $\frac{b}{s^2+b^2} \longmapsto \sin bt\, U(t)$, which $\to 0$ as $t \to 0^+$. When the number of poles exceeds the number of zeros, a rational function is called *strictly proper*.

- If the Laplace transform is constant or grows like a polynomial, then the limit will be infinite. When there are terms like this in the transform, the time function contains impulses, for example, $\frac{s+1}{s+2} = 1 - \frac{1}{s+2} \longmapsto \delta(t) - e^{-2t}$.

What is interesting in the last case is that, having performed one step of long division, we are left with a strictly proper rational function for which the initial value theorem gives a finite answer, namely $\lim_{\sigma \to \infty} -\frac{\sigma}{\sigma+2} = -2$, which is the limit as $t \to 0^+$ of the non-impulsive part of the time function, $-e^{-2t}$. Because the impulsive part is supported at the point $t = 0$, we can speak of a limit as $t \to 0^+$ that stops short of the impulses. This suggests a more general version of the initial value theorem.[10]

Theorem 9.12 (Generalized initial value theorem). Let f be continuous for $t > 0$ but possibly with a jump and impulsive functions at the origin. Its Laplace transform $F_L(s)$ can have the form:

$$F_L(s) = c_0 + c_1 s + \cdots + c_p s^p + F_p(s),$$

where F_p falls off as $1/|s|^k$, $k \geq 1$. (If F_p is rational, the number of poles is at least one more than the number of zeros.) Then

$$\lim_{\sigma \to \infty} \sigma F_p(\sigma) = f(0^+),$$

where the limit is taken along the real axis, $s = \sigma + i0$.

Theorem 9.13 (Final value theorem). If f is differentiable and has Laplace transform F_L, $\mathcal{R}e\, s > 0$,

$$\lim_{t \to \infty} f(t) = \lim_{\sigma \to 0} \sigma F_L(\sigma), \qquad (9.37)$$

when the limits exist.

Proof: The initial and final values of f are connected through the derivative, $\int_0^\infty f'(t)\, dt = f(\infty) - f(0)$. And, by the derivative theorem, $\int_0^\infty f'(t)\, e^{-st}$

---

[10]Kailath (1980, pp. 12–13) and Zemanian (1987, pp. 243–248).

$dt = sF_L(s) - f(0^+)$. Assuming f is of exponential order, the Laplace transform integral converges uniformly and we may take limits under the integral sign:

$$\lim_{\sigma \to 0} \int_0^\infty f'(t) e^{-\sigma t} \, dt = \int_0^\infty \lim_{\sigma \to 0} f'(t) e^{-\sigma t} \, dt = \int_0^\infty f'(t) \, dt = f(\infty) - f(0^+)$$
$$= \lim_{\sigma \to 0} \sigma F_L(\sigma) - f(0^+)$$
$$\Rightarrow f(\infty) = \lim_{\sigma \to 0} \sigma F_L(\sigma).$$

∎

As with the initial value theorem, we may identify a few cases of interest. We will restrict attention to the common case of a rational transform, $F_L(s) = \frac{\Pi_m(s-b_m)}{\Pi_n(s-a_n)}$. The corresponding time function f is a sum of terms of the form $e^{-a_n t} U(t)$.

- The requirement that the transform converge for $\mathcal{R}e\, s > 0$ excludes poles in the right half-plane, which would cause the final value to be unbounded. But there could be poles on the imaginary axis, for example, $\cos bt \longmapsto \frac{s}{s^2+b^2}$, for which the final value theorem would erroneously give $\lim_{t \to \infty} \cos bt = \lim_{\sigma \to 0} \frac{\sigma^2}{\sigma^2+b^2} = 0$. So one must be sure that a final value exists before using the theorem.

- If none of the poles is at the origin (or on the imaginary axis), then $\sigma F_L(\sigma) \to 0$ as $\sigma \to 0$. In the time domain, each of the terms $e^{-a_n t} U(t)$ goes to 0 as $t \to \infty$.

- If one of the poles is at the origin, the factor of s in $sF_L(s)$ cancels the pole, and the limit as $\sigma \to 0$ will be nonzero. In fact, the final value is the residue at that pole. In the time domain, one of the terms will be of the form $e^{0t} U(t) = U(t)$; the time function f is a step plus a sum of decaying exponentials, and the final value is due to the step.

- If there are two or more poles at the origin, the time function contains at least a ramp function, which is unbounded as $t \to \infty$. In the final value theorem, only one of the poles is cancelled in $sF_L(s)$, and the remaining poles blow up as $\sigma \to 0$.

Example 9.16. Consider the Laplace transform pairs $e^{-2t} U(t) \longmapsto \frac{1}{s+2}$ and $(1 - e^{-2t}) U(t) \longmapsto \frac{2}{s(s+2)}$. Both transforms meet the region of convergence requirement. By inspection of the time functions, we see that the initial values are 1 and 0, respectively, and the corresponding final values are 0 and 1. Applying the initial value theorem,

$$\lim_{\sigma \to \infty} \frac{\sigma}{\sigma + 2} = 1 \quad \text{and} \quad \lim_{\sigma \to \infty} \frac{2\sigma}{\sigma(\sigma + 2)} = 0.$$

Applying the final value theorem,

$$\lim_{\sigma \to 0} \frac{\sigma}{\sigma + 2} = 0 \quad \text{and} \quad \lim_{\sigma \to 0} \frac{\sigma}{\sigma(\sigma + 2)} = 1.$$

For an example of a case that fails, consider the growing exponential $e^t U(t) \longmapsto \frac{1}{s-1}$, for which the region of convergence is $\mathcal{R}e\, s > 1$. The initial value is 1, and the final value is unbounded. Applying the theorems,

$$\lim_{\sigma \to \infty} \frac{\sigma}{\sigma - 1} = 1 = \lim_{t \to 0^+} e^t U(t),$$

$$\lim_{\sigma \to 0} \frac{\sigma}{\sigma - 1} = 0 \neq \lim_{t \to \infty} e^t U(t).$$

The initial value is correct, but the final value is not. The final value theorem requires that the Laplace transform converge for $\mathcal{Re}\, s > 0$ so that the limit as $\sigma \to 0$ may be approached within the region of convergence. In this case $\mathcal{Re}\, s > 1$, violating the condition. ∎

When there are two poles at the origin, there is a ramp function in the time domain which, though it has no finite final value, does have a finite derivative (e.g., constant velocity ⟶ linearly increasing position). This "final velocity" can be extracted from the Laplace transform by combining the derivative theorem with the final value theorem. The derivative theorem says $f' \to sF_L(s)$, and then the final value theorem says

$$f'(\infty) = \lim_{\sigma \to 0} \sigma^2 F_L(s) \tag{9.38}$$

when the limit exists.

Example 9.17. For $X_L(s) = \frac{1}{s^2(s+1)} = -\frac{1}{s} + \frac{1}{s^2} + \frac{1}{s+1}$, the time function is $x(t) = \left(-1 + t + e^{-2t}\right) U(t)$. The position clearly grows without bound, as the final value theorem predicts, $\lim_{\sigma \to 0} \frac{\sigma}{\sigma^2(\sigma+1)} \to \infty$. The velocity is $x'(t) = 1 - 2e^{-2t}$, which approaches 1 asymptotically as $t \to \infty$. Using Equation 9.38, $\lim_{\sigma \to 0} \frac{\sigma^2}{\sigma^2(\sigma+1)} = \lim_{\sigma \to 0} \frac{1}{\sigma+1} = 1$. ∎

More about Differential Equations

The great appeal of the Laplace transform for solving initial value problems is that the homogeneous and particular solutions are obtained together by one method, rather than separately. This was illustrated in an earlier example, and we give another example below.

Example 9.18 (System with initial condition and driving function). In this example we use the Laplace transform to solve a first-order differential equation with a nonzero initial condition and a sinusoidal driving function and compare the result with the sinusoidal steady state solution obtained using the Fourier transform. The differential equation is

$$y' + ay = B \cos \omega_0 t, \quad y(0^+) = y_0.$$

9.1 THE LAPLACE TRANSFORM

Apply the Laplace transform to both sides, and solve for $Y_L(s)$:

$$sY_L(s) - y_0 + aY_L(s) = \frac{Bs}{s^2 + \omega_0^2},$$

$$Y_L(s) = \frac{y_0}{s+a} + \frac{Bs}{(s+a)(s^2 + \omega_0^2)}.$$

There are poles at $s = -a, \pm i\omega_0$; the right-hand side is expanded in partial fractions:

$$Y_L(s) = \frac{y_0}{s+a} - \frac{aB/(a^2 + \omega_0^2)}{s+a} + \frac{B}{2}\frac{(a+i\omega_0)/(a^2 + \omega_0^2)}{s+i\omega_0} + \frac{B}{2}\frac{(a-i\omega_0)/(a^2 + \omega_0^2)}{s-i\omega_0}.$$

Then, using the basic transform pair $e^{-\alpha t}U(t) \longleftrightarrow \frac{1}{s+\alpha}$ with $\alpha = a, i\omega_0,$ and $-i\omega_0$ in turn, we obtain

$$y(t) = y_0 e^{-at} - \frac{aB}{a^2 + \omega_0^2}e^{-at} + \frac{B}{2}\frac{a+i\omega_0}{a^2 + \omega_0^2}e^{-i\omega_0 t} + \frac{B}{2}\frac{a-i\omega_0}{a^2 + \omega_0^2}e^{+i\omega_0 t}$$

$$= y_0 e^{-at} - \frac{aB}{a^2 + \omega_0^2}e^{-at} + \frac{aB}{a^2 + \omega_0^2}\cos\omega_0 t + \frac{\omega_0 B}{a^2 + \omega_0^2}\sin\omega_0 t, \quad t > 0.$$

Or, using Equation 1.21,

$$y(t) = y_0 e^{-at} - \frac{aB}{a^2 + \omega_0^2}e^{-at} + \frac{B}{\sqrt{a^2 + \omega_0^2}}\cos(\omega_0 t + \varphi), \quad t > 0, \qquad (9.39)$$

where $\tan\varphi = -\omega_0/a$.

From left to right, the three terms are the response of the system to the initial condition alone, the transient response of the system to the sudden application of the input at $t = 0$, and the steady-state response to the input after the transients have died away. You can easily check that at $t = 0$, the negative transient and the steady-state response cancel so that only the initial condition y_0 is present (Figure 9.4).

In the extreme as $a \to 0$, the differential equation becomes $y' = B\cos\omega_0 t$, for which the solution is obtained by direct integration, $y(t) = y_0 + \frac{B}{\omega_0}\sin\omega_0 t$. As we take $a \to 0$ in Equation 9.39, the effect of the initial condition persists, $y_0 e^{-at} \to y_0$, but the transient excited by the driving function becomes weaker and disappears in the limit. In the steady-state response, the amplitude $B/\sqrt{a^2 + \omega_0^2} \to B/\omega_0$ and the phase $\varphi = \arctan(-\omega_0/a) \to \arctan(-\infty) = -\pi/2$, giving $B\cos(\omega_0 t - \pi/2) = \frac{B}{\omega_0}\sin\omega_0 t$. The amplitudes of the sinusoid and the negative transient vary together with a so that $y(0^+)$ is always y_0.

The steady-state solution alone may be solved for using the Fourier transform. Assume that $t \gg 0$ so that transients from the initial conditions and the one-sided

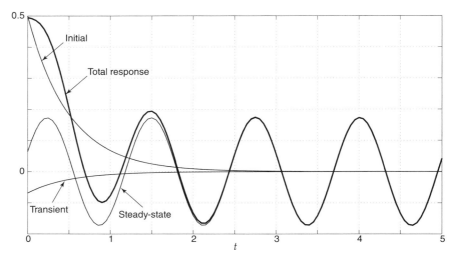

FIGURE 9.4 Cosine-driven response of a first-order system with initial condition, Equation 9.39. Parameters are $y_0 = 0.5$, $B = 1$, $a = 2$, $\omega_0 = 5$. The transient and steady-state responses cancel at $t = 0$. The transient response and the initial condition response decay with increasing t, leaving the steady-state response.

driving function have died out. Apply the Fourier transform to both sides, then solve for $Y(\omega)$:

$$i\omega Y(\omega) + aY(\omega) = \pi B \delta(\omega - \omega_0) + \pi B \delta(\omega + \omega_0)$$

$$Y(\omega) = \frac{\pi B}{a + i\omega}[\delta(\omega - \omega_0) + \delta(\omega + \omega_0)]$$

$$= \frac{\pi B (a - i\omega_0)}{a^2 + \omega_0^2}\delta(\omega - \omega_0) + \frac{\pi B (a + i\omega_0)}{a^2 + \omega_0^2}\delta(\omega + \omega_0).$$

The inverse Fourier transform is

$$y(t) = \frac{B}{2}\frac{a - i\omega_0}{a^2 + \omega_0^2}e^{+i\omega_0 t} + \frac{B}{2}\frac{a + i\omega_0}{a^2 + \omega_0^2}e^{-i\omega_0 t},$$

which matches the steady-state part of the Laplace transform solution. Note that the coefficients of the delta functions, $\frac{1}{a \pm i\omega_0}$, are identical to the transfer function $\frac{1}{s+a}$, evaluated at $s = \pm i\omega_0$. Again, this is because the Fourier transform frequency response is the Laplace transform transfer function evaluated on the imaginary axis. ∎

Sometimes the driving function has an impulse. Consider, for example, the series resistor–inductor circuit driven by a voltage source. The differential equation of the voltage y across the inductor is

$$y' + \frac{R}{L}y = v'.$$

9.1 THE LAPLACE TRANSFORM

When a step change in voltage is applied to the circuit, the driving function becomes an impulse whose area equals the height of the step. According to the equation, this causes an impulse in y', which means there is a jump in y as a result of the jump in v—the familiar "back emf" that opposes a sudden change in the current through the inductor. Laplace transforming both sides of the equation,

$$sY_L(s) + \frac{R}{L}Y_L = \mathcal{L}_-\{\delta(t)\} = 1.$$

Now, this is the identical result one obtains from the equation

$$y' + \frac{R}{L}y = 0, \quad y(0^+) = 1,$$

that is,

$$sY_L(s) - 1 + \frac{R}{L}Y_L = 0.$$

The impulsive driving function causes a jump in y from $t = 0^-$ to $t = 0^+$ and is equivalent to specifying an initial condition $y(0^+)$. When there are impulses present in the driving function, one must be careful to specify whether initial conditions are at $t = 0^-$ (before the impulse) or at $t = 0^+$ (after the impulse). An initial condition specified at $t = 0^-$ will be changed to a new value at $t = 0^+$ by the impulse. On the other hand, an initial condition specified at $t = 0^+$ already includes the effects of any impulses at $t = 0$. It would not make physical sense in this example to have a driving function with an impulse at $t = 0$ and to separately specify an initial condition at $t = 0^+$.

Example 9.19 (Response of an RL circuit). We will follow through with the solution of the first-order differential equation for the series RL circuit. Let the input be $v(t) = V_0 \cos \omega_0 t \, U(t)$, then $v'(t) = V_0 \delta(t) - V_0 \sin \omega_0 t \, U(t)$. We expect to see a jump discontinuity in the response due to the impulse. Laplace transforming both sides,

$$sY_L(s) + \frac{R}{L}Y_L = (s + R/L)Y_L(s) = sV_0 \frac{s}{s^2 + \omega_0^2}.$$

Solve for Y_L and invert the transform:

$$Y_L(s) = \frac{V_0 s^2}{(s + R/L)(s^2 + \omega_0^2)}$$

$$= \underbrace{\frac{V_0 (R/L)^2}{(R/L)^2 + \omega_0^2} \cdot \frac{1}{s + R/L}}_{\longmapsto \exp[-(R/L)t]} + \underbrace{\frac{V_0 \omega_0^2}{(R/L)^2 + \omega_0^2} \cdot \frac{s - R/L}{s^2 + \omega_0^2}}_{\longmapsto \sin \omega_0 t, \cos \omega_0 t}$$

$$\longmapsto \frac{V_0 (R/L)^2}{(R/L)^2 + \omega_0^2} e^{-(R/L)t} + \frac{V_0 \omega_0^2}{(R/L)^2 + \omega_0^2} \cos \omega_0 t - \frac{V_0 \omega_0^2}{(R/L)^2 + \omega_0^2} \sin \omega_0 t, \quad t > 0.$$

The first term represents the transient in y following the impulse produced by the step in the $\cos \omega_0 t$ input. The second and third terms are the sinusoidal response, which also includes a jump from the cosine component. We observe that at $t = 0^+$,

$$y(0^+) = \frac{V_0 (R/L)^2}{(R/L)^2 + \omega_0^2} + \frac{V_0 \omega_0^2}{(R/L)^2 + \omega_0^2} + 0 = V_0.$$

We may obtain the same result using the initial value theorem:

$$y(0^+) = \lim_{\sigma \to \infty} \sigma \frac{V_0 \sigma^2}{(\sigma + R/L)(\sigma^2 + \omega_0^2)} = V_0.$$

■

9.1.4 The Inverse Laplace Transform

The informal Laplace/Fourier relationship $\mathcal{L}\{f(t)\} = \mathcal{F}\{f(t)e^{-\sigma t}U(t)\}$ points the way to a formula for the inverse Laplace transform. Beginning with

$$F_L(s) = \int_{-\infty}^{\infty} [f(t)e^{-\sigma t} U(t)] e^{-i\omega t} \, dt$$

and applying the inverse Fourier transform (Equation 5.2), we have

$$f(t)e^{-\sigma t}U(t) = \frac{1}{2\pi} \int_{-\infty}^{\infty} F_L(\sigma + i\omega) e^{+i\omega t} \, d\omega.$$

Carry the convergence factor to the other side and absorb it into the integral:

$$f(t)U(t) = \frac{1}{2\pi} \int_{-\infty}^{\infty} F_L(\sigma + i\omega) e^{(\sigma + i\omega)t} \, d\omega.$$

Change variables, $s = \sigma + i\omega$, $ds = i d\omega$. We now have a path integral in the complex plane:

$$f(t)U(t) = \frac{1}{i2\pi} \int_{\sigma - i\infty}^{\sigma + i\infty} F_L(s) e^{st} \, ds.$$

The step function on the left-hand side is a reminder that the result is one sided. The improper integral is interpreted as a Cauchy principal value, giving the final result,

$$f(t) = \lim_{\Omega \to \infty} \frac{1}{i2\pi} \int_{\sigma - i\Omega}^{\sigma + i\Omega} F_L(s) e^{st} \, ds, \quad t > 0. \quad (9.40)$$

Like the Fourier transform, if $f(t)$ has a jump discontinuity at t_0, the inverse Laplace transform converges to the average of the right and left limits, $\frac{f(t_0+0^+)+f(t_0+0^-)}{2}$.

This is the *Bromwich inversion formula* for the Laplace transform. The path of integration is a vertical line, called the *Bromwich contour*, abbreviated Br. The value of σ must be chosen to place the Bromwich contour in the region of convergence of $F_L(s)$ (Figure 9.5). It can be shown by a path independence argument that the value of the integral does not depend on the particular value of σ chosen for the contour, as long as σ is within the region of convergence.

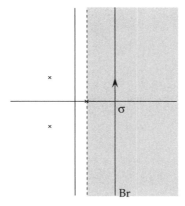

FIGURE 9.5 The path of integration for the inverse Laplace transform is a vertical line in the transform's region of convergence.

Complex Integration

The Laplace inversion integral is approached just like the real integrals we performed via contour integration in the last chapter—embed the Bromwich integral in an integral around a closed contour, calculate the residues at the poles inside the contour, show that the integral on the large semicircular arc goes to zero, so on. A variation on Jordan's lemma takes care of the large arc. Consider the closed contour shown in Figure 9.6.

We require that $|F_L(s)|$ decay sufficiently rapidly, $O(|s|^{-k})$, $k > 0$. In the common case that $F_L(s)$ is a rational function, $F_L(s) = \frac{b(s)}{a(s)}$, this requirement is met if the order of the denominator, $a(s)$, is greater than the order of the numerator, $b(s)$, that is, F_L is strictly proper. Transforms which are not strictly proper may still be inverted by resorting to generalized functions, but that will come later.

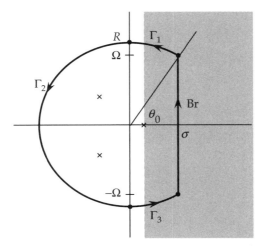

FIGURE 9.6 A closed contour for evaluating the inverse Laplace transform.

We want to show that the integrals along the three arcs, Γ_1, Γ_2, and Γ_3, go to zero as $R \to \infty$. If $\sigma \leq 0$, the integrals on Γ_1 and Γ_3 do not appear. Assume, then, that $\sigma > 0$. On Γ_1, $s = Re^{i\theta}$, $\theta \in (\theta_0, \pi/2)$, where $\cos \theta_0 = \sigma/R$:

$$\oint_{\Gamma_1} = \int_{\theta_0}^{\pi/2} F(Re^{i\theta}) \exp(Re^{i\theta} t) \, iRe^{i\theta} \, d\theta.$$

This integral is bounded above:

$$\left| \oint_{\Gamma_1} \right| \leq \int_{\theta_0}^{\pi/2} |F(Re^{i\theta}) \exp(Rt \cos \theta + iRt \sin \theta) \, iRe^{i\theta}| \, d\theta$$

$$\leq \int_{\theta_0}^{\pi/2} \frac{\mu}{R^k} \exp(Rt \cos \theta) R \, d\theta = \frac{\mu}{R^{k-1}} \int_{\theta_0}^{\pi/2} \exp(Rt \cos \theta) \, d\theta.$$

On the interval of integration $(\theta_0, \pi/2)$, the integrand is decreasing (because $\cos \theta$ is decreasing toward 0 at $\theta = \pi/2$). We may bound it above by $\exp(Rt \cos \theta_0) = \exp(Rt(\sigma/R)) = \exp(\sigma t)$ and use the ML inequality to bound the integral:

$$\left| \oint_{\Gamma_1} \right| \leq \mu \frac{e^{\sigma t}}{R^{k-1}} \left(\frac{\pi}{2} - \theta_0 \right).$$

Now $\pi/2 - \theta_0 = \pi/2 - \arccos(\sigma/R) = \arcsin(\sigma/R)$, and $\arcsin(\sigma/R) < (\pi/2)\sigma/R$. So we have

$$\left| \oint_{\Gamma_1} \right| < \frac{\mu \pi \sigma e^{\sigma t}}{2R^k},$$

and this upper bound goes to zero as $R \to \infty$, so the integral $\oint_{\Gamma_1}$ goes to zero. By a similar derivation, the integral on Γ_3 is also shown to go to zero.

The proof for the integral on Γ_2 closely follows the earlier proof of Jordan's lemma. We parametrize the path by $s = Re^{i\theta}$. If $\sigma \geq 0$, Γ_2 will be a half circle, and $\theta \in (\frac{\pi}{2}, \frac{3\pi}{2})$. If $\sigma < 0$, then the range of θ will be smaller, $\theta \in (\theta_0, 2\pi - \theta_0)$, where again, $\cos \theta_0 = \sigma/R$. On this path, the integral is again bounded above:

$$\left| \oint_{\Gamma_2} \right| \leq \frac{\mu}{R^{k-1}} \int_{\theta_0}^{2\pi - \theta_0} \exp(Rt \cos \theta) \, d\theta < \frac{\mu}{R^{k-1}} \int_{\pi/2}^{3\pi/2} \exp(Rt \cos \theta) \, d\theta,$$

because the integrand is positive. Changing variables, $\theta' = \theta - \pi/2$,

$$\int_{\pi/2}^{3\pi/2} \exp(Rt \cos \theta) \, d\theta = \int_0^{\pi} \exp(-Rt \sin \theta') \, d\theta' = 2 \int_0^{\pi/2} \exp(-Rt \sin \theta') \, d\theta'.$$

Now, just as in the proof of Jordan's lemma, we note that on the interval $(0, \pi/2)$, $\sin \theta' > 2\theta'/\pi$, so

$$\exp(-Rt \sin \theta') < \exp(-2Rt \theta'/\pi)$$

as long as $t > 0$. We then have

$$\int_{\pi/2}^{3\pi/2} \exp(Rt \cos \theta) \, d\theta < 2 \int_0^{\pi/2} \exp(-2Rt\theta'/\pi) \, d\theta' = \frac{\pi}{Rt} t(1 - e^{-Rt})$$

and

$$\left|\int_{\Gamma_2}\right| < \frac{\pi\mu(1-e^{-Rt})}{R^k t}, \quad t > 0.$$

As $R \to \infty$, this upper bound goes to zero, and so the integral on Γ_2 goes to zero.

We made a crucial assumption that $t > 0$. For $t < 0$, the bound $\exp(-Rt\sin\theta') < \exp(-2Rt\theta'\pi)$ is not true; in fact, the integral is unbounded. On the other hand, if the contour is closed to the right, rather than the left, the integral on the semicircular arc goes to zero for $t < 0$. This is just like Jordan's lemma—close the contour in one direction for $t > 0$, and in the other direction for $t < 0$.

Finally, we know that the integral around the entire closed contour is equal to $2\pi i$ times the sum of the residues at the poles encircled by the contour. With right-sided functions f, the region of convergence is a half-plane $\mathcal{R}e\, s > \sigma_a$, so there will be no singularities to the right of the Bromwich contour. Closing the contour to the right gives zero for $t < 0$, as expected. For the transform of a two-sided function, the region of convergence is a strip rather than a half-plane, and there are singularities on both sides of the Bromwich contour, resulting in nonzero contributions for both $t < 0$ and $t > 0$.

This derivation is summarized in the following theorem.

Theorem 9.14 (Laplace inversion). Let $F_L(s) = \mathcal{L}\{f(t)\}$, the Laplace transform of $f(t)$ with region of convergence $\sigma_- > \mathcal{R}e\, s > \sigma_+$, be analytic everywhere in the complex plane except for a finite number of isolated singular points $\{s_n\}$. Further, let $F_L(s)$ be $O(|s|^{-k})$, $k > 0$. Then, for any $\sigma \in (\sigma_+, \sigma_-)$, the inverse Laplace transform of F is given by

$$f(t) = \lim_{\Omega \to \infty} \frac{1}{2\pi i} \int_{\sigma-i\Omega}^{\sigma+i\Omega} F_L(s) e^{st}\, ds$$

$$= \begin{cases} \lim_{R\to\infty} \dfrac{1}{2\pi i} \oint_{\Gamma_L} F_L(s) e^{st}\, ds & t > 0 \\[6pt] \lim_{R\to\infty} -\dfrac{1}{2\pi i} \oint_{\Gamma_R} F_L(s) e^{st}\, ds & t < 0 \end{cases}$$

$$= \begin{cases} \sum\limits_{s_n \text{ inside } \Gamma_L} \text{Res}\left[F_L(s)e^{st}, s = s_n\right], & t > 0 \\[6pt] \sum\limits_{s_n \text{ inside } \Gamma_R} -\text{Res}\left[F_L(s)e^{st}, s = s_n\right], & t < 0 \end{cases}, \quad (9.41)$$

where Γ_L is the combination of the Bromwich contour and a circular path, centered at the origin, closing the contour to the left and enclosing all the singularities to the left of the Bromwich contour, and Γ_R is the combination of the Bromwich contour and a circular path, centered at the origin, closing the contour to the right (and traversed in the negative direction), enclosing all the singularities to the right of Bromwich contour.

Example 9.20. $F_L(s) = \frac{1}{s+2}$, $\mathcal{R}e\ s > -2$. The Bromwich contour must be placed in the region of convergence. Any vertical path to the right of $\mathcal{R}e\ s = -2$ will do, so for convenience take $\sigma = 0$ (run the Bromwich contour up the imaginary axis). Because F is $O(|s|^{-1})$, the previous theorem applies. Closing the contour to the right encloses no poles, so $f(t) = 0$, $t < 0$. Closing the contour to the left ($t > 0$) encloses the pole at $s = -2$. The residue at this pole is $\lim_{s \to -2} \frac{e^{st}}{1} = e^{-2t}$. Therefore, $f(t) = e^{-2t}$, $t > 0$. ∎

Example 9.21. $F_L(s) = \frac{1}{s^2+\sqrt{2}s+1}$. The poles are at $s = \frac{-1\pm i}{\sqrt{2}}$. Again we may place the Bromwich contour along the imaginary axis, closing it with a semicircle to the left. The order of the denominator (2) is greater than the order of the numerator (0), so $|F_L(s)|$ is $O(|s|^{-2})$ which guarantees that the integral on the semicircle will go to zero. The residues at the poles are

$$\text{Res}\left[F_L(s)e^{st},\ s = \frac{-1\pm i}{\sqrt{2}}\right] = \lim_{s \to \frac{-1\pm i}{\sqrt{2}}} \frac{e^{st}}{2s+\sqrt{2}} = \frac{\exp\frac{-1\pm i}{\sqrt{2}}t}{\pm i\sqrt{2}},$$

and their sum is the inverse transform:

$$f(t) = \frac{\exp\frac{-1+i}{\sqrt{2}}t}{i\sqrt{2}} - \frac{\exp\frac{-1-i}{\sqrt{2}}t}{i\sqrt{2}}$$

$$= \sqrt{2}\,e^{-t/\sqrt{2}}\sin\frac{t}{\sqrt{2}}.$$ ∎

When the transform is not strictly proper, it does not fall off as $|s|$ increases and so the integrals on the semicircular arcs do not go to zero as required. To deal with this, perform long division until a strictly proper remainder is obtained, whence $F_L(s) = r(s) + F_p(s)$, where $r(s)$ is a polynomial. Then invert F_p by integration and invert $r(s)$ term-by-term using the basic result $\delta^{(n)}(t) \longleftrightarrow s^n$.

Example 9.22 (Inverse Laplace transform of s^2). The function s^2 is not strictly proper and cannot be inverted by integration. However, $s^2 = s^3 \cdot \frac{1}{s}$, and $\frac{1}{s} \longmapsto U(t)$. By the derivative theorem, $s^2 = s^3 \frac{1}{s} \longmapsto \frac{d^3}{dt^3}U(t) = \delta''(t)$. ∎

Example 9.23 (Inverse transform of an improper rational function). Invert the improper rational transform $F_L(s) = \frac{s^3+2}{s^2+4s+3}$, $\mathcal{R}e\ s > -1$. Divide the denominator into the numerator until a strictly proper remainder is obtained:

$$\frac{s^3+2}{s^2+4s+3} = s - \frac{4s^2+3s-2}{s^2+4s+3} = s - 4 + \frac{13s+14}{s^2+4s+3}.$$

9.1 THE LAPLACE TRANSFORM

Now invert the rational function by integration, using Equations 9.41:

$$\mathcal{L}^{-1}\left\{\frac{13s+14}{s^2+4s+3}\right\} = \mathcal{L}^{-1}\left\{\frac{13s+14}{(s+3)(s+1)}\right\}$$

$$= \text{Res}\left[\frac{(13s+14)e^{st}}{(s+3)(s+1)}, s=-3\right] + \text{Res}\left[\frac{(13s+14)e^{st}}{(s+3)(s+1)}, s=-1\right]$$

$$= \frac{25}{2}e^{-3t}U(t) + \frac{1}{2}e^{-t}U(t).$$

Then, adding the contributions from the polynomial, $s - 4 \longmapsto \delta'(t) - 4\delta(t)$,

$$f(t) = \delta'(t) - 4\delta(t) + \frac{25}{2}e^{-3t}U(t) + \frac{1}{2}e^{-t}U(t). \qquad \blacksquare$$

Partial Fraction Expansion

The method of partial fractions is a well-known way to invert Laplace transforms. One learns in introductory calculus that any strictly proper rational function $Y(s) = \frac{P(s)}{Q(s)}$ can be expressed as a finite sum of simpler fractions of the form

$$\frac{A}{(s+a)^m} \quad \text{and} \quad \frac{Bs+C}{((s+a)^2+b^2)^n}.$$

In our case, if Y has resulted from the Laplace transform analysis of an ordinary linear differential equation with real, constant coefficients (i.e., a real LTI system), then the polynomials P and Q have real coefficients. The roots of Q are either real or occur in complex conjugate pairs and the parameters a and b in the partial fractions are real, as are the coefficients A, B, and C. The respective inverse Laplace transforms are

$$\frac{A}{(m-1)!}t^{m-1}e^{-at}U(t) \quad \text{and} \quad c(t)e^{-at}\cos(bt)U(t) + s(t)e^{-at}\sin(bt)U(t),$$

where $c(t)$ and $s(t)$ are $(n-1)$th-order polynomials in t whose coefficients, all real, depend on a, b, B, and C. These functions are fundamental solutions of constant-coefficient linear ODEs. The general time response of an LTI system is a linear combination of these fundamental modes. The partial fraction expansion is an expression, in the transform domain, of a modal decomposition of the system's response.

The hard part of the method is obtaining the partial fraction expansion. The most basic approach, suitable for simple functions, is algebraic. You have probably seen this before, and we will review it with a couple of examples, then go on to a better approach based on complex analysis.

Example 9.24. $F_L(s) = \frac{2}{s^2+3s+2}$. The first thing to do is identify the roots of the denominator. The quadratic formula, a root-finding program, or some good guessing yields $s = -1, -2$. The partial fraction expansion is of the form

$$\frac{A}{s+1} + \frac{B}{s+2}.$$

Place the two terms over a common denominator and equate the result to the original function:

$$\frac{A(s+2) + B(s+1)}{(s+1)(s+2)} = \frac{(A+B)s + (2A+B)}{s^2 + 3s + 2} = \frac{2}{s^2 + 3s + 2}.$$

This gives two equations in two unknowns:

$$A + B = 0,$$
$$2A + B = 2,$$

which solve to $A = 2$ and $B = -2$. Therefore, we have

$$F_L(s) = \frac{2}{s+1} + \frac{-2}{s+2}.$$ ∎

Example 9.25. $F_L(s) = \frac{s-1}{s^3 + 7s^2 + 16s + 12} = \frac{s-1}{(s+2)^2(s+3)}$. We have a single pole and a double pole. The rules one learns in calculus dictate that a double pole contributes two terms to the expansion, $\frac{A}{s+a} + \frac{B}{(s+a)^2}$. For this function, the partial fraction has the form

$$F_L(s) = \frac{A}{(s+2)^2} + \frac{B}{s+2} + \frac{C}{s+3}.$$

Placing all three terms over a common denominator, the numerator works out to

$$(s+3)A + (s+2)(s+3)B + (s+2)^2 C$$
$$= (B+C)s^2 + (A + 5B + 4C)s + (3A + 6B + 4C) = s - 1,$$

leading to three equations in three unknowns:

$$B + C = 0,$$
$$A + 5B + 4C = 1,$$
$$3A + 6B + 4C = -1.$$

The solution is $A = -3$, $B = 4$, and $C = -4$. The partial fraction expansion is

$$\frac{s-1}{(s+2)^2(s+3)} = -\frac{3}{(s+2)^2} + \frac{4}{s+2} - \frac{4}{s+3}.$$ ∎

For anything higher than second order, partial fraction expansion by the algebraic method becomes tedious. We will now develop a more direct method, building up to the final result by steps.

In the first example, $F_L(s) = \frac{2}{s^2 + 3s + 2} = \frac{2}{s+1} + \frac{-2}{s+2}$. Observe that the coefficients in the partial fraction expansion are the residues of F_L at the simple poles $s = -1, -2$.

In the next example, $F_L(s) = \frac{s-1}{(s+2)^2(s+3)} = \frac{-4}{s+3} + \frac{-3}{(s+2)^2} + \frac{4}{s+2}$, there is a simple pole at $s = -3$ and a double pole at $s = -2$. The coefficient of $1/(s+3)$ in the expansion is, again, the residue at the pole:

$$c_{-1} = \lim_{s \to -3} (s+3) F_L(s) = \lim_{s \to -3} \frac{s-1}{(s+2)^2} = -4.$$

Also, the coefficient of $\frac{1}{s+2}$ is the residue at that pole:

$$c_{-1} = \lim_{s \to -2} \frac{d}{ds}(s+2)^2 F_L(s) = \lim_{s \to -2} \frac{d}{ds}\frac{s-1}{s+3} = \lim_{s \to -2}\frac{4}{(s+3)^2} = 4.$$

So it appears that the partial fraction expansion is closely related to the Laurent series. If so, then perhaps the coefficient of $1/(s+2)^2$ is the Laurent coefficient c_{-2}. We can isolate this term by multiplying by $(s+2)^2$ and evaluating at $s = -2$:

$$c_{-2} = \lim_{s \to -2}(s+2)^2 F_L(s) = \lim_{s \to -2}\frac{s-1}{s+3} = -3.$$

It does indeed seem to be the case that the the partial fraction expansion is the sum of the negative-power terms, that is, the principal parts, $P(s, a)$, of the Laurent expansions around each of the poles:

$$F(s) = \sum_{j=1}^{N} P(s, a_j).$$

If a function is proper, but not strictly proper, for example, $F_L(s) = \frac{s^2+2s+2}{s^2+3s+2} = \frac{s^2+2s+2}{(s+1)(s+2)}$, we can perform one step of long division, obtaining $F_L(s) = 1 - \frac{s}{(s+1)(s+2)}$, and then calculate the principal parts of the remainder:

$$F_L(s) = 1 + \frac{1}{s+1} - \frac{2}{s+2}.$$

But observe that if we calculate the principal parts directly, without doing the long division first, we obtain (using the simple pole residue formula)

$$\lim_{s \to -1}(s+1)F_L(s) = \lim_{s \to -1}\frac{s^2+2s+2}{s+2} = 1,$$

$$\lim_{s \to -2}(s+2)F_L(s) = \lim_{s \to -1}\frac{s^2+2s+2}{s+1} = 2.$$

The principal parts are the same, with or without the long division. This makes sense, since the the constant obtained by long division corresponds to the c_0 term of the Laurent series, which is separate from the terms in the principal parts. Knowing this, we have an alternative method for obtaining the constant out front once the principal parts are calculated. In the expansion

$$F(s) = C + \sum_{j=1}^{N} P(s, a_j),$$

if F is finite at $s = 0$, we can calculate $C = F(0) - \sum_{j=1}^{N} P(0, a_j)$. In the current example,

$$F(s) = \frac{s^2+2s+2}{s^2+3s+2} = C + \frac{1}{s+1} - \frac{2}{s+2},$$

$$F(0) = 1 = C + 1 - 1,$$

$$\Rightarrow C = 1.$$

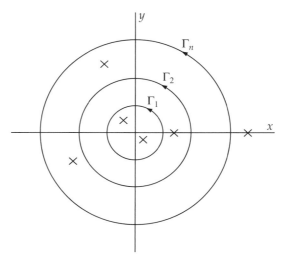

FIGURE 9.7 Sequence of contours for applying the Mittag–Leffler expansion. The meromorphic function F must be uniformly bounded on each of the Γ_n.

We have worked our way, by examples, to the following form for the partial fraction expansion, based not on matching coefficients algebraically, but on the properties of the complex functions:

$$F(s) = F(0) + \sum_{j=1}^{N} [P(s, a_j) - P(0, a_j)].$$

Of course, this is only a promising hypothesis at this point, but it turns out to be correct, and is not limited just to rational functions.

A function which is analytic except for a possibly infinite number of poles is called *meromorphic*. Rational functions, proper and improper, are meromorphic. So are ratios of analytic functions and polynomials, such as $\frac{e^s}{s^2+2s+1}$, and functions which have no polynomial component, like $\frac{1}{\sin \pi s}$. If a function has an essential singularity or a branch point, it is not meromorphic. The general result for partial fraction expansions of meromorphic functions is stated in the following theorem.[11]

Theorem 9.15 (Mittag–Leffler expansion). Let $F : \mathbb{C} \to \mathbb{C}$ be meromorphic, analytic at $s = 0$, and uniformly bounded on a set of simple closed curves Γ_n, as shown in Figure 9.7. Then F can be represented by the expansion

$$F(s) = F(0) + \sum_{j=1}^{\infty} [P(s; a_j) - P(0; a_j)], \qquad (9.42)$$

where $\{a_j\}$ are the poles of F and $\{P(s, a_j)\}$ are the respective principal parts of F.

---

[11] See LePage (1980, pp. 153–162) for a complete derivation.

This general result for meromorphic functions has a few important particular cases.

1. If F is a rational function, then the boundedness requirement is the same as saying that F is proper. If it is rational and improper, a partial fraction expansion may still be possible, in the form $F(s) = r(s) + F_p(s)$, where r is entire and F_p is a partial fraction expansion. The principal parts of F are calculated in the usual way to get F_p, then r is obtained by subtracting F_p from F.

2. If F is a strictly proper rational function, then the constant term $F(0) - \sum_{j=1}^{N} P(0, a_j)$ is identically zero and the partial fraction expansion simplifies to

$$F(s) = \sum_{j=1}^{N} P(s, a_j). \tag{9.43}$$

3. If F is singular at the origin, calculate the principal part there and subtract it from F to obtain a function G which is analytic at the origin: $G(s) = F(s) - P(s, 0)$. Then expand G in a partial fraction expansion. Because $P(s, 0)$ is analytic at all the other pole locations, the principal parts of G are the same as those of F (except for $s = 0$). The partial fraction expansion is

$$F(s) = G(0) + P(s, 0) + \sum_{j=1}^{\infty} [P(s, a_j) - P(0, a_j)], \tag{9.44}$$

where $\{a_j\}$ are the poles of F other than the ones at the origin.

4. If the individual series $\sum_{j=1}^{\infty} P(s, a_j)$ and $\sum_{j=1}^{\infty} P(0, a_j)$ are convergent, the sum may be split and the partial fraction expansion may be written

$$F(s) = \left[F(0) - \sum_{j=1}^{\infty} P(0, a_j) \right] + \sum_{j=1}^{\infty} P(s, a_j). \tag{9.45}$$

The problem of finding a partial fraction expansion thus boils down to calculating particular coefficients of Laurent series. For a general nth-order pole at $s = a$,

$$F(s) = c_{-n}(s-a)^{-n} + \cdots + c_{-2}(s-a)^{-2} + c_{-1}(s-a)^{-1} + c_0 + c_1(s-a) + \cdots,$$

multiplying $F(s)$ by $(s-a)^n$ cancels the singular terms out of the series, giving

$$(s-a)^n F(s) = c_{-n} + c_{-(n-1)}(s-a) + \cdots + c_{-2}(s-a)^{n-2} + c_{-1}(s-a)^{n-1} + O((s-a)^n).$$

We see immediately that $c_{-n} = \lim_{s \to a} (s-a)^n F(s)$. To get the next term, differentiate once to clear out c_{-n}:

$$c_{-(n-1)} = \lim_{s \to a} \frac{d}{ds} (s-a)^n F(s).$$

A general formula is obtained by following the pattern. For an nth order pole at $s = a$, the principal part is

$$P(s, a) = \sum_{k=1}^{n} \frac{c_{-k}}{(s-a)^k} \tag{9.46a}$$

$$c_{-k} = \frac{1}{(n-k)!} \lim_{s \to a} \frac{d^{n-k}}{ds^{n-k}} \left[(s-a)^n F(s)\right] \tag{9.46b}$$

For n any larger than 2 or 3, the multiple derivatives can get nasty. Fortunately, in engineering practice one rarely encounters poles of such high order. First-order poles are the norm, and so the particular formulas for calculating residues at first-order poles are probably the most widely used for hand calculation. But when a multiple pole must be dealt with, this general approach, which always works, is available.

Example 9.26 (Coverup method). $F(s) = \frac{1}{s^2 - 3s + 2}$. The poles are at $s = 1, 2$. Because F is strictly proper, we can use Equation 9.43. The partial fraction expansion is of the form

$$\frac{1}{s^2 - 3s + 2} = \frac{1}{(s-1)(s-2)} = \frac{1}{s-1} + \frac{1}{s-2},$$

The coefficients of the partial fractions are residues at the poles.

$$\text{Res}[F, s = 1] = \frac{s-1}{(s-1)(s-2)}\bigg|_{s=1} = \frac{1}{s-2}\bigg|_{s=1} = -1$$

$$\text{Res}[F, s = 2] = \frac{s-2}{(s-1)(s-2)}\bigg|_{s=2} = \frac{1}{s-1}\bigg|_{s=2} = 1$$

This is the origin of the "coverup" method (attributed to Heaviside) for finding partial fraction coefficients: one "covers up" the pole (i.e., cancels it in the residue formula) and evaluates what is left at the pole location. The partial fraction expansion is

$$\frac{1}{s^2 - 3s + 2} = \frac{-1}{s-1} + \frac{1}{s-2}. \qquad \blacksquare$$

Example 9.27. $F(s) = \frac{s-1}{s^3 + 7s^2 + 16s + 12} = \frac{s-1}{(s+2)^2(s+3)}$. We have a single pole and a double pole. The single pole can be handled by a residue calculation:

$$P(s, -3) = \frac{\text{Res}[F, s = -3]}{s + 3} = \frac{-4}{s + 3}.$$

For the double pole, the principal part is, using Equation 9.46,

$$P(s, -2) = \frac{c_{-2}}{(s+2)^2} + \frac{c_{-1}}{s+2},$$

where $c_{-2} = \lim_{s \to -2} \frac{s-1}{s+3} = -3$,

$$c_{-1} = \lim_{s \to -2} \frac{d}{ds} \frac{s-1}{s+3} = \lim_{s \to -2} \frac{4}{(s+3)^2} = 4$$

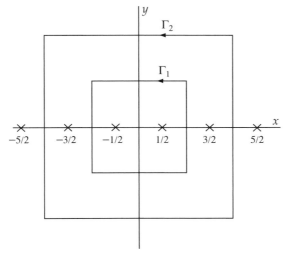

FIGURE 9.8 Sequence of contours for applying the Mittag–Leffler expansion to $F(s) = \sec \pi s$.

Therefore,

$$F(s) = \frac{-4}{s+3} + \frac{-3}{(s+2)^2} + \frac{4}{s+2}$$

■

Example 9.28. $F(s) = \frac{s^3+s^2+1}{s^2-4}$. This function is not proper, so we begin with long division:

$$F(s) = \frac{s^3+s^2+1}{s^2-4} = s+1+\frac{4s+5}{s^2-4}.$$

Now the partial fraction expansion is applied to the remainder, which has poles at $s = -2, 2$. The coverup method then gives

$$F(s) = s+1 + \frac{13/4}{s-2} + \frac{3/4}{s+2}$$

■

Example 9.29. $F(s) = \sec \pi s = \frac{1}{\cos \pi s}$. There are first-order poles at $s = \pm(\frac{1}{2} + k)$, $k = 0, 1, 2, \ldots$. To check the boundedness of F, we use the system of square contours shown in Figure 9.8. With $s = x + iy$,

$$|\cos \pi s| = (\cos^2 \pi x + \sinh^2 \pi y)^{1/2}.$$

On the sides of contour Γ_n, $s = \pm n + iy$, $n \geq y \geq -n$, so

$$|\cos \pi s| = (1 + \sinh^2 \pi y)^{1/2} \geq 1,$$

and on the top and bottom, $s = x \pm in$, $n \geq x \geq -n$, so

$$|\cos \pi s| = \left(\cos^2 \pi x + \sinh^2 \pi n\right)^{1/2} \geq |\sinh \pi n| > 1, \quad n > 0.$$

Thus, everywhere on each of the Γ_n, $|\cos \pi s| \geq 1$, so $|F| \leq 1$, proving that F is uniformly bounded.

The poles are first order, so the principal parts are obtained by calculating residues at the poles:

$$\text{Res}\left[\frac{1}{\cos \pi s}, s = \frac{1}{2} + k\right] = \lim_{s \to \frac{1}{2}+k} \frac{1}{-\pi \sin \pi s} = -\frac{(-1)^k}{\pi},$$

$$\text{Res}\left[\frac{1}{\cos \pi s}, s = -\frac{1}{2} - k\right] = \lim_{s \to -\frac{1}{2}-k} \frac{1}{-\pi \sin \pi s} = \frac{(-1)^k}{\pi}.$$

The expansion, therefore, is

$$\frac{1}{\cos \pi s} = 1 + \sum_{k=0}^{\infty} \frac{(-1)^k}{\pi} \left[\left(\frac{-1}{s-(\frac{1}{2}+k)} + \frac{1}{s+(\frac{1}{2}+k)}\right) - \left(\frac{1}{\frac{1}{2}+k} + \frac{1}{\frac{1}{2}+k}\right)\right]$$

$$= 1 - \sum_{k=0}^{\infty} \frac{(-1)^k}{\pi} \left[\frac{2k+1}{s^2 - (k+\frac{1}{2})^2} + \frac{2}{k+\frac{1}{2}}\right].$$

If the individual series

$$\sum_{k=0}^{\infty} \frac{(-1)^k}{\pi} \frac{2k+1}{s^2 - (k+\frac{1}{2})^2}$$

and

$$\sum_{k=0}^{\infty} \frac{(-1)^k}{\pi} \frac{2}{k+\frac{1}{2}}$$

are convergent, then some simplification is possible by writing the expansion as the sum of the individual series. The second series is known:

$$\sum_{k=0}^{\infty} \frac{(-1)^k}{\pi} \frac{2}{k+\frac{1}{2}} = \frac{2}{\pi} \sum_{k=0}^{\infty} \frac{(-1)^k}{k+\frac{1}{2}} = \frac{2}{\pi} \cdot \frac{\pi}{2} = 1.$$

The first series converges if its real and imaginary parts individually converge. Separating them,

$$\mathcal{R}e\left\{\sum_{k=0}^{\infty} \frac{(-1)^k}{\pi} \frac{2k+1}{s^2 - (k+\frac{1}{2})^2}\right\} = \sum_{k=0}^{\infty} \frac{(-1)^k}{\pi} \frac{(2k+1)\left((x^2 - y^2) + (k+\frac{1}{2})^2\right)}{\left((x^2 - y^2) + (k+\frac{1}{2})^2\right)^2 + 2x^2y^2},$$

$$\mathcal{I}m\left\{\sum_{k=0}^{\infty} \frac{(-1)^k}{\pi} \frac{2k+1}{s^2 - (k+\frac{1}{2})^2}\right\} = \sum_{k=0}^{\infty} \frac{(-1)^k}{\pi} \frac{-(2k+1)(2xy)}{\left((x^2 - y^2) + (k+\frac{1}{2})^2\right)^2 + 2x^2y^2}.$$

The imaginary part converges because the terms are $O(k^{-3})$. As for the real part, the terms are $O(k^{-1})$, so convergence is not obvious. For any s we may divide the series:

$$\sum_{k=0}^{\infty} = \sum_{k=0}^{N} + \sum_{k=N+1}^{\infty},$$

where $N > |x^2 - y^2|^{1/2} - \frac{1}{2}$. Then the first sum is finite, and the second is an alternating series with decreasing terms, hence convergent. We therefore have the final result

$$\frac{1}{\cos \pi s} = 1 - \sum_{k=0}^{\infty} \frac{(-1)^k}{\pi} \frac{2k+1}{s^2 - (k+\frac{1}{2})^2} - 1 = \sum_{k=0}^{\infty} \frac{(-1)^{k+1}}{\pi} \frac{2k+1}{s^2 - (k+\frac{1}{2})^2}. \quad \blacksquare$$

Example 9.30. Consider $F(s) = \csc \pi s = \frac{1}{\sin \pi s}$. There are first-order poles at $s = k$, $k = 0, \pm 1, \pm 2, \ldots$. The principal part at $s = 0$ is

$$P(s, 0) = \text{Res}\left[\frac{1}{\sin \pi s}, s=0\right] \cdot \frac{1}{s} = \frac{1}{\pi \cos 0} \cdot \frac{1}{s} = \frac{1}{\pi s}$$

and

$$g(s) = \frac{1}{\sin \pi s} - \frac{1}{\pi s} = \frac{\pi s - \sin \pi s}{\pi s \sin \pi s},$$

with

$$g(0) = \lim_{s \to 0} \frac{\pi s - \sin \pi s}{\pi s \sin \pi s} = 0.$$

Following steps similar to the previous example, one arrives at the partial fraction expansion:

$$\frac{1}{\sin \pi s} = \frac{1}{\pi s} + \frac{1}{\pi} \sum_{k=1}^{\infty} \frac{(-1)^k 2s}{s^2 - k^2} = \frac{1}{\pi} \sum_{k=-\infty}^{\infty} \frac{(-1)^k}{s - k}. \quad \blacksquare$$

Finally, we note that MATLAB provides tools for performing partial fraction expansion of rational functions. MATLAB's approach is numeric—given a rational function with numeric coefficients, the `residue` command provides the coefficients of a partial fraction expansion. Complex pole pairs are kept separate.

9.1.5 Laplace Transform of Sampled Functions

Many modern feedback control systems are combinations of continuous-time and discrete-time subsystems, connected with analog-to-digital and digital-to-analog converters (Figure 9.9).[12] The continuous-time portions are often modeled by rational transfer functions derived from a Laplace transform of a differential equation. The discrete-time portions are often modeled by the Z transform, which is introduced

---
[12] For introductions to sampled data control systems, see Franklin, et al. (1998) and Kuo (1963).

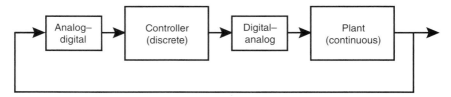

FIGURE 9.9 A sampled-data feedback control loop. The system to be controlled (the plant) operates in continuous time, and the controller operates in discrete time. They are connected with analog–digital and digital–analog converters.

in the next section. The linkage between the continuous and discrete domains is provided by the Laplace transform of a sampled function. While the application to control theory is beyond the scope of this text, it will be beneficial to consider here what happens when the Laplace transform is applied to a sampled function.

As in Section 6.5.3, we wish to model a sampled function using the comb function:

$$f_s(t) = f(t) \cdot \frac{1}{\Delta t} \text{III}\left(\frac{t}{\Delta t}\right),$$

where Δt is the sampling interval (Equation 6.49). We assume that f is one sided, with a possible jump discontinuity at $t = 0$, so $f(t) = 0$ for $t < 0$ and $f(0^+) \ne f(0^-)$. We denote the Laplace transform of a sampled function by F_{L*} and seek a relationship between F_{L*} and F_L. Prior experience with sampling and the Fourier transform leads us to expect that F_{L*} will, somehow, be a periodic replication of F_L.

Using the product theorem (9.34), the Laplace transform of $f(t) \cdot \frac{1}{\Delta t} \text{III}\left(\frac{t}{\Delta t}\right)$ is

$$F_{L*}(s) = \frac{1}{2\pi i} \int_{c-i\infty}^{c+i\infty} F_L(s-z) G_L(z) \, dz,$$

where

$$G_L(z) = \mathcal{L}\left\{\frac{1}{\Delta t} \text{III}\left(\frac{t}{\Delta t}\right)\right\} = \int_{0^-}^{\infty} \sum_{n=0}^{\infty} \delta(t - n\Delta t) e^{-zt} \, dt$$

$$= \sum_{n=0}^{\infty} e^{-nz\Delta t} = \frac{1}{1 - e^{-z\Delta t}}.$$

Thus,

$$F_{L*}(s) = \frac{1}{2\pi i} \int_{c-i\infty}^{c+i\infty} \frac{F_L(s-z)}{1 - e^{-z\Delta t}} \, dz. \tag{9.47}$$

The denominator, $1 - \exp(-z\Delta t)$, has first-order roots at $z = i2\pi k/\Delta t$. The path of integration is positioned between these roots and the poles of $F_L(s - z)$ (Figure 9.10), where the regions of convergence overlap.

The contour may be closed to the left or to the right. On the right, the curved path Γ_R is an arc of radius $R_k = 2\pi(k + 1/2)/\Delta t$. On the left, the path Γ_L consists of two short horizontal segments and an arc of radius R_k. Initially, assume that $F_L(s)$ is

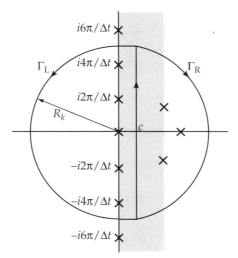

FIGURE 9.10 Contours for calculating the Laplace transform of a sampled function (Equation 9.47). The path of integration is located in the strip between the poles of $\frac{1}{1-e^{-z\Delta t}}$ (on the imaginary axis) and the poles of $F(s-z)$ (in the right half-plane). The contour may be closed to the left or right with an arc of radius $R_k = 2\pi(k+1/2)/\Delta t$.

rational, with two more poles than zeros, guaranteeing that the integrals on Γ_L and Γ_R go to zero as $R_k \to \infty$ ($k \to \infty$). With the contour closed to the right, the Laplace transform that results is

$$F_{L^*}(s) = \sum_{n=0}^{\infty} f(n\Delta t)e^{-sn\Delta t},$$

and closing the contour to the left, we obtain

$$F_{L^*}(s) = \frac{1}{\Delta t}\sum_{n=-\infty}^{\infty} F_L(s - i2\pi n/\Delta t).$$

Details of these calculations are left to the problems.

In the event that $F_L(s)$ has only one more pole than zero (e.g., a step function or a single exponential), it is more difficult to show that the integral on Γ_L goes to zero, and an additional term appears from the integral around Γ_R[13] giving the final form

$$F_{L^*}(s) = \sum_{n=0}^{\infty} f(n\Delta t)e^{-sn\Delta t} \tag{9.48a}$$

$$= \frac{1}{2}f(0^+) + \frac{1}{\Delta t}\lim_{N\to\infty}\sum_{n=-N}^{N} F_L(s - i2\pi n/\Delta t). \tag{9.48b}$$

[13] Wilts (1960, pp. 197–200, 261–265) and Kuo (1963, pp. 731–737).

606 CHAPTER 9 LAPLACE, Z, AND HILBERT TRANSFORMS

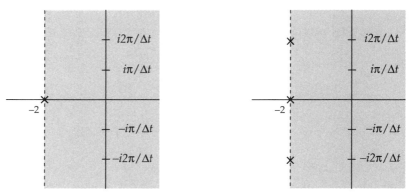

FIGURE 9.11 Effect of sampling on the Laplace transform. *Left*: The function $F_L(s) = \frac{1}{s+2}$ has a pole at $s = -2$. *Right*: Sampling causes the pole to repeat at locations $s = -2 + i2\pi n/\Delta t$.

This added term, $\frac{1}{2}f(0^+)$, is identically zero when $F_L(s)$ has two more poles than zeros (consider the initial value theorem). The form of the infinite series in Equation 9.48b, analogous to a Cauchy principal value integral, is necessary to obtain convergence when there is only one more pole than zero.

We saw in Section 6.5.3 that sampling in the time domain produces a periodic replication at $\Delta \nu = 1/\Delta t$ in the frequency domain (Equation 6.51). Here, in Equation 9.48b, $F_L(s - i2\pi/\Delta t)$ is a vertical shift of F_L in the complex plane. A pole at $s = -2$, $\frac{1}{s+2}$, becomes $\frac{1}{s-i2\pi/\Delta t+2} = \frac{1}{s+(2-i2\pi/\Delta t)}$, a pole at $s = -2 + i2\pi/\Delta t$. Sampling replicates the Laplace transform F_L at intervals of $\Delta \omega = 2\pi/\Delta t$ up and down the complex plane. The original and sampled transforms, F_L and F_{L^*}, have the same region of convergence (Figure 9.11).

Example 9.31. $f(t) = e^{-at}U(t)$ $(a \geq 0)$. The Laplace transform of f is

$$F_L(s) = \frac{1}{s+a}, \quad \mathcal{R}e\, s > -a.$$

The samples of f are $(f(n\Delta t))_{n \geq 0} = (1, e^{-a}, e^{-2a}, \ldots)$. So, Equation 9.48a gives

$$F_{L^*}(s) = \sum_{n=0}^{\infty} e^{-na\Delta t} e^{-sn\Delta t} = \sum_{n=0}^{\infty} (e^{-(s+a)\Delta t})^n$$
$$= \frac{1}{1 - e^{-(s+a)\Delta t}}, \quad \mathcal{R}e\, s > -a.$$

The other form, Equation 9.48b, produces

$$F_{L^*}(s) = \frac{1}{2} + \frac{1}{\Delta t} \sum_{n=-\infty}^{\infty} \frac{1}{s + \left(a - i\frac{2\pi n}{\Delta t}\right)}.$$

The equivalence of these two expressions for F_{L*} is confirmed by calculating the partial fraction expansion of $\frac{1}{1-e^{-(s+a)\Delta t}}$ using Theorem 9.15. Begin by subtracting $\frac{1}{2}$ from both sides, giving

$$\frac{1}{1-e^{-(s+a)\Delta t}} - \frac{1}{2} = \frac{1}{2}\frac{\cosh\left(\frac{(s+a)\Delta t}{2}\right)}{\sinh\left(\frac{(s+a)\Delta t}{2}\right)} = \frac{1}{2}\coth\left(\frac{(s+a)\Delta t}{2}\right).$$

The function has simple poles at $s = -a + i2\pi n/\Delta t$. The residues are

$$\frac{1}{2}\frac{\cosh\left(\frac{(s+a)\Delta t}{2}\right)}{\frac{d}{ds}\sinh\left(\frac{(s+a)\Delta t}{2}\right)}\Bigg|_{s=-a+\frac{i2\pi n}{\Delta t}} = \frac{1}{2}\frac{\cosh\left(\frac{(s+a)\Delta t}{2}\right)}{\frac{\Delta t}{2}\cosh\left(\frac{(s+a)\Delta t}{2}\right)}\Bigg|_{s=-a+\frac{i2\pi n}{\Delta t}} = \frac{1}{\Delta t}.$$

This gives the partial fraction expansion:

$$\frac{1}{1-e^{-(s+a)\Delta t}} - \frac{1}{2} = \sum_{n=-\infty}^{\infty} \frac{1/\Delta t}{s + (a - i\frac{2\pi n}{\Delta t})},$$

which is what we sought to show. ∎

The sampled Laplace transform is subject to aliasing. If f is bandlimited to $|\nu| < B$, its Laplace transform F_L is confined to the strip $2\pi B \geq \mathcal{I}m\, s \geq -2\pi B$. Sampling replicates the strips at $\Delta\omega = 2\pi/\Delta t = 4\pi B$. If $\Delta t < 1/2B$, when F_L is replicated, the strips will not overlap and aliasing will not occur. It can be shown, however, that no one-sided function can be bandlimited (cf. the Paley–Wiener theorem), so some aliasing will always occur. Figure 9.12 illustrates the point. As the sampling interval Δt is decreased, the replicated poles move to higher positions in the complex plane, leaving the single pole at $s = -2$ dominant at lower frequencies. And in the other form of the sampled transform, $\frac{1}{1-e^{-(s+2)\Delta t}}$, we have $e^{-(s+2)\Delta t} \approx 1 - (s+2)\Delta t$, so

$$\frac{1}{1-e^{-(s+2)\Delta t}} \rightarrow \frac{1}{\Delta t}\frac{1}{s+2}.$$

9.2 THE Z TRANSFORM

9.2.1 Definition

A different interpretation of the sampled Laplace transform leads to the *Z transform*, which applies to discrete-time systems. Begin with Equation 9.48a,

$$F_{L*}(s) = \sum_{n=0}^{\infty} f(n\Delta t)e^{-sn\Delta t}.$$

Define the sequence $f[n] = f(n\Delta t)$, make the change of variable $z = e^{s\Delta t}$, and write the sum as a geometric series, $\sum_{n=0}^{\infty} f[n]z^{-n}$. The result is the Z transform.

608 CHAPTER 9 LAPLACE, Z, AND HILBERT TRANSFORMS

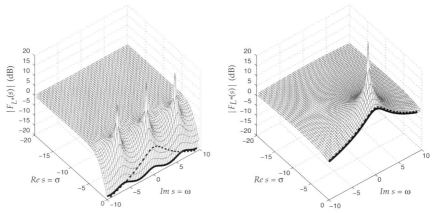

FIGURE 9.12 The Laplace transform $e^{-2t}U(t) \longmapsto \frac{1}{s+2}$, after sampling (compare Figure 9.11). The dashed line is the unsampled Fourier magnitude normalized by the sampling interval, $\frac{1}{\Delta t}\left|\frac{1}{i\omega+2}\right|$. *Left:* with $\Delta t = 1.0$, the transform is replicated at $\Delta\omega = 2\pi/\Delta t = 2\pi$. The magnitude along $s = i\omega$ exhibits substantial aliasing. *Right:* with $\Delta t = 0.2$, the transform is replicated at $\Delta\omega = 10\pi$. Aliasing is reduced as the separation between the replicated poles is increased by a factor of five.

Definition 9.3 (Z transform). Let f be a right-sided sequence, real or complex valued. If $f[n]\alpha^n \in \ell^1$ for some real α, then the Z transform of f is defined:

$$F_Z(z) = \sum_{n=0}^{\infty} f[n] z^{-n}, \tag{9.49}$$

where $z = re^{i\theta}$. The operator notation for the Z transform is $F_Z = \mathcal{Z}\{f\}$.

Here we will compare the Laplace and Z transforms in order to build a bridge from your understanding of the Laplace transform to the Z transform. Eventually, though, you should think of the Z transform in its own right as a transform for sequences which may or may not have resulted from sampling a continuous time function.

The sampled Laplace transform and the Z transform are connected through the change of variable $z = e^{s\Delta t}$:

$$F_{L^*}(s) = F_Z(e^{s\Delta t}). \tag{9.50}$$

The properties of the mapping $z = e^{s\Delta t}$ were considered in Problem 7.2 of Chapter 7. They are summarized in Table 9.1.

Writing $s = \sigma + i\omega = \sigma + i2\pi\nu$, a point in the S-plane maps to the point $z = e^{\sigma\Delta t}e^{i2\pi\nu\Delta t} = re^{i\theta}$. Polar coordinates are the natural choice for the Z-plane, and just as we write $e^{st} = e^{\sigma t}e^{i\omega t}$, we shall write $z^n = r^n e^{i\theta n}$. The S-plane coordinate σ represents the rate of growth or decay of the complex exponential e^{st}, and the other coordinate, ω, is the frequency of oscillation. On the Z-plane, the radius $r = e^{\sigma\Delta t}$ represents the rate of growth or decay, and the angle, $\theta = \omega\Delta t = 2\pi\nu\Delta t$, corresponds to the

TABLE 9.1 Properties of the mapping $z = e^{s\Delta t}$.

S-plane	Image in Z-plane
Point, $s = \sigma + i\omega$	$z = e^{\sigma \Delta t} e^{i\omega \Delta t} = re^{i\theta}$
Origin, $s = 0$	$z = 1$
$s = \pm i\pi$	$z = -1$
Infinity, $s \to -\infty$	$z \to 0$
Imaginary axis	Unit circle
Positive real axis, $s \in (0, \infty)$	Real axis, $z \in (1, \infty)$
Negative real axis, $s \in (-\infty, 0)$	Real axis, $z \in [0, 1)$
Left half-plane, $\mathcal{R}e\, s < 0$	Inside unit circle, $|z| < 1$
Right half-plane, $\mathcal{R}e\, s > 0$	Outside unit circle, $|z| > 1$
Horizontal line (constant frequency ω)	Ray at angle $\theta = \omega \Delta t$
Vertical line (constant σ)	Circle of radius $r = e^{\sigma \Delta t}$

frequency of oscillation. It is the same as the digital frequency introduced in Section 3.1. The sequence (r^n) is geometric. If $r < 1$ (z inside the unit circle), then r^n decays as $n \to \infty$, and if $r > 1$ (z outside the unit circle), then r^n grows.

The Laplace transform exists if $e^{-\sigma t} f(t) U(t) \in L^1$ for some real σ, and we say that $f(t)$ is of exponential order if $f(t) < Ce^{-\sigma t}$ for some real $C > 0$ and σ, and $t > T > 0$. Analogously, the Z transform exists if $\alpha^n f[n] U[n] \in \ell^1$ for some real α, and $f[n]$ is of exponential order if $f[n] < C\alpha^n$ for some real $C > 0$, $\alpha > 0$, and $n > N > 0$.

Example 9.32. Let $f(t) = e^{at} U(t)$, for which $f[n] = e^{an\Delta t} U[n]$. The Z transform is

$$F_Z(z) = \sum_{n=0}^{\infty} e^{an\Delta t} z^{-n} = \sum_{n=0}^{\infty} (e^{a\Delta t} z^{-1})^n.$$

This is a geometric series, and

$$\sum_{n=0}^{\infty} (e^{a\Delta t} z^{-1})^n = \lim_{N \to \infty} \frac{1 - (e^{a\Delta t} z^{-1})^N}{1 - e^{a\Delta t} z^{-1}}.$$

The series will converge if $|e^{a\Delta t} z^{-1}| < 1$ or $|z| > e^{a\Delta t}$. The Z transform, then, is

$$F_Z(z) = \frac{1}{1 - e^{a\Delta t} z^{-1}}, \quad |z| > e^{a\Delta t}. \quad \blacksquare$$

Compare this Z transform with the Laplace transform, which we know to be $F_L(s) = \frac{1}{s-a}$ (Figure 9.13). When $a > 0$, f is a growing exponential. The pole at $s = a$ is in the right half of the complex S-plane. When $a < 0$, the exponential is decaying and the pole is in the left half-plane. The Laplace transform goes to zero as $|s| \to \infty$ (there is a "zero at infinity"). The Z transform,

$$F_Z(z) = \frac{1}{1 - e^{a\Delta t} z^{-1}} = \frac{z}{z - e^{a\Delta t}},$$

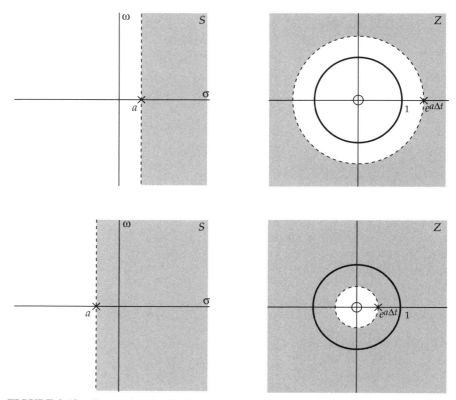

FIGURE 9.13 Comparing the Laplace transform of $f(t) = e^{at}U(t)$ with the Z transform of $f[n] = e^{an\Delta t}U[n]$. The S and Z planes are related through the mapping $z = e^{s\Delta t}$ described in Table 9.1. *Top*: When $a > 0$, the S-plane pole is in the right half-plane and the corresponding Z-plane pole at $z = e^{a\Delta t}$ is outside the unit circle. *Bottom*: When $a < 0$, the S-plane pole is in the left half-plane and the corresponding Z-plane pole is inside the unit circle. The Laplace transform's region of convergence is the half-plane $\mathcal{R}e\ s > a$. The Z transform converges outside a circle of radius $e^{a\Delta t}$.

has a pole at $z = e^{a\Delta t}$ and a zero at the origin. When $a > 0$, the pole is outside the unit circle and when $a < 0$, it is inside the unit circle. The region of convergence of the Z transform is the region outside a circle of radius $e^{a\Delta t}$, the image of the half-plane $\mathcal{R}e\ s > a$ under the mapping $z = e^{s\Delta t}$. The zero at $z = 0$ corresponds to the zero at infinity in the left-half S-plane.

Also compare the Z transform to the sampled Laplace transform (Figures 9.11 and 9.12):

$$F_Z(z) = \frac{z}{z - e^{a\Delta t}}, \quad F_{L^*}(s) = \frac{1}{1 - e^{-(s+a)\Delta t}}.$$

The vertical line $s = -a + i\omega$, where the poles of F_{L^*} are located, is mapped to the circle $z = e^{-a\Delta t}e^{i\omega\Delta t}$ in the Z-plane, where the single pole of F_Z is found. Moving up the line in the S-plane encounters each of the poles of F_{L^*} in turn, at $s = -a$,

$-a + 2\pi/\Delta t, -a + 4\pi/\Delta t, \ldots$. Likewise, moving around the circle in the Z-plane, the pole at $z = e^{-a\Delta t}$ is revisited with every cycle, at $\theta = 0, 2\pi, 4\pi, \ldots$. The replicated poles of F_{L*} are all represented by the single pole in F_Z.

The example showed that the exponential sequence $f[n] = \alpha^n U[n]$ is the discrete-time version of the one-sided exponential function $e^{at}U(t)$. Its Z transform is

$$\alpha^n U[n] \longmapsto \frac{1}{1 - \alpha z^{-1}}, \quad |z| > |\alpha| \tag{9.51}$$

for real or complex α. From this basic transform pair we may construct other useful pairs. If we let $\alpha = 1$, we obtain the Z transform of the step sequence,

$$U[n] \longmapsto \frac{1}{1 - z^{-1}}, \quad |z| > 1, \tag{9.52}$$

which has a pole on the unit circle at $z = 1$. If $\alpha = e^{i\beta}$, then the pole is on the unit circle at angle β. Combining this sequence with its complex conjugate yields Z transforms for cosine and sine sequences:

$$\cos \beta n \, U[n] \longmapsto \sum_{n=0}^{\infty} \frac{1}{2}\left(e^{i\beta n} + e^{-i\beta n}\right) z^{-n} = \frac{1}{2}\frac{1}{1 - e^{i\beta}z^{-1}} + \frac{1}{2}\frac{1}{1 - e^{-i\beta}z^{-1}}$$

$$= \frac{1 - \cos \beta \, z^{-1}}{1 - 2\cos \beta \, z^{-1} + z^{-2}}, \quad |z| > 1, \tag{9.53}$$

$$\sin \beta n \, U[n] \longmapsto \frac{\sin \beta \, z^{-1}}{1 - 2\cos \beta \, z^{-1} + z^{-2}}, \quad |z| > 1. \tag{9.54}$$

The Z transforms for exponentially damped sinusoids are simple extensions of these results:

$$\alpha^n \cos \beta n \, U[n] \longmapsto \frac{1 - \alpha \cos \beta \, z^{-1}}{1 - 2\alpha \cos \beta \, z^{-1} + \alpha^2 z^{-2}}, \quad |z| > |\alpha|, \tag{9.55}$$

$$\alpha^n \sin \beta n \, U[n] \longmapsto \frac{\alpha \sin \beta \, z^{-1}}{1 - 2\alpha \cos \beta \, z^{-1} + \alpha^2 z^{-2}}, \quad |z| > |\alpha|. \tag{9.56}$$

The Z transform of the discrete impulse, or unit sample sequence, is

$$\delta[n] \longmapsto \sum_{n=0}^{\infty} \delta[n] \, z^{-n} = 1, \quad |z| > 0. \tag{9.57}$$

Example 9.33. Let $f(t) = e^{at}\sin bt$, for which $f[n] = e^{na\Delta t}\sin(nb\Delta t)$. Let $\alpha = a\Delta t$ and $\beta = b\Delta t$. The Z transform is, using Equation 9.56,

$$F_Z(z) = \frac{\alpha \sin \beta \, z^{-1}}{1 - 2\alpha \cos \beta \, z^{-1} + \alpha^2 z^{-2}} = \frac{\alpha \sin \beta \, z}{z^2 - 2\alpha \cos \beta \, z + \alpha^2}.$$

There are two poles, at $z = \alpha e^{i\beta}$ and $z = \alpha e^{-i\beta}$. The magnitude, or radius, of the poles is $\alpha = e^{a\Delta t}$. For a damped sinusoid, $a < 0$ and $\alpha < 1$, placing the poles inside the unit circle. Increasing the damping moves the poles closer to the origin, and decreasing the

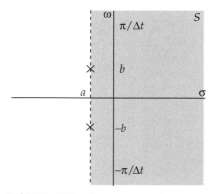 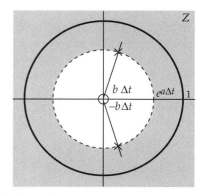

FIGURE 9.14 Comparing the Laplace transform of $f(t) = e^{at} \sin bt\, U(t)$ with the Z transform of $f[n] = e^{an\Delta t} \sin bn\Delta t\, U[n]$. The Laplace transform's region of convergence is the half-plane $\mathcal{R}e\, s > a$. The Z transform converges outside a circle of radius $e^{a\Delta t}$. The Laplace transform poles are located on a vertical line. Increasing the damping moves the line and the poles to the left; increasing the frequency moves the poles along the line, farther from the origin. The Z transform poles are located on a circle of radius $e^{a\Delta t}$, at angles $\pm b\Delta t$. Increasing the damping moves the poles toward the origin, and increasing the frequency moves them to higher angles around the circle. When $b\Delta t = \pi$, the Nyquist frequency has been reached and the poles meet at the back of the circle.

damping moves them closer to the unit circle. The angles of the poles are $\pm\beta = \pm b\Delta t$. Increasing the frequency increases β, moving the poles in opposite directions around a circle of radius α. When $\beta = \pi$, corresponding to the Nyquist frequency $b = \pi/\Delta t$, the poles meet. If the frequency is increased beyond this value, the poles cross the negative real axis and continue around the circle. Their locations reproduce the pole pattern corresponding to a lower frequency (aliasing) (Figure 9.14).

The possibility of aliasing is also observable in the mathematical form of the Z transform:

$$F_Z(z) = \frac{\alpha \sin \beta\, z}{z^2 - 2\alpha \cos \beta\, z + \alpha^2}.$$

The denominator polynomial, whose roots give the pole locations, depends on $\cos \beta$. Since the cosine has period 2π, the polynomial is unchanged if β is replaced by $2\pi - \beta$. The poles for a sinusoid of frequency β are indistinguishable from the poles for a sinusoid of frequency $2\pi - \beta$, just as, according to sampling theory, the samples of sinusoids with frequencies β and $2\pi - \beta$ are indistinguishable. ∎

As in continuous time, we may define a Z transform for two-sided sequences:

$$F_Z(z) = \sum_{n=-\infty}^{\infty} f[n] z^{-n}, \qquad (9.58)$$

with special cases for right-sided and left-sided sequences. The behavior of the Z transform for one-sided and two-sided sequences is analogous to that for the Laplace transform of one-sided and two-sided functions. The Z transform of a right-sided sequence converges outside a circle of some radius, and the Z transform of a left-sided sequence converges inside a circle. If a two-sided sequence decays to zero as $|n| \to \infty$, the region of convergence will be an annulus that includes the unit circle. There are no singularities in the region of convergence. A causal system has a right-sided impulse response, and an anticausal system has a left-sided impulse response. The poles of a stable, causal system are inside the unit circle, and the poles of a stable, anticausal system are outside the unit circle.[14]

Example 9.34 (A left-sided sequence). The sequence $f[n] = 2^n U[-n-1]$ decays exponentially as $n \to -\infty$. Its Z transform is

$$F_Z(z) = \sum_{n=-\infty}^{-1} 2^n z^{-n} = \sum_{n=1}^{\infty} 2^{-n} z^n = -1 + \sum_{n=0}^{\infty} (z/2)^n$$

$$= -1 + \frac{1}{1 - z/2}, \quad |z/2| < 1$$

$$= -\frac{1}{1 - 2z^{-1}}, \quad |z| < 2.$$

There is a pole at $z = 2$ and a zero at $z = 0$. The region of convergence is *inside* a circle of radius 2. If $f[n]$ were the impulse response of a noncausal system, the fact that the region of convergence includes the unit circle would indicate stability. ∎

If the region of convergence of the Z transform includes the unit circle, then the transform may be evaluated on the unit circle ($z = e^{i\theta}$), obtaining

$$F_Z(e^{i\theta}) = \sum_{n=0}^{\infty} f[n] e^{-in\theta} = F_d(\theta), \tag{9.59}$$

which is the discrete-time Fourier transform (Section 4.9).

Example 9.35. In an earlier chapter we calculated the discrete-time Fourier transform of the decaying exponential sequence $f[n] = \alpha^n U[n]$, $|\alpha| < 1$ (Example 4.23) and found $F_d(\theta) = \frac{1}{1 - ae^{-i\theta}}$. The Z transform of this sequence is $F_Z(z) = \frac{1}{1 - \alpha z^{-1}}$, with region of convergence $|z| > |\alpha|$. The region of convergence includes the unit circle if $|\alpha| < 1$. We may calculate $F_Z(e^{i\theta})$ and obtain

$$F_Z(e^{i\theta}) = \frac{1}{1 - ae^{-i\theta}} = F_d(\theta).$$

These functions are plotted, for $a = 0.85$, in Figure 9.15.

You are invited to compare this figure with Figure 9.12, the sampled Laplace transform. There it was observed that increasing the sampling interval, Δt, placed the replicated poles closer together and increased the aliasing effect. In the Z transform

---

[14] See Oppenheim and Schafer (2010, Chapter 3) for an in-depth discussion of two-sided Z transforms.

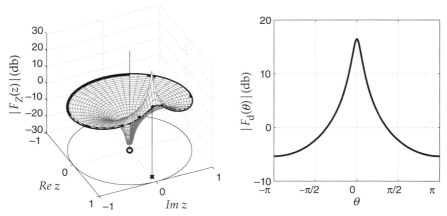

FIGURE 9.15 The Z transform, evaluated on the unit circle, gives the discrete-time Fourier transform. *Left*: Magnitude of the Z transform $F_Z(z) = \frac{1}{1-0.85z^{-1}}$. The Z-plane, showing the pole at $z = 0.85$ and zero at $z = 0$, is plotted below the surface. *Right*: Magnitude of the discrete-time Fourier transform $F_d(\theta) = F_Z(e^{i\theta}) = \frac{1}{1-0.85e^{-i\theta}}$.

of a sampled function, increasing Δt moves the poles closer to the origin. How does this correspond to the effect on the poles in the sampled Laplace transform? ∎

9.2.2 Z transform Theorems

The theorems for the Z transform closely resemble those for the Laplace and discrete-time Fourier transforms (see Section 4.9.2). The derivations of most are straightforward and will be omitted. Unless otherwise indicated, the theorems are restricted to the one-sided transform, particularly as this bears on the region of convergence.

Linearity

Theorem 9.16. If f has Z transform F_Z, and g has Z transform G_Z, then

$$af + bg \longmapsto aF_Z + bG_Z. \tag{9.60}$$

The region of convergence is the intersection of the regions of convergence of f and g.

Example 9.36 (Saturating exponential). The Z transforms of $U[n]$ and $\alpha^n U[n]$ are, respectively, $\frac{1}{1-z^{-1}}$, $|z| > 1$, and $\frac{1}{1-\alpha z^{-1}}$, $|z| > |\alpha|$. By linearity,

$$1 - \alpha^n \longmapsto \frac{1}{1-z^{-1}} - \frac{1}{1-\alpha z^{-1}} = \frac{(1-\alpha)z^{-1}}{(1-z^{-1})(1-\alpha z^{-1})} \tag{9.61}$$

$$|z| > \max(1, |\alpha|).$$

∎

Symmetry

Theorem 9.17. If f has Z transform F_Z, then

$$f^*[n] \longmapsto F_Z^*(z^*), \qquad (9.62)$$

and if f is real, so $f = f^*$,

$$F_Z^*(z) = F_Z(z^*). \qquad (9.63)$$

The region of convergence for F^* is the same as that of F_Z.

When $z = e^{i\theta}$, the latter relationship becomes $F_Z^*(e^{i\theta}) = F_Z(e^{-i\theta})$, that is, $F_Z(e^{i\theta})$ is Hermitian.

Shift

Theorem 9.18. Let $f \longmapsto F_Z$. For the two-sided Z transform,

$$f[n-k] \longmapsto z^{-k} F_Z(z), \qquad (9.64)$$

and for the one-sided transform,

$$f[n-k]\,U[n] \longmapsto f[-k] + z^{-1}f[-k+1] + \cdots + z^{-k+1}f[-1] + z^{-k}F_Z(z) \qquad (9.65)$$

$$\text{and } f[n+k]\,U[n] \longmapsto -z^k f[0] - z^{k-1}f[1] - \cdots - zf[k] + z^k F_Z(z), \qquad (9.66)$$

where $k > 0$. The region of convergence is the same as that of F_Z, excluding the origin for the case of a right shift.

Proof: First, for the two-sided transform,

$$f[n-k] \longmapsto \sum_{n=-\infty}^{\infty} f[n-k]\,z^{-n} = \sum_{n'=-\infty}^{\infty} f[n']\,z^{-(n'+k)} = z^{-k} F_Z(z).$$

For the one-sided transform, we must think about values that are shifted across the origin. With a right shift,

$$f[n-1]U[n] \longmapsto \sum_{n=0}^{\infty} f[n-1]\,z^{-n} = f[-1] + \sum_{n=1}^{\infty} f[n-1]\,z^{-n}$$

$$= f[-1] + \sum_{n'=0}^{\infty} f[n']\,z^{-(n'+1)}$$

$$= f[-1] + z^{-1} F_Z(z).$$

Iterating this process leads to Equation 9.65. If $f[n] = 0$ for $n < 0$, then the extra terms do not appear. With a left shift, the one-sided transform clips off values of f for

$n < k$ that must then be deducted from F_Z:

$$f[n+1]U[n] \longmapsto z \cdot \mathcal{Z}\{f[n]U[n-1]\} = z\sum_{n=1}^{\infty} f[n]z^{-n}$$

$$= -zf[0] + z\sum_{n=0}^{\infty} f[n]z^{-n} = -zf[0] + zF_Z(z).$$

Iterating this process leads to Equation 9.66. ∎

The physical significance of the samples $f[-1], f[-2]$, etc., is brought out by the next theorem.

Finite Difference

The application of the Laplace transform to linear differential equations rests on the derivative theorem, $f(t) \longmapsto sF_L(s) - f(0+)$, with $f(0+)$ understood to be an initial condition. A first-order accurate approximation to the first derivative, sampled at $t = n\Delta t$, is

$$f'(n\Delta t) = \frac{f(n\Delta t) - f((n-1)\Delta t)}{\Delta t} = \frac{f[n] - f[n-1]}{\Delta t} = \frac{\Delta_1 f[n]}{\Delta t},$$

where $f[n] = f(n\Delta t)$ and $\Delta_1 f$ is the the *finite difference*:

$$\Delta_1 f[n] = f[n] - f[n-1]. \tag{9.67}$$

The shift theorem then leads directly to a result for the Z transform that resembles the derivative theorem for the Laplace transform.

Theorem 9.19. The Z transform of $\Delta_1 f$, the finite difference of f, is

$$\mathcal{Z}\{\Delta_1 f\} = (1 - z^{-1})F_Z(z) \tag{9.68}$$

for the two-sided transform and

$$\mathcal{Z}\{\Delta_1 f\} = (1 - z^{-1})F_Z(z) - f[-1] \tag{9.69}$$

for the one-sided transform. The region of convergence is same as the region of convergence of F_Z, excluding the origin.

The value $f[-1]$ is interpreted here as an initial condition, just like $f(0)$ in continuous time. If f is a right-sided sequence, then $f[-1] = 0$.

Example 9.37. In continuous time, the derivative of the step is the delta function, and $\mathcal{L}\{\delta(t)\} = s \cdot \mathcal{L}_-\{U(t)\} = s \cdot \frac{1}{s} = 1$. Here, in discrete time, $\Delta_1 U[n] = \delta[n]$, and

$$\mathcal{Z}\{\Delta_1 U[n]\} = (1 - z^{-1})\frac{1}{1 - z^{-1}} - U[-1] = 1 - 0 = 1,$$

which is also the Z transform of $\delta[n]$. ∎

Example 9.38 (Euler's method). A simple approach to numerically solving the initial value problem $y' + ay = 0$, $y(0^-) = y_0$, known as *Euler's method*, approximates the derivative by a first difference. The unknown function $y(t)$ is represented by its samples on a grid with spacing Δt, and the derivative is approximated by a *forward difference*:

$$y'(n\Delta t) \approx \frac{y((n+1)\Delta t) - y(n\Delta t)}{\Delta t}.$$

Substituting this and collecting terms, we obtain

$$y((n+1)\Delta t) - (1 - a\Delta t) y(n\Delta t) = 0.$$

Writing $y(n\Delta t) = y[n]$, the approximation to the differential equation is a *difference equation* (Section 4.9.3):

$$y[n+1] - (1 - a\Delta t) y[n] = 0, \quad y[0] = y_0.$$

All discretization schemes for differential equations result in some difference equation, and the Z transform is useful for analyzing their stability properties. To solve the equation, take Z transforms (using the shift theorem) and solve for Y_Z:

$$-zy_0 + zY_Z(z) - (1 - a\Delta t) Y_Z(z) = 0$$
$$\Rightarrow (z - (1 - a\Delta t))Y_Z(z) = zy_0$$
$$\Rightarrow Y_Z(z) = \frac{y_0}{1 - (1 - a\Delta t) z^{-1}}.$$

Let us assume that $a > 0$, which corresponds to a differential equation having a stable solution, $y(t) = y_0 e^{-at} U(t)$. The numerical solution, represented by the Z transform $Y_Z(z)$, should also be stable. It has a single real pole, at $z = 1 - a\Delta t$, which must be inside the unit circle, $|1 - a\Delta t| < 1$.

Using the known Z transform pair $\alpha^n U[n] \longleftrightarrow \frac{1}{1 - \alpha z^{-1}}$, we can invert the transform and return to the time domain:

$$y[n] = y_0 (1 - a\Delta t)^n U[n].$$

Observe that if $|1 - a\Delta t| < 1$, the solution will decay with n, as desired. We also see that $1 - a\Delta t$ must be positive, otherwise the solution will change sign at every time step, which is not the correct behavior for a first-order differential equation. In the Z transform domain, this means the pole must be on the positive real axis, between 0 and 1; a pole between -1 and 0 corresponds to the oscillatory time-domain behavior.

We conclude that the numerical solution by Euler's method will at least be stable (accuracy is another issue) for $1 > a\Delta t > 0$. Given a value of a, the sampling interval Δt must be chosen so that $\Delta t < 1/a$. It can be shown that as Δt becomes very small, the discrete-time solution approaches the sampled continuous time solution, that is, $y_0 (1 - a\Delta t)^n U[n] \to y_0 e^{-an\Delta t} U[n]$. ∎

Integration

Theorem 9.20. Let f have Z transform F_Z, $|z| > r$, and define Σ_f to be the cumulative sum:

$$\Sigma_f[n] = \sum_{k=0}^{n} f[k].$$

Then the Z transform is

$$\Sigma_f \longmapsto \frac{F_z(z)}{1 - z^{-1}}, \quad |z| > \max(1, r). \tag{9.70}$$

Multiplication

Theorem 9.21. Let f have Z transform F_Z, $|z| > r$. Then

$$\alpha^{-n} f[n] \longmapsto F_Z(\alpha z), \quad |z| > |\alpha r|, \tag{9.71}$$
$$n f[n] \longmapsto -z F'_Z(z), \quad |z| > r. \tag{9.72}$$

The factor α^{-n}, when $|\alpha| > 1$, increases the decay of f and dilates the Z transform—a pole at $z = z_0$ is scaled to a pole at z_0/α, closer to the origin. This is analogous to what we observed in the Laplace transform, where increasing the decay of a function pushed the poles of its transform deeper into the left half-plane. When $\alpha = e^{i\beta}$ the Z transform is rotated by angle β around the unit circle. The second form provides the Z transforms of polynomials in n, for example, for a ramp,

$$n U[n] \longmapsto -z \frac{d}{dz} \frac{1}{1 - z^{-1}} = -z \cdot -\frac{-z^{-2}}{(1 - z^{-1})^2} = \frac{z^{-1}}{(1 - z^{-1})^2}, \quad |z| > 1. \tag{9.73}$$

Compare the Laplace transform $tU(t) \longmapsto \frac{1}{s^2}$, which has a double pole at the origin.

Dilation

In discrete time, dilation is upsampling (inserting zeros between samples) and downsampling (removing samples). Earlier (Section 4.9.2) we saw that the relationships between dilated sequences and their discrete-time Fourier transforms are

Upsampling:

$$f_{\uparrow P}[n] = \begin{cases} f[n/P], & n = rP, \ r \in \mathbb{Z} \\ 0, & \text{otherwise} \end{cases} \tag{4.58}$$

$$F_{\uparrow P}(\theta) = F_d(P\theta). \tag{4.59}$$

Downsampling:

$$f_{\downarrow P}[n] = f[nP], \tag{4.57}$$

$$F_{\downarrow P}(\theta) = \frac{1}{P} \sum_{m=0}^{P-1} F_d\left(\frac{\theta - 2\pi m}{P}\right). \tag{4.60}$$

9.2 THE Z TRANSFORM

Because of the relationship between the Z transform and the discrete-time Fourier transform, $F_d(\theta) = F_Z(e^{i\theta})$, we should expect similar forms for the Z transforms of dilated sequences.

Theorem 9.22. Let $f \longmapsto F_Z$, with radius of convergence r_f. The Z transforms of the upsampled sequence $f_{\uparrow P}$ and the downsampled sequence $f_{\downarrow P}$ are:

$$f_{\uparrow P}[n] \longmapsto F_Z(z^P), \quad |z| > r_f^{1/P} \tag{9.74}$$

$$f_{\downarrow P}[n] \longmapsto \frac{1}{P} \sum_{m=0}^{P-1} F_Z(e^{-i2\pi m/P} z^{1/P}), \quad |z| > r_f^P \tag{9.75}$$

Plugging in $z = e^{i\theta}$ easily reproduces the Fourier transforms. The downsampling result is difficult to interpret on the complex plane, because the $z^{1/P}$ implies branch points and branch cuts. However, for the typical case of a rational Z transform, intuitive results are possible, as illustrated by the following example.

Example 9.39 (Dilated exponential sequence). The exponential sequence $\alpha^n U[n]$ with $|\alpha| < 1$ has Z transform $\frac{1}{1-\alpha z^{-1}}$, $|z| > |\alpha|$. There is a pole at $z = \alpha$ and a zero at the origin.

First consider upsampling, which inserts $P - 1$ zeros between each sample. This stretches the sequence out, causing it to decay more slowly and, if α is complex, oscillate more slowly. We expect the Z transform to have several poles, because the sequence is no longer a pure exponential. With a longer decay, the poles should be closer than $|\alpha|$ to the unit circle, and with a lower frequency, we expect them to rotate to lower angles. So now consider the Z transform according to the theorem:

$$(\alpha^n U[n])_{\uparrow P} \longmapsto \frac{1}{1 - \alpha z^{-P}}.$$

This function does in fact have P poles, at the Pth roots of α,

$$z = \alpha^{1/P} = |\alpha|^{1/P} \exp\left(i \frac{\text{Arg } \alpha + 2\pi m}{P}\right), \quad m = 0, 1, \ldots P - 1.$$

The poles are equally spaced around a circle of radius $|\alpha|^{1/P}$, and the Z transform converges outside this circle, $|z| > |\alpha|^{1/P}$.

Downsampling removes samples, changing the sequence to $(\alpha^P)^n U[n]$. This is still a pure exponential with parameter α^P and so we expect the Z transform to have only one pole. The sequence decays faster than $|\alpha|^n U[n]$ and oscillates at a higher frequency, thus we expect the new pole to be closer to the origin and at a higher angle. If P Arg α exceeds π, we may expect the transform to show effects of aliasing. Rather than using Equation 9.75, the Z transform may be obtained by inspection:

$$(\alpha^n U[n])_{\downarrow P} = (\alpha^P)^n U[n] \longmapsto \frac{1}{1 - \alpha^P z^{-1}}.$$

Indeed, there is one pole, at $z = \alpha^P = |\alpha|^P \exp(iP \text{ Arg } \alpha)$, and the transform converges outside that pole, $|z| > |\alpha|^P$. ∎

Convolution

Theorem 9.23. Let f and g have Z transforms F_Z and G_Z, with radii of convergence r_f and r_g, respectively. The convolution of f and g is

$$f * g = \sum_{k=0}^{n} f[k]g[n-k],$$

and

$$f * g \longleftrightarrow F_Z G_Z, \quad |z| > \max(r_f, r_g). \tag{9.76}$$

Product Theorem

Theorem 9.24. Let f and g have Z transforms F_Z and G_Z, with radii of convergence r_f and r_g, respectively. The Z transform of the product fg is

$$fg \longmapsto \frac{1}{2\pi i} \oint_\Gamma F_Z(\xi) G_Z\left(\frac{z}{\xi}\right) \frac{d\xi}{\xi} = \frac{1}{2\pi i} \oint_\Gamma F_Z\left(\frac{z}{\xi}\right) G_Z(\xi) \frac{d\xi}{\xi}, \quad |z| \geq r_f r_g, \tag{9.77}$$

where the contour of integration Γ is a circle located in the common region of convergence of $F_Z(\xi)$ and $G_Z\left(\frac{z}{\xi}\right)$, or of $F_Z\left(\frac{z}{\xi}\right)$ and $G_Z(\xi)$.

The derivation follows the same approach used to obtain the product theorem for the Laplace transform.[15] If the path of integration is taken to be a circle of radius r, then the integral can be written

$$\frac{1}{2\pi} \int_{-\pi}^{\pi} F_Z(re^{i(\theta-\varphi)}) G_Z(re^{i\varphi}) \, d\varphi,$$

revealing that the right-hand side of Equation 9.77 is a convolution of transforms.

Parseval's Theorem

Theorem 9.25. Let f and g have Z transforms F_Z and G_Z, with radii of convergence r_f and r_g, respectively. Then,

$$\sum_{n=0}^{\infty} f[n]g^*[n] = \frac{1}{2\pi i} \oint_\Gamma F_Z(z) G_Z^*\left(\frac{1}{z^*}\right) \frac{dz}{z} = \frac{1}{2\pi i} \oint_\Gamma F_Z\left(\frac{1}{z}\right) G_Z^*(z^*) \frac{dz}{z}, \tag{9.78}$$

where the contour of integration Γ is a circle located in the common region of convergence of $F_Z(z)$ and $G_Z(z^{-1})$ or of $F_Z(z^{-1})$ and $G_Z(z)$. When the contour of integration is the unit circle, $z = e^{i\theta}$, we obtain Parseval's theorem for the discrete-time Fourier transform (Equation 4.49):

$$\sum_{n=-\infty}^{\infty} f[n]g^*[n] = \frac{1}{2\pi} \int_{-\pi}^{\pi} F_d(\theta) G_d^*(\theta) \, d\theta.$$

---
[15] Also see Oppenheim and Schafer (2010, pp. 63–66).

Initial and Final Values

Theorem 9.26. If f has Z transform F_Z, $|z| > 1$, then

$$f[0] = \lim_{|z| \to \infty} F_Z(z), \qquad (9.79)$$

$$\lim_{n \to \infty} f[n] = \lim_{z \to 1} (z-1) F_Z(z). \qquad (9.80)$$

In the initial value theorem, because $F_Z(z) = f[0] + O(z^{-1})$, taking $|z| \to \infty$ forces all the terms to zero except $f[0]$. The idea of the final value theorem is the same as for the Laplace transform. In order for f to have a finite final value, it must contain a step equal in height to the final value. This step contributes a single pole at $z = 1$ to the Z transform, whose residue is the final value.

9.2.3 The Inverse Z Transform

Earlier we found that the Laplace transform may be regarded as a Fourier transform:

$$F_L(\sigma + i\omega) = \mathcal{F}\{f(t) e^{-\sigma t} U(t)\}.$$

An analogous relationship exists between the Z transform and the discrete-time Fourier transform, which is defined (Equation 4.44)

$$F_d(\theta) = \sum_{n=-\infty}^{\infty} f[n] e^{-in\theta},$$

$$f[n] = \frac{1}{2\pi} \int_{-\pi}^{\pi} F_d(\theta) e^{in\theta} \, d\theta.$$

Comparing it to the Z transform with $z = re^{i\theta}$, observe that

$$F_Z(re^{i\theta}) = \sum_{n=0}^{\infty} f[n](re^{i\theta})^{-n} = \sum_{n=-\infty}^{\infty} (f[n] r^{-n} U[n]) e^{-in\theta}$$

$$= \mathcal{F}\{f[n] r^{-n} U[n]\}.$$

The Z transform is the Fourier transform of a one-sided sequence that has been multiplied by a convergence factor. The smallest value of r for which the transform exists is the radius of convergence r_0, and the region of convergence is $|z| > r_0$.

Just as with the Laplace transform, the Fourier-Z relationship inspires an inverse transform formula. Begin with

$$F_Z(re^{i\theta}) = \sum_{n=-\infty}^{\infty} (f[n] r^{-n} U[n]) e^{-in\theta}$$

and apply the inverse Fourier transform:

$$f[n] r^{-n} U[n] = \frac{1}{2\pi} \int_{-\pi}^{\pi} F_Z(re^{i\theta}) e^{in\theta} \, d\theta.$$

Move the convergence factor to the other side and absorb it into the integral:

$$f[n]U[n] = \frac{1}{2\pi}\int_{-\pi}^{\pi} F_Z(re^{i\theta})(re^{i\theta})^n \, d\theta.$$

Change variables to $z = re^{i\theta}$, with $dz = ire^{i\theta}d\theta$. This changes the integral to a complex integral on the circular path $|z| = r$, where r is in the region of convergence of F_Z.

$$f[n] = \frac{1}{i2\pi}\oint_{|z|=r} F_Z(z)z^{n-1}\, dz, \quad n \geq 0.$$

We can also obtain the inverse transform by beginning with the observation that the Z transform is a Laurent series:

$$F_Z(z) = \sum_{n=0}^{\infty} f[n]z^{-n} = f[0] + f[1]z^{-1} + f[2]z^{-2} + \cdots$$

In the region of convergence this series may be integrated term-by-term around a closed contour $|z| = r$:

$$\oint_{|z|=r} F_Z(z)dz = \oint_{|z|=r} f[0]dz + \oint_{|z|=r} f[1]z^{-1}dz + \oint_{|z|=r} f[2]z^{-2}dz + \cdots$$

All the integrals will be zero except one, $\oint_{|z|=r} f[1]z^{-1}dz = i2\pi f[1]$ (Equation 8.3). This recovers one element of the original sequence:

$$f[1] = \frac{1}{i2\pi}\oint_{|z|=r} F_Z(z)dz.$$

For the general form, multiply the Z transform by z^{n-1} and integrate:

$$\oint_{|z|=r} F_Z(z)z^{n-1}\, dz = \oint_{|z|=r} \sum_{k=0}^{\infty} f[k]z^{-k}z^{n-1}\, dz = \sum_{k=0}^{\infty} \oint_{|z|=r} f[k]z^{n-k-1}\, dz.$$

Again using Equation 8.3, we observe that the integral on the right-hand side will be zero unless $n - k - 1 = -1$, or $k = n$. That is,

$$\oint_{|z|=r} z^{n-k-1}\, dz = i2\pi\delta[k-n], \quad k \geq 0.$$

Thus, we have

$$\oint_{|z|=r} F_Z(z)z^{n-1}\, dz = \sum_{k=0}^{\infty} i2\pi f[k]\delta[k-n] = i2\pi f[n], \quad n \geq 0,$$

which establishes the following theorem.

Theorem 9.27 (Inverse Z transform). Let f be a one-sided sequence with Z transform F_Z. Let Γ be a simple closed contour in the region of convergence of F_Z. Then

$$f[n] = \frac{1}{2\pi i}\oint_{\Gamma} F_Z(z)z^{n-1}\, dz, \quad n \geq 0. \tag{9.81}$$

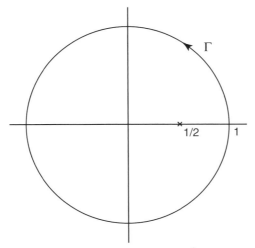

FIGURE 9.16 Inverting the Z transform $F_Z(z) = \frac{1}{1-\frac{1}{2}z^{-1}}$ by complex integration.

Example 9.40. $F_Z(z) = \frac{1}{1-\frac{1}{2}z^{-1}}$ has a simple pole at $z = \frac{1}{2}$. Its region of convergence is $|z| > \frac{1}{2}$. Taking Γ to be the unit circle (Figure 9.16) and using Equation 9.81,

$$f[n] = \frac{1}{2\pi i}\oint_\Gamma \frac{z^{n-1}}{1-\frac{1}{2}z^{-1}}\,dz = \frac{1}{2\pi i}\oint_\Gamma \frac{z^n}{z-\frac{1}{2}}\,dz = \frac{1}{2\pi i} \times 2\pi i \, \text{Res}\left[\frac{z^n}{z-\frac{1}{2}}, \frac{1}{2}\right]$$

$$= \left(\frac{1}{2}\right)^n, \quad n \geq 0.$$

We assume that the inverse transform is zero for $n < 0$ because of the region of convergence. But it is instructive to carry the calculation through for $n \leq -1$. For $n > 0$, the z^n factor in the numerator gives zeros at the origin, but for $n \leq -1$ it becomes an nth-order pole at the origin.

$$f[-1] = \frac{1}{2\pi i} \times 2\pi i \left(\text{Res}\left[\frac{1}{z(z-1/2)}, 0\right] + \text{Res}\left[\frac{1}{z(z-1/2)}, 1/2\right]\right)$$

$$= \frac{1}{-1/2} + \frac{1}{1/2} = 0$$

$$f[-2] = \frac{1}{2\pi i} \times 2\pi i \left(\text{Res}\left[\frac{1}{z^2(z-1/2)}, 0\right] + \text{Res}\left[\frac{1}{z^2(z-1/2)}, 1/2\right]\right)$$

$$= \frac{1}{0!}\left[\frac{d}{dz}\frac{1}{z-1/2}\right]_{z\to 0} + \frac{1}{(1/2)^2} = -\frac{1}{(z-1/2)^2}\bigg|_{z=0} + 4 = 0$$

$$\vdots$$

$$f[-n] = \frac{1}{2\pi i} \times 2\pi i \left(\text{Res} \left[\frac{1}{z^n(z-1/2)}, 0 \right] + \text{Res} \left[\frac{1}{z^n(z-1/2)}, 1/2 \right] \right)$$

$$= \frac{1}{(n-1)!} \left[\frac{d^{n-1}}{dz^{n-1}} \frac{1}{z-1/2} \right]_{z \to 0} + \frac{1}{(1/2)^n}$$

$$= \frac{1}{(n-1)!} \left[\frac{(-1)^{n-1}(n-1)!}{(z-1/2)^n} \right]_{z=0} + \frac{1}{(1/2)^n} = -2^n + 2^n = 0.$$

The contour integration does take care of both the $n \geq 0$ and $n \leq -1$ cases, although we do not need to do the calculation for $n \leq -1$ if we already know the answer is one sided. ∎

Example 9.41. $F_Z(z) = \frac{2z^2-z}{(z+\frac{3}{4})^2}$ has a double pole at $z = -\frac{3}{4}$. Take Γ to be a circle of radius greater than $\frac{3}{4}$, and apply Equation 9.81 with the formula for the residue at a multiple pole (Equation 8.17):

$$f[n] = \frac{1}{2\pi i} \oint_\Gamma \frac{(2z^2-z)z^{n-1}}{(z+\frac{3}{4})^2} dz = \text{Res} \left[\frac{2z^{n+1} - z^n}{(z+\frac{3}{4})^2}, z = -\frac{3}{4} \right]$$

$$= \lim_{z \to -\frac{3}{4}} \frac{d}{dz}(2z^{n+1} - z^n) = \lim_{z \to -\frac{3}{4}} 2(n+1)z^n - n z^{n-1}$$

$$= 2(n+1)\left(-\frac{3}{4}\right)^n - n\left(-\frac{3}{4}\right)^{n-1} = 2n\left(-\frac{3}{4}\right)^n + 2\left(-\frac{3}{4}\right)^n + \frac{4}{3}n\left(-\frac{3}{4}\right)^n$$

$$= \frac{10}{3}n\left(-\frac{3}{4}\right)^n + 2\left(-\frac{3}{4}\right)^n, \quad n \geq 0.$$ ∎

★ Two-Sided Inverse Transform

The Laplace transform for a two-sided function is inverted by closing the contour of integration to the left for $t > 0$, encircling poles in the left half-plane, and closing the contour of integration to the right for $t < 0$, encircling poles in the right half-plane. A similar procedure works for two-sided Z transforms. Poles inside the contour of integration (e.g., the unit circle) contribute to the inverse transform for $n \geq 0$, and poles outside the contour produce the inverse transform for $n < 0$. If the path of integration is understood to encircle the poles outside in the *clockwise* direction, then Equation 9.81 works also for the left-sided part, with a minus sign applied to the integral to account for the reversal of direction.

Example 9.42. Let $F_Z(z) = \frac{-1}{1-2z^{-1}}$, $|z| < 2$, which we calculated earlier from the sequence $f[n] = 2^n U[-n-1]$ (Example 9.34). From the region of convergence, we understand F_Z to be the transform of a left-sided sequence. It has a simple pole at $z = 2$. The inverse transform is calculated:

$$f[n] = -\frac{1}{2\pi i} \int_{\Gamma(\text{ext})} \frac{-z^{n-1}}{1-2z^{-1}} dz = \frac{1}{2\pi i} \times 2\pi i \, \text{Res}\left[\frac{z^n}{z-2}, 2\right] = 2^n, \quad n \leq -1.$$

Alternatively, to calculate the left-sided part we may apply the change of variable $\xi = 1/z$ to the integrand, which reflects the poles outside the unit circle to the inside, and the poles inside to the outside. The path of integration is reflected from $|z| = r$ to $|\xi| = 1/r$. The interior of the contour is traversed clockwise rather than counterclockwise, so a minus sign is applied to the integral:

$$f[n] = -\frac{1}{2\pi i} \oint_{\xi=1/r} F_Z(1/\xi) \, (1/\xi)^{n-1} \, d(1/\xi)$$

$$= \frac{1}{2\pi i} \oint_{\xi=1/r} F_Z(1/\xi) \, \xi^{-n-1} \, d\xi, \quad n \leq -1. \tag{9.82}$$

With this approach, the calculation in the previous example is

$$f[n] = \frac{1}{2\pi i} \oint_{|\xi|=1/r} \frac{-\xi^{-n-1}}{1-2\xi} \, d\xi = \frac{1}{2\pi i} \oint_{|\xi|=1/r} \frac{1}{2} \frac{\xi^{-n-1}}{\xi - 1/2} \, d\xi$$

$$= \operatorname{Res}\left[\frac{1}{2} \frac{\xi^{-n-1}}{\xi - 1/2}, \xi = 1/2\right] = (1/2)^{-n} = 2^n, n \leq -1.$$

For $n > 0$, the factor ξ^{-n-1} is an $n+1$-order pole at the origin, and a calculation like the one in Example 9.40 shows that the right-sided terms in the sequence are zero. ∎

Partial Fraction Expansion

The Mittag–Leffler expansion 9.42 can be used to break a Z transform into a sum of partial fractions, which are then easily inverted using standard transform pairs. The approach is illustrated by the following example.

Example 9.43. Consider the Z transform:

$$F_Z(z) = \frac{4z^2 - \frac{5}{4}z}{z^2 - \frac{3}{4}z + \frac{1}{8}}.$$

It is a proper rational function with two simple poles, located at $z = \frac{1}{2}$ and $z = \frac{1}{4}$. The function is therefore analytic at the origin and the conditions for the Mittag–Leffler expansion are met. The residues at the poles are $\frac{3}{2}$ and $\frac{1}{4}$, respectively. The principal parts are

$$P\left(z, \frac{1}{2}\right) = \frac{\frac{3}{2}}{z - \frac{1}{2}},$$

$$P\left(z, \frac{1}{4}\right) = \frac{\frac{1}{4}}{z - \frac{1}{4}}.$$

The partial fraction expansion is

$$F_Z(z) = F_Z(0) + \left[P\left(z, \frac{1}{2}\right) - P\left(0, \frac{1}{2}\right)\right] + \left[P\left(z, \frac{1}{4}\right) - P\left(0, \frac{1}{4}\right)\right]$$

$$= 0 + \left(\frac{\frac{3}{2}}{z - \frac{1}{2}} + 3\right) + \left(\frac{\frac{1}{4}}{z - \frac{1}{4}} + 1\right) = \frac{3z}{z - \frac{1}{2}} + \frac{z}{z - \frac{1}{4}}$$

$$= \frac{3}{1 - \frac{1}{2}z^{-1}} + \frac{1}{1 - \frac{1}{4}z^{-1}}.$$

The last step is performed to make the partial fractions match up with the basic Z transform pair, $a^n U[n] \longleftrightarrow \frac{1}{1 - az^{-1}}$. Using this pair, the inverse transform is

$$f[n] = 3\left(\frac{1}{2}\right)^n U[n] + \left(\frac{1}{4}\right)^n U[n].$$

∎

Example 9.44. Consider again the Z transform

$$F_Z(z) = \frac{2z^2 - z}{\left(z + \frac{3}{4}\right)^2}.$$

The form of the principal part is

$$P(z, a) = \frac{c_{-2}}{\left(z + \frac{3}{4}\right)^2} + \frac{c_{-1}}{z + \frac{3}{4}}.$$

The coefficients are calculated using Equation 9.45b:

$$c_{-2} = \lim_{z \to -\frac{3}{4}} \left(z + \frac{3}{4}\right)^2 \frac{2z^2 - z}{\left(z + \frac{3}{4}\right)^2} = \lim_{z \to -\frac{3}{4}} (2z^2 - z) = \frac{15}{8},$$

$$c_{-1} = \lim_{z \to -\frac{3}{4}} \frac{d}{dz}(2z^2 - z) = -4.$$

The partial fraction expansion is

$$F_Z(z) = F_Z(0) + P(z, a) - P(0, a) = 0 + \frac{\frac{15}{8}}{\left(z + \frac{3}{4}\right)^2} + \frac{-4}{z + \frac{3}{4}} - \left(\frac{\frac{15}{8}}{\frac{9}{16}} + \frac{-4}{\frac{3}{4}}\right)$$

$$= \frac{\frac{15}{8}}{\left(z + \frac{3}{4}\right)^2} + \frac{-4}{z + \frac{3}{4}} + 2.$$

9.2 THE Z TRANSFORM

The constant term is split across the two partial fractions to get them into standard forms:

$$F_Z(z) = \frac{\frac{15}{8}}{\left(z+\frac{3}{4}\right)^2} + \frac{-4}{z+\frac{3}{4}} + 2 = \frac{\frac{15}{8}}{\left(z+\frac{3}{4}\right)^2} + \frac{-4+2\left(z+\frac{3}{4}\right)}{z+\frac{3}{4}}$$

$$= \frac{\frac{15}{8}}{\left(z+\frac{3}{4}\right)^2} - \frac{\frac{5}{2}}{z+\frac{3}{4}} + \frac{2z}{z+\frac{3}{4}} = \frac{\frac{15}{8} - \frac{5}{2}\left(z+\frac{3}{4}\right)}{\left(z+\frac{3}{4}\right)^2} + \frac{2z}{z+\frac{3}{4}}$$

$$= \frac{-\frac{5}{2}z}{\left(z+\frac{3}{4}\right)^2} + \frac{2z}{\left(z+\frac{3}{4}\right)} = \frac{\frac{10}{3}\left(-\frac{3}{4}\right)z^{-1}}{\left(1+\frac{3}{4}z^{-1}\right)^2} + \frac{2}{\left(1+\frac{3}{4}z^{-1}\right)}.$$

Using the table at the end of the chapter, the inverse Z transform is

$$f[n] = \frac{10}{3}n\left(-\frac{3}{4}\right)^n + 2\left(-\frac{3}{4}\right)^n, \quad n \geq 0.$$

The manipulation of the constant term may seem a bit tricky and *ad hoc*. Certainly, one is aided by having an idea, based on experience, of what forms to look for in the partial fractions. If we were to leave the constant term alone and proceed with the table lookups, we would still get the right answer. Write

$$F_Z(z) = \frac{\frac{15}{8}}{\left(z+\frac{3}{4}\right)^2} + \frac{-4}{z+\frac{3}{4}} + 2$$

and use the table, together with the shift theorem:

$$F_Z(z) = -\frac{5}{2}z^{-1}\frac{-\frac{3}{4}z^{-1}}{(1-\frac{3}{4}z^{-1})^2} - 4z^{-1}\frac{1}{1+\frac{3}{4}z^{-1}} + 2,$$

$$f[n] = -\frac{5}{2}\cdot(n-1)\left(-\frac{3}{4}\right)^{n-1}U[n-1] - 4\cdot\left(-\frac{3}{4}\right)^{n-1}U[n-1] + 2\delta[n]$$

$$= -\frac{5}{2}n\left(-\frac{3}{4}\right)^{n-1}U[n-1] - \frac{3}{2}\left(-\frac{3}{4}\right)^{n-1}U[n-1] + 2\delta[n]$$

$$= \frac{10}{3}n\left(-\frac{3}{4}\right)^n U[n-1] + 2\left(-\frac{3}{4}\right)^n U[n-1] + 2\delta[n].$$

Now substitute $U[n-1] = U[n] - \delta[n]$:

$$f[n] = \underbrace{\frac{10}{3}n\left(-\frac{3}{4}\right)^n U[n] - \frac{10}{3}n\left(-\frac{3}{4}\right)^n\delta[n]}_{=0} + 2\left(-\frac{3}{4}\right)^n U[n] \underbrace{- 2\left(-\frac{3}{4}\right)^n\delta[n] + 2\delta[n]}_{=2\delta[n]}$$

$$= \frac{10}{3}n\left(-\frac{3}{4}\right)^n U[n] + 2\left(-\frac{3}{4}\right)^n U[n].$$

In this example, complex integration provides a more direct path to the inverse transform than the partial fraction expansion. ∎

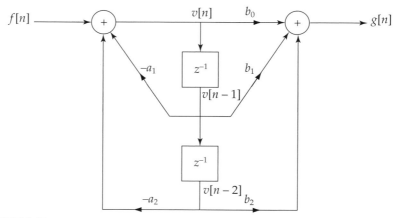

FIGURE 9.17 A second-order discrete-time LTI system. The boxes labelled "z^{-1}" represent unit time delays. An arrow with a coefficient symbolizes multiplication by that coefficient.

9.2.4 Discrete-time Systems

A causal linear, time-invariant discrete-time system (introduced in Section 4.9.3) is described by a linear difference equation:

$$g[n] + a_1 g[n-1] + \cdots + a_N g[n-N] = b_0 f[n] + b_1 f[n-1] + \cdots + b_M f[n-M].$$

where f is the input and g is the output. The current value of the output, $g[n]$, is recursively computed from the past values of the output and the current and past values of the input. There may be initial conditions specifying the values $g[-1], g[-2], \ldots, g[-N]$, but in many applications the initial conditions are assumed to be zero.

Figure 9.17 shows a representative second-order system. It is described by a pair of difference equations:

$$g[n] = b_0 v[n] + b_1 v[n-1] + b_2 v[n-2],$$
$$v[n] = f[n] - a_1 v[n-1] - a_2 v[n-2].$$

To eliminate v between the equations, take the Z transform of both sides:

$$G_Z(z) = (b_0 + b_1 z^{-1} + b_2 z^{-2}) V_Z(z),$$
$$V_Z(z) = F_Z(z) - a_1 z^{-1} V_Z(z) - a_2 z^{-2} V_Z(z),$$
$$\Rightarrow V_Z(z) = \frac{1}{1 + a_1 z^{-1} + a_2 z^{-2}} F_Z(z).$$

then substitute into the equation for G_Z:

$$G_Z(z) = \frac{b_0 + b_1 z^{-1} + b_2 z^{-2}}{1 + a_1 z^{-1} + a_2 z^{-2}} F_Z(z).$$

The Z transforms of input and output are connected by a rational function of z, which is the transfer function, $H_Z(z)$. Assuming that the poles of H_Z are inside the unit circle, $H_Z(e^{i\theta})$ gives the frequency response $H_d(\theta)$, a discrete-time Fourier transform. The inverse Z transform of $H_Z(z)$ is the impulse response $h[n]$, and by the convolution theorem,

$$g[n] = h * f[n].$$

When the coefficients a_n are zero, the output depends solely on the current and past values of the input. The impulse response of the second-order system is

$$h[n] = \mathcal{Z}^{-1}\{b_0 + b_1 z^{-1} + b_2 z^{-2}\} = b_0 \delta[n] + b_1 \delta[n-1] + b_2 \delta[n-2].$$

The impulse response has a finite duration of three time steps. Once an impulse input has passed through the two time delays, it has no further effect on the output. This type of system, with no feedback and a finite-duration impulse response, is called a *finite impulse response* (FIR) system or, when used as a frequency-selective filter, an *FIR filter*.

When one or more of the feedback coefficients a are nonzero, an input impulse continues to re-enter the delay elements and, consequently, its influence on the output never completely ends. This type of system has an infinite-duration impulse response and is called an *infinite impulse response* (IIR) system, or *IIR filter*. A particularly simple, but illustrative, case results when $a_2 = 0$ and $b_1 = b_2 = 0$. The transfer function is

$$H_Z(z) = \frac{b_0}{1 + a_1 z^{-1}}$$

and the impulse response is exponential:

$$h[n] = b_0 \left(-a_1\right)^n U[n].$$

In the time domain, with every cycle around the feedback loop, the intermediate term $v[n]$ is multiplied by another factor of $-a_1$. If $|a_1| < 1$, this results in an exponential decay of the impulse response. Both FIR and IIR systems are important in discrete-time control and signal processing.[16]

9.3 THE HILBERT TRANSFORM

9.3.1 The Fourier Transform of One-sided Functions

One-sided functions, which are nonzero only for $t \geq 0$, represent several real phenomena. A signal that turns on at $t = 0$, for example, $f(t) = \cos 2\pi t \, U(t)$, is one sided. The impulse response of a causal linear system (one which does not produce an output before the input is applied) is one sided, for example, $h(t) = e^{-t}U(t)$. Causality is an important constraint on the design of systems that operate in real time on streams of data.

---

[16]For applications of discrete-time systems in feedback control, see Franklin *et al.* (1998) and Kuo (1963). For applications to signal processing (digital filters), see Oppenheim and Schafer (2010), and also Porat (1997).

The theme of this section is that the real and imaginary parts of the Fourier transforms of one-sided functions are not independent, but can be calculated one from the other. To see how this works, break the one-sided function f into its even and odd parts, f_e and f_o. Because $f(t) = 0$ for $t < 0$,

$$f_e(t) = \frac{f(t) + f(-t)}{2} = \begin{cases} \frac{1}{2}f(t), & t \geq 0 \\ \frac{1}{2}f(-t), & t < 0 \end{cases},$$

$$f_o(t) = \frac{f(t) - f(-t)}{2} = \begin{cases} \frac{1}{2}f(t), & t \geq 0 \\ -\frac{1}{2}f(-t), & t < 0 \end{cases}.$$

Thus, the even and odd parts are connected,

$$f_o(t) = f_e(t) \operatorname{sgn} t.$$

Now take the Fourier transform of both sides of this expression. The Fourier transform of a real and even function is real, and the transform of a real and odd function is imaginary. So, with $F = F_r + iF_i$, $f_o \longmapsto iF_i$ and $f_e \longmapsto F_r$, and F_i is related to F_r by

$$iF_i(\nu) = F_r(\nu) * \mathcal{F}\{\operatorname{sgn} t\} = F_r(\nu) * \frac{1}{i\pi\nu}$$

or,

$$F_i(\nu) = F_r(\nu) * -\frac{1}{\pi\nu} = \frac{1}{\pi} \int_{-\infty}^{\infty} \frac{F_r(\eta)\, d\eta}{\eta - \nu}. \tag{9.83a}$$

It is also true that $f_e(t) = f_o(t) \operatorname{sgn} t$ (except at the origin, where $f_o(0) = 0$ but $f_e(0)$ need not be), so

$$F_r(\nu) = iF_i(\nu) * \frac{1}{i\pi\nu} = -\frac{1}{\pi} \int_{-\infty}^{\infty} \frac{F_i(\eta)\, d\eta}{\eta - \nu}. \tag{9.83b}$$

The integral in both of these equations is known as the *Hilbert transform*.[17] The Hilbert transform of a function $f : \mathbb{R} \to \mathbb{C}$ or a sequence $f : \mathbb{Z} \to \mathbb{C}$ is denoted $\mathcal{H}i\{f\}$ or $f_{\mathcal{H}i}$. We begin with the continuous case.

Definition 9.4 (Continuous Hilbert transform). Let $f : \mathbb{R} \to \mathbb{C}$ be a function. The Hilbert transform of f is a function $f_{\mathcal{H}i} : \mathbb{R} \to \mathbb{C}$ defined

$$f_{\mathcal{H}i}(x) = \mathcal{H}i\{f\} = -\frac{1}{\pi x} * f(x) = \frac{1}{\pi} \mathcal{P} \int_{-\infty}^{\infty} \frac{f(y)\, dy}{y - x}, \tag{9.84}$$

when the integral exists. The integral is taken as a Cauchy principal value because of the singularity at $y = x$.

---

[17] For historical notes on the Hilbert transform, see King (2009, Vol. 1, pp. 3–8). A comprehensive table of transforms, with many graphs of transform pairs, may be found in Vol. 2, pp. 453–546.

9.3 THE HILBERT TRANSFORM

With this notation, the real and imaginary parts of the Fourier transform of a one-sided function $f(t)$ (Equation 9.83) are compactly written

$$F_i = \mathcal{H}i\{F_r\} \quad \text{and} \quad F_r = -\mathcal{H}i\{F_i\}.$$

Example 9.45. The Fourier transform of the step function is $\frac{1}{2}\delta(v) + \frac{1}{i2\pi v} = \frac{1}{2}\delta(v) - i\frac{1}{2\pi v}$. We will show that the real and imaginary parts follow Equations (9.83), that is,

$$-\frac{1}{2\pi v} = \mathcal{H}i\left\{\frac{1}{2}\delta(v)\right\}$$

and

$$\frac{1}{2}\delta(v) = -\mathcal{H}i\left\{\frac{1}{2\pi v}\right\}.$$

The Hilbert transform of $\frac{1}{2}\delta(v)$ is easy, because of the sifting property of the delta function:

$$-\frac{1}{\pi}\mathcal{P}\int_{-\infty}^{\infty}\frac{\frac{1}{2}\delta(\eta)\,d\eta}{v-\eta} = -\frac{1}{2\pi v}.$$

For the imaginary part, use the convolution theorem:

$$-\mathcal{H}i\left\{\frac{-1}{2\pi v}\right\} = \frac{1}{\pi v} * -\frac{1}{2\pi v}$$

$$= \mathcal{F}\left\{i\,\text{sgn}\,t \cdot -\frac{1}{2}i\,\text{sgn}\,t\right\} = \frac{1}{2}\mathcal{F}\{1\} = \frac{1}{2}\delta(v). \quad \blacksquare$$

Example 9.46. The Fourier transform of the one-sided exponential, $e^{-t}U(t)$, is $\frac{1}{1+i2\pi v} = \frac{1}{1+(2\pi v)^2} + i\frac{-2\pi v}{1+(2\pi v)^2}$. We will calculate the Hilbert transform of the real part by direct integration. The verification for the imaginary part is similar and is left to the problems:

$$\mathcal{H}i\left\{\frac{1}{1+(2\pi v)^2}\right\} = -\frac{1}{\pi}\mathcal{P}\int_{-\infty}^{\infty}\frac{d\eta}{(v-\eta)(1+(2\pi\eta)^2)}.$$

This appears well suited to calculation with a contour integral. Consider the complex integral

$$\frac{1}{2\pi^3}\oint_{\Gamma}\frac{dz}{(z-v)(z^2+(1/2\pi)^2)}$$

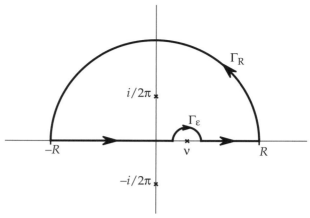

FIGURE 9.18 Contour for calculating the Hilbert transform of $\frac{1}{1+(2\pi v)^2}$.

on an indented contour, as shown in Figure 9.18. There are simple poles at $z = v$, $\frac{i}{2\pi}$, and $-\frac{i}{2\pi}$. The pole at $z = \frac{i}{2\pi}$ is within the contour, and its residue is

$$\text{Res}\left[\frac{1}{(z-v)\left(z^2+(1/2\pi)^2\right)}, \frac{i}{2\pi}\right] = \frac{-2\pi^2}{1+i2\pi v}.$$

So,

$$\oint_\Gamma = 2\pi i \frac{-2\pi^2}{1+i2\pi v} = \frac{-i4\pi^3}{1+i2\pi v}.$$

Using the ML inequality, the integral on Γ_R is seen to go to zero as $R \to \infty$. The contribution of the indentation Γ_ϵ is

$$\oint_{\Gamma_\epsilon} = -i\pi \,\text{Res}\left[\frac{1}{(z-v)\left(z^2+(1/2\pi)^2\right)}, v\right] = \frac{-i4\pi^3}{1+(2\pi v)^2}.$$

Thus,

$$\mathcal{H}i\left\{\frac{1}{1+(2\pi v)^2}\right\} = \frac{1}{2\pi^3}\left[\oint_\Gamma - \oint_{\Gamma_\epsilon}\right] = \frac{1}{2\pi^3}\left[\frac{-i4\pi^3}{1+i2\pi v} + \frac{i4\pi^3}{1+(2\pi v)^2}\right]$$
$$= 2\frac{-i(1-i2\pi v)+i}{1+(2\pi v)^2} = \frac{-2\pi v}{1+(2\pi v)^2},$$

which is the imaginary part of $\frac{1}{1+i2\pi v}$. ∎

One-Sided Infinite Sequences

Now consider the even and odd parts of a one-sided sequence $(f[n])_{n=0}^{\infty}$:

$$f_e[n] = \frac{f[n] + f[-n]}{2} = \begin{cases} \frac{1}{2}f[n], & n > 0 \\ f[0], & n = 0 \\ \frac{1}{2}f[-n], & n < 0 \end{cases}$$

$$f_o[n] = \frac{f[n] - f[-n]}{2} = \begin{cases} \frac{1}{2}f[n], & n > 0 \\ 0, & n = 0 \\ -\frac{1}{2}f[-n], & n < 0 \end{cases}$$

Again the even and odd parts are connected:

$$f_e[n] = f[0]\,\delta[n] + f_o[n]\,\mathrm{sgn}\,[n],$$
$$f_o[n] = f_e[n]\,\mathrm{sgn}\,[n],$$

where the discrete-time signum is defined:

$$\mathrm{sgn}\,[n] = \begin{cases} 1, & n > 0 \\ 0, & n = 0 \\ -1, & n < 0 \end{cases} \qquad (9.85)$$

The real and imaginary parts of the discrete-time Fourier transform, F_{dr} and F_{di}, are calculated from f_e and f_o, respectively, using the convolution theorem 4.16.

$$F_{\mathrm{dr}} = f[0] + iF_{\mathrm{di}} \circledast \mathcal{F}\{\mathrm{sgn}\},$$
$$iF_{\mathrm{di}} = F_{\mathrm{dr}} \circledast \mathcal{F}\{\mathrm{sgn}\}.$$

Now, the Fourier transform of the signum sequence is

$$\mathcal{F}\{\mathrm{sgn}\} = \sum_{n=-\infty}^{\infty} \mathrm{sgn}\,[n]\,e^{-in\theta}.$$

The signum is not absolutely summable. We will apply a convergence factor $a^{|n|}$, $0 < a < 1$, and calculate the transform in the limit. The steps should all be familiar by now:

$$\mathcal{F}\{\text{sgn}\} = \lim_{a \to 1} \sum_{n=-\infty}^{\infty} a^{|n|} \text{sgn}[n] e^{-in\theta}$$

$$= \lim_{a \to 1} \left[\sum_{n=1}^{\infty} (ae^{-i\theta})^n - \sum_{n=-\infty}^{-1} (ae^{+i\theta})^{-n} \right] = \lim_{a \to 1} \sum_{n=1}^{\infty} [(ae^{-i\theta})^n - (ae^{+i\theta})^n]$$

$$= \lim_{a \to 1} \lim_{N \to \infty} \left[\frac{1 - (ae^{-i\theta})^N}{1 - ae^{-i\theta}} - \frac{1 - (ae^{+i\theta})^N}{1 - ae^{+i\theta}} \right] = \lim_{a \to 1} \left[\frac{1}{1 - ae^{-i\theta}} - \frac{1}{1 - ae^{+i\theta}} \right]$$

$$= \frac{-2i \sin \theta}{2 - 2 \cos \theta} = -i \cot \left(\frac{\theta}{2} \right). \tag{9.86}$$

With this, the real and imaginary parts of F_d are connected through a Hilbert transform relationship:

$$F_{\text{dr}}(\theta) = f[0] + i F_{\text{di}}(\theta) \circledast -i \cot \left(\frac{\theta}{2} \right)$$

$$= f[0] - \frac{1}{2\pi} \mathcal{P} \int_{-\pi}^{\pi} F_{\text{di}}(\varphi) \cot \left(\frac{\varphi - \theta}{2} \right) d\varphi, \tag{9.87a}$$

$$F_{\text{di}}(\theta) = -i F_{\text{dr}} \circledast -i \cot \left(\frac{\theta}{2} \right)$$

$$= \frac{1}{2\pi} \mathcal{P} \int_{-\pi}^{\pi} F_{\text{di}}(\varphi) \cot \left(\frac{\varphi - \theta}{2} \right) d\varphi. \tag{9.87b}$$

The integrals are taken as Cauchy principal values because the cotangent blows up at $\varphi = \theta$.

The discrete-time Fourier transform is periodic with period 2π. Equations 9.87 are a particular case of the following theorem.

Theorem 9.28 (Hilbert transform, periodic functions). Let $f : \mathbb{R} \to \mathbb{C}$ be periodic with period L. The Hilbert transform of f is

$$\mathcal{H}i\{f\} = \frac{1}{\pi} \mathcal{P} \int_{-\infty}^{\infty} \frac{f(y) \, dy}{y - x}$$

$$= \frac{1}{L} \mathcal{P} \int_{-L/2}^{L/2} f(y) \cot \left(\frac{\pi}{L}(y - x) \right) dy. \tag{9.88}$$

★ *Hilbert Transform and Sampling*

The Hilbert transform relationships for the discrete-time Fourier transform of a one-sided sequence, Equations 9.87, are connected with the Hilbert transform of a continuous time function, Equations 9.83, through sampling. It will be instructive to derive them again from this point of view. A proof of Theorem 9.28 can be constructed along these lines.

Consider the function $f(x)$, sampled at $x = n\Delta x$. From sampling theory, we know that $f_s(x) = f(x) \frac{1}{\Delta x} \text{III} \left(\frac{x}{\Delta x} \right)$ has Fourier transform $F_s(\nu) = F(\nu) * \text{III}(\Delta x \, \nu)$. If f is one sided, then so is f_s, and the real and imaginary parts of F_s, which we denote

F_{sr} and F_{si}, must also be connected by the Hilbert transform. Calculate[18]

$$F_{si}(v) = \mathcal{H}i\{F_r(v) * III(\Delta xv)\}$$
$$= -\frac{1}{\pi v} * [F_r(v) * III(\Delta xv)] = F_r(v) * \left[-\frac{1}{\pi v} * III(\Delta xv)\right]$$
$$= F_r(v) * \sum_{k=-\infty}^{\infty} -\frac{1}{\pi \Delta x} \frac{1}{v - k/\Delta x}.$$

It can be shown (see the problems) that the periodic replication of $1/\pi v$ is the cotangent function, $\cot(\pi \Delta x\, v)$, a Mittag–Leffler expansion:

$$\sum_{k=-\infty}^{\infty} \frac{1}{\pi \Delta x} \frac{1}{v - k/\Delta x} = \cot(\pi \Delta x\, v).$$

With this result, we have

$$F_{si}(v) = F_r(v) * -\cot(\pi \Delta x\, v)$$
$$= \int_{-\infty}^{\infty} F_r(\eta) \cot[\pi \Delta x (\eta - v)]\, d\eta, \qquad (9.89a)$$

and similarly,

$$F_{sr}(v) = F_i(v) * -\cot(\pi \Delta x\, v)$$
$$= \int_{-\infty}^{\infty} F_r(\eta) \cot[\pi \Delta x (\eta - v)]\, d\eta. \qquad (9.89b)$$

The imaginary (real) part of the replicated F is equal to the convolution of the real (imaginary) part of the original F with the periodic cotangent function.

Because $\cot(\pi \Delta x\, v)$ is periodic with period $1/\Delta x$, it can be written as the convolution of one period (isolated with a rectangle function) and a comb:

$$\cot(\pi \Delta x\, v) = [\cot(\pi \Delta x\, v)\, \text{rect}(\Delta x\, v)] * \Delta x\, III(\Delta x\, v),$$

which enables F_{si} to be written again as a (different) triple convolution:

$$F_{si}(v) = -F_r(v) * ([\cot(\pi \Delta x\, v)\, \text{rect}(\Delta x\, v)] * \Delta x\, III(\Delta x\, v))$$
$$= -\Delta x \underbrace{(F_r(v) * III(\Delta x\, v))}_{F_{sr}(v)} * [\cot(\pi \Delta x\, v)\, \text{rect}(\Delta x\, v)]$$
$$= \int_{-1/2\Delta x}^{1/2\Delta x} \Delta x\, F_{sr}(\eta - v) \cot(\pi \Delta x\, \eta)\, d\eta.$$

Finally, recall that the discrete-time Fourier transform $F_d(\theta)$ of the sequence $f[n] = f(n\Delta x)$ is related to the Fourier transform $F_s(v)$ by the analog-to-digital

---

[18] See Champeney (1987, pp. 139–144) for a discussion of the associativity of convolution for generalized functions. Briefly, $f * (g * h) = (f * g) * h$ if $f * \varphi$ is a good function and the remaining convolution, $g * h$, is defined. Here, $F_r * \varphi$ is good because f is assumed bandlimited and square integrable, and, by direct calculation, the convolution of $1/\pi v$ and $III(\Delta x\, v)$ exists.

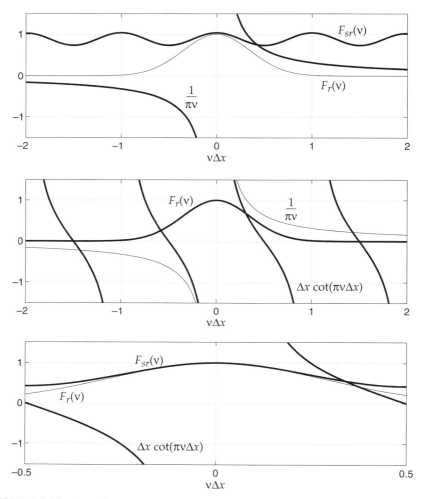

FIGURE 9.19 *Top*: The Hilbert transform of the replicated function F_{sr} is a convolution with $\frac{1}{\pi v}$. *Middle*: It is equivalent to the convolution of the unreplicated F_r with $\cot(\pi \Delta x v)$. *Bottom*: Finally, it is equivalent to a convolution of F_{sr} with $\Delta x \cot(\pi \Delta x v)$ over the interval $(-\frac{1}{2\Delta x}, \frac{1}{2\Delta x})$.

frequency mapping, $\theta = 2\pi \Delta x \, v$ (Section 6.6.2),

$$F_d(\theta) = F_s\left(\frac{\theta}{2\pi \Delta x}\right).$$

Make this change of variable, and also let $\phi = 2\pi \Delta x \eta$, $d\eta = d\phi/2\pi \Delta x$; then,

$$F_{di}(\theta) = \frac{1}{2\pi} \mathcal{P} \int_{-\pi}^{\pi} F_{dr}(\phi - \theta) \cot\left(\frac{\phi}{2}\right) d\phi = \frac{1}{2\pi} \mathcal{P} \int_{-\pi}^{\pi} F_{dr}(\phi) \cot\left(\frac{\phi - \theta}{2}\right) d\phi,$$

which is, again, Equation 9.87. These convolutions are illustrated in Figure 9.19.

One-Sided Finite Sequences

To derive the analogous Hilbert transform relationships for the DFT, we need to define a finite one-sided sequence. Recall that the DFT implicitly regards a vector as one period of an infinite periodic sequence, for which evenness and oddness are defined (Equation 3.2):

$$\text{Even: } f_e[n] = f_e[N - n],$$
$$\text{Odd: } f_o[n] = -f_o[N - n].$$

With N even, an odd periodic sequence must have $f_o[0] = -f_o[N] = 0$ and $f_o[\frac{N}{2}] = -f_o[\frac{N}{2}] = 0$. We define a finite signum sequence by

$$\text{sgn}[n] = \begin{cases} 0, & n = 0, \frac{N}{2} \\ 1, & n = 1, 2, \ldots, \frac{N}{2} - 1 \\ -1, & n = \frac{N}{2} + 1, \ldots, N - 2, N - 1 \end{cases} \quad (9.90)$$

Then, for a one-sided sequence $f = f_e + f_o$, the odd part is connected to the even part in this way:

$$f_o[n] = f_e[n] \, \text{sgn}[n],$$

and the even part is connected to the odd part by

$$f_e[n] = f[0]\,\delta[n] + f\left[\frac{N}{2}\right]\delta\left[n - \frac{N}{2}\right] + f_o[n]\,\text{sgn}[n].$$

If f is considered as one period of an infinite periodic sequence, the entire sequence cannot be one sided. Rather, a one-sided periodic sequence is zero for the latter half of each period ($f[\frac{N}{2} + 1]$ through $f[N - 1]$). This is what we observe if we add f_e and f_o:

$$f_e[n] + f_o[n] = f_e[n](1 + \text{sgn}[n]) = \begin{cases} f_e[n], & n = 0, N/2 \\ 2f_e[n], & n = 1, \ldots N/2 - 1 \\ 0, & n = N/2 + 1 \ldots N - 1 \end{cases}.$$

With these definitions and the convolution theorem (Equation 3.31), we calculate the DFTs:

$$iF_i[m] = \frac{1}{N} F_r[m] \circledast \text{DFT}\{\text{sgn}\},$$

$$F_r[m] = f[0] + f\left[\frac{N}{2}\right] e^{-i\pi m} + \frac{1}{N} iF_i[m] \circledast \text{DFT}\{\text{sgn}\}.$$

For the DFT of the signum sequence,

$$\text{DFT}\{\text{sgn}\} = \sum_{n=0}^{N-1} \text{sgn}[n] e^{-i2\pi mn/N} = \sum_{n=1}^{\frac{N}{2}-1} e^{-i2\pi mn/N} - \sum_{n=\frac{N}{2}+1}^{N-1} e^{-i2\pi mn/N}$$

$$= \sum_{n=1}^{\frac{N}{2}-1} e^{-i2\pi mn/N} - e^{-i\pi m} \sum_{n=1}^{\frac{N}{2}-1} e^{-i2\pi mn/N}$$

$$= (1 - e^{-i\pi m}) \left(-1 + \frac{1 - e^{-i\pi m}}{1 - e^{-i2\pi m/N}} \right).$$

The factor $(1 - e^{-i\pi m})$ is 0 for even m and 2 for odd m. The other factor simplifies to

$$-1 + \frac{1 - e^{-i\pi m}}{1 - e^{-i2\pi m/N}} = \frac{e^{-i2\pi m/N} - e^{-i\pi m}}{1 - e^{-i2\pi m/N}}$$

$$= \frac{e^{-i\pi m/N} - e^{-i\pi m} e^{+i\pi m/N}}{2i \sin\left(\frac{\pi m}{N}\right)}.$$

The factor of $e^{-i\pi m}$ in the numerator is $+1$ for even m and -1 for odd m. But only odd-indexed terms will be nonzero, so we replace $e^{-i\pi m}$ by -1 and the numerator becomes $2\cos\left(\frac{\pi m}{N}\right)$. Putting the pieces together, we have

$$\text{DFT}\{\text{sgn}\} = \begin{cases} 0, & m \text{ even} \\ -i2 \cot\left(\frac{\pi m}{N}\right), & m \text{ odd} \end{cases}. \tag{9.91}$$

(You are invited to compare this result to Equation 9.86 in light of the relationship between the DFT and the discrete-time Fourier transform.)

Therefore, the Hilbert transform relationships are

$$F_i[m] = \sum_{k=0}^{N-1} F_r[k] S[k-m], \tag{9.92a}$$

$$F_r[m] = f[0] + (-1)^m f\left[\frac{N}{2}\right] - \sum_{k=0}^{N-1} F_i[k] S[k-m], \tag{9.92b}$$

where

$$S[m] = \begin{cases} 0, & m \text{ even} \\ \frac{2}{N} \cot\left(\frac{\pi m}{N}\right), & m \text{ odd} \end{cases}. \tag{9.92c}$$

Like Theorem 9.28, these relationships for a periodic sequence are a particular case of a more general Hilbert transform pair for sequences.

9.3 THE HILBERT TRANSFORM

Definition 9.5 (Discrete Hilbert transform). Let $f : \mathbb{Z} \to \mathbb{C}$ be a sequence. The Hilbert transform of f is a sequence $f_{\mathcal{H}i} : \mathbb{Z} \to \mathbb{C}$ defined

$$f_{\mathcal{H}i}[n] = \sum_{k=-\infty}^{\infty} f[k]s[k-n],$$

$$\text{where } s[n] = \begin{cases} 0, & n \text{ even} \\ \dfrac{2}{\pi n}, & n \text{ odd} \end{cases}, \quad (9.93)$$

when the series converges. When f is periodic, the kernel $S[m]$ in Equation 9.92 can be shown to be the periodic replication of the kernel $s[n]$ in Equation 9.93.

Minimum Phase Systems

A real, causal, stable LTI system has a right-sided impulse response $h(t)$ or $h[n]$. The poles of the transfer function $H_L(s)$ are in the left half-plane for a continuous time system, and the poles of $H_Z(z)$ are inside the unit circle for a discrete-time system. The system has a frequency response, $H(\nu)$ or $H_d(\theta)$, and the real and imaginary parts of the transfer function are connected by the Hilbert transform, Equation 9.83 or 9.87.

While causality and stability require the poles of the transfer function to be in the left half-plane or inside the unit circle, they leave the zeros unconstrained. However, if the zeros as well as the poles are in the left half-plane or inside the unit circle, so that the region of convergence of $H_L(s)$ or $H_Z(z)$ is free of zeros as well as poles, the system has the additional property of *minimum phase*.

How this works may be seen with the aid of Figure 9.20. An S-plane frequency response is the ratio of polynomials, which for simplicity we assume have factors of multiplicity one. That is,

$$H_L(i\omega) = A \frac{\prod_m (i\omega - c_m)}{\prod_n (i\omega - d_n)}.$$

where the $\{c_m\}$ are the zeros and $\{d_m\}$ are the poles. A single zero, $i\omega - c$, is illustrated in the figure. (The same geometry pertains to the contributions of poles in the left half-plane.) The magnitude $|i\omega - c|$ is its contribution to the magnitude response, and the indicated angle $\phi = \arg(i\omega - c)$ is its contribution to the phase response. There is a symmetric point in the right half-plane, $c' = -c^*$, which gives the same value of magnitude, that is, $|i\omega - c| = |i\omega - c'|$, for all frequencies ω. The magnitude response of the system is the same whether the zero is in the left half-plane at $s = c$ or in the right half-plane at $s = c'$. However, $\arg(i\omega - c')$ is greater than $\arg(i\omega - c)$ at all frequencies. The right half-plane zero contributes more phase shift to the frequency response than the left half-plane zero. Thus, system with all its zeros in the left half-plane will have the minimum phase shift of all systems with the same magnitude response.

Similarly, a discrete-time frequency response is the ratio of polynomials in $e^{i\theta}$, which again we assume have factors of multiplicity one:

$$H_d(\theta) = H_Z(e^{i\theta}) = A \frac{\prod_m (e^{i\theta} - c_m)}{\prod_n (e^{i\theta} - d_n)}.$$

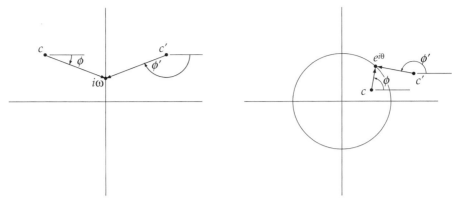

FIGURE 9.20 *Left*: Two symmetric points $c = -a + ib$ and $c' = -c^* = a + ib$ are the same distance from a point $i\omega$. The vectors drawn from c to $i\omega$ and from c' to $i\omega$ represent the complex factors $i\omega - c$ and $i\omega - c'$. The contribution of a zero at $s = c$ to the magnitude of frequency response $H_L(i\omega)$ is the same as the contribution of a zero at $s = c'$ for all frequencies, but the right half-plane zero always contributes more phase shift, $\phi' > \phi$. *Right*: Two symmetric points $c = re^{i\beta}$ and $c' = 1/c^* = r^{-1}e^{i\beta}$. The vectors drawn from c to $e^{i\theta}$ and from c' to $e^{i\theta}$ represent the complex factors $e^{i\theta} - c$ and $e^{i\theta} - c'$. The magnitude of a transfer function $H_d(\theta) = H_Z(e^{i\theta})$ with a zero at c' differs from the magnitude with a zero at c by the constant factor r for all frequencies, but the zero outside the unit circle always contributes more phase shift, $\phi' > \phi$.

A single zero is illustrated in the figure. The magnitude $|e^{i\theta} - c|$ is the contribution of c to the magnitude response, and the indicated angle $\phi = \arg(e^{i\theta} - c)$ is its contribution to the phase response. There is a symmetric point outside the unit circle, at $c' = 1/c^*$. The magnitude $|e^{i\theta} - c'|$ is not the same as $|e^{i\theta} - c|$, but their ratio is a constant for all frequencies. Whether a zero is inside the unit circle or outside only scales the magnitude response. However, as the figure shows, the angle ϕ' is always greater than ϕ, so the zero outside the unit circle contributes a larger phase shift than the zero inside the unit circle. A discrete-time system with all its zeros inside the unit circle will have the minimum phase shift of all systems having the same (within a constant factor) magnitude response.

If a system is minimum phase, there will be no zeros or poles in the right half-plane or outside the unit circle; both H and $1/H$ are analytic in the right half-plane (outside the unit circle). As a result, the logarithm of the transfer function, $\log H_L(s)$ or $\log H_Z(z)$, will be singularity-free in those regions—no poles, zeros, logarithmic singularities, or branch points. Moreover, because the original impulse response $h(t)$ or $h[n]$ is real valued, the frequency response $H_L(i\omega)$ or $H_Z(e^{i\theta})$ is Hermitian. Now, as we have seen before, a Hermitian function has an even real part and an odd imaginary part; it therefore has an even magnitude and an odd phase. So, $\log H_L(i\omega) = \log |H_L(i\omega)| + i \arg H_L(i\omega)$ is Hermitian as well. This is similarly true for $\log H_Z(e^{i\theta})$. Consequently, in a minimum phase system, the (log) magnitude response and the phase response are also connected by the Hilbert transform.

Hilbert transform relationships for magnitude and phase are presented below and considered further in the problems. In continuous time,[19] one form is (recall $H(\nu) = H_L(i2\pi\nu)$)

$$\phi(\nu) = -\frac{2}{\pi}\mathcal{P}\int_0^\infty \frac{\log|H(\eta)|\,d\eta}{\nu(1-\eta^2/\nu^2)}, \qquad (9.94a)$$

$$\log|H(\nu)| - \log|H(0)| = \frac{2}{\pi}\mathcal{P}\int_0^\infty \frac{\phi(\eta)\,d\eta}{\eta(1-\eta^2/\nu^2)}. \qquad (9.94b)$$

For discrete time,[20] a straightforward application of Equations 9.92 (recalling $H_d(\theta) = H_Z(e^{j\theta})$) gives

$$\arg(H_d(\theta)) = -\frac{1}{2\pi}\mathcal{P}\int_{-\pi}^\pi \log|H_d(\theta)|\cot\left(\frac{\theta-\eta}{2}\right)d\eta, \qquad (9.95a)$$

$$\log|H_d(\theta)| - \hat{h}[0] = \frac{1}{2\pi}\mathcal{P}\int_{-\pi}^\pi \arg(H_d(\theta))\cot\left(\frac{\theta-\eta}{2}\right)d\eta, \qquad (9.95b)$$

where $\hat{h}[0] = \frac{1}{2\pi}\int_{-\pi}^\pi \log|H_d(\theta)|\,d\theta$.

In practice, numerical integration would be employed to compute an appropriate phase given an experimentally measured magnitude.

⋆ Signal Reconstruction Problems

We have now seen a few examples of connections between constraints in one domain and dependencies in the other domain.

- If a function is bandlimited, it cannot simultaneously be time limited, and vice versa (Paley–Wiener).
- If a function is causal, the real and imaginary parts of its Fourier transform are connected by the Hilbert transform. Likewise, the magnitude and phase of its Fourier transform are connected by the Hilbert transform.

Similar problems of practical importance have been and continue to be studied. A few of these are listed below. Space does not permit a detailed discussion, but references are given for the interested reader.

- Suppose it is desired to construct a causal LTI system with a specified magnitude response $|H(\nu)| \in L^2$. The desired causal impulse response $h(t)$ can be found if and only if

$$\int_{-\infty}^\infty \frac{|\log|H(\nu)||}{1+\nu^2}\,d\nu < \infty \qquad (9.96)$$

---

[19] King (2009, Vol. 2, pp. 257–263, 437–443); Several magnitude–phase relationships are developed in the seminal work by Bode (1945, Chapters 13 and 14). Connections between Bode's results developed for electrical engineering, and the Kramers–Kronig equations, familiar in physics, are explored in Bechhoefer (2011).

[20] Oppenheim and Schafer (2010, Chapter 11).

(this is called the *Paley–Wiener criterion*).[21] A corollary of this result is the fact that a function cannot simultaneously be causal and bandlimited (take, for example, the Fourier pair sinc $t \longleftrightarrow$ rect v in Equation 9.96).

- Suppose a function f is known to be bandlimited, and only partial measurements of f, for example, $f_T(t) = f(t)\operatorname{rect}(t/T)$, are available. The *bandlimited extrapolation* problem is to reconstruct all or part of f outside $(-T, T)$. In principle, because a bandlimited function is entire, a Taylor series developed from the known measurements of f_T can be used to extrapolate to f. However, noisy measurements prohibit accurate knowledge of the series coefficients, and other methods have been studied.[22] In image processing, the *superresolution* problem is to computationally extrapolate the frequency spectrum of an image beyond the natural bandlimit imposed by the physics of image formation and acquisition.

- Suppose a function f is known to have finite support (time or space limited), and measurements of its Fourier magnitude $|F|$ are available, but the Fourier phase is unknown. With the additional knowledge that f is real and nonnegative, calculate the phase of F and so recover f. Unlike the Hilbert transform, there is no simple integral transform solution to this problem, known as *phase retrieval*. Rather, iterative solutions have been developed.[23]

9.3.2 Hilbert Transform Properties

We begin by listing four forms of the Hilbert transform, for different kinds of functions. Three of these were derived in the previous section. The derivation of the Hilbert transform for discrete time is left to the problems.

The Hilbert transforms $f_{Hi} = \mathcal{H}i\{f\}$ for continuous and discrete-time functions f are defined as follows:

$$\text{Continuous time: } f_{Hi}(x) = -\frac{1}{\pi x} * f(x) = \frac{1}{\pi} \mathcal{P} \int_{-\infty}^{\infty} \frac{f(y)\, dy}{y - x}, \tag{9.84}$$

$$\text{Periodic function: } f_{Hi}(x) = -\frac{1}{L} \cot\left(\frac{\pi x}{L}\right) * f(x) \operatorname{rect}\left(\frac{x}{L}\right)$$

$$= \frac{1}{L} \mathcal{P} \int_{-L/2}^{L/2} f(y) \cot\left(\frac{\pi}{L}(y - x)\right) dy, \tag{9.88}$$

$$\text{Discrete time: } f_{Hi}[n] = \sum_{k=-\infty}^{\infty} f[k]s[k - n],$$

$$s[n] = \begin{cases} 0, & n \text{ even} \\ \dfrac{2}{\pi n}, & n \text{ odd} \end{cases}, \tag{9.93}$$

---

[21] Paley and Wiener (1934, pp. 16–19); see Papoulis (1977, pp. 227–231) for a discrete-time version.
[22] Papoulis (1977, pp. 243–251).
[23] A classic reference on phase retrieval is Fienup (1982). A recent perspective, with references to a wide variety of applications, is Fienup (2013).

9.3 THE HILBERT TRANSFORM

Periodic sequence: $f_{Hi}[n] = \sum_{k=0}^{N-1} f[k] s[(k-n) \mod N]$,

$$s[n] = \begin{cases} 0, & n \text{ even} \\ \dfrac{2}{N} \cot\left(\dfrac{\pi n}{N}\right), & n \text{ odd} \end{cases}. \qquad (9.92)$$

The integrals (and sums) are taken as Cauchy principal values (or the analogous form for sums) because of the singularities at $y = x$ ($k = n$).

Existence of the Hilbert Transform

The Hilbert transform exists for many of the functions we have worked with before, but not all of them, owing to the singular nature of the transform kernel, $-1/\pi x$.

- If a function (sequence) is L^2 (ℓ^2), its Hilbert transform exists and is also L^2 (ℓ^2).
- If a function (sequence) is L^1 (ℓ^1), its Hilbert transform exists, but is not necessarily also L^1 (ℓ^1).[24] In Example 9.46, $1/(1+x^2) \in L^1$, its Hilbert transform, $-x/(1+x^2)$, is not L^1, and both functions are L^2.
- The delta function and its derivatives have Hilbert transforms. The sine and cosine functions have Hilbert transforms. However, slowly growing functions like step, signum, and polynomials, do not have Hilbert transforms.[25]

Hilbert Transform as a Linear Filter

Unlike the Fourier transform, both the function f and its Hilbert transform f_{Hi} are in the same domain, time or frequency. The convolutional form of the Hilbert transform gives it an interpretation as a linear filter. In continuous time,

$$f_{Hi}(t) = h(t) * f(t).$$

The impulse response of the filter is

$$h(t) = -\frac{1}{\pi t} \qquad (9.97a)$$

and the transfer function is

$$H(\nu) = \mathcal{F}\left\{-\frac{1}{\pi t}\right\} = i \operatorname{sgn} \nu. \qquad (9.97b)$$

Example 9.47 (Cosine and sine are a Hilbert transform pair). The Hilbert transform of the cosine function $\cos 2\pi b t$ ($b > 0$) is the integral

$$\mathcal{H}i\{\cos 2\pi b t\} = \frac{1}{\pi} \mathcal{P} \int_{-\infty}^{\infty} \frac{\cos 2\pi b \tau \, d\tau}{\tau - t},$$

---

[24] See King (2009, Vol. 1, pp. 96–99, 203–215) for proofs for L^p functions.

[25] The generalized functions that have Hilbert transforms belong to a narrower class than tempered distributions, with different testing functions. See King (2009, Vol. 1, Chapter 10); also, Pandey (1996, Chapters 3 and 4).

which can be evaluated by calculating

$$\mathcal{P}\int_{-\infty}^{\infty}\frac{e^{-i2\pi b\tau}}{\pi(\tau-t)}\,d\tau$$

and taking the real part of the result. This is almost an integral we have done before (Example 8.25). Make the change of variable $\xi = \tau - t$,

$$\mathcal{P}\int_{-\infty}^{\infty}\frac{e^{-i2\pi b\tau}}{\pi(\tau-t)}\,d\tau = \mathcal{P}\int_{-\infty}^{\infty}\frac{e^{-i2\pi b(t+\xi)}}{\pi\xi}\,d\xi$$
$$= e^{-i2\pi bt}\,\mathcal{P}\int_{-\infty}^{\infty}\frac{e^{-i2\pi b\xi}}{\pi\xi}\,d\xi.$$

The integral is the Fourier transform of $\frac{1}{\pi x}$, with x replaced by ξ and ν replaced by b. We know that $\frac{1}{x} \longmapsto -i\pi\,\mathrm{sgn}\,\nu$, so

$$e^{-i2\pi bt}\,\mathcal{P}\int_{-\infty}^{\infty}\frac{e^{-i2\pi b\xi}}{\pi\xi}\,d\xi = e^{-i2\pi bt}\cdot(-i\,\mathrm{sgn}\,b) = -ie^{-i2\pi bt}$$

($\mathrm{sgn}\,b = 1$ because $b > 0$). Taking the real part,

$$\mathcal{H}i\{\cos 2\pi bt\} = \mathcal{R}e\{-ie^{-i2\pi bt}\} = -\sin 2\pi bt.$$

Calculating in the frequency domain instead,

$$\cos 2\pi bt \longmapsto \frac{1}{2}\delta(\nu-b) + \frac{1}{2}\delta(\nu+b).$$

Multiply this by the Hilbert transform transfer function $i\,\mathrm{sgn}\,\nu$:

$$\left[\frac{1}{2}\delta(\nu-b) + \frac{1}{2}\delta(\nu+b)\right]\cdot i\,\mathrm{sgn}\,\nu = \frac{1}{2}i\,\mathrm{sgn}(b)\,\delta(\nu-b) + \frac{1}{2}i\,\mathrm{sgn}(-b)\,\delta(\nu+b)$$
$$= -\frac{1}{2i}\delta(\nu-b) + \frac{1}{2i}\delta(\nu+b).$$

This is the Fourier transform of $-\sin 2\pi bt$. By similar calculations one can show $\mathcal{H}i\,\{\sin 2\pi bt\} = \cos 2\pi bt$. ∎

The example suggests that the Hilbert transform has the effect on a sinusoid of shifting its phase by $\frac{\pi}{2}$. This is apparent when we look at the transfer function of the Hilbert transform filter (Equation 9.97b) in polar form:

$$|H(\nu)| = 1,$$
$$\arg H(\nu) = \begin{cases} \frac{\pi}{2}, & \nu > 0 \\ 0, & \nu = 0 \\ -\frac{\pi}{2}, & \nu < 0 \end{cases}.$$

The phase response of this filter is a constant $\frac{\pi}{2}$ (90°), positive for positive frequency and negative for negative frequency. The complex exponential $e^{i2\pi bt}$ is shifted to $e^{i(2\pi bt+\pi/2)}$ and its complex conjugate $e^{-i2\pi bt}$ is shifted to $e^{-i2\pi bt-i\pi/2} = e^{-i(2\pi bt+\pi/2)}$. Thus, $\cos 2\pi bt$ is shifted to $\cos(2\pi bt + \frac{\pi}{2}) = -\sin 2\pi bt$ and $\sin 2\pi bt$ is shifted to

$\sin(2\pi bt + \frac{\pi}{2}) = \cos 2\pi bt$. The Hilbert transform of an arbitrary function may be interpreted in the same way, as the result of passing all the Fourier components of the function through a 90° phase shifter. The fact that the Hilbert transform can be interpreted as a linear filter leads to a variety of ways to compute the Hilbert transform as the output of a discrete-time system.[26]

The following properties of the Hilbert transform are straightforwardly established. Their proofs are left to the problems. Unless otherwise indicated, each theorem applies to all four forms of the Hilbert transform.

Linearity

Because the Hilbert transform is a convolution, we easily have linearity:

Theorem 9.29. Let f and g have Hilbert transforms $f_{\mathcal{H}i}$ and $g_{\mathcal{H}i}$, and a and b be constants. Then

$$\mathcal{H}i\{af + bg\} = af_{\mathcal{H}i} + bg_{\mathcal{H}i}. \tag{9.98}$$

Inverse Transform

Theorem 9.30. Applying the Hilbert transform twice in succession gives

$$\mathcal{H}i\{\mathcal{H}i\{f\}\} = -f. \tag{9.99}$$

Thus, the inverse Hilbert transform is

$$\mathcal{H}i^{-1}\{f_{\mathcal{H}i}\} = -\mathcal{H}i\{f_{\mathcal{H}i}\} \tag{9.100}$$

Shift

Theorem 9.31. Let f have Hilbert transform $f_{\mathcal{H}i}$. Then

$$\mathcal{H}i\{f(x-b)\} = f_{\mathcal{H}i}(x-b), \tag{9.101a}$$
$$\mathcal{H}i\{f[n-k]\} = f_{\mathcal{H}i}[n-k]. \tag{9.101b}$$

Dilation

Theorem 9.32. Let $f : \mathbb{R} \to \mathbb{C}$ (i.e., continuous time only) have Hilbert transform $f_{\mathcal{H}i}$. Then

$$\mathcal{H}i\{f(ax)\} = f_{\mathcal{H}i}(ax). \tag{9.102}$$

---

[26] See Oppenheim and Schafer (2010, pp. 361–363), Bracewell (2000, pp. 364–367), Hahn (1996, Chapter 5); also see King (2009, Vol. 1, Chapter 14) for computational methods based on numerical integration.

Inner Product and Norm

The Hilbert transform preserves inner product and norm for square-integrable functions.

Theorem 9.33. Let $f, g \in L^2$ (ℓ^2). Then their Hilbert transforms $f_{\mathcal{H}i}, g_{\mathcal{H}i}$ are also in L^2 (ℓ^2), and

$$\|f\|^2 = \|f_{\mathcal{H}i}\|^2, \tag{9.103a}$$
$$\langle f, g \rangle = \langle f_{\mathcal{H}i}, g_{\mathcal{H}i} \rangle, \tag{9.103b}$$
$$\langle f, g_{\mathcal{H}i} \rangle = -\langle f_{\mathcal{H}i}, g \rangle, \tag{9.103c}$$
$$\text{and} \quad \langle f, f_{\mathcal{H}i} \rangle = 0 \tag{9.103d}$$

(a function and its Hilbert transform are orthogonal).

Convolution

Theorem 9.34. Let $f, f_{\mathcal{H}i}$ and $g, g_{\mathcal{H}i}$ be Hilbert transform pairs. Then

$$\mathcal{H}i\{f * g\} = f_{\mathcal{H}i} * g = f * g_{\mathcal{H}i} \tag{9.104a}$$
$$\text{and} \quad f * g = -f_{\mathcal{H}i} * g_{\mathcal{H}i}, \tag{9.104b}$$

when the integrals (sums) exist.

Product

Theorem 9.35 (Bedrosian's theorem[27]). Let f and g be functions with lowpass and highpass spectra, respectively; that is, F is zero for $|\nu| > B$ ($\pi \geq |\theta| > B$) and G is zero for $|\nu| < B$ ($B > |\theta| \geq 0$). Then,

$$\mathcal{H}i\{fg\} = fg_{\mathcal{H}i}. \tag{9.105}$$

Modulation

This is a special case of the product theorem.

Theorem 9.36. Let f be bandlimited to ν $(\theta) \in (-B, B)$. Then, for all $\nu_0 > B$ ($\pi > \theta_0 > B$),

$$\mathcal{H}i\{f(t) \cos 2\pi \nu_0 t\} = -f(t) \sin 2\pi \nu_0 t, \tag{9.106a}$$
$$\mathcal{H}i\{f[n] \cos \theta_0 n\} = -f[n] \sin \theta_0 n. \tag{9.106b}$$

9.3.3 The Analytic Signal

It is often convenient to use the complex exponential representation of a sinusoid instead of the real signal. The cosine function $A \cos(2\pi bt + \phi)$ is the real part of the

---

[27] Bedrosian (1963).

complex exponential $Ae^{i\phi}e^{i2\pi bt}$, where the complex factor $Ae^{i\phi}$ is called a *phasor* (Figure 1.14). The Fourier transforms of the real function and its phasor form are

$$A\cos(2\pi bt + \phi) \longmapsto \frac{1}{2}Ae^{i\phi}\delta(\nu - b) + \frac{1}{2}Ae^{-i\phi}\delta(\nu + b),$$
$$Ae^{i\phi}e^{i2\pi bt} \longmapsto Ae^{i\phi}\delta(\nu - b).$$

The phasor form has a one-sided transform. The generalization of a phasor is frequently useful, in which an arbitrary real function f is the real part of a complex function $\tilde{f}$ whose Fourier transform is one sided. The resulting complex function $\tilde{f}$ is called the *analytic signal* associated with f.

We write $\tilde{f} = f + ig$, where g is real valued, and seek to identify what g must be in order to obtain a one-sided spectrum for $\tilde{f}$. Let $\tilde{F} = \mathcal{F}\{\tilde{f}\}$ and write

$$\tilde{F} = \mathcal{F}\{f + ig\} = F + iG.$$

Because f and g are real, their transforms F and G are Hermitian: F_e and G_e are real, and F_o and G_o are imaginary. Hence, expressing F and G in terms of their even and odd parts,

$$\tilde{F} = (F_e + F_o) + i(G_e + G_o) = \underbrace{(F_e + iG_o)}_{\mathcal{R}e\,\tilde{F}} + \underbrace{(F_o + iG_e)}_{i\,\mathcal{I}m\,\tilde{F}}.$$

We require the real and imaginary parts of $\tilde{F}$ to be one sided. Neither F nor G is one sided, nor are their respective even and odd parts. But it is possible for $F_e + iG_o$ and $F_o + iG_e$ to be one sided, because each is the sum of an even and an odd function, and cancellation could occur for $\nu < 0$. This will happen if $iG_o = F_e \operatorname{sgn} \nu$ and $iG_e = F_o \operatorname{sgn} \nu$. Adding these two equations,

$$i(G_e + G_o) = (F_e + F_o) \operatorname{sgn} \nu,$$

from which we obtain

$$G(\nu) = -iF(\nu) \operatorname{sgn} \nu$$

and the simple result

$$\tilde{F}(\nu) = F(\nu) + i(-iF(\nu)\operatorname{sgn}\nu) = F(\nu)(1 + \operatorname{sgn}\nu)$$
$$= 2F(\nu)\,U(\nu). \tag{9.107}$$

The analytic signal is obtained, in the Fourier domain, simply by deleting the negative frequency components and doubling the positive frequency components. Calculating the inverse Fourier transform of $\tilde{F} = F + \operatorname{sgn}(\nu)F = F - i(i\operatorname{sgn}(\nu)F)$ we obtain a time-domain expression for the analytic signal:

$$\tilde{f}(t) = f(t) - if_{\mathcal{H}i}(t). \tag{9.108}$$

The connection between the analytic signal and the theory of analytic functions is taken up in the problems.

Example 9.48. Check that the phasor form $Ae^{i\phi}e^{i2\pi bt}$ is the analytic signal for $A\cos(2\pi bt + \phi)$. We will do this in two ways: in the time domain with the Hilbert transform, and in the frequency domain using Equation 9.107:

$$\mathcal{H}i\{A\cos(2\pi bt + \phi)\} = \mathcal{H}i\{A\cos(2\pi bt)\cos\phi - A\sin(2\pi bt)\sin\phi\}$$
$$= A\cos\phi\,\mathcal{H}i\{\cos(2\pi bt)\} - A\sin\phi\,\mathcal{H}i\{\sin(2\pi bt)\}.$$

We know from Example 9.47 that $\mathcal{H}i\{\cos 2\pi bt\} = -\sin 2\pi bt$ and $\mathcal{H}i\{\sin 2\pi bt\} = \cos 2\pi bt$. Thus,

$$\mathcal{H}i\{A\cos(2\pi bt + \phi)\} = -A\cos\phi\sin(2\pi bt) + A\sin\phi\cos(2\pi bt)$$
$$= -A\sin(2\pi bt + \phi).$$

The analytic signal is

$$A\cos(2\pi bt + \phi) - i(-A\sin(2\pi bt + \phi)) = A(\cos(2\pi bt + \phi) + i\sin(2\pi bt + \phi))$$
$$= Ae^{i(2\pi bt + \phi)},$$

as expected. Following the frequency-domain approach, the Fourier transform of $A\cos(2\pi bt + \phi)$ is $\frac{Ae^{-i\phi}}{2}\delta(v + b) + \frac{Ae^{i\phi}}{2}\delta(v - b)$. We drop the component at $v = -b$ and double the component at $v = b$, giving for the analytic signal

$$\mathcal{F}^{-1}\{Ae^{i\phi}\delta(v - b)\} = Ae^{i\phi}e^{i2\pi bt}. \qquad \blacksquare$$

Bandpass Signals

A *bandpass* signal is a function whose Fourier transform is nonzero only for $|v| \in (v_1, v_2)$, where $v_2 > v_1 > 0$ (Figure 9.21). The difference $v_2 - v_1$ is the bandwidth, B.

When $B \ll v_1$, the signal is also called *narrowband*. This is the case in a communication system where a voice signal $m(t)$, with $B \approx 4000$ Hz, modulates a carrier wave $\cos 2\pi v_0 t$, with v_0 ranging from under 1 MHz to over 1 GHz for radio, and even higher, on the order of 10^5 GHz for lightwave communications. It is also the case when a radio telescope is used to scan a portion of the electromagnetic spectrum, searching for signals from astronomical objects.

From the point of view of a signal analyst, the portion of the spectrum outside the bandpass range (v_1, v_2) is unimportant. It is therefore of interest to develop a model for the bandpass signal that isolates the information-bearing portion of the spectrum

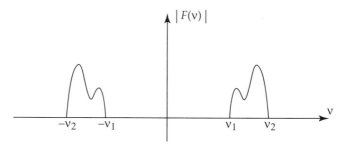

FIGURE 9.21 Fourier spectrum $|F(v)|$ of a bandpass signal.

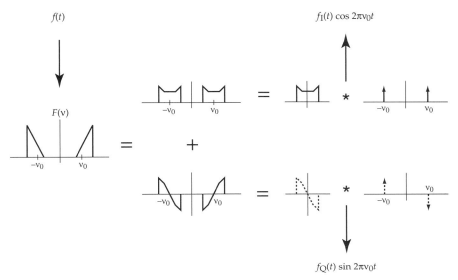

FIGURE 9.22 Decomposing a bandpass signal into in-phase and quadrature components. For convenience of illustration, only the real part of the bandpass spectrum F is shown.

in a convenient form. We begin with the assumption that f is real valued, as is the case for physical signals. The Fourier transform of f is Hermitian. For convenience of illustration, only the real part of F is shown in Figure 9.22 (the treatment of the imaginary part is left to the problems).

Referring to the figure, the frequency v_0 defines a reference point in the band of interest, which may be chosen to be the mean frequency $\frac{v_1+v_2}{2}$, but need not be. Divide F into two components by breaking the spectral islands into even and odd parts with respect to the reference points $\pm v_0$, as shown. The upper graph is further expressed as the convolution of a *lowpass*, or *baseband* spectrum with a pair of impulses at $\pm v_0$. The lower graph may be likewise expressed, but the odd symmetry of the spectral islands requires that one of the replicas of the lowpass spectrum be inverted as well as translated. In this case the lowpass spectrum and the impulses are taken to be imaginary. This makes the lowpass spectrum imaginary and odd, which is the Fourier transform of a real and odd function. Finally, the inverse Fourier transform of the two convolutions gives the following representation for the bandpass signal:

$$f(t) = f_I(t) \cos 2\pi v_0 t + f_Q(t) \sin 2\pi v_0 t. \tag{9.109}$$

The functions f_I and f_Q are called the *in-phase* and *quadrature* components, respectively, of the bandpass signal f. (In communications terminology, two signals are said to be in quadrature if their relative phase difference is 90°. In this context, f_I modulates a cosine, while f_Q modulates a sine. The sine is in quadrature with the cosine.) Equation 9.109 is often called the IQ form of the bandpass signal.

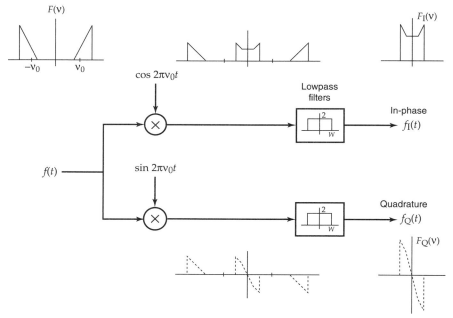

FIGURE 9.23 Mixing and filtering scheme for transforming the bandpass signal $f(t)$ into lowpass signals $f_I(t)$ and $f_Q(t)$. For convenience of illustration, only the real part of $F(\nu)$ is shown. Mixing places the in-phase and quadrature spectra at baseband, with images at $\nu = \pm 2\nu_0$. The lowpass filters remove the images, leaving f_I and f_Q.

The process for deriving the in-phase and quadrature components from the bandpass signal is shown in Figure 9.23. Mathematically, the process is easy to understand. Consider the upper signal path, in which the bandpass signal is multiplied by a cosine at the reference frequency ν_0. The result of this process, also called *mixing*, is

$$f(t) \cos 2\pi\nu_0 t = f_I(t) \cos^2 2\pi\nu_0 t + f_Q(t) \sin 2\pi\nu_0 t \cos 2\pi\nu_0 t$$
$$= \frac{1}{2} f_I(t) + \frac{1}{2} f_I(t) \cos 4\pi\nu_0 t + \frac{1}{2} f_Q(t) \sin 4\pi\nu_0 t.$$

The mixing process produces the in-phase component at baseband, but also a pair of spectral islands, or images, centered at $\pm 2\nu_0$. The mixer is followed by a lowpass filter whose cutoff frequency, W, is sufficient to just pass f_I and reject the high-frequency images. The filter also amplifies the signal to restore it to the proper amplitude. Mixing the bandpass signal with a sine function in the lower signal path gives

$$f(t) \sin 2\pi\nu_0 t = f_I(t) \sin 2\pi\nu_0 t \cos 2\pi\nu_0 t + f_Q(t) \sin^2 2\pi\nu_0 t$$
$$= \frac{1}{2} f_I(t) \sin 4\pi\nu_0 t + \frac{1}{2} f_Q(t) + \frac{1}{2} f_Q(t) \cos 4\pi\nu_0 t.$$

9.3 THE HILBERT TRANSFORM

This time, it is the quadrature component appearing at baseband, and images again appear at $\pm 2v_0$. The lowpass filter removes these images and amplifies the signal, giving f_Q at its output.

There is a close connection between the in-phase and quadrature components and the analytic signal representation of the bandpass signal. Apply the modulation theorem for the Hilbert transform (Equation 9.106) and calculate the Hilbert transform of the bandpass signal in IQ form:

$$f_{\mathcal{H}i}(t) = \mathcal{H}i\{f_I(t) \cos 2\pi v_0 t + f_Q(t) \sin 2\pi v_0 t\} = -f_I(t) \sin 2\pi v_0 t + f_Q(t) \cos 2\pi v_0 t,$$

whence

$$\begin{aligned}\tilde{f}(t) &= f(t) - if_{\mathcal{H}i}(t) \\ &= [f_I(t) \cos 2\pi v_0 t + f_Q(t) \sin 2\pi v_0 t] - i[f_Q(t) \cos 2\pi v_0 t - f_I(t) \sin 2\pi v_0 t] \quad (9.110) \\ &= (f_I(t) - if_{Q(t)})(\cos 2\pi v_0 t + i \sin \sin 2\pi v_0 t) \\ &= (f_I(t) - if_Q(t))e^{i2\pi v_0 t}. \quad (9.111)\end{aligned}$$

Taking the real part of Equation 9.111 recovers the original signal (Equation 9.109). In this form, the bandpass signal is represented as a phasor with a slowly-varying complex amplitude, $\tilde{f}(t) = \tilde{A}(t)e^{i2\pi v_0 t}$. The magnitude of the phasor, also called the *envelope* of the analytic signal, is

$$A(t) = |\tilde{A}(t)| = \sqrt{f_I^2(t) + f_Q^2(t)} \quad (9.112a)$$

and the argument, or phase, is

$$\theta(t) = \arg(f_I(t) - if_Q(t)) = \arctan\left(\frac{-f_Q(t)}{f_I(t)}\right). \quad (9.112b)$$

In terms of envelope and phase, the bandpass signal is

$$f(t) = \mathcal{R}e\, A(t)e^{i\theta(t)}e^{i2\pi v_0 t} = A(t)\cos(2\pi v_0 t + \theta(t)). \quad (9.113)$$

Further properties of bandpass signals are taken up in the problems.[28]

---

[28] For further reading, see Papoulis (1977, Chapters 10 and 11). With slight modification, the results of this section apply to discrete-time signals as well. See Oppenheim and Schafer (2010, Chapter 12).

9.4 SUMMARY

Some Laplace and Z Transform Pairs

$f(t),\ t \geq 0,$	$F_L(s)$	Equation	$f[n],\ n \geq 0$	$F_Z(z)$	Equation
$\delta(t)$	1	9.17	$\delta[n]$	1	9.57
$\delta'(t)$	s	9.18			
1	$\dfrac{1}{s},\ \mathcal{R}e\,s > 0$	9.15	1	$\dfrac{1}{1-z^{-1}},\ |z| > 1$	9.52
t	$\dfrac{1}{s^2},\ \mathcal{R}e\,s > 0$	9.31	n	$\dfrac{z^{-1}}{(1-z^{-1})^2},\ |z| > 1$	9.73
e^{at}	$\dfrac{1}{s-a},\ \mathcal{R}e\,s > \mathcal{R}e\,a$	9.3	α^n	$\dfrac{1}{1-\alpha z^{-1}},\ |z| > |\alpha|$	9.51
$1 - e^{at}$	$\dfrac{-a}{s(s-a)}$ $\mathcal{R}e\,s > \max(0, \mathcal{R}e\,a)$	9.20	$1 - \alpha^n$	$\dfrac{(1-\alpha)z^{-1}}{(1-z^{-1})(1-\alpha z^{-1})}$ $|z| > \max(1, |\alpha|)$	9.61
te^{at}	$\dfrac{1}{(s-a)^2},\ \mathcal{R}e\,s > \mathcal{R}e\,a$	9.7	$n\alpha^n$	$\dfrac{\alpha z^{-1}}{(1-\alpha z^{-1})^2},\ |z| > |\alpha|$	9.72
$\cos bt$	$\dfrac{s}{s^2+b^2},\ \mathcal{R}e\,s > 0$	9.4	$\cos \beta n$	$\dfrac{1-(\cos\beta)z^{-1}}{1-2(\cos\beta)z^{-1}+z^{-2}},\ |z| > 1$	9.53
$\sin bt$	$\dfrac{b}{s^2+b^2},\ \mathcal{R}e\,s > 0$	9.21	$\sin \beta n$	$\dfrac{(\sin\beta)z^{-1}}{1-2(\cos\beta)z^{-1}+z^{-2}},\ |z| > 1$	9.54
$e^{at}\cos bt$	$\dfrac{s-a}{(s-a)^2+b^2},\ \mathcal{R}e\,s > a$	9.8	$\alpha^n \cos \beta n$	$\dfrac{1-(\alpha\cos\beta)z^{-1}}{1-2(\alpha\cos\beta)z^{-1}+\alpha^2 z^{-2}},$ $|z| > |\alpha|$	9.55
$e^{at}\sin bt$	$\dfrac{b}{(s-a)^2+b^2},\ \mathcal{R}e\,s > 0$	9.8	$\alpha^n \sin \beta n$	$\dfrac{(\alpha\sin\beta)z^{-1}}{1-2(\alpha\cos\beta)z^{-1}+\alpha^2 z^{-2}},$ $|z| > |\alpha|$	9.56

(Refer to the text at the indicated references for details of regions of convergence.)

Theorem	Laplace	Equation	Z	Equation
Linearity	$\alpha f + \beta g \longleftrightarrow \alpha F_L + \beta G_L$	9.19	$\alpha f + \beta g \longleftrightarrow \alpha F_Z + \beta G_Z$	9.60
Symmetries	$f^*(t) \longleftrightarrow F_L^*(s^*)$	9.22	$f^*[n] \longleftrightarrow F_Z^*(z^*)$	9.62
	f real $\Longrightarrow F_L^*(s) = F_L(s^*)$	9.23	f real $\Longrightarrow F_Z^*(z) = F_Z(z^*)$	9.63
Shift	$f(t-\tau) \longleftrightarrow e^{-s\tau}F_L(s),\ \tau > 0$	9.24	$f[n-k] \longleftrightarrow z^{-k}F_Z(z)$	9.64
			(also see Equations 9.65 and 9.66)	
Modulation	$e^{-at}f(t) \longleftrightarrow F_L(s+a),\ a \in \mathbb{C}$	9.25	$\alpha^{-n}f[n] \longleftrightarrow F_Z(\alpha z),\ \alpha \in \mathbb{C}$	9.71
Dilation	$f(at) \longleftrightarrow \dfrac{1}{a}F_L\left(\dfrac{s}{a}\right),\ a > 0$	9.26	$f_{\uparrow P}[n] \longleftrightarrow F_Z(z^P)$	9.74
			$f_{\downarrow P}[n] \longleftrightarrow \dfrac{1}{P}\sum_{m=0}^{P-1} F_Z\left(e^{-i2\pi m/P}z^{1/P}\right)$	9.75
Derivative	$f'(t) \longleftrightarrow sF_L(s) - f(0^+)$	9.28	$f[n] - f[n-1] \longleftrightarrow (1-z^{-1})F_Z(z) - f[-1]$	9.69
	(also see Equations 9.27 and 9.29)			
	$tf(t) \longleftrightarrow -F_L'(s)$	9.30	$nf[n] \longleftrightarrow -zF_Z'(z)$	9.72
Integral	$\displaystyle\int_0^t f(\tau)d\tau \longleftrightarrow \dfrac{1}{s}F_L(s)$	9.32	$\displaystyle\sum_{k=0}^n f[k] \longleftrightarrow \dfrac{1}{1-z^{-1}}F_Z(z)$	9.70
Convolution	$f*g \longleftrightarrow F_L G_L$	9.33	$f*g \longleftrightarrow F_Z G_Z$	9.76
Product	$fg \longleftrightarrow \dfrac{1}{2\pi i}\displaystyle\int_{\sigma-i\infty}^{\sigma+i\infty} F_L(s-\xi)G_L(\xi)d\xi$	9.34	$fg \longleftrightarrow \dfrac{1}{2\pi i}\displaystyle\oint_{|z|=r} F_Z\left(\dfrac{z}{\xi}\right)G_Z(\xi)\dfrac{d\xi}{\xi}$	9.77
	σ in ROC of G_L		r in ROC of G_Z	
Parseval	$\displaystyle\int_0^\infty f(t)g^*(t)dt = \int_{\sigma-i\infty}^{\sigma+i\infty} F_L(s)G_L^*(-s^*)\,ds,$	9.35	$\displaystyle\sum_0^\infty f[n]g^*[n] = \oint_{|z|=r} F_Z(z)G_Z^*\left(\dfrac{1}{z^*}\right)\dfrac{dz}{z}$	9.78
	σ in common ROC of $F_L(s)$ and $G_L(-s)$		r in common ROC of $F_Z(z)$ and $G_Z(z^{-1})$	
Initial value	$\displaystyle\lim_{t\to 0^+} f(t) = \lim_{s\to\infty} sF_L(s)$	9.36	$f[0] = \displaystyle\lim_{|z|\to\infty} F_Z(z)$	9.79
Final value	$\displaystyle\lim_{t\to\infty} f(t) = \lim_{s\to 0} sF_L(s)$	9.37	$\displaystyle\lim_{n\to\infty} f[n] = \lim_{z\to 1}(z-1)F_Z(z)$	9.80
Inversion	$f(t) = \dfrac{1}{2\pi i}\displaystyle\int_{\sigma-i\infty}^{\sigma+i\infty} F_L(s)e^{st}ds$	9.40	$f[n] = \dfrac{1}{2\pi i}\displaystyle\oint_{|z|=r} F_Z(z)z^{n-1}dz$	9.81
	σ in ROC of F_L		r in ROC of F_Z	
Fourier	$F(\nu) = F_L(i2\pi\nu)$	9.14	$F_d(\theta) = F_Z(e^{i\theta})$	9.59
	imaginary axis in ROC		unit circle in ROC	

Hilbert Transform Theorems

Theorem	Formula	Equation
Linearity	$af + bg \longleftrightarrow af_{\mathcal{H}i} + bf_{\mathcal{H}i}$	9.98
Double transform	$\mathcal{H}i\{f_{\mathcal{H}i}\} = -f$	9.99
Inverse	$\mathcal{H}i^{-1}\{f_{\mathcal{H}i}\} = -\mathcal{H}i\{f_{\mathcal{H}i}\} = f$	9.100
Norm	$\|f(x)\|^2 = \|f_{\mathcal{H}i}(x)\|^2$	9.103a
Inner product	$\langle f, g \rangle = \langle f_{\mathcal{H}i}, g_{\mathcal{H}i} \rangle$	9.103b
Orthogonality	$\langle f, f_{\mathcal{H}i} \rangle = 0$	9.103d
Shift	$f(x-a) \longleftrightarrow f_{\mathcal{H}i}(x-a)$ $f[n-k] \longleftrightarrow f_{\mathcal{H}i}[n-k]$	9.101
Dilation	$f(ax) \longleftrightarrow f_{\mathcal{H}i}(ax)$	9.102
Convolution	$f * g \longleftrightarrow f_{\mathcal{H}i} * g = f * g_{\mathcal{H}i}$	9.104a
Product (Bedrosian)	If f is lowpass and g is highpass, $fg \longleftrightarrow fg_{\mathcal{H}i}$	9.105
Modulation	If f is bandlimited to $(-B, B)$ and $\nu_0 > B$ ($\pi > \theta_0 > B$), $f(t) \cos 2\pi\nu_0 t \longleftrightarrow -f(t) \sin 2\pi\nu_0 t$ $f[n] \cos \theta_0 n \longleftrightarrow -f[n] \sin \theta_0 n$	9.106
Analytic signal	$\tilde{f} = f - if_{\mathcal{H}i}$	9.108

PROBLEMS

9.1. Suppose the Laplace integral $\int_0^\infty f(t)e^{-st}\,dt$ converges for some $\mathcal{R}e\,s = \sigma_0$. Show that it also converges for any $\sigma > \sigma_0$.

9.2. The M test for uniform convergence of an integral[29] states: If there is a positive function $M(x)$ that is integrable and dominates f on $[c, \infty)$,

$$|f(x,y)| \leq M(x), \quad x \in [c, \infty), \, y \in (a,b),$$

$$\int_c^\infty M(x)\,dx < \infty,$$

then $\int_c^\infty f(x,y)\,dy$ converges absolutely and uniformly for $y \in (a,b)$. Show: If f is of exponential order, then its Laplace transform converges absolutely and uniformly.

9.3. Let f be a function of exponential order, with Laplace transform $F_L(s)$. Show that $|F_L(s)| \to 0$ as $|s| \to \infty$ along paths in the s plane such $\sigma \to \infty$ (i.e., not vertically).

9.4. Show that the inverse Laplace transform does not depend on the position of the Bromwich contour, as long as it is in the region of convergence. To do this, consider the contour shown in Figure 9.24. Because the Laplace transform is analytic in its region of convergence, the integral around the closed contour is zero, and by path invariance, the integral along the vertical path from $\sigma - i\Omega$ to $\sigma + i\Omega$ is equal to the integral along the other

---

[29] Folland (2002, pp. 336 ff).

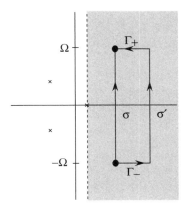

FIGURE 9.24 For Problem 9.4.

three sides. Assume that the Laplace transform $F_L(s) < \mu |s|^{-k}$, $k > 0$, for large $|s|$, and show that the integrals along the short segments Γ_+ and Γ_- go to zero as $\Omega \to \infty$.

9.5. Consider the Laplace transform $F_L(s) = \frac{2}{(s+2)(s-1)}$ and calculate the inverse transform $f(t)$ for each of the following regions of convergence. Comment on the causality and stability (boundedness) of your results.

(a) $\mathcal{R}e\, s > 1$.
(b) $1 > \mathcal{R}e\, s > -2$.
(c) $\mathcal{R}e\, s < -2$.

9.6. Extend the analysis in Example 9.7 to harmonizing the Laplace and Fourier transforms of the one-sided cosine function, $f(t) = \cos(2\pi bt)U(t)$. The Laplace transform is

$$F_L(s) = \frac{1}{2}\frac{1}{s - i2\pi b} + \frac{1}{2}\frac{1}{s + i2\pi b}, \quad \mathcal{R}e\, s > 0.$$

(a) By direct calculation, show that the Fourier transform is

$$\frac{1}{4}\delta(v + b) + \frac{1}{4}\delta(v - b) + \frac{1}{i4\pi}\frac{1}{v + b} + \frac{1}{i4\pi}\frac{1}{v - b}.$$

(b) Take the limit of the Laplace transform as the imaginary axis is approached through the region of convergence and obtain the result you calculated in (a).

9.7. Derive the s-domain derivative theorem (Equation 9.30):

$$t^n f \longmapsto (-1)^n \frac{d^n F_L(s)}{ds^n}. \tag{9.114}$$

9.8. *Parseval's theorem*

(a) Using the product theorem (Equation 9.34), derive Parseval's theorem (Equation 9.35).

(b) Show that Parseval's theorem for the Laplace transform reduces to Parseval's theorem for the Fourier transform when the path of integration is chosen to the imaginary axis ($c = 0$). Under what conditions can this choice be made?

9.9. Prove that if f and g are of exponential order, so is their convolution, $f * g$.

9.10. *More about Parseval's theorem*

Previously we interpreted the magnitude squared of the Fourier transform of a signal, $|F(\nu)|^2$, as a power density. Also, the autocorrelation theorem says

$$|F(\nu)|^2 = F^*(\nu)F(\nu) = \mathcal{F}\left\{\int_{-\infty}^{\infty} f^*(t)f(t+\tau)\,dt\right\} = \mathcal{F}\{f^*(-t) * f(t)\}.$$

In this problem you will generalize this result to the Laplace transform. That is, if $F_L(s)$ is the Laplace transform of $f(t)$, then:

(a) What is the Laplace transform of $f^*(-t)$ and what is its region of convergence? Note that if $f(t)$ is right sided, then $f(-t)$ is left sided. If $s = a + ib$, $a < 0$, is a pole of the transform of $f(t)$, where is the corresponding pole in the transform of $f^*(-t)$?

(b) Using your result for (a) with the convolution theorem, show that the correct generalization is

$$[F_L^*(-s^*)F_L(s)]_{s=i\omega} = |F(\omega)|^2.$$

(c) This result is relevant to the "spectral factorization" problem. For example, what is the transfer function $H_L(s)$ for a causal linear filter whose squared magnitude response is $|H(\omega)|^2 = \frac{1}{1+(\omega/\omega_c)^6}$?

(d) Use the approach taken in (a) and (b) to derive Parseval's theorem for the Laplace transform (Equation 9.35).

9.11. *Initial value theorem*

Here is a simple interpretation of the initial value theorem.[30] In the integral

$$\int_{0^-}^{\infty} f(t)\,\sigma\,e^{-\sigma t}\,dt,$$

graph $\sigma e^{-\sigma t} U(t)$ for an increasing sequence of σ values and observe that it becomes narrower and higher as σ increases. Show that the function also has unit area. That is, $\sigma e^{-\sigma t} U(t)$ behaves like a "right-sided impulse" in the limit $\sigma \to \infty$, sifting out $f(0^+)$.

9.12. *Final value theorem*

It is not difficult to prove the final value theorem if it assumed that F_L has a partial fraction expansion. This is the case for LTI system analysis, where the theorem is particularly useful. So, assume

$$F_L(s) = \sum_{k=1}^{K} \frac{A_k}{s - s_k}$$

with $\mathrm{Re}\,s_k \leq 0$ for all k.

(a) Consider the behavior of $sF_L(s)$ as $s \to 0$ for (1) poles in the left half-plane, (2) poles on the imaginary axis, and (3) a pole at the origin. Compare with the behavior of $f(t)$ as $t \to \infty$ for each of these three cases. Thus, establish the final value theorem.

(b) How do your results for Part (a) change if the poles are not simple, that is, if

$$F_L(s) = \sum_{k=1}^{K}\sum_{n=1}^{N_k} \frac{A_{kn}}{(s - s_k)^n}?$$

---

[30] Kailath (1980, p. 12).

PROBLEMS

9.13. *Final value theorem*

(a) A more general proof of the final value theorem begins with a generalization of the area theorem for the Fourier transform; namely, for a transform pair $g \longleftrightarrow G_L$,

$$\int_0^\infty g(t)dt = \lim_{s \to 0} G_L(s). \qquad (9.115)$$

Under what conditions is this true?

(b) Combine this result with the derivative theorem to derive the final value theorem.

9.14. For the Laplace transform,

$$F_L(s) = \frac{s-2}{(s+1)^2(s^2+4)}.$$

(a) What is the region of convergence (assuming $f(t)$ is one sided)?

(b) Calculate the inverse transform using complex integration and the Laplace inversion formula.

(c) Find a partial fraction expansion using Theorem 9.15 with Equations 9.46.

(d) If you have access to MATLAB, find a partial fraction expansion using the `residue` command.

(e) Calculate the inverse transform using the partial fraction expansion.

9.15. Repeat the calculations in Example 9.18 with $\sin \omega_0 t$ driving function instead of $\cos \omega_0 t$. Comment on the nature of the transient response in this case.

9.16. Calculate the inverse Laplace transforms of the following functions, using both complex integration and partial fraction expansion. If you have access to MATLAB, try using the `residue` command for the partial fraction expansion, in addition to by-hand analysis.

(a) $F_L(s) = \dfrac{1}{s^2 + 2s + 5}$

(b) $F_L(s) = \dfrac{s}{(s+a)^2}, a > 0$

9.17. Calculate the convolution $U(t) * \exp(-t)$, which represents the output of a first-order LTI system driven by a step function:

(a) By direct integration.

(b) Using the Laplace transform and convolution theorem.

9.18. Consider a linear, time-invariant system whose input–output behavior is described by the Laplace transform relationship:

$$\frac{G_L(s)}{F_L(s)} = \frac{s-1}{s^2 + 2s + 2},$$

where F_L and G_L are the Laplace transforms of the input f and output g, respectively. The *step response* is the output when the input is a unit step function, $f(t) = U(t)$. Using complex integration, calculate the step response of the system.

9.19. *System stability*

A linear, time-invariant system is said to be "bounded-input, bounded-output" (BIBO) stable if a bounded input ($\|f\|_\infty < \infty$) produces a bounded output ($\|h * f\|_\infty < \infty$), where h is the impulse response.

(a) Derive a condition on the impulse response which ensures BIBO stability. Your answer should be in the form of some kind of norm.

(b) A common way to determine if a system is stable is to look at the poles of the transfer function $H_L(s)$. For a causal system, if the poles are in the left half of the complex S-plane, then the system is stable. Show that this condition is sufficient to guarantee BIBO stability.

9.20. When a function is not rational but has a finite number of poles, we may also do a partial fraction expansion. Consider $F_L(s) = \dfrac{\cos \pi s}{(s+1)(s+2)}$. It has two poles, at $s = -1$ and $s = -2$.

(a) Verify that F is not bounded as $|s| \to \infty$.

(b) Show that the principal parts of F at the two poles are
$$P(s, -1) = \frac{-1}{s+1},$$
$$P(s, -2) = \frac{-1}{s+2}.$$

(c) Subtract the principal parts from F, obtaining a residual function $r(s)$, and show that this function is entire. Thus, the partial fraction expansion is
$$F_L(s) = r(s) - \frac{1}{s+1} - \frac{1}{s+2}.$$

9.21. Fill in the missing steps in Example 9.30.

9.22. Derive a general expression for the Laplace transform of a one-sided (zero for $t < 0$) periodic function.

9.23. *Sampled functions*

Fill in the details of the calculations leading to Equation 9.48a. Assume that $F_L(s)$ is rational with two more poles than zeros.

(a) Begin with Equation 9.47 and argue that
$$F_{L^*}(s) = \sum_{n=0}^{\infty} \frac{1}{2\pi i} \int_{c-i\infty}^{c+i\infty} F_L(s-z) \, e^{-nz\Delta t} \, dz.$$

Then, considering just the integral, with the contour shown in Figure 9.10, show that the integral on Γ_R goes to zero as its radius goes to infinity.

(b) Perform the usual residue calculation to complete the integration, obtaining
$$F_{L^*}(s) = \sum_{n=0}^{\infty} f(n\Delta t) e^{-sn\Delta t}.$$

(c) Now suppose that $F_L(s)$ has only one more pole than zero. Show that the integral on Γ_R does not go to zero, but rather has the value $-f(0^+)/2$. Thus,
$$F_{L^*}(s) = \frac{1}{2} f(0^+) + \sum_{n=0}^{\infty} f(n\Delta t) e^{-sn\Delta t}.$$

Using the initial value theorem, show that $f(0^+) = 0$ if the number of poles in $F_L(s)$ exceeds the number of zeros by more than one.

9.24. *Sampled functions*
Fill in the details of the calculations leading to Equation 9.48b. Assume that $F_L(s)$ is rational with two more poles than zeros.
(a) Show that the integral along Γ_L goes to zero as $k \to \infty$. It will help to show first that $\left|\frac{1}{1-\exp(-z\Delta t)}\right|$ is bounded on Γ_L.
(b) Perform the usual residue calculation for fixed k, then take $k \to \infty$ to complete the integration, obtaining

$$F_{L^*}(s) = \lim_{k\to\infty} \frac{1}{\Delta t} \sum_{n=-k}^{k} F_L(s - i2\pi n/\Delta t).$$

9.25. *Sampled functions*
(a) Let $f(t) = \cos(2\pi t) U(t)$ and $\Delta t = \frac{1}{4}$. Find and sketch accurately the pole-zero locations of the Laplace transform $F_{L^*}(s)$.
(b) Repeat with $f(t) = \cos 4\pi t U(t)$ and $f(t) = \cos 6\pi t U(t)$, $\Delta t = \frac{1}{4}$. Explain.

9.26. The Laplace transform is useful in problems not related to ordinary differential equations with constant coefficients. In many such problems, the transform to be inverted is not a ratio of polynomials, and partial fraction methods do not apply. Here is an example.
Show that the inverse Laplace transform of $F_L(s) = \frac{\exp(-as^{1/2})}{s^{1/2}}$ is $f(t) = \frac{\exp(-a^2/4t)}{\sqrt{\pi t}} U(t)$. Follow these steps.

(a) This transform does not have any poles, but there is a branch point at $s = 0$. Hence, there is a branch cut which the contour of integration must avoid. The choice of branch cut and contour shown in Figure 9.25 will keep the Bromwich path in the clear: On Γ_1 and Γ_5 your task, as usual, is to show that the integrals go to zero as $R \to \infty$ (adapt Jordan's lemma). Likewise, you integrate around the branch point, taking the limit as $\epsilon \to 0$.

(b) On Γ_2, note that arg $z = \pi$, while on Γ_4, arg $z = -\pi$. So even though the paths are going opposite directions, do not expect the integrals to cancel. In fact, you should obtain the following:

$$\int_{\Gamma_2} + \int_{\Gamma_4} = -i\int_0^R \frac{\exp(-ia\sqrt{r})\exp(-rt)dr}{\sqrt{r}} - i\int_0^R \frac{\exp(+ia\sqrt{r})\exp(-rt)dr}{\sqrt{r}}.$$

Then, make the change of variable $r = x^2$ to get rid of the square roots:

$$\int_{\Gamma_2} + \int_{\Gamma_4} = -2i\int_0^{R^2}\exp(-tx^2)\exp(-iax)dx - 2i\int_0^{R^2}\exp(-tx^2)\exp(+iax)dx,$$

which you can manipulate (via another change of variable) into

$$\lim_{R\to\infty}\int_{\Gamma_2} + \int_{\Gamma_4} = -2i\int_{-\infty}^{\infty}\exp(-tx^2)\exp(-iax)dx,$$

and from here it should be easy (can you recognize this integral?).

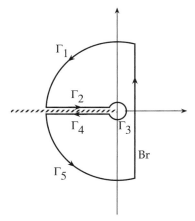

FIGURE 9.25 For Problem 9.26. Contour for calculating the inverse Laplace transform of $\frac{\exp(-as^{1/2})}{s^{1/2}}$.

(c) Using the inverse transform you just derived, obtain the *Fourier* transform of $\frac{1}{\sqrt{t}}U(t)$, and from this derive the Fourier transforms of $|x|^{-1/2}$ and $|x|^{-1/2}\,\mathrm{sgn}\,x$. *Hint*: Think about even and odd symmetries in the Fourier transform.

9.27. Prove the following Z transform theorems:

(a) $f[n+1] \longmapsto zF_Z(z) - zf[0]$

(b) $nf[n] \longmapsto -zF'_Z(z)$

9.28. *Parseval's theorem*

(a) Using the product theorem (Equation 9.77), derive Parseval's theorem (Equation 9.78).

(b) Show that Parseval's theorem for the Z transform reduces to Parseval's theorem for the discrete-time Fourier transform when the path of integration Γ is chosen to be the unit circle. Under what conditions can this choice be made?

9.29. Consider two discrete-time functions (sequences) that are related through the operation $g[n] = (-1)^n f[n]$, that is, every other sample of f is multiplied by -1.

(a) Derive an expression for the Z transform G_Z in terms of the Z transform F_Z. Specify the region of convergence.

(b) Suppose that f is the impulse response of a stable discrete-time system. Is g also stable?

(c) Suppose that f is the impulse response of a discrete-time system having a lowpass frequency response, that is, the discrete-time Fourier magnitude $|F_d(\theta)|$ is a maximum at $\theta = 0$ and decreases monotonically with increasing frequency (up to $\theta = \pi$). Describe the frequency response of the system whose impulse response is g.

Hint: Consider the particular case $f[n] = a^n U[n]$ to gain insight before attempting a general solution.

9.30. Consider the discrete-time system shown in Figure 9.26. The input f is multiplied by the sequence $(-1)^n$ (this amounts to flipping the sign of every other sample) and passed

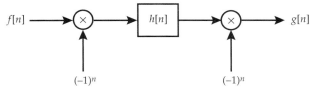

FIGURE 9.26 For Problem 9.30.

through a filter with impulse response h. The output of the filter is multiplied by $(-1)^n$, producing the system output g.

(a) Show that the system is LTI and calculate the transfer function $\frac{G(z)}{F(z)}$. Compare it to the filter's transfer function $H(z)$.

(b) Derive the impulse response for an equivalent system h' such that $g = h' * f$.

9.31. (a) Derive the Laplace transform $te^{at}U(t) \longmapsto \frac{1}{(s-a)^2}$.

(b) Derive the Z transform $na^n U[n] \longmapsto \frac{az^{-1}}{(1-az^{-1})^2}$.

(c) Generalize your derivation in Part (a) to calculate the Laplace transform of $t^2 e^{at}U(t)$. Derive a similar relationship for the Z transform.

9.32. *Upsampling and downsampling*

(a) Suppose a discrete-time signal $f[n]$ is upsampled, and then immediately downsampled, giving $g[n] = (f_{\uparrow P}[n])_{\downarrow P}$. Show, using the relationships in Theorem 9.22, that $G_Z(z) = F_Z(z)$. Explain.

(b) Now suppose the operations are reversed, so $g[n] = (f_{\downarrow P}[n])_{\uparrow P}$. Show that $G_Z(z) = \frac{1}{P} \sum_{m=1}^{P-1} F_Z(e^{-i2\pi m/P} z)$. Interpret this expression (perhaps make a sketch) and explain what happens if f is not bandlimited to $\pi/P > \theta > -\pi/P$.

9.33. *Upsampling and downsampling*
Consider a discrete-time signal $f[n]$ and a discrete LTI system with transfer function $H(z)$. Derive the following identities, which are important in so-called multirate signal processing.

(a) Downsampling: Downsampling f by a factor of P to $f_{\downarrow P}$, then filtering the result with $H(z)$, is equivalent to first filtering f with $H(z^P)$ and then downsampling the output of the filter by a factor of P.

(b) Upsampling: Filtering f with $H(z)$ and then upsampling the result by a factor of P is equivalent to first upsampling f to $f_{\uparrow P}$ and then filtering it with $H(z^P)$.

9.34. (a) Using contour integration, calculate the sample sequence $\{f_n\}$ corresponding to the Z transform $F(z) = \frac{z^2}{z^2+1/4}$. Your result should be:

$$f_n = \left(\frac{1}{2}\right)^n \cos\left(\frac{\pi n}{2}\right), n \geq 0,$$

that is, a damped cosine.

(b) Using MATLAB, compute values for $f_0, \ldots, f_7$ and compute the DFT of this vector. Also compute eight values of $F(z)$ equally spaced on the unit circle ($\theta = 0, \pi/8, \ldots, 7\pi/8$). Plot the Z transform and the DFT in such a way that you can compare their values, and observe that the DFT is identical to the Z transform evaluated on the unit circle.

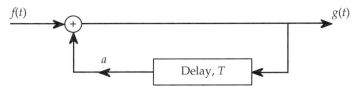

FIGURE 9.27 For Problem 9.35. Model of a reverberant environment.

9.35. Derive the following alternative forms for the Z transforms of $\alpha^n \cos \beta n$ and $\alpha^n \sin \beta n$:

$$\alpha^n \cos \beta n \longmapsto \frac{z(z - \alpha \cos \beta)}{(z - \alpha \cos \beta)^2 + (\alpha \sin \beta)^2}$$

$$\alpha^n \sin \beta n \longmapsto \frac{\alpha z \sin \beta}{(z - \alpha \cos \beta)^2 + (\alpha \sin \beta)^2}.$$

9.36. *Reverberation*

A simple model for a reverberant room was presented in Problem 6.51 of Chapter 6. The block diagram of this system is reproduced in Figure 9.27.

The input–output behavior of this system is described by the *difference* equation:

$$g(t) - ag(t - T) = f(t),$$

where T is the delay time and the factor a models whether the room is "lively" or "dead." This factor will be between 0 and 1, unless there is a PA system—then, a fraction of the sound could be picked up by the microphone, amplified, and sent back out again.

(a) Apply the Laplace transform to the input–output equation and obtain an expression for the transfer function $H_L(s) = \frac{G_L(s)}{F_L(s)}$.

(b) Find all the poles of $H_L(s)$ and classify them (single, double, etc). For what value(s) of a is the system stable? Give your mathematical result a physical interpretation using the block diagram.

(c) Make the change of variable $z = e^{sT}$, converting $H_L(s)$ to a Z transform $H_Z(z)$. Calculate the inverse Z transform, $h[n]$, and use the block diagram to explain how this discrete sequence is related to the system impulse response $h(t)$.

9.37. Carry out a Mittag–Leffler expansion of the function $\cot(\pi \Delta x \nu)$ and show that it is the periodic replication of $\frac{1}{\pi \Delta x \nu}$, that is,

$$\sum_{k=-\infty}^{\infty} \frac{1}{\pi \Delta x} \frac{1}{(\nu - k/\Delta x)} = \cot(\pi \Delta x \nu).$$

9.38. Derive the Hilbert transform for periodic functions, Equation 9.88.

9.39. Prove the following Hilbert transform relationships:

(a) $\mathcal{H}i\{\mathcal{H}i\{f\}\} = -f$

(b) $\mathcal{H}i^{-1}\{f_{\mathcal{H}i}\} = -\frac{1}{\pi} P \int_{-\infty}^{\infty} \frac{f_{\mathcal{H}i}(\tau) d\tau}{\tau - t}$.

9.40. In the Fourier domain, the Hilbert transform is represented as a linear filter with transfer function $i \, \text{sgn} \, \nu$. Because signum is zero at $\nu = 0$, it appears that the Hilbert transform will remove any constant (DC) component from its input. This can also be seen in the time domain, by calculating the convolution $-\frac{1}{\pi x} * c$, where c is a constant.

(a) Show that $-\frac{1}{\pi x} * c = 0$, and hence that the Hilbert transform of a DC signal is zero.

(b) Problem 9.38 showed that applying the Hilbert transform twice in succession returns the negative of the original function, that is, $-\frac{1}{\pi x} * -\frac{1}{\pi x} = -\delta(x)$ (see also Example 6.29). This leads to an apparent contradiction, for:

$$-\frac{1}{\pi x} * \left(-\frac{1}{\pi x} * c\right) = -\frac{1}{\pi x} * 0 = 0,$$

but

$$\left(-\frac{1}{\pi x} * -\frac{1}{\pi x}\right) * c = -\delta(x) * c = -c.$$

Explain.

9.41. Derive the Hilbert transform for discrete time by calculating the inverse Fourier transform:

$$\frac{1}{2\pi} \int_{-\pi}^{\pi} F_d(\theta) \, i \, \text{sgn}(\theta) \, e^{+in\theta} \, d\theta.$$

9.42. Prove the shift and dilation relationships for the Hilbert transform:
  (a) $\mathcal{H}i\{f(t-b)\} = f_{\mathcal{H}i}(t-b)$
  (b) $\mathcal{H}i\{f(at)\} = f_{\mathcal{H}i}(at)$

9.43. Prove the following Hilbert transform relationships for $f \in L^2$:
  (a) The Hilbert transform $f_{\mathcal{H}i}$ is also in L^2, and $\|f\|^2 = \|f_{\mathcal{H}i}\|^2$
  (b) $\langle f, g_{\mathcal{H}i}\rangle = -\langle f_{\mathcal{H}i}, g\rangle$
  (c) $\langle f, g\rangle = \langle f_{\mathcal{H}i}, g_{\mathcal{H}i}\rangle$
  (d) $\langle f, f_{\mathcal{H}i}\rangle = 0$

9.44. Complete Example 9.46 by showing that

$$-\mathcal{H}i\left\{\frac{-2\pi v}{1+(2\pi v)^2}\right\} = \frac{1}{1+(2\pi v)^2}.$$

9.45. Prove the convolution relationships for the Hilbert transform:
  (a) $\mathcal{H}i\{f * g\} = f_{\mathcal{H}i} * g = f * g_{\mathcal{H}i}$
  (b) $f * g = -f_{\mathcal{H}i} * g_{\mathcal{H}i}$

9.46. Prove the product theorem for the Hilbert transform (Theorem 9.35). Let f and g be functions with lowpass and highpass spectra, respectively; that is, F is zero for $|v| > B$ and G is zero for $|v| < B$. First, show that

$$\mathcal{H}i\{fg\} = \int_{-\infty}^{\infty}\int_{-\infty}^{\infty} F(u) \, G(v) \, \mathcal{H}i\left\{e^{i2\pi(u+v)x}\right\} du \, dv$$

$$= \int_{-\infty}^{\infty}\int_{-\infty}^{\infty} F(u) \, G(v) \, i \, \text{sgn}(u+v) \, e^{i2\pi(u+v)x} \, du \, dv.$$

Now, consider the region in the uv plane where the product $F(u)G(v)$ is nonzero (Figure 9.28), and show that

$$\mathcal{H}i\{fg\} = \int_{-\infty}^{\infty}\int_{-\infty}^{\infty} F(u) \, G(v) \, i \, \text{sgn } v \, e^{i2\pi(u+v)x} \, du \, dv.$$

From here, complete the calculation and show $\mathcal{H}i\{fg\} = fg_{\mathcal{H}i}$.

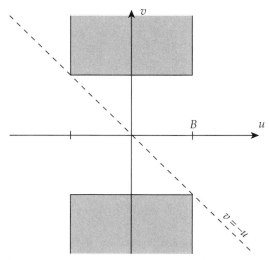

FIGURE 9.28 For Problem 9.45. The product $F(u)G(v)$ is nonzero in the shaded region.

9.47. Prove the modulation theorem for the Hilbert transform (Equations 9.106). Let f be bandlimited to $v \in (-B, B)$. For $v_0 > B$, show that

$$\mathcal{H}i\{f(t) \cos 2\pi v_0 t\} = -f(t) \sin 2\pi v_0 t.$$

Work in the Fourier domain.

9.48. *Kramers–Kronig relationships*
Let f be real and one sided, with Fourier transform $F = F_r + iF_i$. Derive the following alternative form for the Hilbert transform relationships between F_r and F_i:

$$F_r(v) = -\frac{2}{\pi} \int_0^\infty \frac{\eta F_i(\eta) \, d\eta}{\eta^2 - v^2}, \tag{9.116a}$$

$$F_i(v) = \frac{2}{\pi} \int_0^\infty \frac{v F_r(\eta) \, d\eta}{\eta^2 - v^2}. \tag{9.116b}$$

These are called the *Kramers–Kronig relationships*. In a linear optical material, the real and imaginary parts of the refractive index have this relationship. Given spectral measurements of the real or imaginary part, the other component can be computed using the appropriate Kramers–Kronig relationship.

9.49. Verify the following Hilbert transform pairs:
(a) $\sin^2 2\pi bt \longleftrightarrow \frac{1}{2} \cos 4\pi bt$
(b) $\cos^2 2\pi bt \longleftrightarrow -\frac{1}{2} \sin 4\pi bt$
(c) $\text{rect } t \longleftrightarrow \frac{1}{\pi} \log \left| \frac{t - \frac{1}{2}}{t + \frac{1}{2}} \right|$
(d) $\text{sinc } t \longleftrightarrow \left(\frac{\pi t}{2}\right) \text{sinc}^2(t/2) = \frac{\cos \pi t - 1}{\pi^2 t}$

9.50. Calculate the Hilbert transform of $tf(t)$.

9.51. The delta function and the Hilbert transform have an interesting relationship. Derive the following transforms:

(a) $\mathcal{Hi}\{\delta(t)\} = -\frac{1}{\pi t}$

(b) $\mathcal{Hi}\left\{\frac{1}{\pi t}\right\} = \delta(t)$

(c) $\mathcal{Hi}\{\delta'(t)\} = \frac{1}{\pi t^2}$

(d) $\mathcal{Hi}\left\{\frac{1}{\pi t^2}\right\} = -\delta'(t)$

9.52. Show that the Hilbert transform of the odd square wave with period L (see Figure 4.3a) is

$$f_{\mathcal{Hi}}(y) = \frac{2}{\pi}\log\left|\tan\left(\frac{\pi y}{L}\right)\right|.$$

9.53. *Magnitude-phase coupling in minimum phase systems*

The real and imaginary parts of a causal LTI system's transfer function are connected by a Hilbert transform. The objective of this problem is to derive analogous relationships for the magnitude and phase of the transfer function, $H(\nu) = |H(\nu)|e^{i\phi(\nu)}$:

$$\phi(\nu) = -\frac{2}{\pi}\mathcal{P}\int_0^\infty \frac{\log|H(\eta)|\,d\eta}{\nu\left(1 - \eta^2/\nu^2\right)},$$

$$\log|H(\nu)| - \log|H(0)| = \frac{2}{\pi}\mathcal{P}\int_0^\infty \frac{\phi(\eta)\,d\eta}{\eta\left(1 - \eta^2/\nu^2\right)}.$$

Begin with the ratio $H(\nu)/H(0)$, which is the frequency response relative to the response at zero frequency (so-called DC gain). The logarithm of this relative response, using the Laplace transform and writing $\omega = 2\pi\nu$ for convenience, is

$$\log\left(\frac{H_L(i\omega)}{H_L(0)}\right) = \log H_L(i\omega) - \log H_L(0).$$

Because the system is causal and stable, $H_L(s)$ is analytic in the right half-plane. Moreover, assuming the system is minimum phase, there are no zeros in the right half-plane, either. Further, we assume that the number of poles is equal to or greater than the number of zeros.

(a) Show that we may write

$$\log H_L(i\omega) - \log H_L(0) = -\frac{1}{2\pi i}\oint_\Gamma \log H_L(s)\left(\frac{1}{s - i\omega} - \frac{1}{s}\right)ds,$$

where Γ is the contour shown in Figure 9.29.

(b) Calculate the contributions of the segments of the contour and show that

$$-\mathcal{P}\int_{-\infty}^\infty \log H_L(i\omega')\left(\frac{1}{\omega' - \omega} - \frac{1}{\omega'}\right)d\omega' = i\pi[\log H_L(i\omega) - \log H_L(0)].$$

or, changing back from Laplace to Fourier,

$$-\mathcal{P}\int_{-\infty}^\infty \log H(\eta)\left(\frac{1}{\eta - \nu} - \frac{1}{\eta}\right)d\eta = i\pi[\log H(\nu) - \log H(0)].$$

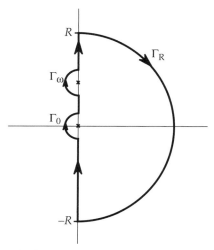

FIGURE 9.29 Contour of integration For Problem 9.52.

(c) Now substitute H in magnitude-phase form, separate real and imaginary parts, and show that the desired relationships result. Note that because h is real, H is Hermitian, so the magnitude is even and the phase is odd.

9.54. *Analytic signals and analytic functions*

Let f be an analytic function. By Cauchy's integral formula,

$$f(z_0) = \frac{1}{2\pi i} \oint_\Gamma \frac{f(z)}{z - z_0} dz.$$

In particular, let the contour Γ be as shown in Figure 9.30.

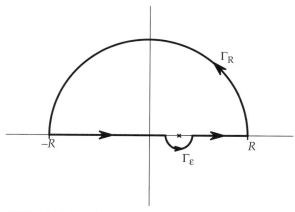

FIGURE 9.30 Contour of integration For Problem 9.53.

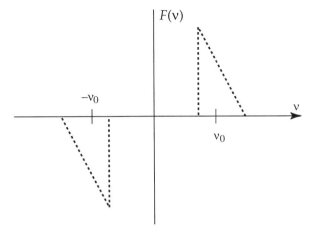

FIGURE 9.31 For Problem 9.55. The Fourier spectrum of a bandpass signal.

Carry out the integration and show

$$f(x_0) = \frac{1}{i\pi} P \int_{-\infty}^{\infty} \frac{f(x)dx}{x - x_0}.$$

Now separate the real and imaginary parts of f, $f = u + iv$, and show

$$u(x_0) = \frac{1}{\pi} P \int_{-\infty}^{\infty} \frac{v(x)dx}{x - x_0},$$

$$v(x_0) = -\frac{1}{\pi} P \int_{-\infty}^{\infty} \frac{u(x)dx}{x - x_0}.$$

That is, u and v are a Hilbert transform pair. In other words, an analytic signal $\tilde{f}(t)$, continued to complex t, is an analytic function, and an analytic function, restricted to the real axis, is an analytic signal.

9.55. Let $\tilde{f}$ and $\tilde{g}$ be analytic signals. Show the following:
   (a) $\mathcal{H}i\{\tilde{f}\} = -i\tilde{f}$.
   (b) (Product theorem): $\tilde{f}\tilde{g}$ is an analytic signal, and $\mathcal{H}i\{\tilde{f}\tilde{g}\} = \tilde{f}\mathcal{H}i\{\tilde{g}\} = \mathcal{H}i\{\tilde{f}\}\tilde{g}$.
   (c) The convolution $\tilde{f} * \tilde{g}$ is an analytic signal.

9.56. Consider the bandpass signal $f(t)$ whose spectrum is shown in Figure 9.31. Derive and sketch the spectra for the in-phase and quadrature components, and verify that $f_I(t)$ and $f_Q(t)$ are real-valued functions.

9.57. Show that the in-phase and quadrature components of the bandpass signal conserve energy, that is,

$$\|f\|^2 = \|f_I\|^2 + \|f_Q\|^2.$$

9.58. Consider the "cartesian" form for the bandpass analytic signal (Equation 9.110):

$$\tilde{f}(t) = [f_I(t)\cos 2\pi v_0 t + f_Q(t)\sin 2\pi v_0 t] - i[f_Q(t)\cos 2\pi v_0 t - f_I(t)\sin 2\pi v_0 t].$$

   (a) Show that the real and imaginary parts of $\tilde{f}$ are orthogonal.
   (b) Show that $\tilde{f}$ has a one-sided spectrum.

9.59. Consider a bandpass signal $f(t)$, as shown in Figure 9.21.
  (a) According to sampling theory, what is the minimum sampling rate for this signal?
  (b) Based on the bandwidth of the interesting part of the spectrum, $B = v_2 - v_1$, it seems that a much lower sampling rate should suffice, that is, that a lot of samples are wasted on the frequencies below v_1. Instead of sampling f, consider sampling the I and Q components of f, with $v_0 = \frac{v_1 + v_2}{2}$. What is the total sampling rate if f_I and f_Q are each sampled at their respective Nyquist rates?

9.60. Consider the problem of passing a bandpass signal f through a bandpass filter with impulse response h. Derive an expression for the filter output, $g = h * f$, in terms of
  (a) The analytic signal representations for f and h.
  (b) The I and Q components of f and h. Given the transfer function H, show how to calculate $H_I = \mathcal{F}\{h_I\}$ and $H_Q = \mathcal{F}\{h_Q\}$.

Assume the same reference frequency, v_0, for both f and h.

9.61. What is the analytic signal corresponding to the AM waveform $f(t) = (1 + M \cos \Omega t) \cos \omega t$? Assume $\Omega \ll \omega$ and $M < 1$.

9.62. In communications, double sideband modulation (DSB) is a method described by the equation $f(t) = Am(t) \cos(2\pi v_c t)$, where $m(t)$ is the modulating signal (message), and v_c is the frequency of the carrier wave. In the simple case of a pure tone message, $m(t) = \cos(2\pi v_m t)$ (in practice, $v_m \ll v_c$), the DSB waveform is

$$f(t) = A \cos(2\pi v_m t) \cos(2\pi v_c t) = \frac{A}{2} \cos(2\pi(v_c + v_m)t) + \frac{A}{2} \cos(2\pi(v_c - v_m)t),$$

and its spectrum consists of impulses at $v = \pm(v_c + v_m)$ (called the upper sideband) and $v = \pm(v_c - v_m)$ (the lower sideband).

A modification to the DSB scheme is described by the equation

$$g(t) = A \left[m(t) \cos(2\pi v_c t) + m_{Hi}(t) \sin(2\pi v_c t) \right],$$

where m_{Hi} is the Hilbert transform of m.
  (a) Calculate a general expression for the Fourier transform G in terms of M, the transform of the message, and other functions.
  (b) Sketch accurately the Fourier transform G, when $m(t) = \cos(2\pi v_m t)$. Contrast this result with the DSB spectrum.
  (c) This modulation method is called *single sideband*. Based on your calculations, explain why.

CHAPTER 10

FOURIER TRANSFORMS IN TWO AND THREE DIMENSIONS

This chapter introduces Fourier and related transforms in two and three dimensions. We shall see that much of the mathematics extends straightforwardly from the one-dimensional transforms developed in Chapters 3–6. Applications include the Fourier analysis of images, which are treated as two- and three-dimensional signals, and more realistic models of wave propagation at radio and optical frequencies, including the diffraction analysis of crystals. We shall encounter special cases of the Fourier transform when the functions under consideration have radial and spherical symmetry, and also the Radon transform, which is the mathematical basis for tomographic (cross-sectional) X-ray imaging.

10.1 TWO-DIMENSIONAL FOURIER TRANSFORM

10.1.1 Definition and Interpretation

The Fourier transform in two dimensions is defined:

$$F(\nu_1, \nu_2) = \int_{-\infty}^{\infty} \int_{-\infty}^{\infty} f(x_1, x_2) \, e^{-i2\pi(\nu_1 x_1 + \nu_2 x_2)} \, dx_1 \, dx_2 \quad (10.1a)$$

$$f(x_1, x_2) = \int_{-\infty}^{\infty} \int_{-\infty}^{\infty} F(\nu_1, \nu_2) \, e^{+i2\pi(\nu_1 x_1 + \nu_2 x_2)} \, d\nu_1 \, d\nu_2 \quad (10.1b)$$

The transform kernel $e^{-i2\pi(\nu_1 x_1 + \nu_2 x_2)}$ is just the product of two one-dimensional kernels, $e^{-i2\pi\nu_1 x_1} e^{-i2\pi\nu_2 x_2}$. The extension to three dimensions is straightforward: just add another factor, for example, $e^{-i2\pi\nu_3 x_3}$. In physical applications, the coordinates (x_1, x_2) are often written (x, y), and the corresponding frequency variables are (ν_x, ν_y). The combination of x and ν coordinates in the exponent has the form of a dot product, $\boldsymbol{\nu} \cdot \mathbf{x}$, so the Fourier transform may be more compactly and extensibly written using

---

Fourier Transforms: Principles and Applications, First Edition. Eric W. Hansen.
© 2014 John Wiley & Sons, Inc. Published 2014 by John Wiley & Sons, Inc.

the notation

$$F(\mathbf{v}) = \int_{-\infty}^{\infty} f(\mathbf{x}) e^{-i2\pi \mathbf{v}\cdot\mathbf{x}} d\mathbf{x}, \qquad (10.2a)$$

$$f(\mathbf{x}) = \int_{-\infty}^{\infty} F(\mathbf{v}) e^{+i2\pi \mathbf{v}\cdot\mathbf{x}} d\mathbf{v}, \qquad (10.2b)$$

where $\mathbf{x} = (x_1, x_2)$, $d\mathbf{x} = dx_1 dx_2$, etc.,

and the one integral stands for integration over all variables. We will use this notation for developing the familiar Fourier theorems and go back to component-wise notation as needed for clarity in particular applications.

The definitions for the Fourier transform in L^1 and L^2 carry over to higher dimensions, for example, a function f belongs to the space $L^2(\mathbb{R}^2)$ if

$$\int_{-\infty}^{\infty} |f(\mathbf{x})|^2 \, d\mathbf{x} = \int_{-\infty}^{\infty}\int_{-\infty}^{\infty} |f(x_1, x_2)|^2 \, dx_1 \, dx_2 < \infty.$$

The forward and inverse transforms in L^2 are interpreted as limits of sequences in the L^2 norm:

$$F(\mathbf{v}) = \lim_{n\to\infty} \int_{-\infty}^{\infty} f(\mathbf{x}) e^{-\pi(\|\mathbf{x}\|/n)^2} e^{-i2\pi \mathbf{v}\cdot\mathbf{x}} d\mathbf{x},$$

$$f(\mathbf{x}) = \lim_{n\to\infty} \int_{-\infty}^{\infty} F(\mathbf{v}) e^{-\pi(\|\mathbf{v}\|/n)^2} e^{i2\pi \mathbf{v}\cdot\mathbf{x}} d\mathbf{v},$$

where $e^{-\pi(\|\mathbf{x}\|/n)^2}$ denotes a two-dimensional Gaussian (extendible to higher dimensions):

$$e^{-\pi(\|\mathbf{x}\|/n)^2} = e^{-\pi(x_1^2 + x_2^2)/n^2}.$$

Moreover, the delta function extends to two dimensions, with sifting property:

$$\iint \delta(x_1 - a_1, x_2 - a_2) \, \varphi(x_1, x_2) \, dx_1 \, dx_2 = \varphi(a_1, a_2) \quad (10.3a)$$

or, compactly,
$$\int \delta(\mathbf{x} - \mathbf{a}) \, \varphi(\mathbf{x}) \, d\mathbf{x} = \varphi(\mathbf{a}). \qquad (10.3b)$$

The testing functions φ are good functions of two variables, which means that they are infinitely continuously differentiable with respect to both variables, including mixed partial derivatives, and rapidly decreasing as $\|\mathbf{x}\| \to \infty$ in all directions. Conveniently, in Cartesian coordinates the two-dimensional delta function is separable into the product of two one-dimensional delta functions, $\delta(x_1 - a_1, x_2 - a_2) = \delta(x_1 - a_1)\delta(x_2 - a_2)$. Equivalence is demonstrated in the usual way:

$$\iint \delta(x_1 - a_1)\delta(x_2 - a_2) \, \varphi(x_1, x_2) \, dx_1 dx_2 = \int \delta(x_1 - a_1) \, \varphi(x_1, a_2) \, dx_1 = \varphi(a_1, a_2).$$

As with the one-dimensional delta function, the sifting property also holds for any function that is continuous at $\mathbf{x} = \mathbf{a}$, for example,

$$\iint \delta(x - 2, y + 3) \, xy \, dx \, dy = (2)(-3) = -6.$$

10.1 TWO-DIMENSIONAL FOURIER TRANSFORM

The inverse Fourier transform integral says that $f(\mathbf{x})$ is a superposition of complex exponentials $e^{+i2\pi \mathbf{v}\cdot\mathbf{x}}$ according the "recipe" specified by $F(\mathbf{v})$. The simplest case of a pure sinusoid follows from an impulse and the sifting property:

$$\delta(\mathbf{v}-\mathbf{b}) \longmapsto \int \delta(\mathbf{v}-\mathbf{b})\,e^{i2\pi\mathbf{v}\cdot\mathbf{x}}\,d\mathbf{v} = e^{i2\pi\mathbf{b}\cdot\mathbf{x}},$$

$$\frac{1}{2}\delta(\mathbf{v}-\mathbf{b}) + \frac{1}{2}\delta(\mathbf{v}+\mathbf{b}) \longmapsto \frac{1}{2}e^{i2\pi\mathbf{b}\cdot\mathbf{x}} + \frac{1}{2}e^{-i2\pi\mathbf{b}\cdot\mathbf{x}} = \cos(2\pi\mathbf{b}\cdot\mathbf{x}).$$

This cosine is drawn in Figure 10.1.

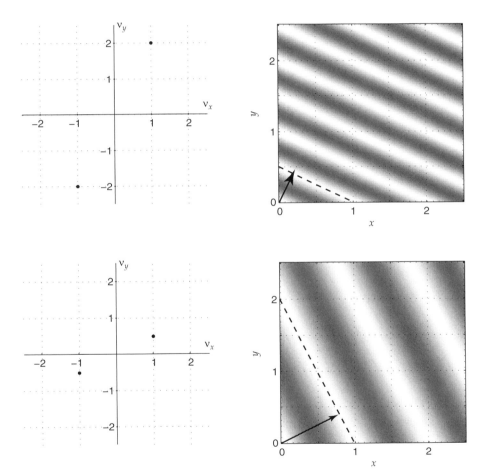

FIGURE 10.1 A two-dimensional cosine $\cos(2\pi\mathbf{b}\cdot\mathbf{x})$ is the inverse Fourier transform of a pair of impulses in the (v_x, v_y) plane. *Top:* $v_x = 1$, $v_y = 2$. The equiphase line shown is $x + 2y = 1$. *Bottom:* $v_x = 1$, $v_y = 0.5$. The equiphase line shown is $x + y/2 = 1$. The vectors shown perpendicular to the equiphase lines are $\mathbf{b}/\|\mathbf{b}\|^2$; their lengths, $1/\|\mathbf{b}\|$, are the periods of the cosines.

The phase $2\pi \mathbf{b} \cdot \mathbf{x} = 2\pi(b_x x + b_y y)$ varies with position in the plane. For those (x, y) such that the phase is an integer multiple of 2π, we are on an equiphase of the cosine function, $2\pi(b_x x + b_y y) = 2\pi m$, or $\mathbf{b} \cdot \mathbf{x} = m$. From the figure, the slope of the equiphase is $\frac{-1/b_y}{1/b_x} = -b_x/b_y$, and the slope of the frequency vector $\mathbf{b}$ is b_y/b_x; $\mathbf{b}$ is perpendicular to the equiphase. The vector drawn perpendicular to the equiphase can be shown to be $\mathbf{b}/\|\mathbf{b}\|^2$. It is proportional to the frequency vector, and its length, $1/\|\mathbf{b}\|$, is the period of the cosine, $\frac{1}{\sqrt{b_x^2+b_y^2}}$.

For every frequency vector $\mathbf{v}$ there is a complex exponential, or complex sine–cosine pair, of some frequency and orientation in the $\mathbf{x}$ plane. The inverse Fourier transform integral superposes these complex exponentials, with magnitude and phase specified by $F(\mathbf{v})$, to construct the function $f(\mathbf{x})$.

In several cases of practical interest, the function to be transformed is separable, $f(x, y) = f_x(x) f_y(y)$. In this case the Fourier transform also separates:

$$F(v_x, v_y) = \int_{-\infty}^{\infty} \int_{-\infty}^{\infty} f_x(x) f_y(y) \, e^{-i2\pi(v_x x + v_y y)} \, dx \, dy$$

$$= \int_{-\infty}^{\infty} f_x(x) \, e^{-i2\pi(v_x x)} \, dx \int_{-\infty}^{\infty} f_y(y) \, e^{-i2\pi v_y y} \, dy = F_x(v_x) F_y(v_y). \quad (10.4)$$

Example 10.1. The two-dimensional rectangle function $\text{rect}(x/X) \, \text{rect}(y/Y)$ models truncation in two dimensions, such as an optical aperture. Its Fourier transform is, using Equation 10.4 with the one-dimensional dilation theorem,

$$\text{rect}(x/X) \, \text{rect}(y/Y) \longmapsto X \, \text{sinc}(X v_x) \, Y \, \text{sinc}(Y v_y). \quad (10.5)$$

This function is shown in Figure 10.2 for the particular case $X = 2$, $Y = 1$. ∎

Example 10.2 (Plotting two-dimensional functions in MATLAB). The graphs in Figure 10.2 were created in MATLAB in the following way. A two-dimensional function is represented by a matrix of function values:

$$\begin{bmatrix} f(x_1, y_1) & \cdots & f(x_M, y_1) \\ \vdots & \ddots & \vdots \\ f(x_1, y_N) & \cdots & f(x_M, y_N) \end{bmatrix}.$$

A separable function, $f(x, y) = f_x(x) f_y(y)$, can be computed as the outer product of two vectors:

$$\begin{bmatrix} f_x(x_1) f_y(y_1) & \cdots & f_x(x_M) f_y(y_1) \\ \vdots & \ddots & \vdots \\ f_x(x_1) f_y(y_N) & \cdots & f_x(x_M) f_y(y_N) \end{bmatrix} = \begin{bmatrix} f_y(y_1) \\ \vdots \\ f_y(y_N) \end{bmatrix} \begin{bmatrix} f_x(x_1) & \cdots & f_x(x_M) \end{bmatrix}$$

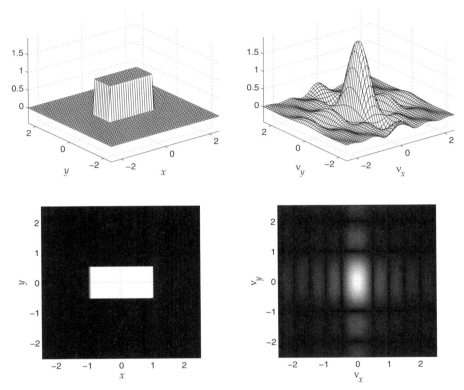

FIGURE 10.2 A two-dimensional rectangle function, rect($x/2$) rect(y) (*left*) and its Fourier transform, 2 sinc($2v_x$) sinc(v_y) (*right*).

and similarly for the Fourier transform $F(v_x, v_y) = F_x(v_x)F_y(v_y)$. For example, to compute the matrix for the Fourier transform $F(v_x, v_y) = 2\,\text{sinc}(2v_x)\,\text{sinc}(v_y)$,

```
vx = linspace(-2.5, 2.5, M);     % Coordinate vectors
vy = linspace(-2.5, 2.5, N);
Ff = 2 * sinc(2*vy') * sinc(vx); % Fourier transform of f
```

The mesh function is called to make the altitude plot:

```
mesh(vx, vy, Ff)
```

To view the function as an image instead, apply the imagesc command to the magnitude of the transform:

```
imagesc(vx, vy, abs(Ff))
axis xy; axis square;
```

The two axis commands orient the image so that its axes match a cartesian coordinate system and have a 1:1 aspect ratio. Additional commands, not shown, label the axes and change the orientation of the mesh plot. ∎

674 CHAPTER 10 FOURIER TRANSFORMS IN TWO AND THREE DIMENSIONS

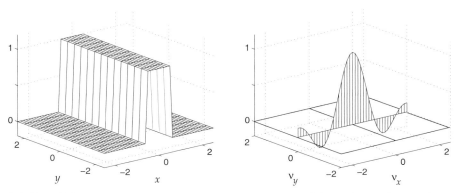

FIGURE 10.3 A one-dimensional rectangle function, rect x (*left*) and its Fourier transform, $\text{sinc}(\nu_x)\,\delta(\nu_y)$ (*right*).

Example 10.3. The function $f(x, y) = \text{rect}\, x$ is constant in y and a rectangle in x. The two-dimensional Fourier transform is

$$F(\nu_x, \nu_y) = \iint_{-\infty}^{\infty} \text{rect}\, x \, e^{-i2\pi(\nu_x x + \nu_y y)}\, dx\, dy$$

$$= \int_{-\infty}^{\infty} \text{rect}\, x \, e^{-i2\pi \nu_x x}\, dx \times \int_{-\infty}^{\infty} 1\, e^{-i2\pi \nu_y y}\, dy$$

$$= \text{sinc}\left(\nu_x\right)\, \delta(\nu_y).$$

This is a blade running along the line $\nu_y = 0$, modulated by the sinc function sinc ν_x (Figure 10.3). ∎

10.1.2 Fourier Transform Theorems

Most of the one-dimensional Fourier transform theorems generalize to multiple dimensions in straightforward ways.

Linearity

Of course, the multidimensional Fourier transform is linear, since multiple integrals are linear:

$$af(\mathbf{x}) + bg(\mathbf{x}) \longleftrightarrow aF(\mathbf{\nu}) + bG(\mathbf{\nu}). \tag{10.6}$$

Symmetry

The familiar symmetry relationships hold for multidimensional Fourier transforms, *coordinate-by-coordinate*. That is, if we have a function $f(x, y)$ that is even with respect to x, $f(-x, y) = f(x, y)$, but odd with respect to y, $f(x, -y) = -f(x, y)$, the Fourier transform will be even in ν_x and odd in ν_y.

Parseval's Formula

The derivation of Parseval's formula in two dimensions and higher is the same as the one-dimensional derivation, just with more integrals. The result is, for functions in $L^2(\mathbb{R}^n)$,

$$\int_{-\infty}^{\infty} f(\mathbf{x})g^*(\mathbf{x})d\mathbf{x} = \int_{-\infty}^{\infty} F(\mathbf{v})G^*(\mathbf{v})d\mathbf{v}, \tag{10.7a}$$

$$\int_{-\infty}^{\infty} |f(\mathbf{x})|^2 \, d\mathbf{x} = \int_{-\infty}^{\infty} |F(\mathbf{v})|^2 \, d\mathbf{v}. \tag{10.7a}$$

Shift

The difference here is that more than one coordinate can be shifted, so shifts can occur along diagonals as well as along axes:

$$f(\mathbf{x} - \mathbf{a}) \longleftrightarrow e^{-i2\pi \mathbf{v} \cdot \mathbf{a}} F(\mathbf{v}), \tag{10.8a}$$

$$e^{i2\pi \mathbf{b} \cdot \mathbf{x}} f(\mathbf{x}) \longleftrightarrow F(\mathbf{v} - \mathbf{b}). \tag{10.8a}$$

The phase factor $e^{-i2\pi \mathbf{v} \cdot \mathbf{a}} = e^{-i2\pi v_1 a_1} e^{-i2\pi v_2 a_2}$ is the product of phases due to the shifts in each coordinate.

Dilation

Dilation poses a few surprises. The generalization from a scalar dilation $x \longmapsto ax$ is a linear coordinate transformation, $\mathbf{x} \longmapsto \mathbf{Ax}$.[1] Particular cases of interest are (Figure 10.4)

- Coordinate scaling: $\mathbf{A} = \begin{bmatrix} a_1 & 0 \\ 0 & a_2 \end{bmatrix}$.
- (Horizontal) shear: $\mathbf{A} = \begin{bmatrix} 1 & \sigma \\ 0 & 1 \end{bmatrix}$.
- Rotation: $\mathbf{A} = \begin{bmatrix} \cos\theta & \sin\theta \\ -\sin\theta & \cos\theta \end{bmatrix}$.

Transformations may be combined, for example, a vertical shear can be made by a 90° rotation, followed by horizontal shear, followed by a −90° rotation. An arbitrary transformation may be decomposed into a product of these basic operations (see the problems).

We are interested in how a linear coordinate transformation affects the Fourier transform:

$$f(\mathbf{Ax}) \longmapsto \int_{-\infty}^{\infty} f(\mathbf{Ax}) \, e^{-i2\pi \mathbf{v} \cdot \mathbf{x}} \, d\mathbf{x}.$$

---

[1] When we write $\mathbf{v} \cdot \mathbf{x}$, it is not necessary to say whether $\mathbf{v}$ or $\mathbf{x}$ is a row or column vector. But when we operate on a vector, for example, $\mathbf{Ax}$, we follow linear algebra convention and assume that $\mathbf{x}$ is a column vector. Similarly, for the dot product $\mathbf{v} \cdot \mathbf{Ax}$ we may equivalently write $\mathbf{v}^T \mathbf{Ax}$ to be explicit about how the algebra is done.

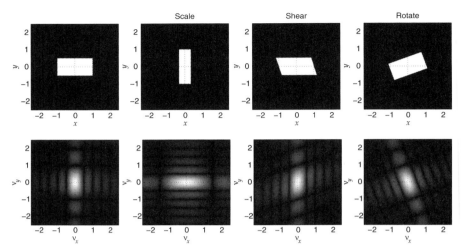

FIGURE 10.4 The function $\text{rect}(x/2)\text{rect}(y)$ (*top*) and its Fourier transform, $2\,\text{sinc}(2v_x)\,\text{sinc}(v_y)$ (*bottom*), are transformed by scaling, shearing, and rotation. See Example 10.4.

This is just a change of variable in a multiple integral. Write $\boldsymbol{\xi} = \mathbf{A}\mathbf{x}$. Then, assuming that $\mathbf{A}$ is invertible (which it must be in order to be physically useful),

$$\int_{-\infty}^{\infty} f(\mathbf{A}\mathbf{x})\, e^{-i2\pi \mathbf{v}\cdot\mathbf{x}}\, d\mathbf{x} = \int_{-\infty}^{\infty} f(\boldsymbol{\xi})\, e^{-i2\pi \mathbf{v}\cdot \mathbf{A}^{-1}\boldsymbol{\xi}}\, \frac{d\boldsymbol{\xi}}{|\det \mathbf{A}|}.$$

Now, we need to rewrite the dot product in the exponent to put the transformation on $\mathbf{v}$. Writing $\mathbf{v}$ and $\mathbf{x}$ as column vectors,

$$\mathbf{v}\cdot\left(\mathbf{A}^{-1}\boldsymbol{\xi}\right) = \mathbf{v}^T \mathbf{A}^{-1}\boldsymbol{\xi} = \left(\mathbf{v}^T \mathbf{A}^{-1}\right)\boldsymbol{\xi} = \left((\mathbf{A}^{-1})^T \mathbf{v}\right)^T \boldsymbol{\xi} = \left(\mathbf{A}^{-T}\mathbf{v}\right)\cdot\boldsymbol{\xi}.$$

Then the Fourier integral becomes

$$\int_{-\infty}^{\infty} f(\boldsymbol{\xi})\, e^{-i2\pi(\mathbf{A}^{-T}\mathbf{v})\cdot\boldsymbol{\xi}}\, \frac{d\boldsymbol{\xi}}{|\det \mathbf{A}|} = \frac{1}{|\det \mathbf{A}|}\, F\left(\mathbf{A}^{-T}\mathbf{v}\right).$$

Thus, for an invertible coordinate transformation $\mathbf{A}$, we have the multidimensional dilation theorem:

$$f(\mathbf{A}\mathbf{x}) \longleftrightarrow \frac{1}{|\det \mathbf{A}|}\, F\left(\mathbf{A}^{-T}\mathbf{v}\right). \tag{10.9}$$

In one dimension, $\mathbf{A}^{-T}$ is $1/a$, $|\det \mathbf{A}|$ is $|a|$, and we recover the familiar result, $f(ax) \longleftrightarrow \frac{1}{|a|} F(v/a)$.

We will also need a formula for the dilation of a delta function (we will include a shift), $\delta(\mathbf{Ax} - \mathbf{b})$. Let $\boldsymbol{\xi} = \mathbf{Ax} - \mathbf{b}$, then $\mathbf{x} = \mathbf{A}^{-1}(\boldsymbol{\xi} + \mathbf{b})$, and

$$\int \delta(\mathbf{Ax} - \mathbf{b})\, \varphi(\mathbf{x})\, d\mathbf{x} = \int \delta(\boldsymbol{\xi})\, \varphi\left(\mathbf{A}^{-1}(\boldsymbol{\xi} + \mathbf{b})\right) \frac{1}{|\det \mathbf{A}|}\, d\boldsymbol{\xi}$$

$$= \frac{1}{|\det \mathbf{A}|}\, \varphi\left(\mathbf{A}^{-1}\mathbf{b}\right)$$

$$\Rightarrow \delta(\mathbf{Ax} - \mathbf{b}) = \frac{1}{|\det \mathbf{A}|}\, \delta\left(\mathbf{x} - \mathbf{A}^{-1}\mathbf{b}\right). \tag{10.10}$$

Let us see how the dilation theorem operates in the three particular cases of scale, shear, and rotation. Because all of these operations are invertible, each of the operations will result in a transform pair:

- Coordinate scaling: $\mathbf{A} = \begin{bmatrix} a_1 & 0 \\ 0 & a_2 \end{bmatrix}$. The determinant is $\det \mathbf{A} = a_1 a_2$. The transpose of the inverse is $\mathbf{A}^{-T} = \begin{bmatrix} 1/a_1 & 0 \\ 0 & 1/a_2 \end{bmatrix}$. So, $\mathbf{A}^{-T}\mathbf{v} = (v_1/a_1, v_2/a_2)^T$ and we have

$$f(a_1 x_1, a_2 x_2) \longleftrightarrow \frac{1}{|a_1 a_2|}\, F\left(\frac{v_1}{a_1}, \frac{v_2}{a_2}\right).$$

This has the same form as the one-dimensional dilation theorem, with each coordinate separately scaled.

- (Horizontal) shear: $\mathbf{A} = \begin{bmatrix} 1 & \sigma \\ 0 & 1 \end{bmatrix}$. The determinant is 1, the transpose of the inverse is $\mathbf{A}^{-T} = \begin{bmatrix} 1 & 0 \\ -\sigma & 1 \end{bmatrix}$. Thus,

$$f\left(x_1 + \sigma x_2, x_2\right) \longleftrightarrow F\left(v_1, v_2 - \sigma v_1\right).$$

A horizontal shear of f transforms to a vertical shear of F.

- Rotation: $\mathbf{A} = \begin{bmatrix} \cos\theta & \sin\theta \\ -\sin\theta & \cos\theta \end{bmatrix}$. A rotation is an orthogonal matrix. The determinant is $\det \mathbf{A} = 1$ and the inverse is $\mathbf{A}^{-1} = \mathbf{A}^T$. Thus the transpose of the inverse is $\mathbf{A}$ itself, and we have

$$f(\mathbf{Ax}) \longleftrightarrow F(\mathbf{Av}).$$

A rotation of f transforms to an identical rotation of F.

Example 10.4 (Transformation of rect $\longleftrightarrow$ sinc**).** The three operations are illustrated in Figure 10.4 for a rectangle, $f(x, y) = \text{rect}(x/2)\,\text{rect}(y)$:

- Scaling, $\begin{bmatrix} 3 & 0 \\ 0 & 1/2 \end{bmatrix}$.

$$\text{rect}(3x/2)\,\text{rect}(y/2) \longmapsto \frac{4}{3}\,\text{sinc}(2v_x/3)\,\text{sinc}(2v_y).$$

The transformed rectangle is $\frac{2}{3} \times 2$, and the distances from the origin to the first zero crossings (dark bands) of its Fourier transform are $3/2$ and $1/2$.

- Shearing, $\begin{bmatrix} 1 & 1/3 \\ 0 & 1 \end{bmatrix}$.

$$\text{rect}\left(\frac{x+y/3}{2}\right) \text{rect}(y) \longmapsto 2 \, \text{sinc}(v_x/2) \, \text{sinc}(v_y - v_x/3).$$

The transformed rectangle is sheared horizontally into a parallelogram, but has the same dimensions measured along the axes. Its Fourier transform is sheared vertically, but the horizontal and vertical distances between zero crossings are unchanged.

- Rotation, $\begin{bmatrix} \cos\theta & \sin\theta \\ -\sin\theta & \cos\theta \end{bmatrix}$, $\theta = 20°$.

$$\text{rect}\left(\frac{x\cos\theta + y\sin\theta}{2}\right) \text{rect}(-x\sin\theta + y\cos\theta)$$
$$\longmapsto 2 \, \text{sinc}\left(2(v_x\cos\theta + v_y\sin\theta)\right) \text{sinc}\left(-v_x\sin\theta + v_y\cos\theta\right).$$

Both the rectangle and its Fourier transform are rotated by $20°$. The dimensions of both are unchanged. ∎

Note that the sheared and rotated rectangles are not separable. Calculating their Fourier transforms by integration would be difficult, but they are not hard to do using the dilation theorem, because there is a transformation that connects them to a separable function.

Example 10.5 (Plotting two-dimensional functions in MATLAB, II). Here is how to compute and plot, in MATLAB, a nonseparable function like the sheared sinc function in the previous example:

$$f(x, y) = 2 \, \text{sinc}(x/2) \, \text{sinc}(y - x/3).$$

An input array of points is mapped into an output array of function values, for example,

$$\begin{bmatrix} (x_1, y_1) & \cdots & (x_M, y_1) \\ \vdots & \ddots & \vdots \\ (x_1, y_N) & \cdots & (x_M, y_N) \end{bmatrix} \longmapsto \begin{bmatrix} f(x_1, y_1) & \cdots & f(x_M, y_1) \\ \vdots & \ddots & \vdots \\ f(x_1, y_N) & \cdots & f(x_M, y_N) \end{bmatrix}.$$

An array of points is actually represented by two $M \times N$ arrays of coordinates, one for the x values and one for the y values:

$$X = \begin{bmatrix} x_1 & \cdots & x_M \\ \vdots & \ddots & \vdots \\ x_1 & \cdots & x_M \end{bmatrix}, \quad Y = \begin{bmatrix} y_1 & \cdots & y_1 \\ \vdots & \ddots & \vdots \\ y_N & \cdots & y_N \end{bmatrix}.$$

Note that the pair $(X_{mn}, Y_{mn}) = (x_n, y_m)$. MATLAB supplies a function `meshgrid` to create the arrays, which are called `xx` and `yy` in MATLAB (it is bad practice to have different variables called x and X). For equally spaced x and y coordinates, with $x, y \in [-2, 2]$, `meshgrid` is called like this:

```
Npts = 51;
xmax = 2;
[xx, yy] = meshgrid(linspace(-xmax, xmax, Npts));
```

The arrays `xx` and `yy` are each 51×51.

The `xx` and `yy` arrays are passed to the function that computes f, for example,

```
f = 2 * sinc(xx/2) .* sinc(yy - xx/3);
```

From here, `mesh` or `imagesc` (or some other plotting function) may be called for display. ∎

Example 10.6 (Transformation of a cosine). The same transformations are performed for a cosine, $f(x, y) = \cos(2\pi v_0 x)$, whose Fourier transform is $F(v_x, v_y) = \frac{1}{2}\delta(v_x - v_0)\delta(v_y) + \frac{1}{2}\delta(v_x + v_0)\delta(v_y)$. For simplicity, we separate the cosine into the sum of complex exponentials and do the calculation for one of the exponentials. Write the complex exponential as $e^{i2\pi \mathbf{b} \cdot \mathbf{x}}$, with $\mathbf{x} = (x, y)$ and $\mathbf{b} = (v_0, 0)$, and the delta function as $\delta(v_x - v_0)\delta(v_y) = \delta(\mathbf{v} - \mathbf{b})$, with $\mathbf{v} = (v_x, v_y)$. Then, by the dilation theorems (Equations 10.9 and 10.10),

$$e^{i2\pi \mathbf{b} \cdot \mathbf{A}\mathbf{x}} \longmapsto \frac{1}{|\det \mathbf{A}|} \delta\left(\mathbf{A}^{-T}\mathbf{v} - \mathbf{b}\right) = \frac{1}{|\det \mathbf{A}|} \left|\det \mathbf{A}^{T}\right| \delta\left(\mathbf{v} - \mathbf{A}^{T}\mathbf{b}\right)$$
$$= \delta\left(\mathbf{v} - \mathbf{A}^{T}\mathbf{b}\right)$$

(because $\det \mathbf{A}^T = \det \mathbf{A}$), and for the full cosine,

$$\cos(2\pi \mathbf{b} \cdot \mathbf{A}\mathbf{x}) \longmapsto \frac{1}{2}\delta\left(\mathbf{v} - \mathbf{A}^{T}\mathbf{b}\right) + \frac{1}{2}\delta\left(\mathbf{v} + \mathbf{A}^{T}\mathbf{b}\right).$$

We will consider only a rotation here and leave the other transformations to the problems. For a rotation, $\det \mathbf{A} = 1$. The delta functions at $\mathbf{v} = \pm \mathbf{b}$ are transformed to delta functions at

$$\mathbf{v} = \pm \mathbf{A}^T \mathbf{b} = \pm \begin{bmatrix} \cos\theta & -\sin\theta \\ \sin\theta & \cos\theta \end{bmatrix} \begin{bmatrix} v_0 \\ 0 \end{bmatrix} = \pm \begin{bmatrix} v_0 \cos\theta \\ v_0 \sin\theta \end{bmatrix}.$$

These transformations are shown in Figure 10.5 for a cosine with $\mathbf{b} = (2, 0)$. ∎

Derivatives

The essence of the derivative theorem is that differentiation in one domain becomes multiplication in the other domain (Equation 5.24):

$$f'(x) \longmapsto i2\pi v F(v).$$

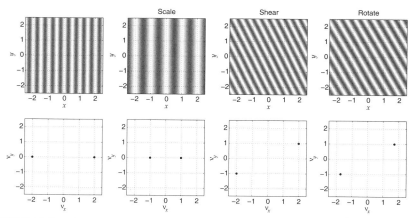

FIGURE 10.5 The function $\cos(2\pi \cdot 2x)$ (*top*) and its Fourier transform, $\frac{1}{2}\delta(v_x - 2)\delta(v_y) + \frac{1}{2}\delta(v_x + 2)\delta(v_y)$ (*bottom*), are transformed by scaling ($a_x = 0.5$), shearing ($\sigma = 0.5$), and rotation (30°). See Example 10.6. Only the locations of the delta functions are shown, not their strengths. While the sheared and rotated cosines are similar, note that the horizontal frequency is unchanged by shearing, while it is reduced by rotation.

Now consider the extension to a first derivative in each of two dimensions (a mixed partial):

$$\frac{\partial^2}{\partial x \partial y}f(x,y) = \frac{\partial^2}{\partial x \partial y}\int_{-\infty}^{\infty} F(v_x, v_y)e^{i2\pi(v_x x + v_y y)}\,dv_x dv_y$$

$$= \int_{-\infty}^{\infty} F(v_x, v_y)\left(\frac{\partial^2}{\partial x \partial y}e^{i2\pi(v_x x + v_y y)}\right)dv_x dv_y$$

$$= \int_{-\infty}^{\infty} (i2\pi v_x)(i2\pi v_y)F(v_x, v_y)e^{i2\pi(v_x x + v_y y)}\,dv_x dv_y$$

$$\Rightarrow \frac{\partial^2}{\partial x \partial y}f(x,y) \longmapsto (i2\pi v_x)(i2\pi v_y)F(v_x, v_y).$$

We see that the one-dimensional result extends straightforwardly to higher dimensions. For each derivative with respect to coordinate x_k, multiply F by a factor of $i2\pi v_k$.

To keep track of this in general, we introduce the so-called multi-index notation. A multi-index is a "vector" of nonnegative integers, $\boldsymbol{\alpha} = (\alpha_1, \alpha_2, \ldots, \alpha_N)$. This enables us to define, compactly,

$$|\boldsymbol{\alpha}| = \alpha_1 + \alpha_2 + \cdots + \alpha_N,$$

$$\mathbf{x}^{\boldsymbol{\alpha}} = x_1^{\alpha_1} x_2^{\alpha_2} \cdots x_N^{\alpha_N},$$

$$\text{and } \partial^{\boldsymbol{\alpha}} = \frac{\partial^{\alpha_1}}{\partial x_1^{\alpha_1}} \frac{\partial^{\alpha_2}}{\partial x_2^{\alpha_2}} \cdots \frac{\partial^{\alpha_N}}{\partial x_N^{\alpha_N}}.$$

Then using this notation, the multidimensional derivative theorem is, simply,

$$\partial^\alpha f(\mathbf{x}) \longleftrightarrow (i2\pi\mathbf{v})^\alpha F(\mathbf{v}), \tag{10.11a}$$

$$(-i2\pi\mathbf{x})^\alpha f(\mathbf{x}) \longleftrightarrow \partial^\alpha F(\mathbf{v}). \tag{10.11b}$$

Another derivative that is important in partial differential equations is the Laplacian:

$$\nabla^2 f(\mathbf{x}) = \left(\frac{\partial^2}{\partial x^2} + \frac{\partial^2}{\partial y^2}\right) f(x,y) \longmapsto (i2\pi)^2 \left(v_x^2 + v_y^2\right) F(v_x, v_y) \tag{10.12a}$$

or, in a form that covers all dimensions:

$$\nabla^2 f(\mathbf{x}) \longmapsto -4\pi^2 \|\mathbf{v}\|^2 F(\mathbf{v}). \tag{10.11b}$$

Example 10.7 (Diffusion on an unbounded domain). Solution of the diffusion (heat) equation on a bounded one-dimensional domain was discussed in Section 4.5. Here we revisit the problem for an unbounded two-dimensional domain. Previously, the heat equation was written

$$\nabla^2 u(\mathbf{r}, t) = \frac{1}{k} \frac{\partial u(\mathbf{r}, t)}{\partial t}, \tag{4.27}$$

where the function u is the temperature at a point $\mathbf{r} = (x, y)$ and time t, and k is a constant. The initial temperature distribution is $u(\mathbf{r}, 0)$. Because u is defined on an unbounded spatial domain, we may Fourier transform both sides, using the Laplacian relationship (Equation 10.12b),

$$-4\pi^2 \|\mathbf{v}\|^2 U(\mathbf{v}, t) = \frac{1}{k} \frac{\partial U(\mathbf{v}, t)}{\partial t}.$$

This is a first-order ordinary differential equation:

$$\frac{\partial U(\mathbf{v}, t)}{\partial t} + 4\pi^2 k \|\mathbf{v}\|^2 U(\mathbf{v}, t) = 0,$$

with initial condition $U(\mathbf{v}, 0)$. For now, consider an initial point distribution, modeled by a delta function, $u(\mathbf{r}, 0) = C\delta(\mathbf{r})$. Then $U(\mathbf{v}, 0) = C$. The solution is

$$U(\mathbf{v}, t) = C e^{-4\pi^2 k \|\mathbf{v}\|^2 t}, \quad t \geq 0.$$

To calculate the inverse Fourier transform, rewrite the solution in the form

$$U(\mathbf{v}, t) = C e^{-\pi(4\pi kt) \|\mathbf{v}\|^2},$$

a Gaussian. Using the dilation theorem, we obtain the final result:

$$u(\mathbf{r}, t) = \frac{C}{4\pi kt} e^{-\pi \left(\|\mathbf{r}\|/\sqrt{4\pi kt}\right)^2}. \tag{10.13}$$

These solutions are graphed in Figure 10.6 for $C = 1$, $k = 0.025$. The peak value of the distribution is $\frac{C}{4\pi kt}$ and the half width when it is $1/e$ of its peak value is $\sqrt{4kt}$. The cross-sectional area at this point is $4\pi kt$, and not surprisingly, the product of the peak

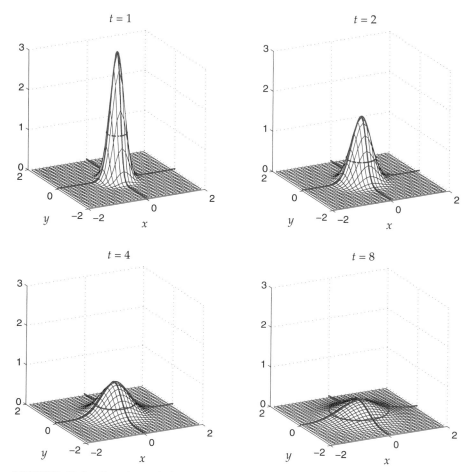

FIGURE 10.6 Gaussian solutions of the two-dimensional diffusion equation (Equation 10.13) with $C = 1, k = 0.025$, for times $t = 1, 2, 4, 8$. The peak value of the distribution is $C/4\pi kt$. The area of the $1/e$ circle shown is $4\pi kt$. To conserve the mass (energy) of the initial distribution, the distribution becomes shorter and wider as t increases.

value and this area is constant, C. That is, the initial mass (energy) of the distribution is conserved as it diffuses. ∎

Moment Theorems
The area theorem carries over to the multidimensional transform:

$$\int f(\mathbf{x})\, d\mathbf{x} = F(\mathbf{0}), \qquad (10.14\text{a})$$

$$f(\mathbf{0}) = \int F(\mathbf{v})\, d\mathbf{v}. \qquad (10.14\text{b})$$

10.1 TWO-DIMENSIONAL FOURIER TRANSFORM

Moments of multidimensional functions are taken with respect to each coordinate. In two dimensions,

$$\mu_x^{(n)} = \int_{-\infty}^{\infty}\int_{-\infty}^{\infty} x^n f(x,y)\,dxdy = \frac{1}{(-i2\pi)^n}\frac{\partial^n F(v_x,v_y)}{\partial v_x^n}\bigg|_{v=0},$$

$$\mu_y^{(n)} = \int_{-\infty}^{\infty}\int_{-\infty}^{\infty} y^n f(x,y)\,dxdy = \frac{1}{(-i2\pi)^n}\frac{\partial^n F(v_x,v_y)}{\partial v_y^n}\bigg|_{v=0}. \quad (10.15)$$

All the $n=0$ moments are the same and are equal to the area. The centroid of a multidimensional distribution is a vector:

$$\bar{x} = \frac{1}{\mu^{(0)}}\left(\mu_1^{(1)},\mu_2^{(1)},\ldots,\mu_N^{(1)}\right) = \frac{\nabla F(v)}{-i2\pi F(0)}\bigg|_{v=0}, \quad (10.16)$$

where ∇ is the gradient operator. Mixed moments are also defined in higher dimensions:

$$\mu^\alpha = \int \mathbf{x}^\alpha f(\mathbf{x})\,d\mathbf{x} = \frac{1}{(-i2\pi)^{|\alpha|}}\partial^\alpha F(v)\bigg|_{v=0}, \quad (10.17)$$

where α is a multi-index. In fact, this is the most general form of the moment theorem, for example, in three dimensions, $\mu_x^{(2)} = \mu^{(2,0,0)}$, $\mu_{xy}^{(1)} = \mu^{(1,1,0)}$.

Example 10.8 (Moments in statistics). In probability and statistics, the Fourier transform F of a probability density function f is called the *moment generating function* because of the connection between moments and derivatives of the Fourier transform. For a bivariate distribution $f(x,y)$, the expected values, or means, of the random variables x and y are the first moments:

$$\bar{x} = \mu_x^{(1)} = \int_{-\infty}^{\infty}\int_{-\infty}^{\infty} xf(x,y)\,dxdy = \frac{1}{-i2\pi}\frac{\partial F}{\partial v_x}\bigg|_{(0,0)},$$

$$\bar{y} = \mu_y^{(1)} = \int_{-\infty}^{\infty}\int_{-\infty}^{\infty} yf(x,y)\,dxdy = \frac{1}{-i2\pi}\frac{\partial F}{\partial v_y}\bigg|_{(0,0)}.$$

It is not necessary to divide by the area because all probability density functions have unit area (total probability $=1$). The variances are computed from the first and second moments:

$$\sigma_x^2 = \int_{-\infty}^{\infty}\int_{-\infty}^{\infty}(x-\bar{x})^2 f(x,y)\,dxdy = \mu_x^{(2)} - \left(\mu_x^{(1)}\right)^2 = \frac{1}{(-i2\pi)^2}\left[\frac{\partial^2 F}{\partial v_x^2} - \left(\frac{\partial F}{\partial v_x}\right)^2\right]_{(0,0)},$$

$$\sigma_y^2 = \int_{-\infty}^{\infty}\int_{-\infty}^{\infty}(y-\bar{y})^2 f(x,y)\,dxdy = \mu_y^{(2)} - \left(\mu_y^{(1)}\right)^2 = \frac{1}{(-i2\pi)^2}\left[\frac{\partial^2 F}{\partial v_y^2} - \left(\frac{\partial F}{\partial v_y}\right)^2\right]_{(0,0)}.$$

In addition, the correlation, which measures the linear dependence between two random variables, is a mixed moment:

$$\rho_{xy} = \int_{-\infty}^{\infty}\int_{-\infty}^{\infty} xyf(x,y)\,dxdy = \frac{1}{(-i2\pi)^2}\frac{\partial^2 F}{\partial v_x \partial v_y}\bigg|_{(0,0)}.$$

Specific examples are in the problems. ∎

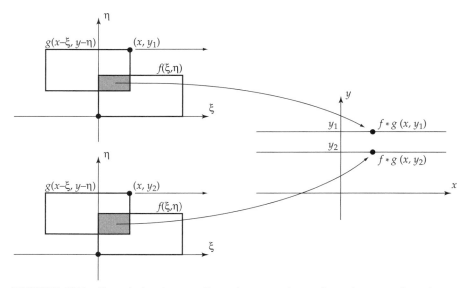

FIGURE 10.7 Convolution in two dimensions may be performed as a series of one-dimensional convolutions. Here, two functions with rectangular support in the first quadrant are convolved. Fixing $y = y_1$, sliding $g(x - \xi, y_1 - \xi)$ to the right, and integrating over the shaded region yields $f * g(x, y_1)$. Likewise, fixing $y = y_2$, sliding $g(x - \xi, y_2 - \xi)$ to the right, and integrating over the shaded region yields $f * g(x, y_2)$. Repeating for all y produces the full convolution, $f * g(x, y)$.

Convolution

The two-dimensional convolution and convolution theorem are exactly as one would expect:

$$f * g(\mathbf{x}) = \int f(\boldsymbol{\xi}) g(\mathbf{x} - \boldsymbol{\xi}) d\boldsymbol{\xi}, \tag{10.18a}$$

$$f * g(\mathbf{x}) \longleftrightarrow F(\boldsymbol{\nu})G(\boldsymbol{\nu}). \tag{10.18b}$$

The integration can often be done one dimension at a time, for example, in two dimensions (Figure 10.7),

$$f * g(x, y) = \int_{-\infty}^{\infty} \left[\int_{-\infty}^{\infty} f(\xi, \eta) g(x - \xi, y - \eta) d\xi \right] d\eta.$$

10.2 FOURIER TRANSFORMS IN POLAR COORDINATES

10.2.1 Circular Symmetry: Hankel Transform

When the function being Fourier transformed has radial symmetry, it is more natural to describe it in polar coordinates and to develop a Fourier transform in polar coordinates.

10.2 FOURIER TRANSFORMS IN POLAR COORDINATES

Consider, then, a function $f(r)$, where $r = \sqrt{x^2 + y^2}$. In the two-dimensional Fourier transform, Equation 10.1a, make the changes of variable:

$$x = r\cos\theta, \quad y = r\sin\theta$$
$$v_x = \rho\cos\phi, \quad v_y = \rho\sin\phi,$$

then

$$F(\rho) = \int_0^{2\pi} \int_0^\infty f(r) e^{-i2\pi(r\rho\cos\theta\cos\phi + r\rho\sin\theta\sin\phi)} \, r\, dr\, d\theta$$

$$= \int_0^\infty f(r) \left[\int_0^{2\pi} e^{-i2\pi r\rho\cos(\theta-\phi)} \, d\phi \right] r\, dr.$$

Now, the *Bessel functions* of the first kind, J_n, are defined by the integral[2]

$$J_n(x) = \frac{1}{2\pi} \int_{-\pi}^{\pi} e^{-i(nu - x\sin u)} \, du. \tag{10.17}$$

The first three Bessel functions, J_0 through J_2, are graphed in Figure 10.8.

Some useful Bessel function identities, which may be derived from the integral definition, are

$$2nJ_n(x) = xJ_{n-1}(x) + xJ_{n+1}(x), \tag{10.20a}$$
$$J_n'(x) = J_{n-1}(x) - \frac{n}{x} J_n(x), \tag{10.20b}$$
$$J_{-n}(x) = (-1)^n J_n(x). \tag{10.20c}$$

In addition, the Bessel functions have unit area:

$$\int_0^\infty J_n(x) \, dx = 1, \tag{10.21}$$

and an orthogonality property:

$$2\pi\rho \int_0^\infty J_n(2\pi r\rho) J_n(2\pi r\rho') \, r\, dr = \delta\left(2\pi(\rho - \rho')\right). \tag{10.22}$$

Using Equation 10.19 with $n = 0$, $x = 2\pi r\rho$, and $u = \frac{\pi}{2} + \theta - \phi$, we have

$$\int_0^{2\pi} e^{-i2\pi r\rho\cos(\theta-\phi)} \, d\phi = 2\pi J_0(2\pi r\rho),$$

and so,

$$F(\rho) = 2\pi \int_0^\infty f(r) J_0(2\pi r\rho) \, r\, dr. \tag{10.23}$$

---

[2] There are other ways to define a Bessel function, but this form is convenient for our purposes. Bessel's own research into this function began with this integral. See Watson (1995, pp. 19ff). The following Bessel function identities, and many more, may be found in a standard mathematics reference such as Abramowitz and Stegun (1972, Chapter 9), or the newer *NIST Digital Library of Mathematical Functions*, http://dlmf.nist.gov/, Chapter 10 (accessed February 12, 2013). Also see Churchill and Brown (1987, Chapter 8), which includes applications to the solution of partial differential equations.

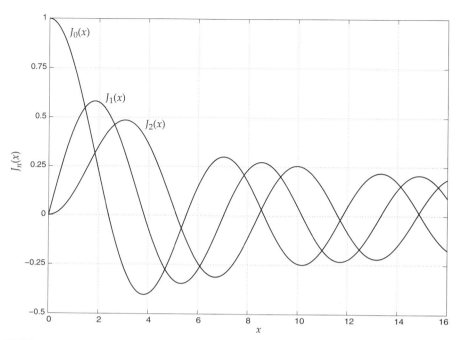

FIGURE 10.8 Bessel functions of the first kind, J_0, J_1, J_2 (Equation 10.19). The zero crossings are not uniformly spaced, as with sine and cosine. Of all Bessel functions J_n, only J_0 is nonzero at the origin.

This transform is also known as the *Hankel transform*.[3] The inverse Hankel transform is identical to the forward:

$$f(r) = 2\pi \int_0^\infty F(\rho) J_0(2\pi r \rho) \, \rho \, d\rho. \qquad (10.24)$$

The derivation of this is left as an exercise for the reader.

The Circle Function
A useful example of a radially symmetric function is the circle function, which can be used to model a circular aperture in an optical system:

$$\text{circ}(r) = \begin{cases} 1, & r < 1 \\ \frac{1}{2}, & r = 1 \\ 0, & \text{otherwise} \end{cases}. \qquad (10.25)$$

---

[3] More precisely, it is the zeroth-order Hankel transform. Higher order transforms are made using the *n*th order Bessel functions $J_n(\cdot)$. See the problems for an example.

The circle function is the polar coordinate analog of the two-dimensional rectangle function. To calculate its Hankel transform, begin by writing down the integral,

$$\mathrm{circ}(r) \longmapsto 2\pi \int_0^\infty \mathrm{circ}(r) J_0(2\pi r\rho) r\, dr = 2\pi \int_0^1 J_0(2\pi r\rho) r\, dr.$$

Make the change of variable $\xi = 2\pi r\rho$, and the integral becomes

$$2\pi \int_0^1 J_0(2\pi r\rho) r\, dr \to \frac{1}{2\pi\rho^2} \int_0^{2\pi\rho} J_0(\xi)\, \xi\, d\xi.$$

Now, using Equation 10.20b,

$$\xi J_0(\xi) = \xi J_1'(\xi) + J_1(\xi) = \frac{d}{d\xi}\left(\xi J_1(\xi)\right)$$

and

$$\int_0^{2\pi\rho} J_0(\xi)\, \xi\, d\xi = \xi J_1(\xi)\Big|_0^{2\pi\rho} = 2\pi\rho J_1(2\pi\rho).$$

Therefore, we have the result

$$\mathrm{circ}(r) \longmapsto \frac{J_1(2\pi\rho)}{\rho}. \tag{10.26}$$

By the two-dimensional area theorem, because the area (volume) under the unit circle function is π, the peak value of the Fourier transform, $\frac{J_1(2\pi\rho)}{\rho}$, is also π. Dividing by π so that it has unit height, $\frac{J_1(2\pi\rho)}{\pi\rho}$, it has a form analogous to the sinc function, $\frac{\sin \pi x}{\pi x}$. This normalized function is often called the "besinc," "jinc," or even "sombrero" function. We prefer "jinc." So with the definition

$$\mathrm{jinc}(\rho) = \frac{J_1(2\pi\rho)}{\pi\rho}, \tag{10.27}$$

the Fourier transform pair for the circle function is

$$\mathrm{circ}(r) \longleftrightarrow \pi\, \mathrm{jinc}(\rho). \tag{10.28}$$

The jinc function is graphed in Figure 10.9. It has the general appearance of a sinc, but notably, the zero crossings are not uniformly spaced.[4]

Delta Functions: Ring and Point

In one dimension, a pair of impulses transforms to a cosine:

$$\delta(x-a) + \delta(x+a) \longmapsto 2\cos 2\pi a\nu.$$

One way to extend this to two dimensions is to consider a pair of vertical impulse segments of length $2b$:

$$\left[\delta(x-a) + \delta(x+a)\right]\mathrm{rect}(y/2b) \longmapsto 4b\cos\left(2\pi a\nu_x\right)\mathrm{sinc}(2b\nu_y).$$

---

[4] Zero crossings of several Bessel functions, including J_0 and J_1, are tabulated in Abramowitz and Stegun (1972, Table 9.5).

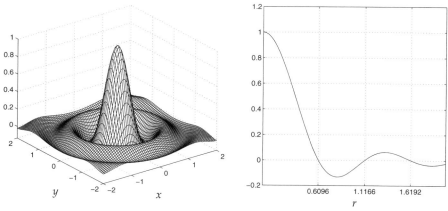

FIGURE 10.9 The jinc function, jinc $r = \frac{J_1(2\pi r)}{\pi r}$, as a function of $r = \sqrt{x^2 + y^2}$ (*left*) and in radial profile (*right*). The profile shows the first three zero crossings, at $r = 0.6096, 1.1166, 1.6192$. Unlike sine, cosine, and sinc, the zero crossings of the Bessel functions and jinc are not uniformly spaced.

Observe that as $b \to \infty$, $\text{rect}(y/2b) \to 1$, $2b \,\text{sinc}(2bv_y) \to \delta(v_y)$, and we recover $\delta(x - a) + \delta(x + a) \longmapsto 2\cos(2\pi a v_x)\,\delta(v_y)$. From two such sets of impulse segments at right angles we can construct a thin hollow box, or rectangular ring, and its Fourier transform (Figure 10.10):

$$[\delta(x-a) + \delta(x+a)]\,\text{rect}(y/2b) + \text{rect}(x/2a)[\delta(y-b) + \delta(y+b)]$$
$$\longmapsto 4b\cos(2\pi av_x)\,\text{sinc}(2bv_y) + 4a\,\text{sinc}(2av_x)\cos(2\pi bv_y). \tag{10.29}$$

Now, in polar coordinates, we define the ring delta, $\delta(r - a)$, which is supported on a circle of radius a centered at the origin. It has the sifting property

$$\int_0^{2\pi}\int_0^\infty \delta(r-a)f(r,\theta)\,r\,dr\,d\theta = \int_0^{2\pi} af(a,\theta)\,d\theta$$
$$= 2\pi a \cdot \frac{1}{2\pi}\int_0^{2\pi} f(a,\theta)\,d\theta. \tag{10.30}$$

The last expression is the circumference of a circle of radius a times the average of f on the circle. The area of the ring delta is also its circumference, $2\pi a$:

$$\int_0^{2\pi}\int_0^\infty \delta(r-a)\,r\,dr\,d\theta = 2\pi a.$$

A unit-area ring delta is $\delta(r-a)/2\pi a$ which is also $\delta(r-a)/2\pi r$ by the sifting property. The Hankel transform of the normalized ring delta is

$$2\pi\int_0^\infty \frac{\delta(r-a)}{2\pi r} J_0(2\pi r\rho)\,r\,dr = J_0(2\pi a\rho).$$

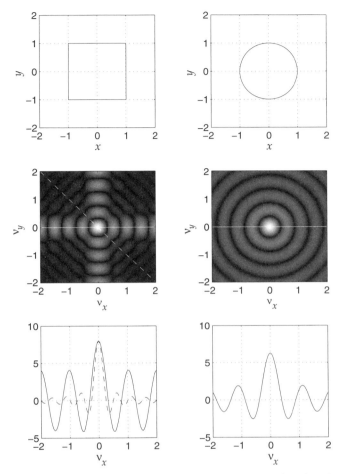

FIGURE 10.10 Hollow unit square (*left*) and unit ring (*right*) delta functions (*top*) and their Fourier transforms (*middle, bottom*), according to Equations 10.29 and 10.31. The images (*middle*) are of the Fourier magnitude, and the profiles (*bottom*) are cross-sections taken through the center of the transform, at $v_y = 0$ (solid) and $v_y = -v_x$ (dashed).

So we have another Hankel transform pair:

$$\frac{\delta(r-a)}{2\pi r} \longleftrightarrow J_0(2\pi a\rho) \quad \text{or} \quad \delta(r-a) \longleftrightarrow 2\pi a\, J_0(2\pi a\rho). \tag{10.31}$$

The square ($a = b$) and circular ring deltas, and their transforms, are compared in Figure 10.10.

In cartesian coordinates, a delta function at the origin is $\delta(x, y)$, with sifting property

$$\int_{-\infty}^{\infty}\int_{-\infty}^{\infty} f(x,y)\delta(x,y)\,dx\,dy = f(0,0).$$

When we take this to polar coordinates, it is straightforward enough to define a point delta away from the origin as the intersection of a ring delta, $\delta(r-a)/r$, and a "ray delta," $\delta(\theta-\theta_0)$:

$$\int_0^{2\pi}\int_0^\infty f(r,\theta)\frac{\delta(r-a)\delta(\theta-\theta_0)}{r}\,r\,dr\,d\theta = f(a,\theta_0).$$

When the point delta is at the origin, we expect to get $f(0,\cdot)$ (the value of θ is immaterial at the origin). We seek a generalized function $g(r)$ such that

$$\int_0^{2\pi}\int_0^\infty f(r,\theta)\,g(r)\,r\,dr\,d\theta = \int_0^\infty f(r,\cdot)\,[2\pi g(r)\,r]\,dr = f(0,\cdot).$$

Now recall the development of the delta function in Chapter 6 as a sequence of pulses:

$$\int_0^\infty f(r,\cdot)\,\delta(r)\,dr = \lim_{n\to\infty}\int_0^\infty f(r,\cdot)\,n\varphi(nr)\,dr.$$

When the integral runs only from 0 to ∞, it picks up only half of the delta pulse. Following the same approach taken in Section 6.2, we see that[5]

$$\int_0^\infty f(r,\cdot)\,\delta(r)\,dr = \frac{1}{2}f(0,\cdot).$$

Therefore, $2\pi g(r)\,r = 2\delta(r)$, and we define the unit impulse at the origin to be $\delta(r)/\pi r$:[6]

$$\int_0^{2\pi}\int_0^\infty f(r)\frac{\delta(r)}{\pi r}\,r\,dr\,d\theta = f(0,\cdot). \tag{10.32}$$

Its Hankel transform is

$$\frac{\delta(r)}{\pi r} \longmapsto 2\pi\int_0^\infty \frac{\delta(r)}{\pi r}J_0(2\pi r\rho)\,r\,dr = 2\int_0^\infty \delta(r)J_0(2\pi r\rho)\,dr = 2\times\frac{1}{2}J_0(0) = 1,$$

thus

$$\frac{\delta(r)}{\pi r} \longleftrightarrow 1. \tag{10.33}$$

---

[5] Instead of integrating $\int_0^{2\pi}\int_0^\infty \cdots r\,dr\,d\theta$, one may consider r to be a bilateral variable and integrate $\int_0^\pi \int_{-\infty}^\infty \cdots |r|\,dr\,d\theta$. Then

$$\int_0^\pi \int_{-\infty}^\infty f(r,\theta)g(r)\,|r|\,dr\,d\theta = f(0,\cdot)$$

implies $\pi g(r)|r| = \delta(r)$ and (but see the next footnote) the impulse at the origin is $\delta(r)/\pi|r|$. This approach is sometimes taken to avoid the "half delta" problem at the origin.

[6] The combination of $\delta(r)$ with $1/\pi r$ is problematic, since both are singular at the origin. In practice, $\delta(r)/\pi r$ always appears in an integral with a differential area $r\,dr\,d\theta$. Formally "cancelling" the $1/r$ and r yields correct results, but is suspect. Alternatively, one may use Equation 6.28 to write $g(r) = -\delta'(r)/\pi$ (see the problems). For a more rigorous approach to polar coordinate transformations of generalized functions that avoids difficulties at the origin see Jones (1982, pp. 300–306).

Hankel Transform Theorems

Theorems for the two-dimensional Fourier transform apply to the Hankel transform, as one would expect. One must be careful with the dilation and shift theorems, since these operations can spoil the circular symmetry. Moreover, convolution is difficult. We will comment on these in detail here, and leave the rest to a table at the end of the chapter.

Recall that the dilation theorem in two dimensions is

$$f(\mathbf{Ax}) \longleftrightarrow \frac{1}{|\det \mathbf{A}|} F(\mathbf{A}^{-T}\mathbf{v}),$$

where $\mathbf{A}$ is an invertible 2×2 matrix. Consider a function $f(r)$, where $r = \sqrt{x^2 + y^2} = \|\mathbf{x}\|$, and its Hankel transform $F(\rho)$, where $\rho = \sqrt{v_x^2 + v_y^2} = \|\mathbf{v}\|$. If $\mathbf{A}$ represents a rotation, then circular symmetry is preserved:

$$f(\|\mathbf{Ax}\|) = f(\|\mathbf{x}\|) \longleftrightarrow F(\|\mathbf{v}\|). \tag{10.34}$$

If $\mathbf{A}$ represents an isotropic scaling, $\mathbf{Ax} = a\mathbf{x}$, then $\det \mathbf{A} = a^2$ and circular symmetry is preserved,

$$f(ar) \longleftrightarrow \frac{1}{a^2} F(\rho/a). \tag{10.35}$$

Other transformations, for example, anisotropic scaling or shearing, break the circular symmetry.

As for the shift theorem, translating a circularly symmetric function breaks the symmetry, and the Fourier transform pair can no longer be written in terms of the radial variables r and ρ. As above, let $f(\|\mathbf{x}\|)$ and $F(\|\mathbf{v}\|)$ be a Hankel transform pair. A translated version of f is $f(\|\mathbf{x} - \mathbf{b}\|)$, and by the shift theorem, its Fourier transform is

$$f(\|\mathbf{x} - \mathbf{b}\|) \longmapsto e^{-i2\pi \mathbf{v} \cdot \mathbf{b}} \mathcal{F}\{f(\|\mathbf{x}\|)\} = e^{-i2\pi \mathbf{v} \cdot \mathbf{b}} F(\|\mathbf{v}\|). \tag{10.36}$$

Example 10.9 (A pair of circular apertures). Consider a pair of circle functions of radius a, centered at $x = \pm b$:

$$f(\mathbf{x}) = \operatorname{circ}\left(\frac{\sqrt{(x+b)^2 + y^2}}{a}\right) + \operatorname{circ}\left(\frac{\sqrt{(x-b)^2 + y^2}}{a}\right).$$

First use the shift theorem:

$$F(\mathbf{v}) = e^{+i2\pi b v_x} \mathcal{F}\left\{\operatorname{circ}\left(\frac{\sqrt{x^2+y^2}}{a}\right)\right\} + e^{-i2\pi b v_x} \mathcal{F}\left\{\operatorname{circ}\left(\frac{\sqrt{x^2+y^2}}{a}\right)\right\}$$

$$= 2\cos(2\pi b v_x) \mathcal{F}\left\{\operatorname{circ}\left(\frac{\sqrt{x^2+y^2}}{a}\right)\right\}.$$

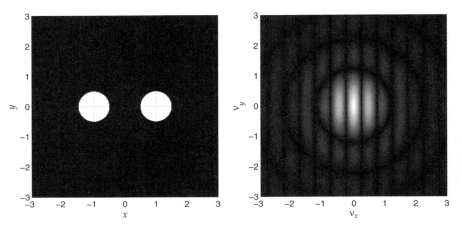

FIGURE 10.11 Two circle functions of radius 0.5, centered at $x = \pm 1$ (*left*) and their Fourier magnitude (*right*). The Fourier transform is the product of a jinc that depends on the aperture radius and a cosine modulation whose frequency depends on the separation of the apertures.

With the shift removed, the remaining Fourier transform is of a circularly symmetric function, and the dilation theorem may be used with the Hankel transform of the circle to obtain

$$F(\mathbf{v}) = 2\pi a^2 \cos(2\pi b v_x) \operatorname{jinc}(a\|\mathbf{v}\|).$$

These functions are shown in Figure 10.11. In optics and antenna theory, this models the interference of waves emanating from two uniformly illuminated circular apertures. As the apertures become smaller and more pointlike, the jinc function spreads so that the cosine becomes the dominant feature, as expected. ∎

As for convolution, the Hankel transform is a Fourier transform, so of course there is a convolution relationship—the Hankel transform of the product $F(\rho)G(\rho)$ is the convolution of $f(r)$ and $g(r)$. Derivation of the convolution integral $f * g(r)$, in polar coordinates, is left to the problems. Here is the result:

$$f * g(r) = \int_0^{2\pi} \int_0^\infty f(u) g\left(\sqrt{r^2 + u^2 - 2ru\cos\theta}\right) u \, du \, d\theta. \qquad (10.37)$$

(You may recognize the Law of Cosines in the argument for g.) Actually performing the integration is difficult, except in some very special cases. Here is one.

Example 10.10 (Self convolution of a circle function). The setup for calculating the convolution of $\operatorname{circ}(r/a)$ with itself is shown in Figure 10.12. Because the circle function is either 1 or 0 and is confined to a bounded area, the integration reduces to calculating the area of the lens-shaped region of overlap, which by symmetry is four times the area of the smaller region shown in the figure. Covering the area with

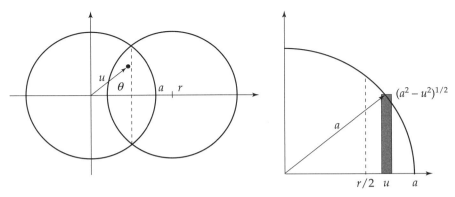

FIGURE 10.12 Convolution of two circle functions of radius a. *Left*: Integration is with respect to u and θ over the lens-shaped area of overlap, for $r \leq 2a$. The convolution is zero for $r > 2a$. *Right*: Detail of calculation for a quarter of the area.

strips of height $\sqrt{a^2 - u^2}$ and width du and integrating from $u = r/2$ to $u = a$ gives (for $r \leq 2a$)

$$\begin{aligned}
\text{circ}(r/a) * \text{circ}(r/a) &= 4 \int_{r/2}^{a} (a^2 - u^2)^{1/2} \, du \\
&= 2 \left[u\sqrt{a^2 - u^2} + a^2 \arctan\left(\frac{u}{\sqrt{a^2 - u^2}}\right) \right]_{r/2}^{a} \\
&= 2 \left[a^2 \frac{\pi}{2} - \frac{ar}{2}\sqrt{1 - (r/2a)^2} - a^2 \arctan\left(\frac{r/2a}{\sqrt{1 - (r/2a)^2}}\right) \right] \\
&= 2a^2 \left[\arccos\left(\frac{r}{2a}\right) - \frac{r}{2a}\sqrt{1 - \left(\frac{r}{2a}\right)^2} \right], \quad r \leq 2a.
\end{aligned}$$
(10.38)

The convolution is zero for $r > 2a$. When $r = 0$, the expression evaluates to πa^2, the area of the circle. As we have seen before with convolution, the support of the result is larger than the support of the individual functions (Figure 10.13). ∎

10.2.2 Spherical Symmetry

To calculate the three-dimensional Fourier transform of a function with spherical symmetry, begin again with the cartesian form and make the changes of variable from (x, y, z) to (r, φ, θ) (Figure 10.14):

$$x = r \sin\theta \cos\varphi, \quad y = r \sin\theta \sin\varphi, \quad z = r \cos\theta$$
$$\nu_x = \rho \sin\zeta \cos\xi, \quad \nu_y = \rho \sin\zeta \sin\xi, \quad \nu_z = \rho \cos\zeta$$

then

$$\mathbf{v} \cdot \mathbf{x} = r\rho \sin\theta \sin\zeta \underbrace{(\cos\varphi \cos\xi + \sin\varphi \sin\xi)}_{\cos(\varphi-\xi)} + r\rho \cos\theta \cos\zeta$$

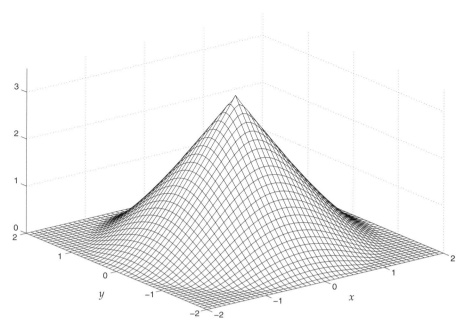

FIGURE 10.13 Self-convolution of a circle function (Equation 10.38). The convolution is radially symmetric and supported on a circle of radius 2. The peak value is π. For a cross-sectional view see Figure 10.24.

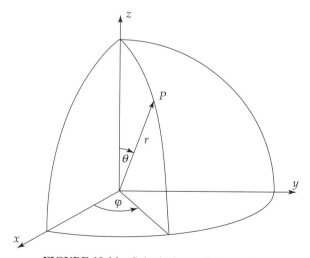

FIGURE 10.14 Spherical coordinate system.

and

$$F(\rho) = \int_0^\pi \int_0^{2\pi} \int_0^\infty f(r) e^{-i2\pi r\rho(\sin\theta \sin\zeta \cos(\varphi-\xi)+\cos\theta \cos\zeta)} r^2 \sin\theta \, dr \, d\varphi \, d\theta.$$

This is a formidable integral, but can be simplified greatly by the assumption of spherical symmetry. Because the Fourier transform is independent of angle, we may choose values of ζ and ξ that are convenient. In particular, with $\zeta = 0$, the integral becomes

$$F(\rho) = \int_0^\pi \int_0^{2\pi} \int_0^\infty f(r) e^{-i2\pi r\rho \cos\theta} r^2 \sin\theta \, dr \, d\varphi \, d\theta$$

$$= 2\pi \int_0^\pi \int_0^\infty f(r) e^{-i2\pi r\rho \cos\theta} r^2 \sin\theta \, dr \, d\theta.$$

Let $u = \cos\theta$, $du = -\sin\theta d\theta$, and

$$F(\rho) = 2\pi \int_0^\infty f(r) \left[\int_{-1}^1 e^{-i2\pi r\rho u} \, du \right] r^2 \, dr$$

$$= 4\pi \int_0^\infty f(r) \, \text{sinc}(2r\rho) \, r^2 \, dr = \frac{2}{\rho} \int_0^\infty f(r) \sin(2\pi r\rho) \, r \, dr. \qquad (10.39)$$

The inverse is identical in form to the forward transform:

$$f(r) = 4\pi \int_0^\infty F(\rho) \, \text{sinc}(2r\rho) \, \rho^2 \, d\rho = \frac{2}{r} \int_0^\infty F(\rho) \sin(2\pi r\rho) \, \rho \, d\rho. \qquad (10.40)$$

Example 10.11 (Unit ball). A unit ball is the spherical analog of the circle function. Its Fourier transform is

$$F(\rho) = \frac{2}{\rho} \int_0^1 \sin(2\pi r\rho) \, r \, dr$$

$$= \frac{2}{\rho} \left. \frac{\sin(2\pi r\rho) - 2\pi r\rho \cos(2\pi r\rho)}{(2\pi\rho)^2} \right|_0^1 = \frac{\sin(2\pi\rho) - 2\pi\rho \cos(2\pi\rho)}{2\pi^2 \rho^3}.$$

Letting $\rho = 0$ in this result should recover the volume of the ball (generalizing the area theorem to three dimensions). Using L'Hospital's rule,

$$F(0) = \lim_{\rho \to 0} \frac{\sin(2\pi\rho) - 2\pi\rho \cos(2\pi\rho)}{2\pi^2 \rho^3}$$

$$= \lim_{\rho \to 0} \frac{2\pi \cos(2\pi\rho) - 2\pi \cos(2\pi\rho) + 4\pi^2 \rho \sin(2\pi\rho)}{6\pi^2 \rho^2}$$

$$= \lim_{\rho \to 0} \frac{2 \sin(2\pi\rho)}{3\rho} = \frac{4\pi}{3},$$

which is indeed the volume of the unit ball. ∎

Example 10.12 (Spherical shell). A thin spherical shell is the three-dimensional analog of the ring delta. To calculate the normalization factor, integrate $\delta(r-a)$ over all space:

$$\int_0^\pi \int_0^{2\pi} \int_0^\infty \delta(r-a)\, r^2 \sin\theta\, dr\, d\varphi\, d\theta = 4\pi a^2,$$

which is, coincidentally, the surface area of the sphere. A spherical unit delta function is defined $\delta(r-a)/4\pi a^2$, or by the sifting property, $\delta(r-a)/4\pi r^2$.

The Fourier transform of the spherical shell is

$$4\pi \int_0^\infty \frac{\delta(r-a)}{4\pi r^2} \operatorname{sinc}(2r\rho)\, r^2\, dr = \int_0^\infty \delta(r-a) \operatorname{sinc}(2r\rho)\, dr = \operatorname{sinc}(2a\rho).$$

Thus,

$$\frac{\delta(r-a)}{4\pi r^2} \longleftrightarrow \operatorname{sinc}(2a\rho) \quad \text{or} \quad \delta(r-a) \longleftrightarrow 4\pi a^2 \operatorname{sinc}(2a\rho). \qquad (10.41)$$

■

An impulse at the origin in spherical coordinates must provide the sifting property. We seek a generalized function g such that

$$\int_0^\pi \int_0^{2\pi} \int_0^\infty f(r,\varphi,\theta)\, g(r)\, r^2 \sin\theta\, dr\, d\varphi\, d\theta = \int_0^\infty f(r,\cdot,\cdot) \left[4\pi r^2 g(r)\right] dr$$
$$= f(0,\cdot,\cdot).$$

(The values of φ and θ are immaterial at the origin.) As with the earlier calculation in polar coordinates, the integral $\int_0^\infty f(r)\delta(r)\, dr = \tfrac{1}{2}f(0)$, so $4\pi r^2 g(r) = 2\delta(r)$, and we define the unit impulse at the origin to be[7] $\delta(r)/2\pi r^2$. Its Fourier transform is

$$4\pi \int_0^\infty \frac{\delta(r)}{2\pi r^2} \operatorname{sinc}(2r\rho)\, r^2\, dr = 2\int_0^\infty \delta(r) \operatorname{sinc}(2r\rho)\, dr = 1;$$

thus,

$$\frac{\delta(r)}{2\pi r^2} \longleftrightarrow 1. \qquad (10.42)$$

Theorems for the three-dimensional transform follow the same pattern as the Hankel transform and are not elaborated upon here.[8]

10.3 WAVE PROPAGATION

10.3.1 Plane Waves

A wave whose complex amplitude is of the form $e^{i(\mathbf{k}\cdot\mathbf{r}-\omega t)}$ is called a *plane wave*. It can be shown to be a solution of the three-dimensional wave equation $\nabla^2 \tilde{u}(\mathbf{r},t) =$

---

[7] With the same caveats that applied to the polar coordinate delta function mentioned in an earlier footnote.
[8] See Bracewell (2000, pp. 329–343) for more examples of two- and three-dimensional radial transforms.

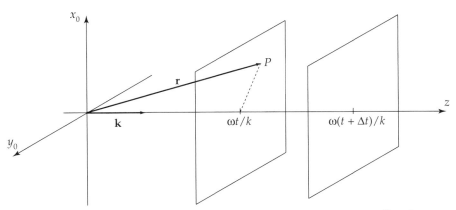

FIGURE 10.15 The equiphase surfaces of the complex wave amplitude $e^{i(\mathbf{k}\cdot\mathbf{r}-\omega t)}$ are the points P such that $\mathbf{k}\cdot\mathbf{r} - \omega t =$ constant. These surfaces are planar, and perpendicular to the $\mathbf{k}$ vector; here, $\mathbf{k}$ is oriented along the z-axis. The equiphase $\mathbf{k}\cdot\mathbf{r} - \omega t = 0$ is located at distance $\omega t / k$ from the origin $z = 0$. After a time Δt has elapsed, the equiphase has propagated a distance $\Delta z = \omega \Delta t / k$ farther. The distance corresponding to a phase change $\omega \Delta t = 2\pi$ is the wavelength, λ.

$\frac{1}{v^2}\frac{\partial^2}{\partial t^2}\tilde{u}(\mathbf{r},t)$, with $v^2 = \omega^2/\|\mathbf{k}\|^2$. The wavefronts, or equiphase surfaces, of the wave are defined by $\mathbf{k}\cdot\mathbf{r} - \omega t =$ constant. For a fixed t, these are planes perpendicular to the vector $\mathbf{k}$ (Figure 10.15). The wavelength λ is the distance in space between two planar equiphases differing by 2π at a fixed t, that is, $k\lambda = 2\pi$, which gives $k = \|\mathbf{k}\| = 2\pi/\lambda$.

For any particular equiphase, with $\mathbf{k}\cdot\mathbf{r} - \omega t =$ constant, as time increases $\mathbf{k}\cdot\mathbf{r}$ must also increase to keep the phase constant. This means that $\mathbf{r}$ for a point on the equiphase must also lengthen, that is, the wave propagates in the direction of $\mathbf{k}$. In a time interval Δt, the equiphase moves a distance $\omega \Delta t / k$. The ratio $\omega / k = \lambda \omega / 2\pi = \lambda \nu$ is the propagation speed, v, of the wave. For electromagnetic waves in vacuum, v is the speed of light, c. Unlike the spherical wave (Equation 4.36), whose amplitude falls off with distance from the point of origin, the plane wave's amplitude ideally remains constant as it propagates.

The direction of propagation of a plane wave is defined by the vector $\mathbf{k} = (k_x, k_y, k_z)$. A common way of specifying the orientation of the $\mathbf{k}$ vector is via the angles a and b that $\mathbf{k}$ makes with the z-axis in the xz and yz planes, respectively (Figure 10.16). The sines of these direction angles are called the *direction sines* of $\mathbf{k}$, $\alpha = \sin a$ and $\beta = \sin b$. In terms of the direction sines, we have

$$\mathbf{k} = (k_x, k_y, k_z) = k\left(\alpha, \beta, \sqrt{1-\alpha^2-\beta^2}\right). \tag{10.43}$$

All the components of $\mathbf{k}$ must be real in order for $e^{i(\mathbf{k}\cdot\mathbf{r}-\omega t)}$ to be a propagating wave. An imaginary component causes the complex exponential to decay and the wave does not propagate (a so-called evanescent wave). Individually, of course, α^2 and β^2 are no greater than 1, being sines. Then, in order for k_z to be real, we have the additional constraint $\alpha^2 + \beta^2 \leq 1$. A plane wave propagating in any direction may

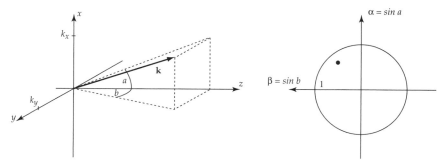

FIGURE 10.16 The orientation of the **k** vector for a plane wave may be specified by the angles a and b that **k** makes with the z-axis in the xz and yz planes, respectively (*left*). For a propagating wave, the pair (α, β) of direction sines $\alpha = \sin a = k_x/k$ and $\beta = \sin b = k_y/k$ must be within the unit circle, $\alpha^2 + \beta^2 \leq 1$ (*right*).

be located by a pair of direction sines (α, β) within the unit circle (Figure 10.16). For example, $(0, 0)$ gives a wave propagating down the z-axis; $(1, 0)$ gives a wave propagating up the x-axis. Orientations falling outside the unit circle correspond to evanescent (non-propagating) waves.

Mathematically, the complex exponential $e^{i\mathbf{k}\cdot\mathbf{r}}$ is of the same form as the multidimensional Fourier transform kernel $e^{i2\pi\mathbf{v}\cdot\mathbf{x}}$, and we will see that plane waves have the same "basis" property for propagating waves that sinusoids have for one-dimensional signals.

10.3.2 Fraunhofer Diffraction

One of the most important applications of the two-dimensional Fourier transform is wave propagation—antennas, optics, acoustics, and more. The basic result connects the complex field amplitude at a plane, denoted $\tilde{u}(x_0, y_0)$, with the field amplitude at a distant observation point—*the far field amplitude is the spatial Fourier transform of the input field amplitude*. This may be arrived at in different ways. The approach taken here will be to extend the results on antenna arrays in Section 4.7. Recall that the complex field at an observation point P_1, due to a point source at P_0, is a spherical wave (Figure 10.17):

$$\tilde{u}_{P_1} = \frac{A e^{i(kr_{01} - \omega t)}}{r_{01}},$$

where $k = 2\pi/\lambda = \omega/c$. In what follows we will drop the sinusoidal time dependence and concern ourselves only with the spatial variations of the field. Regarding an arbitrary source distribution $\tilde{u}_0(\mathbf{r}_0)$ as a superposition of point sources, the field at P_1 can be shown to be a superposition of spherical waves:[9]

$$\tilde{u}_1 = \iint_{\text{source}} \tilde{u}_0(\mathbf{r}_0) \frac{e^{ikr_{01}}}{i\lambda r_{01}} d\mathbf{r}_0, \tag{10.44}$$

---

[9]Goodman (1968, Chapters 3 and 4) gives an excellent historical overview and a derivation from first principles.

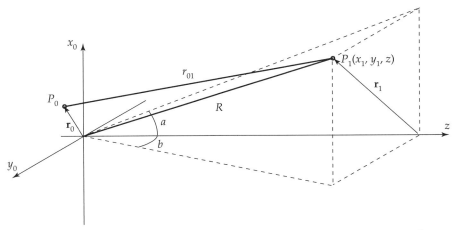

FIGURE 10.17 Geometry for wave propagation between two points. Vectors $\mathbf{r}_0$ and $\mathbf{r}_1$ are the transverse coordinates of the source and observation points, respectively. The longitudinal coordinate of the observation point is z. The angles a and b orient the observation point P_1 with respect to the z-axis in the xz and yz planes, respectively.

under the conditions that the direction angles a and b are not too large—the so-called paraxial approximation—and that the observation point is many wavelengths away from the source, $R \gg \lambda$.

To make this integral practical, we need a convenient expression for the distance r_{01} in terms of R, the direction angles a and b, and the coordinates (x_0, y_0). We begin by adding the y_0 dimension to the previous derivation:

$$r_{01} = \left(R^2 + (x_0^2 + y_0^2) - 2R(x_0 \sin a + y_0 \sin b)\right)^{1/2}$$

$$= R\sqrt{1 - \frac{2(x_0 \sin a + y_0 \sin b)}{R} + \frac{r_0^2}{R^2}},$$

where $r_0^2 = x_0^2 + y_0^2$. Assuming that P_1 is very far from P_0, so that $R \gg r_0$, we approximate the square root by a binomial expansion:

$$r_{01} \approx R\left(1 - \frac{(x_0 \sin a + y_0 \sin b)}{R} + \frac{r_0^2}{2R^2} + \text{higher order terms}\right),$$

and keep only the first-order term, resulting in the so-called Fraunhofer approximation:[10]

$$r_{01} \approx R - (x_0 \sin a + y_0 \sin b).$$

---

[10] See Goodman (1968, pp. 58–62) for a discussion of the physical validity of this approximation.

We may satisfactorily approximate the distance r_{01} in the denominator by R, simplifying the spherical wave to

$$\frac{e^{ikr_{01}}}{r_{01}} \approx \frac{e^{ikR}}{R} e^{-ik(x_0 \sin a + y_0 \sin b)}.$$

This field is a combination of two effects: a spherical wave e^{ikR}/R, and a modulation of the complex amplitude of the spherical wave that depends on the coordinates of the source point, (x_0, y_0), and the direction angles of the observation point, (a, b). This much is just a two-dimensional version of the results in Chapter 4.

Inserting this approximation into the superposition, Equation 10.44, and letting $k = 2\pi/\lambda$, gives the field at P_1:

$$\tilde{u}_1(a, b) \approx \frac{e^{ikR}}{i\lambda R} \iint_{-\infty}^{\infty} \tilde{u}_0(x_0, y_0) \exp\left[-i2\pi \left(x_0 \left(\frac{\sin a}{\lambda}\right) + y_0 \left(\frac{\sin b}{\lambda}\right)\right)\right] dx_0 dy_0$$

$$= \frac{e^{ikR}}{i\lambda R} \tilde{U}_0 \left(\frac{\sin a}{\lambda}, \frac{\sin b}{\lambda}\right). \tag{10.45}$$

Wave propagation under the Fraunhofer approximation is commonly called *Fraunhofer diffraction*.[11] The complex amplitude of the diffracted field, in the Fraunhofer approximation, is proportional to the two-dimensional Fourier transform of the source distribution, $\tilde{U}_0(\nu_x, \nu_y)$, evaluated at $\nu_x = (\sin a)/\lambda$, $\nu_y = (\sin b)/\lambda$. Another version is obtained when the direction angles a and b are small, the so-called *paraxial approximation*. Then, $\sin a \approx \tan a = x_1/z$, $\sin b \approx \tan b = y_1/z$, and $R \approx z + \frac{x_1^2 + y_1^2}{2z}$. We have

$$\tilde{u}_1(x_1, y_1)$$

$$\approx \frac{e^{ikz}}{i\lambda z} e^{ik(x_1^2 + y_1^2)/2z} \iint_{-\infty}^{\infty} \tilde{u}_0(x_0, y_0) \exp\left[-i2\pi \left(x_0 \left(\frac{x_1}{\lambda z}\right) + y_0 \left(\frac{y_1}{\lambda z}\right)\right)\right] dx_0 dy_0$$

$$= \frac{e^{ikz}}{i\lambda z} e^{ik(x_1^2 + y_1^2)/2z} \tilde{U}_0 \left(\frac{x_1}{\lambda z}, \frac{y_1}{\lambda z}\right). \tag{10.46}$$

The integral is the two-dimensional Fourier transform of the source distribution, $\tilde{U}_0(\nu_x, \nu_y)$, evaluated at $\nu_x = x_1/\lambda z$, $\nu_y = y_1/\lambda z$. For small angles, the sphere of radius R in Equation 10.45 is approximately a plane at a distance z from the origin. The radiation patterns of antennas, as seen in Chapter 4, are commonly plotted as functions of the direction sines (Equation 10.45). In systems of lenses, where angles are small enough to justify the paraxial approximation and propagation is between planes, the second form (Equation 10.46) is used.

---

[11] Physically, diffraction is observed when a wave encounters an opaque obstacle such as a sharp edge or aperture, or an object whose transparency or thickness varies spatially on scales comparable to a wavelength.

The Fourier transform $\widetilde{U}_0$ has an interesting physical interpretation using plane waves. The initial field $\tilde{u}_0$ may be written as the inverse transform of $\widetilde{U}_0$:

$$\tilde{u}_0(x_0, y_0) = \iint_{-\infty}^{\infty} \widetilde{U}_0(v_x, v_y) e^{+i2\pi(v_x x + v_y y)} \, dv_x \, dv_y$$

$$= \iint_{-\infty}^{\infty} \widetilde{U}_0\left(\frac{\alpha}{\lambda}, \frac{\beta}{\lambda}\right) e^{+ik(\alpha x + \beta y)} \frac{d\alpha}{\lambda} \frac{d\beta}{\lambda},$$

where α and β are direction sines. Now recall the plane wave:

$$e^{i\mathbf{k}\cdot\mathbf{r}} = \exp\left[i\left(k_x x + k_y y + k_z z\right)\right],$$

where $k_x = k\alpha$, $k_y = k\beta$, and $k_z = k\sqrt{1 - (\alpha^2 + \beta^2)}$ with $\alpha^2 + \beta^2 < 1$. Then

$$e^{i\mathbf{k}\cdot\mathbf{r}} = \exp\left[ik\sqrt{1 - (\alpha^2 + \beta^2)}\,z\right] \exp[ik(\alpha x + \beta y)].$$

At the initial plane, $z = 0$ and the Fourier kernel is identical in form to a plane wave propagating with direction sines α and β. Thus, the field $\tilde{u}_0(x_0, y_0)$, whose equiphases may have arbitrary shape, is expressed as a linear superposition of plane waves propagating with direction sines α and β. The Fourier transform $\widetilde{U}_0(\alpha/\lambda, \beta/\lambda)$, which is constrained to the circular support $\alpha^2 + \beta^2 \leq \lambda^2$, is interpreted as the angular spectrum of plane waves making up the wave $\tilde{u}_0(x_0, y_0)$.[12]

Example 10.13 (Sinusoidal amplitude grating). A *diffraction grating* is an optical element with a fine periodic variation of density (magnitude) or thickness (phase). Consider, for example, a glass plate upon which has been deposited a film with a sinusoidal variation of density. Its transmittance is $t(x, y) = \left[\frac{1}{2} + \frac{1}{2}\cos 2\pi bx\right] \text{rect}(x/L)\text{rect}(y/L)$, where $1/b$ is on the order of a wavelength of light. If this plate is illuminated with a normally incident, unit-amplitude plane wave, the resulting far-field complex amplitude is, dropping the leading phase factors,

$$\tilde{u}_1(x_1, y_1)$$

$$= T(v_x, v_y)\Big|_{(x_1/\lambda z, y_1/\lambda z)}$$

$$= \left[\frac{1}{2}\delta(v_x) + \frac{1}{4}\delta(v_x - b) + \frac{1}{4}\delta(v_x + b)\right]\delta(v_y) * L^2 \text{sinc}(Lv_x)\text{sinc}(Lv_y)\Big|_{(x_1/\lambda z, y_1/\lambda z)}$$

$$= \frac{L^2}{4}\left[\text{sinc}\left(L(v_x + b)\right) + 2\,\text{sinc}(Lv_x) + \text{sinc}(L(v_x - b))\right]\text{sinc}(Lv_y)\Big|_{(x_1/\lambda z, y_1/\lambda z)}.$$
(10.47)

Typically, $L \gg 1/b$ and the observed optical intensity, $|\tilde{u}_1|^2$, consists of three well-separated sinc²-shaped peaks lined up along the x_1-axis, at $x_1 = 0$ and $x_1 = \pm\lambda zb$

---

[12] Propagation and diffraction may correspondingly be interpreted as linear filtering operations carried out on the angular spectrum. See Goodman (1968, Chapter 3) for details.

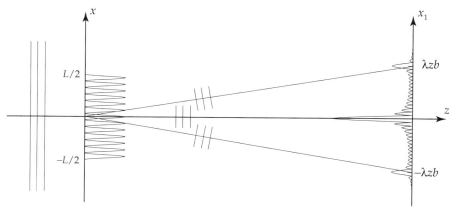

FIGURE 10.18 Diffraction pattern from a sinusoidal amplitude grating, sectioned along the xz plane. Plane wave illumination is diffracted by the grating into an on-axis and two off-axis spots. The width of each spot is inversely proportional to the aperture width, L, and the position of the off-axis spots is proportional to the wavelength, λ.

(Figure 10.18). The width of each spot, measured by the distance between the first zero crossings, is $\lambda z/L$. The central peak represents light that is not diffracted by the grating but only by the bounding aperture. The two off-axis peaks, at $x_1 = \pm \lambda zb$, are the ± 1 *diffracted orders* of the grating. The angle at which they diffract off the grating, which is approximately λb, increases with both the spatial frequency of the grating, b, and the wavelength of the light, λ. Shorter wavelengths, for example, blue light, diffract less than longer wavelengths, for example, red light. Thus, a diffraction grating can be used to separate polychromatic light into its component wavelengths. Gratings are important components in spectrometers and other optical instruments.

The *diffraction efficiency* of a grating is the fraction of the incident power that goes into the desired diffracted order. For this grating, the incident intensity is proportional to the aperture area, L^2, and the intensity of the +1 diffracted order is $L^2/16$. Thus, the diffraction efficiency of the grating is only $1/16$. Such low efficiency is characteristic of amplitude gratings that absorb much of the incident light and have a large undiffracted component. ∎

Example 10.14 (Diffraction grating with tilted input wave). Suppose the grating in the previous example is tilted with respect to the incident light, so that the incident beam makes an angle θ with respect to the normal to the grating, which is still taken to be the z-axis. A tilted plane wave is described by $e^{i\mathbf{k}\cdot\mathbf{r}}$ with $\mathbf{k} = (k\sin\theta, 0, k\cos\theta)$, so the optical field just behind (after) the grating is

$$\tilde{u}_0(x, y) = e^{ikz\cos\theta} e^{ikx\sin\theta} t(x, y).$$

The diffracted field is the Fourier transform of $\tilde{u}_0$, as in the previous example. What is new in this calculation is the phase term $e^{i(k\sin\theta)x} = e^{i2\pi(\sin\theta/\lambda)x}$. Applying the shift

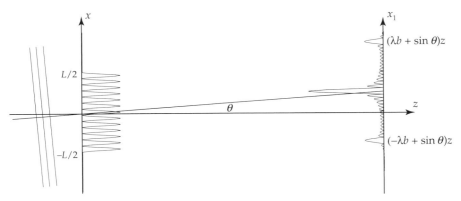

FIGURE 10.19 Diffraction pattern from a sinusoidal amplitude grating with tilted illumination. The tilt θ in the illumination causes the diffraction pattern to shift an amount proportional to $\sin\theta$. When $\sin\theta = \lambda b$, the -1 diffracted order is shifted to the origin of the observation plane. A plot of the on-axis intensity as a function of $\sin\theta$ can be converted to a spectrum of the light as a function of λ.

theorem with Equation 10.46 and dropping the leading phase factors,

$$\tilde{u}_1(x_1, y_1) = e^{ikz\cos\theta} \mathcal{F}\left\{e^{i2\pi(\sin\theta/\lambda)x} t(x,y)\right\}\Big|_{(x_1/\lambda z, y_1/\lambda z)}$$

$$= e^{ikz\cos\theta} T\left(v_x - \frac{\sin\theta}{\lambda}, v_y\right)\Big|_{(x_1/\lambda z, y_1/\lambda z)}$$

$$= e^{ikz\cos\theta} T\left(\frac{x_1 - z\sin\theta}{\lambda z}, \frac{y_1}{\lambda z}\right). \tag{10.48}$$

The complex exponential factor in front has unity magnitude and does not affect the intensity. The diffraction pattern is the same as Equation 10.47, shifted along the x_1-axis in the observation plane by $z\sin\theta$ (Figure 10.19). In this context, the shift theorem connects translation at one plane with the direction (tilt) of a diffracted wave at another plane. ∎

The tilt angle θ in the diffraction pattern, Equation 10.48, enables the grating to be used to determine the wavelength spectrum of a light source. Such an instrument is called a *spectrometer*. If the light is collimated (made into a plane wave) by a system of lenses and used to illuminate a tilted grating, the -1 diffracted order will appear at location $x_1 = (\lambda b - \sin\theta)z$. By placing a photodetector on-axis, $x_1 = 0$, and adjusting θ, different wavelengths are swept across the detector, with the relationship $\lambda = \sin\theta/b$. A plot of the on-axis intensity as a function of $\sin\theta$ can be converted to a spectrum of the light as a function of λ.

The finite aperture limits the ability of the spectrometer to separate, or *resolve*, closely-spaced spectral lines. This can be seen in Equation 10.47, where the ideal diffraction pattern from the unobstructed grating (three delta functions) is convolved with the sinc-function diffraction pattern of the aperture. Two closely spaced lines produce two closely spaced diffraction patterns, which overlap and may be indistinguishable. The larger the aperture, the narrower the sinc functions and the better the resolution. Two spectral lines are said to be "just resolved" when the peak of one overlaps the first zero crossing of the other.

Not just the aperture, but the grating itself affects the performance of the spectrometer. The grating spreads the spectrum of the source in the vicinity of the -1 order, and varying the tilt of the grating sweeps the spectrum across the detector. The grating can usefully be tilted no farther than the point where undiffracted light begins to be detected. Increasing the grating's frequency, b, spreads the diffracted orders farther away from the undiffracted light, creating more room for the source spectrum and improving the spectral bandwidth of the instrument.

Example 10.15 (Sinusoidal phase grating). An amplitude grating loses light to absorption, which decreases the diffraction efficiency. By varying the thickness of the grating rather than its absorption, spatial *phase* variations may be imposed on the incident light that also result in diffraction. With normally incident, unit-amplitude plane wave illumination, the complex amplitude of the field immediately behind a sinusoidal phase grating is (ignoring the finite aperture for now) $\tilde{u}_0(x, y) = e^{im \sin(2\pi b x)}$, where m is a scaling factor that determines the "depth" of the phase modulation. The Fourier transform of this function is

$$\tilde{U}_0(\nu_x, \nu_y) = \left[\int_{-\infty}^{\infty} e^{im \sin(2\pi b x)} e^{-i2\pi \nu_x x} dx \right] \delta(\nu_y).$$

The complex exponential $e^{im \sin(2\pi b x)}$ is a periodic function in x and has a Fourier series,

$$e^{im \sin(2\pi b x)} = \sum_{n=-\infty}^{\infty} c_n e^{i2\pi n b x},$$

$$\text{where } c_n = \frac{1}{b} \int_{-1/2b}^{1/2b} e^{im \sin(2\pi b x)} e^{-i2\pi n b x} dx.$$

Change variables, $\varphi = 2\pi b x$,

$$c_n = \frac{1}{2\pi} \int_{-\pi}^{\pi} e^{-i(n\varphi - m \sin \varphi)} d\varphi = J_n(m),$$

where $J_n(.)$ is the nth order Bessel function of the first kind (Equation 10.19). We then have

$$e^{im \sin(2\pi b x)} = \sum_{n=-\infty}^{\infty} J_n(m) e^{i2\pi n b x} \longmapsto \sum_{n=-\infty}^{\infty} J_n(m) \delta(\nu_x - nb)$$

$$\Rightarrow \tilde{U}_0(\nu_x, \nu_y) = \sum_{n=-\infty}^{\infty} J_n(m) \delta(\nu_x - nb) \delta(\nu_y).$$

Unlike the sinusoidal amplitude grating, which has only the ± 1 orders plus undiffracted light, the sinusoidal phase grating has numerous diffracted orders. But, if the modulation factor m is chosen to be a zero of the Bessel function J_0, there is no undiffracted light. As with the amplitude grating, a finite-sized aperture around the grating causes the impulses in the spectrum to "melt" into sincs. ∎

Phase gratings may also be made with nonsinusoidal profiles, such as a sawtooth shape (see the Chapter 4 problems), with better suppression of undesired diffracted orders.[13]

10.3.3 Antennas

In Section 4.7, we derived the following expression for angular dependence of the far-field radiation pattern of an antenna composed of multiple point sources, measured in the xz plane:

$$u_P(\theta) = AD_{2N+1}\left(\frac{d}{\lambda}\sin\theta\right), \quad (4.40)$$

where D_{2N+1} is the Dirichlet kernel. The same result, in two dimensions, is obtained using Equation 10.45 with source distribution $\tilde{u}_0(x_0, y_0) = A\delta(y_0)\sum_{n=-N}^{N}\delta(x_0 - nd)$. The Fourier transform is

$$\tilde{U}(v_x, v_y) = A\sum_{n=-N}^{N} e^{-i2\pi v_x nd}$$

and, substituting $v_x = (\sin a)/\lambda$,

$$\tilde{u}_1(\sin a, \sin b) \propto A\sum_{n=-N}^{N} e^{-i2\pi n(\sin a)d/\lambda} = AD_{2N+1}\left(\frac{d}{\lambda}\sin a\right).$$

Now consider a more realistic antenna array, with finite-sized elements rather than points. Model each element as a rectangle, $\text{rect}(x_0/X)\,\text{rect}(y_0/Y)$. Then

$$\tilde{u}_0(x_0, y_0) = A\sum_{n=-N}^{N} \text{rect}\left(\frac{x_0 - nd}{X}\right)\text{rect}\left(\frac{y_0}{Y}\right).$$

This is the convolution of the point array with a single rectangular element:

$$\tilde{u}_0(x_0, y_0) = A\delta(y_0)\sum_{n=-N}^{N}\delta(x_0 - nd) * \text{rect}\left(\frac{x_0}{X}\right)\text{rect}\left(\frac{y_0}{Y}\right).$$

Using the convolution theorem, we can immediately write down the far field distribution:

$$\tilde{u}_1(\sin a, \sin b) \propto AD_{2N+1}(v_x d)\, XY\,\text{sinc}(Xv_x)\,\text{sinc}(Yv_y)\Big|_{v_x=\frac{\sin a}{\lambda},\, v_y=\frac{\sin b}{\lambda}}$$

$$= AXY\, D_{2N+1}\left(\frac{d\sin a}{\lambda}\right)\text{sinc}\left(\frac{X\sin a}{\lambda}\right)\text{sinc}\left(\frac{Y\sin b}{\lambda}\right). \quad (10.49)$$

The element dimensions, X and Y, are smaller than the interelement spacing, d. Consequently, the sinc functions in the radiation pattern are wider than the Dirichlet kernel and attenuate the response at higher angles. Problems 5.29 and 5.30 considered this in more detail.

We observed in Equation 10.48 how a tilt of the incident wave, corresponding to a linear phase shift across the aperture, produced a translation of the far-field

---

[13] Born and Wolf (1999, pp. 446–465).

diffraction pattern. The same idea can be used to good effect in an antenna: by imposing a linear phase shift across the antenna aperture, the antenna's beam can be steered in angle. So let the aperture distribution be modified so that the drive to the nth element is delayed by Δt relative to the $n-1$th element. This time delay imposes a phase shift $\omega n \Delta t$ on the wave emanating from the nth element:

$$\tilde{u}_0(x_0, y_0) = A\delta(y_0) \sum_{n=-N}^{N} \left[e^{in\omega\Delta t} \delta(x_0 - nd) \right] * \text{rect}\left(\frac{x_0}{X}\right) \text{rect}\left(\frac{y_0}{Y}\right).$$

The Fourier transform is

$$\tilde{U}_0(\nu_x, \nu_y) = A \sum_{n=-N}^{N} e^{-in(2\pi\nu_x d - \omega\Delta t)} XY \operatorname{sinc}(X\nu_x) \operatorname{sinc}(Y\nu_y),$$

and the far-field distribution is

$$\tilde{u}_1(\sin a, \sin b) \propto AXY D_{2N+1}\left(\frac{d\sin a - c\Delta t}{\lambda}\right) \operatorname{sinc}\left(\frac{X\sin a}{\lambda}\right) \operatorname{sinc}\left(\frac{Y\sin b}{\lambda}\right),$$

where c is the speed of light. The center of the Dirichlet kernel is shifted to $\sin a = c\Delta t/d$. By varying the time delay electronically, the antenna's pattern can be swept through space without physically moving the antenna. Such *phased array* antennas are widely used in radar, communications, and, with acoustic waves, in sonar and medical ultrasonic imaging.

10.3.4 Lenses

You have, no doubt, experienced the focusing property of a convex lens (Figure 10.20). In physical terms, when a plane wave is propagated through a lens, the decrease in the lens' thickness from center to edge imposes a phase curvature on the wave that causes it to converge toward a point behind the lens. This point is called the back *focal point* of the lens. Similarly, if a plane wave is propagated backward through the lens, it converges at a focal point in front of the lens. The distance from the lens to either focal point is called the *focal length* of the lens.

It can be shown, remarkably, that a convex lens modifies an optical field in such a way that the Fraunhofer diffraction pattern appears at the focal plane rather than at a large distance.[14] In particular, the complex amplitude $\tilde{u}_0(\mathbf{r}_0)$ at the front focal plane of the lens is related to the complex amplitude $\tilde{u}_1(\mathbf{r}_1)$ at the back focal plane by the Fourier transform:

$$\tilde{u}_1(\mathbf{r}_1) = \frac{1}{i\lambda f} \tilde{U}_0\left(\frac{\mathbf{r}_1}{\lambda f}\right). \tag{10.50}$$

Example 10.16. Let the incident field be a tilted plane wave, $\tilde{u}_0(x_0, y_0) = e^{ikx_0 \sin\theta}$. The Fourier transform is, using the shift theorem and the relationship $1 \longleftrightarrow \delta(\nu_x, \nu_y)$,

$$\tilde{U}_0(\nu_x, \nu_y) = \delta\left(\nu_x - \frac{\sin\theta}{\lambda}, \nu_y\right).$$

---
[14] Goodman (1968, Chapter 5).

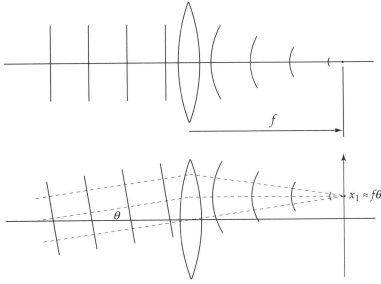

FIGURE 10.20 *Top*: A positive lens causes an incident plane wave to converge toward a focal point, located a distance f behind the lens. *Bottom*: A tilted plane wave is brought to an off-axis focus.

Making the substitutions $v_x = x_1/\lambda f$, $v_y = y_1/\lambda f$, we have the field at the back focal plane:

$$\tilde{u}(x_1, y_1) = \frac{1}{i\lambda f} \delta\left(\frac{x_1}{\lambda f} - \frac{\sin\theta}{\lambda}, \frac{y_1}{\lambda f}\right) = -i\delta(x_1 - f\sin\theta, y_1),$$

that is, a focused spot at $(x_1, y_1) = (f\sin\theta, 0) \approx (f\theta, 0)$ (Figure 10.20). ∎

In reality, the finite size of the lens aperture limits the sharpness of the focus. Consider, for example, a circular aperture of radius a with a normally incident plane wave illumination. Then,

$$\tilde{u}_1(\mathbf{r}_1) = \frac{1}{i\lambda f} \int_{-\infty}^{\infty} \text{circ}(r_0/a)\, e^{-i2\pi \mathbf{r}_0 \cdot (\mathbf{r}_1/\lambda f)}\, d\mathbf{r}_0 = \frac{\pi a^2}{i\lambda f}\, \text{jinc}(ar_1/\lambda f). \quad (10.51)$$

Increasing the size of the aperture narrows the jinc function, making a sharper focus, but physically, the aperture size cannot be increased indefinitely. An infinitely sharp focus is never observed in practice. Of course, a tilted plane wave illuminating a circular aperture produces a shifted jinc function at the focal plane. The details are left to the problems.

When a partially transparent object t (which could be a diffraction grating or a photographic slide) is placed at the front focal plane and illuminated with a

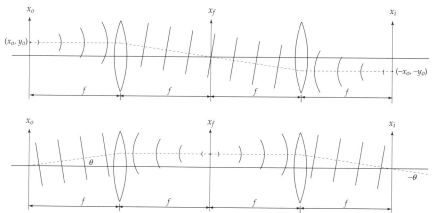

FIGURE 10.21 An optical imaging system formed by cascading two Fourier transform geometries. *Top*: A point at $\mathbf{r}_o = (x_o, y_o)$ in the object plane produces a spherical wave, which is transformed into a tilted plane wave by the first lens. The tilted plane wave is transformed back to a converging spherical wave by the second lens, coming to a focus at the inverted image point, $\mathbf{r}_i = (-x_o, -y_o)$. *Bottom*: A plane wave impinging on the first lens at angle θ is brought to an off-axis focus in the Fourier plane then diverges to the second lens and is transformed to a tilted plane wave at angle $-\theta$.

unit-amplitude normally incident plane wave, the incident field $\tilde{u}_0$ is proportional to t and the result at the back focal plane of the lens is

$$\tilde{u}_1(\mathbf{r}_1) = \frac{1}{i\lambda f} \int_{-\infty}^{\infty} t(\mathbf{r}_0) e^{-i2\pi \mathbf{r}_0 \cdot (\mathbf{r}_1/\lambda f)} \, d\mathbf{r}_0 = \frac{1}{i\lambda f} T\left(\frac{\mathbf{r}_1}{\lambda f}\right). \tag{10.52}$$

The field in the focal plane is proportional to the Fourier transform of the object. It is common to say that the lens Fourier transforms the object, but actually it is diffraction that does the Fourier transform, and the lens merely brings the transform plane in from infinity to a convenient position.

Finally, consider what happens when we place two identical lenses together, separated by $2f$ so that the back focal point of the first lens coincides with the front focal point of the second lens (Figure 10.21). The Fourier transform of the object field $\tilde{u}_o(x_o, y_o)$ is produced at the intermediate focal plane (x_f, y_f). This field propagates through the second half of the system and its Fourier transform is produced at the final plane, (x_i, y_i).[15] This is not a forward transform followed by an inverse transform. It is the cascade of two forward transforms, and by Theorem 5.5,

$$\begin{aligned}
\tilde{u}_i(\mathbf{r}_i) &= \frac{1}{i\lambda f} \mathcal{F}\{\tilde{u}_f(\mathbf{r}_f)\}\Big|_{\boldsymbol{\nu}=\mathbf{r}_i/\lambda f} = \frac{1}{i\lambda f} \mathcal{F}\left\{\frac{1}{i\lambda f} \tilde{U}_o\left(\frac{\mathbf{r}_f}{\lambda f}\right)\right\}\Big|_{\boldsymbol{\nu}=\mathbf{r}_i/\lambda f} \\
&= -\tilde{u}_o(-\lambda f \boldsymbol{\nu})\Big|_{\boldsymbol{\nu}=\mathbf{r}_i/\lambda f} \\
&= -\tilde{u}_o(-\mathbf{r}_i).
\end{aligned} \tag{10.53}$$

---

[15] The subscripts o and i stand for "object" and "image," not "output" and "input."

The output field is a replica of the input field, with a coordinate reversal. This combination of two lenses is an imaging system.[16] The inversion is illustrated in two ways in Figure 10.21.

The accessibility of the Fourier transform at the intermediate focal plane, also called the Fourier plane, makes it possible to process the image by manipulating the transform. For example, a circular aperture is a lowpass filter, and an opaque on-axis dot blocks undiffracted light, which can enhance contrast. These so-called spatial filters have been widely used in optical instruments, including telescopes and microscopes.[17]

10.4 IMAGE FORMATION AND PROCESSING

10.4.1 Fourier Analysis of Imaging Systems

The preceding analysis is idealized in a couple of important respects. First, it assumes that the lenses are infinite in extent, and so capture all the light diffracted from an object. In reality, the finite aperture of the lens truncates the diffracted field (Equation 10.51). Light diffracted at high angles, corresponding to higher spatial frequencies in the object, will "miss" the lens and not reach the image plane. The effect is to lowpass filter the image. Second, real lenses are subject to deviations from ideal behavior, called *aberrations*, that have a deleterious effect on image quality analogous to a nonlinear phase response in an electrical filter. High-performance lenses are built from several elements carefully designed to have countervailing aberrations, resulting in near-perfect imaging. When a lens is perfectly corrected for aberrations and only the finite aperture effects remain, it is said to be *diffraction limited*.

The aperture and aberration effects in a lens are analogous to the frequency-domain magnitude and phase responses of an electrical filter. A larger aperture corresponds to a higher bandwidth, and a lack of aberration corresponds to a linear or flat phase response. The optical analog of the electrical filter's impulse response is called the *point spread function*, and in many practical situations the image may be described by the convolution of the object with the point spread function. There is, as you would expect, a Fourier transform relationship between the aperture function of a lens and its point spread function. To investigate this thoroughly would take us beyond the scope of this text,[18] but we will give one illustrative example.

Example 10.17 (Telescope imaging). Consider a simple telescope consisting of a single circular lens. It is assumed to be diffraction limited, with radius a. A distant star is approximately a point source, and for simplicity we will assume the light it gives off is monochromatic, with wavelength λ. The star is so far away that, also ignoring

---

[16] Inversion of the image is common. For example, the image formed by a camera lens on an image sensor and the image formed by the lens of your eye on your retina are inverted, too.

[17] F. Zernike was awarded the Nobel Prize in physics in 1953 for his invention of the phase contrast microscope, which is based on a particular Fourier plane manipulation.

[18] See, for example, Goodman (1968, Chapter 6). Papoulis (1968, pp. 14–15) shows several analogies between optical and electrical signal processing systems.

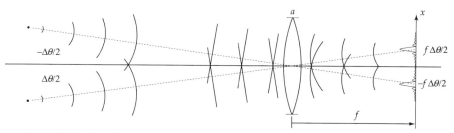

FIGURE 10.22 Imaging a pair of distant point objects with a telescope. The light from each point is spread into a diffraction pattern at the image plane. The diameter of the point image is proportional to the ratio $\lambda f/a$.

the phase-distorting effects of the atmosphere, the wave that reaches the telescope is planar. The setup then matches that shown in Figure 10.20: the plane wave from the star is brought to a focus behind the lens. The focal field is the Fourier transform of the aperture (Equation 10.52), which we know is a jinc:

$$\tilde{u}_f(\mathbf{r}) \propto \mathcal{F}\{\mathrm{circ}\,(r_\ell/a)\}\Big|_{\rho=r/\lambda f} = a^2 \pi \,\mathrm{jinc}\left(\frac{\|\mathbf{r}\|}{\lambda f/a}\right).$$

The image that we observe or measure is the squared magnitude of the complex field amplitude, and this quantity is the point spread function (PSF):

$$h(x,y) = |\tilde{u}_f(x,y)|^2 = \pi^2 a^4 \mathrm{jinc}^2\left(\frac{\sqrt{x^2+y^2}}{\lambda f/a}\right). \tag{10.54}$$

The square of a jinc function is also known as the *Airy disk*. The size of the Airy disk is conventionally taken to be the radius or diameter of the first dark ring, which is $r \approx 0.61 \lambda f/a$. Observe that this dimension is inversely proportional to the radius of the lens aperture; a large telescope not only collects more light (in proportion to πa^2), but it also focuses that light to a finer image.

The image of a distant star is spread by diffraction into a PSF. If the telescope's field of view contains more than one star, the image is the sum of shifted PSFs, weighted by the brightnesses of the stars.[19] In particular, the image of two stars with brightnesses I_1 and I_2, separated horizontally by $\Delta\theta$ about the optical axis, is (Figure 10.22)

$$I(x,y) = I_1\,h(x+f\Delta\theta/2, y) + I_2\,h(x-f\Delta\theta/2, y)$$
$$= \pi^2 a^4 \left[I_1 \mathrm{jinc}^2\left(\frac{\sqrt{(x+f\Delta\theta/2)^2+y^2}}{\lambda f/a}\right) + I_2 \mathrm{jinc}^2\left(\frac{\sqrt{(x-f\Delta\theta/2)^2+y^2}}{\lambda f/a}\right) \right].$$

---

[19]The fields from two stars are independent and combine without interference. This is the case of *incoherent* light. If the two fields have a common source so that they maintain a steady phase relationship, they are said to be coherent. For more on coherence and imaging, see Goodman (1968, Chapter 6).

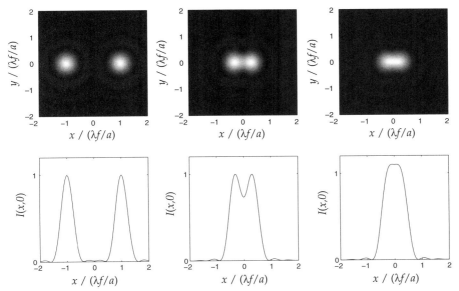

FIGURE 10.23 Images of two closely spaced point objects as would be viewed through a telescope (*top*) and cross sections at $y = 0$ (*bottom*). The image plane coordinates x and y are normalized by $\lambda f/a$. Intensity is in arbitrary units. *Left:* well-separated points, well resolved. *Center:* Rayleigh resolution limit, points separated by the radius of the first dark ring of the point spread function, $\Delta\theta = 0.61\lambda/a$. *Right:* sparrow resolution limit, points barely unresolvable, $\Delta\theta \approx 0.5\lambda/a$.

As the angular separation $\Delta\theta$ decreases, eventually the two images overlap and appear as one object rather than two. How close can they be before they are indistinguishable? With equal brightness stars, the intensity profile along the x-axis is

$$I(x,0) = \pi^2 a^4 \left[\text{jinc}^2\left(\frac{x+f\Delta\theta/2}{\lambda f/a}\right) + \text{jinc}^2\left(\frac{x-f\Delta\theta/2}{\lambda f/a}\right) \right].$$

This function is plotted in Figure 10.23 for a few values of the angular separation $\Delta\theta$. According to the classical Rayleigh criterion, two points are adequately resolved by the naked eye when the peak of one PSF coincides with the first dark ring of the other, that is, they are separated by $0.61\lambda f/a$. This corresponds to an angular separation of the stars, $\Delta\theta$, equal to $0.61\lambda/a$. Again, the resolution of the telescope depends on the size of its aperture. The so-called Sparrow resolution limit is reached when the two points are no longer distinguishable. For a pair of Airy disks, this occurs at $\Delta\theta \approx 0.5\lambda/a$. ∎

The image of an extended object (e.g., something nearby like the Moon) is also a superposition of point spread functions, where each PSF arises from a point in the object. If $g(x, y)$ is the perfect image predicted by geometric optics (an exact replica, though scaled and inverted), then the actual image due to diffraction is the convolution, $g * h$, familiar from linear system theory. Under this model, the blurring due to a poorly corrected optical system is caused by convolution with a broad PSF.

The Fourier transform of the point spread function is called the *optical transfer function* (OTF). By the autocorrelation theorem, because the PSF is the square of the Fourier transform of the aperture function, the OTF is the autocorrelation of the aperture function. By convention, the OTF is normalized to unity at zero frequency. Calculating the OTF for a circular aperture function (see Example 10.10),[20] yields

$$H(\rho) = \frac{2}{\pi}\left[\arccos\left(\frac{\rho}{a/\lambda f}\right) - \frac{\rho}{a/\lambda f}\sqrt{1 - \left(\frac{\rho}{a/\lambda f}\right)^2}\right]\operatorname{circ}\left(\frac{\rho}{a/\lambda f}\right). \quad (10.55)$$

Note that the OTF has a sharp bandlimit, unlike the transfer function of any practical electrical filter. The maximum spatial frequency that passes the OTF is proportional to the aperture radius a. A large aperture gives the system a higher spatial bandwidth and a narrower point spread function. Aberrations (phase errors) change the magnitude and phase of the OTF. The magnitude of the OTF is called the *modulation transfer function* (MTF).

The MTF has an interpretation in terms of the contrast of the Fourier components of an image. The *contrast* of an image is defined:

$$C = \frac{\text{maximum brightness} - \text{minimum brightness}}{\text{maximum brightness} + \text{minimum brightness}}. \quad (10.56)$$

For example, a simple sinusoidal object with brightness between 0 and 1 is $f(x) = \frac{1}{2}(1 + \cos 2\pi v x)$. The contrast of this object is $C = \frac{1-0}{1+0} = 1$, or 100%. When this object is passed through an imaging system with MTF $M(v)$, the Fourier components are weighted by the MTF. Ignoring any shifts due to phase response, the image is $g(x) = \frac{1}{2}(1 + M(v)\cos 2\pi v x)$. The contrast is

$$C = \frac{\left(\frac{1}{2} + \frac{1}{2}M(v)\right) - \left(\frac{1}{2} - \frac{1}{2}M(v)\right)}{\left(\frac{1}{2} + \frac{1}{2}M(v)\right) + \left(\frac{1}{2} - \frac{1}{2}M(v)\right)} = M(v).$$

Loss of contrast at high frequencies (Figure 10.24) shows up as a blurring of fine detail in an image.

The attempt to improve the sharpness of images by computer processing is an important topic of research, complicated by the absolute loss of spatial frequencies above the cutoff frequency $\rho_c = a/\lambda f$. Research into systems capable of *superresolution*, resolving below the classical limits, remains active at the time of this writing.[21]

---

[20] See Goodman (1968, Chapter 6) for other examples of OTF calculations.
[21] Park et al. (2003).

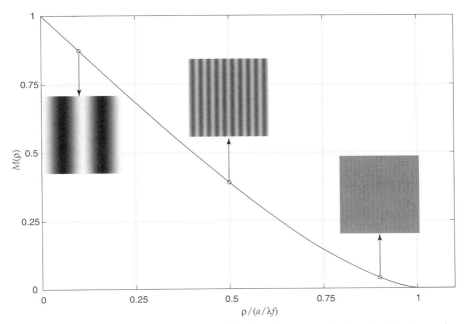

FIGURE 10.24 The frequency response of an imaging system is described by the modulation transfer function, $M(\rho) = |H(\rho)|$. This is the graph of the MTF of a diffraction-limited system with circular pupil (Equation 10.55). Sinusoidal test images are shown for normalized frequencies 0.1, 0.5, and 0.9. As spatial frequency increases, the contrast decreases and the images appear to "wash out."

10.4.2 Image Reconstruction from Projections

The integral of a function $f(x, y)$ along the vertical line $x = x'$ is

$$\int_{-\infty}^{\infty} f(x', y)\, dy.$$

Using the delta function, this can also be written as an integral over the plane:

$$\int_{-\infty}^{\infty} \int_{-\infty}^{\infty} f(x, y)\, \delta(x - x')\, dx\, dy.$$

Likewise, an integral along the horizontal line $y = y'$ is

$$\int_{-\infty}^{\infty} \int_{-\infty}^{\infty} f(x, y)\, \delta(y - y')\, dx\, dy.$$

We will call these line integrals, which are functions of x' and y', *projections*. A projection of a function f at an arbitrary angle θ, denoted $p_\theta(x')$, is shown in Figure 10.25.

The path of integration is the line $\mathbf{e}_\theta \cdot \mathbf{x} = x'$, where $\mathbf{e}_\theta = (\cos\theta, \sin\theta)$ is a unit vector at angle θ to the x-axis, and $\mathbf{x}$ is the vector to the point (x, y). This line may be

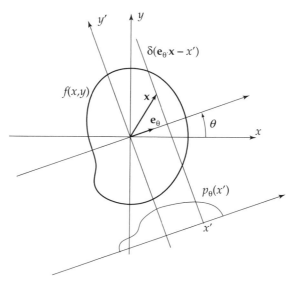

FIGURE 10.25 Geometry for cross-sectional projection. $p_\theta(x')$ is the integral of $f(x, y)$ along the path described by the line delta $\delta(\mathbf{e}_\theta \cdot \mathbf{x} - x')$. Projections are gathered for $\pi > \theta \geq 0$.

represented by the delta function $\delta(\mathbf{e}_\theta \cdot \mathbf{x} - x')$, and so the projection is the integral

$$p_\theta(x') = \int_{-\infty}^{\infty} f(\mathbf{x})\, \delta(\mathbf{e}_\theta \cdot \mathbf{x} - x')\, d\mathbf{x} \tag{10.57a}$$

$$= \int_{-\infty}^{\infty} \int_{-\infty}^{\infty} f(x, y)\, \delta(x \cos\theta + y \sin\theta - x')\, dx\, dy. \tag{10.57b}$$

The projection angle θ ranges from 0 to π. The other angles, from π to 2π (or, equivalently, from $-\pi$ to 0), reproduce the projections, but flipped end-for-end, that is, $p_{-\theta}(x') = p_\theta(-x')$. By convention, the full set of projections $p_\theta(x')$ is considered a function of modified polar coordinates with a signed radius, $(x', \theta) \in \mathbb{R} \times [0, \pi)$.

The mapping (Equation 10.57) that takes the function $f(x, y)$ to its projections $p_\theta(x')$ is called the *Radon transform*.[22] In order for the Radon transform of a function to exist, the function must be integrable along all lines in the plane. Classes of functions possessing Radon transforms include the good functions $\mathcal{S}(\mathbb{R}^2)$, absolutely integrable functions, $L^1(\mathbb{R}^2)$, and square-integrable functions with bounded support, for example, $L^2(\text{circ})$. The Radon transforms of functions in these classes may be shown to also be good, absolutely integrable, and square integrable, respectively.[23] Significantly for the problem of inverting the Radon transform, projections in these classes are Fourier transformable.

---

[22] J. Radon first described this operation in a seminal 1917 paper, available in English translation (1986).

[23] See Natterer (1986, Chapter 2) for more about the properties of the Radon transform. The Radon transform may be posed and inverted in spaces of dimension higher than two. In three dimensions, the integrals are taken over planes rather than lines, and in higher dimensions the integrals are over hyperplanes. Here we shall restrict attention to functions in the plane.

We will write $p_\theta = \mathcal{R}\{f\}$ or $f \xmapsto{\mathcal{R}} p_\theta$ to denote the Radon transform. The Radon transform has a few simple properties that we list here. Being a line integral, it is clearly linear:

$$\mathcal{R}\{af + bg\} = a\mathcal{R}\{f\} + b\mathcal{R}\{g\}. \tag{10.58}$$

Shifting f results in a shift of the projection:

$$f(\mathbf{x} - \mathbf{b}) \xmapsto{\mathcal{R}} p_\theta(x' - \mathbf{e}_\theta \cdot \mathbf{b}). \tag{10.59}$$

A simple dilation of f results in a corresponding dilation of the projection:

$$f(a\mathbf{x}) \xmapsto{\mathcal{R}} p_\theta(ax'), \tag{10.60}$$

and a rotation of f results in a corresponding rotation of the Radon transform. With

$$\mathbf{R}_\phi = \begin{bmatrix} \cos\phi & \sin\phi \\ -\sin\phi & \cos\phi \end{bmatrix},$$

$$f(\mathbf{R}_\phi \mathbf{x}) \xmapsto{\mathcal{R}} p_{\theta-\phi}(x'). \tag{10.61}$$

The Radon transform of a (two-dimensional) convolution is the (one-dimensional) convolution of Radon transforms. Denoting the Radon transforms of f and g by f_θ and g_θ,

$$f * g(\mathbf{x}) \xmapsto{\mathcal{R}} f_\theta * g_\theta(x'). \tag{10.62}$$

Proofs are left to the problems.

Reconstruction

We shall now show that the Radon transform can be inverted, that is, a function f can be reconstructed from its projections. This remarkable result is the basis for computed tomography (also called CT, or CAT scanning), which, since it came into common use in the 1970s, has revolutionized medical imaging and found application in a host of other fields such as nondestructive testing.[24]

If we calculate the Fourier transform of the projection, $P_\theta(u)$, we have

$$P_\theta(u) = \int_{-\infty}^{\infty} \int_{-\infty}^{\infty} f(\mathbf{x}) \delta(\mathbf{e}_\theta \cdot \mathbf{x} - x') e^{-i2\pi u x'} d\mathbf{x}\, dx' = \int_{-\infty}^{\infty} f(\mathbf{x}) e^{-i2\pi u \mathbf{e}_\theta \cdot \mathbf{x}} d\mathbf{x}.$$

This is the Fourier transform of f, with $\mathbf{v}$ replaced by $u\mathbf{e}_\theta$. This establishes the key result, the *projection-slice theorem*.

Theorem 10.1 (Projection-slice). Let $p_\theta(x')$ be the Radon transform of $f(\mathbf{x})$, and $P_\theta(u)$ and $F(\mathbf{v})$ be their respective Fourier transforms. Then

$$P_\theta(u) = F(u\mathbf{e}_\theta) = F(u\cos\theta, u\sin\theta). \tag{10.63}$$

---

[24] In computed tomography, the projections are line integrals of the attenuation of an X-ray beam as it passes through the object. The detected X-ray intensity is proportional to $\exp(-p_\theta(x'))$, and taking the logarithm of the detected signal gives the projection set to be reconstructed.

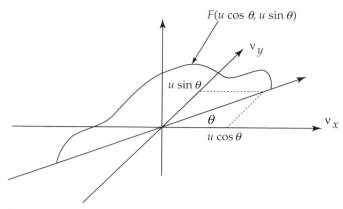

FIGURE 10.26 The Fourier transform of a projection, $P_\theta(u) = F(u\cos\theta, u\sin\theta)$, is a slice through the Fourier transform $F(v_x, v_y)$, at angle θ.

The theorem gets its name from the fact that the set of points $\mathbf{v} = u\mathbf{e}_\theta = (u\cos\theta, u\sin\theta)$ is a line through the origin of the (v_x, v_y) plane at angle θ. Thus, $P_\theta(u) = F(u\cos\theta, u\sin\theta)$ is a slice of the Fourier transform F along that line (Figure 10.26).

You can imagine from the figure that if projections for all angles $\theta \in [0, \pi)$ are collected, the slices will fill the Fourier plane, reconstructing the Fourier transform $F(v_x, v_y)$. An inverse Fourier transform will then recover the original object $f(x, y)$. Using coordinates $(u, \theta) \in \mathbb{R} \times [0, \pi)$ in the Fourier plane as in the Radon transform domain, the differential area is $|u|\, du\, d\theta$, and the inverse Fourier transform is

$$f(x,y) = \int_0^\pi \int_{-\infty}^\infty F(u\cos\theta, u\sin\theta)\, e^{+i2\pi(xu\cos\theta + yu\sin\theta)}\, |u|\, du\, d\theta;$$

then recalling $x\cos\theta + y\sin\theta = \mathbf{e}_\theta \cdot \mathbf{x}$,

$$f(x,y) = \int_0^\pi \underbrace{\left[\int_{-\infty}^\infty |u| P_\theta(u)\, e^{+i2\pi u\, \mathbf{e}_\theta \cdot \mathbf{x}}\, du\right]}_{\tilde{p}_\theta(\mathbf{e}_\theta \cdot \mathbf{x})} d\theta. \tag{10.64}$$

The function $\tilde{p}(\mathbf{e}_\theta \cdot \mathbf{x})$ resulting from the inner integral is the projection p_θ taken at angle θ, filtered with a frequency response $|u|$. Unlike the original projection $p_\theta(x')$, it is not a function of one variable x', but of two coordinates $\mathbf{x} = (x, y)$ in the xy plane. It is constant along lines $\mathbf{e}_\theta \cdot \mathbf{x} = \text{const}$. We say that the function $\tilde{p}_\theta$ has been *backprojected* at the angle θ across the xy plane. A backprojection is like a corrugated surface with profile equal to $\tilde{p}_\theta$ (Figure 10.27).

The outer integral is the superposition of all the filtered backprojections. Thus, the function $f(x, y)$ is reconstructed from its projections by filtering and backprojection. In practice, projections at a discrete set of angles $\{\theta_k\}$ are filtered, then interpolated onto a rectangular grid. The inverse transform is computed with a two-dimensional DFT.

10.4 IMAGE FORMATION AND PROCESSING   717

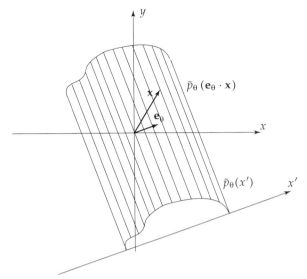

FIGURE 10.27 Backprojection. The function $\tilde{p}_\theta(\mathbf{e}_\theta \cdot \mathbf{x})$ is equal, everywhere along the line $\mathbf{e}_\theta \cdot \mathbf{x} = x'$, to the filtered projection $\tilde{p}_\theta(x')$.

An alternate interpretation of filtering and backprojection is obtained by applying the convolution theorem in Equation 10.64:

$$f(x, y) = \int_0^\pi \left[p_\theta(x') * \left(\int_{-\infty}^\infty |u|\, e^{+i2\pi u x'}\, du \right) \right]_{x'=\mathbf{e}_\theta \cdot \mathbf{x}} d\theta. \tag{10.65}$$

The projections p_θ are convolved with the inverse Fourier transform of $|u|$, again yielding $\tilde{p}_\theta$, then backprojected. This method of reconstruction is called, appropriately enough, *convolution-backprojection*. The frequency response $|u|$ is a function of slow growth, so the impulse response for the filter is a generalized function. We did this calculation in Example 6.33 and found

$$|u| \longmapsto -\frac{2}{(2\pi x)^2}.$$

This is a quite singular function, and in practice the bandwidth of the frequency response must be restricted for computational tractability, for example, multiplying $|u|$ by a rectangle function gives

$$|u|\, \text{rect}\left(\frac{u}{2B}\right) = \text{rect}\left(\frac{u}{2B}\right) - \Lambda\left(\frac{u}{B}\right) \longmapsto 2B\, \text{sinc}(2Bx) - B\text{sinc}^2(Bx).$$

It is interesting to ask what is the purpose of the filter, that is, what would be the result of superposing *unfiltered* backprojections? Consider a delta function object, $f(x, y) = \delta(x, y)$. The projections are identical for all angles, $p_\theta(x') = \delta(x')$. The

Fourier transform is $P_\theta(u) = 1$. If we backproject without filtering, the result is

$$\int_0^\pi \int_{-\infty}^\infty 1\, e^{i2\pi u \mathbf{e}_\theta \cdot \mathbf{x}} \, du\, d\theta = \int_0^\pi \delta(\mathbf{e}_\theta \cdot \mathbf{x})\, d\theta.$$

Change variables, $x = r\cos\varphi$, $y = r\sin\varphi$,

$$= \int_0^\pi \delta(r(\cos\varphi \cos\theta + \sin\varphi \sin\theta))\, d\theta$$

$$= \int_0^\pi \frac{1}{r} \delta(\cos(\theta - \varphi))\, d\theta.$$

Using Equation 6.8,

$$\delta(\cos(\theta - \varphi)) = \sum_{k=-\infty}^\infty \delta(\theta - [\phi + (k + 1/2)\pi]).$$

These impulses are spaced π apart, so only one of them is located between the limits of integration. The area is unity, and the final result is

$$\int_0^\pi \int_{-\infty}^\infty 1\, e^{i2\pi u \mathbf{e}_\theta \cdot \mathbf{x}}\, du\, d\theta = \frac{1}{r}.$$

Without backprojection, an impulse object reconstructs to the function $1/r$, which is singular at the origin but is not localized like the impulse; rather, it decays slowly away from the origin. We may consider $1/r$ to be the impulse response associated with backprojection. Its Fourier transform is the frequency response of backprojection. Because $1/r$ is circularly symmetric, the Fourier transform is given by a Hankel transform:

$$\frac{1}{r} \xmapsto{\mathcal{H}} 2\pi \int_0^\infty \frac{1}{r} J_0(2\pi r\rho)\, r\, dr = 2\pi \int_0^\infty J_0(2\pi r\rho)\, dr = \frac{1}{\rho}. \quad (10.66)$$

Ideally, the operations of projection and reconstruction should cancel, giving a unit net frequency response. However, we see here that the frequency response is $1/\rho$. This necessitates a filtering of the projections before backprojection to cancel the $1/\rho$ response. The factor $|u|$ in the filtered backprojection formula, Equation 10.64, does precisely this.

Circular Symmetry: The Abel Transform

A special case of the Radon transform occurs for circularly symmetric functions, $f(r)$. All projections are equal, and one projection suffices for reconstruction. We begin with the Radon transform formula 10.57:

$$p_\theta(x') = \int_{-\infty}^\infty f(\mathbf{x}) \delta(\mathbf{e}_\theta \cdot \mathbf{x} - x')\, d\mathbf{x}.$$

Because all projections will be the same, we can set $\theta = 0$. Then $\mathbf{e}_\theta \cdot \mathbf{x} = (1,0) \cdot (x,y) = x$. Also, replace $f(\mathbf{x})$ with $f(r)$ and write $x = r\cos\phi$, $dx = r\,dr\,d\phi$:

$$p(x') = \int_{-\pi}^{\pi}\int_0^\infty f(r)\,\delta(r\cos\phi - x')\,r\,dr\,d\phi$$

$$= \int_0^\infty f(r)\left[\int_{-\pi}^{\pi}\delta(\cos\phi - x'/r)\,d\phi\right]dr$$

$$= \int_0^\infty f(r)\left[2\int_0^{\pi}\delta(\cos\phi - x'/r)\,d\phi\right]dr.$$

The delta function is located at $\phi = \cos^{-1}(x'/r)$, and because $|\cos| \le 1$, we must restrict $r \ge x'$. Then, using Equation 6.8,

$$\delta(\cos\phi - x'/r) = \frac{\delta(\phi - \cos^{-1}(x'/r))}{\left|-\sin\left[\cos^{-1}(x'/r)\right]\right|} = \frac{\delta(\phi - \cos^{-1}(x'/r))}{\sqrt{1-(x'/r)^2}},$$

and so,

$$\int_0^\pi \delta(\cos\phi - x'/r)\,d\phi = \frac{1}{\sqrt{1-(x'/r)^2}}.$$

We therefore have the result, valid for positive and negative x,

$$p(x) = \int_{|x|}^\infty \frac{2f(r)\,dr}{\sqrt{1-(x/r)^2}} = \int_{|x|}^\infty \frac{2f(r)\,r\,dr}{\sqrt{r^2-x^2}}, \tag{10.67}$$

where for convenience we have dropped the prime $(')$ from x. This mapping from f to p is called the *Abel transform*.

Example 10.18 (Abel transform of the circle function). The circle function provides a particularly simple example of the Abel transform. With $f(r) = \mathrm{circ}(r/a)$, the transform integral becomes

$$p(x) = 2\int_{|x|}^a \frac{r\,dr}{\sqrt{r^2-x^2}} = 2\sqrt{r^2-x^2}\Big|_{|x|}^a = 2\sqrt{a^2-x^2}\,\mathrm{rect}(x/2a).$$

The projection is the length of a chord of the circle at a distance $|x|$ from the center. This is illustrated in Figure 10.28. ∎

To invert the Abel transform, we mimic the filtered backprojection method and calculate the Fourier transform of the projection:

$$P(u) = \mathcal{F}\left\{\int_{|x|}^\infty \frac{2f(r)\,dr}{\sqrt{1-(x/r)^2}}\right\} = \int_0^\infty 2f(r)\,\mathcal{F}\left\{\frac{U(1-(x/r)^2)}{\sqrt{1-(x/r)^2}}\right\}dr,$$

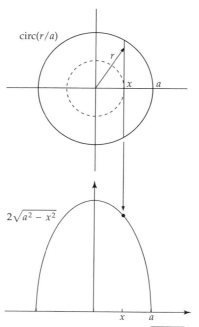

FIGURE 10.28 The Abel transform of circ(r/a) is $2\sqrt{a^2 - x^2}$ rect($x/2a$). The Abel transform at x depends only on points outside a circle of radius $|x|$. The projection of the circle at x is the length of the chord located $|x|$ from the center of the circle.

where $U(\cdot)$ is the unit step function used to pull x out of the limit of integration and into the integrand. The Fourier transform of the Abel kernel:

$$\mathcal{F}\left\{\frac{U(1-(x/r)^2)}{\sqrt{1-(x/r)^2}}\right\} = \int_{-\infty}^{\infty} \frac{U(1-(x/r)^2)\,e^{-i2\pi ux}}{\sqrt{1-(x/r)^2}}\,dx = \int_{-r}^{r} \frac{e^{-i2\pi ux}}{\sqrt{1-(x/r)^2}}\,dx$$

$$= \mathcal{F}\left\{\frac{\text{rect}(x/2r)}{\sqrt{1-(x/r)^2}}\right\} = \pi r J_0(2\pi r u) \qquad (10.68)$$

(for a derivation, see the problems) and therefore, the Fourier transform of the projection is

$$P(u) = 2\pi \int_0^{\infty} f(r)\, J_0(2\pi r u)\, r\, dr. \qquad (10.69)$$

The Fourier transform of the Abel transform is the Hankel transform.[25] To recover $f(r)$, then, calculate the inverse Hankel transform of $P(u)$:

$$f(r) = 2\pi \int_0^{\infty} P(u)\, J_0(2\pi r u)\, u\, du.$$

[25] Bracewell (1956).

10.4 IMAGE FORMATION AND PROCESSING

The transform $P(u)$ is real and even, because p and J_0 are real and even. We may use the even symmetry to write

$$f(r) = 2\pi \times \frac{1}{2}\int_{-\infty}^{\infty} P(u) J_0(2\pi ru) |u|\, du$$

and then, writing $|u| = u\,\text{sgn}\,u$, we have

$$f(r) = \frac{1}{2}\int_{-\infty}^{\infty} [2\pi u P(u)] J_0(2\pi ru)\,\text{sgn}\,u\, du.$$

Now, $i2\pi u P(u)$ is the Fourier transform of $p'(x)$, by the derivative theorem. Therefore, substituting the Fourier transform integral of $p(x)$ for $P(u)$,

$$f(r) = \frac{1}{2}\int_{-\infty}^{\infty}\left[-i\int_{-\infty}^{\infty} p'(x) e^{-i2\pi ux}\, dx\right] J_0(2\pi ru)\,\text{sgn}\,u\, du$$

$$= \frac{1}{2i}\int_{-\infty}^{\infty} p'(x)\left[\int_{-\infty}^{\infty} J_0(2\pi ru)\,\text{sgn}\,u\, e^{-i2\pi ux}\, du\right] dx.$$

The inner integral is the Fourier transform of $J_0(2\pi ru)\,\text{sgn}\,u$. The derivation of this transform is left to the problems. The result is

$$J_0(2\pi ru)\,\text{sgn}\,u \longmapsto \begin{cases} \dfrac{\text{sgn}\,x}{i\pi\sqrt{x^2 - r^2}}, & |x| > r \\ 0, & \text{otherwise} \end{cases} \qquad (10.70)$$

and inserting this into the outer integral,

$$f(r) = \frac{1}{2i}\left[\int_{-\infty}^{-r} p'(x) \frac{-1}{i\pi\sqrt{x^2 - r^2}}\, dx + \int_{r}^{\infty} p'(x) \frac{1}{i\pi\sqrt{x^2 - r^2}}\, dx\right].$$

Because $p(x)$ is even, $p'(x)$ is odd, and the two integrals are in fact the same. We have

$$f(r) = \frac{1}{2i}\times 2\int_{r}^{\infty} \frac{p'(x)}{i\pi\sqrt{x^2 - r^2}}\, dx = -\frac{1}{\pi}\int_{r}^{\infty} \frac{p'(x)}{\sqrt{x^2 - r^2}}\, dx,$$

giving the final result for the Abel transform pair:

$$p(x) = \int_{|x|}^{\infty} \frac{2f(r)\,r\,dr}{\sqrt{r^2 - x^2}}, \qquad (10.67)$$

$$f(r) = -\frac{1}{\pi}\int_{r}^{\infty} \frac{p'(x)}{\sqrt{x^2 - r^2}}\, dx. \qquad (10.71)$$

Example 10.19 (Abel transform of the circle function, continued). Previously, we saw

$$\text{circ}(r/a) \stackrel{A}{\longmapsto} 2\sqrt{a^2 - x^2}\,\text{rect}(x/2a).$$

To illustrate the Hankel–Fourier–Abel cycle, we calculate the Fourier transform:

$$\int_{-a}^{a} 2\sqrt{a^2 - x^2}\, e^{-i2\pi v x}\, dx = 2a^2 \int_{-1}^{1} \sqrt{1 - \xi^2}\, e^{-i2\pi(av)\xi}\, d\xi.$$

The integral can be looked up (or for a derivation, see the problems) and is equal to $J_1(2\pi av)/2av$. So,

$$2\sqrt{a^2 - x^2}\, \text{rect}(x/2a) \xrightarrow{\mathcal{F}} \frac{2a^2 J_1(2\pi av)}{2av} = \pi a^2\, \text{jinc}(av),$$

which we know is the Hankel transform of $\text{circ}(r/a)$. This completes the cycle,

$$\text{circ}(r/a) \xrightarrow{\mathcal{A}} 2\sqrt{a^2 - x^2}\, \text{rect}(x/2a) \xrightarrow{\mathcal{F}} \pi a^2\, \text{jinc}(a\rho) \xrightarrow{\mathcal{H}} \text{circ}(r/a).$$ ∎

10.5 FOURIER TRANSFORM OF A LATTICE

In an earlier chapter we introduced the comb function and its Fourier transform and used its sampling/replicating property in several applications. In this section we generalize the comb to two and three dimensions and discuss some applications.

Recall the definition of the comb in one dimension for impulses spaced by a distance a:

$$\sum_{n=-\infty}^{\infty} \delta(x - na) = \frac{1}{|a|} \sum_{n=-\infty}^{\infty} \delta(x/a - n) = \frac{1}{|a|} \text{III}(x/a).$$

Its Fourier transform is also a comb:

$$\frac{1}{|a|} \text{III}(x/a) \longmapsto \text{III}(av) = \sum_{n=-\infty}^{\infty} \delta(av - n) = \sum_{n=-\infty}^{\infty} \frac{1}{|a|} \delta(v - n/a).$$

In two dimensions, $\text{III}(x)$ is an array of blades extending to $\pm\infty$ in the y direction. Multiplying two combs, $\text{III}(x)\,\text{III}(y)$ gives a square array of impulses with unit spacing. In three dimensions, $\text{III}(x)\,\text{III}(y)$ is an array of lines extending to $\pm\infty$ in the z direction, and multiplying by a third comb, $\text{III}(x)\,\text{III}(y)\,\text{III}(z)$, produces a cubic array of impulses (e.g., point sources) with unit spacing. By analogy with crystallography, we will call these multidimensional impulse arrays *lattices*.

For a compact notation, write

$$\text{III}(x)\,\text{III}(y)\,\text{III}(z) = \sum_{\ell}\sum_{m}\sum_{n} \delta(x-\ell, y-m, z-n) = \sum_{\mathbf{n}} \delta(\mathbf{x} - \mathbf{n}) = \text{III}(\mathbf{x}),$$

where $\mathbf{n} = (\ell, m, n)$ is a multi-index and the sum over $\mathbf{n}$ denotes a triple sum over the indices ℓ, m, and n. It is easy to see, using separability and the one-dimensional Fourier pair $\text{III}(x) \longleftrightarrow \text{III}(v)$, that the Fourier transform of $\text{III}(\mathbf{x})$ is $\text{III}(\mathbf{v})$. The cubic lattice may be made to have non-unit spacings (a, b, c) by a diagonal scaling matrix:

$$\sum_{\ell,m,n} \delta(x - a\ell)\,\delta(y - bm)\,\delta(z - cn) = \sum_{\mathbf{n}} \delta(\mathbf{x} - \mathbf{A}\mathbf{n}),$$

where

$$\mathbf{A} = \begin{bmatrix} a & 0 & 0 \\ 0 & b & 0 \\ 0 & 0 & c \end{bmatrix}.$$

In comb notation,

$$\sum_{\mathbf{n}} \delta(\mathbf{x} - \mathbf{A}\mathbf{n}) = \frac{1}{|\det \mathbf{A}|} \mathrm{III}(\mathbf{A}^{-1}\mathbf{x}), \qquad (10.72)$$

then using Equation 10.9, we calculate the Fourier transform:

$$\frac{1}{|\det \mathbf{A}|} \mathrm{III}(\mathbf{A}^{-1}\mathbf{x}) \longmapsto \mathrm{III}(\mathbf{A}\nu) = \frac{1}{|\det \mathbf{A}|} \sum_{\mathbf{n}} \delta(\nu - \mathbf{A}^{-1}\mathbf{n}). \qquad (10.73)$$

As we would expect from the one-dimensional case, the lattice spacings in the Fourier domain are $(1/a, 1/b, 1/c)$.

10.5.1 Nonorthogonal Lattices and the Reciprocal Lattice

Nonorthogonal lattices are encountered in crystallography and also in certain sampling schemes. Using matrices for scale, shear, and rotation, a cubic lattice may be transformed into a nonorthogonal lattice, but here we will take a different approach. Let a point in the lattice be represented by a delta function located at a position $\mathbf{x} = \ell\mathbf{a} + m\mathbf{b} + n\mathbf{c}$, that is, $\delta(\mathbf{x} - (\ell\mathbf{a} + m\mathbf{b} + n\mathbf{c}))$, where $\{\mathbf{a}, \mathbf{b}, \mathbf{c}\}$ are three linearly independent *lattice vectors*. The entire lattice is a sum of these impulses over all indices. If we gather the three lattice vectors (expressed in terms of their (x, y, z) components) into a matrix $\mathbf{V} = [\mathbf{a}|\mathbf{b}|\mathbf{c}]$, we may write this impulse compactly:

$$\delta(\mathbf{x} - (\ell\mathbf{a} + m\mathbf{b} + n\mathbf{c})) = \delta(\mathbf{x} - \mathbf{V}\mathbf{n}),$$

and then write the lattice in terms of the comb function:

$$\sum_{\mathbf{n}} \delta(\mathbf{x} - \mathbf{V}\mathbf{n}) = \sum_{\mathbf{n}} \frac{1}{|\det \mathbf{V}|} \delta(\mathbf{V}^{-1}\mathbf{x} - \mathbf{n}) = \frac{1}{|\det \mathbf{V}|} \mathrm{III}(\mathbf{V}^{-1}\mathbf{x}). \qquad (10.74)$$

The matrix $\mathbf{V}$ is guaranteed to be nonsingular because the lattice vectors are, by definition, linearly independent.

The Fourier transform of the nonorthogonal lattice is straightforward, using Equations 10.9 and 10.72:

$$\frac{1}{|\det \mathbf{V}|} \mathrm{III}(\mathbf{V}^{-1}\mathbf{x}) \longmapsto \mathrm{III}(\mathbf{V}^T \nu) = \frac{1}{|\det \mathbf{V}|} \sum_{\mathbf{n}} \delta(\nu - \mathbf{V}^{-T}\mathbf{n}). \qquad (10.75)$$

In the special case of a rectangular lattice, where $\{\mathbf{a}, \mathbf{b}, \mathbf{c}\} = \{a\mathbf{e}_x, b\mathbf{e}_y, c\mathbf{e}_z\}$, this expression reduces to Equation 10.73. To interpret the general case, note that the

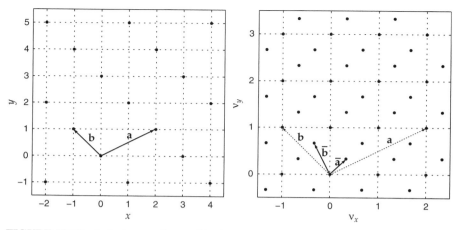

FIGURE 10.29 *Left*: A nonorthogonal lattice generated by lattice vectors $\mathbf{a} = (2, 1, 0)$, $\mathbf{b} = (-1, 1, 0)$, and $\mathbf{c} = (0, 0, 1)$. Only the $z = 0$ plane is shown, for simplicity. *Right*: The reciprocal lattice is the Fourier transform of the lattice and is generated by reciprocal lattice vectors $\bar{\mathbf{a}} = (1/3, 1/3, 0)$, $\bar{\mathbf{b}} = (-1/3, 2/3, 0)$, and $\bar{\mathbf{c}} = (0, 0, 1)$. Only the $v_z = 0$ plane is shown, for simplicity. Note that $\bar{\mathbf{a}} \perp \mathbf{b}$ and $\bar{\mathbf{b}} \perp \mathbf{a}$.

columns of $\mathbf{V}^{-T}$ are the lattice vectors for the Fourier transform, which we denote $\{\bar{\mathbf{a}}, \bar{\mathbf{b}}, \bar{\mathbf{c}}\}$. But $\mathbf{V}^{-1}\mathbf{V} = I$, that is,

$$\begin{bmatrix} \bar{\mathbf{a}}^T \\ \bar{\mathbf{b}}^T \\ \bar{\mathbf{c}}^T \end{bmatrix} [\mathbf{a}|\mathbf{b}|\mathbf{c}] = \begin{bmatrix} 1 & 0 & 0 \\ 0 & 1 & 0 \\ 0 & 0 & 1 \end{bmatrix},$$

and we see that $\bar{\mathbf{a}} \cdot \mathbf{a} = 1$ and $\bar{\mathbf{a}}$ is orthogonal to $\mathbf{b}$ and $\mathbf{c}$, and similarly for $\bar{\mathbf{b}}$ and $\bar{\mathbf{c}}$. This is illustrated in Figure 10.29. Again taking a cue from crystallography, the lattice generated by the vectors $\{\bar{\mathbf{a}}, \bar{\mathbf{b}}, \bar{\mathbf{c}}\}$ is called the *reciprocal lattice* and the vectors that generate it are called reciprocal lattice vectors.[26]

Given the lattice vectors, the reciprocal lattice vectors may be found by forming and inverting the matrix $\mathbf{V}$. Another way follows from the orthogonality relationships between the vector sets. Since $\bar{\mathbf{a}}$ is orthogonal to both $\mathbf{b}$ and $\mathbf{c}$, it is proportional to their cross product:

$$\bar{\mathbf{a}} = k\,\mathbf{b} \times \mathbf{c}.$$

Using the other relationship, $\mathbf{a} \cdot \bar{\mathbf{a}} = 1$, we can determine the constant:

$$\mathbf{a} \cdot \bar{\mathbf{a}} = k\,\mathbf{a} \cdot (\mathbf{b} \times \mathbf{c}) = 1$$

$$\Rightarrow k = \frac{1}{\mathbf{a} \cdot (\mathbf{b} \times \mathbf{c})}.$$

[26]For a discussion of the reciprocal lattice as it applies to crystallography, see Cullity and Stock (2001, pp. 519–614).

Performing the same calculation for $\bar{\mathbf{b}}$ and $\bar{\mathbf{c}}$ leads to these expressions for the reciprocal lattice vectors:

$$\bar{\mathbf{a}} = \frac{\mathbf{b} \times \mathbf{c}}{\mathbf{a} \cdot (\mathbf{b} \times \mathbf{c})} \qquad \bar{\mathbf{b}} = \frac{\mathbf{c} \times \mathbf{a}}{\mathbf{b} \cdot (\mathbf{c} \times \mathbf{a})} \qquad \bar{\mathbf{c}} = \frac{\mathbf{a} \times \mathbf{b}}{\mathbf{c} \cdot (\mathbf{a} \times \mathbf{b})}. \tag{10.76}$$

Example 10.20 (Reciprocal lattice calculation). Consider the lattice shown in Figure 10.29, with $\mathbf{a} = (2, 1, 0)$, $\mathbf{b} = (-1, 1, 0)$, and $\mathbf{c} = (0, 0, 1)$. The $\mathbf{V}$ matrix is

$$\mathbf{V} = \begin{bmatrix} 2 & -1 & 0 \\ 1 & 1 & 0 \\ 0 & 0 & 1 \end{bmatrix},$$

and its transpose inverse is

$$\mathbf{V}^{-T} = \begin{bmatrix} 1/3 & -1/3 & 0 \\ 1/3 & 2/3 & 0 \\ 0 & 0 & 1 \end{bmatrix},$$

from which we have $\bar{\mathbf{a}} = (1/3, 1/3, 0)$, $\bar{\mathbf{b}} = (-1/3, 2/3, 0)$, $\bar{\mathbf{c}} = (0, 0, 1)$. Using Equation 10.76 instead,

$$\mathbf{b} \times \mathbf{c} = (-1, 1, 0) \times (0, 0, 1) = (1, 1, 0)$$
$$\mathbf{a} \cdot (\mathbf{b} \times \mathbf{c}) = (2, 1, 0) \cdot (1, 1, 0) = 3$$
$$\Rightarrow \bar{\mathbf{a}} = \frac{1}{3}(1, 1, 0) = (1/3, 1/3, 0)$$

and similarly for $\bar{\mathbf{b}}$ and $\bar{\mathbf{c}}$. ∎

In one dimension, there is a reciprocal relationship between impulse spacings in the time and frequency domains:

$$\sum_n \delta(x - an) \longleftrightarrow \frac{1}{|a|} \sum_n \delta(\nu - n/a).$$

On a rectangular lattice, a similar relationship is obtained by the separability of the lattice into the product of two comb functions:

$$\sum_{m,n} \delta(x - am, y - bn) \longleftrightarrow \frac{1}{|ab|} \sum_{m,n} \delta(\nu_x - n/a, \nu_y - m/b).$$

The spacings in ν_x and ν_y are the reciprocals of the spacings in x and y, respectively. But for a general lattice, it is not true that the corresponding sample spacings in the two domains, for example, $\|\mathbf{a}\|$ and $\|\bar{\mathbf{a}}\|$, are reciprocals. Rather, because $\mathbf{a} \cdot \bar{\mathbf{a}} = 1$, we have $\|\bar{\mathbf{a}}\| = \frac{1}{\|\mathbf{a}\| \cos(\mathbf{a},\bar{\mathbf{a}})}$, where $\cos(\mathbf{a}, \bar{\mathbf{a}})$ is the cosine of the angle between $\mathbf{a}$ and $\bar{\mathbf{a}}$. In general, $\|\bar{\mathbf{a}}\| \geq 1/\|\mathbf{a}\|$, and only if $\mathbf{a}$ is parallel to $\bar{\mathbf{a}}$ (thus, perpendicular to $\mathbf{b}$ and $\mathbf{c}$) do we have a reciprocal relationship.

However, there is a reciprocal relationship between the *densities* of lattice points in the two domains. Consider the parallelopiped in the space domain whose vertices are lattice points. From analytic geometry, we know the volume of this box is the triple scalar product $\mathbf{a} \cdot (\mathbf{b} \times \mathbf{c})$, which is identical to the determinant of the matrix $[\mathbf{a}|\mathbf{b}|\mathbf{c}] = \mathbf{V}$. The density of points in the space domain is the reciprocal of this volume, $1/\det \mathbf{V}$. The volume of the corresponding box in the Fourier domain (the reciprocal lattice) is the determinant of the matrix $[\bar{\mathbf{a}}|\bar{\mathbf{b}}|\bar{\mathbf{c}}] = \mathbf{V}^{-T}$, and the density of points in the Fourier domain is the reciprocal of this volume. But $\det(\mathbf{V}^{-T}) = 1/\det \mathbf{V}$. So, the density of points in the reciprocal lattice, $\det \mathbf{V}$, is the reciprocal of the density of points in the space lattice, $1/\det \mathbf{V}$.

Example 10.21 (Lattice densities). Consider again the lattice shown in Figure 10.29, with $\mathbf{a} = (2, 1, 0)$, $\mathbf{b} = (-1, 1, 0)$, $\mathbf{c} = (0, 0, 1)$, and $\bar{\mathbf{a}} = (1/3, 1/3, 0)$, $\bar{\mathbf{b}} = (-1/3, 2/3, 0)$, $\bar{\mathbf{c}} = (0, 0, 1)$. The lengths of $\mathbf{a}$, $\mathbf{b}$, and $\mathbf{c}$ are $\sqrt{5}$, $\sqrt{2}$, and 1, respectively. The respective lengths of $\bar{\mathbf{a}}$, $\bar{\mathbf{b}}$, and $\bar{\mathbf{c}}$ are not reciprocals: $\sqrt{2}/3$, $\sqrt{5}/3$, and 1 (as expected, $\sqrt{2}/3 > 1/\sqrt{5}$ and $\sqrt{5}/3 > 1/\sqrt{2}$). But the density of points per unit volume in the space domain is

$$\left(\det \begin{bmatrix} 2 & -1 & 0 \\ 1 & 1 & 0 \\ 0 & 0 & 1 \end{bmatrix}\right)^{-1} = 1/3$$

and the density of points in the Fourier domain is

$$\left(\det \begin{bmatrix} 1/3 & -1/3 & 0 \\ 1/3 & 2/3 & 0 \\ 0 & 0 & 1 \end{bmatrix}\right)^{-1} = 3.$$

There is a reciprocal relationship between the densities of points in the two lattices. ∎

10.5.2 Sampling Theory

The results of the preceding section are straightforwardly applied to the problem of sampling a two- or three-dimensional function, for example, an image. We will restrict attention to the two-dimensional case and leave the three-dimensional case to the reader's imagination. A two-dimensional function $f(\mathbf{x})$ is sampled on an orthogonal grid with sampling intervals Δx and Δy; for purposes of illustration, we take $\Delta x = 1$ and $\Delta y = 2$. The sampling lattice vectors are $\mathbf{a} = (\Delta x, 0) = (1, 0)$ and $\mathbf{b} = (0, \Delta y) = (0, 2)$. The sampled function is modeled by multiplying f with the sampling lattice (Equation 10.74):

$$f_s(\mathbf{x}) = f(\mathbf{x}) \sum_{m,n} \delta(\mathbf{x} - m\mathbf{a} - n\mathbf{b}) = f(\mathbf{x}) \frac{1}{|\det \mathbf{V}|} \operatorname{III}(\mathbf{V}^{-1}\mathbf{x}), \tag{10.77}$$

where

$$\mathbf{V} = [\mathbf{a}|\mathbf{b}] = \begin{bmatrix} 1 & 0 \\ 0 & 2 \end{bmatrix}.$$

10.5 FOURIER TRANSFORM OF A LATTICE

In the Fourier domain, the product becomes a convolution, and (Equation 10.75)

$$F_s(v) = F(v) * \text{III}(V^T v) = F(v) * \frac{1}{|\det V|} \sum_n \delta(v - V^{-T} n)$$

$$= \frac{1}{|\det V|} \sum_n F(v - V^{-T} n). \tag{10.78}$$

In our example,

$$F_s(v) = \sum_{m,n} \frac{1}{|\det V|} F(v - m\bar{a} - n\bar{b}),$$

where $\bar{a}$ and $\bar{b}$ are the reciprocal lattice vectors:

$$[\bar{a}|\bar{b}] = V^{-T} = \begin{bmatrix} 1 & 0 \\ 0 & \frac{1}{2} \end{bmatrix},$$

that is,

$$F_s(v_x, v_y) = \sum_{m,n} \frac{1}{2} F(v_x - m, v_y - n/2).$$

The Fourier domain is shown in Figure 10.30.

Sampling on the lattice generated by the vectors $\{a, b\}$ replicates the Fourier transform on the reciprocal lattice generated by the vectors $\{\bar{a}, \bar{b}\}$. When the lattice is orthogonal, the reciprocal lattice is also orthogonal. Replicating a rectangle of

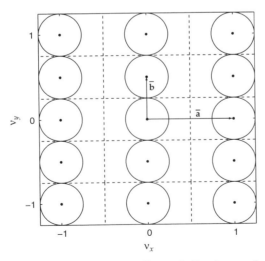

FIGURE 10.30 Two-dimensional sampling replicates the Fourier transform on the reciprocal lattice of the sampling lattice. If the function's Fourier transform is bandlimited to a $\frac{1}{\Delta x} \times \frac{1}{\Delta y}$ rectangle, the function is recoverable from its samples by sinc function interpolation. If the function is bandlimited to a disk, as shown here, there are large gaps between the spectral replicas and coarser, more efficient sampling is possible.

728 CHAPTER 10 FOURIER TRANSFORMS IN TWO AND THREE DIMENSIONS

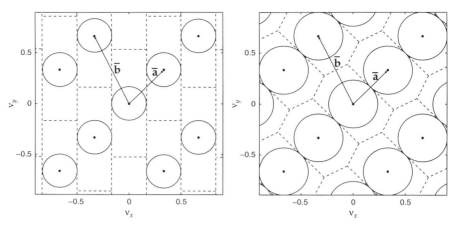

FIGURE 10.31 Two-dimensional sampling replicates the Fourier transform on the reciprocal lattice of the sampling lattice. When the sampling lattice is nonorthogonal, the Fourier domain can be covered by different periodic tilings, with different implications for bandlimiting. *Left*: Rectangular tiling. If the function's Fourier transform is bandlimited to the unit tile, it is recoverable from its samples by sinc function interpolation. The rectangular tiling leads to a pessimistic estimate of the bandlimit for a circularly bandlimited function. *Right*: Hexagonal tiling allows the spectrum to have a different support and shows that a higher circular bandlimit is possible. If the function is bandlimited to the disk shown, it will not be aliased and can be reconstructed by interpolation with jinc functions.

dimensions $\frac{1}{\Delta x} \times \frac{1}{\Delta y}$ with this reciprocal lattice covers the Fourier domain with non-overlapping rectangular tiles. If the function f is bandlimited such that its Fourier transform is zero outside the rectangle $|\nu_x| < \frac{1}{2\Delta x}, |\nu_y| < \frac{1}{2\Delta y}$, then it is not aliased and can be recovered from its samples. The ideal lowpass filter is $2\,\text{rect}(\nu_x)\,\text{rect}(2\nu_y)$ and the corresponding interpolation function is $\text{sinc}(x)\,\text{sinc}(y/2)$. In general,

$$\hat{f}(x,y) = \sum_{m,n} f(x - m\Delta x, y - n\Delta y)\,\text{sinc}(x/\Delta x)\,\text{sinc}(y/\Delta y).$$

Sampling on an orthogonal lattice is thus a straightforward extension of one-dimensional sampling.

A two-dimensional function is not always bandlimited to a rectangular support. We saw earlier that an image formed by an optical instrument with a circular aperture is bandlimited to a circular support. If such a function is sampled on a rectangular grid, the replication in the Fourier domain will produce gaps between the circular spectral islands. The scheme shown in Figure 10.30 with $\Delta x \neq \Delta y$ results in an inefficient coverage of the Fourier domain. Sampling with $\Delta x = \Delta y$ is preferable in this case.

If the function is sampled on a nonorthogonal lattice, then the replication of the Fourier transform on the reciprocal lattice is also nonorthogonal. Two different tilings of the Fourier domain are shown in Figure 10.31 (and there are many other possible tilings besides these, with oddly shaped tiles).[27] In both cases, the unit

---

[27]Dudgeon and Mersereau (1984, Chapter 1); Marks (1991, Chapter 6); Barrett and Myers (2004, pp. 149ff).

tile centered at the origin specifies maximum bandlimits for the function. For the rectangular tiling, interpolation with sinc functions is possible if the support of the Fourier transform matches or fits within the unit tile. But again, if a function is bandlimited to some other support, for example, circular, then keeping the support inside the rectangular tile may be an inefficient use of samples. A Fourier transform with circular support, for example, fits better within the hexagonal tiling shown in the figure. In fact, for a circularly bandlimited function, the optimum sampling scheme occurs when the spectral islands are close packed. Then the central spectral island is isolated with an ideal circular lowpass filter, and the function is recovered from its samples by interpolating with jinc functions. This special case is considered further in the problems.

10.5.3 X-ray Diffraction

A crystal is a periodic replication of a molecule in three dimensions. The replicated molecule may be a single atom, like silicon, a simple diatomic structure like water (ice) or sodium chloride, or a very complex molecule, like a protein. In all cases, the crystal can be modeled by the convolution of the molecule with a lattice.

X-rays are scattered by the electrons in an atom; approximately, the atom behaves like a point and scatters in an amount proportional to its number of electrons. The x-ray wavelengths, on the order of 10^{-10} m (1 Å), are of the same scale as the interatomic distances in the molecule and the molecular spacings in the lattice. Consequently, the scattered x-rays can interfere and produce diffraction patterns. X-ray crystallography works backward from these observed diffraction patterns to infer the electron densities in the crystal, giving information about the structure of the molecule and also of the lattice.

To develop the Fourier transform relationships for diffraction by a crystal, consider the geometry shown in Figure 10.32. An incident plane wave is scattered at point P, for example, by an atom, into a spherical wave that is approximately planar in the far field. Relative to the origin O, the scattered wave experiences a phase shift $\varphi = \frac{2\pi}{\lambda}(\mathbf{r} \cdot \mathbf{e} - \mathbf{r} \cdot \mathbf{e}_0)$, where $\mathbf{e}_0$ and $\mathbf{e}$ are unit vectors in the incident and scattered

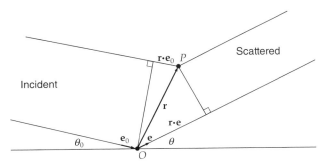

FIGURE 10.32 A plane wave scatters from a point P, for example, X-ray scattering from an atom. The point scatters a spherical wave, which becomes a plane wave in the far field. Relative to the origin O, the scattered wave has phase shift $\varphi = \frac{2\pi}{\lambda}(\mathbf{r} \cdot \mathbf{e} - \mathbf{r} \cdot \mathbf{e}_0) = \mathbf{r} \cdot (\mathbf{k} - \mathbf{k}_0)$.

directions, respectively, and vector **r** locates P relative to O. Writing $\mathbf{k} = \frac{2\pi}{\lambda}\mathbf{e}$ and $\mathbf{k}_0 = \frac{2\pi}{\lambda}\mathbf{e}_0$, the phase shift is $\varphi = \mathbf{r} \cdot (\mathbf{k} - \mathbf{k}_0)$.

If there is also a scatterer at the origin, then the wave in the far field is (suppressing a common plane wave factor $e^{i(\omega t - \mathbf{k} \cdot \mathbf{x})}$)

$$\widetilde{U}(\mathbf{k}) = 1 + e^{-i(\mathbf{k}-\mathbf{k}_0)\cdot\mathbf{r}} = 2e^{-i(\mathbf{k}-\mathbf{k}_0)\cdot\mathbf{r}/2} \cos[(\mathbf{k}-\mathbf{k}_0)\cdot\mathbf{r}/2]. \qquad (10.79)$$

The cosine expresses the interference between the two scattered waves. For a wave with incident direction given by the vector $\mathbf{k}_0$, intensity maxima occur in directions $\mathbf{k}$ such that $(\mathbf{k} - \mathbf{k}_0) \cdot \mathbf{r} = 2\pi n$. This is, of course, just a more general formulation of the two-point interference used earlier to describe antenna arrays. For example, if $\mathbf{r}$ is taken along the horizontal direction with length d, $\mathbf{r} = d\mathbf{e}_x$, then the maxima appear at

$$\frac{2\pi d}{\lambda}(\mathbf{e} \cdot \mathbf{e}_x - \mathbf{e}_0 \cdot \mathbf{e}_x) = \frac{2\pi d}{\lambda}\left(\cos\theta_0 - \cos\theta\right) = 2\pi n$$

$$\Rightarrow \cos\theta = \cos\theta_0 + \frac{n\lambda}{d},$$

a familiar result (compare Equation 4.39).

The two-point model may easily be extended to an arbitrary number of points, assuming that the total scattered field is the superposition of fields resulting from scattering the incident field off each point. Contributions from multiple scatterings are assumed to be too weak to contribute significantly to the total.[28] When the points are in a lattice, their position vectors are $\mathbf{r} = \mathbf{V}\mathbf{n}$ where, as before, $\mathbf{V} = [\mathbf{a}|\mathbf{b}|\mathbf{c}]$ are the lattice vectors and $\mathbf{n}$ is a multi-index. Then the scattered wave is

$$\widetilde{U}(\mathbf{k}) = \sum_{\mathbf{n}} \exp[-i(\mathbf{k} - \mathbf{k}_0) \cdot \mathbf{V}\mathbf{n}],$$

which can be written in terms of the comb function model of a lattice:

$$\widetilde{U}(\mathbf{k}) = \sum_{\mathbf{n}} \int \exp\left[-i(\mathbf{k} - \mathbf{k}_0) \cdot \mathbf{x}\right] \delta(\mathbf{x} - \mathbf{V}\mathbf{n}) \, d\mathbf{x}$$

$$= \int \frac{1}{|\det \mathbf{V}|} \mathrm{III}\left(\mathbf{V}^{-1}\mathbf{x}\right) \exp\left[-i(\mathbf{k} - \mathbf{k}_0) \cdot \mathbf{x}\right] d\mathbf{x}.$$

Now this is the Fourier transform of a comb—the diffracted field is the Fourier transform of the lattice. Calculating the transform, we have

$$\widetilde{U}(\mathbf{k}) = \mathcal{F}\left\{\frac{1}{|\det \mathbf{V}|} \mathrm{III}\left(\mathbf{V}^{-1}\mathbf{x}\right)\right\}\bigg|_{\boldsymbol{\nu}=(\mathbf{k}-\mathbf{k}_0)/2\pi} = \mathrm{III}\left(\mathbf{V}^T(\mathbf{k}-\mathbf{k}_0)/2\pi\right)$$

$$= \sum_{\mathbf{n}} \frac{2\pi}{|\det \mathbf{V}|} \delta\left[(\mathbf{k}-\mathbf{k}_0) - 2\pi\mathbf{V}^{-T}\mathbf{n}\right]. \qquad (10.80)$$

---

[28]The assumption of single scattering is sometimes referred to as the first Born approximation—Born and Wolf (1999, pp. 699–708); Barrett and Myers (2004, pp. 542–547).

Recall that $\mathbf{V}^{-T} = [\bar{\mathbf{a}}|\bar{\mathbf{b}}|\bar{\mathbf{c}}]$ are the reciprocal lattice vectors. The peaks in the diffraction pattern, where the scattered waves constructively interfere, are given by the locations of the delta functions in reciprocal space:

$$\mathbf{k} - \mathbf{k}_0 = 2\pi[\bar{\mathbf{a}}|\bar{\mathbf{b}}|\bar{\mathbf{c}}]\mathbf{n}, \qquad (10.81)$$

that is, when $\mathbf{k} - \mathbf{k}_0$ is directed to a point in the reciprocal lattice. Thus the reciprocal lattice has a physical interpretation as the directions in which the incident radiation is scattered by the crystal. Experimentally, the reciprocal lattice may be inferred by measuring the diffraction directions for several incident directions, and from this the crystal structure may be determined.

A real crystal is a periodic repetition of a basic arrangement of atoms or unit cell. Mathematically, it is modeled by the convolution of the unit cell with the lattice. The diffraction pattern, then, is related to the product of the reciprocal lattice with the Fourier transform of the unit cell. In some experiments, not only the lattice but also the structure of the unit cell, for example, a protein, is determined from diffraction measurements.[29]

10.6 DISCRETE MULTIDIMENSIONAL FOURIER TRANSFORMS

The Fourier series, discrete-time Fourier transform, and discrete Fourier transform all have useful multidimensional versions. Two-, three- and even four-dimensional (three spatial dimensions plus time) Fourier series appear in the solutions to certain partial differential equations.[30] Multidimensional DTFTs and DFTs are applicable in signal and image processing.[31] The Z transform also extends to two dimensions, with some interesting complications owing to the complex nature of the transform variables.

We will develop the multidimensional discrete-time Fourier transform and DFT by beginning with a sampled continuous-time function and using the relationships among the members of the Fourier family (Figure 6.21). A multidimensional function $f(\mathbf{x})$, sampled on a nonorthogonal lattice, results in a multidimensional sampled function (Equation 10.77):

$$f_s(\mathbf{x}) = f(\mathbf{x}) \frac{1}{|\det \mathbf{V}|} \mathrm{III}\left(\mathbf{V}^{-1}\mathbf{x}\right) = \sum_{\mathbf{n}} f(\mathbf{V}\mathbf{n}) \delta(\mathbf{x} - \mathbf{V}\mathbf{n}),$$

where, as usual, $\mathbf{V} = [\mathbf{a}|\mathbf{b}]$ in two dimensions. The Fourier transform of f_s is

$$F_s(\mathbf{\nu}) = \sum_{\mathbf{n}} f(\mathbf{V}\mathbf{n}) \exp(-i2\pi\mathbf{\nu} \cdot \mathbf{V}\mathbf{n}) = \frac{1}{|\det \mathbf{V}|} \sum_{\mathbf{n}} F\left(\mathbf{\nu} - \mathbf{V}^{-T}\mathbf{n}\right).$$

The Fourier transform is periodic in $\mathbf{\nu}$, along the reciprocal lattice. Moving integer numbers of reciprocal lattice periods in $\mathbf{\nu}$ (replace $\mathbf{\nu}$ by $\mathbf{\nu} + r_1\bar{\mathbf{a}} + r_2\bar{\mathbf{b}} = \mathbf{\nu} + \mathbf{V}^{-T}\mathbf{r}$)

---

[29] For more on X-ray crystallography, see McPherson (2009), Cullity and Stock (2001), and Ramachandran and Srinivasan (1970).

[30] Churchill and Brown (1987).

[31] Dudgeon and Mersereau (1984).

yields

$$\exp\left[-i2\pi\left(\nu + \mathbf{V}^{-T}\mathbf{r}\right)\cdot\mathbf{Vn}\right] = \exp\left[-i2\pi\nu\cdot\mathbf{Vn}\right]\exp\left[-i2\pi\left(\mathbf{V}^{-T}\mathbf{r}\right)^{T}\mathbf{Vn}\right]$$
$$= \exp\left[-i2\pi\nu\cdot\mathbf{Vn}\right]\exp\left[-i2\pi\mathbf{r}^{T}\mathbf{V}^{-1}\mathbf{Vn}\right]$$
$$= \exp\left[-i2\pi\nu\cdot\mathbf{Vn}\right]\underbrace{\exp\left[-i2\pi\mathbf{r}^{T}\mathbf{n}\right]}_{=1}.$$

Thus,

$$F_s(\nu + \mathbf{V}^{-T}\mathbf{r}) = F_s(\nu) \tag{10.82}$$

as we saw earlier in Figure 10.30.

The discrete-time Fourier transform of the multidimensional sequence $f[\mathbf{n}] = f(\mathbf{Vn})$ is defined:

$$F_d(\boldsymbol{\theta}) = \sum_{\mathbf{n}} f[\mathbf{n}]\exp(-i\boldsymbol{\theta}\cdot\mathbf{n}). \tag{10.83}$$

This DTFT is periodic in $\boldsymbol{\theta}$ with period 2π in any component, that is, $F_d(\boldsymbol{\theta} + 2\pi\mathbf{k}) = F_d(\boldsymbol{\theta})$. The connection with the Fourier transform $F_s(\nu)$ is made by comparing the Fourier kernels, $\exp(-i2\pi\nu\cdot\mathbf{Vn})$ and $\exp(-i\boldsymbol{\theta}\cdot\mathbf{n})$. They are equivalent for $\nu = \mathbf{V}^{-T}\boldsymbol{\theta}/2\pi$, thus

$$F_d(\boldsymbol{\theta}) = F_s\left(\mathbf{V}^{-T}\boldsymbol{\theta}/2\pi\right) \quad \text{and} \quad F_s(\nu) = F(2\pi\mathbf{V}^T\nu). \tag{10.84}$$

In one dimension, the relationship between digital frequency and continuous frequency is $\theta = 2\pi\nu T$, where T is the sampling interval. Increasing θ by 2π corresponds to an increase in ν of $1/T$, the sampling frequency, and moves to the identical value of F_s in an adjacent period. In two dimensions with orthogonal sampling, the lattice vectors are $\mathbf{a} = T_x\mathbf{e}_x$ and $\mathbf{b} = T_y\mathbf{e}_y$, where T_x and T_y are the respective sampling intervals. The matrix $\mathbf{V}$ is $\begin{bmatrix} T_x & 0 \\ 0 & T_y \end{bmatrix}$. The mapping from ν to $\boldsymbol{\theta}$ is $\boldsymbol{\theta} = 2\pi\mathbf{V}^T\nu = (2\pi\nu_x T_x, 2\pi\nu_y T_y)$, a straightforward extension of the one-dimensional result $\theta = 2\pi\nu T$. Increasing θ_x or θ_y by 2π corresponds to an increase of ν_x by $1/T_x$ or ν_y by $1/T_y$, again moving to adjacent periods of F_s. When the sampling is on a nonorthogonal lattice, the relationship between ν and $\boldsymbol{\theta}$ is $\nu = \mathbf{V}^{-T}\boldsymbol{\theta}/2\pi = (\theta_1/2\pi)\bar{\mathbf{a}} + (\theta_2/2\pi)\bar{\mathbf{b}}$. As θ_1 goes from 0 to 2π, the frequency increases by one unit of $\bar{\mathbf{a}}$, which is one period of F_s in the $\bar{\mathbf{a}}$ direction. The elements of $\boldsymbol{\theta}$ are components of frequency *in the directions specified by the reciprocal lattice vectors* $\bar{\mathbf{a}}$ and $\bar{\mathbf{b}}$, not the orthogonal frequency axes ν_x and ν_y (Figure 10.33).

The definition of the multidimensional DTFT in Equation 10.83 is independent of any consideration of the sampling lattice. Consequently, the inverse DTFT in M dimensions is also lattice independent:

$$f[\mathbf{n}] = \frac{1}{(2\pi)^M}\int_{-\pi}^{\pi} F_d(\boldsymbol{\theta})\exp(+i\boldsymbol{\theta}\cdot\mathbf{n})\,d\boldsymbol{\theta}. \tag{10.85}$$

So, the sampling lattice connects $f(\mathbf{x})$ with $f[\mathbf{n}]$, the reciprocal lattice connects $F_d(\boldsymbol{\theta})$ with $F_s(\nu)$, and the transform from $f[\mathbf{n}]$ to $F_d(\boldsymbol{\theta})$ is independent of the lattice.

10.6 DISCRETE MULTIDIMENSIONAL FOURIER TRANSFORMS

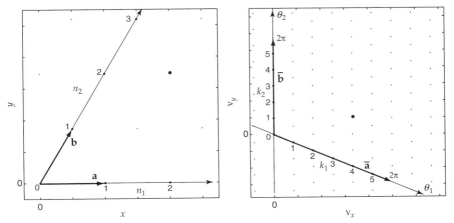

FIGURE 10.33 *Left:* sampling a two-dimensional function on a hexagonal lattice, $f[\mathbf{n}] = f(n_1 \mathbf{a} + n_2 \mathbf{b})$. The heavy dot indicates $f[1,2] = f(\mathbf{a} + 2\mathbf{b})$. *Right:* in the Fourier domain, the discrete frequency axes θ_1 and θ_2 are aligned with the reciprocal lattice vectors $\bar{\mathbf{a}}$ and $\bar{\mathbf{b}}$, and $\mathbf{v} = (\theta_1/2\pi)\bar{\mathbf{a}} + (\theta_2/2\pi)\bar{\mathbf{b}}$. Since the Fourier transform is periodic on the reciprocal lattice, as θ_1 increases from 0 to 2π, the frequency $\mathbf{v}$ increases by one unit of $\bar{\mathbf{a}}$, and similarly for θ_2. The DFT samples $F_d(\theta)$ along θ_1 and θ_2; here, $N_1 = N_2 = 6$. The heavy dot at $\mathbf{k} = (4,3)$ corresponds to discrete frequency $\theta = \left(\frac{4}{6} \cdot 2\pi, \frac{3}{6} \cdot 2\pi\right) = \left(\frac{4}{3}\pi, \pi\right)$ and continuous frequency $\mathbf{v} = \frac{2}{3}\bar{\mathbf{a}} + \frac{1}{2}\bar{\mathbf{b}}$.

In one dimension, the discrete Fourier transform is obtained from the discrete-time Fourier transform of an N-sample vector by taking N samples in θ, uniformly from 0 to 2π: $\theta = 2\pi k/N$, $k = 0, 1, \ldots, N-1$. Thus, the DFT is

$$F[k] = F(2\pi k/N) = F_s(k/NT) = \sum_{n=0}^{N-1} f(nT) \exp(-i2\pi kn/N).$$

When we generalize to the multidimensional case, we first truncate the multidimensional sequence $f[\mathbf{n}]$ to finite size, $N_1 \times N_2$ (again, without loss of generality, we restrict attention to two dimensions). Then $\mathbf{n} \in \{0, 1, \ldots N_1 - 1\} \times \{0, 1, \ldots N_2 - 1\}$. We sample $F_d(\theta)$ with $\theta = (2\pi k_1/N_1, 2\pi k_2/N_2)$, or

$$\theta = 2\pi \mathbf{N}^{-1} \mathbf{k},$$

where $\mathbf{N} = \begin{bmatrix} N_1 & 0 \\ 0 & N_2 \end{bmatrix}$. Substitute this into Equation 10.83,

$$\begin{aligned} F[\mathbf{k}] = F\left(2\pi \mathbf{N}^{-1} \mathbf{k}\right) &= \sum_{\mathbf{n}} f[\mathbf{n}] \exp\left[-i\left(2\pi \mathbf{N}^{-1} \mathbf{k}\right) \cdot \mathbf{n}\right] \\ &= \sum_{\mathbf{n}} f[\mathbf{n}] \exp\left[-i2\pi \mathbf{k}^T \mathbf{N}^{-1} \mathbf{n}\right]. \end{aligned} \quad (10.86)$$

To invert, multiply both sides by $\exp\left[+i2\pi\mathbf{k}^T\mathbf{N}^{-1}\mathbf{n}'\right]$ and sum over $\mathbf{k}$:

$$\sum_{\mathbf{k}} F[\mathbf{k}] \exp\left[+i2\pi\mathbf{k}^T\mathbf{N}^{-1}\mathbf{n}'\right] = \sum_{\mathbf{n}} f[\mathbf{n}] \sum_{\mathbf{k}} \exp[-i2\pi\mathbf{k}^T\mathbf{N}^{-1}(\mathbf{n}-\mathbf{n}')].$$

The sum over $\mathbf{k}$ expands into the product of two sums, $\sum_{k_1=0}^{N_1-1} \exp[-i2\pi k_1(n_1-n'_1)/N_1]$ and $\sum_{k_2=0}^{N_2-1} \exp[-i2\pi k_2(n_2-n'_2)/N_2]$, which we know from Chapter 3 are $N_1\delta[n_1-n'_1]$ and $N_2\delta[n_2-n'_2]$. Thus,

$$\sum_{\mathbf{k}} \exp\left[-i2\pi\mathbf{k}^T\mathbf{N}^{-1}(\mathbf{n}-\mathbf{n}')\right] = N_1 N_2 \delta[\mathbf{n}-\mathbf{n}'] = \det \mathbf{N}\, \delta[\mathbf{n}-\mathbf{n}'],$$

and we have the inverse DFT:

$$f[\mathbf{n}] = \frac{1}{\det \mathbf{N}} \sum_{\mathbf{k}} F[\mathbf{k}] \exp\left[+i2\pi\mathbf{k}^T\mathbf{N}^{-1}\mathbf{n}\right]. \tag{10.87}$$

The frequency-domain sampling that generates the DFT produces a periodic replication of $f[\mathbf{n}]$, so that $f[\mathbf{n}+\mathbf{Nr}] = f[\mathbf{n}]$ (where $\mathbf{r}$ is a vector of integers). So, just as in the one-dimensional case, the time and frequency domains are both sampled and periodic.

The form of the DFT is independent of the sampling lattice that gave rise to the sequence $f[\mathbf{n}]$. The sampling lattice connects the original function $f(\mathbf{x})$ to the sample sequence $f[\mathbf{n}]$, and the reciprocal lattice connects the DFT $F[\mathbf{k}]$ back to the Fourier transform $F_s(\mathbf{v})$.[32] In two dimensions, the sample values at $\mathbf{n} = (0,0), (1,0), \ldots, (N-1,0)$ are equally spaced at multiples of $\mathbf{a}$: $f(\mathbf{0}), f(\mathbf{a}), \ldots, f((N-1)\mathbf{a})$. In the Fourier domain, the DFT samples at $\mathbf{k} = (0,0), (1,0), \ldots, (N-1,0)$ divide the reciprocal lattice vector $\bar{\mathbf{a}}$, representing one period of the sampled Fourier transform, into N equal intervals, yielding values $F_s(\mathbf{0}), F_s\left(\frac{1}{N_1}\bar{\mathbf{a}}\right),\ldots, F_s\left(\frac{N_1-1}{N_1}\bar{\mathbf{a}}\right)$ (Figure 10.33). This is illustrated for the transform pair $\mathrm{circ}(r/8) \longleftrightarrow 64\pi\,\mathrm{jinc}(8\rho)$ in Figure 10.34.

Between $f[\mathbf{n}]$ and $F[\mathbf{k}]$, the DFT is always the same, and one multidimensional DFT algorithm will suffice for any sampling scheme. The simplest approach follows from the fact that the DFT kernel $\exp[-i2\pi\mathbf{k}^T\mathbf{N}^{-1}\mathbf{n}]$ separates into the product of one-dimensional kernels:

$$\exp\left[-i2\pi\mathbf{k}^T\mathbf{N}^{-1}\mathbf{n}\right] = \exp(-i2\pi k_1 n_1/N_1)\exp(-i2\pi k_2 n_2/N_2),$$

and so the two-dimensional DFT may be computed by a succession of one-dimensional DFTs:

$$F[\mathbf{k}] = \sum_{n_2=0}^{N_2-1} \left[\sum_{n_1=0}^{N_1-1} f[\mathbf{n}] \exp(-i2\pi k_1 n_1/N_1)\right] \exp(-i2\pi k_2 n_2/N_2).$$

One simply computes the DFTs (by FFT, say) of the rows of $f[\mathbf{n}]$ to produce an intermediate array. The DFT is completed by computing the DFTs of the columns

---

[32] Barrett and Myers (2004, pp. 173ff).

10.6 DISCRETE MULTIDIMENSIONAL FOURIER TRANSFORMS

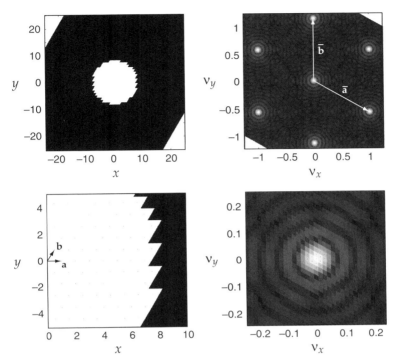

FIGURE 10.34 Discrete Fourier transform of a circle function, circ($r/8$), sampled on a hexagonal lattice defined by the vectors, $\mathbf{a} = (1, 0)$, $\mathbf{b} = (\frac{1}{2}, \frac{\sqrt{3}}{2})$ (*left*). The Fourier transform is replicated on the reciprocal lattice, $\bar{\mathbf{a}} = (1, -\frac{1}{\sqrt{3}})$, $\bar{\mathbf{b}} = (0, \frac{2}{\sqrt{3}})$ (*right*). Details (*bottom*) show the rectilinear pixels shaped by the lattice vectors. Dots are the sample points. The first zero crossing (dark ring) of the jinc is at $\rho = 0.6096/8 = 0.0762$.

of this intermediate array. So-called vector-radix FFT algorithms also exist, which attack the multidimensional DFT directly, breaking an $N \times N$ DFT into a combination of four $N/2 \times N/2$ DFTs, working down to a base case of 2×2 DFTs.[33]

---

[33] Dudgeon and Mersereau (1984, pp. 76–86).

10.7 SUMMARY

Multidimensional Fourier Transform Theorems

Theorem	Formula	Equation
Parseval	$\int_{-\infty}^{\infty} f(\mathbf{x})g^*(\mathbf{x})d\mathbf{x} = \int_{-\infty}^{\infty} F(\mathbf{v})G^*(\mathbf{v})d\mathbf{v}$	10.7a
	$\int_{-\infty}^{\infty} |f(\mathbf{x})|^2 \, d\mathbf{x} = \int_{-\infty}^{\infty} |F(\mathbf{v})|^2 \, d\mathbf{v}$	10.7b
Area	$\int_{-\infty}^{\infty} f(\mathbf{x}) \, d\mathbf{x} = F(\mathbf{0}), f(\mathbf{0}) = \int_{-\infty}^{\infty} F(\mathbf{v}) \, d\mathbf{v}$	10.14
Moment	$\mu^\alpha = \int_{-\infty}^{\infty} \mathbf{x}^\alpha f(\mathbf{x}) \, d\mathbf{x} = \left. \frac{1}{(-i2\pi)^{|\alpha|}} \partial^\alpha F(\mathbf{v}) \right|_{\mathbf{v}=0}$ $\alpha = (\alpha_1, \alpha_2, \ldots, \alpha_N)$	10.17
Linearity	$af(\mathbf{x}) + bg(\mathbf{x}) \longleftrightarrow aF(\mathbf{v}) + bG(\mathbf{v})$	10.6
Shift	$f(\mathbf{x} - \mathbf{b}) \longleftrightarrow e^{-i2\pi \mathbf{v} \cdot \mathbf{b}} F(\mathbf{v})$ $e^{i2\pi \mathbf{b} \cdot \mathbf{x}} f(\mathbf{x}) \longleftrightarrow F(\mathbf{v} - \mathbf{b})$	10.8
Dilation	$f(\mathbf{A}\mathbf{x}) \longleftrightarrow \frac{1}{|\det \mathbf{A}|} F\left(\mathbf{A}^{-T}\mathbf{v}\right)$	10.9
Derivative	$\partial^\alpha f(\mathbf{x}) \longleftrightarrow (i2\pi \mathbf{v})^\alpha F(\mathbf{v})$ $(-i2\pi \mathbf{x})^\alpha f(\mathbf{x}) \longleftrightarrow \partial^\alpha F(\mathbf{v})$	10.11
Laplacian	$\nabla^2 f(\mathbf{x}) \longmapsto -4\pi^2 \|\mathbf{v}\|^2 F(\mathbf{v})$	10.12
Convolution	$f * g(\mathbf{x}) \longleftrightarrow F(\mathbf{v})G(\mathbf{v})$	10.18

Hankel Transform Theorems

Theorem	Formula	Equation
Parseval	$\int_0^\infty f(r)g^*(r) \, r \, dr = \int_0^\infty F(\rho)G^*(\rho) \, \rho \, d\rho$	
	$\int_0^\infty |f(r)|^2 \, r \, dr = \int_0^\infty |F(\rho)|^2 \, \rho \, d\rho$	
Area	$2\pi \int_0^\infty f(r) \, r \, dr = F(0), f(0) = 2\pi \int_0^\infty F(\rho) \, \rho \, d\rho$	
Linearity	$af(r) + bg(r) \longleftrightarrow aF(\rho) + bG(\rho)$	
Shift	$f(\|\mathbf{x} - \mathbf{b}\|) \longleftrightarrow e^{-i2\pi \mathbf{v} \cdot \mathbf{b}} F(\|\mathbf{v}\|)$	10.36
Dilation	$f(ar) \longleftrightarrow \frac{1}{a^2} F(\rho/a)$	10.35
Laplacian	$\nabla^2 f(r) \longmapsto -4\pi^2 \rho^2 F(\rho)$	10.90
Convolution	$f * g(r) \longleftrightarrow F(\rho)G(\rho)$	10.37

PROBLEMS

10.1. Carry through the geometric calculation to show that the vectors perpendicular to the equiphases in Figure 10.1, whose lengths are $1/\|b\|$, have x and y components $b_x/(b_x^2 + b_y^2)$ and $b_y/(b_x^2 + b_y^2)$, respectively.

10.2. *Decomposing coordinate transformations*
A 2×2 coordinate transformation matrix $\mathbf{A}$ can be decomposed into a product of elementary transformations. One of these is the QR factorization, $\mathbf{A} = \mathbf{QR}$, where $\mathbf{Q}$ is unitary and $\mathbf{R}$ is upper triangular. The QR factorization is performed numerically by the MATLAB function `qr`.

(a) Because $\mathbf{Q}$ is unitary, its determinant is either 1 or -1. Show that if $\det \mathbf{Q} = 1$, then $\mathbf{Q}$ is a rotation by θ, and if $\det \mathbf{Q} = -1$, then $\mathbf{Q}$ is a reflection about a line at angle θ. Give expressions for θ in terms of the elements of $\mathbf{Q}$.

(b) Show that $\mathbf{R}$ is the composition of a coordinate scaling with a horizontal shear, and derive expressions for the scale factors and shear factor in terms of the elements of $\mathbf{Q}$.

10.3. *Decomposing coordinate transformations*
A 2×2 coordinate transformation matrix $\mathbf{A}$ can be decomposed into a product of elementary transformations. One of these is the Schur factorization, $\mathbf{A} = \mathbf{UTU}^T$, where $\mathbf{U}$ is unitary and $\mathbf{T}$ is upper triangular. The Schur factorization is performed numerically by the MATLAB function `schur`.

(a) The Schur factorization algorithm always returns a matrix $\mathbf{U}$ with determinant $+1$. Show that this is a rotation about an angle θ and give an expression for θ in terms of the elements of $\mathbf{U}$.

(b) Show that $\mathbf{T}$ is the composition of a coordinate scaling with a horizontal shear, and derive expressions for the scale factors and shear factor in terms of the elements of $\mathbf{T}$.

10.4. Carry out the calculations for scale and shear in Example 10.6, and compare with Figure 10.5.

10.5. *Anisotropic diffusion*
When the diffusion constant k is anisotropic, the diffusion equation has the form

$$\left(k_x \frac{\partial^2}{\partial x^2} + k_y \frac{\partial^2}{\partial y^2} \right) u(\mathbf{r}, t) = \frac{\partial u(\mathbf{r}, t)}{\partial t}.$$

Derive the solution to this equation with a point initial distribution, $u(x, y, 0) = C\delta(x, y)$.

10.6. *Bessel functions*
Show that

$$J_0'(x) = -J_1(x). \qquad (10.88)$$

10.7. *Bessel functions*
Derive and graph these Fourier transform pairs:

$$J_0(2\pi x) \longleftrightarrow \frac{\text{rect}(v/2)}{\pi \sqrt{1 - v^2}}, \qquad (10.68)$$

$$J_0(2\pi x) \, \text{sgn} \, x \longleftrightarrow \frac{[1 - \text{rect}(v/2)] \, \text{sgn}(v)}{i\pi \sqrt{v^2 - 1}}. \qquad (10.70)$$

Hint: Begin by substituting the Bessel function definition, Equation 10.19, into the Fourier integral. Performing the integration with respect to x, obtain, for Equation 10.68,

$$J_0(2\pi x) \longmapsto \frac{1}{\pi} \int_0^\pi \delta(v + \cos\phi)\, d\phi$$

and, for Equation 10.70,

$$J_0(2\pi x)\,\mathrm{sgn}\,x \longmapsto \frac{1}{\pi}\int_0^\pi \frac{d\phi}{i\pi(v+\cos\phi)}.$$

Alternatively, for Equation 10.68, make the trigonometric substitution $v = \cos\phi$ on the right-hand side and use the Bessel function definition.

10.8. *Bessel functions*
Derive the Fourier transform pair for the jinc function:

$$\sqrt{1-x^2}\,\mathrm{rect}\left(\frac{x}{2}\right) \longleftrightarrow \frac{J_1(2\pi v)}{2v} = \frac{\pi}{2}\,\mathrm{jinc}(v). \tag{10.89}$$

Hint: Use Equations 10.68 and 10.88.

10.9. *Hankel transforms*
A two-dimensional function $f(r,\theta)$ may be expressed as a Fourier series in θ:

$$f(r,\theta) = \sum_{n=-\infty}^{\infty} f_n(r)\,e^{in\theta},$$

$$\text{where } f_n(r) = \frac{1}{2\pi}\int_{-\pi}^{\pi} f(r,\theta)\,e^{-in\theta}.$$

Derive the following expression for the Fourier transform, $F(\rho,\phi)$:

$$F(\rho,\phi) = \sum_{n=-\infty}^{\infty} F_n(r)\,i^{-n}\,e^{in\phi},$$

$$\text{where } F_n(r) = 2\pi \int_0^\infty f_n(r)\,J_n(2\pi r\rho)\,r\,dr,$$

and $J_n(.)$ is the nth-order Bessel function of the first kind. *Hint:* Use the Bessel function integral identity:

$$J_n(x) = \int_{-\pi}^{\pi} e^{-i(nu - x\sin u)}\,du.$$

10.10. *Inverse Hankel transform*
(a) Show that the inverse Hankel transform (of zeroth order) is identical in form to the forward transform.

(b) Derive an expression for the inverse Hankel transform of arbitrary order.

10.11. *Ring delta*
The ring delta function may be defined as a sequence of functions. The difference of two circle functions of radius $1 + 1/2n$ and $1 - 1/2n$ yields a cylindrical shell with wall thickness $1/n$. The volume of the cylindrical shell is $2\pi/n$. Thus, we have a sequence of functions:

$$f_n(r) = \frac{n}{2\pi}\left[\mathrm{circ}\left(\frac{r}{1+1/2n}\right) - \mathrm{circ}\left(\frac{r}{1-1/2n}\right)\right].$$

Calculate the sequence of Hankel transforms $F_n(\rho)$ and show that, as $n \to \infty$, $F_n(\rho) \to J_0(2\pi\rho)$.

10.12. *Point delta in polar coordinates*

Show that $\int_0^{2\pi}\int_0^\infty -\frac{1}{\pi}\delta'(r)f(r,\theta)\,r\,dr\,d\theta = f(0,\cdot)$ and show that the Hankel transform of $-\frac{1}{\pi}\delta'(r)$ is unity.

10.13. *Derivatives of the multidimensional delta function*

In one dimension, $\delta^{(n)}(x)$ is the generalized function that maps $f(x)$ to $(-1)^n f^{(n)}(0)$. This can be extended to the multidimensional delta function $\delta(\mathbf{x})$. For example, in two dimensions, $\partial_x \delta(x,y)$ is approached in the expected way:

$$\iint [\partial_x\delta(x,y)]\,\varphi(x,y)\,dx\,dy = -\iint \delta(x,y)\,[\partial_x\varphi(x,y)]\,dx\,dy = -\varphi_x(0,0),$$

where $\varphi_x = \partial_x\varphi$. With this in mind, show that the following are consistent definitions:

(a) $\int (\nabla\delta(\mathbf{x}))\,\varphi(\mathbf{x})\,d\mathbf{x} = -\nabla\varphi|_0$

(b) $\int (\nabla^2\delta(\mathbf{x}))\,\varphi(\mathbf{x})\,d\mathbf{x} = \nabla^2\varphi|_0$

10.14. *Ellipse function*

The equation of an ellipse is $\frac{x^2}{a^2} + \frac{y^2}{b^2} = 1$. Show that an "ellipse function" analogous to the circle function can be created by a linear transformation $\mathbf{Ax}$ applied to the circle function, and use the dilation theorem to calculate its Fourier transform in terms of the jinc function.

10.15. *Hankel transform*

Derive the following Hankel transform pairs:

(a) $e^{-\pi r^2} \longleftrightarrow e^{-\pi\rho^2}$ (this is very easy)

(b) $\dfrac{1}{r} \longleftrightarrow \dfrac{1}{\rho}$

(c) $2\pi a\,\text{sinc}(2ar) \longleftrightarrow \dfrac{\text{circ}(\rho/a)}{\sqrt{a^2 - \rho^2}}$

(d) $\dfrac{e^{-ar}}{r} \longleftrightarrow \dfrac{2\pi}{\sqrt{a^2 + (2\pi\rho)^2}}$

10.16. *Hankel transform*

By integrating in polar coordinates, derive the formula for the Hankel transform of the Laplacian:

$$\nabla^2 f(r) = -4\pi^2\rho^2 F(\rho). \tag{10.90}$$

The Laplacian in polar coordinates is $\nabla^2 = \frac{1}{r}\frac{\partial}{\partial r}\left(r\frac{\partial}{\partial r}\right)$.

10.17. *Hankel transform*

Use the Laplacian relationship and other Fourier theorems, as necessary, to calculate the Hankel transform of $r^2 e^{-\pi r^2}$.

10.18. *Convolution in polar coordinates*
Derive Equation 10.37,

$$f * g(r) = \int_0^{2\pi} \int_0^\infty f(u) g\left(\sqrt{r^2 + u^2 - 2ru\cos\theta}\right) u\, du\, d\theta.$$

10.19. *Fourier transform in cylindrical coordinates*
Show that the Fourier transform of a cylindrically symmetric function $f(r, z)$ is

$$F(\rho, v_z) = \int_{-\infty}^\infty \int_0^\infty f(r, z) J_0(2\pi r\rho) e^{-i2\pi z v_z} r\, dr\, dz, \qquad (10.91a)$$

$$f(r, z) = \int_{-\infty}^\infty \int_0^\infty F(\rho, v_z) J_0(2\pi r\rho) e^{+i2\pi z v_z} \rho\, d\rho\, dv_z. \qquad (10.91b)$$

Then, as an example, calculate the Fourier transform of a disk, $\text{circ}(r)\,\text{rect}(z/a)$. Consider the special cases of a very thin disk and a very thick disk.

10.20. *Circular symmetry in spherical coordinates*
When a three-dimensional function is independent of the azimuthal angle φ but not the polar angle θ, it has circular symmetry. Show that the Fourier transform in this case is

$$F(\rho, \zeta) = 2\pi \int_0^\pi \int_0^\infty f(r, \theta) J_0(2\pi r\rho \sin\zeta \sin\theta) e^{-i2\pi r\rho \cos\zeta \cos\theta} r^2 \sin\theta\, dr\, d\theta, \qquad (10.92)$$

then show that this reduces to the spherically symmetric Fourier transform under the additional condition that f is independent of θ.

10.21. *Dilation under spherical symmetry*
By direct integration in spherical coordinates, show that $f(r/a) \longmapsto |a|^3 F(a\rho)$.

10.22. *Spectral resolution*
The diffracted order from a grating with a square aperture is of the form (Equation 10.47) $\text{sinc}\,(L(v_x + b))\,\text{sinc}\,(Lv_y)$. Two closely spaced spectral lines are said to be "just resolved" when the peak of one overlaps the first zero crossing of the other.

(a) Graph the function $\text{sinc}^2\,(L(v_x - \Delta v_x/2 + b)) + \text{sinc}^2\,(L(v_x + \Delta v_x/2 + b))$, varying Δv_x to see what "just resolved" looks like.

(b) Derive an expression for the minimum resolvable line spacing $\Delta\lambda = |\lambda_1 - \lambda_2|$ in terms of the grating size L, grating frequency b, and mean wavelength $\bar\lambda$, which may be taken to be the average of λ_1 and λ_2. This may also be written in terms of $N = Lb$, the number of grating cycles across the aperture.

10.23. *Tilted illumination*
Carry out the calculation to show that a tilted plane wave illumination of a circular aperture results in a shifted version of the jinc function pattern (Equation 10.51).

10.24. *Radon transform*
Show that the Radon transform of an impulse located at polar coordinates (a, ϕ) is $\delta(x' - a\cos(\theta - \phi))$, and sketch it accurately as a function of x' and θ. What is the effect of a and ϕ on the graph? The Radon transform of an impulse is called the *sinogram*.[34]

---

[34] Barrett and Myers (2004, pp. 204ff).

10.25. *Radon transform theorems*
Prove the basic theorems for the Radon transform.
  (a) Shift (Equation 10.59)
  (b) Dilation (Equation 10.60)
  (c) Rotation (Equation 10.61)
  (d) Convolution (Equation 10.62)

10.26. *Radon transform*
If p_θ is the projection of f at angle θ, show that $\int_{-\infty}^{\infty} p_\theta(x')\,dx' = \int_{-\infty}^{\infty} f(\mathbf{x})\,d\mathbf{x}$, for all angles θ.

10.27. *Abel–Fourier–Hankel cycle*
Calculate the following transforms:
  (a) $p(x) = \mathcal{A}\{\delta(r - a)\}$
  (b) $P(u) = \mathcal{F}\{p(x)\}$
  (c) $f(r) = \mathcal{H}\{P(u)\} = \delta(r - a)$

10.28. *Line spread function*
A line source in the object plane of an imaging system is modeled by $\delta(x)$ (it has infinite extent in the y direction). The system is assumed to have a radially symmetric point spread function, $h(\mathbf{x}) = h(r)$. The image of the line source is the convolution of the point spread function with the line source, $h(r) * \delta(x)$. It is, appropriately, called the *line spread function*, $\ell(x)$.

  (a) Derive an expression for the line spread function in terms of the point spread function, and show how the point spread function may be calculated from the line spread function.

  (b) The *edge spread function* $E(x)$ is the image of a unit step, $U(x)$. Relate the edge spread function to the line spread function and show how to recover the point spread function from the edge spread function.

10.29. *Two-dimensional sampling*
A two-dimensional function $f(\mathbf{x})$ is bandlimited to $\|\mathbf{v}\| < B$.

  (a) Find the coarsest rectangular sampling lattice that avoids aliasing. Sketch the Fourier plane showing the replicated spectral islands. What is the sample density?

  (b) It is argued that a more efficient sampling scheme uses a nonorthogonal lattice such that the circular spectral islands are close packed in the Fourier domain. Find the sampling lattice, and sketch the Fourier plane, showing the replicated spectral islands. What is the sample density, and how does it compare with the rectangular lattice in part (a)?

10.30. *Two-dimensional sampling*
Consider a sampling pattern analogous to sampling on the black squares of chessboard, as an approximation to the hexagonal sampling pattern in the previous problem.

  (a) What are the lattice vectors and the reciprocal lattice vectors?

  (b) What is the sampling density, and how does it compare to hexagonal sampling and rectangular sampling for a signal bandlimited to a circle of radius B?

BIBLIOGRAPHY

Abramowitz, M. and Stegun, I. (editors). *Handbook of Mathematical Functions, with Formulas, Graphs, and Mathematical Tables*. Dover, New York, 1972.

Ahmed, N., Natarajan, T., and Rao, K.R. Discrete cosine transform. *IEEE Transactions on Computers*, C-23:90–93, 1974.

Andrews, H.C. and Hunt, B.R. *Digital Image Restoration*. Prentice-Hall, Englewood Cliffs, NJ, 1977.

Arfken, G. *Mathematical Methods for Physicists*, 3rd edition. Academic Press, Orlando, FL, 1985.

Axler, S. *Linear Algebra Done Right*, 2nd edition. Springer-Verlag, New York, 1997.

Barrett, H.H. and Myers, K.J. *Foundations of Image Science*. John Wiley & Sons, Inc., Hoboken, NJ, 2004.

Bechhoefer, J. Kramers-Kronig, Bode, and the meaning of zero. *American Journal of Physics*, 79(10):1053–1059, 2011.

Bedrosian, E. A product theorem for Hilbert transforms. *Proceedings of the IEEE*, 51(5):868–869, 1963.

Beerends, R.J., ter Morsche, H.G., van den Berg, J.C., and van de Vrie, E.M. *Fourier and Laplace Transforms*. Cambridge University Press, New York, 2003.

Boas, R.P. *Entire Functions*, Vol. 5: *Pure and Applied Mathematics*. Academic Press, New York, 1954.

Bode, H.W. *Network Analysis and Feedback Amplifier Design*. Van Nostrand, New York, 1945.

Boggess, A. and Narcowich, F.J. *A First Course in Wavelets with Fourier Analysis*. Prentice-Hall, Englewood Cliffs, NJ, 2001.

Born, M. and Wolf, E. *Principles of Optics*, 7th edition (expanded). Cambridge University Press, New York, 1999.

Bracewell, R.N. Strip integration in radio astronomy. *Australian Journal of Physics*, 9:198–217, 1956.

Bracewell, R.N. *The Hartley Transform*. Oxford University Press, New York, 1986.

Bracewell, R.N. *The Fourier Transform and Its Applications*, 3rd edition. McGraw-Hill, New York, 2000.

Brigham, E.O. *The Fast Fourier Transform and Its Applications*. Prentice-Hall, Englewood Cliffs, NJ, 1988.

Cannon, R.H. *Dynamics of Physical Systems*. McGraw-Hill, New York, 1967.

Carleson, L. On convergence and growth of partial sums of Fourier series. *Acta Mathematica*, 116:135–157, 1966.

Carslaw, H.S. and Jaeger, J.C. *Operational Methods in Applied Mathematics*. Clarendon Press, Oxford, 1941.

Carson, J.R. *Electric Circuit Theory and the Operational Calculus*. McGraw-Hill, New York, 1926.

Champeney, D.C. *A Handbook of Fourier Theorems*. Cambridge University Press, New York, 1987.

Churchill, R.V. and Brown, J.W. *Fourier Series and Boundary Value Problems*, 4th edition. McGraw-Hill, New York, 1987.

Cooley, J.W. The re-discovery of the fast Fourier transform algorithm. *Mikrochimica Acta*, 111:33–45, 1987.

Cooley, J.W. and Tukey, J.W. An algorithm for the machine calculation of complex Fourier series. *Mathematics of Computation*, 19:297–301, 1965.

Cullity, B.D. and Stock, S.R. *Elements of X-Ray Diffraction*, 3rd edition. Prentice-Hall, Englewood Cliffs, NJ, 2001.

Doetsch, G. *Introduction to the Theory and Application of the Laplace Transform*. Springer-Verlag, New York, 1974.

Fourier Transforms: Principles and Applications, First Edition. Eric W. Hansen.
© 2014 John Wiley & Sons, Inc. Published 2014 by John Wiley & Sons, Inc.

Dudgeon, D.E. and Mersereau, R.M. *Multidimensional Digital Signal Processing*. Prentice-Hall, Englewood Cliffs, NJ, 1984.

Dym, H. and McKean, H.P. *Fourier Series and Integrals*. Academic Press, Boston, 1972.

Elmore, W.C. and Heald, M.A. *Physics of Waves*. McGraw-Hill, New York, 1969.

Fienup, J.R. Phase retrieval algorithms: a comparison. *Applied Optics*, 21(15):2758–2769, 1982.

Fienup, J.R. Phase retrieval algorithms: a personal tour. *Applied Optics*, 52(1):45–56, 2013.

Flanigan, F.J. *Complex Variables: Harmonic and Analytic Functions*. Dover, New York, 1983.

Folland, G.B. *Fourier Analysis and Its Applications*. Brooks/Cole, Boston, 1992.

Folland, G.B. *Real Analysis: Modern Techniques and Their Applications*, 2nd edition. John Wiley & Sons, Inc., New York, 1999.

Folland, G.B. *Advanced Calculus*. Prentice-Hall, Englewood Cliffs, NJ, 2002.

Fourier, J. *The Analytic Theory of Heat*. Dover, New York, 2003.

Franklin, G.F., Powell, J.D., and Workman, M.L. *Digital Control of Dynamic Systems*, 3rd edition. Addison-Wesley, Menlo Park, CA, 1998.

Gabor, D. Theory of communication. *Journal of the IEE*, 93:429–457, 1946.

Gasquet, C. and Witomski, P. *Fourier Analysis and Applications: Filtering, Numerical Computation, Wavelets*. Springer-Verlag, New York, 1999.

Goodman, J.W. *Introduction to Fourier Optics*. McGraw-Hill, New York, 1968.

Grafakos, L. *Classical and Modern Fourier Analysis*. Pearson, Upper Saddle River, NJ, 2004.

Gray, R.M. and Goodman, J.W. *Fourier Transforms: An Introduction for Engineers*. Kluwer, Boston, 1995.

Hahn, L.-S. and Epstein, B. *Classical Complex Analysis*. Jones and Bartlett, Boston, 1996.

Hahn, S. *Hilbert Transforms in Signal Processing*. Artech House, Boston, 1996.

Hartley, R.V.L. Transmission of information. *Bell System Technical Journal*, 7:535–563, 1928.

Heaviside, O. *Electromagnetic Theory*. Dover, New York, 1950.

Howell, K.B. *Principles of Fourier Analysis*. CRC Press, Boca Raton, FL, 2001.

Jahnke, E., Emde, F., and Lösch, F. *Tables of Higher Functions*, 6th edition. McGraw-Hill, New York, 1960.

Jones, D.S. *The Theory of Generalised Functions*, 2nd edition. Cambridge University Press, New York, 1982.

Kailath, T. *Linear Systems*. Prentice-Hall, Englewood Cliffs, NJ, 1980.

Kammler, D.W. *A First Course in Fourier Analysis*. Prentice-Hall, Englewood Cliffs, NJ, 2000.

Kay, S.M. *Modern Spectral Estimation*. Prentice-Hall, Englewood Cliffs, NJ, 1988.

King, F.W. *Hilbert Transforms*, 2 vols. Cambridge University Press, New York, 2009.

Kolmogorov, A.N. and Fomin, S.V. *Introductory Real Analysis*. Dover, New York, 1975.

Kovačević, J. and Chebira, A. Life beyond bases: the advent of frames (part I). *IEEE Signal Processing Magazine*, 24(4):86–104, July 2007.

Kovačević, J. and Chebira, A. Life beyond bases: the advent of frames (part II). *IEEE Signal Processing Magazine*, 24(5):115–125, September 2007.

Kuo, B.C. *Analysis and Synthesis of Sampled-Data Control Systems*. Prentice-Hall, Englewood Cliffs, NJ, 1963.

LePage, W.R. *Complex Variables and the Laplace Transform for Engineers*. Dover, New York, 1980.

Levanon, N. and Mozeson, E. *Radar Signals*. John Wiley & Sons, Inc., Hoboken, NJ, 2004.

Lighthill, M.J. *Fourier Analysis and Generalized Functions*. Cambridge University Press, New York, 1958.

Mallat, S. *A Wavelet Tour of Signal Processing*, 2nd edition. Academic Press, New York, 1999.

Marks, R.J. *Introduction to Shannon Sampling and Interpolation Theory*. Springer-Verlag, New York, 1991.

Marsden, J.E. and Hoffman, M.J. *Basic Complex Analysis*, 3rd edition. W.H. Freeman, New York, 1998.

McPherson, A. *Introduction to Macromolecular Crystallography*. Wiley-Blackwell, Hoboken, NJ, 2009.

Natterer, F. *The Mathematics of Computerized Tomography*. John Wiley & Sons, New York, 1986.

Nyquist, H. Certain factors affecting telegraph speed. *Bell System Technical Journal*, 3:324–346, April 1924.

Nyquist, H. Certain topics in telegraph transmission theory. *Transactions of the AIEE* 47(2):617–644 (1928). Reprinted in: *Proceedings of the IEEE*, 90(2):280–305, 2002.

Oden, J.T. and Demkowicz, L.F. *Applied Functional Analysis*. CRC Press, New York, 1996.

Oppenheim, A.V. and Schafer, R.W. *Discrete-Time Signal Processing*, 3rd edition. Prentice-Hall, Englewood Cliffs, NJ, 2010.

Paley, R.E.A.C. and Wiener, N. *Fourier Transforms in the Complex Domain*, Vol. 19: Colloquium Publications. American Mathematical Society, New York, 1934.

Pandey, J.N. *The Hilbert Transform of Schwartz Distributions and Applications*. John Wiley & Sons, Inc., New York, 1996.

Papoulis, A. *Systems and Transforms with Applications in Optics*. McGraw-Hill, 1968.

Papoulis, A. *Signal Analysis*. McGraw-Hill, New York, 1977.

Park, S.C., Park, M.K., and Kang, M.G. Super-resolution image reconstruction: a technical overview. *IEEE Signal Processing Magazine*, pp. 21–36, May 2003.

Pennebaker, W.B. and Mitchell, J.L. *JPEG Still Image Data Compression Standard*. Van Nostrand Reinhold, New York, 1993.

Percival, D.B. and Walden, A.T. *Spectral Analysis for Physical Applications*. Cambridge University Press, New York, 1993.

Percival, D.B. and Walden, A.T. *Wavelet Methods for Time Series Analysis*. Cambridge University Press, New York, 2000.

Porat, B. *A Course in Digital Signal Processing*. John Wiley & Sons, Inc., New York, 1997.

Pratt, W.K. *Digital Image Processing: PIKS Scientific Inside*, 4th edition. Wiley-Interscience, New York, 2007.

Radon, J. Über die Bestimmung von Funktionen durch ihre Integralwerte längs gewisser Mannigfaltigkeiten. *Berichte der Sächsischen Akadamie der Wissenschaft*, 69:262–277, 1917. Reprinted: Radon, J. and Parks, P.C. (trans). On the determination of functions from their integral values along certain manifolds. *IEEE Transactions on Medical Imaging*, MI-5(4):170–176, December 1986.

Ramachandran, G.N. and Srinivasan, R. *Fourier Methods in Crystallography*. John Wiley & Sons, Inc., New York, 1970.

Rosenlicht, M. *Introduction to Analysis*. Scott, Foresman and Co, Glenview, IL, 1968.

Saff, E.B. and Snider, A.D. *Fundamentals of Complex Analysis for Mathematics, Science, and Engineering*. Prentice-Hall, Englewood Cliffs, NJ, 1976.

Schwartz, L. *Théorie des Distributions*. Hermann, Paris, 1951.

Stade, E. *Fourier Analysis*. John Wiley & Sons, Inc., Hoboken, NJ, 2005.

Strichartz, R.S. *A Guide to Distribution Theory and Fourier Transforms*. CRC Press, Boca Raton, FL, 1994.

Unser, M. On the approximation of the discrete Karhunen-Loeve transform for stationary processes. *Signal Processing*, 7:231–249, 1984.

Unser, M. Sampling — 50 years after Shannon. *Proceedings of the IEEE*, 88(4):569–587, 2000.

Watson, G.N. *A Treatise on the Theory of Bessel Functions*, 2nd edition. Cambridge University Press, New York, 1995.

Wilcox, H.J. and Myers, D.L. *An Introduction to Lebesgue Integration and Fourier Series*. Dover, New York, 1978.

Wilts, C.H. *Principles of Feedback Control*. Addison-Wesley, Reading, MA, 1960.

Wunsch, A.D. *Complex Variables with Applications*, 2nd edition. Addison-Wesley, Reading, MA, 1994.

Young, N. *An Introduction to Hilbert Space*. Cambridge University Press, New York, 1988.

Zemanian, A.H. *Distribution Theory and Transform Analysis*. Dover, New York, 1987.

INDEX

Abel transform, 718–722
Absolute convergence, *see* Convergence
Absolutely integrable, 29, 69. *See also* L^1
Absolutely summable, 67. *See also* ℓ^1; Sequence
Airy disk, 710–711. *See also* Jinc function
Aliasing, 110, 123, 423–424, 728
   and computing Fourier series, 235, 249
   and computing Fourier transform, 323
   and Laplace transform, 607–608, 619
Almost everywhere, a.e., 73, 190
Amplitude, 20, 229, 251, 696, 698, 701
Amplitude modulation (AM), 441, 443
Analytic function, 467, 505, 512. *See also* Complex function; Cauchy–Riemann equations
Analytic signal, 647
Antenna array, 227–233, 357, 449, 705
Anticausal, *see* System, anticausal
Aperiodic function, 112, 274, 430
Application examples
   antennas and arrays, 227, 267, 357–358, 449, 705
   audio and acoustics, 286, 445, 662
   bandpass signals, 648
   communications, 286, 355, 441, 443, 668
   diffraction gratings, 263, 701–705
   diffusion, 681, 737
   electronics, 263–265, 295, 443, 446–447, 589
   field theory, 491
   filters, 295, 319, 352, 356, 439, 559
   harmonic distortion, 214, 265, 270
   heat conduction, 215, 266, 361
   JPEG, 60, 157
   lasers, 444, 445
   optics and imaging, 108, 353, 441, 447–449, 664, 691, 701, 706ff, 741
   probability and statistics, 97, 351, 683
   quantum mechanics, 362
   radar, 47, 364
   risetime and bandwidth, 288
   sample rate conversion, 253, 272
   seismology, 450
   signal processing, 124–125, 250, 326, 440, 452
   spectroscopy, 363, 703, 740
   spectrum analysis, 173–175, 352, 356
   system theory, 104, 354
   tomography, 713ff
   vibrating string, 223, 266–267
   X-ray diffraction, 729
Area theorem
   for DFT, 128
   for discrete time Fourier transform, 241
   for Fourier series, 206
   for Fourier transform, 248
Autocorrelation, *see* Correlation

Backprojection, 716
Banach space, 79. *See also* Vector space
Bandlimited function, 235, 249, 423, 555, 607, 641, 646, 727
Bandpass signal, 648
   IQ model, 649
Bandwidth, 288, 318, 648, 704, 712
Baseband signal, 649
Basis
   complex exponential, 52, 112, 150, 429–430
   Haar, 52, 57
   JPEG, 60, 164
   orthogonal, 5
   orthonormal, 50, 55, 94, 426
   standard, 51
   trigonometric, 162, 182, 263
   wavelet, 348
Bedrosian's theorem, 646

Fourier Transforms: Principles and Applications, First Edition. Eric W. Hansen.
© 2014 John Wiley & Sons, Inc. Published 2014 by John Wiley & Sons, Inc.

748 INDEX

Bessel function, 685, 704
Bessel's inequality, 83
Big-O notation, 25
Bijective, *see* Mapping
Bin, *see* discrete Fourier transform
Bound
 essential supremum, 75
 ML inequality, 496
 upper and lower (supremum and infimum), 17
Bounded function, 17, 29, 69, 74, 179
Bounded interval, 28, 188
Bounded operator, 88
Bounded sequence, 67
Bounded support, *see* Support
Branch, 17, 476
Branch cut 476, 511, 541
Branch point, 478
Bromwich contour, 571

Carrier wave, 286, 364, 441, 648
Cauchy principal value, 29, 396, 407, 533, 537, 631
Cauchy sequence, 65, 66
Cauchy's integral formula, 512, 516
Cauchy's (Cauchy–Goursat) integral theorem, 505
Cauchy–Riemann equations, 464, 466
Cauchy–Schwarz inequality, 46
Causality, *see* System, causal
Centroid, 297, 683
Chirp function, 294, 338ff
Circle function, 686
Comb function, 418, 420, 722
Comb sequence, 114, 247, 429
Complete set of functions, 84
Complete vector space, 66, 79
Complex function, 455. *See also* Singularity
 analytic, 467, 512, 525
 entire, 467
 exponential, 470
 logarithm, 473
 meromorphic, 598
 multivalued, 473ff
 power, 484ff
 square root, 479, 484
 trigonometric, 471–472, 489
Complex integral, 497ff
Complex number, 8
Complex variable, 455

Continuous function, *see* Function
Continuous operator, *see* Operator
Continuous time signal, 37
Contour of integration, 496ff
 Bromwich, 590
 deformed, 506
 indented, 537ff
Contrast, *see* Image processing
Convergence, *see also* Region of convergence; Sequence
 absolute, 190
 of discrete-time Fourier transform, 240
 dominated, 403
 of Fourier series, 190–196, 416
 of Fourier transform, 311ff, 406, 427ff, 543, 564, 621
 in norm (mean, mean square), 190
 uniform, 190, 544, 567
 weak, 389
Convergence factor, *see* Convergence, of Fourier transform
Convolution, 105, 210, 300. *See also* System, linear time-invariant
 bounds on, 212, 301
 computing with the DFT, 330ff
 cyclic, circular, 137, 140
 of generalized functions, 401
 in imaging systems, 709
 multidimensional, 684, 692
 of periodic functions, 210
 of sequences, 212, 243, 245
Convolution theorem
 for DFT, 136, 142
 for discrete-time Fourier transform, 245
 for Fourier series, 213
 for Fourier transform, 305, 316, 684
 generalized, 412
 for Hankel transform, 692
 for Hilbert transform, 646
 for Laplace transform, 581
 multidimensional, 684
 for Radon transform, 715
 for Z transform, 620
Correlation
 autocorrelation, 310, 712
 crosscorrelation, 48, 309
 lag, 310
 receiver (matched filter), 47, 364
 theorem, 310

Cyclic convolution, *see* Convolution, cyclic
Cyclic shift, 134, 140, 324. *See also*
    Convolution; Shift theorem

Delta function, 371. *See also* Generalized
    functions
  multidimensional, 670, 687, 696
  operational properties, 383
  sifting property, 374, 399, 670, 688, 696
Derivative theorem
  for discrete time Fourier transform, 242
  for Fourier series, 207
  for Fourier transform, 290, 409, 681
  for Laplace transform, 578
  for Z transform, 616
Difference equation, 249, 628. *See also*
    System, discrete time
Diffraction
  Fraunhofer, 700
  grating, 701–704
  X-ray, 729ff
Diffraction limited, 709
Diffusion equation, 215, 681. *See also* Heat
    equation
Digital filter, *see* Filter, digital
Digital (discrete) frequency, 110, 239, 732
Dilation, 246, 288, 374, 675. *See also*
    Downsampling; Upsampling
Dilation theorem
  for discrete time Fourier transform, 249
  for Fourier transform, 288, 676
  for Hankel transform, 691
  for Hilbert transform, 645
  for Laplace transform, 578
  for Radon transform, 715
  for Z transform, 619
Dimensionality, *see* Vector space
Direct sum, *see* Vector space
Direction angle, sine, 232, 697
Dirichlet kernel, 120, 196, 233
Discrete cosine transform (DCT), 162
Discrete Fourier transform (DFT), 115
  and discrete time Fourier transform, 255ff
  and Fourier series, 233ff
  and Fourier transform, 323ff
  frequency bins, 118, 124
  tables, 164–165
  theorems, 126ff

Discrete-time Fourier transform, 238, 427
  and Hilbert transform, 634
  multidimensional, 732
  tables, 435
  theorems, 259
  and Z transform, 613
Discrete time signal, 37. *See also* Sequence
Distance, *see* Metric; Norm
Domain
  of analyticity, 467
  in the complex plane, 466, 503
  of a mapping, 15
  of an operator, 87
  simply-connected, 503
  time and frequency, 273
Dot product, 4, 44, 92, 669. *See also* Inner
    product
Double sideband modulation, 286
Downsampling, 246, 619

Eigenfunction, 24, 178, 216, 309
Eigenvalue, 14, 24, 216, 309
Energy, *see* Norm; Parseval's formula
Entire function, *see* Complex function
Envelope, 364, 443, 651
Equiphase, 671, 697
Equivalent width, 318
Essential singularity, 468
Euclidean space, 40
Euler equations, 22
Euler's method, 617
Evanescent wave, 697
Even symmetry, 17, 112, 131, 158, 202, 282, 392, 674
Exponential function, 278, 281, 470, 565
  complex, 22, 112
  two-sided, 281, 570
Exponential order, 567
Exponential type, 556

Fairly good function, 386
Fast Fourier transform (FFT), 152ff
Father wavelet, *see* Scaling function
Filter, 53, 87, 171, 643
  with backprojection, 716
  bandpass, 353
  Butterworth, 559, 572
  digital, 251, 452
  finite impulse response (FIR), 251, 629
  highpass, 355

Filter (*Continued*)
  infinite impulse response (IIR), 253, 629
  inverse, 441
  lowpass, 252, 288, 424, 709
  matched, *see* Matched filter
  with sampling, 424, 728
  with upsampling, 253
Final value theorem, 585, 621
Finite difference, 242, 616
Focal length, *see* Lens
Fourier series, 177ff
  classical, 182
  complex, 180
  computing, with DFT, 234ff
  convergence, 190ff
  existence, 179
  and Fourier transform, 416
  and generalized functions, 415ff
  line spectrum, 183
  and partial differential equations, 215ff
  tables, 258
  theorems, 200ff
Fourier transform, 273ff, 669ff
  and complex integration, 543ff
  computing, with DFT, 324ff
  existence and invertibility, 275–276, 311ff
  generalized, 404
  in the limit (convergence factor), 311ff, 405
  and Laplace transform, 572
  multidimensional, 670
  tables, 349, 435
  theorems, 350
Frame, 150, 345
Frequency response, 251, 286, 309. *See also* Transfer function
Fubini's theorem, 32
Function, 17, 188. *See also* Complex function; Generalized function; Sequence
  continuous, 18, 187
  continuously differentiable, 189
  differentiable, 19
  even and odd parts, 17
  Hermitian, 17
  integrable, 28, 69
  piecewise continuous, 18, 187
  piecewise smooth, 19, 187
  real and imaginary parts, 10

Function space, 69. *See also* Vector space
Functional, 386
Fundamental frequency, in Fourier series, 180
Fundamental theorem of calculus (complex functions), 510

Gaussian function, 293
  complex, *see* Chirp function
  as convergence factor, 311
  delta sequence, 373
  in diffusion, 681
  multidimensional, 670
  in short time Fourier transform, 338
  uncertainty, 321
Generalized derivative, 393
Generalized function, 388. *See also* Comb function; Delta function; Good function
  calculus, 389
  equivalent, 389
  Fourier transform, 404ff
  and ordinary functions, 385
  periodic, 418
  products and convolutions, 396ff
  regular vs. singular, 388
  sequences, 388–389
Gibbs phenomenon (overshoot), 199
Good function (Schwartz function), 386–387
Gram–Schmidt process, 94. *See also* Basis, orthonormal
Green's theorem, 505

Haar wavelet, 346ff
Hamming window, 174, 342, 358
Hankel–Fourier–Abel transform cycle, 722
Hankel transform, 675
  theorems, 691, 736
Harmonic frequency, in Fourier series, 180
Heat equation, 215ff. *See also* Diffusion equation
Hermite–Gaussian wavefunction, 353
Hermitian function, 17, 112, 130, 201, 252, 577, 615, 640
Hilbert space, 79. *See also* Vector space
Hilbert transform, 642
  and analytic signal, 647
  and bandpass signal, 651

INDEX   751

and causality, 630
theorems, 654
Hölder's inequality, 79. *See also*
        Cauchy–Schwarz inequality

Identity operator, 89
Image point, *see* Mapping, image and
        preimage
Image processing
    compression, 60, 156ff
    contrast, 712
    Fourier analysis, 709
    reconstruction from projections,
        713ff
    restoration, 53
Impulse, 307. *See also* Delta function
Impulse response, 249, 307, 629, 641.
        See also Filter; Point spread
        function; System; Transfer function
Inequality
    Bessel, 83
    Cauchy–Schwarz, 46
    Hölder, 79
    Minkowski, 104
    Schwarz, 41
    triangle, 40
    uncertainty principle, 320
Infinite impulse response (IIR), *see* Filter
Initial value theorem, 583, 584, 621
Injective, *see* Mapping
Inner product, 45, 50, 56, 64. *See also*
        Norm; Vector space
Instantaneous frequency, 295
Integral
    complex (contour), 497ff
    double, 32
    Fourier, *see* Fourier transform
    improper, 29, 533. *See also* Cauchy
        principal value
    Laplace, *see* Laplace transform
    Lebesgue, 79
    line, 494–496
    multivalued function, 540
    performed via complex integration,
        531ff
    principal value, *see* Cauchy principal
        value
    Riemann, 308
    singular, *see* Cauchy principal value
    trigonometric, 531

Integral theorem
    Cauchy, *see* Cauchy's integral theorem
    for discrete time Fourier transform, 242
    for Fourier series, 208
    for Fourier transform, 295
    for Laplace transform, 580
    for Z transform, 618
IQ model, *see* Bandpass signal
Isolated singularity, 468. *See also* Pole
Isometry, 92
Isomorphic spaces, 92

Jinc function, 687, 692, 707, 722, 728
Jordan's lemma, 544, 593
JPEG, 60, 157

Kronecker delta, 114

$\mathcal{L}_+$, $\mathcal{L}_-$, *see* Laplace transform
L^1, 69, 79
    Fourier series, 179, 188, 192, 212
    Fourier transform, 281, 311, 316, 423
    Hilbert transform, 643
    Laplace transform, 564
    Radon transform, 714
L^2, 69, 72, 79
    Fourier series, 179, 188, 192, 212
    Fourier transform, 314, 316, 423
    Hilbert transform, 643
    Radon transform, 714
L^p, 69
L^∞, 69, 188, 212
ℓ^0, 68, 245
ℓ^1, 67, 71
    discrete time Fourier transform, 240,
        245
    Fourier series, 191
    Hilbert transform, 643
    Z transform, 608
ℓ^2, 67, 79, 93
    discrete time Fourier transform, 240,
        245
    Fourier series, 192
    Hilbert transform, 643
ℓ^p, 68, 79, 244
ℓ^∞, 68, 245
Laplace transform, 563ff
    abcissa of convergence, 565
    and Fourier transform, 572
    initial value problems, 564, 586

Laplace transform (*Continued*)
  inversion, 593ff
    by complex integration, 593
    by partial fractions, 598
  $\mathcal{L}_+, \mathcal{L}_-$, 575
  of generalized function, 574
  region of convergence, 565, 569
  of sampled function, 605
  table, 652
  theorems, 576ff, 653
  two-sided, 569
Lattice, 722
  crystal, 729
  nonorthogonal, 723
  reciprocal, 724
Laurent series, 523, 597, 622
Lens, 706ff
  as a Fourier transformer, 708
  in image formation, 708, 710
Line spectrum, *see* Spectrum
Linear combination, 5, 50, 81, 177, 426, 595
Linear dependence and independence, 15, 50, 723
Linear phase, 286, 705
Linear space, *see* Vector space
Linear system, *see* System, linear
Linearity theorem
  for DFT, 126
  for discrete time Fourier transform, 241
  for Fourier series, 200
  for Fourier transform, 282, 407, 674
  for Hilbert transform, 645
  for Laplace transform, 576
  for Z transform, 614

Magnitude response, 251, 286, 573, 639, 641
Mapping, 15, 86, 386, 455
  bijective (one-to-one and onto), 16, 87
  image and preimage, 15, 90
  injective (one-to-one), 15, 87
  multivalued, 17, 473ff, 540
  surjective (onto), 15, 87
Matched filter, 48, 364
Matrix, 11
  adjoint, 13, 93
  circulant, 140
  DFT, 116
  identity, 12
  inverse, 15, 90
  orthogonal, 15, 93
  singular, 15
  transpose, 12
  unitary, 15, 93, 117
Mean-square width, 319
Measure of a set, 73
Meromorphic function, *see* Complex function
Metric, 42ff. *See also* Norm
Metric space, *see* Vector space
Minimum phase, *see* System, minimum phase
Minkowski's inequality, 104. *See also* Triangle inequality
Mittag–Leffler expansion, 598. *See also* Partial fraction expansion
Mixer, 286, 650. *See also* Modulation theorem
ML inequality, 496, 534, 542, 543, 549, 592
Mode, 225
Modulation theorem
  for Fourier transform, 286
  for Hilbert transform, 646
  for Laplace transform, 577
  for Z transform, 618
Modulation transfer function (MTF), 712
Moment, 297, 683
  generating function, 351, 683
  theorems, for Fourier transform, 297–298, 682
Morera's theorem, 512
Mother wavelet, 346
Moving average, 250, 252
Multi-index, 680, 730
Multivalued function, *see* Mapping, multivalued

Negative frequency, 124, 182, 324, 647
Neighborhood, 44, 466, 478
Noise equivalent bandwidth, 319, 559
Noncausal, *see* System
Norm, 3, 40
  absolute value (1-norm), 41
  essential supremum (L^∞ norm), 75
  Euclidean (2-norm), 40, 67, 69
  and inner product, 45
  and metric, 42
  p-norm, 68, 69
  of an operator, 88
  supremum (uniform, ℓ^∞ norm), 67–69
Normed space, *see* Vector space

Null set, *see* Set
Null space, 87
Nyquist rate, 111. *See also* Aliasing; Sampling, Nyquist

Odd symmetry, 17, 130, 201–202, 282, 392, 630
One-sided function, 243, 301, 308, 324, 403, 574, 607. *See also* Hilbert transform; System, causal
One-sided spectrum, *see* Analytic signal
One-to-one (injective), *see* Mapping
Onto (surjective), *see* Mapping
Operator, 86, 306. *See also* Mapping
  bounded, 88
  invertible, 90
  isometric (norm-preserving), 95
  linear, 86
  projection, 94
  unitary, 92
Operator norm, 88
Optical transfer function (OTF), 712
Orthogonal complement, 53
Orthogonal decomposition, expansion, 51, 57, 85, 178, 182, 425
Orthogonal functions, 76, 96
Orthogonal matrix, 15
Orthogonal vectors, 50. *See also* Basis
Orthogonality principle, 56
Orthonormal vectors, 50. *See also* Basis

Paley–Wiener criterion, 642
Paley–Wiener theorem, 556
Parallelogram law, 46, 68
Parametric curve, 494
Paraxial approximation, 699, 700
Parseval's theorem (formula), 56, 85
  conservation of energy, 57
  for DFT, 127
  for discrete time Fourier transform, 241, 620
  for Fourier series, 205
  for Fourier transform, 285, 675
  and generalized Fourier transforms, 404
  for Hilbert transform, 646
  for Laplace transform, 583
  for short-time Fourier transform, 341, 344
  for Z transform, 620
Partial differential equations, *see* Diffusion equation; Heat equation; Wave equation

Partial fraction expansion, 595
  Heaviside (coverup) method, 600
  Mittag–Leffler, 598
Periodic function, 112, 180, 415, 418, 634
Periodic replication, 151, 421, 430, 606, 734
Phase, 20, 182, 229, 251, 286, 294, 639, 642, 651, 704, 706
Phase response, 251, 286, 639
Phase retrieval, 642
Phasor, 24, 647
Plancherel, *see* Parseval's theorem
Plane wave, 696, 729
Point spread function, 289, 308, 709
Poisson sum formula, 417
Polarization identity, 46
Pole, 468, 525. *See also* Residue
  determining order of, 529
  of a transfer function, 573, 585, 599, 614, 640
Power spectrum, *see* Spectrum
Preimage, *see* Mapping, image and preimage
Principal part of Laurent series, 524
Principal value
  Cauchy, *see* Cauchy principal value
  of a function, 9, 470
Product theorem
  for analytic signals, 667
  for Fourier transform, 316
  for Hilbert transform (Bedrosian's theorem), 646
  for Laplace transform, 582
  for Z transform, 620
Projection-slice theorem, 715

Quadrature, 649, 651. *See also* Analytic signal; Bandpass signal

Radius of convergence, 521, 610, 621. *See also* Region of convergence
Radon transform, 714ff
Range, of a mapping, 15, 87, 475
Rapid descent (rapidly decreasing), 191, 222, 245, 386, 411
Rational function, 469
  strictly proper, 584, 595
Rayleigh's theorem, *see* Parseval's theorem
Reciprocal lattice, 724, 727, 731
Rectangle function, 168, 239, 277, 672

Region of convergence
  Laplace transform, 565, 569, 572
  Laurent series, 524
  Taylor series, 521
  Z transform, 613
Regression, 97
Regular part of Laurent series, 524
Regular sequence, *see* Generalized function
Residue, 526ff
Resolution
  frequency, 328
  Rayleigh and Sparrow, 686
  spectral, 356, 704, 740
Riemann–Lebesgue lemma, 183, 241, 277, 569
Risetime, 288, 368

Sampling, 110. *See also* Aliasing; Periodic replication
  and discrete time Fourier transform, 247, 255
  and Fourier series, 235ff, 417
  and Fourier transform, 323ff
  and Hilbert transform, 634
  and Laplace transform, 604
  multidimensional, 726
  Nyquist, 111, 235, 612
  rate, 110
  and replication, 423, 434
  theorem, 423ff
Scaling function, 346
Schwartz function, *see* Good function
Schwarz inequality, 41. *See also* Cauchy-Schwarz inequality
Separable function, 216, 672
Separation of variables, 216
Sequence, 15, 37
  Cauchy, 65
  convergence, 64
  of partial sums, 64, 179
  summabililty, 67
Set, 1–2
  closed, 466
  connected, 466–467, 503
  measure of, 73
  null, 73
  open, 466
Shift theorem
  for DFT, 134
  in diffraction, 703
  for discrete time Fourier transform, 242
  for Fourier series, 206
  for Fourier transform, 285, 408, 675
  for Hilbert transform, 645
  for Laplace transform, 577
  for Z transform, 615
Short-time DFT, 342
Short-time Fourier transform, 336
Sifting property, *see* Delta function
Signal, 36–37
Signum function, 20, 405, 427, 633, 638
Sinc function, 69, 277, 423, 672
Singularity, 27, 456, 521, 571. *See also* Branch point; Pole
  essential, 468, 526
  isolated, 468, 479
  removable, 27, 525
Sinusoid, 20, 109
Slow growth (slowly growing), 386–387, 402, 411, 427, 643
Space-bandwidth product, 326
Span, 5, 50
Spectral leakage, 119, 185
Spectrogram, 337
Spectrometer, 363, 703
Spectrum
  bandpass, 649
  line, 183, 430
  one-sided, 647
  plane wave, 701
  power, 125, 183, 300, 339
Square integrable, 69. *See also* L^2
Square summable, 67. *See also* ℓ^2; Sequence
Square wave, 186, 192, 196ff, 210, 416
Stability, *see* System
Step function, 18, 407, 428, 566, 573
Strictly proper, *see* Rational function
Subspace, 53, 60, 87, 179, 348, 426
Superresolution, 642, 712
Support, of a function, 301, 431, 642, 714, 728
Surjective (onto), *see* Mapping
Symmetry
  in DFT, 130
  in discrete-time Fourier transform, 202, 238
  even and odd, *see* Even symmetry; Odd symmetry
  in Fourier series, 202
  in Fourier transform, 282, 284, 408, 674, 684, 693

Hermitian, *see* Hermitian function
  in Laplace transform, 577
  in Z transform, 615
System, 86, 306. *See also* Filter; Frequency response; Impulse response; Transfer function
  anticausal, 572, 613
  causal, 249, 306, 572, 582, 628, 641
  discrete time, 249, 628
  linear, 86, 306
  linear time-invariant (LTI), 249, 306, 309, 582, 595, 628
  minimum phase, 639
  noncausal, 249
  stability, 572, 613

Tables
  convolutions and products, 212, 245, 316
  delta function properties, 383
  DFT, 164
  discrete time Fourier transform, 259, 435
  Fourier series, 192, 258
  Fourier transform, 349, 350, 411, 435
  Fourier transform, multidimensional, 736
  generalized functions, 383, 387, 433–434
  Hankel transform, 736
  Hilbert transform, 654
  Laplace and Z transforms, 609, 652, 653
Taylor series, 26, 521
Temperate convergence, *see* Convergence, weak
Tempered distribution, 378. *See also* Generalized function
Testing function, 385. *See also* Generalized function
Time-bandwidth product, 326
Tomography, 715
Tonelli's theorem, 32
Transfer function, 251, 293, 309, 572, 628, 640
Transformation, *see* Operator
Triangle function, 280
Triangle inequality, 3, 40, 42
Twiddle factor, 153

Uncertainty principle, 320, 363
Uniform convergence, *see* Convergence
Unit impulse, *see* Delta function

Unitary matrix, *see* Matrix, unitary
Unitary operator, *see* Operator, unitary
Upsampling, 246ff, 618

Variance, 99, 297, 683
Vector space, 38. *See also* L^1; L^2; L^p; L^∞; ℓ^1; ℓ^2; ℓ^p; ℓ^∞; Subspace
  Banach, 79, 80
  $\mathbb{C}^n$, 43, 79, 80, 93, 109
  $\mathbb{C}^\infty$, 67
  $C^{(p)}$, 188ff, 192, 241, 387, 411, 521
  complete, 66
  dimensionality (finite, infinite), 32, 50
  direct sum, 53, 348
  Hilbert, 79, 80
  inner product, 45
  metric, 42
  normed, 40
  $\mathbb{R}^n$, 38, 79
  $\mathbb{R}^\infty$, 67
Vibrating string, 223ff. *See also* Wave equation

Wave
  plane, 696–697
  spherical, 229ff, 698ff, 729
Wave equation, 178, 223, 696
Wavefront, 229, 697
Wavelength, 225, 229, 697
Wavelet, 345ff
Weak convergence, *see* Convergence
White noise, 319
Width, of a function, 318, 319

X-ray diffraction, 729

Z transform, 608ff
  and discrete-time Fourier transform, 613
  inversion
    by complex integration, 623
    by partial fractions, 625
  and Laplace transform, 607, 610, 612
  region of convergence, 613
  table, 652
  theorems, 614ff, 653
  two-sided, 612
Zero packing, 150. *See also* Upsampling
Zero padding, 143, 332
Zeros, of a transfer function, 584, 639–640